Young Scientist 2021 (YS21)

IOP Conference Series: Materials Science and Engineering
Volume 1209

High Tatras, Slovakia
13 - 15 October 2021

Editors:

Peter Mesaros
Vinayakam Jothiprakash
Kamila Kotrasova
Martina Zelenakova

ISBN: 978-1-7138-4668-0
ISSN: 1757-8981

Printed from e-media with permission by:

Curran Associates, Inc.
57 Morehouse Lane
Red Hook, NY 12571

Some format issues inherent in the e-media version may also appear in this print version.

Institute of Physics
Dirac House, Temple Back
Bristol BS1 6BE UK

Phone: 44 1 17 929 7481
Fax: 44 1 17 920 0979

techtracking@iop.org

Additional copies of this publication are available from:

Curran Associates, Inc.
57 Morehouse Lane
Red Hook, NY 12571 USA
Phone: 845-758-0400
Fax: 845-758-2633
Email: curran@proceedings.com
Web: www.proceedings.com

TABLE OF CONTENTS

Preface

Peer review declaration

A data structures for purpose of the BIM-based Life Cycle Assessment: A review and theoretical background ... 1
M Brandtner, V Venkrbec

The level of detail for 4D BIM modeling ... 7
Y Nechyporchuk, R Baskova

Insulation of historic buildings and case study simulation ... 13
K Hutkai, D Katunský

Thermophysical parameters of the Kežmarok sandstone ... 23
P Jaroš, M Vertal

Reverberation measurement set for the interrupted noise method ... 28
D Jun, O Nespěšný

Analysis of technical solutions of stacks in high-rise buildings ... 34
M Sokol, J Peráčková

Advanced building materials - prefabricated and modular designs as part of Construction 4.0 with a focus on transport constructions ... 43
P Krupík

Opportunities for rehabilitation of facades of historic buildings in Moldova with limestone elements ... 48
D C Albu, A Lesan

IoT and 5G enabled construction detection model ... 59
H Begić

Comparison of spreadsheet-based optimization tools applied to construction site layout problem ... 65
B Dasović, U Klanšek

Comparison of the planned and actual course of construction of reinforced concrete monolithic structures ... 71
R Doubek

Analysis of time consumption for extensive green roof realization ... 76
M Hejl, M Mohapl, J Jílek

Management of health and safety on building site using smart technologies ... 82
A Husovská, Z Struková

Sustainability of material solutions of modern methods of construction ... 88
D Mokrenko, M Kozlovská

Carbon footprint of selected building structures ... 98
J Budajová

Removal of organic micropollutants in the biological units of a Swedish wastewater treatment plant............ 104
C Burzio, E Nivert, A Mattsson, O Svahn, F Persson, O Modin, B M Wilén

Alternatives how to mitigate urban flooding and CSO in urbanized areas used in European cities and towns.......... 108
M Csóka, G Rózsa, I Marko, Š Stanko

Analysis of CO2 emissions in municipality Most pri Bratislave.......... 114
M Garaj, M Zeleňaková

Combined sewer overflow chamber analysis based on CFD modelling.......... 119
J Hrudka, M Šutúš, M Csóka, A Raczková, I Škultétyová

The proposals for intensification of a non-capacity wastewater treatment plant.......... 126
T Kubínová, M Kyncl

The effectiveness of SuDS measures in reducing the rainwater runoff from urbanized areas 132
I Marko, R Wittmanová, J Hrudka, Š Stanko

Assessment of streamflows in the Hornád river basin and analysis of trends 138
P Nagy

Single-step and double-step heath recuperation of waste water from thermal baths.......... 142
A Predajnianska, J Takacs

Development of the level of sorting of municipal waste in the selected locality.......... 150
G Rózsa, R Wittmanová, J Hrudka, I Škultétyová

Flood risk assessment methodology.......... 156
M Šugareková

Hydrodynamic assessment of object on the sewer network in the city of Trnava.......... 162
M Šutúš, G Rózsa, M Csóka, Š Stanko, I Škultétyová

Life-cycle assessment of a photovoltaic panel: Assessment of energy intensity of production and environmental impacts.......... 167
M Vácha, J Kodymová, V Lapčík

Development of design rainfalls: A review.......... 175
R Wittmanová, I Marko, M Šutúš, Š Stanko

Parasitic Architectural Forms (PAF) S01.E02 "Methodology and Ontology".......... 181
T Baroš, D Katunský

Biaxial testing of textile material - methodology and instrument assembly procedure.......... 188
N Freiherrova, M Hornakova, D Juracka, L Stulerova, L Kapolka

Theoretical Analysis of Structural I-beams Using Various Types of Transparent Wood 194
V Karla, V Bajzecerová, J Kanócz

Methods for identifying non-value-adding activities in construction processes.......... 201
D Klosova, M Kozlovská

Experimental verification referring to the new shape of the perforated steel strip.......... 209
A Petrík

Measurement of the runoff coefficient of extensive greenroof 215
J Vystrčil, O Nespešný, K Šuhajda, D Bečkovský, P Selník

Silicate conductive composites with graphite-based fillers 221
Š Baránek, V Černý, G Yakovlev, R Drochytka

Impact of dampness to changes in the mechanical properties of solid fired bricks 228
V Bartoň, J Šnédar, P Cikrle

Study of the properties of chemically resistant repair mortar with the use of secondary raw materials 234
P Figala, R Drochytka, R Hermann

Study of polymer-based adhesive mortar with higher durability 241
R Hermann, P Figala, L Mészárosová, R Drochytka

Overview of the presence of predominant microorganisms on the interface of cementitious materials and soil from selected locations 248
M Hološová, A Eštoková, A Luptáková

The use of construction and rubber recycled material for the production of cement composites with thermal insulation function 255
J Charvát, J Svoboda, K Máčalová, L Chlebková, R Papesch

Analysis of the behaviour of 3D samples printed by FFF/FDM technologies under bending stress with a focus on infill 263
D Juračka, M Kawulok

Mechanical fracture parameters of concrete drill-core specimens supported by a slenderness ratio study 270
D Lisztwan, I Kumpova, P Danek, P Frantík, Z Kersner

Self-sensing performance of metakaolin geopolymer with carbon nanotubes subjected to repeated compressive loading 285
C Mizerová, P Rovnaník, I Kusák, P Schmid

Characterization of materials used for earth architecture in Eastern Croatia 291
A Perić

Preparation of triclinic alite with short soakings in a small experimental high-temperature furnace 298
S Ravaszová

Influence of fly ash addition on physical-mechanical properties of clay bricks – literature review 304
I Vrdoljak

Methods of calculation of cold-formed structures 312
E Hodovanets, M Al Ali, V Kvocak

Numerical analysis of asymmetrical using of actuators to membrane structure 318
L Kapolka, L Stulerova

Crack behaviour at the interface of a surface layer applied on a steel substrate by laser cladding 324
L Malíková, P Doubek, P Miarka, S Seitl

Evaluation of loading methods for determination of material and elastic constants of cement fiber board 330
O Nespesny, J Pencik, J Vystrcil, D Beckovsky

Computational modelling of mechanical response of glulam beams due to moisture inducted delamination 335
J Pełczyński, P A Król

Modal analysis of a membrane structure with actuators 339
L Stulerova, L Kapolka, S Kmet

Evaluation of dynamic behaviour of an experimentally tested model on a shaking table 344
D Taušan, A Brandis, I Kraus

Comparison of the level of required prestressing of road bridges according to the historical development of technical standards in Slovakia 356
J Baran, V Borzovič

Prepared experimental study for durability verification of FRP reinforced concrete members 363
S Blaho, K Gajdošová

Strengthening flat slabs without shear reinforcement against punching shear by concrete overlay 368
D Čereš, K Gajdošová

Theoretical research of dimensional optimization and choice of material strength in steel-concrete composite beams 373
K Hrabovska, J Brecka

Effect of the number and arrangement of inclusions on the strength of concrete specimens 380
M Kovac, S Kusnir

Influence of time intervals of microwave drying of masonry on the final result of drying 386
V Kvapilová

Application of critical shear crack theory on punching of flat slabs 390
S Sarvaicova, V Borzovic

Comparison of the effect of GFRP and steel reinforcement in concrete members subjected to compressive stress 396
Ž Šenšelová, V Borzovič

The influence of signal noise to measurements at bridge structure 402
R Štecák

Construction types and typologies of existing masonry arch bridges in Croatia 411
D Trajber, D Penava, F Rinaudo, V Sarhosis, L Abrahamczyk

Push-out tests of shear dowels 421
P Vanova, D Dubecky, V Kvocak

Material characteristics of push-out tests 425
P Vanova, P Orolin, D Dubecky

Determination of temperature gradient of slender structure 431
M Velešík, M Požár, R Nečas

Experimental analysis of shear connection effect composite beams with concrete 441
M Weissova, D Dubecky, V Kvocak

Sustainable concept of energy management of buildings as an effective tool for green building 450
J Domanický, Z Vranayová, F Vranay

The trend of using solar energy of a green intelligent building and thermal energy storage to reduce the energy intensity of the building 460
M Gorás, Z Vranayová, F Vranay

Sustainability of green walls 466
I Halaszova, M Kozlovska

A method of shaping an energy-efficient building envelope based on natural patterns in order to achieve save of energy 473
I Wawrek, R Staňa

Assessment of the effect of shapes of a shaft intake structure using the PIV method 479
L Bytčanková

Managed Aquifer Recharge (MAR) in areas used for agricultural purposes: the Rye Island pilot study 485
M Červeňanská, J Mydla, A Šoltész, Z Danáčová, E Kullman

Drought assessment and prediction for Gidra river, Slovakia 495
L Čubanová, W Almikaeel

Analysis of anomalies in the control probes of the Rozgrund dam 506
F Koval

Designing the sewerage network of the city in purpose to reduce impact to the environment and water bodies 517
M Portnov, A Popov, J Hrudka, S Stanko

Restoration of a wetland Čéne in eastern Slovakia 522
A Repel

Effect of percent air voids content on the deformation properties of the asphalt mixture 527
J Bokomlasko, J Mandula

Analysis of available shared transport solutions in Krakow 533
A Grabarz, W Sawicki

Introduction of cold-in-place recycling technology in practice as an innovative tool 540
R Kovács, J Mandula

Evaluation of the mechanical properties of fluvial coarsegrained soils from dynamic penetration test 549
E Kučová, M Kuvik

Author Index

IOP Conf. Series: Materials Science and Engineering 1209 (2021) 011001 doi:10.1088/1757-899X/1209/1/011001

PREFACE

The 13th International Scientific Conference of Civil and Environmental Engineering for the PhD. students and young scientists below 35 years - Young Scientists 2021 (YS21) under auspices of

- assoc. prof. Peter Mésároš, the dean of the Civil Engineering Faculty of the Technical University of Košice,
- Ing. Andrej Doležal, the Minister of Transport and Construction of the Slovak Republic,
- Mgr. Branislav Gröhling, the Minister of Education, Science, Research and Sport of the Slovak Republic, and
- Ján Budaj, the Minister of Environment of the Slovak Republic,

was held in Štrbské Pleso, the High Tatras, Slovakia on 13th - 15th October 2021. The traditional meeting of PhD. students, their supervisors and young scientists working in the field of civil and environmental engineering was organised by the Faculty of Civil Engineering, the Technical University of Košice, Slovakia.

The Young Scientist 2021 offered an international platform for the dissemination of the original research results. It provides a pleasant environment to present a new, great ideas of young researchers and discovered advances in the field of civil and environmental engineering and related interdisciplinary topics. The conference provided an opportunity for PhD. students and young researchers to discussed scientific topics and research results, to share their knowledge and experiences, to collaborate in innovative designs and proposals.

The Conference assembled more than 99 delegates including PhD. students and other representatives from the universities from

- Technical University of Košice, Slovakia
- Slovak University of Technology in Bratislava, Slovakia
- Czech Technical University in Prague, Czech Republic
- Brno University of Technology, Czech Republic
- VSB-Technical University of Ostrava, Czech Republic
- Warsaw University of Technology, Poland
- Cracow University of Technology, Poland
- Chalmers University of Technology, Sweden
- University of Maribor, Slovenia
- Josip Juraj Strossmayer University of Osijek, Croatia
- "Gheorghe Asachi" Technical University of Iasi, Romania.

The scientific contributions involved a wide range of research fields:

- Building Physics and Services
- Sustainable Architecture and Energy Efficiency
- Statics, Dynamics and Modeling
- Structural Engineering and Bridges
- Transport and Geotechnical Engineering

- Construction Technology, Organization and Management
- Construction Economics, Marketing and Management
- Building Information Modeling
- Innovations in Construction Design and Technology
- Environmental Engineering
- Healthy Indoor Environment
- Material Engineering and Recycling
- Sustainable Water Management.

Eighty-seven contributions have been selected to be published in the volume of IOP Conference Series: Materials Science and Engineering. The organizing committee appreciates the enormous work of the both the editorial board and reviewers in the preparation of this volume. Many thanks also to all authors for their flexibility and meeting deadlines.

Issue editors
Peter Mésároš
Vinayakam Jothiprakash
Kamila Kotrasová
Martina Zeleňáková

EDITORS

Peter MÉSÁROŠ, Technical University of Košice, Slovakia
Vinayakam JOTHIPRAKASH, Indian Institute of Technology in Mumbai, India
Kamila KOTRASOVÁ, Technical University of Košice, Slovakia
Martina ZELEŇÁKOVÁ, Technical University of Košice, Slovakia

COMMITTEES

SCIENTIFIC COMMITTEE

Miroslav BAJER, Brno University of Technology, Czech Republic
Radim ČAJKA, Technical University of Ostrava, Czech Republic
Marián DRUSA, University of Žilina, Slovakia
Vinayakam JOTHIPRAKASH, Indian Institute of Technology in Mumbai, India
Dušan KATUNSKÝ, Technical University of Košice, Slovakia
Stanislav KMEŤ, Technical University of Košice, Slovakia
Mária KOZLOVSKÁ, Technical University of Košice, Slovakia
Katarína KRAJNÍKOVÁ, Technical University of Košice, Slovakia
Lech LICHOŁAI, Rzesow University of Technology, Poland
Jiří MÁCA, Czech Technical University in Prague, Czech Republic
Maciej MAJOR, Czestochowa University of Technology, Poland
Peter MÉSÁROŠ, Technical University of Košice, Slovakia
Abdelazim NEGM, Zagazig University, Egypt
Maria Manuela PORTELA, University of Lisbon, Portugal
Andrzej SZARATA, Cracow University of Technology, Poland
Michal TOMKO, Technical University of Košice, Slovakia
Stanislav UNČÍK, Slovak University of Technology, Slovakia
Vojtěch VÁCLAVÍK, Technical University of Ostrava, Czech Republic
Damir VAREVAC, Josip Juraj Strossmayer University of Osjek, Croatia
Silvia VILČEKOVÁ, Technical University of Košice, Slovakia
Zuzana VRANAYOVÁ, Technical University of Košice, Slovakia
Martina ZELEŇÁKOVÁ, Technical University of Košice, Slovakia

ORGANISING GUARANTEE

Kamila KOTRASOVÁ, Technical University of Košice, Slovakia

ORGANISING COMMITTEE

Jana BUDAJOVÁ, Technical University of Košice, Slovakia
Viktor KARĽA, Technical University of Košice, Slovakia
Daria KLOSOVA, Technical University of Košice, Slovakia
Martin LAVKO, Technical University of Košice, Slovakia
Lenka ŠTULEROVÁ, Technical University of Košice, Slovakia
Patrícia VAŇOVÁ, Technical University of Košice, Slovakia

REVIEW STATEMENT

All papers published in this volume of IOP Conference Series: Materials Science and Engineering have been peer reviewed per two reviewers through processes administered by the Editors. Reviews were conducted by expert referees to the professional and scientific standards expected of a proceedings journal published by IOP Publishing.

Peer review declaration

- All papers published in this volume of *IOP Conference Series: Materials Science and Engineering* have been peer reviewed through processes administered by the Editors. Reviews were conducted by expert referees to the professional and scientific standards expected of a proceedings journal published by IOP Publishing.

  Type of peer review:
  The review process was of an open peer review type. Each article was reviewed by two reviewers. First review was provided by the author, who addressed the reviewer himself and made the revisions in the article before submitting both article and the review form. The first reviewer was asked to be someone close to their research, who could provide helpful advice. Second review was provided mostly by employees of Technical University of Košice. The reviewers reviewed articles close to their research area. They were provided with full articles including the authors names and the authors then received the review form which included the reviewer's name. The authors had then an opportunity to revise the papers and the papers were accepted only after the authors made the changes asked for by the second reviewer in their articles.
  In the review form the reviewers were asked to judge the quality of the paper, choosing from options: excellent, good, average, or poor; provide some comments to support their argument and give the authors some notes to help them improve the quality of their paper. For the conclusion of the review form the reviewers had to choose the status of acceptance of the paper, choosing from following options: accepted without revisions, accepted with minor revisions, accepted with major revisions, or rejected. If the paper was accepted without revisions, the authors were then not required to provide any revisions. If the second reviewer decided to choose the reject option, the authors did not have an option to revise their paper and the paper was rejected even if the first reviewer accepted the paper.

- **Conference submission management system:** CaptainForm

- **Number of submissions received:** 88

- **Number of submissions sent for review:** 88

- **Number of submissions accepted:** 87

- **Acceptance Rate (Number of Submissions Accepted / Number of Submissions Received X 100):** 98.86

- **Average number of reviews per paper:** 2

- **Total number of reviewers involved:** 99

- **Any additional info on review process:**

IOP Conf. Series: Materials Science and Engineering 1209 (2021) 011002 doi:10.1088/1757-899X/1209/1/011002

- **Contact person for queries:** Kamila Kotrasová, kamila.kotrasova@tuke.sk

IOP Conf. Series: Materials Science and Engineering 1209 (2021) 012001 doi:10.1088/1757-899X/1209/1/012001

A data structures for purpose of the BIM-based Life Cycle Assessment: A review and theoretical background

M Brandtner[1] and V Venkrbec[1]

[1] Faculty of Civil Engineering, Brno University of Technology, Veveří 331/95, 602 00 Brno, Czech Republic

E-mail: 159111@vutbr.cz

Abstract. The article deals with the data structure for the purpose of Life Cycle Assessment (LCA) of buildings using the Building Information Model (BIM). Construction industry produces a significant amount of waste and on the other hand the capacities of landfills are almost filled. It is necessary to deal with the effective use of materials that have already been used and have potential to be reused again. LCA is a method that can be used to demonstrate the suitability of proposed materials, structures or buildings in terms of their whole life cycle and its environmental impact. BIM includes, in addition to geometry, the information part. This data can be used for life cycle inventory (LCI) and then for the assessment itself. The aim of the article is to analyse previous approaches and define which data structure is necessary to be obtained from the BIM model for the LCI purpose of a specific material. The proposed methodology of the data recognition and selection is based on data structure of non-graphical database called SNIM, which was developed for the Czech construction environment. The article is also focused on the theoretical background of the newly developed classification system Construction Classification International (CCI).

1. Introduction

The Life Cycle Assessment (LCA) of construction products represents a current issue related to the requirement for sustainable construction production, resulting from Directive No. 305/2011 of the European Parliament and the European Commission [1]. This new requirement for both construction products and the entire construction structure consists in the sustainable use of natural resources in order to guarantee that the buildings are environmentally friendly. Sustainable use of natural resources is related to the possibility of reusing or recycling already used construction materials, however, also to recycle entire constructions or to use environmentally friendly materials and raw materials already during the construction process. The LCA method [2] represents one possibility how to assess this new requirement.

LCA is an analytical method based on the assessment of environmental impacts of products, services or technologies. Kočí [2] describes the approach of the LCA method to the assessment of environmental impacts with regard to the entire life cycle of products, i.e. from the stage of acquisition and manufacturing the primary raw materials, through the stage of manufacturing products, their use until the actual disposal or reuse of products via recycling.

In the past, the method used for obtaining data for the life cycle assessment of various products was described. The authors described the data entering the LCA. This article focuses on a summary of past, present and also expected forthcoming methods of obtaining data for the LCA assessment. Furthermore, the research described in the article aims to determine how the information from the SNIM database is sorted out for its application into the LCA study.

2. Current state of the researched issue

In 1990, for the first time, a comprehensive environmental analysis of loads and models of full-fledged life cycle assessment and life cycle calculation appeared. At the beginning of the 21st century, the Life Cycle Initiative was created with the aim to introduce life cycle ideas into real life [3].

Many authors have already dealt with the interconnection of the LCA method and the possibility of using the data structure of the BIM model. Antón and Díaz [4] presented two approaches to integration between the LCA and the BIM at the end of her article. One of them is the assessment of the entire life cycle of the construction and the other approach is material oriented. In the article [5], the authors evaluated the latest studies that focus on the BIM-based LCA. At the same time, they focus on how BIM could contribute to simplifying data entry, optimizing both the data output and the results during the LCA application at specific structures.

Finnveden et al. [6] described the development of the LCA. This article provides an overview of recent LCA developments. The authors discuss areas where progress has been made. These areas may include: the difference between attributive and the subsequent LCA, the hybrid development of the LCA database and the categories of impacts (such as spatial differentiation, toxicity, indoor air pollution and impacts resulting from land, resource and water use). The authors highlight the greatest successes. Nowadays we should be able to better understand the difference between the attribute and the subsequent LCA, as well as methods for hybrid Input–Output Analysis (IOA). However, the authors emphasize the fact that there are many areas to focus on. These areas include: the development of LCA tools and methods for assessing the impacts on ecosystems that result from land use and water use. Databases should be the main area of development.

Another article on the LCA [7] summarizes all life cycle assessment research that was applied in the construction industry. The authors state that more than ever before, the construction industry is concerned with improving social, economic and natural indicators of sustainability. If the LCA method was used more, it would be possible to achieve better optimization of these aspects and thus better control of the building life cycle (from the extraction of natural resources to the final liquidation phase and disposal of building materials). Thus, it can be concluded that the authors try to analyse the previous research of the LCA and compare it with the current research of the LCA and thus highlight the main challenges for the development of the LCA and the construction industry. The study therefore provides a basic description of the LCA and its methodology. It also deals with the ways and differences the LCA uses Building Materials and Component Combinations (BMCC) compared to the use of the LCA in the LCA of the Whole Process of the Construction (WPC). The study confirms that the LCA represents a really innovative method that improves the sustainability of the life cycle of a building through all its phases.

In his study, Marzouk [8] describes connecting the LCA with the BIM. This study presents the BIM approach, which allows for estimates of 6 types of emissions. These include the greenhouse gases, sulphur dioxide, particular matter, eutrophication particles, ozone depleting particles and smog. Thanks to this tool, it is possible to calculate the production of direct and indirect emissions during the entire construction process, i.e. from the production phase through the transport, construction, operation, and maintenance phases to the demolition phase. The authors used Athena Impact Estimator and Autodesk Revit applications for the calculations.

The authors of the article [9] came up with a method of automatic LCA calculations at an early stage, i.e. already at the first level of the BIM detail – LOD100. This should lead to easier adjustments of the calculations during the further development of the BIM model. To achieve this goal, new layers of data and their formats are needed. This new data layer should fill the information gaps between the extracted BIM data and the already existing LCA data provided by the common LCA databases. The result of the article was therefore an approximation of this method. The next step is to demonstrate the benefits of this method through further scientific research.

The article [10] concerned the integration of environmental impacts within the BIM. The secondary goal was to find solutions on how designers could mitigate these environmental impacts. The authors designed a BIM-integrated assessment as well as a visualization of the environmental impacts. The results show that it would be possible to use this approach to determine what environmental impacts the building material shall have in the initial stages of designing. Visualization of a geometric model

IOP Conf. Series: Materials Science and Engineering 1209 (2021) 012001 doi:10.1088/1757-899X/1209/1/012001

of a building can be used as a representation of a design that can lead to a better decision-making process on material choices for individual building elements.

The efficiency of the LCA was addressed by the article [11]. The authors of this article dealt with the connection between a tool based on the BIM and the LCA tool Gabi 6. The research in the article demonstrated that the BIM plug-in can provide necessary data for each design phase. This means that the designer can perform a full life cycle analysis more efficiently.

Based on the above-stated references, it can be concluded that the authors in the past dealt with the possibility of linking the LCA studies with the information contained in the 3D model in the BIM environment. In their publications, the authors used the division of product life cycle stages using impact categories as well as verified the chosen methodology for calculating the LCA studies using various software tools such as Autodesk Revit, Gabi 6, etc.

3. Theoretical background

Product and its environmental impact. In his publication, Kočí describes the life cycle of products, which have various impacts on the environment, in four main stages. The first stage comprises obtaining raw materials to produce the necessary materials for the manufacturing of the product, which involves the extraction of crude oil, iron ore or wood and their transport from the place where they are obtained to the place of their further processing. The second stage describes the manufacturing of the product from already obtained materials and raw materials in the first stage. The stage involves the transformation of the obtained raw materials into materials that are usable in further industrial production. However, the transformation of raw materials includes, for example, the consumption of electricity or various fuels. The third stage of using a product involves consuming the product or using it for the function for which it was manufactured. An integral part of this stage is energy and raw material requirements, which serve for the operation of the product itself, it's possible repairs, storage or maintenance. The fourth, last stage, describes the final disposal of the product, including energy and material requirements for the removal of the product or its recycling [2].

Life cycle impact assessment using environmental impact categories. The life cycle of the products and evaluation of their impact on the environment using the so-called impact categories is further divided by the author into six specific areas. These include global warming and climate change, stratospheric ozone depletion, acidification, eutrophication, photooxidant formation and depletion of raw materials. Each impact category includes both the processes and consequences that adversely affect the environment. Global warming involves an increase in the average temperature of the atmosphere due to capturing more energy in the atmosphere, which leads to climate change at the global level or to the so-called greenhouse effect. The loss of stratospheric ozone leads to UV radiation reaching the surface of the Earth, which leads to an adverse effect on human health or the quality of the environment. Acidification is the process of acidifying the soil or water environment caused by the release of acid-forming substances into the atmosphere, water and soil. Acidified precipitations condense on solid surfaces or dissolve in water, which leads to soil, water and building materials acidification. Such substances are also harmful to plants and trees. Eutrophication is associated with nutrient enrichment of the environment, which leads to overgrowth of surface water, lack of oxygen in water or deterioration of the drinking water sources quality. Formation of photooxidants involves the adverse effects of ozone or other substances in the ground layer of the atmosphere, which leads to a negative impact on the environment resulting in various diseases and the influence on the health of living organisms. The decrease in the amount of raw materials is associated with the excessive extraction of non-renewable resources by humans, which leads to the disruption of ecosystems and affects the amount of raw materials available in the future [2].

Standardization of Non-Graphical 3D Model information (SNIM) is a set of information and standards that were developed especially for designing using the BIM model in the conditions of the Czech Republic. Parameters (non-graphic information) were assigned to individual elements of the 3D model, which were further divided according to the requirements for individual stages of project documentation in connection with the given construction phase. SNIM contains a list of parameters that are assigned to specific structures and a classification tool of categories of building structures, which divides individual types of structures according to the technical and parametric information.

The classification tool of building structures was created in order to strive for better orientation in the project using a database search.

Construction Classification International (CCI) is an international classification system, which represents a promising solution to the categorization and description of buildings [12]. This classification system is particularly suitable for digital processing. The construction industry involves an inexhaustible amount of information exchange between a large number of entities, each with different information requirements at different stages of the construction life cycle. The success of the whole project depends on the possibility of passing on complete and relevant information at the right time. The Czech construction industry does not have a respected uniform classification system, nor a common "language" between the individual participants of the construction process. The CCI classification system offers a general common "language" to suit all phases and roles of the participants throughout the entire construction life cycle. CCI offers solutions to the classification of both the entire constructions and more detailed parts of the information model of constructions. In order to adopt a new classification, it is necessary to unambiguously identify information for machine data processing with the highest possible degree of automation using a facet classification system. The solution consists of the classification in the form of partial facets for built spaces, technical systems, construction systems and individual detailed construction elements. This information forms an interconnected system enabling clarity and unambiguous classification of information about the construction for subsequent use by various software tools.

The subject of information construction modelling is the exchange of all kinds of information during the project between its participants and the applications used. For the successful exchange of data during the project, it is necessary to require a comprehensive and consistent method for the classification of buildings in the individual projects. Types of information exchanged include geometric data, functionality data, technical data and cost and maintenance data with a focus on the project timeline running from the project plan to the actual disposal of the construction. The basic idea of sorting information in the CCI is specified by the user, who classifies the individual elements according to their purpose of use or according to their way of use. Project information can also be classified according to the CCI classification system, which shall lead to easier determination of the parameters and specific specifications necessary for the calculation of LCA studies. The advantage of information classified in such way can be its easy identification for machine data processing and its easy use for the creation of various types of automated software. Thanks to the newly introduced common "language" for all participants of the entire life cycle of the construction, it shall be possible to use the results of the calculated LCA studies in all phases of the project, including the phase of use of the newly built object.

4. Discussion and proposal of the research direction

A database of standards and parameters serving for civil engineering activities in connection with the use of information from the BIM environment, especially for the environment of the Czech Republic, was created in the BIM environment. The above-mentioned parameters, i.e. non-graphical information, were classified by the authors according to the newly created classification tool of building structures and assigned to individual elements of the 3D model in the BIM environment. The classification tool also includes the recommended format and form of entering individual information. This is the so-called standardization of non-graphical information in the model [13].

It is possible to use categorized information from this environment to simplify the calculations of the LCA studies in connection with the new introduction of the BIM environment for the construction industry in the Czech Republic. This new method uses knowledge of the life cycle stages of products in relation to possible environmental impacts. Furthermore, the division of impact categories into six specific areas was considered. Data sorted according to the parameters assigned to individual elements of the 3D model in the BIM environment of the SNIM database are considered to be further selected for selecting for the LCA calculations and handled according to precisely defined criteria.

To assess the life cycle of products with the possibility of evaluating their impact on the environment using the LCA method, it is necessary to determine the values of individual indicators of impact categories with regard to the individual stages of the product life cycle. These values, which form a set of results, must have, in addition to specific values, also clearly defined units. The individual sets of results are then assigned to specific types of impact categories that have different environmental impacts. The parameters that can be used for LCA evaluation are clearly defined in the graphic and information part of the BIM model of the construction work. In connection with the standardization of non-graphical information of the model and the emergence of the above-mentioned standards and parameters, it is possible to supplement the BIM model with specific information from SNIM for a comprehensive assessment of the construction work in terms of its entire life cycle.

Methodology of data selection from the SNIM environment for the LCI purposes. When selecting data from the SNIM environment, data at the level of study detail are used. The reason is a sufficient amount of information in this phase of construction for the basic purposes of the LCI and also an effort to deal with the real situation as much as possible. In the construction practice, only some information is available in the study phase, however, even in this construction phase, there is an increasing effort to pre-calculate the LCI studies based on the LCA in order to reduce the environmental impact of the planned construction.

The considered methodology is based on the presumption that the volumes of individual constructions as well as data that contain structural markings are taken over from the SNIM database. Together, this is data that contains the name of the element, its volume, marking according to the SNIM, specification or other specific description. This data is used for the purposes of the basic structure of the LCI (LCA study). An example of the data structure scheme for one material category is given in Table 1.

Table 1. Example of data structure scheme for one material category.

	Marking according to SNIM	Sub-category	Description	MIN	MAX
Type of construction	XX000	Construction specification	....	000	000

Unfortunately, the data taken from the SNIM database does not contain indicators of impact categories (emissions). Therefore, this data must be assigned to the calculation from external sources, such as existing databases or specific data obtained directly from the manufacturers of the material.

It is possible to assign a specific LCI value to each specific material based on the specification. The LCI value assigned to a specific material is always only for one basic unit of the amount of the given material (i.e. m^3, m^2, m, t, etc.). Therefore, the LCI value is then multiplied by the volume value of the element. A specific LCI value of all materials of each element in a specific structure is generated using the MATLAB matrix solver based on this information. The final result is the percentage distribution of individual emission factors across the overall material characteristics in the construction.

5. Conclusion

This article presents scientific research and an elementary concept of the methodology of using data from the structure information model for the purposes of calculating LCA studies, with the emphasis on the appropriately selected classification system used in the construction sector. Only one of the two possible variants for the environment of the Czech Republic are presented, using the SNIM standard; on the contrary, the use of the CCI system is not possible to precisely define at the time of this article's publication due to the ongoing development of this system. It is essential to point out that the presented methodology, whether it works on the basis of SNIM or CCI, must be verified by a sufficient number of case studies of specific building structures. This verification will be the subject of further research into this issue. The aim of this article was to summarize the theory, the current state of the issue research and also to outline the framework procedure by which the calculation of environmental impacts can be approached in the future. In the next phase of the research, it will be possible to choose a specific case study, the results of which will be compared with the already presented results of other authors, using various types of software and parameters. In the future, it would also be appropriate to create specialized software dealing with the LCA and the LCI calculations or the development of a plug-in for the Czech environment, which will be easily usable in construction practice.

6. References

[1] European Council 2011 *Regulation (EU) No 305/2011 of the European Parliament and of the Council: laying down harmonized conditions for the marketing of construction products and repealing Council Directive 89/106/EEC* (Strasbourg: European Council)

[2] Kočí V 2012 *LCA a EPD stavebních výrobků: posuzování životního cyklu a environmentální prohlášení o produktu jako cesta k udržitelnému stavebnictví* (Prague: Czech Council for Green Buildings)

[3] Guinée J B et al. 2011 Life Cycle Assessment: Past, Present, and Future *Environ. Sci. Technol* **45** 90–96

[4] Antón L A and Díaz J 2014 Integration of Life Cycle Assessment in a BIM Environment *Procedia Engineering* **85** 26–32

[5] Soust-Verdaguer B, Llatas C and García-Martínez 2017 A Critical review of bim-based LCA method to buildings *Energy and Buildings* **136** 110–120

[6] Finnveden G, Hauschild M Z, Ekvall T, et al. 2009 Recent developments in Life Cycle Assessment *Journal of Environmental Management* **91** 1 1–21

[7] Ortiz O, Castells F and Sonnemann G 2009 Sustainability in the construction industry: A review of recent developments based on LCA *Construction and Building Materials* **23** 1 28–39

[8] Marzouk M, Abdelkader E M and Al-Gahtani K 2017 Building information modeling-based model for calculating direct and indirect emissions in construction projects *Journal of Cleaner Production* **152** 351–363

[9] Dupuis M, April A, Lesage P and Forgues D 2017 Method to Enable LCA Analysis through Each Level of Development of a BIM Model *Procedia Engineering* **196** 857–863

[10] Röck M, Hollberg A, Habert G and Passer A 2018 LCA and BIM: Visualization of environmental potentials in building construction at early design stages *Building and Environment* **140** 153–161

[11] Bueno C and Fabricio M M 2018 Comparative analysis between a complete LCA study and results from a BIM-LCA plug-in *Automation in Construction* **90** 188–200

[12] BIM Koncepce *Klasifikační systém CCI* [Accessed 2020-09-25] Available online https://www.koncepcebim.cz/634-klasifikacni-systemcci?fbclid=IwAR0SvP407GaKz7P cFAbKceGt43pdjfM3M -aAZ16_IR8A3HOoD8kdFNagdI

[13] BIM cz *Standardizace negrafických informací 3D modelu* [Accessed 2020-06-05] Available online http://www.czbim.org/nodes/nodes/view/type:stranka/slug:standardizace-negrafickych informaci-3d-modelu

Acknowledgments
The article was published with support from a Internal research grant of the Brno University of Technology with the registration number FAST-J-21-7288.

The level of detail for 4D BIM modeling

Y Nechyporchuk[1] and R Baskova[1]

[1] Department of Technology, Economics and Management in Civil Engineering, Faculty of Civil Engineering, Technical university of Kosice, Vysokoskolska 4, Kosice 042 00, Slovakia

E-mail: yaroslav.nechyporchuk@tuke.sk

Abstract. 4D modeling has been actively developing over the past decade along with the progress of BIM implementation. 4D model can provide enhanced early decisions about the space-temporal criticality of work elements. This models is a collection of graphical and scheduling information about an object. These inputs can have different levels of detail (LOD). In creating and using BIM projects, the LOD of datasets is an important aspect. However, to date there is limited research thoroughly investigating the issue of LOD within 4D models. The article provides an overview of studies related to the level of detail for 4D models, and also describes the impact of LOD on the final 4D model.

1. Introduction

Construction is one of the oldest and still demanded branches of industrial human activity. The world is developing very rapidly today. Construction cannot stand still and it also has to constantly progress and introduce new technologies and approaches. One of the technologies that helps construction keep up with the fast-paced world is Building Information Modeling (BIM). The concept of BIM technology is not a novelty, it was known about it at the end of the last century. The BIM concept was proposed by Charles Eastman back in 1974. In his work, he described the Building Description System (BDS) which then developed into BIM [1]. Since the early 2000s, the concept of BIM has become widespread in the architecture, design and construction (AEC) industry when parametric 3D modeling was developed [2]. Further, the technology was actively developed and today is an integral part of modern construction. Building information modeling (BIM) is defined as the process of creating, storing, managing, exchanging and sharing information about objects [3]. BIM can also be described as a virtual process that encompasses all aspects, disciplines and systems of an object within a single virtual model, allowing all members of the design team (owners, architects, engineers, contractors, subcontractors and suppliers) to collaborate more accurately and efficiently than using traditional processes [4].

2. Dimensions of BIM

BIM technology assumes the possibility of accumulating a large amount of different information. Depending on the nature of the information that contains BIM models, there are several of their dimensions.

The basis of BIM projects are 3D models of objects. A 3D model is a set of graphical data about an object presented in a three-dimensional coordinate system (x, y, z). Unlike 3D CAD models, BIM uses parametric elements. The 3D dimension is considered one of the most important, as it allows a clear idea and visualization of the result of the work performed. In addition, many project participants work with information in 3D: architects, designers, structural engineers and more. In addition to project visualization, which helps you better understand and present your project and avoid possible graphic collisions, there are many other benefits. 3D BIM helps participants more effectively manage their multidisciplinary collaboration in modeling and analyzing complex structural problems. Furthermore, it is possible to collect accurate data throughout the project life cycle and store it in the building information model - new value scan be added to predictive models that allow proactive problem solving [2].

The 4D dimension refers to the time aspect of the project. The timing associated with the BIM project processes helps determine how long it will take to complete the project and how the project will evolve over time. The 4D visualization provides an illustration of how the structure will evolve during construction. In addition to planning the phases and critical points / milestones of construction, logistics of supply and installation of construction products and components, risks or subsequent use of the building, this also includes the simulation of the BIM model in terms of structural durability. The use of 4D-BIM technology can lead to improved control over conflict detection or the complexity of changes that occur during a construction project. 4D BIM provides approaches to managing and anticipating site status information, the impact of changes, as well as informing stakeholders or warning of potential threats. The integration of BIM with 4D CAD simulation models brings participants benefits in terms of optimizing team planning, construction and coordination [14].

Model 5D is created by adding more information about the construction project - costs. This aspect of the BIM project analyzes the prices of components and services, helps to create accurate calculations and reports, helps plan and control costs, analysis of the price impacts of project variants and the financial impact of project changes. Incorporating the budget into the 4D model allows you to monitor financial flows during construction and dynamically evaluate various analyzes. Visualization of the progress of activities of project participants allows to estimate the total costs associated with them, which leads to higher accuracy and feasibility of the project. In addition, 5D BIM allows architects to show clients the relationship between alternative design choices and costs [2].

6D BIM (also known as Integrated BIM) contributes to the sustainability and energy efficiency of buildings. This process estimates energy consumption, which makes it possible to analyze the energy use of a building / structure already at the design stage. Unlike the traditional approach, which focuses on start-up costs, 6D BIM provides detailed project information that can improve the management of facilities and the operation of the building and its technical equipment in terms of their energy performance. This dimension essentially contains information about the manufacturer of the building component, the date of installation, the maintenance schedule, configuration information on the best performance, energy requirements and decommissioning information [2].

7D BIM is primarily about the operation and management of facilities by building managers and owners. The parameters of this dimension are intended to monitor important property data, such as its condition, service / operating manuals, warranty information, technical data, etc., that are used in the use phase of the building / construction. 7D BIM is a unique approach in which everything related to the asset management process is gathered in one place in the building information model. This tactic helps to improve the quality of service provision throughout the project life cycle. The use of 7D BIM ensures that everything in the project remains in the best condition from the first day of use until the day the structure is demolished.

3. LOD for BIM

Level of Detail (LOD) was the original name of a concept that originated in the USA at VicoSoftware. The AIA (American Institute of Architects) took it over and renamed it the Level of Development (LOD), which better describes this concept and contains both graphical and informative detail of the model (levels: LOD 100, LOD 200, LOD 300, LOD 350, LOD 4OO, LOD 500). The level of detail is most often interpreted only as a graphical detail of the model. In contrast, in England the NBS (National Building Specification) set up its own system and called it the Level of Definition (LOD), which is very similar but more complex (levels: LOD 1 to LOD 7) [7].

The LOD shows with what accuracy and how detailed the data provided in the project are elaborated. It represents the overall state of the information model in a specific phase of the design process, which includes not only graphic objects, but also data associated with objects.

Table 1. LOD definition[5][6].

LOD	Description
LOD100	LOD 100 - this is the beginning of the project, and it can not be a part of many projects. The elements of the model are graphically represented with a symbol or other way of generic representation. This is information attached to other model elements or symbols to indicate the presence of a component, but not its shape, size, or exact location. Data exchange between project participants takes place on paper or electronic media.
LOD200	Model elements are graphically represented as an overall system, object, or assembly with approximate quantities, sizes, shapes, locations, and orientations. Information obtained from LOD 200 items is approximate. At this point, the conceptual mass model and its associated data will be transformed into a model from which construction documents can be created. This is the stage at which preliminary coordination at a high level should take place.
LOD300	The stage at which the project begins to closely resemble what will be installed. LOD 300 can already be compared with construction documentation. Model elements are graphically represented in the model as a specific system or object. The model contains specific elements, their number, size, location and relationships. Non-graphic information can also be attached to the model element.
LOD 400	In the LOD 400 phase, the model contains a specific design of structures and the internal environment. Each element contains technical data from manufacturers and suppliers. The equivalent of documentation is workshop documentation. The key focus of the LOD 400 is cost planning and scheduling. At levels from 100 to 300, different professions complete the stages at different times. In LOD 400, this will change and instead all professions will be connected to the physical space of the BIM model.
LOD 500	The LOD 500 level model contains specific implemented constructions according to the actual state. The equivalent of the documentation is the documentation of the actual construction. The model at the LOD 7 level is used to optimize the costs associated with the operation and maintenance of the object. The graphic data no longer changes and the FM (facility management) model is used for viewing, additional data or individual documents (revisions, photographs, technical documents,) are assigned to it.

The American Institute of Architects has created the LOD specification BIMForum, which is a detailed interpretation of the LOD schema. It is a reference tool that explains the content of the model at different stages of the design and construction process, and also aims to ensure a high level of communication between BIM users. BIMForum also provides a LOD specification table that describes the requirements for building elements to achieve the required LOD [5].

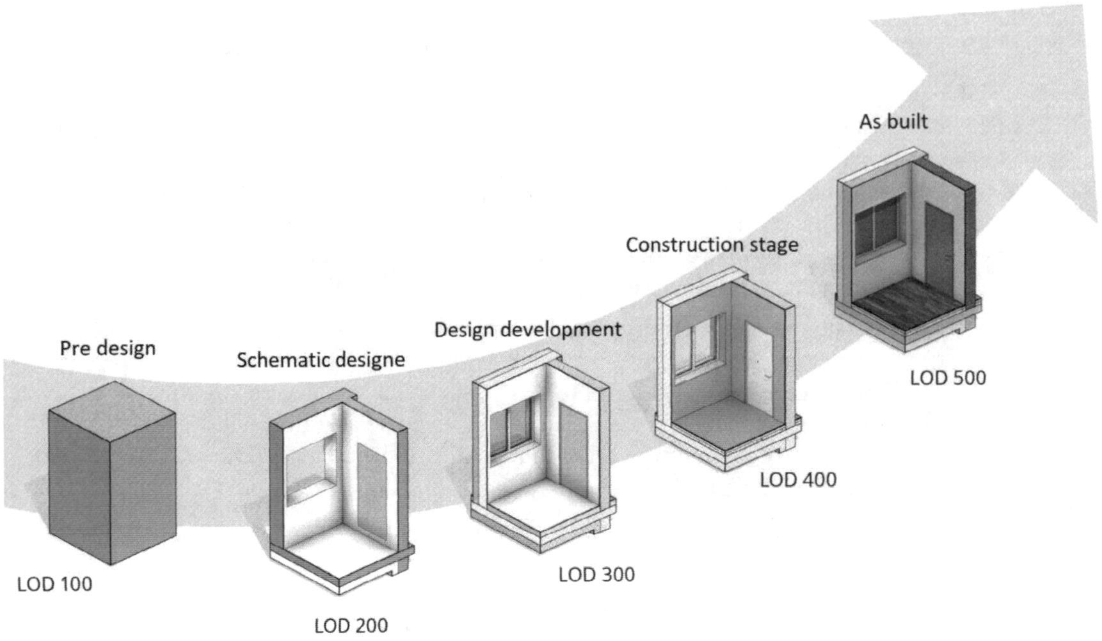

Figure 1. Example of LOD.

4. LOD and 4D BIM

4.1. 4D Model

To create a 4D model, temporal characteristics are added to the 3D model, which allow the construction process to be displayed using animation. The 4D model allows users to simulate the construction process, but also provides additional opportunities for improving and managing construction progress. The BIM concept is created visually as each component task has a start and end time. Construction management level, logistics and planning can be supported with 4D BIM [8]. Timing information for a specific item can include information about delivery time, installation / disassembly time, time required for commissioning, order in which components are to be installed, and dependencies on other areas of the project [11]. The four-dimensional model has been recognized as an approach that improves construction planning and management methods [9]. 4D adoption is increasingly used in construction and construction managers are beginning to realize the potential benefits of its use. Using the 4D model, it is possible to perform an analysis of the entire sequence of project work, as well as the search for possible collisions in design solutions. In addition, it makes it possible to detect space-time collisions that may occur during construction. The use of 4D models helps to analyze and prevent many problems in advance, even before construction begins.

4.2. Level of Detail for 4D modeling

LOD and 3D geometry issues during design and construction have attracted the attention of both practitioners and researchers [12][11]. The LOD protocols for the design model provide guidance for those interacting with the 3D geometric aspect of the final design. There is not enough research to fully understand and define the exact level of graphical and temporal detail required for 4D modeling throughout the construction process. Thus, the level of detail for 4D is still not well defined [12].

In order to facilitate more dynamic 4D, 3D modelers need to approach the geometry model in such a way that it can support 4D actions. 3D models must be organized into work components to match the level of detail on the schedule. However, in many cases, the original design model is often not created with the intention of using it for planning or control purposes. This often results in the contractor creating a new model for the purpose of developing a 4D simulation [13].

In 4D models, the LOD specification must manage graphical levels of detail and time level of information to provide realistic and more accurate 4D simulations. There is still a clear link between BIM Geometry Level of Detail and 4D LOD requirements.

4.2.1. LOD100
LOD100 is a primitive model. The building is roughly located and is missing a large number of dimensions. Objects may be depicted schematically and may not correspond to their size and location. In this regard, the creation of 4D models with LOD100 is often impractical.

4.2.2. LOD200
LOD200 models are more detailed. But still the model will not be accurate, the number of elements, their sizes and relationships will be approximate. At this stage, larger tasks will be solved at the expense of small details (for example, allocation of space for utility networks, rather than for pipe-to-pipe collisions) With LOD200, a 4D model can be created, but at this stage its benefits will be minimal.

4.2.3. LOD300
LOD 300 is similar to construction documents, the stage at which the project begins to closely resemble what will be installed. Upon completion of level 300, the model will contain the exact number, size, location and systematic relationships of all the objects that will eventually be established. The disadvantage is the different level of detail of individual professions (for example, architectural elements at a level higher than MEP). This level of detail is acceptable for its creation and further effective use.

4.2.4. LOD400
LOD 400 is the exact model of the object to be built. The key difference between the LOD 400 is the same level of all disciplines and professions. This level is probably the best for creating a four-dimensional model, as it creates the maximum approximation to real conditions. But this model is quite voluminous and working with it is more time consuming and requires equipment that will allow you to work with it.

4.2.5. LOD 500
The LOD 500 is the post-construction phase of the project. This is the final stage of the first cost of any project. At LOD 500, the model will contain all the building elements. They will be geometrically accurate, but without some of the data needed for implementation. At this stage, usually the creation of a 4D model is not relevant.

5. Conclusion
In the study, the problems of the level of detail of BIM models were considered, and specifically the requirements for 4D modeling. After analyzing the literature, we can conclude that the most suitable levels of detail for the purposes considered are LOD 300 and LOD 400. LOD 400 is the most suitable, but also more difficult to implement. If you need to create a 4D model that may not include some information, you can use LOD 300 to reduce labor costs. LOD 100 and LOD 200 often do not have enough information to create a 4D model that can be used effectively. In addition, you need to consider the level of detail of the hourly plan.

6. References

[1] Eastman C M, Fisher D, Lafue G, Lividini J, Stoker D and Yessios C 1974 An outline of the building description system *Research report no. 50* (Pittsburgh: Carnegie-Mellon University)

[2] Eastman C M, Teicholz P, Sacks R and Liston K 2018 *BIM handbook: A guide to building information modelling for owners, managers, designers, engineers and contractors* (New Jersey: John Wiley & Sons, Inc.)

[3] Azhar S, Hein M, and Sketo B 2008 Building information modeling (BIM): benefits, risks and challenges *Proc. of the 44th ASC (Auburn)*

[4] Azhar S 2011 Building information modeling (BIM): Trends, benefits, risks, and challenges for the AEC industry *Leadership and Manag. in Eng.* **11** 241-52

[5] BIMForum 2017 *Level of Development Specification* **1** p 231

[6] U.S General Services Administration 2019 *Document Guides: Level of Detail* p 289

[7] Špalek M 2020 Co znamená pojem LOD v BIM? [online] https://www.tzb-info.cz/bim-informacni-model-budovy/20352-co-znamena-pojem-lod-v-bim

[8] Zhang J and Hu Z 2011 BIM- and 4D-Based Integrated Solution of Analysis and Management for Conflicts and Structural Safety Problems during Construction: 1. Principles and Methodologies *Automation in Construction* **20** 155-166

[9] Butkovic B and Heesom D 2017 Towards A Framework For Multi-LOD 4D BIM Simulations *Proc. Int. Research Conf. Shaping Tomorrow's Built Environment (Salford)* p 578

[10] Gledson B 2015 Investigating the diffusion of 4D BIM Innovation *Proc. 31st Annual ARCOM Conf. vol.1 (Lincoln)* p 641

[11] Grytting I, Svalestuen F, Lohne J, Sommerseth H, Augdal S and Lædre O 2017 Use of LoD decision plan in BIM-projects *Proc. Engineering* **196** 407-14

[12] Guevremont M and Hammad A 2018 Multi-LOD 4D Simulation in Phased Rehabilitation Projects *Proc. of the 17th ICCCBE (Tampere)* pp 724-731

[13] Aouad G, Lee A, Wu S and Onyenobi T 2012 Computer Aided Design Guide for Architecture *Engineering and Construction* (New York: Spon Press)

[14] Koutamanis A 2020 Dimensionality in BIM: Why BIM cannot have more than four dimensions? *Automation in Construction* **114** 1-5

[15] Beirer M 2020 LoG, LoI, LoD – Was sind die Unterschiede? *Build Informed* [online] https://www.buildinformed.com/log-lod-loi/

Acknowledgments

The paper presents a partial research results of project VEGA 1/0557/18 and project KEGA 038TUKE-4/2020.

Insulation of historic buildings and case study simulation

K Hutkai[1] and D Katunský[1]

[1] Institute of Architectural Engineering, Faculty of Civil Engineering, Technical university of Kosice, Vysokoškolská 4, 042 00 Kosice, Slovakia

E-mail: kristian.hutkai@tuke.sk

Abstract. The article deals with the historic buildings and their insulation. It is generally known that classical thermal insulation of walls from the outside is not permissible due to the cultural value of historic facades. Compared to external insulation solutions, the implementation of internal insulation is more technically complex and faces several risks. In the article is devoted the chapter which deals with the aim of the application of internal insulation in historic buildings. There are mentioned risks associated with the application of internal insulation, which are related to the thermal-humidity behaviour of the structure. One chapter is devoted to faults that may occur after the application of internal insulation. These disorders have a building-physical character and also affect the hygienic criteria of the building. The larger chapter deals with insulating materials suitable for the purposes of interior insulation of historic buildings. The results of a case study from Dublin and Košice are also presented, where the thermal-humidity behaviour of selected thermal insulation materials was investigated. The results of moisture content and surface temperature before and after application of insulating materials are compared.

1. Introduction

Historic buildings built in the first half of the 20th century form a significant part of the European building stock. Many of these buildings have a high architectural or cultural value, but they also account for almost 40% of the total energy consumption of buildings and are responsible for 36% of CO2 emissions from buildings in Europe. For this reason, it is essential to focus on the renovation of historic buildings in order to achieve the EU's climate and energy goals. By using internal insulation of the perimeter walls, it is possible to save potentially 15-20% of the total energy consumption of the building. Insulation improves the thermal insulation capacity of the wall and improves the level of thermal comfort. The most important are the thermal conductivity λ (W/(mK)), the capillary moisture content w_{cap} (kg/m^3) and water absorption coefficient A (kg/(m^{2}s$^{0.5}$)) which are important for water transport analysis and capillary activity and also the diffusion resistance factor μ (-) [1] [2].

Thermal insulation from the outside of the wall is the most effective way to increase the thermal insulation properties of the wall and it is also the safest in terms of moisture behaviour of the structure [3-8]. However, it is impossible to use external insulation in historic buildings due to the high cultural value of historic facades and their elements. The only way to reduce energy consumption is to use thermal insulation from the inside. However, this solution is accompanied by risks such as condensation between the existing structure and the thermal insulation, reduction of drying of the structure inwards, the resulting increased humidity in the structure and thus the risk of mould growth [9-14]. Damage to wooden beams due to wood rot can occur when the wooden beam ends are no longer located in the warm part of the wall [15]. On the market are available various types of insulation that are suitable for the internal insulation of historic buildings [16]. These are capillary active and hydrophilic insulations [17], [18] and commonly used hydrophobic insulations with a vapour barrier. Researchers from the Faculty of Civil Engineering of the Technical University in Košice also dealt with the topic of historic buildings in various publications [19-25].

2. Objectives of internal insulation applications

Increasing the level of energy efficiency is a complex process. It includes various technical solutions, heritage values, internal environment, cost-effectiveness, etc. The application of internal insulation is a complex process because it consists of many variables that can be independent or interconnected. These are the thermal-humidity properties of existing and used building materials, indoor and outdoor climate, energy costs, operating and maintenance costs, human behaviour, technical systems in the building, environmental impact, value of cultural heritage, building location, etc. The application of interior insulation in historic buildings can be achieved on the basis of several objectives: reducing energy consumption or savings, economic objectives, improving the indoor climate, reducing floor space loss and reducing environmental impact [26].

2.1. Reduction of energy consumption

The reduction of energy consumption is related to the value of the heat transfer coefficient U (W/(m^2K)), but the standard requirements do not apply to listed buildings, so it is not assessed in terms of U value. In real conditions, we cannot even effectively reach the target values due to the thick massive walls. However, with the application of internal insulation, we can reduce the energy consumption for heating to a certain extent [1].

2.2. Reducing the impact on the environment

The aim is to reduce the impact on the environment and also to reduce CO2 emissions [1].

2.3. Reduction of energy and other costs

This objective is set to minimize all costs of applying internal insulation in historic buildings, in particular energy, material and installation costs, as well as life-cycle costs [1].

2.4. Improving the indoor environment

This goal is set to improve thermal comfort in interiors. Thermal discomfort is related to the low surface temperature, which causes thermal discomfort [1].

3. Disorders associated with the application of internal insulation

3.1. Moisture damage

The application of internal insulation increases the risk of moisture damage, as the internal insulation reduces the drying potential of the existing wall as well as its temperature. For this reason, it is necessary to examine and remove all signs of excessive moisture in the walls before applying the internal insulation [26].

3.2. Mold growth

By adding internal insulation, the original wall construction will be cooler and the drying potential will be reduced. This in turn can locally increase the relative humidity (RH) and thus the risk of mold growth and condensation. The risk of high humidity levels can be further increased if the internal insulation allows moist indoor air to pass into the wall structure, especially if there is unbalanced ventilation (internal overpressure). If the humidity rises, rot problems can also occur over time [26].

3.3. Frost damage

The addition of internal insulation to the masonry can affect the conditions of frost damage (damp and colder existing wall). The wall becomes colder and the conditions for freezing the water are becoming more and more intense. Next, the freezing will take place deeper into the wall. Consequently, the walls will be more difficult to dry due to internal insulation, especially when using a vapor-tight insulation system, which increases the risk of a critical level of humidity [26].

3.4. Wood rot

If the wall is insulated from the inside, the wooden ends of the beams are no longer located in the warm part of the wall. As a result, the ends of the wooden beam could be in adverse environmental conditions during the cold part of the year. Prolonged exposure of the ends of wooden beams to high humidity could lead to their disintegration. In one case study by Professor Tywoniak, the temperature in the pockets of wooden beams before insulation is 4-6 ° C higher than after insulation [15].

3.5. Thermal bridges

Thermal bridges are a problem in all possible solutions for thermal wall insulation applications. During application, the continuity of the added insulation may be disturbed, at the protrusions from the surface to be insulated, or at the connections with other building elements. If the insulation is done from the inside, common walls, partitions or interior floors are risk places for the formation of thermal bridges. The problem can also occur with windows, roofs or foundation structures [2], [27].

4. Thermal insulation materials suitable for historic buildings

According to [1], in order to choose the right insulation solution, it is important to know the thermal-moisture properties of materials in existing structures and insulation systems. The most important are the thermal conductivity λ (W/(mK)), capillary activity and the water vapor diffusion resistance factor μ (-). Various internal insulation systems and products are available on the market. They differ mainly in terms of material properties, installation method, environmental impact, durability and cost. Based on the thermal-technical properties of the insulation material, internal insulation systems can be divided into two main groups: vapor-tight and vapor-open systems [1].

4.1. Vapour-tight internal insulation system

Prevent the penetration of warm and humid indoor air into the insulation. This can be achieved by using a vapor barrier insulation material or alternatively by attaching a vapor barrier to the vapor open material [1].

4.2. Vapour open internal insulation system

In this type of insulation system, a vapor-open insulation material is generally used, which also allows capillary suction (capillary-active insulation material). The capillary activity of the material in fact allows the transfer of moisture through the insulating material towards the internal environment, if the inner surface of the existing wall becomes wet, due to condensation between the wall and the insulation [1].

Capillary activity is the ability of materials to absorb liquid water due to capillary suction and redistribute it through the material. The pores in capillary active materials allow the absorption of liquid water and its redistribution into the room in the vapor and liquid phase according to the internal capillary pressure gradient [6]. The best known capillary active materials include calcium silicate plates (CaSi) and fibreboard (WFB) [1].

Researchers from the University of Dublin, R. Walker and S. Pavía, studied the thermal-humidity behaviour of seven insulation systems on a historic 840 mm brick wall. When measuring in-situ, they compared insulation materials such as aerogels, cork lime, hemp lime, calcium silicate board, fibreboard and PIR boards with traditional lime plaster. The initial U-value of the historic wall was 1.32 W/(m^{2}K) and the mentioned insulating materials reduced this value in the range of 34 to 61%. As the U-value of historic buildings is not assessed by Slovak standards, it would be more appropriate to point out the difference between surface temperatures and moisture content before and after the application of insulation systems [10].

Table 1. Surface temperatures and moisture content of the historic wall before and after insulation [7].

Insulation material	Temperature						Moisture content	
	Before insulation		After insulation		Difference at the wall	Difference under the window	Before	After
	Wall	Under window	Wall	Under window				
Lime plaster	16.7	14.5	17.5+0.5	15.2+0.5	1.3	1.2	10.2	11.7
Aerogel	17.6	14.6	18.2+0.5	17.1+0.5	1.1	3.0	10.8	11.2
Lime and cork	17.7	14.6	17.9+1	16+1	1.2	2.4	9.7	11.1
Lime and hemp	17.5	14.6	17.9+1	16.2+1	1.4	2.6	10.0	10.4
Calcium-silic.board	17.4	16.0	19.1+0.5	17.2+0.5	2.2	1.7	10.7	11.9
Timber fibre board	18.1	16.0	19.7+0.5	18.5+0.5	2.1	3.0	10.7	10.8
PIR	17.4	14.7	19.5	17.8	2.1	3.1	10.4	13.2

From Table 1, it is clear that the best results in terms of surface temperature and moisture content of the historic wall are the fiberboard. It is generally known that PIR board excels in its thermal insulation properties and nevertheless fails to provide such thermal comfort in terms of surface temperature as capillary active wood fiber board. There is an even bigger difference in the moisture content of the wall, which can be attributed to the properties of the capillary activity of the fibreboard relative to the vapor-proof PIR board. The results can be seen shown in the attached graphs in figures 1 and 2

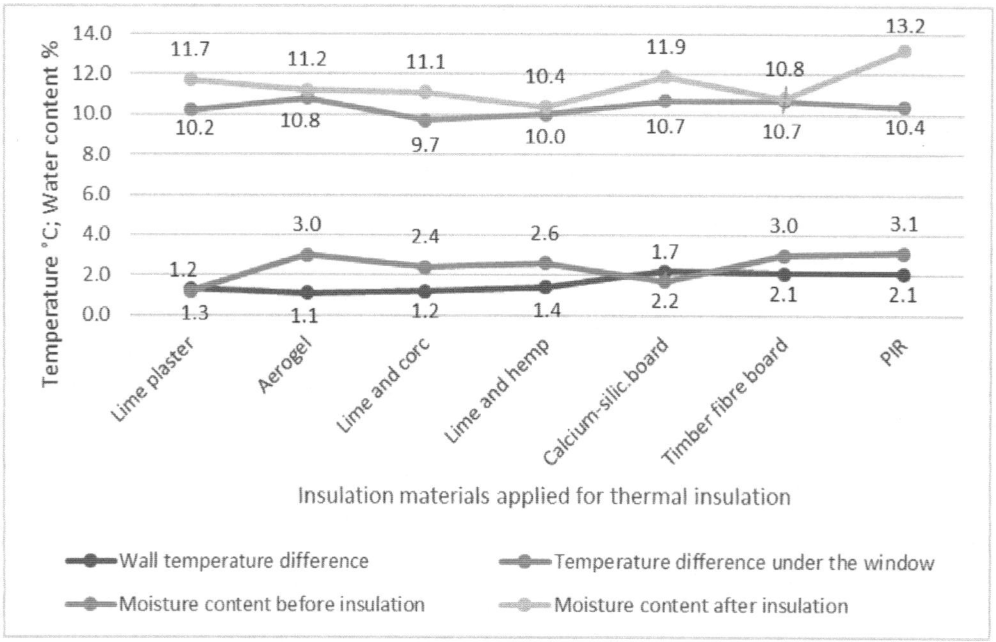

Figure 1. Differences in temperature and moisture content before and after insulation.

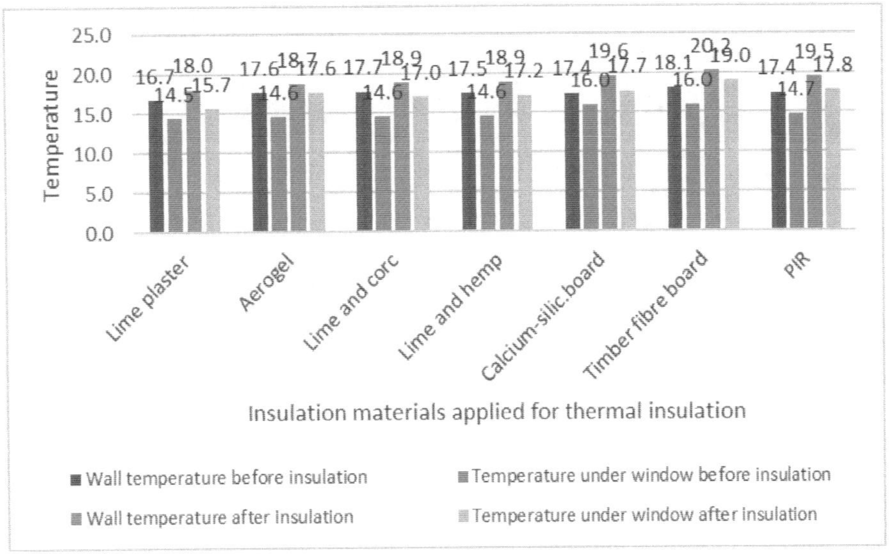

Figure 2. Surface temperatures of the historic wall before and after insulation.

5. Case study simulation in Košice

The simulation of the internal thermal insulation was performed on a historic building in Košice. The building is currently used as a puppet theater. The effect of calcium-silicate plate 50 and 100 mm thick in a slightly sunken technical room was investigated.

5.1. Boundary and initial conditions

In HAM (heat-air-moisture) simulations dynamic external boundary conditions for the city of Košice were used according to [28] (Table 2, Figure 3). For the internal boundary condition, an average temperature of 21.2 ° C with an amplitude of 1K with a daily maximum of July 3 of 22.2 ° C was considered. The internal relative humidity was considered an average value of 60% with an amplitude of 5% with a daily maximum of 16 September. Based on the moisture survey of the perimeter masonry, a value of 67.07 kg/m^3 was selected as the initial condition for the water content for the lime-cement exterior plaster.

Table 2. Boundary conditions.

Parameter	Value
Location	Košice
Latitude (°)	48.67 North
Longitude (°)	21.24 East
Altitude (m)	231
Mean Temperature (°C)	10.2
Max. Temperature (°C)	31.8
Min. Temperature (°C)	-12.2
Mean Relative Humidity (%)	73.3
Max. Relative Humidity (%)	100
Mean Relative Humidity (%)	22
Mean Wind Speed (m/s)	3.1
Normal Rain Sum (mm/a)	469.8

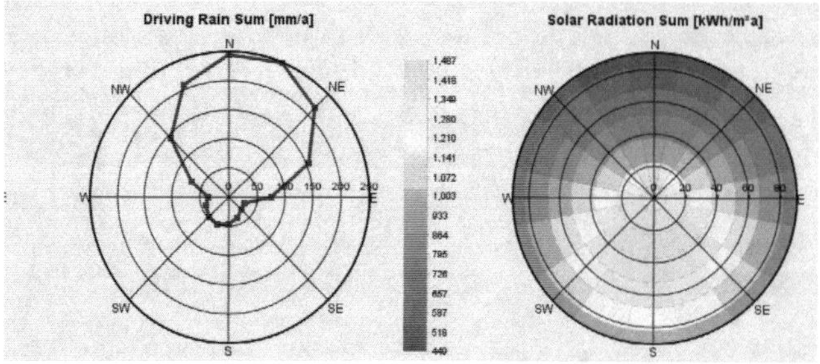

Figure 3. Solar radiation and driving rain sum for Košice (from Wufi database).

5.2. Material properties

The monitored wall is made of quarry tuff stone, connected by a light-grey mortar with large pieces of quicklime about 1.4 m thick. From the inside, it is plastered with lime plaster with the final surface covering of the white paint. The wall is recessed partially below the ground level and external lime-cement plaster was considered. The material properties of existing wall are presented in Table 3., and material properties of the wall after adding insulation material are presented in Table 4,5.

Table 3. Material properties of existing solid wall.

	Thick. (m)	Bulk density (kg/m3)	Porosity (m3/m3)	Specific heat capacity (J/kgK)	Thermal Conductivity (W/mK)	Water vapour diffusion resistance factor (-)
Cement lime plaster	0.05	1900	0.24	850	0.8	19
Historical solid brick	1.4	1800	0.31	850	0.6	15
Lime plaster	0.02	1600	0.3	850	0.7	7

Table 4. Material properties of existing solid wall with calcium-silicate insulation th. 50mm.

	Thick. (m)	Bulk density (kg/m3)	Porosity (m3/m3)	Specific heat capacity (J/kgK)	Thermal Conductivity (W/mK)	Water vapour diffusion resistance factor (-)
Cement lime plaster	0.05	1900	0.24	850	0.8	19
Historical solid brick	1.4	1800	0.31	850	0.6	15
Cement lime plaster	0.02	1900	0.24	850	0.8	19
Ytong Multipor	0.05	115	0.96	850	0.04	4.1
Ytong Multipor Adhesive	0.006	833	0.686	850	0.155	15.1

Table 5. Material properties of existing solid wall with calcium-silicate insulation th. 100mm.

	Thick. (m)	Bulk density (kg/m3)	Porosity (m3/m3)	Specific heat capacity (J/kgK)	Thermal Conductivity (W/mK)	Water vapour diffusion resistance factor (-)
Cement lime plaster	0.05	1900	0.24	850	0.8	19
Historical solid brick	1.4	1800	0.31	850	0.6	15
Cement lime plaster	0.02	1900	0.24	850	0.8	19
Ytong Multipor	0.1	115	0.96	850	0.04	4.1
Ytong Multipor Adhesive	0.006	833	0.686	850	0.155	15.1

6. Results and discussions

In the Figure 4 we see the course of water content in the historic wall for 4 years. The initial water content is approximately 4.5 kg/m^3. With the original composition over 4 years, we see a declining trend in the water content, in winter the volume of water increases, in summer it gradually dries out. After applying calcium silicate board thickness 50mm to the inner surface, there is a clear increase in water content over 4 years. In the summer months it tends to dry out to its original content, but we can observe an increase in the water content in the wall structure. After the application of calcium silicate board thickness 100mm, we can already see a more significant increase in the water content compared to the insulation board thickness 50mm. During the summer months, it no longer dries to its original volume and the water content in the structure increases every year. Data from 24.4.2023 are marked, where for 4 years we can see the maximum value of the water content in the structure at 50mm and 100mm insulation.

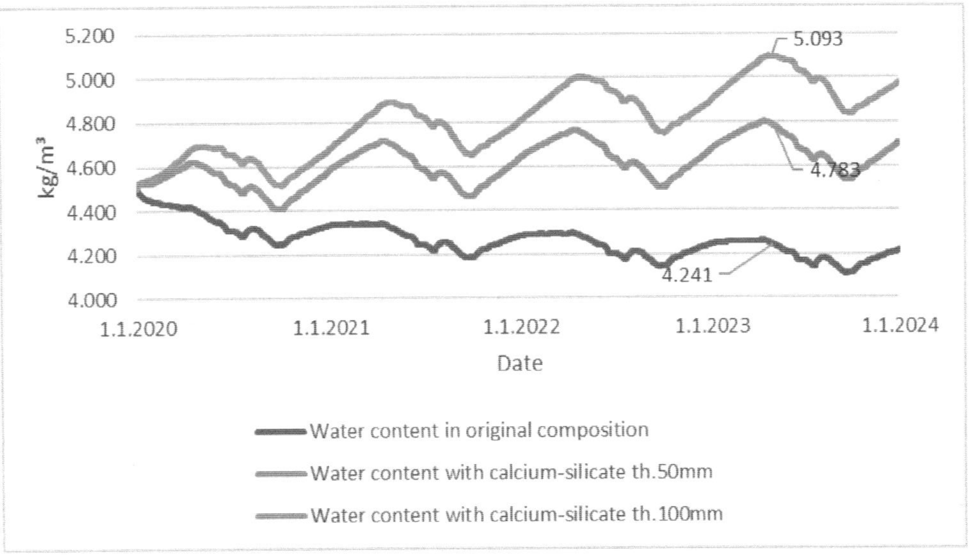

Figure 4. Water content in historic masonry layer in original composition, after using 50mm calcium-silicate board and after using 100mm calcium-silicate board.

6.1. Discussion

By applying the measured values of mass moisture as the initial condition of the long-term simulation, the state of the structure at the beginning of the heat transfer process became realistic. The influence of water content in the structure as an initial condition in the simulation is well known [24]. We can therefore consider the results of long-term simulations to be reliable. The result of the simulation is the water content in the structure after the application of two different thicknesses of internal insulation, resp. without insulation of the perimeter masonry wall. The water content increases with increasing thickness of the thermal insulation and its long-term accumulation occurs. With a thermal insulation thickness of 100 mm, the inability of the drying structure in summer is especially visible. The structure slowly gets rid of the accumulated amount of water in the winter months. In contrast, with a thermal insulation thickness of 50 mm, the structure is able to reduce its percentage water content from 0.28 to 0.25%. Findings from the results of long-term simulations indicate that it is necessary to optimize the thickness of thermal insulation with respect to the accumulation of water in the structure.

The original wall with a thickness of 1.47 m has a heat transfer coefficient U = 0.398 W / (m^{2}K). After insulation with a calcium-silicate capillary-active insulation board 50 mm thick was used, the heat transfer coefficient decreased to U = 0.274 W / (m^{2}K). When 100mm thick insulation was used, this value was reduced to 0.21 W / (m^{2}K). (Figure 5)

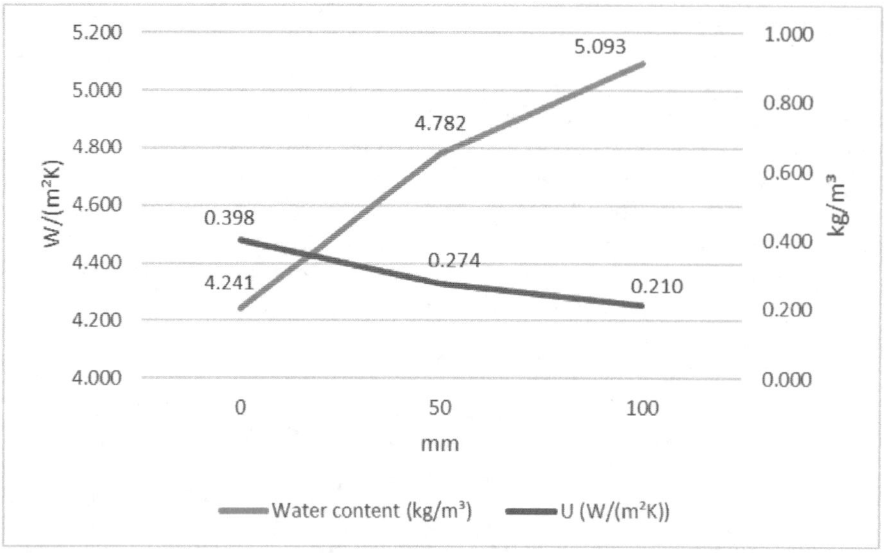

Figure 5. Influence of insulation thickness on water content and thermal conductivity of the historic masonry.

7. Conclusions

Historical masonry of massive thickness is able to accumulate heat but also water to a large extent. The long-term simulation results of water content accumulation in the structure after insulation confirm the fact that internal insulation prevents drying into the interior in summer period. Due to the internal insulation, the temperature in the masonry decreases when the structure is subcooled and slightly heated. With a decrease in the temperature in the masonry and a reduced ability of the masonry to dry into the interior in the summer, the accumulation of water in the structure will increase after its insulation. The percentages of masonry water content after insulation and after two years do not exceed the level - very low humidity. We can conclude that even when the thickness of 100 mm from the inside is insulated the construction will remain functional in heat-air-moisture terms of view. However, it should be noted that the assessment is carried out in a one-dimensional hygrothermal transport. With the multidimensional details of structures and various structural conditions of historic buildings undesirable effects of internal insulation can occur locally. The task for the future work is experimentally in-situ verification of the performance of the exterior wall structures of the building before and after the insulation. The obtained data from further research can be used to achieve a comprehensive picture of the hygrothermal behaviour of the masonry of historic buildings on which the internal insulation was carried out.

8. References

[1] RiBuild.eu 2021 Available online: https://www.ribuild.eu
[2] Blumberga A, Kašs K, Kamendere E, Žogla G, Kamenders A, Blumberga D and de Place Hansen E J 2019 *Robust Internal Thermal Insulation of Historic Buildings-State of the art on historic building insulation materials and retrofit strategies*
[3] Kossecka E and Kosny J 2002 Influence of insulation configuration on heating and cooling loads in a continuously used building *Energy and Buildings - ENERG BLDG* **34** 321-31
[4] Bjarløv SP, Finken G R and Odgaard T 2015 Retrofit with interior insulation on solid masonry walls in cool temperate climates–An evaluation of the influence of interior insulation materials on moisture condition in the building envelope *Energy Procedia* **78** 1461-6
[5] Munch-Andersen J 2008 *SBi-anvisning 221 Efterisolering af etageboliger* (Hørsholm: Danish Building Research Institute, Aalborg University)
[6] Brandt E, Bunch-Nielsen T, Christensen G, Gudum C, Hansen M H and Møller E B 2009 *SBi-anvisning 224: Fugt i bygninger* (Hørsholm: Statens Byggeforskningsinstitut)

[7] Toman J, Vimmrova A and Černý R 2009 Long-term on-site assessment of hygrothermal performance of interior thermal insulation system without water vapour barrier *Energy and Buildings* **41** 1 51-5.

[8] Jorgensen G J, Kim H M and Wendelin T J 1996 Durability studies of solar reflector materials exposed to environmental stresses. In Durability Testing of Nonmetallic Materials *ASTM International.*

[9] Zhao J, Grunewald J, Ruisinger U and Feng S 2017 Evaluation of capillary-active mineral insulation systems for interior retrofit solution Building and Environment **115** 215-227

[10] Walker R and Pavía S 2015 Thermal performance of a selection of insulation materials suitable for historic buildings *Building and environment* **94** 155-165.

[11] Kolaitis D I, Malliotakis E, Kontogeorgos D A, Mandilaras I, Katsourinis D I and Founti M A 2013 Comparative assessment of internal and external thermal insulation systems for energy efficient retrofitting of residential buildings *Energy and Buildings* **64** 123-131

[12] Klõšeiko P, Arumägi E and Kalamees T 2013 Hygrothermal performance of internally insulated brick wall in cold climate: field measurement and model calibration *Central European Symposium on Building Physics.*

[13] Straube J, Schumacher C 2007 Interior insulation retrofits of load-bearing masonry walls in cold climates *Journal of Green Building* **2** 2 42-50

[14] Vereecken E, Van Gelder L, Jansse H and Roels S2015 Interior insulation for wall retrofitting–A probabilistic analysis of energy savings and hygrothermal risks *Energy and Buildings* **89** 231-244

[15] Kopecký P, Staněk K and Tywoniak J 2020 Hygro-thermal performance of wooden beam ends: experimental investigations of uninsulated and internally insulated brick wall *E3S Web of Conferences* **172** 01009

[16] Hanse T K et al. 2018 Long term in situ measurements of hygrothermal conditions at critical points in four cases of internally insulated historic solid masonry walls *Energy and Buildings* **172** 235-248

[17] Finken G R, Bjarløv S P and Peuhkuri R H 2016 Effect of façade impregnation on feasibility of capillary active thermal internal insulation for a historic dormitory–A hygrothermal simulation study *Construction and Building Materials* **113** 202-214

[18] Pavlík Z and Černý R 2009 Hygrothermal performance study of an innovative interior thermal insulation system *Applied Thermal Engineering* **29** 10 1941-1946

[19] Katunská J, Katunský D and Labovská V 2019 Selected problems of thermal insulation of historical buildings *Selected Scientific Papers-Journal of Civil Engineering* **14** 1 67-74

[20] Katunský D, Nemec M Kamenský M 2013 Airtightness of Buildings in Slovakia *Advanced Materials Research* **649** 3–6

[21] Katunská J and Katunský D 2015 Application of Chemical Grouting as an Option of Removing Soil Moisture-a Case Study in the Reconstruction of the Church *Selected Scientific Papers-Journal of Civil Engineering* **10** 2 93-102

[22] Labovská V and Katunský D 2016 In situ monitoring of internal surface temperature of the historic building envelope *Selected Scientific Papers-Journal of Civil Engineering* **11** 1 77-84

[23] Vertal M, Vaskova A, Korjenic A and Katunsky D 2016 Case study investigations on drying process cellular concrete external walls with ETCIS *Bauphysik* **38** 6 378-88

[24] Zozulák M, Vertaľ M and Katunský D 2016 Hygrothermal Initial Conditions-Case Studies. In Applied mechanics and materials **820** 273-8

[25] Katunsky D and Katunska J 2020 Assessment of Thermal Insulation Properties of Envelope Structures of a Burgher House in Kosice *International Scientific Conference EcoComfort and Current Issues of Civil Engineering* 166-73

[26] Blumberga A, Freimanis R, Blumberga D, Veidenbergs I, de Place Hansen E J, Hansen T.K and Wagner G 2020 Written guidelines for decision making concerning the possible use of internal insulation in historic buildings

[27] Orlik-Kożdoń B 2019 Interior insulation of masonry walls—selected problems in the

design *Energies* **12** 20 3895
[28] Available online:
http://climate.onebuilding.org/WMO_Region_6_Europe/SVK_Slovakia/index.html

Acknowledgments
This research is supported by VEGA 1/0674/18 Theoretical and experimental analysis of architectural-structural shapes and fragments of building envelope structures designed for harsh conditions.

IOP Conf. Series: Materials Science and Engineering 1209 (2021) 012004 doi:10.1088/1757-899X/1209/1/012004

Thermophysical parameters of the Kežmarok sandstone

P Jaroš[1] and M Vertal'[1]

[1] Technical university of Kosice, Faculty of Civil Engineering, Institute of Architectural Engineering, Slovakia

E-mail: pavol.jaros@tuke.sk, marian.vertal@tuke.sk

Abstract. Thermophysical parameters of building materials are required for calculating the complex heat and water transfer in building structures. It can be performed by modern simulation software such as Wufi, Delphin, Math, Comsol Multiphysics and other. This software is suitable for evaluation of water and heat transport in construction of historical buildings, because it can include the impact of water on material properties, driven rain, ground water, heat and water accumulation and other. The material properties of historical building materials are required for the use of this software. In Slovakia, the most used building material was sandstone. Sandstone from Kežmarok was chosen for this paper, which was used in the construction of historic buildings such as churches and town houses. The method of dynamic impulse transition by thermophysical tester RTB was used to determine the thermal properties of sandstone.

1. Introduction

Thermal and hygric evaluation of building structures of historic buildings is a difficult process. These structures are built mainly of natural material, specifically stone. In Slovakia, sandstone was the most used building material in Gothic architectural style from the 13th century 0. It is not appropriate to use a normative procedure 0 to evaluate the thermal and hygric behavior, because it does not consider the influence of water on the properties of the material. There is simulation software available, which can include influence of water to material properties. To use this software, it is necessary to know heat and water transport parameters of historical buildings. In this paper, Kežmarok sandstone was selected for analysis of thermal parameters and influence of water to them. Thermal conductivity, specific heat capacity and thermal diffusivity was measured by transient pulse method with laboratory instrument Thermophysical Tester RTB 1.02.

2. Selected material

Kežmarok sandstone was selected for the analysis of thermal properties. This type of historic building material was used for load-bearing and non-bearing structures in the Gothic and Renaissance periods. It was used in construction of the Basilica of the Holy Cross (15th century), the church of the Visitation of the Virgin Mary (18th century), the historic bell tower in city, but also for smaller chapels 0. Sandstone samples were taken from an abandon quarry on the eastern side of the town of Kežmarok. It has a light to gray color and is located in the area of the flysh zone. A European Union program called Historic quarries was used to locate the quarry, in which Comenius University in Bratislava cooperates and is accessible at 0,0.

3. Experimental method

To determine thermal conductivity λ(W/mK), specific heat capacity c(J/kgK) and thermal diffusivity α(m^2/s), non-stationary measuring method was used. The principle of the transient pulse method is based on the analysis of the temperature response to a heat impulse, which is generated by the passing of the electrical current trough the plane electrical resistor made of thin nickel foil. If a planar heat source is used, the process can be described by the following equation 0

$$T(h,t) = T_0 + \frac{Q}{c_p\rho(\pi a t)^{\frac{1}{2}}} \exp\left(-\frac{h^2}{4at}\right) \quad (1)$$

where T_0 is the initial temperature (K), Q heat source (W/m^2), a thermal diffusivity (m^2/s), λ thermal conductivity (W/mK), c specific heat capacity (J/kgK), t time (s), h is the distance of the temperature maximum point (m)

The principle of the method is shown in Figure 1. Thermophysical parameters were determined from the maximum of the temperature T_m at time t_m. Non stationary method was published in0,0,0,0. As is shown in Figure 1, three samples are needed for the measurement and planar heat source is installed between the first and the second sample and thermocouple is installed between the second and the third sample. Block or cylindrical samples can be used for measurement. Compared to other methods, the advantage of the method is the need for smaller samples.

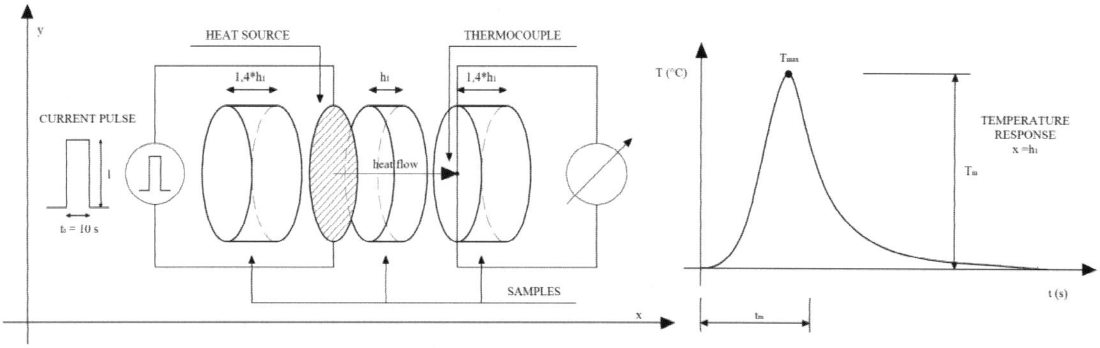

Figure 1. Scheme of the transient pulse method according to 0.

One cylindrical piece of sandstone with diameter 40 mm was made for laboratory measurement, which was divided into 5 parts, three with a thickness h_1 (measured samples) and two with thickness 1,4* h_1. Because water is an important factor in historic buildings, thermophysical parameters were measured with three different moisture contents. Before the test, samples were dried in an electrical oven at 105°C, then were placed in desiccator with salt solution sodium chloride, that provides 75% relative humidity and then were placed in a tank of distilled water (w_{sat}). The moisture content was determined by the gravimetric method from dry and saturated samples. During the test, temperature of the samples was maintained at 20°C. Measuring device Thermophysical Tester RTB 1.02 consist of isothermal cover, vacuum cover, cooper heat exchangers, system for specimen fixing, clamping plate for fixing the reference junction of the thermocouple, miniature RTD sensor, vacuum pump, thermostat and datalogger, which is connected to the PC. It is shown in Figure 2. To include material variability, a minimum three samples of Kežmarok sandstone were used for measuring and each sample was measured 5 times to account for the effect of measurement noise. For this experiment, the bulk density ρ (kg/m^3) is needed 0. It was determined as the ratio of the weight of the dry sample to the volume from 5 cylindrical samples with a diameter of 50 mm and height of 50 mm. One measuring cycle took place 30 minutes, the temperature response was recorded for 20 minutes with thermocouple, and the temperature stabilization took place 10 minutes. The width of the thermal pulse was 10 seconds at a current I_p = 2,5 (A), a voltage U_p = 3,5 (V). The heat source was ensured with a foil (Kapton (25μm)-

glue (3μm)- Ni foil (20μm)-glue (3μm)-Kapton (25μm). Thermocouple was Chromel-Alumel with Teflon wall (70μm).

Figure 2. Thermophysical Tester RTB 1.02 and sample preparation.

From the temperature response and from the maximum temperature T_m at time t_m, thermal diffusivity (m^2/s), specific heat capacity (J/kgK), thermal conductivity λ(W/mK) can be calculated according following equations 0:

The thermal diffusivity can be calculated by equation

$$\alpha = \frac{{h_1}^2}{2t_m} \tag{2}$$

The specific diffusivity can be calculated by equation

$$c = \frac{\varrho}{(2\pi e)^{1/2} h_1 \rho T_m} \tag{3}$$

The thermal conductivity can be calculated by equation

$$\lambda = \alpha c \rho \tag{4}$$

where $\varrho = RI^2 t_0$, α (m^2/s) thermal diffusivity, λ(W/mK) thermal conductivity, c specific heat capacity (J/kgK), h_1 is the thickness of the measured sample, T_m is the temperature maximum of the response, t_m is the time, when response reach the temperature maximum.

4. Results and Discussion

The temperature responses of Kežmarok sandstone in the dry state are shown in Figure 3. Similar temperature curves can be observed for sample 1 and 3, the resulting is similar thermal conductivity values given in Table 1. Sample 2 has a slightly higher temperature maximum and thus a lower value of thermal conductivity. The parameter values of each sample were calculated as the average of at least five consecutive measurements.

IOP Conf. Series: Materials Science and Engineering 1209 (2021) 012004 doi:10.1088/1757-899X/1209/1/012004

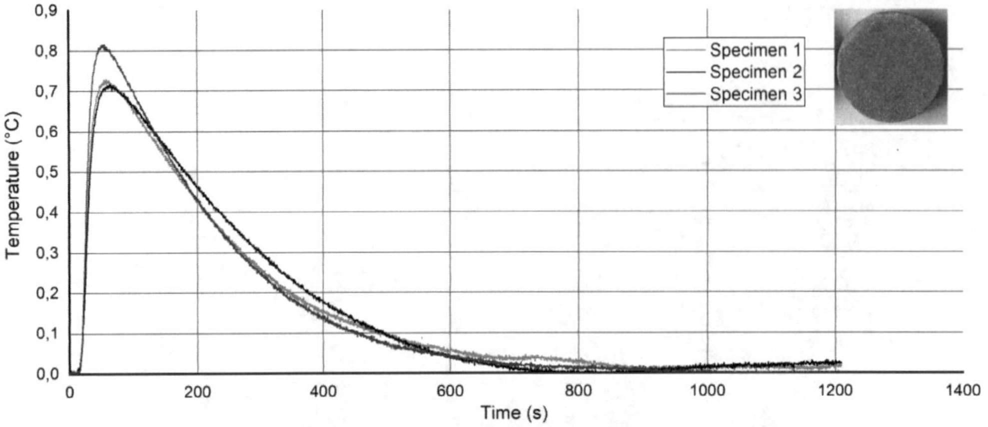

Figure 3. Temperature response of sandstone in dry state.

Table 1. Thermal conductivity of Kežmarok sandstone at different water content.

	Thermal conductivity λ(W/mK) depended on water content		
Specimen	dry state	75% Relative humidity	(w_{sat})
K1	2.61	2.71	3.03
K2	2.58	2.80	3.26
K3	2.62	2.75	3.07

The values of specific heat capacity of selected sandstone are given in Table 2. As can be seen, the water content does not have significant effect on this parameter, because the sandstone has a low open porosity of 10 to 12 percent. From the above values it follows that the noise of the measurement is higher than the total effect of water. However, according to 0, rough values of the specific heat capacity in the dry state is sufficient for simulations, because it does not have a great influence on the hygrothermal behaviour of building structures.

Table 2. Thermal conductivity of Kežmarok sandstone at different water content.

	Specific heat capacity c (J/kgK) depended on water content		
Specimen	dry state	75% Relative humidity	w_{sat}
K1	907.5	852.3	829.1
K2	918.3	948.9	812.5
K3	778.2	874.1	783.4

The dependence of thermal conductivity on the water content of Kežmarok sandstone of three samples is shown in Figure 4.

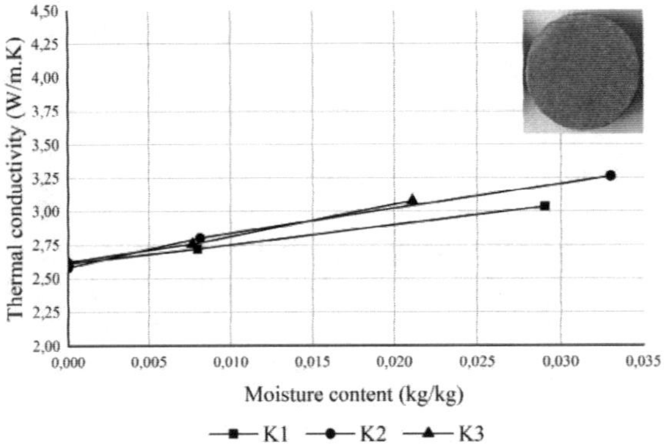

Figure 4. Thermal conductivity depended on water content.

5. References

[1] Čabalová D 2013 *The beauty of a stone in a person life* (Bratislava: Veda) p 374
[2] Kubicar L, Boháč V and Vretenár V 2002 *Transient methods for the measurement of thermophysical properties: The pulse transient method.* High Temperatures-High Pressures
[3] Vertaľ M, Katunský D and Šimíček J 2014 *Transient Pulse Method for Measuring of Heat Conductivity of Autoclaved Aerated Concrete.* Advanced Materials Research 899 pp 466-469
[4] Boháč V, Gustavsson M, Kubicar L and Vretenár V 2003 *Measurements of building materials by transient methods*
[5] Stefkova P and Zmeškal O 2006 *Thermophysical parameters of PERSPEX determined from the temperature responses measured by pulse transient method* Thermophysics
[6] STN 73 0540-2: *Thermal protection of buildings. Thermal performance of buildings and components. Part 2: Functional requirements*
[7] Available: http://hq.chc.sbg.ac.at/
[8] Available: http://apl.geology.sk/igatlasg/
[9] Software WUFI Pro 6.0, Manual

Acknowledgments
This work was supported by the VEGA: 1/0674/18 Theoretical and experimental analysis of architectural-structural shapes and fragments of building envelope structures designed for harsh conditions.

IOP Conf. Series: Materials Science and Engineering 1209 (2021) 012005 doi:10.1088/1757-899X/1209/1/012005

Reverberation measurement set for the interrupted noise method

D Jun[1] and O Nespěšný[1]

[1] Faculty of Civil Engineering, Brno University of Technology, Veveří 331/95, 602 00 Brno, Czech Republic

E-mail: David.Jun@vutbr.cz

Abstract. This paper describes the methodology of measuring the sound absorption coefficient in a reverberation room. For its purpose, a measuring set was created combining professional with widely available hardware suitable for acoustic reverberation measurements. For the measurement itself, the interrupted noise method was chosen based on the standard ČSN EN ISO 354. Measurement and evaluation of data take place using scripts written in the Python programming language. An effective measuring set was developed both in terms of operation during measurement and in terms of subsequent data processing, accessing and graphical presentation.

1. Introduction

Several methods can be used to determine the effectiveness of acoustically absorbing structures. The sound absorption coefficient by which this efficiency is most often quantified can be measured in an impedance tube, in a free field or in a reverberation room. These measurement variants are described in more detail and compared by Cox and D'Antonio [1].

Each of the mentioned methods is suitable for a different purpose, however, for the needs of room acoustics, the most commonly used measurement is in a reverberant room, the output of which is Sabin sound absorption coefficient α_S [-]. The main reason is probably a certain similarity between boundary conditions of this measurement and the placement in a real room. There, the normal incidence of sound (measured in an impedance tube) or a specific incidence angle (measured in a free field) rarely occurs. The situation in a room is more often close to a partially diffuse acoustic field, where the direction of sound impact on adjacent structures is more or less random. It is possible to simulate this situation in the free field, but it is a method demanding instrumentation and accuracy of the experiment.

The reverberation room method also has some disadvantages, which stem from the difficult-to-achieve fully diffuse acoustic field and thus the problem of non-uniform boundary conditions across different laboratories. Jeong and Chang, for example, have researched this problem [2].

The reverberation room method is based on measuring the reverberation time T [s], ie the time during which the sound pressure level drops by 60 dB after the source is switched off [3]. However, this quantity is seldom determined from the whole decrease. The most used variants and evaluated ranges of reverberation curves are summarized in table 1. It is possible to use several excitation signals for the measurement itself, but most often it is an impulse or broadband noise. These two variants are described in the ČSN EN ISO 354 standard [4], the national equivalent of the ISO 354 standard [5].

This contribution presents a measuring set and a software part of the measurement developed especially for measuring the sound absorption coefficient by the interrupted noise method. The aim was to develop an inexpensive set that could be operated from a computer throughout the measurement and that would provide data from all steps necessary to evaluate the sound absorption coefficient α_S [-].

Table 1. The most common types of reverberation times. The level of 0 dB represents the level immediately after switching off the source, or the maximum of the impulse.

Variable [s]	Upper limit [dB]	Lower limit [dB]
T_{30} [6]	–5	–35
T_{20} [6]	–5	–25
EDT [7]	0	–10

2. Measuring set and used software

A set consisting of a Brüel & Kjær OmniPower 4296 omnidirectional power supply, a Brüel & Kjær Power Amplifier Type 2716 amplifier, a Beyerdynamic MM1 measurement microphone, a Steinberg UR22 USB audio interface and an HP ProBook 470 G0 notebook was used for the measurements. The scheme of the measuring set is shown in Figure 1.

While on the one hand the professional Brüel & Kjær hardware was used, on the other hand, the omnidirectional microphone and the audio interface of a rather lower category were used. This did not have a limiting effect on the reverberation measurements, as level differences are evaluated and it is not necessary to know the exact sound pressure level.

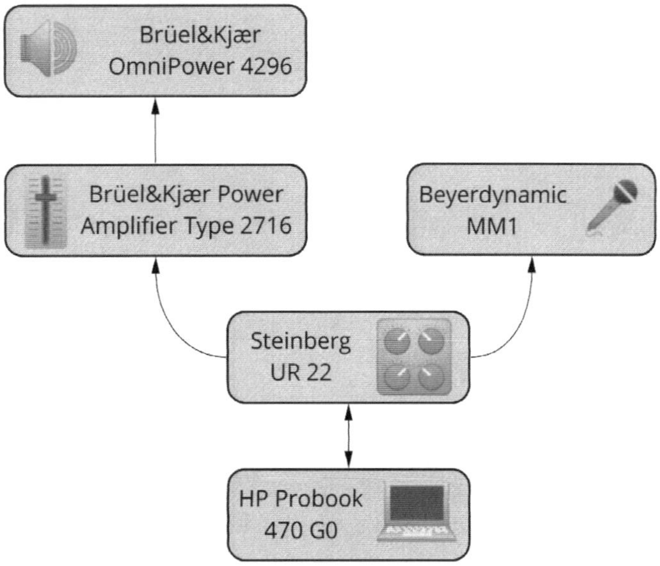

Figure 1. Measuring set scheme.

The software part of the measurement used both existing software and own scripts written in the Python programming language [8] and it was operated from the interactive Jupyter Notebook interface [9].

The first software used is the JACK Audio Connection Kit [10], which is an extension of the audio driver that allows to connect virtual and physical audio inputs and outputs of the computer with low latency. In this case, JACK was supplemented by modules from the jack-tools package [11], which allows basic operations such as playback and recording from the command line environment.

The second software used is the Room EQ Wizard (REW) [12], which is an application that allows the measurement and evaluation of various parameters of room acoustics. In this case, only its pink noise generator was used.

The Python libraries used summarises in table 2. They were used to measure and evaluate the measured data and to create graphical outputs.

Table 2. Python libraries used for measurement and data processing.

Name	Usage
NumPy [13]	Efficient work with large vectors and matrices.
pandas [14]	Data manipulation and analysis.
SciPy [15]	Extensive library for data, image and signal processing. In this work, it was used mainly for digital signal filtering.
Matplotlib [16]	Graphical output creation.
python-acoustics [17]	Acoustic measurements and calculations. In this work, the room module was mainly used, containing a function for reverberation time determination from the impulse response of the room. This function has been modified to suit the interrupted noise method and to return a larger number of outputs. Above all, it was necessary to obtain reverberation curves for further averaging.

3. Measurement procedure and obtained data

According to the standard, the measurement of absorption in a reverberation room is measured in at least 12 position-independent configurations of the source and microphone. In addition, in the case of the interrupted noise method, at least 3 submeasurements have to be made in each configuration to reduce the statistical error [4]. This leads to a minimum of 36 submeasurements for each specimen. Because of a large number of submeasurements, effective measurement requires maintaining a good overview of source-microphone configurations. The second requirement is the near real-time unwanted externality detection during the measurement and its repetition only for configurations where the problem occurred.

In this work, these requirements were included in a measurement algorithm consisting of the following steps, which are summarized in Figure 2:

1. Selection of the number of measuring points of the microphone and the source, or the number of surveys in one configuration, if it is to be greater than 3. Based on this selection, the process of measurement is defined.
2. Saving of boundary conditions of measurement (relative humidity and temperature).
3. Entry into the measuring cycle.
4. Prompt for readiness to perform measurements for the first configuration and then prompt to leave the room.
5. Loading a noise file previously exported from REW software. For the presented work, it was pink noise with a spectrum slope of –6 dB/octave.
6. Measurement in a notified configuration with a predefined number of submeasurements. (Figure 2 - Inner cycle)
    a. Cut out a noise segment lasting 5 s at a random time.
    b. Play noise and start recording 0.5 s before the end of playback. Recording takes 10 seconds from the start.
    c. Power-off detection and trimming of the recorded file.
7. Audible signalization of the end of the measurement in the given configuration, which allows entering the room without the risk of disturbing the measurement.
8. Output of the measurement to CSV files.
9. Display the reverberation time T [s] graph for the current 3 submeasurements to evaluate the need for the measurement repetition.
10. Prompt for repetition of the submeasurements in the same position.
11. Repeat steps 4 through 10 for other configurations. (Figure 2 - Outer cycle)

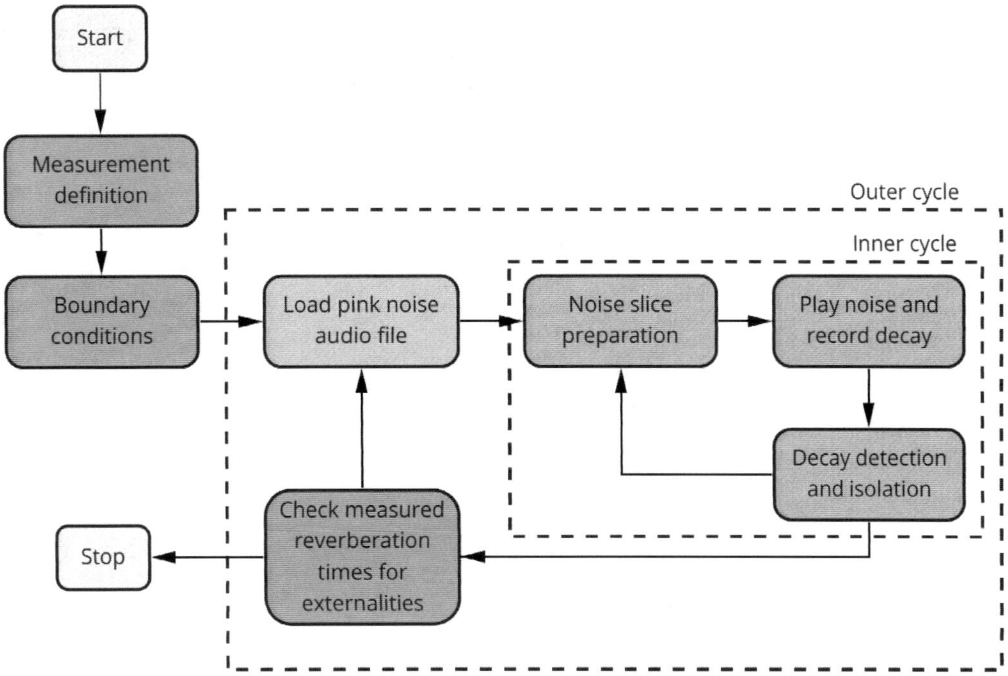

Figure 2. The measurement process. Outer cycle – measuring in different configurations, Inner cycle – performing pre-set number of submeasurements in a specific configuration.

Using the described procedure, reverberation recordings in individual configurations are obtained. Subsequent evaluation of this audio data is facilitated by the previously mentioned uniform naming of source-microphone configurations. Thanks to this, it was also possible to create a unified output system in the form of CSV files. The data processing procedure was as follows:

- Calculation of decay curves for each submeasurement separately using the modified room module from the python-acoustics library. Made in 1/3-octave resolution. (Figure 3a)
- Averaging of the decay curves according to individual configurations. (Figure 3b)
- Reverberation time calculation in individual configurations. (Figure 3c)
- Total absorption calculation of the room with sample A_2 [m^2] according to ČSN EN ISO 354 [4]. (Figure 3d)
- Loading the total absorption of the room without sample A_1 [m^2].
- Calculation of the sound absorption coefficient α [-] according to the formula $\alpha = (A_2 - A_1)/S$ where S [m^2] is the actual area of the measured specimen. (Figure 3e)
- Calculation of the average sound absorption coefficient course from all available configurations. (Figure 3f)

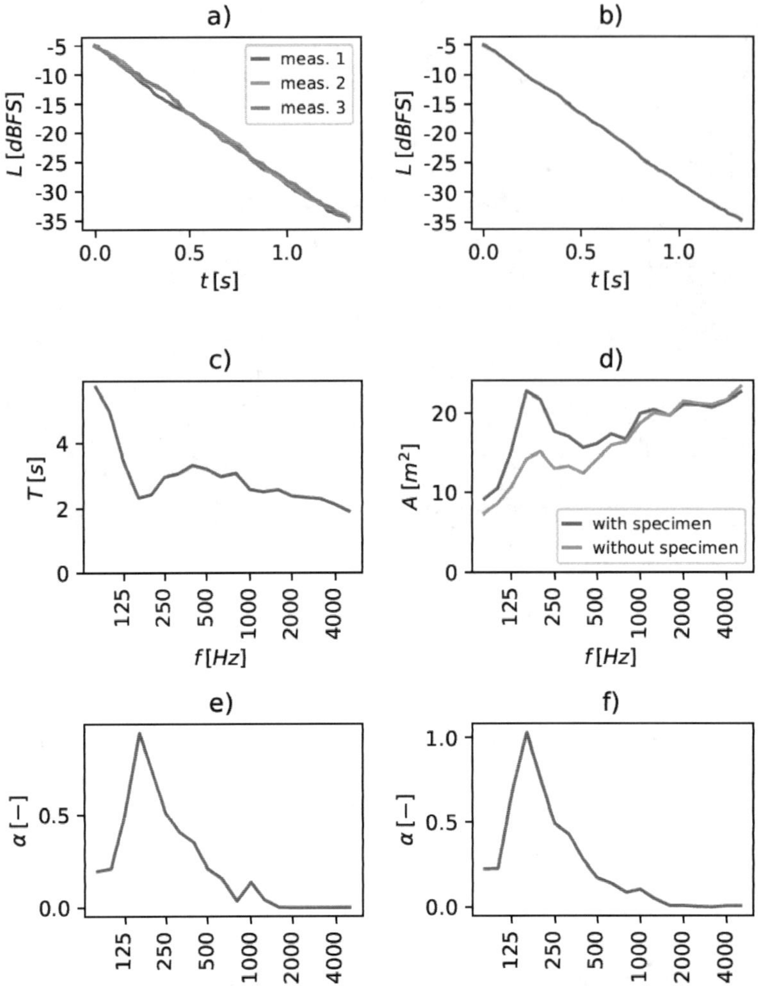

Figure 3. Data processing. a) decay curves in one configuration for one third octave band, b) average decay curve in one configuration for one third octave band, c) reverberation time in one configuration, d) total absorption in one configuration, e) sound absorption coefficient in one configuration, f) sound absorption coefficient averaged from all configurations.

4. Conclusion

This paper described a possible approach to measure the sound absorption coefficient in a reverberant room using the interrupted noise method. Own measuring set and measurement software solution were introduced, which enable time-efficient and operator-friendly measurement with the possibility of above-standard control of the obtained data. Freely available software and libraries and the Python programming language were used for this purpose. The measuring set included both a professional power amplifier and omnidirectional source from the manufacturer Brüel & Kjær, as well as a relatively low-cost measurement microphone and a USB audio interface. The measuring set has already been used for measuring acoustic resonators formed by perforated cement fiber boards [18,19]. One of the main advantages of the measurement is the back access to audio files and CSV files with data stored in important calculation steps. Further developments should focus on the use of alternative measurement signals and methods to obtain the impulse response of the room and to eliminate the error in the processing of the measured reverberation.

5. References

[1] Cox T J and D'Antonio P 2017 *Acoustic Absorbers and Diffusers: Theory, Design and*

Application (New York: Taylor)
[2] Jeong C H and Chang J H 2015 Reproducibility of the random incidence absorption coefficient converted from the sabine absorption coefficient *Acta Acustica united with Acustica* **101**
[3] Fišarová Z 2014 *Stavební fyzika - stavební akustika v teorii a praxi* (Brno: Vysoké učení technické v Brně)
[4] ČSN EN ISO 354 (730535) 2003 *Akustika – Měření zvukové pohltivosti v dozvukové místnosti* **354** 1–24
[5] ISO 354 2003 *Acoustics — Measurement of sound absorption in a reverberation room*
[6] ISO 3382-1: 2009 Acoustics - Measurement of Room Acoustic Parameters - Part 1: Performance Spaces
[7] Jordan V L 1970 Acoustical Criteria for Auditoriums and Their Relation to Model Techniques *The Journal of the Acoustical Society of America* **47** van Rossum G and Drake F L 2009 *Python 3 Reference Manual* (Scotts Valley: CreateSpace)
[9] Pérez F and Granger B E 2007 IPython: A system for interactive scientific computing *Computing in Science and Engineering* **9** 21–9
[10] JACK Audio Connection Kit Available online https://jackaudio.org/
[11] Details of source package jack-tools in buster Available online https://packages.debian.org/source/stable/jack-tools
[12] REW - Room EQ Wizard Room Acoustics Software Available online https://www.roomeqwizard.com/
[13] Harris C R and et al. 2020 Array programming with NumPy *Nature* **585** 357–62
[14] McKinney W 2010 Data Structures for Statistical Computing in Python *Proceedings of the 9th Python in Science Conference* ed S van der Walt and J Millman pp 56–61
[15] Virtanen P and et al. 2020 SciPy 1.0: fundamental algorithms for scientific computing in Python *Nature Methods* **17** 261–72
[16] Hunter J D 2007 Matplotlib: A 2D graphics environment *Computing in Science and Engineering* **9** 90–5
[17] Python-acoustics: A Python library aimed at acousticians Available online https://github.com/python-acoustics/python-acoustics
[18] Jun D, Nespesny O, Pencik J, Fisarova Z and Rubina A 2021 Optimized method for Helmholtz resonator design formed by perforated boards *Applied Acoustics* **184**
[19] Nespěšný O 2020 *Rozbor vlastností materiálu kompozitních desek na bázi cementu a organických vláken v návaznosti na jeho využití v rámci návrhu interiérových schodišť*

Acknowledgments
The contribution was produced with support for specific university research at Brno University of Technology, FAST-J-20-6458 (2020) and the Technological Agency of the Czech Republic within programme EPSILON, TH04020263 (2019-2021).

IOP Conf. Series: Materials Science and Engineering 1209 (2021) 012006 doi:10.1088/1757-899X/1209/1/012006

Analysis of technical solutions of stacks in high-rise buildings

M Sokol[1] and J Peráčková[1]

[1]Faculty of Civil Engineering, Slovak University of Technology in Bratislava, Radlinského 11, 810 05 Bratislava, Slovak Republic

E-mail: xsokolm@stuba.sk, jana.perackova@stuba.sk

Abstract. Nowadays, due to the maximum use of building lands, high-rise buildings in the cities are coming to the forefront of construction. A high-rise building is a building that, due to its height, requires a different technical, technological and structural design. Due to the increasing floors of buildings, more and more demands are placed on sanitary installations in terms of hygiene and quality of distribution. With the drainage in a high-rise building, the biggest problem arises in the design of stacks, which must be given individual attention. Foul water stacks must be designed in such a way that the overpressure and negative pressure created in them do not cause the extraction of water from traps and thus the spread of unpleasant smell in the building. Fluctuations in the pressure in stacks are influenced by many other factors, such as the method of venting of the stack, the fitting used on the stack, or the transition of the stack to the drain. When dimensioning stacks in addition to the correct design, it is also necessary to use a suitable system, which significantly affects the investment costs. A properly chosen system can save a lot of money that could arise from excessive dimensions, unnecessary fire protection systems, or anchorages. The paper deals with a theoretical analysis of modern technical solutions that positively affect pressure fluctuations in waste water pipes in high-rise buildings.

1. Introduction

The drainage issue in high-rise buildings is a relatively extensive and demanding topic due to complicated hydraulic conditions in the flow of foul water in drainage pipes. The biggest problem are stacks, in which high overpressure and negative pressure is created. Exceeding the critical values leads to the suction of water from traps and thus the spread of unpleasant smells in the building. Other problematic areas include changes of the direction of stacks, which cause excessive vibrations and noise. Traditional stacks systems are insufficient for high-rise buildings, and their shortcomings are as follows:

- stack with direct vent – large dimensions that enlarge the shafts and take up the usable area of the building, Figure 1a,
- stack with additional vent – stack supplemented by extra ventilation pipe, additional costs for fire protection systems and anchoring, Figure 1b.

Special systems and fittings have been designed for high-rise buildings, which, when properly designed, ensure optimal water flow to avoid undesired consequences. System solutions for high-rise buildings are as follows:

- stack with flow-optimized Sovent inlet fitting – a suitable solution for buildings with more than 5 ground floors, Figure 1c,
- stack supplemented with positive air pressure attenuator – a suitable solution for buildings with more than 10 ground floors, Figure 1d.

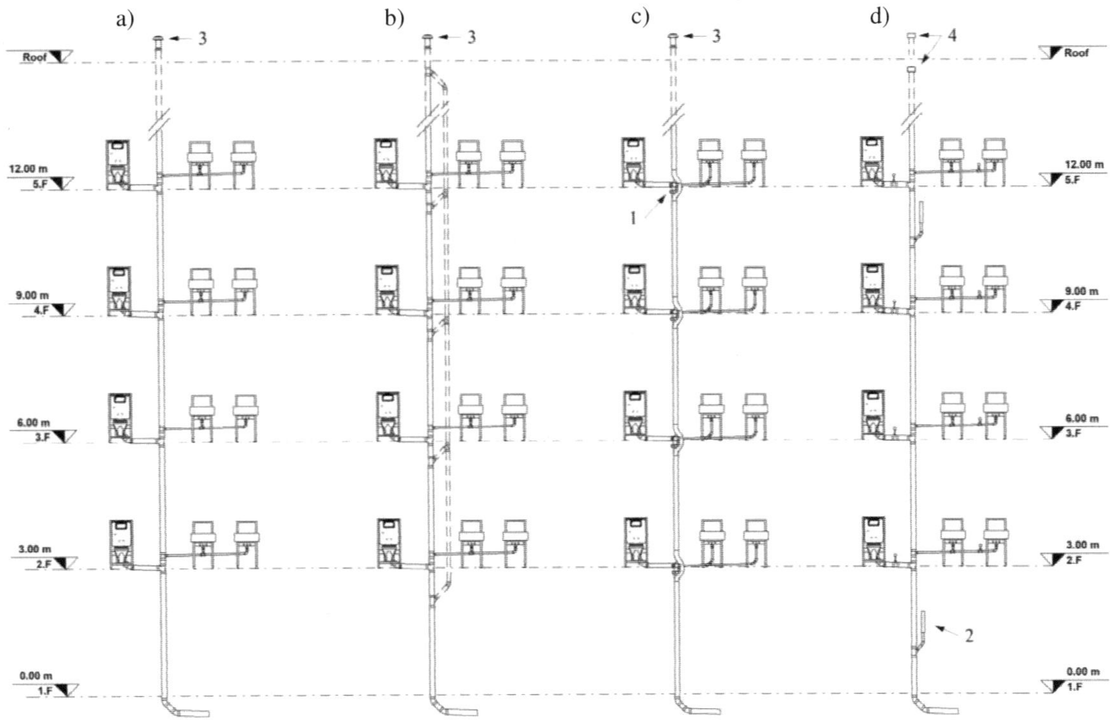

Figure 1. Overview of technical solutions for stacks in high-rise buildings: a) stack with direct vent, b) stack with additional vent, c) stack with Sovent fitting, d) stack supplemented with valves damping positive pressure, 1 – Sovent fitting, 2 – positive air pressure attenuator, 3 – vent head, 4 – air admittance valve (above or below over roof).

2. Water flow in foul water stacks

The water in the foul water stack flows around the pipe's internal walls and creates the so-called hollow cylinder with an air core in the middle. At the connection points of branch pipes with a larger flow, the air core closes, and an negative pressure occurs, which results in the suction of several times the amount of air from the vent pipe (piston effect), Figure 3c.

If there is little or no flow in a pipe, air flows upwards through the stack. The pressure loss that occurs during this flow is another cause of the negative pressure in the stack.

Problematic areas also include changes in the direction of the stacks, such as bounce, bending, or passage into the drain. If the direction changes with a bend with the angle bigger than 45°, a hydraulic jump occurs due to a change in the flow rate. The water stops flowing around the internal walls of the pipe, and the pipe starts to fill. There is an overpressure above the change of direction and an negative pressure below the change of direction. Among other things, water that strikes the wall of the arch when changing direction will cause excessive noise and vibration [6].

2.1. The velocity of water flow in foul water stacks

Figure 2 shows two water flow rates in the stack. According to Torricelli's formula, curve one represents the theoretical falling velocity. Curve two represents falling velocity in stacks.

Curve two shows that after 20 meters is falling velocity 12 m/s, the increase in velocity is minimal from this limit. After approximately 35 meters, it reaches its maximum velocity of 13 m/s, which it no longer exceeds due to air resistance and friction against the internal surface [9].

The velocity of water flow in the stack is negatively affected by bends in the stack by more than 45°, which causes a sudden decrease in velocity. A sharp reduction in velocity results a hydraulic jump that causes excessive vibration and noise.

Optimization of the water flow rate in the stack is possible with the Sovent fitting, which limits the kinetic pressure by directing the flow of water around the connection points of the branch pipes.

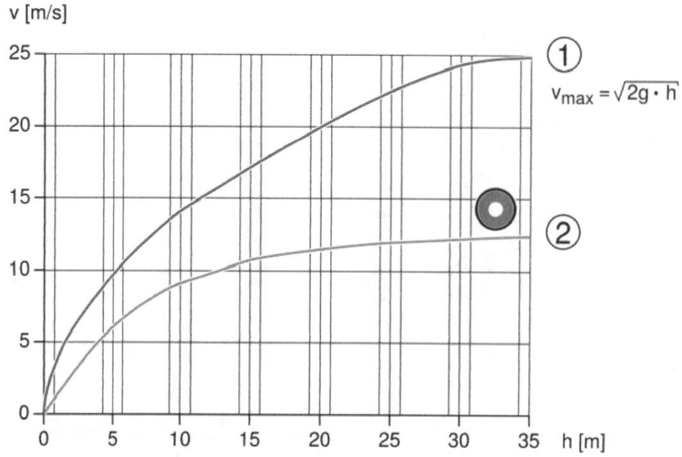

Figure 2. Theoretical falling velocity and falling velocity in stacks [9]: 1 – theoretical falling velocity, 2 – falling velocity in stacks (water film with air column), v – falling velocity (m/s), h – height of fall (m).

2.2. Hydraulic conditions in foul water stacks

Stack with direct vent, where the sum of the lengths of the stack and direct vent pipe exceeding 70 m, must be assessed. The assessment of stacks is not applied at some systems, such as stacks with Sovent fittings, additional vent or positive pressure dampers.

There are several ways to assess stacks based on Dobromyslov and Wyly-Eaton formulas. The stacks are assessed for the critical length of the stack L_{cr}, the maximum negative pressure in the stack Δp_{max}, and the maximum pressure loss Δp_{op}.

According to Dobromyslov formula, the maximum negative pressure in the stack Δp_{max} (Pa) is calculated:

$$\Delta p_{max} = \frac{3590 \, x \left[\frac{Q_{tot}}{(1 + \cos \alpha) \, x \, d_{op}^2}\right]^{1,667}}{\left(\frac{d_{op}}{d_{pp}}\right)^{0,71}} \quad \text{(Pa)} \qquad (1)$$

where:

Q_{tot} – total foul water flow rate in a stack (m^3/s),
d_{op} – internal diameter stack (m),
d_{pp} – internal diameter branch pipe (m),
α – connecting angle to a stack (°).

According to Wyly-Eaton formula, the flow of air sucked into the stack Qa (m^3/s) is calculated:

$$Q_a = 1,5 \, x \, Q_{tot} \, x \, \frac{1-f}{f} \quad (\text{m}^3/\text{s}) \qquad (2)$$

where:

Q_{tot} – total foul water flow rate in the stack (m^3/s),
f – degree of filling of the stack (-).

The pressure loss when air flows from the atmosphere Δp_{op} (Pa) is calculated:

$$\Delta p_{op} = 2240 \, x \, Q_a^{1,85} x \, \frac{L}{d_{op}^5 \cdot p_a} \quad \text{(Pa)} \qquad (3)$$

where:

Q_a – the flow of air sucked into the stack (m^3/s),
L – the sum of the heights of stack and direct vent pipe (m),
d_{op} – internal diameter of a stack (m),
p_a – atmospheric pressure (100 000 Pa).

The following two criteria are taken into account when assessing stacks:

- stacks from 70 to 100 m: $\Delta p_{max} < 464$ Pa, $\Delta p_{op} < 250$ Pa,
- stacks above 100 m: $\Delta p_{max} + \Delta p_{op} < 464$ Pa.

The value of 464 Pa represents the height of the water in the trap after 14 days of not using the sanitary appliance [5] [6].

3. Design of foul water stacks

The water from the branch pipe striking on the opposite wall of the stack and then descends downwards. The fitting used to connect the branch pipes to the stack significantly affects the hydraulic conditions in the stack, Figure 3. The factors that affect the pressure fluctuations are following:

- the angle of connection of the branch pipe to the stack, Figure 3,
- dimensions of pipes (DN),
- foul water flow rate from branch pipes Q_{ww}.

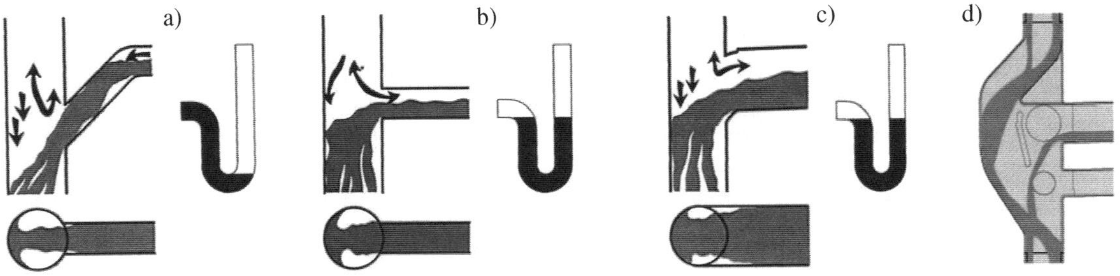

Figure 3. Fittings on the stack [9]: a) simple Y-branch 45°, b) simple Y-branch 88,5°, c) simple Y-branch 88,5° with internal angle 45°, d) Sovent fitting.

Experimental measurements have shown that when connecting the branch pipes to the stack by a simple Y-branch 45°, the water from the trap can be sucked out due to the negative pressure, Figure 3a. It is not recommended to install such branches to stack.

Better flow occurs in:

- simple Y-branch 88,5°, where the dimension of the branch pipe is smaller than the dimension of the stack, Figure 3b,
- simple Y-branch 88,5° with internal arch 45°, where the dimension of branch pipe is DN 100, Figure 3c,
- Sovent fitting, Figure 3d.

With a simple Y-branch 88,5°, water from the trap does not suck out when properly designed and achieves sufficient ventilation of the branch pipes, Figure 3b, c. For a simple Y-branch 88,5° with an internal angle of 45°, it is possible to design stacks for higher maximum flows, Figure 3c. The best flow occurs in the Sovent fitting, Figure 3d.

3.1. Sovent fitting for high-rise buildings

The Sovent fitting is designed to prevent pipes from becoming clogged and creating a piston effect in stack, Figure 3d. The swirling of the flow creates a rotation that allows the flow of foul water to move along the walls of the stack, thus creating a continuous air column, Figure 4. Because the main flow in the stack is always directed around the connection points of branch pipes, the foul water following from the branch pipe has "time" to be included in the main flow, Figure 3d. This system minimalizes collisional turbulences between the two foul water flows, resulting is reduced pressure fluctuations in the stack, Figure 6. This solution has achieved a velocity reduction that limits the kinetic pressure in the stack pipe [9].

Another advantage of this fitting is compatibility with other special fittings from Supertube series, such as BottomTurn and BackFlip. By combining these fitting, it is possible to bend stacks in a particularly economical way, and like Sovent, they prevent the stack pipe from being filled.

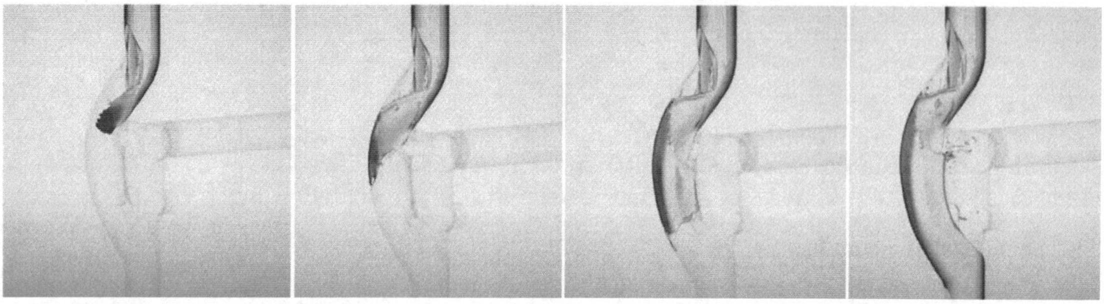

Figure 4. Simulation of the swirling character of water flow in the Sovent Fitting [10].

3.2. Positive air pressure valves for high-rise buildings

This is a foul water stack, which is supplemented by elements of active protection, which are following:

- positive air pressure attenuator installed at the stack, Figure 5a, Figure 1d, tab. 1,
- air admittance valves on each branch pipe, Figure 5b,
- air admittance valve on the top of the stack, Figure 5c.

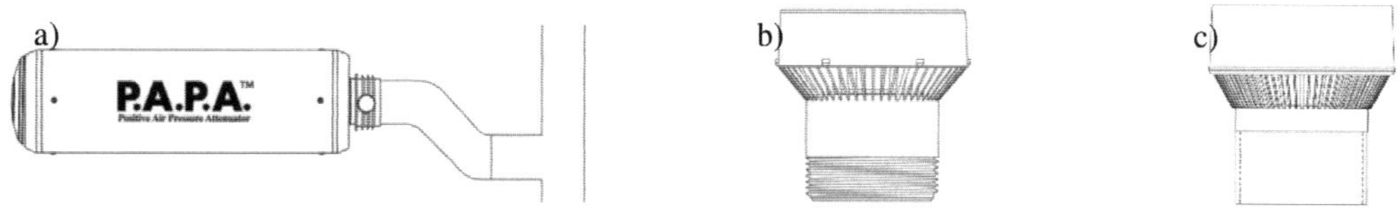

Figure 5. Elements of active protection of stacks [7]: a) positive air pressure attenuator, b) air admittance valve for a branch pipe, c) air admittance valve for a stack.

An active stack system prevents high negative pressure and overpressure values when properly designed, Figure 1d. Air admittance valves prevent the suction of water from traps, and a positive air pressure attenuator prevents water from being ejected from sanitary appliance. For the correct function of the system is necessary to follow the requirements given in the table 1.

As with the Sovent fitting system, it is possible to bend the stacks in an economical way.

Table 1. Requirements for stacks equipped with positive air pressure attenuators [7].

The height of the object	Location of positive air pressure attenuator
5 – 10 floors	one unit on the base
11 – 15 floors	one unit on the base and one halfway
16 – 25 floors	one unit on the base, one on floor 5, one halfway between remaining floors above floor 5
26 – 50 floors	two units in series on the base, then one unit on every 5th floor to the 25th floor, then every 10th floor
50 + floors	a necessary consultation with the supplier

4. Experimental measurements of hydraulic conditions in stacks

The following part of the work describes experimental measurements aimed at assessing negative pressure and overpressure in stacks in various systems for high-rise buildings. Experimental measurements were performed by two foreign companies:

- Geberit in Switzerland comparing pressure fluctuations in stacks using Sovent fittings and simple Y-branch 88,5°,
- Studor in China comparing pressure fluctuations in stacks with direct vent, without and with active pressure protection.

4.1. Pressure fluctuations in stacks when using Sovent and simple 88,5° fitting

Two experimental measurements were performed at Geberit Switzerland:

- the first measurement consisted of comparing the pressure fluctuations in the stacks in the 6-floor building using Sovent fitting and a simple Y-branch 88,5° at two flows in a pipe, Figure 6,
- the second measurement consisted of monitoring of pressure fluctuations in the stack in the 20-floor building using Sovent fitting, Figure 7.

In the first measurement, two modes were compared in the stack of DN 100. In the first mode was a toilet on the 5th floor flushed, Figure 6a. In the second mode was a toiled on the 5th floor flushed, and water flowed with a steady flow of 0,5l/s from the 6th floor.

Stack with a simple Y-branch 88,5° exceeded maximum negative pressure 464 Pa on 2nd – 5th floors at first mode. In contrast, pressure in the stack with Sovent did not fluctuate significantly, Figure 5a. Exceeding the maximum negative pressure means that no water did not remain in traps, which would prevent spreading an unpleasant smell.

During the second mode, negative pressure reached to 900 Pa in the stack with a simple Y-branch 88,5°. Maximum negative pressure was exceeded on each floor. The negative pressure in the stack with Sovent fitting was not significantly increased, Figure 5b.

The stack achieved the best results in this measurement with the Sovent Fitting, which significantly eliminated the risk of suction water from the traps. The maximum value of negative pressure that is 250 Pa.

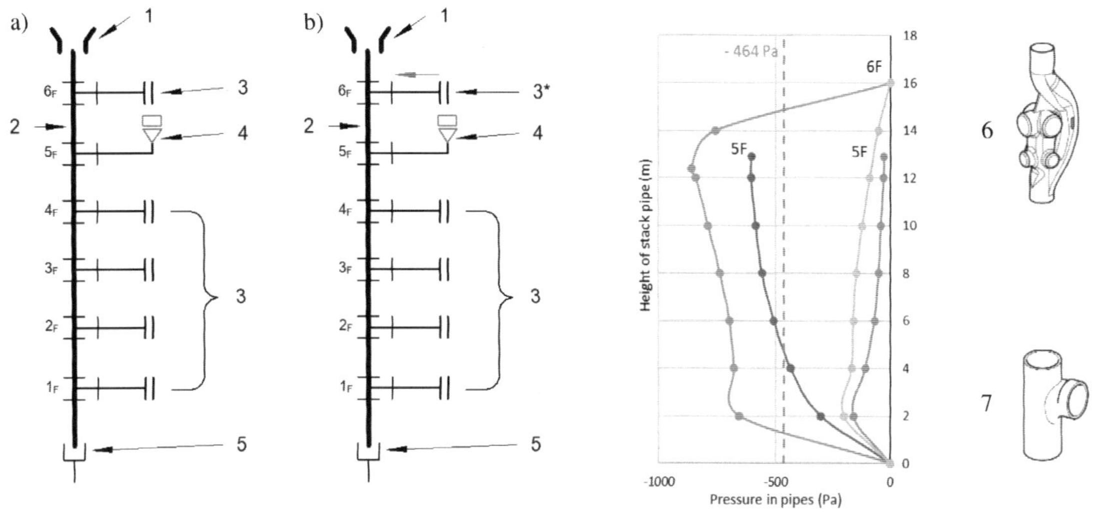

Figure 6. Pressure fluctuations in the stack with Sovent fitting and simple Y-branch 88,5° [11]:
a) flushing the toilet on the 5th floor: — simple Y-branch 88,5°, — Sovent fitting,
b) flushing the toilet on the 5th floor, steady flow 0,5 l/s from 6th floor: — simple Y-branch 88,5°, — Sovent fitting, 1 – air supply, 2 – stack DN 100, 3 – branch pipe without flow, 3* – branch pipe with a steady flow, 4 – branch pipe with toilet, 5 – free discharge, 6 – Sovent DN 100, 7 – simple Y-branch 88,5°.

IOP Conf. Series: Materials Science and Engineering 1209 (2021) 012006 doi:10.1088/1757-899X/1209/1/012006

Figure 7 shows pressure fluctuation in the stack with Sovent on each floor. Toilets with toilet paper were flushed on the 20th, 14th, and 8th randomly. The area graph, which was created when the individual curves overlap, follows that the limit value of maximum negative pressure has not been exceeded. The pressure in the stack fluctuated approximately in a range from -250 to +180 Pa. The stack is suitable in terms of maximum negative pressure.

If a simple Y-branch were used on the stack constructed in this way, the stack's dimension with a direct vent would be around the nominal size DN 150. This solution would be uneconomical compared with the current dimension DN 100.

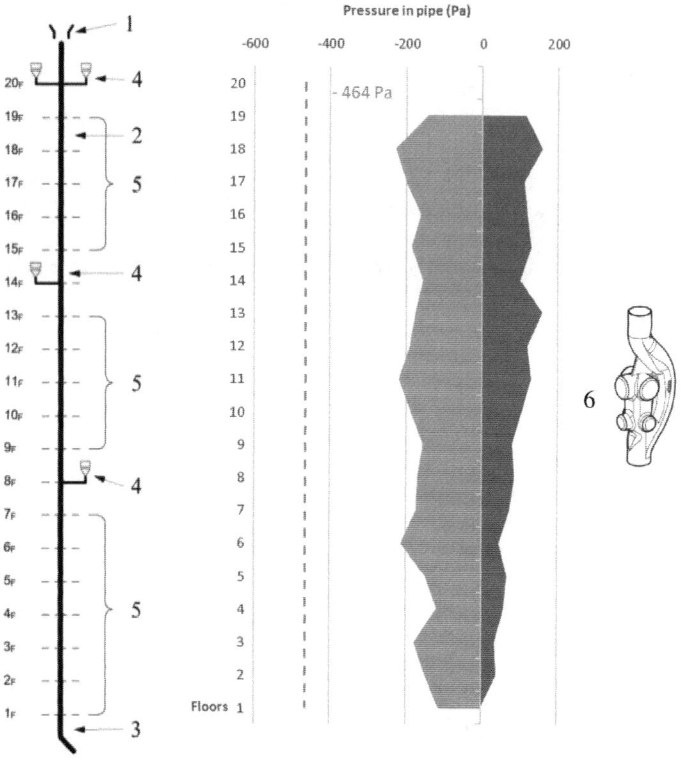

Figure 7. Pressure fluctuations in the stack with Sovent fitting [11]: 1 – air supply, 2 – stack DN 100, 3 – transition to drainage, 4 – branch pipes with toilets, 5 – branch pipes without flow, 6 – Sovent DN 100.

4.2. Pressure fluctuations in stacks without and when using active protection

In the company, Studor was an experimental measurement on the stack with DN 100 on 33rd floors building without equipment and with equipment for active protection compared. Stack structurally corresponded to the description in the chapter 3.2. Water flowed from branch pipe on 30th – 33rd floors. The maximum flow rate in the stack was 6 l/s.

Stack with direct vent exceeded maximum negative pressure in 15th – 29th floor, as shown on the Figure 8. The maximum achieved value of underpressure was 1050 Pa. The stack with active protection does not exceed maximum negative pressure.

Stack with positive air pressure attenuators, which eliminates suction of water from traps, reached the best results in measurements. The maximum value of negative pressure, which has been achieved, was 350 Pa.

Dimension of the stack should be around nominal size DN 200, to the stack with direct vent was complied at maximum negative pressure. This solution would be uneconomical compared with the current dimension DN 100.

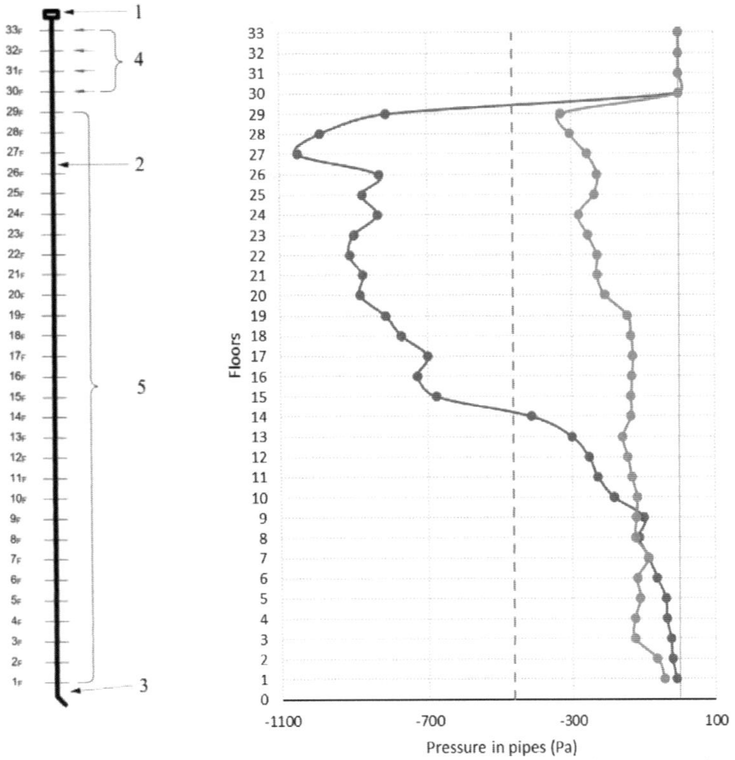

Figure 8. Compare the stack without equipment and with equipment of active protection [7]:
— stack with direct vent, — stack with active protection,
1 – air admittance valve, 2 – stack DN 100, 3 – transition to drainage, 4 – branch pipes with the steady flow, 5 – branch pipes without flow.

5. Conclusion

By analyzing the hydraulic conditions of waste water flow in stacks, and systems for high-rise buildings, it is possible to state:

- water in stack reaches approximately 35 m its maximum velocity 13 m/s, which does not exceed due to air resistance and friction on the internal walls, Figure 2,
- with Sovent fitting it is possible to decrease the velocity of water flow in a stack, Figure 4,
- when using special fittings and systems for high-rise buildings, it is possible to significantly reduce pressure fluctuations in a stack, Figure 6, 7, 8.

Issues of hydraulic conditions water flow stacks is a demanding and extensive topic, especially in high-rise buildings. It is necessary to observance the principles arising from standards, legislation, and assembly constructions to avoid high negative pressure and overpressure in the stack, spreading noise and vibration.

There are more and more demands on drainage. Due to the increasing number of the floors of a building, new technical and material solutions will also be needed to improve the quality of these drainage systems.

6. References

[1] Cheng L, Ho C and Lu H 2004 Current Desiggn of High-Rise Building Drainage System in Taiwan *CIB W062 Symposium 2004 (*France)
[2] Öngören A, Materna R and Tanner M 2005 Investigation of the effect of swept entry configuration on the air entertaiment and self-siphoning behaviour in gravity drainage systems *CIB W062 Symposium 2005* (Brussels) pp 2 - 4

[3] Öngören A and Weiss R 2018 Principles of a genuine high flow waste water fitting with an in-pipe integrated ventilation *CIB W062 Symposium 2018* (Portugal) pp 17 – 24
[4] Pomogaeva V, Matechko L, Prokofiev D and Narezhnaya T 2018 Investigation of the motion processes of wastewater in sewerage of high-rise buildings *E3S Web of Conferences 2018 pp 1 – 9*
[5] Valášek J 1986 Vnútorná kanalizácia. Komentár k ČSN 73 6760. Dimenzovanie potrubia vnútornej kanalizácie (Praha: *Vydavatelství úradú pro normalizaci a měrení) pp 21-30 ISBN 59-313-83*
[6] Vrána J 2004 *Vnitřní kanalizace ve vysokých budovách* (Brno:TZB info)
[7] While S and Hill T 2015 High-rise Testing of Active Drainage System Compared with Conventional Venting on Two Test Wowers *CIB W062 Symposium 2015* (Beijing) pp 11 - 29
[8] Yabe S, Otsuka M, Kawaguchi T and Sugimoto R 2015 An Experimental Investigation of the Influence of Different Offset Piping Methods on the Drainage Performance of a Drainage Stack. *CIB W062 Symposium 2015*(Beijing) pp 30 - 55
[9] GEBERIT – Planning manual – Geberit Supertube. Available on the internet: https://www.geberit.com.au/local-media/files/planning-manual-geberit-supertube-geberitsovent-australia-final.pdf
[10] GEBERIT – Functionality – Geberit Supertube. Available on the internet: https://www.youtube.com/watch?v=_oxfRwf3whc
[11] GEBERIT – Geberit Sovent. Available on the internet: https://www.yumpu.com/en/document/view/53150715/geberit-sovent
[12] Vyhláška č. 549/2007 Z. z. ktorou sa ustanovujú podrobnosti o prípustných hodnotách hluku, infrazvuku a vibrácií a o požiadavkách na objektivizáciu hluku, infrazvuku a vibrácií v životnom prostredí (Slovakia)
[13] STN EN 12056 – 2002: Gravitačné kanalizačné systémy vnútri budov (Slovakia)
- STN EN 12056 – 1: Všeobecné a funkčné požiadavky.
- STN EN 12056 – 2: Potrubia pre splaškové odpadové vody. Navrhovanie a výpočet.
- STN EN 12056 – 5: Inštalácia a skúšanie, pokyny na prevádzku, údržbu a použitie.
[14] STN 73 6760 – 2009: Kanalizácia v budovách (Slovakia)

Acknowledgments
This work was supported by the Ministry of education, science, research, and the sport of the Slovak republic through the Scientific Grant Agency VEGA 1/0303/21 and KEGA n.005STU-4/2021.

IOP Conf. Series: Materials Science and Engineering 1209 (2021) 012007 doi:10.1088/1757-899X/1209/1/012007

Advanced building materials - prefabricated and modular designs as part of Construction 4.0 with a focus on transport constructions

P Krupík[1]

[1] Department of Construction Management and Economics, Faculty of Civil Engineering, CTU Prague, Thakurova 7/2077, 166 29 Prague 6 Dejvice, Czech Republic

E-mail: pavel.krupik@fsv.cvut.cz

Abstract. Digital and technological innovations that have grown rapidly in recent years have been referred to as the 4th Industrial Revolution or Industry 4.0. The paper comprehensively deals with the description of the Advanced building materials, Prefabricated and modular designs as part of Construction 4.0 with a focus on transport construction. The aim of the article is description of this issue comprehensively and with the help of SWOT analysis identify weaknesses, strengths, opportunities and threats with focus on the transport constructions. The content is complemented by an assessment of the current situation, indicating the possibility of use in practice.

1. Introduction

Cost-effectiveness, exhibition speed and improved environmental performance are critical variables that are part of modern construction. Advanced building materials, prefabricated and modular constructions as part of the 4.0 construction are one of the ways to achieve better values in these variables.
Advanced building materials on the one hand and prefabricated and modular constructions on the other are two different things that can achieve synergies within the Construction 4.0 concept. The use of innovative prefabricated modular structures has the potential to contribute to the prefabrication of the environment, embodied in the areas of embodied energy disputes arising from waste reduction and improved waste. Such use of better materials is more reliable and stable in production halls than directly on the building, and it can also be assumed that stable qualities have been met [1].
This article provides an expert overview of advanced building materials, prefabricated and modular designs within construction 4.0 with a focus on traffic structures.

2. Advanced building materials

Given the limited supply of convection materials, the use of advanced materials is one of the possible ways to prevent the complete depletion of these resources.
Due to the number of innovative materials below, the author only lists some of them, which can also be used in traffic construction.

2.1. Ultra high performance concrete

They are modern composite materials based on cements, fine aggregates and the most modern chemical additives, with a very low water content. They achieve very high values of physical and mechanical properties [2].

These concretes are a challenge for designers and architects, they open up the possibility of subtle and tough constructions with a very reduced dead weight, but on the other hand with a long durability and, above all, with high shape variability. They are suitable for environments with high stress from chlorides, frost cycles or chemically aggressive substances. They are extremely stable in appearance [2]. This material is characterised by its excellent properties, exceptional strength and excellent durability achieved through improved homogeneity and packing density. However, it is slowly encountering obstacles that limit its use. High initial costs, limited regulations, design difficulties and complex manufacturing techniques, together with the limited resources available, seriously hinder its commercial development and use in modern construction [6].

For example, the use of full-depth precast waffle panels made of ultra-high-strength concrete achieved accelerated bridge construction and increased bridge deck life [3].

2.2. Fiber reinforced polymers

Fibre-reinforced polymers are already used in the rehabilitation/strengthening of built-up infrastructure made of concrete and masonry [4].

These polymers include epoxy resins that are capable of curing and hardening at normal temperatures and humidity. These resins show several advantages over traditional materials in terms of high strength-to-weight and stiffness-to-weight ratios, great versatility, shorter intervention times, and, as a result, interruption of activities, thus reducing overall costs. Maintenance work is also reduced when polymer composites are used to replace traditional construction materials [4].

2.3. Corrosion of steel in concrete

Investigating the interface between steel and concrete and understanding the role of many different local characteristics plays an important role in corrosion initiation. Utilizing effects such as steel concentration polarization, pore plugging, and self-generated cracks in concrete can slow the rate of onset of corrosion on engineered structures and prestressed and post-tensioned reinforced concrete structures, thereby extending their service life and reducing the life cycle cost of the transportation structure [5].

3. Prefabricated and modular designs

3.1. Prefabricated steel structure system

The key advantage of a prefabricated steel structure system in a case study over a conventional concrete structure and prefabricated wooden building is the ability to reuse a significant portion of the file structure at the end of the life of the building. This can result in a significant reduction in waste landfill. The study also clearly demonstrated indicated that the prefabricated structure is able to provide a better environment performance compared to conventional design methods if they were originally designed to be repeatedly, either adaptively or through disassembly [1].

3.2. Modular designs

Modular construction arises from the prefabrication of large-scale modules in production and their assembly on the construction site. It is fast, environmentally friendly and, thanks to its linear production and quality control, achieves a high standard in the construction industry. The main advantages are the speed of assembly, disassembly and the associated flexibility. Modular construction can respond to the current needs of the owner and can be expanded. However, the main advantage is the extreme environmental friendliness of the construction site. The living space is not subject to the long-term noise and dust pollution that is characteristic of standard construction [8].

Its use in transport construction is mainly in the construction of site facilities. Modular living and sanitary containers can quickly be turned into offices, changing rooms, sanitary facilities or even dormitories. In addition, it can be permanent or temporary modular fuel stations as a fuel storage facility. Today, fuel stations are multi-functional, where in addition to fuel, they offer a mixture for rest, cleaning and refreshment, but also for the purchase of food or gifts. In addition, the modular solution provides the possibility of quick expansion of the station with additional modules by minimizing the time when the station is out of operation [8].

Figure 1. Cell site made of modules [7].

3.3. Prefabricated preloaded beams

Prefabricated preloaded beams made of high dense concrete with evenly distributed reinforcement from short steel wires. The beams are reinforced with preloaded reinforcement anchored by cohesion with the absence of conventional concrete reinforcement. These beams can be used in all areas of road and railway construction, as well as in civil and industrial construction. The advantage of these beams is their lighter weight, lower requirements for transport and assembly mechanisms [2].

Figure 2. Prefabricated preloaded beam [2].

IOP Conf. Series: Materials Science and Engineering 1209 (2021) 012007 doi:10.1088/1757-899X/1209/1/012007

Figure 3. Prefabricated preloaded beam [2].

3.4. Bridge abutments anchored with OMO steel anchor

The abutment in Figure 4 has a New Jersey-shaped leading edge, with steel supports and a steel handrail mounted on the abutment every 2 metres. On the reverse, outer side, the abutment may be fitted with a 'nose' to replace the bridge ledge. The pieces are connected by two safety cables. They can be connected to steel type NH4 bollards. The bollard can be equipped with a transparent noise barrier wall with a height of 4000mm from the road level [2].

Figure 4. Bridge abutments [2].

The advantages of this type of anchors are high fixing strength at a lower anchor depth, immediate effectiveness after tightening, no temperature limitations and the possibility of reusing the anchors after an impact. The disadvantage is that the anchors are more expensive than other types of anchors [9].

4. SWOT analysis

Due to the very different uses of Advanced building materials and Prefabricated and modular designs, the possibilities and possible problems of these parts of Construction 4.0 in relation to the swot analysis are not presented in summary, but the most important ones are outlined for each of the selected cases. In general, the common main disadvantage is the novelty of these technologies and thus the lesser known possibilities for practical application. However, a review of the available published literature has shown the interconnectedness of these two elements and thus the appropriateness of the choice of the topic.

5. Conclusion

At present, advanced building materials and prefabricated and modular structures are in the research and development phase, but some possibilities are already used in practice. The author believes that after removing the current shortcomings and obstacles, these possibilities are used to expand much more and thus become a commonly used tool for use in construction. As such, however, it should play an important role in Construction 4.0.

6. References

[1] Gunawardena D, Mendis P, Ngo D, Aye L and Alfano J 2014 Sustainable Prefabricated Modular Buildings *5 th International Conference on Sustainable Built Environment (Kandy)* ed R Dissanayake (Washington: ICSB) pp 1–7

[2] KŠ prefa 2021 *Dopravní stavby: Mostní nosníky, římsy, svodidla, kotvená svodidla, opěrné a protihlukové stěny a další* Available online https://ksprefa.cz/produkty/dopravni-stavby/

[3] Ebadollah H, Sritharan S, Rouse J M and Aaleti S 2016 Bridge Decks with Precast UHPC Waffle Panels: A Field Evaluation and Design Optimization *Journal of Bridge Engineering* **21** 1 1–32

[4] Frigione M and Lettieri M 2018 Durability issues and challenges for material advancements in FRP employed in the construction industry *Polymers* **10** 3 247

[5] Angst U M 2018 Challenges and opportunities in corrosion of steel in concrete *Mater. Struct.* **51** 4

[6] Azmee N M and Shafiq N 2018 Ultra-high performance concrete: From fundamental to applications *Case Studies in Construction Materials* **9** e00197

[7] WAREX: modulové stavby 2021 *Cell site made of modules* Available online https://warex.cz/reference/modulove-stavby/

[8] KOMA modular *Modularita* Available online https://www.koma-modular.cz/modularita

[9] OMO 2021 *Kotvení Mostních Svodidel* Available online http://www.svodidla-omo.cz/cz/kotveni-mostnich-svodidel

Acknowledgments

This work was supported by the Grant Agency of the Czech Technical University in Prague, grant No. SGS20/100/OHK1/2T/11.

IOP Conf. Series: Materials Science and Engineering 1209 (2021) 012008 doi:10.1088/1757-899X/1209/1/012008

Opportunities for rehabilitation of facades of historic buildings in Moldova with limestone elements

D C Albu[1] and A Lesan[2]

[1] Faculty of Civil Engineering and Building Services, "Gheorghe Asachi" Technical University of Iasi, 1 Profesor Dimitrie Mangeron Avenue, 700050, Iasi, Romania.
[2] Faculty of Constructions, Geodesy and Cadastre, Technical University of Moldova, 41 Dacia Avenue, block 10, MD-2060, Chisinau, Republic of Moldova.

E-mail: doina-cezara.albu@student.tuiasi.ro

Abstract. Most buildings of historical and architectural value located in the Republic of Moldova are built of limestone or brick. The facades are decorated with columns, semi-columns and pilasters, balusters, bas-relief, and consoles. The degradation of historic buildings over time leads to the need for rehabilitation and restoration. Often privately owned buildings are repaired with the application of modern materials without taking into account the original materials, without researching the original elements and materials. Authentic decorative elements are replaced by polyurethane foam elements, which due to their easy installation and relatively low cost, can significantly reduce the financial costs of finishing and decorating facades. The interior walls of the brick are covered with plasterboard. The exterior walls of limestone are plastered with cement-based mortar instead of lime-based mortar. In some cases, the walls were covered with polystyrene. All these replacements reduce the architectural and historical value of the buildings lead to the loss of their authenticity. In order to preserve the value of historic buildings, it is necessary to use local materials. This research examines the possibilities of using limestone both for the rehabilitation of masonry and for the restoration of the decorative elements of buildings built in the XVIII-XX centuries.

1. Introduction

Historical buildings have an important cultural significance and impact on the development of the younger generation.

A current problem for the whole world is the maintenance, consolidation, and rehabilitation of heritage buildings. Historical buildings are highly vulnerable due to the way they are built and the way they are constructed.

In Chisinau, the capital of the Republic of Moldova, there are many historical buildings in an inadequate condition, which require reconstruction or complete rehabilitation [1]. A series of historical buildings in Chisinau have been destroyed and others are in ruins or in an advanced state of decay [2]. Since the early 1990s, architectural monuments have been destroyed to be replaced by hotels, shopping centres and residential buildings.

The purpose of this work is to highlight the limestone elements that can be used to restore the exterior walls of historic buildings.

In this article the author examines the diversity of limestone elements that can be used in the rehabilitation of historic building facades. The technical characteristics of the architectural elements and the technologies for the rehabilitation of facades of historical buildings in Chisinau are examined.

The authors use a variety of research methods such as historical methods, analysis, and synthesis to make it possible to highlight the historical building elements restored with limestone; inspection and technical evaluation make it possible to determine the restoration project to be carried out; comparative analysis can choose the most appropriate method. To repair the building under study.

2. The diversity of limestone elements for the rehabilitation of historic building facades

The historical buildings in Chisinau in 18th-20th centuries were built in an eclectic style with rococo elements.

The majority of the facades of the historical buildings were clad with natural stone and limestone elements. Over the years, the facades become deplorable for several reasons.

New trends in the rehabilitation of historical facades by replacing stone elements with decorative polystyrene elements lead to the loss of the architectural and historical value of the building.

The Republic of Moldova has sufficient resources of limestone, which allows the rehabilitation of historical buildings with original elements. The buildings rehabilitation with limestone elements allows to preserve not only their shape but also their unique architectural value.

From an architectural point of view, various elements shown in Figure 1 can be rehabilitated using limestone.

Figure 1. Limestone elements for façade.

Figure 1 shows the basic limestone elements of a facade, which can be custom-made depending on the architectural style of the historical building.

The natural stone *columns* are based on limestone, which is extracted from mines from the Republic of Moldova on demand. Their properties are unique, so no artificial material can be compared to a natural and environmentally friendly product.

In construction, the limestone used for making columns is of the brand M50 or M75. In the mines of the Republic of Moldova limestone is of 2 types: oolite and of shell. For the precise shape of the columns, oolitic limestone should be used, as it is more compact than the shell limestone.

The limestone columns have a seismic resistance up to 8 degrees on the Richter scale. The density of the columns can reach 1900 kg/m^3. High durability and wear resistance are some of the advantages of using natural elements in reconstruction. The limestone does not destroy as quickly as concrete and is less affected by moisture and temperature changes.

Decorative stone columns serve as load-bearing columns in a building. For example: they support the roof, or the platform of the veranda, the balcony. The use of this material guarantees the durability and safety of the construction.

There are stone facade columns of various shapes shown in Figure 2 - rectangular and cylindrical constructions, including difficult passageways, various decorative elements, balustrades, keystones and consoles.

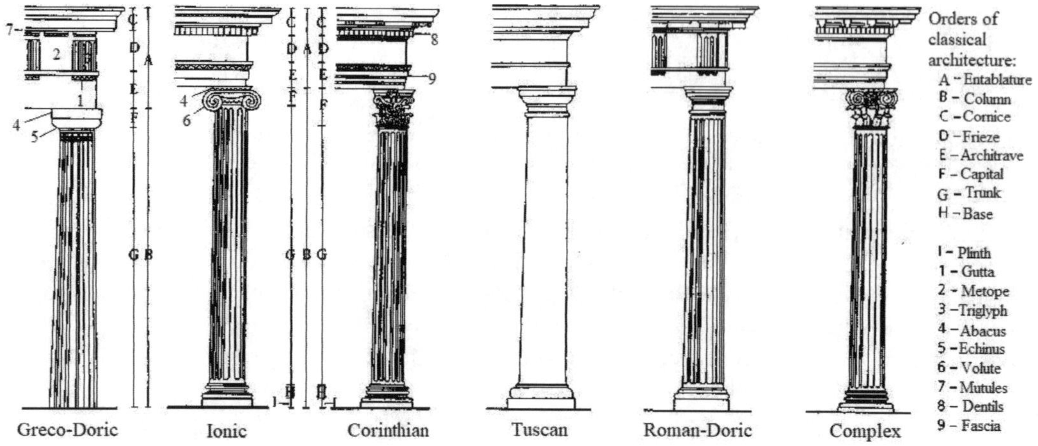

Figure 2. Types of columns and its elements.

The manufacture of limestone *cornices* requires complex production technologies, skills, time, and professional high-precision tools.

The limestone used for the cornices can be of the M25 and M35 marks. This type of limestone has a lower strength compared to the M50 and M75. The compressive resistance of limestone blocks ranges from 0.4 MPa to 300 MPa.

Depending on the size of the desired cornices, limestone blocks are extracted from the Cricova mines to be handed over to the craftsmen, who process the stone, cut out everything superfluous, creating new shapes. The advantages of limestone cornices are as follows:

- Sustainability;
- Longevity;
- Resistance to temperature changes;
- Moisture resistance;
- Ability to create shapes of any difficulty;
- Possibility to create articles of various proportions;
- Choice of colours, shade of material, from which the stone cornices for the building facade should be produced.

Carved stone *consoles* are a special type of decorative construction, which are used for the ornamentation of facade elements such as windows, balconies, doorways and cover plates. The application of consoles is actual in the decoration of buildings of various styles, starting from classical and ending with the "empire" and "baroque" styles. Due to its variety of sizes and oases, limestone consoles can be used for the reconstruction of buildings of historical interest.

In Chisinau, facade consoles can be seen not only on old buildings in classical style, but also on most modern buildings.

Facade consoles have a wide range of other advantages such as:

- High durability and temperature resistance indicators;
- Ability to use a unique style when implementing the design;
- Speed of making stone consoles;
- Simple installation, thanks to which installation costs are significantly reduced;
- Possibility to make the facade look unique;
- Preservation of the full external appearance throughout the entire service life.

Because natural stone is an expensive material, many similar products have appeared on the local market, such as expanded polystyrene, fibre-glass reinforced concrete and artificial stone - all these materials are cheaper, but the high quality of natural stone items allows them to be used for a longer period of time and to serve as a formidable ornament for the entire period of use of the building.
The natural stone *panel* presents a unique decorative element, which is made of limestone and is used to decorate facades. The use of stone elements in the creation of the facades highlights the style of the 18th-20th centuries of buildings, as well as gives the building originality. So natural stone items have a wide range of use.

In most cases, stone panels are made of materials such as limestone, granite, marble, travertine or onyx. The combination of the given stones can bring refinement to any building.

Natural stone panels in Moldova are becoming more and more popular due to the fact that this material has many positive properties, which ensure its distinction from artificial similarities. Among the distinguishing features are:

- Long term of exploitation;
- Minimum care requirements;
- Can be shaped into a wide range of forms and styles;
- If necessary, it can be advantageously combined with other materials;
- The stone is environmentally friendly and doesn't present any danger.

Natural stone is a very durable material, so the panel will last a long time.

The limestone *keystone* from Moldovan mines has a high resistance and a long period of exploitation, it is not vulnerable to low temperatures, high temperatures, snow and rain. Therefore, such an item will serve for a long time. The limestone used to make the keystone resists without signs of damage to a minimum of 6 cycles of freezing at -15° and thawing at +15±5 °C.

High durability is one of the basic factors when choosing raw materials. The keystone for facade vaulting is the supporting detail for arch constructions, passages, cupolas. It is installed during cladding at the end. In most cases it has a pointed shape, which allows the article to be integrated into any construction and fixed with an adhesive.

Natural stone facade anchors in Moldova are used to decorate windows, facing the same direction of the building as the entrance group. The facade is the face of any building and its appearance influences the perception as a whole. If on the outside natural raw material is used for columns, balustrades and other details, then it is reasonable to use window frames and facade anchors made of natural stone to get unique compositions.

The metal or plastic does not have the necessary properties and qualities. Metal over time gets covered with rust deposits, and other materials, such as mould and moss. The resistance of plastic to low temperatures decreases with each passing year. Continuous stretching and compression gradually lead to the destruction of the anchorage, and moisture completes this process.
Stone window anchors can serve decades and even centuries. An example are historic buildings that require rebuilding after centuries, but the resistance to external factors is impressive compared to the longevity of the building. Many of the buildings are still built in the 18th century, but the anchorages have been preserved to the present day.

Instead of repairing and restoring facade every year the elements that have been deteriorating over time, it is safer to use durable and qualitative materials. Natural stone window frames and other items are better than most similar products. It is a beautiful, durable, and environmentally friendly material.
Natural stone *railings* are used for both interior and exterior works. They are used on staircases, balconies, verandas, or to create large decorative elements. On balconies these elements serve as support structures for the balustrade, and on stairs - for the staircase balustrade. In the first case they can be combined with small columns. If limestone balustrades are used in compositions, then railings become component parts of the construction.

Figure 3 shows the types of railings that can be made from limestone.

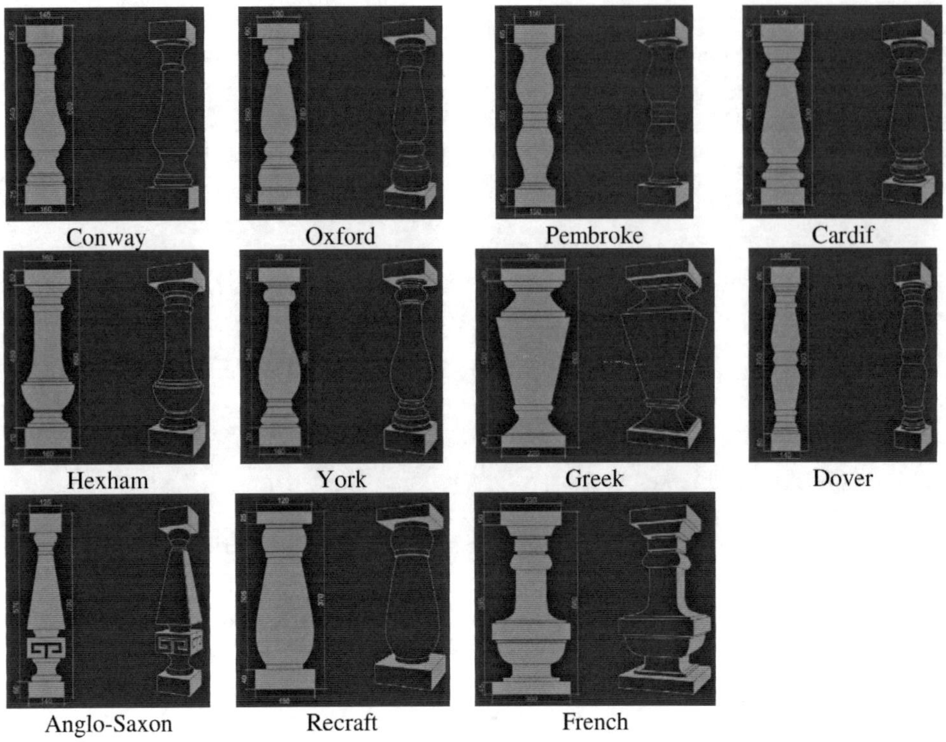

Figure 3. Types of natural stone railings.

The shape of the railings, its size and other parameters are determined individually for each case. The limestone balusters have the following advantages:

- Undeniable aesthetics;
- Long life span;
- Resistance to high humidity;
- Resistance to high/low temperatures and sudden temperature changes.

The decorative element, such as the stone *balustrade*, consists of several component parts: balusters and their supports, stringers, parapet, pillars, spheres, and other elements, which embellish the construction.

Natural raw materials like sandstone, limestone, travertine, and others are used to make balustrades (except those cast from reinforced concrete). They differ in density, processing methods, structure, and other parameters.

The profile is the standard element, representing a plate of well-defined sizes and shapes. However, in most cases the stone facade band can be decorated with ornaments or carvings.

The advantages of a stone panel over similar products made of other materials are:

- Resistance to heat and cold;
- Resistance to atmospheric precipitation;
- Wind resistance;
- Direct sunlight resistance.

Wood, plastic, metal, etc. also have some advantages, but they do not always achieve the desired operating characteristics. Simplicity of processing, the possibility of creating beautiful natural stone articles with difficult shapes, durability, strength is just some of major advantages.

Various types of limestone can also be used as raw material and such an amazing stone as travertine, which is also called Tibur stone or limestone tuff.

Natural stone cladding is a durable material to serve as a stunning decorative element over a long period of time and is an excellent material for the rehabilitation of historic buildings. The facades of natural stone cladding have many advantages, thanks to which they have become popular since the 16th-17th centuries. Among their characteristics are:

- Maintaining the original exterior appearance regardless of weather conditions;
- In high humidity conditions decorative stone for facade collects moisture;
- The material is not vulnerable to the sun's rays;
- The load-bearing structures of the house are not subject to additional weight;
- The stone combines beautifully with other elements of interior and exterior decoration;
- Total resistance against harmful mechanical action.

Decorative stone in Moldova is widely used to finish residential houses, villas, as well as hotels, sanatoriums and tourist complexes, so it is no less important than building stone. Also, the reconstruction of historical monuments and facades, using decorative limestone helps to preserve the historical-architectural appearance of buildings.

The modern market for cladding materials offers consumers, in addition to natural stone and artificial cladding stone, 2 other types:

- limestone cladding, involving the use of limestone;
- cotilet classing, involving the use of cotilet stone;

The rehabilitation of historical buildings in Chisinau is carried out with the help of modern materials, which do not have a long resistance to climatic factors. The author proposes the use of natural stone and limestone elements, which provide resistance to the facade and ensure the preservation of the authentic historical-architectural appearance of the buildings.

3. Rehabilitation of the facades of historical buildings in Chisinau, 18th-20th century

Over the last 30 years, the cultural-historical heritage of the Republic of Moldova has been in a state of continuous degradation. In the period 2003-2006, 977 monuments with the status of protected monument of national and local category were identified in Chisinau and registered in the Register of Monuments of the Republic of Moldova [3]. Already in 2010 it was found that more than 25% of them are either destroyed or in active process of destruction. In Chisinau, 82 monuments have been demolished, of which 44 were demolished between 1993 and 2006 and 38 between 2006 and 2012. 160 of these monuments have undergone illegal interventions. The total number of buildings that have suffered due to non-compliance with the legislation on the protection of monuments is 254 monuments with the status of protected monument of national and local category. And this figure continues to rise, primarily due to existing corruption and non-compliance with legal provisions [1].

The research identified a building of historical value, which requires a complete rehabilitation of the facade.

The following steps should be taken into account before starting to rehabilitate the facades of historic buildings:

1. Identification and/or classification of existing materials;
2. Assessment of existing facade systems, including drainage systems, insulation, vapour barriers and structural supports;
3. Thorough inspection of all items, documenting all signs of damage and defects;
4. Investigation of deterioration of materials by field and laboratory test methods;
5. Identification of potentially hazardous and/or unsafe conditions [4].

Damage to the facades of historical buildings is caused by weathering, invasive plant growth, pollution, and the accumulation of dirt. Other damage can occur through erosion as a result of stone movement, defective or broken anchors and even human error in design, engineering or construction.

Restoration of historical buildings facades can be achieved by the following methods:

- Restoration by stitching involves fixing cracks in the facade and this method can be of 2 types:
  - Mechanical stitching is performed using steel beams;
  - Chemical stitching involves injecting chemical adhesives into cracks.

- Restoration by injection;
- Restoration through reconstruction involves partial or total replacement of architectural elements of the historical building facade;
- Restoration by disassembly and reassembly;
- Restoration by reinforcement with support;
- Restoration by rebalancing trends;
- Restoration through trend stabilisation;
- Restoration by completion involves adding missing elements;
- Restoration by replacement;
- Restoration by casting;
- Restoration by cleaning;
- Restoration by hardening using iron bonds [5].

The historical monument under analysis is the *House of Report with rooms for social functions* on 148 Stefan cel Mare Boulevard, corner with M. G. Bănulescu-Bodoni Street.
In the 1930s Ivan and Annastasia Monastârsky built 2 two-storey houses next to each other in the neighbourhood bordered by Stefan cel Mare Boulevard and Mitropolit Gavriil Bănulescu Bodoni Street. In the 60s of the 19th centuries, these buildings were rebuilt and integrated into one, for the location of Theological Seminary. After 1870 the building was rented by various educational institutions. In 1874 a lease was concluded with Charl Sailladeni, a citizen of Switzerland, who rebuilt it for the hotel "Suisse". During the Second World War the building was damaged and rebuilt in the late 1940s according to the E.R. project. Spirer, who benefited from the consultations of the academician A.V. Shchusev, with the raising of the attic floor. The building occupies the side of a narrow block, has a square plan with facades aligned with the red lines of the side streets, shown in Figure 4. The facades have symmetrical compositions, with side elevations, superimposed by porticos with four columns of the Ionic order, supporting triangular pediments [6].

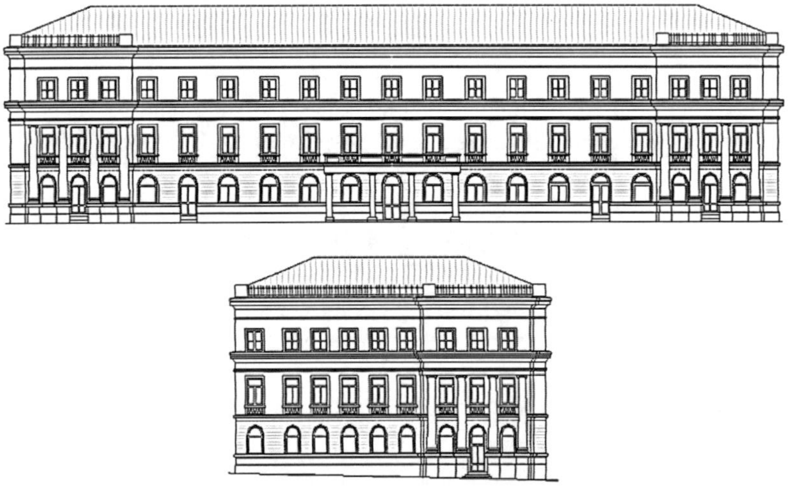

Figure 4. Sketch of the facade of the historical monument on 148 Stefan cel Mare Boulevard, Chisinau, Republic of Moldova [6].

The building has undergone a number of restorations over the years, the last restoration being of the clean-up type. At present the condition of the facade is in a less than satisfactory state, shown in Figure 5. For the restoration of the façade, the method of restoration by reconstruction should be used, with architectural elements made of natural stone, namely limestone.

Figure 5. Facade of the historical monument on 148 Stefan cel Mare Boulevard, Chisinau, Republic of Moldova, 2021.

Before starting the restoration and proposing effective solutions, an on-site inspection of the facade was carried out and is shown in Figure 6.

Figure 6. Inspected facade of the historical monument at 148 Stefan cel Mare Boulevard, Chisinau, Republic of Moldova, 2021.

The main purpose of the reconstruction is to restore the building's architectural-historical exterior appearance, its technical and economic qualities. The main objective is to improve the operational comfort of the building.

During reconstruction, only the load-bearing elements of the building are preserved: solid walls, ceilings, stairs, and platforms. All others should be replaced in whole or in part. The following works are planned here:

- radical reorganisation of the structure of the object, including changes to the technical characteristics and overall dimensions of the structure;
- changing the function of the premises;
- complete re-equipment of utilities;
- reinforcement of load-bearing structures;
- additional measures to improve attics, basements, roofs, exterior finishes.

Before reconstruction begins, a project is drawn up taking into account the technical condition of the object, operational deficiencies and defects detected.

The following steps shall be taken to rehabilitate the facade of the historical monument at 148 Stefan cel Mare Boulevard:

1. Removal of old plaster;
2. Removal of destroyed items such as:
    a. Cornice;
    b. Railings;
    c. Console;
    d. Anchor;
    e. Columns.
3. Prime the walls;
4. Apply plaster over the mesh;
5. Priming walls over plaster;
6. Mounting the decorative limestone elements using adhesive.

The basic technology for applying stone and decorative limestone elements is shown in Figure 7.

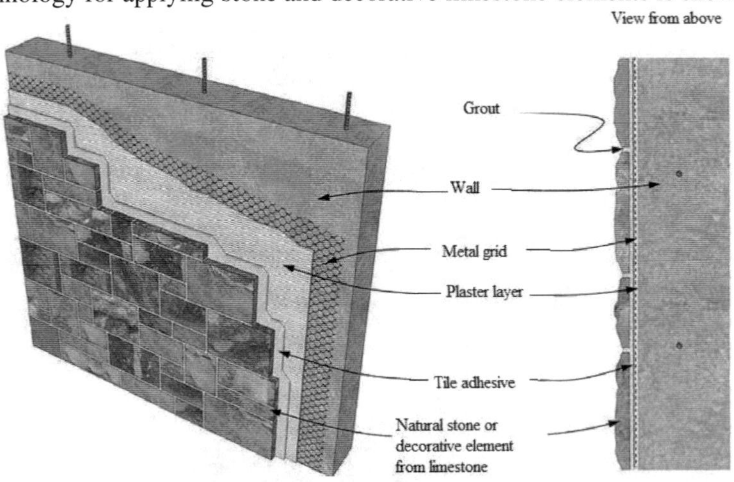

Figure 7. Technology for mounting natural decorative limestone elements on building facades.

Following the full analysis of the facade of the historical building on 148 Stefan cel Mare Boulevard, it was decided to respect the architectural appearance by applying decorative limestone elements described in section 2 and plastering the walls with decorative limestone-based plaster. Figure 8 shows the changes to be applied to the building facade.

IOP Conf. Series: Materials Science and Engineering 1209 (2021) 012008 doi:10.1088/1757-899X/1209/1/012008

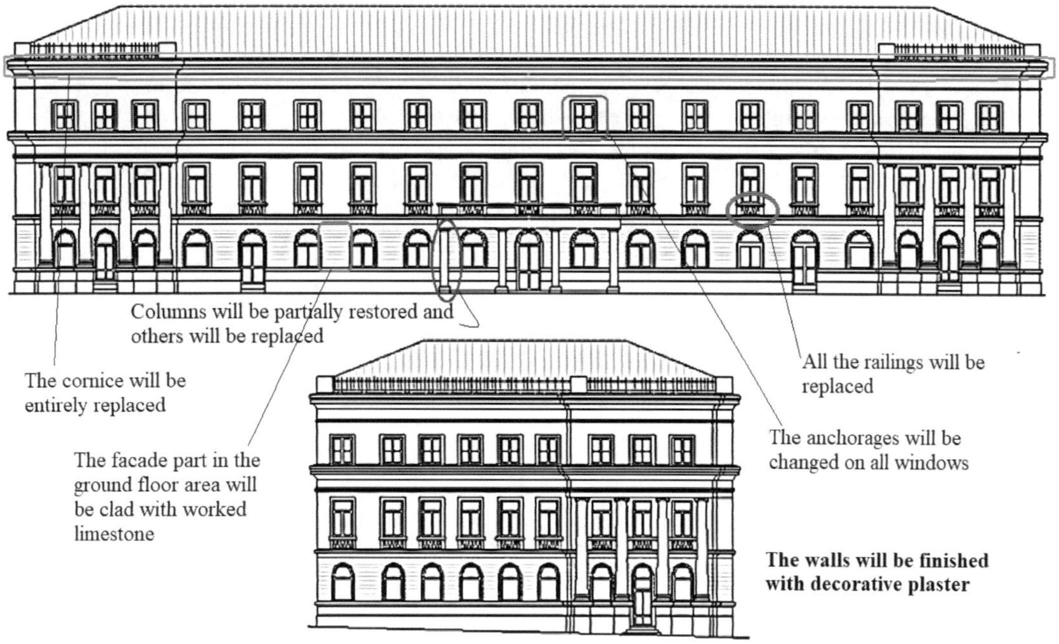

Figure 8. Sketch of the changes that should be made in the restoration of the façade.

Through the example shown above, it was intended to draw attention to the advantages of using limestone elements for the restoration of facades of historical monuments in Chisinau and the use of local materials.

4. Conclusions

In the conclusions we will point out the advantages of decorative limestone elements and the rehabilitation of historic building facades:

One of the main advantages is that limestone elements are natural and environmentally friendly. Other advantages of natural stone elements are frost resistance, of 50 freeze-thaw cycles, and wide range of density, from 800 (shell rock) to 2800 kg/m^3 (crystalline). Crystalline limestone is recommended for facade cladding.

Natural stone has low hardness according to Mohs scale - 3. The material is easy to work, can be cut and carved to form decorative elements in architectural styles from different centuries.

Due to its porous structure, limestone has a low thermal conductivity. The average coefficient is 0.5-0.9 W/m deg, which is comparable to silicate bricks and hollow bricks.

Although deterioration of limescale is inevitable over time, there are ways to slow down the process and prevent severe damages. Routine maintenance and inspections can reveal surface deterioration, displacement, and joint failure, which can indicate more serious underlying conditions. Early detection and prompt, careful attention is essential. Adherence to maintenance of limestone elements leads to longevity of the elements.

The opportunity of preserving the architecture of the facades of the historical heritage of Chisinau should be examined through the prism of facade restoration using decorative elements made of natural stone such as limestone. In our opinion, limestone elements will preserve the architectural authenticity of the buildings for many years compared to decorative elements made of other materials.

5. References

[1] Albu S, Albu I and Ursu V 2020 Patrimoniul arhitectural: aspecte legale, tehnice și economice (Chisinau: MS Logo)

[2] Albu S and A. Lesan 2021 Expropriation as a Way for Protecting the Architectural Heritage Current Urban Studies 9 2 299–314

[3] Jitari L 2021 General Evaluation of the Potential of the Cultural-Historical Heritage Built in the Republic of Moldova Open Journal of Applied Sciences 11 6 633–646
[4] Croitoru G 2021 Approaches Regarding the Functional and Structural Rehabilitation of Historical Monumental Buildings Open Access Library Journal 8 6 1–10
[5] Anwar M 2019 Practical Techniques for Restoration of Architectural Formation Elements in Historical Buildings World Journal of Engineering and Technology 7 1 193–207
[6] Centrul istoric al Chişinăului 2008 Ştefan cel Mare, 148 (colţ str. Bănulescu-Bodoni) Available online http://www.monument.sit.md/stefan-cel-mare/148/ [Accessed 16 June 2021]

Acknowledgments
The research was carried out within the project 20.80009.0807.34 Increasing the value of architectural heritage of the Republic of Moldova.

IoT and 5G enabled construction detection model

H Begić[1]

[1] Faculty of Civil Engineering and Architecture Osijek, Josip Juraj Strossmayer University of Osijek, Street Vladimir Prelog 3, 31000 Osijek, Croatia; hbegic@gfos.hr (H.B.)

E-mail: hbegic@gfos.hr

Abstract. The Fourth Industrial Revolution (4IR) introduced positive changes to some industries, while for most construction industries it is still enthusiastically anticipated. The 4IR focusing on the construction industry in literature is known as Construction 4.0. The Construction 4.0 concept is invoked to transform the current ways the construction industry operates while ensuring benefits, such as reduced overall construction projects' costs and duration, improved quality and work safety, etc. Due to the increasing web usage, it is anticipated that the 4IR technologies will achieve full potentials by the uprising of the fifth generation technology standard for broadband cellular networks (i.e. 5G). One of the most important 4IR technologies is found to be the Internet of Things (IoT). In this perspective, a construction monitoring approach, more precisely a model for construction detection and object spatial/time positioning, is presented in this paper. While still in its initial phase, the model was tested and verified in the laboratory environment for small-scale object detection. It was found that the quality of the model will be significantly improved with the use of the 5G network, while the objects' pool, as big data required for the model's deep learning, is highly dependent on the IoT.

1. Introduction

The fourth industrial revolution (4IR), as the previous ones, is also driven by technological innovations. In the first revolution, it was the printing press and steam power. The second revolution focused on using electrical power for encouraging mass production and the third caused increased use of electronics, telecommunications, personal computers, and above all the Internet i.e. the digital revolution. The term 4IR first appeared in 2011 in Hannover, Germany at an industrial fair and became a German national strategy [1]. The 4IR concept continued to gain more interest in all fields of industry, and consequently the construction industry, where it obtained a new term called Construction 4.0. All industries can benefit from the 4IR, but due to the specificity of the construction industry, such as the change of workplace with every new construction site and the uniqueness of each project, the benefits from 4IR in the construction industry are especially emphasized with the goal of digitalization and automation of the construction site and the whole construction project's lifecycle. The Construction 4.0 concept can deliver many benefits to the construction industry such as cost reduction, time saving, quality improvement, sustainability improvement, etc. Despite the mentioned benefits, the construction industry is known for its sluggishness in adopting new technologies, so many of the technologies and innovations are still in the development phase and lack physical application on the site. The reasons for this can be found in the continuation and concerning lack of skilled workforce, lack of specific knowledge required by the Construction 4.0 concept, cybersecurity problems, etc. Most of the 4IR technologies, i.e. drivers, are also considered to be the drivers of Construction 4.0. Among numerous, the most highlighted are Big Data (BD), the Internet of Things (IoT),

Additive Manufacturing (AM), Cloud Computing, Augmented reality (AR), Robotics, Drones, Radio-frequency identification (RFID), and in the specific context of the construction industry – Building Information Modeling (BIM). While some of the listed technologies are already in a far-off phase of their development and implementation into the construction industry, some are still not so widely used which results in mismatching technologies. Another driver worthy to underline is the Fifth Generation technology standard for broadband cellular networks (5G). 5G can be considered as a Construction 4.0 technology, while also being the driver for other Construction 4.0 technologies by enabling enhanced performances of them. As the 4IR, 5G is also an advancement based on the previous inventions since every new generation of wireless networks brought higher speeds and more functionality to the network. 1G brought the first cellphones and voice calls, 2G enabled texting, 3G enabled internet access and 4G enabled speeds that are still most widespread today and resulted in smartphones. With the continuously increasing number of users of the network, 4G has reached its limits and it is clear that there is a need for improvement. Therefore, the 4G network is not able to provide a quality experience while communicating with drones and robots, provide instant cloud services, and the development of IoT on the desired level [2]. It is expected that 5G can offer many improvements by increasing mobility and capacity of the network, shortening the latency, and increasing network security [3]. Regarding the characteristics of 5G, it is driven by 8 specification requirements as follows:

- Up to 10Gbps data rate → 10 to 100x speed improvement over 4G and 4.5G networks,
- 1-millisecond latency,
- 1000x bandwidth per unit area,
- Up to 100x number of connected devices per unit area (compared with 4G LTE),
- 99.999% availability,
- 100% coverage,
- 90% reduction in network energy usage, and
- Up to 10-year battery life for low-power IoT devices [4].

Ultimately, it is believed that 5G will result in a smart society since it serves not only people but also things and the whole society by driving the intelligent interconnection of everything [5].

2. Methodology

The first part of this paper presented a concise literature review regarding the 4IR and the technologies it brought to the construction industry, the integration of 5G and IoT and object detection in the construction industry. The second part presented an IoT and 5G enabled construction detection model. The model proposes the use of a Raspberry Pi 4 Model B, 4GB, together with a Noir 8MP camera module for Raspberry Pi. It is important to mention that all codes were written in Python. First, the use of the model for time-lapse is presented. The object detection was performed using a TensorFlow object detection API, i.e. the framework which focuses on the training and inference of deep neural networks. To represent the computation in an algorithm and also the state on which it operates, TensorFlow uses a unified data graph [6]. Other prerequisites include Open Source Computer Vision Library (OpenCV) which is an open-source computer vision and machine learning software library that has more than 2500 optimized algorithms [7]. Finally, the object detection is performed using a solid-state drive (SSD) that employs MobileNet V2 since it is fast enough to be used on a real-time video.

3. Short review of the recent findings of the research topics

3.1. 5G and IoT

The main focus of 5G in 4IR technologies is inevitably the impact of 5G on the IoT. IoT usually implies the integration of many identification and tracking technologies to make the things around us communicate with each other [8]. Devices that use IoT can be sensors, actuators, gadgets, appliances, etc. These devices can transmit data through the internet and can also be programmed for various applications to bring intelligence and autonomy to systems and processes [9]. Since IoT devices are often connected with many sensors, there is a huge amount of data to be generated and transmitted [10]. Furthermore, it is expected that in future IoT scenarios, the number of connected devices is going to continuously grow, especially in the comparison with the current 4G scenario. This is supported by the fact that the number of IoT-connected devices has increased from 15,41 billion in 2015 to 23,14 billion in 2018, and is even expected to reach 75.44 billion in 2025 [11]. Due to the huge bandwidth of 5G, the data can be wirelessly transmitted to control or central devices [10]. So, the 5G technologies that will be used for the transmission of data from connected devices will have to maintain multiple network connections with many devices while using limited resources [2]. In 4G LTE, the transmission of high-quality images from the site is difficult and can be significantly delayed and the monitoring system can hardly be real-time. Since the transmission of a high-quality 4K video requires a data rate of 40Mbps per channel, LTE cannot operate in the needed speed. Due to the characteristics needed for high-quality video transmission which are many IoT terminals and 4K cameras, 5G is necessary for the application of IoT to monitoring systems [12]. It is believed that with the help of 5G, IoT will provide a high speed of reporting what's happening on the site. Therefore, it has the potential to reduce the number of human errors and enhance process control [13]. 5G can also be seen as the driver in accelerating the construction industry towards automation and remote operations. Remote technology does currently exist, but the low latency of 5G can ensure enhance speed and precision [14]. One of the examples is the system proposed in the paper [12] which implies a 4K camera installed in a truck, crane car, or other equipment, takes a photo of the surroundings and transmits it to the base station using 5G technology. Then, the data of the surroundings is analyzed and the system replies to the truck or crane car and informs them of a possible collision on the site. Another example is the South Korean conglomerate Doosan, which operated an excavator in South Korea from 8,500km away using 5G [15]. The sensors in the concrete can monitor its temperature, maturity, humidity in real-time and inform engineers using the cloud, so a decision can be made at the right time [16]. Also, some companies use IoT technology for tracking their dispatched trucks on their way to the delivery location by equipping the trucks with temperature and acceleration sensors. That way, they can obtain real-time data on whether the concrete is going to arrive to the destination with the planned properties or something has possibly changed. Furthermore, these sensors can monitor corrosion of the rebar, corrosion potential, electrical resistivity, etc. [16].

3.2. Object detection

An extensive literature search on the use of object detection in the construction industry was done, but it was found that there are not many papers regarding the subject. A method for automated checking of construction machines from video streaming was presented in paper [17], i.e. for the extraction of machine active status to compare it with the project schedule. The method was developed by using a deep learning detection algorithm You Only Look Once (YOLO). Furthermore, they presented a case study on developing an image data set for construction machines of a total of 10000 images that belong to 10 types of construction machines. Additionally, it is stated that the same development method can be used for building other construction data sets such as workers and materials [18]. YOLO was also used by another author [19] who presented three deep learning models for the verification of personal protective equipment of workers such as helmets and vests. In the first model the equipment was detected using the algorithm and then the machine learning model verifies whether it is worn appropriately, the second approach implies the simultaneous detection performed by the algorithm of workers and the equipment compliance using a single convolutional neural network (CNN) framework, and the third approach first detects the workers in images and then the equipment. A similar example is a system presented in [20] where a helmet-wearing detection dataset was made from images of various on-site conditions for automated monitoring of workers wearing helmets. The method is based on CNN and it also identifies the colors of the helmets.

4. Model for Construction Detection and Object Spatial/Time Positioning

4.1. Time-lapse monitoring - photo oriented static

Time-lapse is a photographing technique where the same scene is photographed multiple times in regular time intervals. It is usually used to create the impression of high-speed, i.e. to speed up the photographed process since the photographed processes are often not visible to the human eye due to their slow change, which is possible because the photographs are taken at a frequency that is much lower than the finite frame rate. For the purpose of this paper, images of a construction site in Osijek, Croatia are presented. The images show a paver laying asphalt on a road for 55 seconds. The red line shows the starting line, and the red lines show the distance. Also, the time of each sequence is shown in minutes and seconds. Images 1-10 show only the distance in that moment, while images 11 and 12 show the distance at that moment and all the previous ones.

Figure 1. Time-lapse monitoring of asphalt paver and distances.

The images show a paver laying asphalt on a road for 45 seconds. The red line shows the starting line, and the red lines show the distance. Also, the time of each sequence is shown in minutes and seconds. This kind of monitoring is less convenient due to the low automation level, the amount of time it consumes and is highly dependent on the user. Furthermore, the output of such monitoring is just the speed and performances of the paver.

4.2. Stream monitoring - object dynamic oriented

Perhaps the most efficient type of monitoring is live-streaming through cameras positioned on the construction site. The advances can be found in the possibility of remote controlling the site, and the constant information that can be obtained from anywhere consequently decreasing the need for the engineer's site visits. Live-streaming can be challenging without a proper internet speed for the transmission of data and can therefore decrease the quality of the streamed video. Figure 1 presents the proposed model for construction detection.

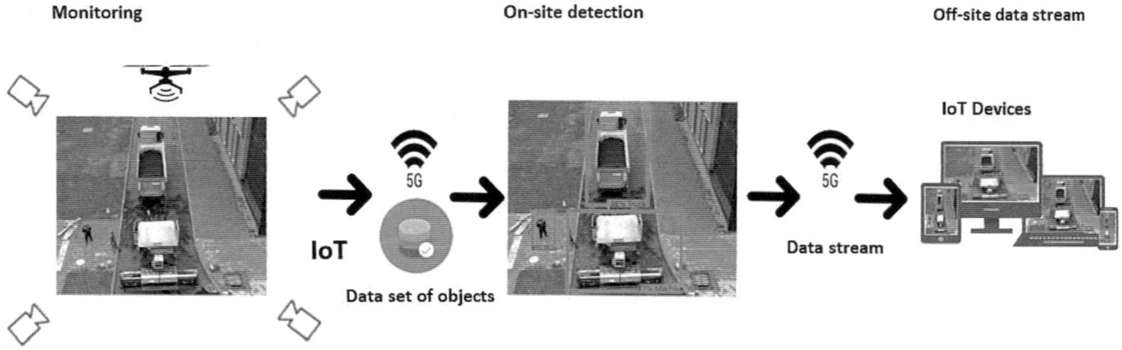

Figure 2. Proposed model for construction detection.

When looking at figure 1 it can be noted that the first phase comprises site monitoring using cameras and drones, i.e. what the construction site manager sees. The next phase begins by using an IoT-based data set of objects and 5G which enables detection performed on the construction site (on-site detection) and it is visible what the camera detects. Using 5G network the data is further transferred to the IoT devices which perform off-site data stream.

5. Conclusions

This paper presented an IoT and 5G enabled construction detection model. Since IoT originates from the 4IR, an introduction was presented with the characteristics of the 4IR. It was found that the 4IR can bring many benefits to the construction industry such as cost reduction, construction duration reduction, quality improvement, sustainability improvement, etc. Also, these benefits are especially emphasized with the goal of digitalization and automation of the construction site and the construction project's lifecycle. Furthermore, the IoT as a 4IR technology and 5G as its enhancer were analyzed and it was concluded that IoT can be beneficial for the IoT, mainly in the context of faster data transmission. Therefore, the quality of real-time monitoring can be significantly improved which consequently increases the project's quality, safety and also the quality of object detection. A short object detection review was provided as an introduction to the model for both time-lapse monitoring and stream monitoring with construction detection. It was found that the model can be improved by using the 5G network since the data from the object data set can be transferred to the IoT devices faster and provide both enhanced on-site detection and off-site data stream.

6. References

[1] Zhou K, Liu T and Zhou L 2015 Industry 4.0: Towards future industrial opportunities and Challenges *12th International conference on fuzzy systems and knowledge discovery (FSKD)* pp 2147–52

[2] Yu H, Lee H and Jeon H 2017 What is 5G? Emerging 5G mobile services and network requirements. *Sustainability* **9** 10 1848

[3] Jovović I, Husnjak S, Forenbacher I and Maček S 2019 Innovative application of 5G and blockchain technology in industry 4.0. *EAI Endorsed Transactions on Industrial Networks and Intelligent Systems* **6** (18)

[4] Troxler P 2013 Making the 3rd industrial revolution: The struggle for polycentric structures and a new peer-production commons in the FabLab community *Fab Lab: Of Machines, Makers and Inventors, Cultural and Media Studies Transcript* (Bielefeld: Bielefeld University Press) pp 181-194

[5] Wang X and Gao L 2020 *When 5G Meets Industry 4.0* (Luxembourg: Springer)
[6] Abadi M, Barham P, Chen J, Chen Z, Davis A, Dean J, et al. 2016 Tensorflow: A system for large-scale machine learning *12th symposium on operating systems design and implementation* pp. 265–83
[7] Petrillo A, De Felice F and Cioffi R Zomparelli F 2018 Fourth industrial revolution: Current practices, challenges, and opportunities *Digital transformation in smart manufacturing* pp 1–20
[8] Li G, Hou Y and Wu A 2017 Fourth Industrial Revolution: technological drivers, impacts and coping methods *Chinese Geographical Science* **27** 4 626–37
[9] Lee J, Bagheri B and Kao H-A 2015 A cyber-physical systems architecture for industry 4.0-based manufacturing systems *Manufacturing letters* **3** 18–23
[10] Khurpade JM, Rao D and Sanghavi PD 2018 A Survey on IOT and 5G Network *International conference on smart city and emerging technology (ICSCET)*
[11] Marchese M, Moheddine A and Patrone F 2019 IoT and UAV integration in 5G hybrid terrestrial satellite networks *Sensors* **19** 17 3704
[12] Nozaki D, Okamoto K, Mochida T, Qi X, Wen Z, Tokuda K, et al. 2018 AI management system to prevent accidents in construction zones using 4K cameras based on 5G network *21st International Symposium on Wireless Personal Multimedia Communications (WPMC) 2018 IEEE* pp 462–6
[13] Ghosh A, Edwards DJ and Hosseini MR 2020 Patterns and trends in Internet of Things (IoT) esearch: future applications in the construction industry *Engineering, Construction and Architectural Management* **28** 2 457–81
[14] Park H-A 2016 Are we ready for the fourth industrial revolution? *Yearbook of medical informatics* **1** 1-3
[15] Schwab K 2017 *The fourth industrial revolution* (Currency)
[16] Paul S, Naik B and Bagal DK 2020 Enabling Technologies of IoT and Challenges in Various Field Of Construction Industry in the 5G Era: A Review *IOP Conf. Ser.: Mater. Sci. Eng.* **970** 012019
[17] Xiao B and Kang S 2019 Deep Learning Detection for Real-time Construction Machine Checking *ISARC Proceedings of the International Symposium on Automation and Robotics in Construction* (Pittsburgh: IAARC Publications)
[18] Xiao B and Kang S-C 2021 Development of an Image Data Set of Construction Machines for Deep Learning Object Detection *Journal of Computing in Civil Engineering* **35** 2 05020005
[19] Nath ND, Behzadan AH and Paal SG 2020 Deep learning for site safety: Real-time detection of personal protective equipment *Automation in Construction* **112** 103085
[20] Wu J, Cai N, Chen W, Wang H and Wang G 2019 Automatic detection of hardhats worn by construction personnel A deep learning approach and benchmark dataset *Automation in Construction* **106** 102894

Comparison of spreadsheet-based optimization tools applied to construction site layout problem

B Dasović[1] and U Klanšek[1]

[1] University of Maribor, Faculty of Civil Engineering, Transportation Engineering and Architecture, Smetanova 17, 2000 Maribor, Slovenia

E-mail: borna.dasovic@um.si; uros.klansek@um.si

Abstract. This paper compares different spreadsheet-based optimization tools applied to a practical example of a construction site layout problem. The objective of the optimization is to minimize the total time of material transportation by optimal positioning of tower crane and work facilities on the construction site with repetitive operations. Computer programs, such as MS Excel, LibreOffice Calc, and Google Sheets can be applied as modeling tools for a variety of construction optimization problems in addition to their usual functionalities. In this study, LibreOffice Calc Solver, MS Excel Solver, along with other MS Excel add-ins, i.e., OpenSolver, Evolver, and What'sBest, are analyzed. The capabilities of optimization tools mentioned above are examined on the problem of optimal positioning of tower crane and work facilities on the construction site. The results obtained by optimization tools are noted and discussed. The paper ends with conclusions and recommendations for further research.

1. Introduction

Selecting the optimal contractor and planning the site layout can be considered important tasks in any construction project [1,2]. The positioning of a tower crane on a construction site is essential for overall safety and project efficiency in terms of energy consumption. Economic efficiency and environmental impact can be rationalized by the efficient placement of tower crane and work facilities [3]. In common practice, such decisions are mainly based on experience and intuition, without any numerical analysis about whether or not the selected locations are optimal for a particular construction site. When planning construction site layout, the optimization can be helpful since all participants in construction project aim to minimize costs [4]. An optimization model can be developed to reduce the tower crane's operating time and reduce overall project costs.

The model for the site layout problem can be developed in a spreadsheet environment. Spreadsheets can be used to calculate the time required for each transport cycle from storage units or facilities to demand points on a building. Although it seems like a simple task to use an optimization model to minimize the total duration of supply cycles, the aforesaid problem is much more complex than that. Numerous spatial constraints must be considered, as well as the load capacity of the tower crane, the speed of the lifting load, and combinations of three different types of tower crane movements. These specifics require a rather complex mathematical model and, most importantly, a robust solving algorithm for this type of problem. The work of Hawarneh et al. [2] provides current state-of-the-art on the subject of construction site layout planning problems and the development of mathematical modeling techniques.

There are two main optimization approaches, exact and heuristic, for solving highly complex site layout problems. The exact techniques, such as branch-and-bound and simplex, can be used to obtain an exact optimal solution but require more processing time and effort than heuristic methods. Heuristic methods can produce result faster, but the result may not be exactly optimal. Heuristic methods can provide near-optimal solutions in a reasonable amount of time. The most commonly applied heuristic methods are genetic algorithms, particle swarm optimization, ant colony optimization, neural networks, and other nature-inspired algorithms.

This paper provides an overview of currently available commercial and open-source spreadsheet-based solvers that can be used to solve construction site layout problems. This study highlights the differences of MS Excel add-ins for optimization purposes, such as Solver, OpenSolver, Evolver, and What'sBest!, as well as the LibreOffice Calc Solver. While most studies focus on developing a mathematical model, this study examines the spreadsheet-based solving algorithms and compares the results obtained on the construction example presented in a previous study [3].

This article is organized as follows: the next section introduces spreadsheet-based optimization tools along with the basic information and functionalities of each tool; Section 3 explores practical usability of considered optimization tools on an example of actual construction project and provides results; Section 4 discusses advantages and limitations of each optimization tool and present recommendations for further research; Section 5 draws conclusions on spreadsheet-based optimization tools for solving construction site layout problems.

2. Spreadsheet-based solvers

A variety of spreadsheet-based optimization tools can be employed to solve different construction problems, for instance scheduling problems, resource allocation problems, routing problems, site layout problems etc. These tools are software integrated into a spreadsheet environment, and they are developed to solve mathematical problems. The problem description is taken in the form of a spreadsheet, and it is solved by applying an appropriate algorithm. In this chapter, some spreadsheet-based optimization tools will be introduced with basic information on their capabilities and limitations in the size of the problem that they can solve.

2.1. MS Excel Solver

MS Excel Solver [5] is probably the most used spreadsheet optimization tool that comes with basic MS Excel installation. Frontline Systems developed it for solving optimization problems in MS Excel workbook. With this solver, it is possible to find an optimal value (minimum or maximum) for the objective function cell, subject to the given constraints or limitations by adjusting the values of decision variable cells. The objective function, conditions, and decision variable cells as well as the formulas interrelating them form a Solver model. The Solver uses various methods, from linear programming and non-linear optimization to genetic and evolutionary algorithms, to find solutions.

The MS Excel Solver tool uses the generalized reduced gradient (GRG) method as implemented in GRG2 code [6] to perform non-linear optimization. For linear programming, simplex and dual simplex methods are used. The branch-and-bound method is used for problems with integer constraints as explained in study [7]. There is also an evolutionary solving method that incorporates genetic algorithms and local search methods for non-smooth optimization, implemented by several individuals at Frontline Systems.

The Solver has a limitation of 200 variables and 100 constraints for optimization model formulation. Suppose the model becomes too complex for a basic Solver. In that case, there is also a paid Premium Solver that does not have any restrictions and is equipped with additional functionalities and additional optional plug-in solver engines, such as KNITRO, MOSEK, Gurobi, XPRESS, and OptQuest.

2.2. LibreOffice Calc Solver

LibreOffice Calc [8] is a powerful and free MS Excel alternative. It is compatible with MS Excel, so it can open .xls and .xlsx files created with MS Office. LibreOffice Calc comes equipped with its Solver tool. As with the Excel Solver, the target is to find those variable values that result in an optimal value of the target cell (minimum or maximum) under certain limiting conditions. To solve optimization problem, a variety of search algorithms can be applied, such as DEPS evolutionary algorithm, SCO evolutionary algorithm, LibreOffice linear solver, and CoinMP linear solver, as well as LibreOffice swarm non-linear solver in an experimental phase.

DEPS consists of two heuristic algorithms, i.e. differential evolution and particle swarm optimization especially suited for non-linear optimization. Social cognitive optimization (SCO) is also a heuristic method that mimics human behavior of learning and sharing information. Linear solvers are two variations of the branch-and-bound algorithm. Swarm non-linear solver is the same as the DEPS solver, but it allows users to choose between differential evolution and particle swarm optimization algorithms.

2.3. Evolver Add-in for MS Excel

The DecisionTools Suite developed by Palisade [9] is an integrated set of programs for risk analysis and decision-making under uncertainty. It is equipped with an Evolver optimization add-in that uses a genetic algorithm, OptQuest, and linear programming technology to solve a wide variety of problems modeled in MS Excel. The developers state that it has a unique ability to find overall global solutions to problems that traditional solvers typically miss. The use of OptQuest as an optimization engine includes metaheuristic optimization algorithms such as evolutionary algorithms, tabu search, neural networks, and scatter search. The tool can be applied to solve non-linear problems, and afterward, the user can choose to keep the best solution or return pre-optimization decision variables. The tool can also generate an optimization report with the log of all calculated combinations and progress.

2.4. OpenSolver Add-in for MS Excel

The OpenSolver [10] is an open-source linear and non-linear optimizer for MS Excel. OpenSolver features and their development are described in the study [11]. It is fully compatible with the existing solver models, so it does not require spreadsheet modifications. There are no limitations on the size of the optimization model but only the restrictions regarding the processing power of a personal computer. The OpenSolver offers various solvers, including the open-source COIN-OR algorithms – CBC (COIN-OR branch-and-cut), Bonmin, and Couenne. To solve the optimization problem there is also possibility of utilization Gurobi software installed on local computer, and also NEOS Server can be used. NEOS Sever is an internet-based service for solving numerical optimization problems and the OpenSolver offers previously mentioned COIN-OR algorithms alongside the CPLEX solver. The online-based solvers can settle the problem of the low processing power of a personal computer since the optimization problem is solved on publicly available supercomputers used by NEOS Server. This add-in transforms the spreadsheet model to algebraic modeling language formulation. Based on the selected solver, either sends the optimization problem to NEOS Server to be solved or uses the personal computer's processing power.

2.5. Lindo What'sBest! Add-in for MS Excel

What'sBest! [12] is an MS Excel add-in that allows building large-scale optimization models in spreadsheet form for linear, integer, and non-linear optimization. It was developed by Lindo Systems and the developers state that the add-in is designed to efficiently solve even the toughest and biggest models. There is no capacity limit for the optimization model, and it was noted that it could solve models of well over 100 000 variables. It offers primal and dual simplex as well as a barrier solver for linear models. A GRG method is used to solve non-linear programming problems with the help of successive linear programming for finding a good feasible solution quickly. Also, What'sBest! is equipped with a global solver that transforms original non-convex and non-linear problems into several convex linear problems. Afterward, the branch-and-bound technique is used to solve sub-problems and find a global solution to the optimization problem. For integer optimization problems, Lindo presents their integer solver that works in conjunction with linear and non-linear solvers.

3. Application example

To test the abovementioned spreadsheet-based solvers, an example from an earlier study [3] was used. The aim of this study was to optimize the position of tower crane (*Cr*) and work facilities – brick storage (S_1) and mortar production plant (S_2), on the construction site with repetitive operations. The objective function was set to calculate the overall duration of supply cycles of the tower crane from facility points to demand points (D_1-D_{16}) within the area of building floor that was to be constructed. The optimization model in the study was developed in MS Excel while the optimization problem was solved by the Solver add-in. The optimization model was presented in detail in the paper [3]. The layout of the construction site can be seen in Figure 1.

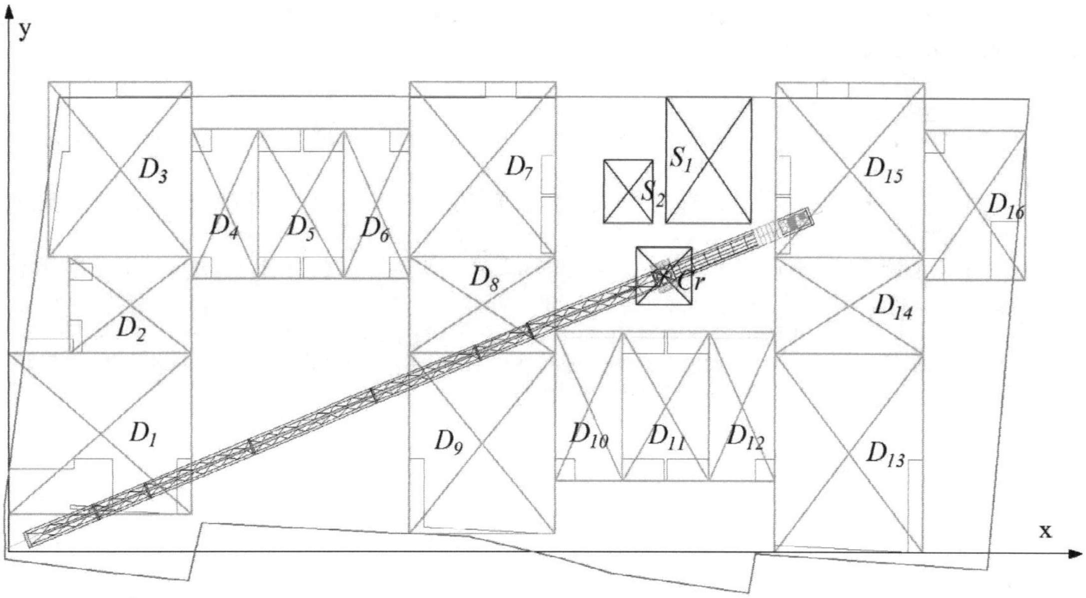

Figure 1. Construction site layout with elements to be optimally positioned.

In the present study, other spreadsheet-based solvers will be applied to solve the optimization problem and compare the results. All optimization models will have the same assumed start position for the tower crane and two of the work facilities on the construction site, as shown in Figure 1. All computer work was accomplished on a 64-bit operating system with a personal computer AMD Ryzen 7 2700 Eight-Core Processor at 3.2 GHz, 64 GB DDR4 RAM, 512 GB NVMe PCIe M.2 SSD, and NVIDIA GeForce GTX 1660 graphics card. The optimization results gained by a particular solver are shown in table 1.

Table 1. Results of optimization gained by spreadsheet-based optimization tools.

	Assumed position	Excel Solver	OfficeLibre Solver	Evolver
Cr	53.78, 78.94	49.23, 27.55	49.78, 25.47	49.23, 27.55
S_1	57.43, 31.13	28.09, 14.21	28.43, 14.51	27.94, 14.48
S_2	50.84, 28.63	47.93, 22.75	43.93, 30.27	47.93, 22.02
Objective function	450.19	290.38	332.70	290.96
Processing time	/	2:17	2:47	>1 h

It can be noted that only three of the previously presented spreadsheet-based solvers offered a solution to this construction site layout problem. Solvers What'sBest! and OpenSolver were also utilized. Still, there are some compliance issues as the model uses some of MS Excel functions that these solvers probably do not process (e.g., IF, SQRT, ABS, MAX, ROUNDUP). Some of these functions could be avoided, but some are essential for the model to be usable and could not be replaced by using different formulas. So, for this study, only MS Excel Solver, Evolver add-in, and LibreOffice Calc Solver will be considered.

The best solution from the assumed position was obtained by MS Excel Solver and GRG non-linear solving method after two consecutive optimization runs. Evolver add-in produced the second best duration of supply cycles with their automatic optimization mode that selects the optimization engine and its settings. This solver required more processing time, presumably because every trial is calculated and recorded to be reviewed afterward. The solver was engaged multiple times since sometimes, it would not produce a better solution for more than 20 minutes than the currently best solution. In OfficeLibre Calc Solver, DEPS and SCO algorithms were each employed to solve the optimization problem. A somewhat better result was obtained by DEPS algorithm.

Even though MS Excel Solver produced the best overall result, the result could no longer be improved with the same solver. Therefore, in a second experiment, other solvers were engaged for possible further improvement from this position. First, the Evolver was applied and improved the objective function by 1.68 minutes. And finally, the OfficeLibre Solver DEPS algorithm was used and produced the overall best recorded total duration of supply cycles. The results can be seen in table 2.

Table 2. Results of optimization from previous best solution obtained by MS Excel Solver.

	Excel Solver	Evolver	OfficeLibre Solver
Cr	49.23, 27.55	49.25, 27.50	49.23, 27.55
S_1	28.09, 14.21	28.39, 15.69	28.43, 15.58
S_2	47.93, 22.75	47.93, 21.12	47.93, 22.75
Objective function	290.38	288.70	287.66

The resulting minimum total duration of supply cycles found in all three examples was better than in the previous study [3] where optimal objective function value was 294.80 minutes. One reason for this could be the newer version of MS Excel or the higher computing power of a personal computer. That way, the initial solution has been improved by an additional 2.42%. It can be noticed that the resulting coordinates given in Table 2 are somewhat similar, and there are no significant differences. Every coordinate was in the 1.5 m range of the original solution. In other words, the result mainly was just fine-tuned to obtain a few minutes better solution.

4. Discussion

The results of this study reveal the dominance of MS Excel Solver in many fields. First of all, it found the best overall solution compared to other optimization tools and in less time. The result needed two optimization re-runs, but the use was straightforward without complications. Its GRG non-linear method proved to be more powerful than the rest of the tested techniques. The software comes together with MS Excel, but it has limitations in the number of decision variables and constraints it can deal with. If the model outgrows the standard Solver edition, there is also a paid Premium Solver option that does not have any restrictions and it can solve the optimization problem through utilization of Frontline System's online server. That way, the end-user does not need a powerful personal computer, and the optimization problem can be solved even on a tablet or a smartphone.

Evolver add-in for MS Excel is a part of the DecisionTools Suite for risk analysis and decision making. It produced an equally good result as previous optimization software, but it needed more processing time. Sometimes the solver would get stuck for quite a time, and after the restart of the optimization process, it would move faster from the previous best solution. The Evolver even improved the best solution obtained by MS Excel Solver. This could mean that maybe the assumed start position for the tower crane and facilities should have been better determined beforehand to achieve optimal solution faster. Evolver is equipped with visualization functionalities, so the improvements in results can be tracked over time.

OpenOffice Calc Solver can be used as an open-source alternative to MS Excel Solver. It can produce satisfactory results within a single run but then falls in local optimum from which it is hard to get out. The tested algorithms were also susceptible to the quality of selected start points for decision variables. After a different set of coordinates was entered into the model, DEPS and SCO algorithms have managed to improve previously found solution and obtain the best recorded result in this study. Given the model complex design, it is possible that the rest of presented optimization tools could not recognize some of the functions used in a spreadsheet.

In further research, more optimization tools could be considered and tested. The spreadsheet-based optimization tools could also be applied to solve other construction problems, such as scheduling problems or routing problems. In further research, online spreadsheet tools as Google Sheets could be considered. Google Sheets also have optimization abilities since there are Solver and OpenSolver add-ins available. Also, the model formulation for application example should be developed in algebraic modeling language in order to find the exact optimal solution that can be compared to those obtained by other optimization tools.

5. Conclusion

This paper has briefly introduced various spreadsheet-based optimization tools that can be applied to construction site layout problems. The results of performed study emphasize the importance of utilized solving algorithms. In the present article, spreadsheet-based optimization tools, i.e. MS Excel Solver, LibreOffice Calc Solver, Evolver, OpenSolver, and Lindo What'sBest!, were introduced with their basic characteristics and capabilities. Mentioned tools were applied to solve practical construction example of site layout optimization.

The results show that the MS Excel Solver obtained better solutions in less time than other mentioned optimization tools for this type of problem. The limitation in the number of constraints and variables allowed in basic Solver was not exceeded in the presented example. Still, it would be interesting to see how well the Premium Solver would solve more complex problems. The rest of the optimization tools also performed well, but the Evolver required more time for processing, and LibreOffice Calc Solver algorithms could not fall into the domain of optimal solutions found by the other two solvers.

Future studies should aim to integrate optimization possibilities within building information modeling tools to optimize all permanent and temporary equipment locations on construction sites. That way, the details of construction site specifics and equipment dimensions can be reviewed in 3D and over the project implementation phases, reducing the possibility of clashes, improving the initial solution, and enhancing project efficiency and overall safety.

6. References

[1] Marović I, Perić M and Hanak T 2021 A multi-criteria decision support concept for selecting the optimal contractor *Appl. Sci.* **11** 1660
[2] Hawarneh A Al, Bendak S and Ghanim F 2021 Construction site layout planning problem: Past, present and future *Expert Syst. Appl.* **168** 114247
[3] Dasović B, Galić M and Klanšek U 2019 Active BIM approach to optimize work facilities and tower crane locations on construction sites with repetitive operations *Buildings* **9** 21
[4] Mésároš P, Mandičák T, Spišáková M, Behúnová A and Behún M 2021 The implementation factors of information and communication technology in the life cycle costs of buildings *Appl. Sci.* **11** 2934
[5] Frontline Systems 2021 Frontline Solvers
[6] Lasdon L S, Waren A D, Jain A and Ratner M 1978 Design and testing of a generalized reduced gradient code for nonlinear programming *ACM Trans. Math. Softw.* **4** 34–50
[7] Fylstra D, Lasdon L, Watson J and Waren A 1998 Design and use of the Microsoft Excel Solver *Interfaces* **28** 29–55
[8] The Document Foundation 2021 LibreOffice - Free Office Suite
[9] Palisade 2021 DecisionTools Suite
[10] Mason A J 2021 OpenSolver for Excel
[11] Mason A J 2012 OpenSolver - An open source add-in to solve linear and integer progammes in Excel pp 401–406
[12] Lindo Systems 2021 What'sBest! and Excel optimization

Acknowledgments

The authors acknowledge the financial support from the Slovenian Research Agency (research core funding No. P2-0129).

IOP Conf. Series: Materials Science and Engineering 1209 (2021) 012011 doi:10.1088/1757-899X/1209/1/012011

Comparison of the planned and actual course of construction of reinforced concrete monolithic structures

R Doubek[1]

[1] Institute of Technology, Mechanization and Construction Management, Faculty of Civil Engineering, Brno University of Technology, Veveří 331/95, 602 00 Brno, Czech republic

E-mail: doubek.r@fce.vutbr.cz

Abstract. During construction, there is often a change in contractual terms, either in advance of construction or delayed construction. This phenomenon is due to many influencing factors such as delayed subcontracting, climatic conditions, the surrounding construction site, the quality of project documentation, the solvency of the investor, the expertise of the general contractor, etc. The aim of the paper is to assess the planned and actual construction of reinforced concrete monolithic structural elements using tower cranes. The subject of the paper is to state the expected method of implementation in time and how the construction proceeded. Milestones for accelerating and slowing down construction, analyzing their causes and proposing future measures will be presented. Data are obtained by measuring real buildings.

1. Introduction

The aim of this paper is to assess the planned and actual course of construction of reinforced concrete monolithic structural elements using tower cranes.

The subject of the paper is to state the expected method of implementation in time and how the construction proceeded. Milestones for accelerating and slowing down construction, analysing their causes and proposing future measures will be presented. Data are obtained by measuring real buildings.

The milestones listed below can be used for quick determination of time requirements for reinforced concrete monolithic buildings of building construction or as a support tool in determining construction plans or pre-production business planning.

These milestones are necessary for quick and correct decision-making in construction technology design with use in supplier planning, design activities and for study purposes.

2. The state-of-the-art

Construction companies commonly use the subjective knowledge of technical staff in scheduling. This allows them to make quick decisions and plan project implementation. At present, there is no known public work or publication directly dealing with the analysis of the completed construction.

The current problem uses software (eg Build Power, Kros, Contec), which within its data (time) base solves time planning using technical and economic indicators [1].

The authors of the source publications deal with the analysis of construction within specific specific topics, such as the performance of mechanization, which is one of the partial factors influencing construction [6].

This contribution is an effort to follow up on the publication [5] describing the specific course of construction from the point of view of the finances of the main and associated construction production.

3. Methodology for the achievement of results

To achieve the results, it is necessary to know the construction system of the building, the technological process of work, the area of items with a standard schedule. Based on this, a tender and selection procedure is performed for the general contractor of the construction.

At this stage, the contractual terms for the construction of reinforced concrete monolithic structural elements are known.

After the start of construction work, construction will begin to be affected by factors that may delay or accelerate the planned implementation dates.

Factors influencing construction are elements entering the course of construction, which can affect it mainly negatively, if the parties involved are not mentally or resource-ready for them [2].

The expertise of the contractor is the factor that can affect the contractor the most. It is mainly about securing professional and motivated employees, especially in the position of construction management. Another necessity is to have a secured cash flow for efficient and timely fulfillment of payment obligations (wages, income invoices, bank guarantees, etc.).

The Contractor may, to a large extent, influence the entire course of a given technological stage, craft or profession by choosing a suitable subcontractor. It is mainly about his expertise, or employees in all positions. Furthermore, the source background (materials, mechanization, finance), thanks to which quality and effective contractual obligations can be fulfilled.

The effect of climatic conditions can significantly affect the course of construction. However, this is an occasional or seasonal effect, which can be reduced by flexible planning and adaptation of work with regard to work safety. However, in the mild climate zone, to which Central Europe largely belongs, these effects are not the key factor of today. The mentality, ie the experience of the contractors and contractors, does not allow the climate to affect the construction to such an extent that it is necessary to take precautionary measures on the grounds that it would not have to be used effectively, thus rendering the building inefficient. Leaving aside the occurrence of isolated tornadoes or millennial floods, it is justified to put aside allocating of funds for measures against climate change.

The principle of force majeure is introduced in contractual relations. But again, it was always a backward factor, which had a rather fictitious weight. It is mainly a state of war in the country where the construction is being carried out, or just a millennial flood. Today, however, the influence of the epidemic has been prioritized within this factor, thanks to which the construction or the contractor may also terminate the construction.

The construction site may be affected mainly due to the daily traffic situation (columns), traffic services (motorways or special-purpose roads), the nature of the terrain (valley or hill ridge). Furthermore, it is the size of the construction site, which will affect the possibility of pre-supply of material on the construction site, the background of workers and mechanization, compliance with occupational safety, etc.

The client is more likely to influence the quality of the project documentation, unless he transfers this obligation to the contractor so that he completes the project documentation. The documentation and its quality depends on the experience and expertise of the designer. Furthermore, on the possibilities and willingness to continue cooperating on the project, which may not always be a matter of course.

The solvency of the investor may affect the course of construction by non-fulfilling the contractual, especially payment obligations towards the contractor, either intentionally or unintentionally. Furthermore, the customer can decide to interrupt or stop construction [3].

The above factors can be ascertained after the completion of the work, especially from construction diaries, inspection days and business negotiations.

IOP Conf. Series: Materials Science and Engineering 1209 (2021) 012011 doi:10.1088/1757-899X/1209/1/012011

After the completion of the construction, it is possible to state the actual dates of the performed works and milestones of the construction and subsequently to determine whether the construction was accelerated or slowed down and for what reasons (factors).

4. Results

The building which is the subject of this contribution is a new administrative building in the existing development.

The building consists of one basement and eight floors, of which the 8th floor is smaller in layout and construction (receding) compared to the lower floor. It is based on drilled piles and a base plate, which is part of the white tub system. The load-bearing construction system is a combined, reinforced concrete monolithic. The cladding is a system of light perimeter cladding (modular facade) and a coating foil covering.

The structural load-bearing elements are columns with extended heads, internal (elevator and staircase cores), perimeter walls, ceiling slabs, landings and intermediate landings are designed for all floors.

The working platoon was composed of: 3 carpenters platoon 1, 3 carpenters platoon 2, 4 carpenters platoon 3, 4 ironworkers platoon 4, 4 ironworkers platoon 5, 4 concrete workers platoon 6, 3 auxiliary workers platoon 7, 2 crane operators platoon 8.

The lifting mechanism used was two Liebherr LTM 1030 mobile cranes, and Liebherr 81 K.1 and 71 K stationary tower cranes.

The sources of materials were formwork from the manufacturer DOKA with elements of column formwork KS Xlife and RS ceiling Dokaflex, reinforced bar bonded B500B diameters 12 to 20 mm, concrete mix C 20/25 XC2 to C 30/37 X0.

The total volume of built-in material is 8 000 tons of reinforcement, 9 500 m3 of concrete mix.

As part of the contractual relationship between the client (builder-investor) and the contractor (general contractor) of the construction, a contractual schedule was created, which the contractor had to follow and fulfill.

Within its corporate structure and the possibility of capacity utilization, the contractor implemented a monolithic reinforced concrete load-bearing structural system with the help of a subcontractor. The obligation to perform was transferred by the contractor to the subcontractor, and thus the fulfillment of the contractual construction schedule.

Milestones have always been considered within the framework of contractual relations, taking into account the implementation of concreting, especially on horizontal structures. Last but not least, the final handover of the work (subcontractor to the contractor), or part of the work (contractor to the investor).

The technological implementation of the construction proceeded as planned. All monolithic structures consisted of the same partial construction processes, namely formwork assembly, reinforcement binding, concreting and formwork disassembly.

The chronological progress of the work was as follows:

1. Base plate in order: assembly of engagement formwork 1 to 4, reinforcement binding 1 to 4 and assembly of engagement formwork 5 to 7, concreting of engagement 1 to 4 and reinforcement bonding engagement 5 to 7, dismantling of formwork engagement 1 to 4 and concreting of engagement 5 to 7, dismantling the engagement formwork 5 to 7.
   The source deployment consisted of platoons one, four, five, six, seven and eight. Two mobile cranes from mechanization.
2. 1st basement in order: binding of reinforcement of columns and walls of engagement 1, assembly of formwork of columns and walls of engagement 1, concreting of columns and walls of engagement 1, disassembly of formwork of columns and walls of engagement 1. Installation of formwork of ceiling of engagement 1, binding of ceiling reinforcement of engagement 1, execution of concreting of the ceiling of engagement 1, dismantling of the formwork of the ceiling of engagement 1. Simultaneously with the execution of partial construction processes of

the ceiling of engagement 1, partial construction processes of columns and walls of engagement 2 took place.

The source deployment consisted of all platoons and mechanization two tower cranes.

3. 1st to 8th floors: the construction process is the same as the 1st basement. [4]

The anticipated contractual schedule envisaged the actual start of implementation on April 15 and the handover of the work on November 30. A total of 25 partial terms (milestones) were contractually binded, according to which the construction took place. 3 milestones were met within the agreed deadline, 12 milestones were reached in advance and 10 milestones were delayed. The initial milestone was completed 16 days in advance, the final milestone was reached with a delay of 8 days. The most favorable milestone was 17 days ahead and the most unfavorable with a delay of 24 days.

Table 1. Overview of meeting contractual deadlines part 1.

construction	contract	reality	days	contract	reality	days
	working shot 1-4			working shot 5-7		
baseplate - concreting	15.04.2018	30.03.2018	16	25.04.2018	08.04.2018	17

Table 2. Overview of meeting contractual deadlines part 2.

construction	contract	reality	days	contract	reality	days
	working shot 1			working shot 2		
ceiling 2nd basement concreting	30.04.2018	25.04.2018	5	10.05.2018	29.04.2018	11
ceiling 1st basement concreting	30.05.2018	19.05.2018	11	10.06.2018	30.05.2018	11
ceiling 1st floor concreting	20.06.2018	28.06.2018	-8	30.06.2018	07.07.2018	-7
ceiling 2nd floor concreting	11.07.2018	18.07.2018	-7	19.07.2018	25.07.2018	-6
ceiling 3th floor concreting	01.08.2018	01.08.2018	0	06.08.2018	10.08.2018	-4
ceiling 4th floor concreting	19.08.2018	18.08.2018	1	25.08.2018	25.08.2018	0
ceiling 5th floor concreting	08.09.2018	02.09.2018	6	15.09.2018	11.09.2018	4
ceiling 6th floor concreting	28.09.2018	21.09.2018	7	01.10.2018	27.09.2018	4
attic 6th floor	08.10.2018	05.10.2018	3	12.10.2018	12.10.2018	0
ceiling 7th floor concreting	14.10.2018	28.10.2018	-14	24.10.2018	06.11.2018	-13
construction site cleaning	31.10.2018	23.11.2018	-23	14.11.2018	08.12.2018	-24
handover of the work	30.11.2018	08.12.2018	-8			

5. Discussion

3 milestones were met within the agreed deadline, 12 milestones were reached in advance and 10 milestones were delayed. The initial milestone was completed 16 days in advance, the final milestone was reached with a delay of 8 days. The most favorable milestone was 17 days ahead and the most unfavorable with a delay of 24 days.

From the above, it is clear that the execution of the rough substructure was in advance, thus creating a time reserve for the execution of the following floors.

The second to fourth floors were made with a relative delay (weekly). If the reserve from the previous floors had not been created, there would have been a significant delay in the implementation of reinforced concrete monolithic structural elements.

Fifth to seventh floors are typical repetitive layouts, which were a little ahead.

For the last three contractual milestones, there was a significant delay in meeting deadlines.

The reason for the delay of milestones in the period from June to August was the repeated recall of platoon 3 and 4 to another building, which delayed the installation of ceiling formwork, binding of ceiling reinforcement, dismantling of ceiling formwork.

The delay in October and November was due to unfavorable weather (rain, wind above 11 m / s). On the contrary, advance was achieved in the period from April to June due to the early start of the subcontractor's work, his knowledge of local and project conditions (completed previous building of the set of buildings) and a strong resource background. The course of August to October was influenced mainly by the repetition of the layout of the floor.

The delay in the construction of the eighth floor in particular is not so critical, although there has been a delay of two weeks, as it is a receding installation floor that does not affect the design of the light perimeter cladding located on the second to seventh floors.

6. Conclusion

According to the above, we can say that the climatic conditions enter our construction noticeably and cannot be fully influenced. The contractor is not able to influence the solvency of the customer and force majeure. The contractor may influence the construction site and the quality of the project documentation on his own initiative, but only partially.

The contractor can most influence the factors of his own professional capacities and the deployment of a quality subcontractor.

An important measure against construction delays is to ensure a sufficient permanent number of workers, operational work planning in case the change of weather.

It is not necessary to influence the advance in principle. However, it is clear from the case that advance will not guarantee the planned completion, but without it there would be a greater delay.

7. References

[1] Motyčka V, Henková S and Kantová R 2014 Considers the Opinions of Construction Companies When Planning Study Programme Content *Conference Proceedings of 3rd Cyprus International Conference on Educational Research (Procedia Social and Behavioral Sciences)* (Amsterdam: Elsevier) pp 423–7

[2] Jarský Č, Musil F, Gašparík J, Motyčka V, Kovářová B, Svoboda P, Václav P and Vyčítal M 2019 *Příprava a realizace staveb* vol 2 (Brno: CERM) p 348

[3] Kantová R and Motyčka V 2014 Construction Site Noise and its Influence on Protected Area of the Existing Buildings *Electronic proceedings of the enviBuild conference (Advanced Materials Research)* (Switzerland: Trans Tech Publications) pp 419–23

[4] Šťastný J, Motyčka V, Brandtner M, Bříza L and Doubek R 2018 Monitoring staveb – sledování nasazení zvedacích prostředků *Czech Journal of Civil Engineering* **1** 85-92

[5] Doubek R 2020 Share of the volume of work typesand financial costs according to the stages of construction of reinforced concrete monolithic buildings *Conference Young Scientist (Štrbské Pleso)*

[6] Štěrba M 2021 *Časové posouzení využití zdvihacích mechanismů v pozemním stavitelství* Dissertation work, Brno

Analysis of time consumption for extensive green roof realization

M Hejl[1], M Mohapl[1] and J Jílek[1]

[1] Brno University of Technology, Faculty of Civil Engineering, Institute of Technology, Mechanization and Construction Management, Czech Republic

E-mail: 167046@vutbr.cz

Abstract. The article deals with the analysis of time consumption for the construction of an extensive green roof. Also, it deals with the construction of several extensive roofs in terms of time and then sets a new standard hour for the construction of an extensive green roof. At the end of the article, it is said how it would be possible to speed up the construction of extensive roofs or in some cases simplify it.

1. Introduction

This article discusses in detail the time required to build a green roof. Due to the collected data, especially from the construction of extensive green formations, these formations will be analyzed and compared in detail. For individual construction processes, a partial standard hour will be determined for each process on selected roof structures, and in the end, a standard hour for the total time of each roof will be determined. If the same processes are included in the construction, the exact standard hour for the implementation of the given activity will then be determined from the data. The total standard hour for the realization of an extensive green roof using the given materials will be determined by calculating the arithmetic mean according to formula No. 1, where x is the value of the given realization time and n is the value of the number of realizations.

$$\bar{x} = \frac{1}{n}\sum_{i=1}^{n} x_i = \frac{x_1+x_2+\cdots x_n}{n} \tag{1}$$

Partial calculations were performed using other formulas, where the time required for the construction of the roof by one person is presented by formula No.2. In this formula, t_l represents the time required for implementation by one person, it represents the total implementation time, and n_l represents the number of employees involved.

$$t_1 = t_c \times n_l. \tag{2}$$

To calculate the time for which one worker manages to construct 1 m^2 of the total area of the composition of the green roof, or 1 m^2 of the technical part was used formula No. 3, where t_{c1} means the total time required to complete the activity by one worker, the letter P total area, for which the activity is performed and t_l the time for which one worker can construct the total area.

$$t_{c_1} = \frac{t_1}{P} \tag{3}$$

IOP Conf. Series: Materials Science and Engineering 1209 (2021) 012012 doi:10.1088/1757-899X/1209/1/012012

To determine the units of work (standard hours), it was necessary to visit several buildings and measure the time of their implementation. At the same time, it was necessary to experiment within the ADMAS research center, where the time of implementation of the green roof and subsequent evaluation took place.

The used compositions of vegetation roofs were based on the research of Ing. Petr Selník and his collaborating researchers presented in articles such as Designing and Testing of the Recycled Vegetative-Retention Layer of the Vegetative Upper Skin of the Pitched Roof [1] and Use of textile recycling in the construction of green roofs. [2]

2. Construction time of extensive green roofs

2.1 The roof of the Brno Žebětín Primary School

The roof structure of the Žebětín primary school was implemented in 2020 and an area of 430 m^2, was decided by the building designer to green it using extensive greenery. The greening of this green roof was taken care of using stonecrop mats. The height of the part of the building on which the construction of the green roof was carried out is 6.5 m. The primary school building is easily accessible, as it is free-standing and not connected to any other building. Thus, there was no major problem with transporting the material to the roof that would make it impossible to transport it quickly using lifting mechanisms or pumps. The composition of this roof was chosen as inverse (with the use of thermal insulation above the waterproofing itself). The green roof was realized by the realization firm only above the layer of thermal insulation. A drainage / hydro accumulation layer was established on the 300 g/m^2 geotextile using retention boards made of recycled polyester fibers with a height of 30 mm. This layer was omitted in the area of 500 mm from the edges of the roof and the penetrations that were led out on the roof to construct a safety zone (gravel) of washed aggregate. Gravel is separated from the substrate using a perforated aluminum ledges and a geotextile of 300 g/m^2, which prevents clogging of the gravel by floating fine particles of the substrate. In the free space of the roof, the substrate was then inflated using a conveyor with a pump and spread to a height of 50 mm above the retention plates. Furthermore, it was necessary to modify all the details of the roof and cut the protruding geotextile in the roof area. Stonecrop mats were then spread on the substrate to form a vegetation layer.

Table 1. Flat roof Brno Žebětín.

	Total time (h)	Total time (min)	Time for 1 worker	Time for 1 m^2 (min)
Laying of retention mats and gravel	12	720	3600	8.37
Laing of substrate (50 mm)	6	360	1800	4.19
Laying of stonecrop mats	4	240	1200	2.79
			Total time	**15.35**

The implementation time of one square meter of green roof, in this case, took one worker 15 minutes and 21 seconds. Laying retention boards and creating gravel from this time makes up the whole 8 minutes and 22 seconds, which makes up more than 55% of the total implementation time. From the remaining activities of pumping (blowing) the substrate using the pump, it took 4 minutes and 11 seconds, and the actual transport of the stonecrop mats to the roof and their distribution took 2 minutes and 47 seconds.

Figure 1. Flat roof ZŠ Žebětín (author).

Figure 2. Flat roof ZŠ Žebětín (author).

2.2 Green roof on the garage, Valašské Meziříčí

Another monitored roof was a vegetation roof realized on a garage in Valašské Meziříčí. The garage building is detached and has a rectangular floor plan measuring 3.5 x 6.5 m. The floor area of the roof itself is 22.75 m^2. The roof is located at a height of 3.5 m above the ground. There is almost no penetration on this roof, so a 500 mm wide gravel safety layer was set up only around the edges of the roof and gutters. The roof itself was designed as an extensive green roof and the composition of the roof layer was chosen similarly to the roof of the primary school in Žebětín. The first layer on the waterproofing made of PVC-P foil is a separating layer of geotextile 500 g/m^2. A layer of retention plates made of recycled polyester with a height of 40 mm was realized on this layer, which is to serve as drainage and hydro accumulation layer. On this layer, gravel of washed aggregate was realized, separated from the substrate layer by aluminum ledges and a geotextile of 300 g/m^2. The substrate chosen for this roof was the substrate for extensive green roofs. The vegetation for the roof was chosen from the stonecrops.

Table 2. Green roof on the garage, Valašské Meziříčí.

	Total time (h)	Total time (min)	Time for 1 worker	Time for 1 m^2 (min)
Laying of retention mats and gravel	1,283	77	154	6.77
Laing of substrate (50 mm)	1,333	80	160	7.03
Laying of stonecrop mats	0,667	40	80	3.52
			Total time	**17.32**

The data from this construction show less complex construction. This can be seen especially in the time of laying retention boards and the formation of gravel when the final time of implementation of this part of the construction would take one worker per 1 m^2 5 minutes and 43 seconds. The realization of the substrate layer using only human forces took 7 minutes on this construction. Planting stonecrops on individual plants took 3 minutes and 30 seconds. The total construction time of this structure would take 17 workers and 19 seconds per worker per 1 square meter.

2.3 Test green roof in the ADMAS research center

As a third roof, a similar construction of an extensive flat green roof was constructed for research purposes in the area of the ADMAS research center of the Faculty of Civil Engineering in Brno. The roof was built on a special platform, which was designed to verify the construction time of the roof structure using various mechanisms and to verify the functionality of entire roof layers and new materials usable in the composition of vegetation roofs.

The first composition realized on this platform was the composition of an extensive flat green roof. On the test platform, concerning the smaller floor plan dimensions, the realization of safety gravels made of washed aggregate along the edges of the roof was abandoned. The composition was therefore realized on the platform when the first geotextile was laid at 300 g/m^2. A hydro accumulation layer made of retention plates made of recycled polyester fibers with a thickness of 40 mm was realized on this filter and separation layer. A substrate for extensive green roofs with a height of 60 mm was then applied to the hydro accumulation layer. Due to the need for testing, the vegetation on this formation was applied only afterward.

Table 3. Test green roof in the ADMAS research center.

	Total time (h)	Total time (min)	Time for 1 worker	Time for 1 m^2 (min)
Laying of retention mats and gravel	0.4	24	48	5.08
Laing of substrate (50 mm)	0.5	36	72	7.63
Laying of stonecrop mats	0.25	15	30	3.18
			Total time	**15.89**

Figure 3. Test green roof ADMAS (author).

It can be read from the data that the laying time of retention boards per 1 square meter by one worker is lower than the time required for the yield and spreading of the substrate, by 2 and a half minutes. The time of laying the retention plates even with the installation of the geotextile is 5 minutes, while the time of spreading the substrate and its installation is 7 minutes and 38 seconds. The planting itself took 3 minutes and 11 seconds. The total realization time of one square meter of this composition, therefore, lasted 15 minutes and 49 seconds.

To determine the standard hour of the green roof structure, the three above-mentioned roofs were selected, which are ideal for these purposes due to their similar type of structure and almost the entire composition. We can therefore responsibly claim that for a given type of composition of an extensive vegetation roof, the implementation time will not be too different.

After analyzing the individual sub-processes in the construction of an extensive green roof, it is necessary to determine the standard hour for the construction of the overall green extensive roof. In the following table, we can see the individual implementation times of the extensive green roof.

Table 4. The setting of the average time for realization

	Technical layers	Substrate	Planting plants	Total time of realization
	Time for 1 m^2 (min)	Time for 1 m^2 (min)	Time for 1 m^2 (min)	Time for 1 m^2 (min)
Zš Žebětín	8,37	4,19	2,79	15,35
ADMAS	5,08	7,63	3,18	15,89
Garage Valašské Meziříčí	5,71	7,74	3,87	17,32
Average time	**6,39**	**6,52**	**3,28**	**16,19**

As can be read from the individual data, the implementation time of the hydro accumulation layer and gravel took the most time on the extensive green roof at the Primary School in Brno Žebětín, namely 8.37 minutes. The implementation on the roof of the garage in Valašské Meziříčí took 2.66 minutes less and the implementation on the platform in ADMAS took 0.63 minutes less. Using the arithmetic mean, a value of 6.39 minutes was subsequently determined for the construction of the hydro accumulation layer, including gravel. We can then correct this value by a 5% possibility of statistical error and determine the range for the applicable standard hour of this construction, which would therefore be in the range of 6.07 minutes to 6.71 minutes. Similarly, we can also determine the values of the time required for laying the substrate and planting the plants.

It can be read from the data that the laying of the substrate and planting of plants at the Žebětín primary school was much faster than in the case of the other two roofs. The fact is that in the case of this roof, the substrate was transported to the roof using a pump and the planting took place using stonecrop carpets. In other cases, the substrate was applied to the roofs by hand and the plants were planted.

It is possible to see from the individual times of the implementation times that the least time-consuming roof of the monitored roofs was the implementation of the most complex variant of the roof at the Primary School in Žebětín, which lasted 15.35 minutes. The roof on the test platform within the ADMAS research center lasted 0.54 minutes longer for one worker per square meter. The extensive green roof of the garage in Valašské Meziříčí was 1.97 minutes longer than the roof in Žebětín.

But the very purpose of this research was the fact that in the Czech Republic the software used to determine the construction time is not currently equipped with the possibility of appropriate pricing and construction time for individual types of green roofs.[3]

3. Conclusion

Based on field research, it was possible to determine the average standard hour for the implementation of one square meter of an extensive flat roof, which is 16.19 minutes. In the case of the implementation of an extensive flat green roof, we can now count on this time. Of course, since this is an average value from a relatively small sample of roofs, the actual implementation time may vary in the case of difficult implementations. However, we can continue to reduce the implementation time by using, for example, pre-grown panels for flat extensive green roofs, as we can see in the article Comparison of the construction time of the extensive green roof constructed from prefabricated panels and classical construction presented at the JUNIORSTAV 2021 conference in the Czech Republic.[4] In this article we can read about possibility to reduce time for construction of the flat green roof to 3,43 minutes per 1 m^2 just by using new technology. With new types of green roof construction and more building with green roof we will see that green roof can be completed much faster now than before.

4. References

[1] Selník P, Nečadová K, Mohapl M and Bedlivá H 2016 Designing and Testing of the Recycled Vegetative-Retention Layer of the Vegetative Upper Skin of the Pitched Roof *Applied Mechanics and Materials* **824** 824.67 67–6
[2] Selník P, Bečkovský D, Rebrova T 2018 Využití textilního recyklátu v konstrukci zelených střech *Konference Hydroizolace* (Mikulov: NOVPRESS)
[3] Hejl M 2018 Process of green roof pricing *Young scientist (Tatranská Lomnica)* p 1–7
[4] Hejl M 2021 Srovnání doby výstavby extenzivní zelené střechy konstruované z předpěstovaných panelů a pomocí klasické konstrukce *JUNIORSTAV 23. odborná konference doktorského studia s mezinárodní účastí Sborník příspěvků* (Brno: ECON publishing) pp 99–104

Acknowledgments

This paper was written as a part of the project No. FAST-J-21-7378 "Analysis of construction-technological processes in the implementation, operation and maintenance of environmentally friendly roofing of buildings" („Analýza stavebně-technologických procesů při realizaci, provozování a údržbě environmentálně šetrného zastřešení budov ").

IOP Conf. Series: Materials Science and Engineering 1209 (2021) 012013 doi:10.1088/1757-899X/1209/1/012013

Management of health and safety on building site using smart technologies

A Husovská[1] and Z Struková[1]

[1] Department of Technology, Economics and Management in Civil Engineering, Faculty of Civil Engineering, Technical university of Kosice, Vysokoskolska 4, Kosice 042 00, Slovakia

E-mail: alzbeta.husovska@gmail.com

Abstract. The construction industry is a dynamic manufacturing sector whose major problem is occupational safety and its management. There are various information and communication technologies or applications that can improve health and safety status. Their use gives every company management a great advantage to become more competitive in the construction trade. The presented article deals with smart phones and related applications as a type of technologies that can be easily used to improve health and safety performance in our conditions. The aim of the article is to present the advantages and benefits of this technology for increasing safety on construction sites. The presented results confirm the trend of efficient processing of the obtained information using ICT technologies. They intensify the safety management process and help staff to handle unpredictable dangerous incidents and contribute to the successful management of the whole construction site.

1. Introduction

The construction industry, together with manufacturing and transport and storage sector, is one of three sectors where the most fatal occupational accidents and serious injuries have been recorded in the last five years (2015-2020) [1]. It is therefore a great challenge for companies in these sectors to manage occupational health and safety as effectively as possible and to monitor and check compliance with the set health and safety rules. In addition to diversity of construction sites, differences in occupations, technology, tools and materials used, make occupational health and safety management very challenging [2]. Several information technologies, which are becoming part of both, our personal as well as working life, can significantly contribute to development and maintenance of a healthy and safe working environment on a construction site.

However, it is demanding and data-intensive for site engineers to collect and share site information in real-time due to ugly and specific conditions of construction sites. The locations of materials, labour, and equipment, along with the current status of progress, are difficult to be understood at sites. Moreover, construction projects are mostly full of various boundaries and distinctions between different participants' knowledge demands, resources and custom practices, related what is and is not safe and how to achieve health and safety goals [3]. These challenges require the development of effective tools equipped with suitable sensing and communication capabilities to acquire and exchange construction information efficiently [4]. Establishment and maintenance of clear communication channels belong to the most important and the simplest ways to improve safety on site.

Mobile smart phones and tablets are part from pieces of ICT technologies that have found a permanent place on construction sites. In addition to enhanced real-time communication and easier coordination of activities, their complete ubiquity makes them a natural part of the core platforms, making applications and tools easily accessible to workers. Their use enables faster and more effective judgments. Decisions and changes can be taken on the fly.

This paper focuses on smart phones and related applications as a part of technologies that can be easily used to improve health and safety performance in our conditions. Benefits and improvements in connection with application of smart phones on construction site are presented. The paper also deals with specific examples of applications intent for management of occupational safety and health in construction site.

2. The role of information and communication technologies in construction

Information and communication technology (ICT) is a concept which contains technologies that provide access to information through telecommunications. They have a huge impact on our daily activities. This includes the internet, wireless networks, cell phones, computers, software, middleware, video-conferencing, social networking, and other media applications and services. They can be accessed, searched, stored, transmitted and manipulated by users in digital form. [5].

Their rapid development and advancement already requires us to be educated in this issue and to be able to work effectively with these technologies. The great advantage of using ICT is their rapid application in practice. The 'construction site of the future' is conceived as a smart working place, where self-adapting sensor networks and decision support systems monitor and complement workers' activities, at every level and in real time: from workers to quality and safety inspectors, from site managers to suppliers [6]. Digitization of the whole process of construction, from design through its implementation to subsequent use and maintenance, is essential to do construction industry more effective. Increased adoption of ICT in the process of construction is related to the overall effort to fully digitize the sector. Concepts such as increased productivity and efficiency, sustainability, increased occupational health and safety and financial savings should become synonymous with digitization and adoption of ICT in construction companies and on sites.

2.1. Adoption of ICT to enhance health and safety

Researchers and engineers continuously search for various strategies and practices that may improve safety performance in the construction industry and have potential to minimize the high number of workplace injuries and fatalities. The publication on the utilization of technology for safety management has increased by approximately three times from the early 2000s to the early 2010s and this rate has been sustained through the last decade (2010 – 2019) as well [7].

ICT technologies are used in many aspects of construction industry, such as pre-construction process, safety training, safety inspections, on-site operational safety issues and improving workers' communication on projects, onsite safety monitoring, on-site hazard identification [8]. The use of ICT applications in general brings about improved health and safety management and reduced number of on-site accidents and near-misses. They can also help project managers as well as clients or investors, who may not have specific information and knowledge about a project, with decision-making processes.

2.2. Possibilities of using ICT to manage health and safety on construction sites

Precise and real-time control over workers, equipment, materials, construction methods and work environment are need for safety risk prevention and emergency response [9]. Information systems have been extensively discussed for managing massive multi-source heterogeneous data generated during construction and health and safety hazard on site was assessed. So, the consistent and timely data can be shared between project participants for effective safety collaborations [10].
Examples of ICT applications that have an impact on overall safety on the construction site are such as follows [7]:

- electronic attendance system,
- dates of employee training,
- regular inspections and revisions, especially of reserved technical equipment,
- different types of reporting.

The quality of a construction project can be improved using the proposed system. To reduce incidents of defects during construction processes, the system provides work task information for site engineers to learn about accurate construction processes [4]. There is a space for improvement mainly in the relatively low usability of various ICT in the Slovak construction industry.

3. Safety management via applications in smart phone

Various construction management software has been around for a while now. For convenience, construction software also already come with their mobile app versions.

Several important functions of on-site H&S management can be identified [4], [6], [11]:

i. preventing workers from accessing predefined danger areas;
ii. keeping workers away from the ground level projection of moving suspended loads;
iii. site monitoring to understand the current status of the construction project - Monitoring of unauthorized intrusion on construction sites is crucial to reducing safety accidents. The existing methods have shortages such as detection error, locating error, alarming delay and difficulties in identifying intruders based on their backgrounds.
iv. to manage construction tasks - to effectively manage construction resources and
v. real-time information sharing to facilitate efficient interaction among construction participants – Real-time decisions can be made and workers can be given new instructions no matter how far they are from their bosses.

However, the full potential of sensing technologies can be achieved only when the information obtained on site is effectively distributed and shared among project participants. The real-time project information can enable a range of project participants, including project managers, site engineers, and construction workers, to make informed decisions [4]. The camera in the smart phone adds another level of information dissemination, because workers can photograph and send pictures of any questions or jobs in progress.

Smartphones are now typically equipped with touch screen, GPS receiver, gyroscope, accelerometer, and wireless communication capability. The strengthened features of smartphones enable a new generation of on-site management processes, such as location-based customized work orders, real time information exchange, and augmented reality (AR)-based site visualization [4].

In addition to management of occupational health and safety, other benefits and improvements in connection with application of smartphones on construction site can be found [12]:

i. construction time reducing - access to the project information database in real time reduces the time needed to visit the office of construction company to obtain information about the project, such as drawings, technical data and plans;
ii. reduction in cost of construction - systems can minimize the possibility of reprocessing through an appropriate work order and efficient information sharing. The system also helps engineers effectively manage project material and equipment costs to avoid wasting project budgets.
iii. increase in predictability;
iv. reduction in waste;
v. increase in productivity - daily tasks can be done with a few swipes and taps. The managers are also able to keep track of worked hours or worker location through GPS.
vi. reduction in operation and maintenance costs.

Smartphones features make them attractive as components of reliable and convenient analytical devices for measuring environmental pollution, either directly by using a smartphone camera as an optical detector [11]. Tablets also fall under the mobile category and there will be more people who prefer to view schedules or documents from a bigger screen. Tablets are then a better alternative than the mobile phone.

If we want to make the most of the potential of smartphones or tablets to our advantage, we need to download and use the right, most suitable application. Mobile applications are the way to facilitate safety management on building site. Especially on the bigger ones.

4. Practical examples of using applications in smartphones

There are more special examples of mobile ICT tools that have been developed to perform safety inspections. Skanska Tekra Oy, the civil engineering division of Skanska in Finland, has introduced an SMS/WAP and MMM based system which allows data to be collected electronically using the supervisor's phone. It then sends problem notifications to the subcontractors who may respond via SMS when the problem is rectified. Remedial actions that previously took days to complete are now done in a few hours, and accidents have decreased since the implementation of the system. The data collected are all stored on a central database enabling the identification of trends to be automated [12].

4.1. A management solution for active tracking of construction equipment

An example of an available solution that is used in the Slovak construction industry is ON!Track program by Hilti. The base of the system, a rugged barcode AI T380, is presented in the Figure 1. Together with bluetooth tags on jobsite equipment, that communicate with cloud-based construction equipment management software that runs on mobile or desktop devices – meaning user can track his tools in an instant [13]. So he can be able to find his assets very quickly.

These robust Bluetooth and barcode tags are designed to withstand the toughest conditions of construction site including abrasion, corrosion, chemicals and extreme temperatures. The Bluetooth technology has a range of 30 m. Data on time, date and location is then transferred to the Hilti Cloud and is accessible anywhere via desktop and mobile devices [13].

Figure 1. The rugged barcode on equipment [13].

Every company that knows exactly where to find its equipment and assets, does not lose valuable time tracking down its tools or waste money buying or hiring replacements. It also quickly gains better control over consumable spend (bits, Personal Protection Equipment etc.) [13].

The most meaningful advantage is that the system is cloud-based. Unlimited users can access the construction equipment management software 24/7 – and from any location – with mobile and desktop apps developed by Hilti. At the touch of a screen or the click of a mouse, „ON!Track equipment management for the construction industry“ tells exactly what equipment company has, where it is and who is using it. When tools need maintenance or calibration it alerts on it. When it's time to renew training or certification the construction equipment management solution tells it in advance [13]. A basic description of the system is presented in Figure 2.

In this application, its users especially appreciate its intuitiveness. Working with it is easy not only for a health and safety or inspection technician, but for anyone who has experience with using smartphones in general. The application is an excellent helper in regular inspections and revisions of hand tools and work aids. Another great advantage is that in the application it is possible to see who is currently using the work tool, but also who has worked with it before. A history of use is very much needed in dealing with work-related injuries or near misses. A necessary benefit, which is perceived by technicians and workers on the construction site, is also that it is possible to store instructions for use for a specific tool.

4.2. Autodesk BIM 360 - Construction application built for the field

To build up an effective construction safety planning, several construction companies in Slovakia, like for example Skanska a.s., use the application Autodesk BIM 360. The construction management software for field execution and project management has a large number of subcategories, therefore only the benefits related to the area of occupational safety and health will be described below.

The principle is in creating and distributing safety checklists and incident reports that provide a project-wide view of safety program conformance. The basic task of health and safety technicians is to use this program to record safety problems and incidents and to monitor project safety performance. The big advantage is that the program completes tasks from anywhere on the jobsite with seamless access to Autodesk Build data, even when you're offline. Technicians or inspectors log the root causes of issues and add photos or comments to help identify and prevent common problems during a site inspection. Use the gallery is possible to quickly find information and generate photo reports. Because it is a BIM program, it is easy to find in the project where a unsafe event or disagreement with system occurred. Then unsafe observations are immediately allocated to the site managers or foremen for quick resolution. In that way all team members can be able to take ownership of site safety.

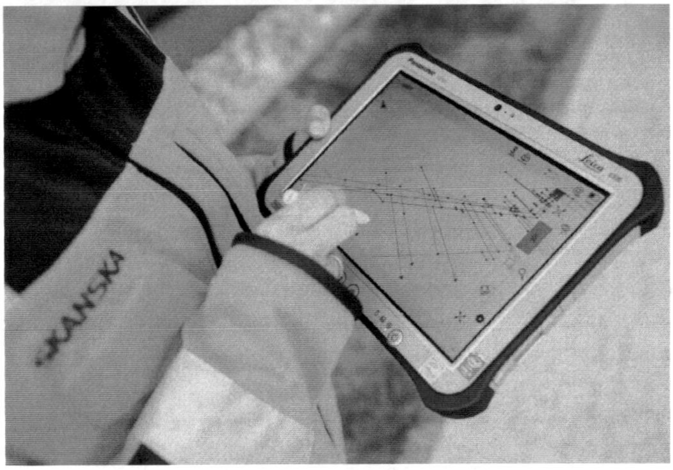

Figure 2. The Skanska CZ uses BIM 360 on tablets [14].

5. Conclusions

Advanced ICT can provide safety professionals as well as entire construction personnel with a useful tool for monitoring and recording data that they can further use to manage safety on the construction site. The functioning of the information system in the company is effective if the manager can find the necessary information in time, pass it on wherever it is needed and where it will be best used. It is important for construction companies to choose the most suitable application that can help them effectively carry out safety management and thus ultimately reduce the number of accidents on construction sites.

If a construction company uses even a simple application, for example to manage its assets and tools, it already increases the level of safety in its workplace. Each use of modern ICT technology speeds up its processes, which would otherwise be lengthy. It is no longer necessary to use papers, wait for a colleague to sign, but I will process everything and send it for reading and acquaintance directly in the application. By being able to see the conclusions and results of the inspection by everyone who has the app in their smartphone, work is faster and more efficient, the security plan is followed and the risks are more predictable. It is possible to say that the smart phones significantly and permanently change the operation way of construction companies.

6. References

[1] Health and Safety Executive Statistics of fatal injuries 2021 *Health and Safety Executive 2015-2020* Available online https://www.ip.gov.sk/bozp/analyzy/?ip=nip

[2] Hallowell M R and Gambatese J A 2009 Construction safety risk mitigation *Journal of Construction Engineering and Management* **135** 12 1316–23
[3] Xu J, Cheung C, Manu P and Ejohwomu O 2021 Safety leading indicators in construction: A systematic review *Safety Science* **139** 105250
[4] Changyoon K, Taeil P, Hyunsu L and Hyoungkwan K 2013 On-site construction management using mobile computing technology *Automation in Construction* **35** 415–23
[5] ICT Definition Tech Terms 2010 Avaialble online https://techterms.com/definition/ict
[6] Carbonari A, Giretti A and Naticchia B 2011 A proactive system for real-time safety management in construction sites *Automation in Construction* **20** 686–98
[7] Mihic M, Vukomanovic M and Završki I 2019 Review of previous applications of innovative information technologies in construction health and safety *Organization Technology and Management in Construction An International Journal* **11** 1 1952–67
[8] Azmy N and Zain A Z M 2016 The application of technology in enhancing safety and health aspects on Malaysian construction projects *ARPN Journal of Engineering and Applied Sciences* **11** 11 7209–13
[9] Skibniewski M J 2014 Information technology Applications in Construction Safety Assurance *Journal of Civil Engineering and Management* **20** 6 778–94
[10] Lee K P, Lee H S, Park M, Kim H and Han S 2014 A realtime location-based construction labor safety management system *Journal of Civil Engineering and Management* **20** 5 724-36
[11] Cerrato-Alvarez M, Frutos-Puerto S, Arroyo P, Miro-Rodriguey C and Pinilla-Gil E 2021 A portable, low-cost, smartphone assisted methodology for on-site measurement of NO2 levels in ambient air by selective chemical reactivity and digital image analysis *Sensors & Actuators: B. Chemical* **338** 129867
[12] Bowden S, Dorr A, Thorpe T and Anumba Ch 2006 Mobile ICT support for construction process improvement *Automation in Construction* **15** 664–76
[13] ON!Track by Hilti, Tool Services, Available online https://www.hilti.group/content/hilti/CP/XX/en/services/tool-services/on-track.html
[14] Internal data of Skanska, a.s.

Acknowledgments
This paper was elaborated with the financial support of the grant no. VEGA 1/0557/18 of the Scientific Grant Agency and the grant no. 038TUKE-4/2020 of the Cultural and Educational Grant Agency of the Ministry of Education, Science, Research and Sport of the Slovak Republic.

IOP Conf. Series: Materials Science and Engineering 1209 (2021) 012014 doi:10.1088/1757-899X/1209/1/012014

Sustainability of material solutions of modern methods of construction

D Mokrenko[1] and M Kozlovská[1]

[1] Institute of Construction Technology, Economic and Management, Faculty of Civil Engineering, Technical University of Kosice, Vysokoskolska 4, 042 00 Kosice, Slovakia

E-mail: daria.mokrenko@tuke.sk

Abstract. In the last decade, the construction industry has been going through a technological revolution known as Construction 4.0. This progress has brought digitization, automation, virtual reality, 3D printing and prefabrication to production. Prefabrication is the basic solution of most Modern Methods of Construction (MMC), thanks to which construction is faster, safer, better quality and with a lower impact on the environment. MMC design solutions can include volumetric, panel, hybrid systems, parts and components bat also on-site innovative technologies. From the point of the sustainability, material solutions of MMC are also very important and open the potential for their innovations. The aim of this article is to explore the possibilities of higher use of composite materials and mixtures, preferably with an emphasis on more environmentally friendly and recycled materials.

1. Introduction

At present, the development of the construction industry is not possible without innovations, which consist in connection with the innovations of the industry as a whole. One such innovation is Industry 4.0, which is being transferred to construction as Construction 4.0. This trend in construction includes digitization, automation, intelligent factories, the Internet of Things, 3D printing, modern methods of construction (MMC), material innovation, etc. [1]. The topic of the article is closely connected with the use of not only traditional but also innovative materials in modern construction methods and provides an overview of currently used materials in MMC.

When choosing a building material or product, it is decisive to what extent it meets the requirements of legislation, technical, economic, subjective, but currently more emphasis is placed on environmental parameters, such as renewable material for the material or product, recyclability, durability, absence of harmful substances [2]. According to Sičáková, there are three main requirements for the material:

- quality,
- reliability,
- environmental properties,

where the term "quality" means the setting and adherence to specific criteria that must improve the quality of building materials. The term "reliability" refers to the fulfilment of certain conditions, such as safety, durability and usability of the material. The term "environmental properties" means the positive impact of the material on human health and also the reduction of the negative impacts of their production on the environment [3].

2. Material base for MMC

Modern methods of construction are a combination of off-site and on-site construction methods, which are mostly carried out in the factory (off-site), but are also those that have to be performed on-site by methods increasing their efficiency of their production. In the case of off-site methods, after completion of work in the factory, the building structure is transported and assembled on site. These methods have many advantages, such as reducing construction time, improving the quality of the final product, addressing labour shortages on the construction site, reducing the environmental impact, and many other benefits [4].

The most commonly used materials for MMC technologies are almost the same as those used in traditional construction. In traditional construction, concrete, steel and wood are generally used because they are widely available and applicable. They have a long-developed production background, norms and standards for the design and construction of structures [5].

2.1. Basic division of MMC

MMC can be divided according to design solutions into five main groups [4]:

- panel systems (open and closed);
- volumetric (volume) systems;
- hybrid systems (semi-volumetric);
- prefabricated parts and components;
- on-site innovative technologies.

Panel systems can be based on concrete, wood or steel. There are several different types of panels, with different materials used. They can be closed, open, filled, insulation panels (SIP) or concrete panels [6].

Closed panels have a wooden or steel frame, usually filled with insulation and lined with building boards (for example, plasterboard, cement or magnesium boards can be used) with wiring and windows already installed in the factory (complexity varies depending on the system type) [7]. They usually use mineral wool and expanded polystyrene as insulation.

Open panels consist only of steel or wooden frame and require the installation of insulation, wiring and cladding boards on the construction site [6, 7].

Filled panels are similar to a closed panel, but usually do not have an internal cladding board (plasterboard or other) installed. Mineral wool, expanded polystyrene or insulating foam can be used as insulation. This panel allows easy installation of wiring on the construction site and eliminates damage to fragile finishing boards [6, 7].

Structural Insulated Panels (SIP) have an insulation core - polymer foam or polystyrene, to which boards are glued on each side. The materials for the board vary depending on the manufacturer, but they are usually plywood, OSB boards (OSB - Oriented Strand Board) or one of the modern composite boards (plasterboard, cement or magnesium board) [6, 7].

Volumetric, semi-parametric, or hybrid systems can be made from a variety of materials, including wood, cold-formed steel, hot-rolled steel, concrete, or a combination of material assemblies. The modules can be supplied with all external and internal surface treatments or only with the basic construction. The first type of module consists only of a frame and a cladding on one side. The second type includes a frame, insulation and cladding boards on both sides without finishing work. The third type is a completely finished type of module in the factory, including finishing work and installed wiring [6, 7].

On-site innovative methods include many designs and technologies that use modern building elements. For example, lost formwork using expanded polystyrene, paper form, filigree slabs, thin-walled slotted steel profiles. Also innovative methods include aerated concrete, light silicate blocks and others can be used to form the main building elements, such as walls, roofs and floors [6, 7, 8].

2.2. Innovative and sustainable MMC products based on concrete

In addition to traditional materials in various innovated forms, new types of materials are also being developed that improve not only the properties of traditional materials throughout the construction life cycle, but also the economic and environmental aspects of construction.

Table 1. Innovative products based on concrete [9 - 15].

Product	Production technology, material base	Specific features and use
The innovation consists in improving the properties of materials		
Concrete, reinforced with carbon fiber [9]	It has a basic structure of plain concrete supplemented with fibers, which strengthen the structure of the composite and thus improve some of its properties.	The advantages of using this material include - high strength, frost and wear resistance, water resistance, lower operating costs, low weight and more creative architectural design. Disadvantages include high start-up costs, lack of knowledge and lack of standards
Concrete, reinforced with basalt fiber [10]	By adding basalt fiber to the concrete mortar, a composite material is formed	Corrosion resistant, has low thermal conductivity, the fibers are firmly adhered to the concrete, which ensures the strength of the reinforcement
Self-healing concrete [11]	Due to the addition of capsules with bacteria and nutrients (calcium lactate) to the concrete solution, the phenomenon occurs that when a crack appears, the nutrients are activated and produce limestone, which clogs it.	The advantages of using this material are extended service life of buildings, reduced costs for repairs and condition monitoring
Light-transmitting concrete [15]	Mixture of optical glass fibers and fine concrete, The proportion of fibers is low, representing 4% of the total volume. Due to their size, they become a structural part of concrete.	The surface of the blocks remains similar to homogeneous concrete. It can be used on interior or exterior walls, illuminated sidewalks
The innovation lies in the use of recyclable materials		
Ercconcrete [12]	Used recycled filler from construction and demolition waste, such as bricks, roofing, sanitary ware or ordinary cement concrete, replaces up to 100% of natural aggregates.	Conserves natural mineral resources. It can achieve the same quality parameters as concrete with natural aggregates. Bulk density is up to 20% lower compared to concrete made of natural aggregate, which brings savings for transport. The life cycle is much longer. At the end of its service life, new concrete mixes or precast units can be produced repeatedly
Polybeton [13]	The binder in the form of coarsely sorted waste plastic material (PET bottles, plastic packaging) is crushed into a fraction of 2 to 20 mm. The filler (sand, aggregates, crushed glass, etc.) is heated, dried	Physical properties better than concrete, Fire resistance - B1

	and homogenized with the binder in a mixing reactor. The bulk mixture is then tempered to the melting point of the thermoplastic with a screw extruder and fed as a melt to the individual molds, where the final shaping and cooling takes place.	
Popbeton [14]	Instead of Portland cement, waste material from large heating plants is used as a binder - fly ash.	Replacement of Portland cement makes it possible to reduce the amount of CO2

In table 1 includes innovative building materials based on concrete, which lead to the sustainability of the building built using MMC. Concretes are being innovated to a greater extent, with the addition of recycled materials. These materials, such as glass, plastic, fly ash, ceramics, no longer have to be stored in landfills and are used in construction, thus saving the environment, freeing up space in landfills, through their use it is possible to shorten the production of various concrete admixtures. At the same time, the essence of material innovation is to improve their properties in favor of longer service life or to reduce operating costs, such as self-healing concrete (figure 1) or light-transmitting concrete (figure 2).

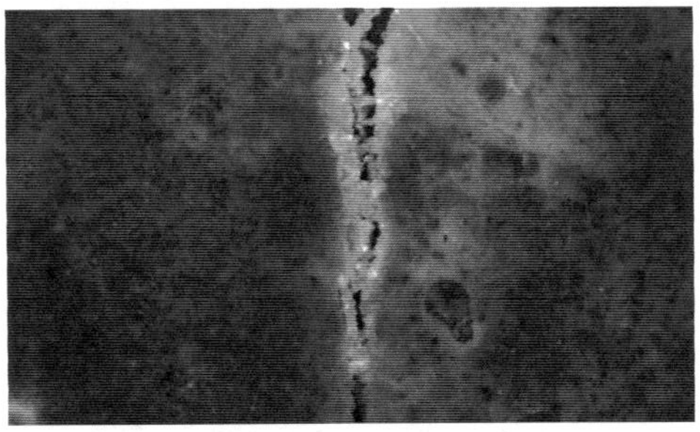

Figure 1. The process of "healing" a crack in self-healing concrete [11].

Figure 2. Light-transmitting concrete [15].

2.3. Innovative and sustainable MMC products based on other materials

In addition to composite materials recently innovated products based on wood, ceramics or organic materials. Last but not least, there are new products that are made from organic materials such as straw, mushrooms or grass and are 100% organic.

Table 2. Products based on other materials [16– 25].

Product	Production technology, material base	Specific features and use
	The innovation consists in improving the properties of materials	
CLT panels [16]	Panels made by the technology of cross-gluing of boards from boards (lamellas).	The risk of shrinkage due to changes in humidity is practically excluded, high resistance to static load. Panels made using CLT technology are 6 times lighter than reinforced concrete. Thanks to this, the load on the foundation is significantly reduced, and transport costs are reduced.
DLT panels [26]	DLT is similar to nail-laminated timber (NLT), without the use of screws or nails. To form DLT members, softwood lumber panels are stacked like NLT and friction-fit together with hardwood dowels. The dowels hold the boards together, and the friction fit—achieved by the differing moisture content of the softwood panels and the hardwood dowels—affords additional dimensional stability.	DLT is best suited for flooring and roofing applications. It also helps improve a project's acoustics and visual appeal through the addition of curves and kerfs.
PSL panels [27]	PSL is part of a family of products, structural composite lumber (SCL), that are made of dried and graded wood veneers, strands or flakes that are layered upon one another and bonded together with a moisture-resistant adhesive into large blocks known as billets.	Given its bending strength, PSL is used as long-span beams, heavily loaded columns and large headers and is well-suited to applications where high bending or compression stress is required. It can be used in both interior and exterior applications.
Luminescent cement [17]	Composite material that has been created by adding fluorescent components to cement that can absorb solar energy during the day and emit light at night. It is made of sand, dust or clay and the only residue during production is water vapour.	This cement is resistant to sunlight, offers a reduction in the cost of street lighting and has an estimated life of 100 years, environmental friendliness.
PROSOLVE370E Decorative architectural modules [18]	The modules are made of a lightweight plastic shell, coated with a layer of superfine titanium dioxide (TiO2), which is a nano photocatalytic version of conventional TiO2 commonly used as a pigment.	This technology effectively reduces air pollutants to harmless amounts of carbon dioxide and water, is activated by ambient daylight and humidity. The modules can be used for facades or their e.g. place in places with high traffic.
	The innovation lies in the use of recyclable materials	

SOLID TEXTILE BOARD Visor boards made of textile waste [19]	Made from cuttings of coating materials and waste from industrial treatment plants. The textile fibres are crushed into a homogeneous mixture and mixed with a special binder to make the boards more durable.	It is used for horizontal applications, meets the standards of the furniture industry, with high density, is directly usable and resistant to wear, offers flame retardant properties
PACKWALL Boards of used beverage cartons [20]	Produced by hot pressing, beverage cartons are made of polyethylene, aluminium and paper	They have 50% lower absorbency than wood-based panels. Used for interior partitions and ceilings, sandwich panels, perimeter walls, base and levelling boards
SILICASTONE [21]	Board formed from recycled waste glass and ceramics by sintering at high temperature and pressure.	UV resistant, does not change colour, standardized dimensions, easy cutting, suitable for interior and exterior
MOGU [22]	Composite material from agricultural by-products, which are made entirely of fungal mycelium and recycled textile residues.	Sustainable, suitable as packaging material, thermal or acoustic insulation
EKOPANELS [23]	It is pressed at high temperature and pressure from straw without the use of binders, glued with recycled cardboard.	Ekopanel is a 100% natural, fully recyclable, solid and diffusely open building material suitable for permanent installation in buildings. Suitable for making partitions, ceilings, interior and exterior cladding. Ensures low operating costs due to thermal insulation properties and high heat accumulation, fast and easy to install, is 100% ecological, sound insulates, is mechanically resistant
ORGANOID [24]	Thin surface materials pressed from locally available grasses, flowers, cereal bran, shells and other natural materials from the vicinity of the Austrian Wildspitze mountain range in North Tyrol	The emphasis is on environmentally friendly production. Wide range of products. In addition to board materials and acoustic panels, Organoid Technologies currently offers the application of their surface layer to an already finished veneer or bent wood molding.
CHIP[S] BOARD [25]	Composite material bonded with an adhesive from potato peels in combination with fibers from potatoes, bamboo, wood or hops and hot-pressed into robust boards.	An environmentally friendly alternative to MDF boards or chipboard from potato chips. Unlike MDF boards, this material is biodegradable and does not contain formaldehyde or other toxic resins or chemicals.

In table 2 innovative building materials were divided into two groups. In the first part the stop is placed on improving the properties of materials, in the second on the use of recycled materials. The first group of materials, which includes CLT panels, Luminescent cement (figure 3) and PROSOLVE370E Decorative Architectural Modules (figure 4), is aimed at improving the specific properties of the material, but more recently greater stop to reduce the impact on the environment or reduce operating costs.The second group includes building boards, made using straw, textile waste, potato peels, grass or glass. The technology of production of these materials is similar, either recycled materials are pressed at high temperatures or crushed and then glued with a special glue. Such materials are an environmentally friendly alternative to traditional materials, but some of them offer much better properties than traditional materials, such as PACKWALL boards.

Figure 3. Luminescent cement [17].

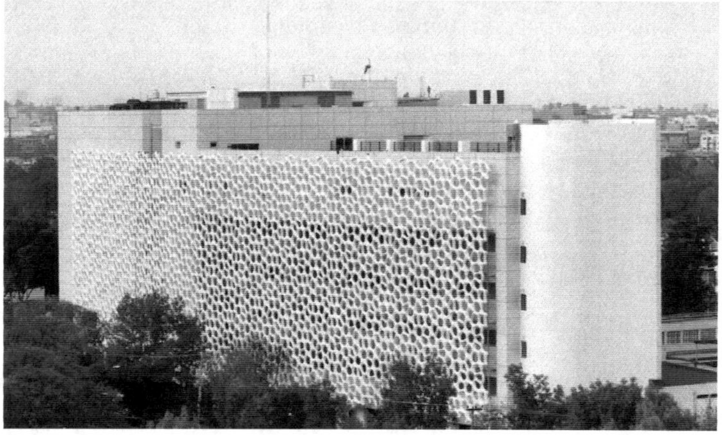

Figure 4. PROSOLVE370E decorative architectural modules used on the facade of Manuel Gea Gonzalesa Hospital in Mexico [18].

3. Discussion and results

Climate change is forcing the construction industry to adopt new technologies and innovate materials to reduce its carbon footprint. There are many new materials that are specially designed to solve many of the problems that construction encounters every day and are environmentally friendly. The analysis of the above sources makes it possible to note that recently in the construction industry they are trying to innovate materials by using recycled waste for the disposal of landfills. At present, when there is an acute problem of increasing the number of landfills that surround buildings in cities and towns, environmental issues are very important. Considering that recycled material is cheaper and the use of recycled materials reduces waste, the benefits are obvious.

Table 3. Innovative products and places of their use in construction

Product / Use	Material base	Wall	Ceiling	Partition	Facade	Surface treatment
Concrete, reinforced with carbon fibre	composite material	+	+		+	
Concrete, reinforced with basalt fibre	composite	+	+		+	
Self-healing concrete	composite	+	+		+	
Light-transmitting concrete	composite	+	+		+	
Ercconcrete	ceramic	+	+		+	
Polybeton	composite +plastic	+	+		+	
Popbeton	composite	+	+		+	
CLT panels	wood	+	+	+		
DLT panels	wood	+	+	+		
PSL panels	wood	+	+	+		
Luminescent cement	composite					+
Prosolve370e	plastic				+	
Solid textile board	coating materials					+
Packwall	paper, cartons			+		+
Silicastone	glass +ceramic					+
Mogu	organic					+
Ekopanels	organic					+
Organoid	organic			+		+
Chip[s] board	organic			+		+

* Products that use recycled materials in their production are marked in green.

In table 3 the above-mentioned materials are listed, broken down by use in construction. The main recycled materials are glass, ceramics, paper, straw, plastic, agricultural by-products, textiles [19-25]. As already mentioned, the advantages of using recycled materials mainly reduce the environmental impact and make it possible to reduce material costs. The second goal of material innovation is to innovate their properties primarily to reduce environmental impact, for example by using PROSOLVE370E [18] decorative architectural modules, which reduce air pollutants to harmless amounts of carbon dioxide and water, or to reduce operating costs as in the case of luminescent cement [17], which offers a reduction in the cost of street lighting.

Some of the above materials have great potential use in construction, but some are already widely implemented in modern construction methods, for example CLT panels. MMC offers many benefits, such as speed of construction, reduction of material used, better productivity, reduction of waste, increase of safety on site, reduction of environmental impact by reduction of noise and dust on the construction site, etc. The use of such materials extends the benefits of MMC mainly to environmental parameters, namely the reduction of CO2 emissions, the absence of toxic substances, recyclability, products are environmentally friendly.

4. Conclusion

At present, building materials are undergoing an innovative way to change the construction industry. Literature analysis shows successes in the implementation of innovative materials at MMC. Building material innovation has been divided into two groups, the first of which aims to focus on improving specific material properties, or to innovate building materials using recycled material.

The use of recyclable materials in combination with the improvement of the structural and thermal insulation properties of the material could thus increase the potential of innovative materials in construction, increase the efficiency of MMC and reduce its impact on the environment. However, it can be stated that at present there are no general requirements and standards in the development of innovative materials and MMC as a whole. However, improving existing materials and their properties creates room for further innovation.

5. References

[1] Craveiro F, Duarte J, Bartolo H and Bartolo P 2019 Additive manufacturing as an enabling technology for digital construction: A perspective on Construction 4.0. *Autom. Constr.* **103** 251–267

[2] Nagy E 2004 Environmental Assessment of Building Materials, Constructions and Buildings *Život. Prostr.* **38** 4 182 – 187

[3] Sičaková A Materiálové inžinierstvo: prednáška 1, úvod do materiálového inžinierstva

[4] Hashemi A 2015 Offsite Manufacturing: A Survey on the Current Status and Risks of Offsite Construction in Iran *Journal of Civil Engineering and Architecture* **9** 141-152.

[5] The Building Societies Association 2016 Laying the foundations for MMC [online] Available at: https://www.bsa.org.uk/BSA/files/20/20cee386-9dc6-4e0e-9ae9-504e3bbe4171.pdf

[6] Saghi H, Chaffari A 2020 Studying the effect of modern construction technologies on time, cost, and quality of Iran mass housing projects *J. Civil Eng. Mater.* **4** 233-242

[7] Modern Methods of Construction (MMC) & Eco-Building *Lenders housebuilding conference*, [online], Available at: https://www.selfbuildportal.org.uk/images/information/modern_methods_of_construction.pdf

[8] Mesároš P and Mandičák T 2015 Factors affecting the use of modern methods and materials in construction *IOP Conf. Series: Materials Science and Engineering* 71 012053

[9] Caulfield L 2020 The world's first building made from carbon-fiber reinforced concrete starts construction in Germany [online], Available at: https://www.bdcnetwork.com/world%E2%80%99s-first-building-made-carbon-fiber-reinforced-concrete-starts-construction-germany

[10] Fathima I I 2014 Strength Aspects of Basalt Fiber Reinforced Concrete *International Journal of Innovative Research in Advanced Engineering (IJIRAE)* **1** 8 ISSN: 2349-2163

[11] Srewart A 2016 The 'living concrete' that can heal itself [online], Available at: https://edition.cnn.com/2015/05/14/tech/bioconcrete-delft-jonkers/index.html

[12] Official website ERC concrete [online] Available at: https://www.erc-tech.eu/cs/ercconcrete-r/#ercconcrete-r

[13] Official website Polybet [online] Available at: http://www.polybet.eu/

[14] Sloboda P, Doležal J, Škvára F, Kopecký L, Lucuk M, Dvořáček K, Myšková L, Pawlasová S, Beksa M and Šulc R Beton bez cementu s názvem popbeton [online] Available at: https://docplayer.cz/105427959-Beton-bez-cementu-s-nazvem-popbeton.html

[15] Desing build network 2021 Litracon – Light Transmitting Concrete [online], Available at: https://www.designbuild-network.com/projects/litracon/

[16] Khodadadi A 2019 Eco-friendly Mid-rise Apartments Using CLT Panels *Proceedings of the IASS Annual Symposium 2019 – Structural Membranes 2019*

[17] Desarrollo 2016 Looking to light highways with light-emitting cement [online] Available at:

https://phys.org/news/2016-05-highways-light-emitting-cement.html
[18] Official website Prosolve370e [online] Available at: http://www.prosolve370e.com/
[19] Official website Kvadrat [online] Available at: https://www.kvadrat.dk/en/really
[20] Official website Packwall [online] Available at: http://www.packwall.cz/
[21] Official website Alusid [online] Available at: https://www.alusid.co.uk/products/silicastone/
[22] Official website MOGU [online] Available at: https://mogu.bio/
[23] Official website Ekopanely [online] Available at: https://www.ekopanely.cz/
[24] Official website Organoids [online] Available at: https://www.organoids.com/en/
[25] Official website Chipsboard [online] Available at: https://www.chipsboard.com/materials
[26] Official website Think wood [online] Available at: https://www.thinkwood.com/mass-timber/dlt
[27] Official website Naturally wood [online] Available at: https://www.naturallywood.com/products/parallel-strand-lumber/

Acknowledgments
The activity presented in the article is part of VEGA research grant No. 1/0557/18 "Research and development of technological and product innovations of modern construction methods in the context of the principles of Industry 4.0"

IOP Conf. Series: Materials Science and Engineering 1209 (2021) 012015 doi:10.1088/1757-899X/1209/1/012015

Carbon footprint of selected building structures

J Budajová[1]

[1] Technical University of Kosice, Faculty of Civil Engineering, Institute of Civil Engineering, Vysokoskolska 4, 042 00 Kosice, Slovakia

E-mail: jana.budajova@tuke.sk

Abstract. In general, we can call the carbon footprint as emissions of gases that affect the Earth's climate, while being used by humans. The impact of construction, building materials, structures, or the overall life cycle of a building on the environment is great. Sustainable architecture is gaining more prominence, using reduced carbon footprint. Today's construction industry is increasingly moving towards sustainable construction, which is constantly being formed. The great weather fluctuations that take place from day to day are forcing us to reduce our greenhouse gas emissions. The global warming potential GWP (global warming potential) caused by these greenhouse gas emissions is increased to carbon dioxide CO_2 and expressed as carbon dioxide equivalent $CO2_{eq}$. Using GWP we can determine the carbon footprint of a product. The aim of this paper is to change the three compositions of the perimeter walls using LCA analysis (life cycle assessment) and to choose the composition that has the best carbon footprint and is therefore more advantageous. The need for a sustainable built environment is urgent due to its positive impact on the environment.

1. Introduction

Construction is an industry with a high share of CO_2 emissions. Carbon dioxide is one of the basic greenhouse gases that cause global warming and climate change. In construction sector, the life cycle of buildings begins with the acquisition of raw materials and building materials through the active life of the building to its demise, demolition and disposal of its waste. The construction and operation of buildings currently requires up to 39% of total CO_2 emissions. Within the European Union, Slovakia and other Member States have committed themselves to achieving carbon neutrality by 2050. The impact of CO_2 emissions on the environment is very large, so it is necessary to reduce it as much as possible. Operation of buildings, materials, building structures during construction form a significant part of total emissions. Low carbon materials, building structures or technologies are made to produce carbon neutrality. The efficient use of renewable energy sources and the elimination of emissions and greenhouse gases caused by climate change reduce the impact on the environment

2. Carbon footprint

Increasing concentrations of gaseous greenhouse gases in the atmosphere are disturbing the environment and causing severe global warming [1]. The carbon footprint as a subset of the ecological footprint is part of the expression of the overall impact of human activities on the environment. In general, the carbon footprint is the amount of emissions of those gases that have an impact on the Earth's climate and are caused by humans [2]. The definition of a "carbon footprint" is surprisingly unclear given the growing use of the term over the last decade [3].

The carbon footprint of a product represents the total amount of CO_2 and other greenhouse gases expressed as kg of CO_2 equivalents released into the environment in relation to the life cycle of that product. This is a simplified output of the LCA cycle focused on the midpoint category of the impact of global warming. When calculating the carbon footprint in the narrower sense, not only carbon dioxide or more carbon-containing gases (e.g. methane) but also carbon-free greenhouse gases (e.g. nitrous oxide) are considered to be greenhouse gases [3]. The calculation of the carbon footprint is calculated in accordance with STN EN ISO 14040 and ISO 14044.

2.1. Sustainable construction

The construction industry has a major impact on global consumption of natural resources, energy worldwide as well as consumption of greenhouse gas emissions [4]. Kibert's definition of sustainable construction is "... the creation and operation of healthy buildings and their environments based on resource efficiency and eco-design". Sustainable construction is therefore the most comprehensive solution to the ecological, social, and economic problems of the building [5]. When designing a sustainable building, it must be considered that the requirements are inseparable and that make sense to exclude in the context of sustainability. If we want to create sustainable buildings, we need to have as few carbon emissions as possible throughout the life cycle of the building. Location and function of the building, its adaptation to changes, service life, orientation, size, shape and construction, materials used, method of heating and ventilation - all this determines the amount of energy needed to build buildings, transport building materials, subsequent maintenance, and operation building, for the transport of equipment to and from the building. We proceed to the final phase of the construction and demolition cycle [6].

2.2. Life cycle of buildings

Environmental life cycle assessments (LCA) of products or buildings have become an important tool for assessing the environmental impact of products in our case of buildings and services from the very beginning [7]. The life cycle of a building includes all stages, from its creation through development during life - operation to its demise. During the life cycle, buildings consume considerable amounts of resources and contribute to the transformation of areas. This can result in significant economic consequences and impacts on the environment and human health. Therefore, we try to reduce the impact of buildings on the environment throughout their life cycle [6]. The complete life cycle analysis of the building is dealt with the LCA method - "Life Cycle Assessment", which assesses the impact of construction from mining, material acquisition, to produce building structures through load during operation and active life of the building to waste disposal after demise, demolition of the building [8]. According to this assessment, the building can be assessed with system boundaries:

- "Cradle to Gate" or "Cradle to Gate with options"
- "Cradle to Grave"
- "Cradle to Cradle" [6].

2.3. Global warming potential (GWP)

The global warming potential is a measure of how much heat (infrared radiation) a unit of greenhouse gas captures in the atmosphere during a given period compared to what is captured in the same amount of carbon dioxide (CO_2). GWP can be estimated in a specific time frame that ranges from several decades to several centuries. Atmospheric lifetime refers to the time that a greenhouse gas remains in the atmosphere before it is decomposed by chemical processes. This means that greenhouse gases with a longer lifetime in the atmosphere will have a higher degree of warming than another gas with a shorter lifetime in the atmosphere, assuming the same GWP. Apart from water vapor, which has a residence time of 9 days, all other greenhouse gases are completely decomposed from the atmosphere for several years [9].

3. Methods

Three compositions of perimeter walls are proposed. Wall 1 is made of Ytong aerated concrete block. Wall 2 is made of reinforced concrete and wall 3 is made of wooden construction. The walls are assessed from the thermal characteristic point of view and meet the thermo-physical requirements for perimeter walls, which have been in force since 2021. The compositions are listed in tables 1, 2 and 3.

Table 1. Composition 1.

Mode	Material	Λ [W/mK]	Thickness [mm]
1	Internal plaster	0.35	5
2	Aerated concrete blocks Ytong	0.105	300
3	Isover EPS 70F	0.034	250
4	Exterior plaster	0,19	15
	Thermal resistance of the structure	$R = 10.30\ m^2K/W$	
	Heat transfer coefficient of the structure	$U = 0.096\ W/m^2K$	

Table 2. Composition 2.

Mode	Material	Λ [W/mK]	Thickness [mm]
1	Internal plaster	0.35	5
2	Reinforced concrete	1.43	400
3	Isover EPS 100S	0.037	100
4	Isover EPS 250S	0.034	250
5	External silicone plaster	0.19	15
	Thermal resistance of the structure	$R = 10.30\ m^2K/W$	
	Heat transfer coefficient of the structure	$U = 0.096\ W/m^2K$	

Table 3. Composition 3.

Mode	Material	Λ [W/mK]	Thickness [mm]
1	Gypsum fiber board	0.35	20
2	Isover Unirol	0.033	60
3	Isover Vario KM Duplex	0.34	2
4	OSB boards	0.13	20
5	Isover Unirol	0.033	160
6	OSB boards	0.13	20
7	Isover Super-Vent	0.031	120
8	CLT panel	0.12	30
	Thermal resistance of the structure	$R = 10.30\ m^2K/W$	
	Heat transfer coefficient of the structure	$U = 0.096\ W/m^2K$	

OneClickLCA software, which works in accordance with STN EN ISO 14040 and ISO 14044, was used to evaluate the LCA analysis.

4. Results

As part of the life cycle assessment, the following phases were assessed: product production (A1 – A3), transport from production to the construction site (A4), use phases (B): replacement (B4), renovation (B5), (C1-C4) end of life phase. The results of the assessment of the carbon footprint through the global warming potential GWP for the evaluated wall compositions are shown in table 4.

Table 4. Global warming potential GWP [kg CO_{2eq}].

Composition 1		Composition 2		Composition 3	
Phase	Global warming potential kg CO_{2eq}	Phase	Global warming potential kg CO_{2eq}	Phase	Global warming potential kg CO_{2eq}
A1-A3	50.26	**A1-A3**	190.44	**A1-A3**	32.1
A4	0.41	**A4**	3.24	**A4**	0.5
B4-B5	2.69	**B4-B5**	85.96	**B4-B5**	13.34
C1-C4	16.11	**C1-C4**	31.05	**C1-C4**	2.55
Total	69.83	**Total**	310.29	**Total**	48.49

The carbon footprint of individual compositions was calculated per $1m^2$ of floor area over a period of 50 years. Phase A1-A3 of the extraction and production of the product is highest for composition 2, where up to 47% of the share is made of reinforced concrete wall. For phase A4 transport and start of construction, the highest value of the global warming potential is for composition 2, also in phase B4-B5 of replacement and renovation and phase C1-C4 at the end of life. The values are also processed graphically in figure 1.

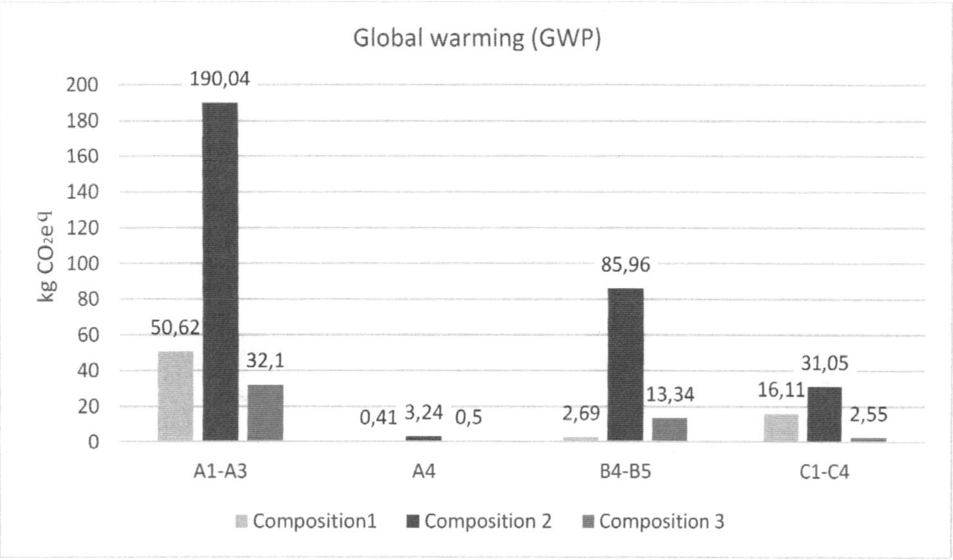

Figure 1. Global warming potential GWP [kg CO_{2eq}].

In composition 1, which is based on Ytong aerated concrete brick, the phase of extraction and production of material contributes the most to the large carbon footprint and the negative impact on the environment of the material extraction and production phase of 72.5%. Modul of transport 0.6%.

In composition 2, which is based on a reinforced concrete wall, the phase of extraction and production of material contributes the most to the large carbon footprint and the negative impact on the environment, as in composition 1. These figures are most influenced by the reinforced concrete wall. At least also the phase of transport 1.0%.

For composition 3, in which the main material is wood, the phase of extraction and production of material contributes the most to the large carbon footprint and the negative impact on the environment, as in the two previous compositions. These numbers are most affected by gypsum fiber board up to 45%. At least also the phase of transport 1.0%.

As can be seen in graph 1, the circumferential wall with composition 2 has the most negative impact on the environment and a large carbon footprint. In all phases, it has the highest values of the global warming potential GWP of the studied compositions. The most advantageous perimeter wall with respect to the GWP global warming indicator is composition 3. Compositions and buildings in general, built of natural materials, are characterized by low CO_{2e} emissions. On the contrary, concrete structures and gypsum fiber boards are the most involved in the creation of GWP.

5. Conclusion

Construction, operation of buildings, the life cycle of the building and the entire construction industry currently account for almost 40% of total CO_2 emissions. From 2021, newly built buildings must belong to category A0 - buildings with almost zero energy consumption, which achieves high energy efficiency. Renewable resources, recyclable materials that significantly reduce the building's carbon footprint are to be used. Different materials have several advantages during their life cycle. Wooden structures serve as a carbon sink during use, but this changes when they are disposed of. They are usually used as energy waste and carbon is released back into the air. Concrete and steel structures produce large carbon emissions during their production. Of the assessed compositions, the most advantageous composition is 3 - a wooden construction of the perimeter wall. The composition has the lowest global warming potential GWP kg CO_{2e}, the lowest carbon footprint and thus a negative impact on the environment. The choice of material is one of the key success factors in achieving carbon neutral buildings and emphasizes the importance of the design and construction phases in reducing carbon. The need for sustainable construction is very important in order to minimize the burden on the environment and destroy the planet.

6. References

[1] Pandey D, Agrawal M and Pandey J S 2011 Carbon footprint: current methods of estimation *Environmental monitoring and assessment* **178** 135-160

[2] Števo S 2008 Uhlíková stopa bývania v globalizovaných domoch *TZB Haustechnik* roč. **25** 16-20

[3] Matthews H S, Hendrickson C T and Weber C L 2008 *The importance of carbon footprint estimation boundaries* 5839-5842

[4] Moňoková A, Vilčeková S 2019 Environmental impact analysis of five family houses in Eastern Slovakia through a life cycle assessment *Selected Scientific Papers - Journal of Civil Engineering* **14** 81-92

[5] Mao X, Lu H and Li Q 2009 A comparison study of mainstream sustainable/green building rating tools in the world *In 2009 International Conference on Management and Service Science* pp 1-5 IEEE

[6] Krajcsovics L, Pifko H and Pifková T Ukážka energetickej efektívnosti a využitie obnoviteľných zdrojov energie na príklade verejných budov *Zručnosti a príklady. Národný školiaci materiál.* (CEC5, 3sCE412P3)

[7] Guinée J B and Heijungs R 2000 *Life cycle assessment* (Kirk-Othmer Encyclopedia of Chemical Technology) doi:10.1002/0471238961.lifeguin.a01

[8] Pifko S, Janches D, Close S, Sparks J, Nakamura T and Nesvorny D 2013 The Meteoroid Input

Function and predictions of mid-latitude meteor observations by the MU radar *Icarus* **223** 444-59
[9] Amoo L M and Fagbenle R L 2020 Climate change in developing nations of the world *In Applications of Heat, Mass and Fluid Boundary Layers* (Woodhead Publishing) pp 437-71
[10] World Green Building Council 2019 *New report: the building and construction sector can reach net zero carbon emissions by 2050,* Available: https://www.worldgbc.org/news- media/WorldGBC-embodied-carbon-report-published

Acknowledgments
This study was financially supported by Grant Agency of Slovak Republic to support project No. 1/0512/20.

IOP Conf. Series: Materials Science and Engineering 1209 (2021) 012016 doi:10.1088/1757-899X/1209/1/012016

Removal of organic micropollutants in the biological units of a Swedish wastewater treatment plant

C Burzio[1], E Nivert[2], A Mattsson[2], O Svahn[3], F Persson[1], O Modin[1] and B M Wilén[1]

[1] Department of Architecture and Civil Engineering, Chalmers University of Technology, SE-412 96 Gothenburg, Sweden
[2] Gryaab AB, Box 8984, Göteborg, Sweden
[3] Department of Environmental Science and Bioscience, Faculty of Natural Science, Kristianstad University, SE-291 88 Kristianstad, Sweden

E-mail: burzio@chalmers.se

Abstract. The present study investigates the presence and removal of target organic micropollutants in a large Swedish wastewater treatment plant designed for nutrient removal including activated sludge, trickling filters, nitrifying moving bed biofilm reactors (MBBRs) and post-denitrifying MBBRs. A total of 28 organic micropollutants were analysed, at concentrations ranging from few ng/L to µg/L, in the influent and effluent of the different biological reactors in two sampling campaigns. The observed micropollutant removal efficiencies of the wastewater treatment plant varied from insignificant (< 20%) to high (> 90%) between compounds. The activated sludge reactor, being the first in line, contributed to most of the removal from the water phase. Additional removal of a few compounds was observed in the biofilm units, but most of the persistent compounds remained stable through all biological treatments.

1. Introduction

The presence of organic hazardous substances in the aquatic environment, such as pharmaceutically active compounds, personal care products, pesticides, synthetic and natural hormones, and industrial chemicals, has become a worldwide issue of increasing environmental concern. Present at concentration of nano- to milligram per liter, they are defined as organic micropollutants (OMPs).

Wastewater treatment plants (WWTPs) have been recognized as the main route of emission of OMPs into the environment and as hotspots for antibiotic resistance. Not being designed for the elimination of micropollutants, the removal is often incomplete, resulting in continuous discharge. The regulation of the release of OMPs from WWTPs into the environment is a topic of discussion and the plants might face in the coming future new treatment objectives concerning the emissions of those substances.

Release of OMPs from WWTPs has been extensively reported and has been found to be significantly influenced by the biological treatment processes. Further optimisation of biological treatment strategies towards OMP removal is possible by applying different redox conditions and biological post-treatment units [1]. Attached biomass has been reported to remove certain OMPs to a higher extent compared to suspended biomass, suggesting that biofilm systems could potentially provide extended biological micropollutant removal [2]. In a real WWTP consisting of a combination of suspended- and attached growth reactors with different redox conditions, it is unclear which process contributes the most to OMP removal. The objective of this study was, thus, to investigate the removal of target OMPs along the biological units of a large Swedish WWTP, which consist of a set of oxic and anoxic processes, with suspended- and attached growth under different substrate concentrations.

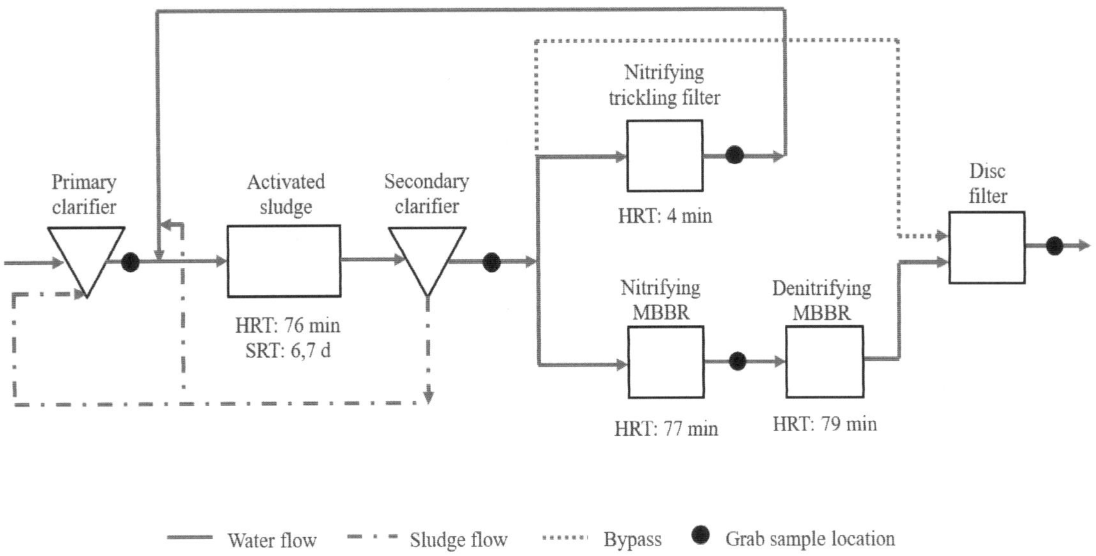

Figure 1. Flow scheme of the investigated WWTP and retention times of the biological units.

2. Materials and methods

The presence and removal of target OMPs were investigated at the Rya WWTP, which treats wastewater from a population of 780 000 people or 970 000 pe (person equivalent) in the Gothenburg region (Sweden). The yearly average incoming wastewater flow is 387 470 m^3 d^{-1} with annual average concentration of 349 mg L^{-1} COD_{Cr}, 160 mg L^{-1} BOD_7, 27 mg L^{-1} N_{tot}, 3.4 mg L^{-1} and P_{tot}, 349 mg L^{-1} COD. Wastewater treatment includes primary settling, chemical removal of phosphorus with ferrous sulphate, and biological nitrogen removal. The WWTP consists of several biological treatment processes, including activated sludge reactor, nitrifying trickling filters, nitrifying moving bed bioreactors (MBBRs) and post-denitrifying MBBRs (figure 1). After primary sedimentation, the wastewater flows to the activated sludge process. The activated sludge tanks are separated into anoxic and aerated zones, where the organic matter and the nitrate recirculated from the nitrifying trickling filter are removed. The sludge is recirculated from the secondary sedimentation back to the activated sludge process to maintain a specific level of biomass concentration. The wastewater effluent from the secondary clarifiers flows either to the nitrifying trickling filter or to post nitrifying MBBRs. A small fraction is led directly to the last treatment step of the plant. The effluent of the nitrifying trickling filter is recirculated back to the activated sludge basins, while the wastewater from the nitrifying MBBRs flows to denitrifying MBBRs. The final step in the plant consists of suspended solid removal by means of disc filters. During the investigation, the activated sludge was operated at an average solid retention time of 6.7 days and a sludge recirculation equal to 0.99 times the inflow. The nitrifying trickling filter flow was 2.17 times the inflow. The average water temperature was 17.4 °C.

A total of 25 pharmaceutical substances, two insecticides and a plasticiser, bisphenol A (BPA), were measured in the wastewater influent and effluent of the different biological treatment processes. The sampling was performed over two separated summer campaigns and duplicate samples were analysed in whole water (unfiltered samples according to instructions in the EU Watchlist) using innovative solid phase extraction - liquid chromatography - tandem mass spectrometry (SPE-UPLC MS/MS) technique and methods [3].

3. Results

All measured compounds were detected in the primary clarified wastewater at average concentrations ranging from 1 ng/L to 43 μg/L (figure 2). These values are in accordance with the concentrations detected and compiled by previous studies investigating pharmaceutical levels in Swedish wastewater [4, 5]. Only three of the investigated substances showed an average influent concentration > 1 μg/L, namely ibuprofen, paracetamol and naproxen, which are frequently used anti-inflammatory and analgesic substances.

The removal efficiencies varied greatly among the compounds (figure 3). The substances with the highest incoming concentrations were readily removed from the water phase, with an average removal between 80% and 100%. High removal (> 70%) was observed for clarithromycin, sulfamethoxazole, bisphenol a, methotrexate, ciprofloxacin and ketoconazole. Venlafaxine, citalopram, sertraline and estrone showed significant removal (40-70%). The high removal of ciprofloxacin, ketoconazole, venlafaxine and citalopram can be explained by adsorption to sludge [6]. Partial or no removal was obtained for the most persistent OMPs.

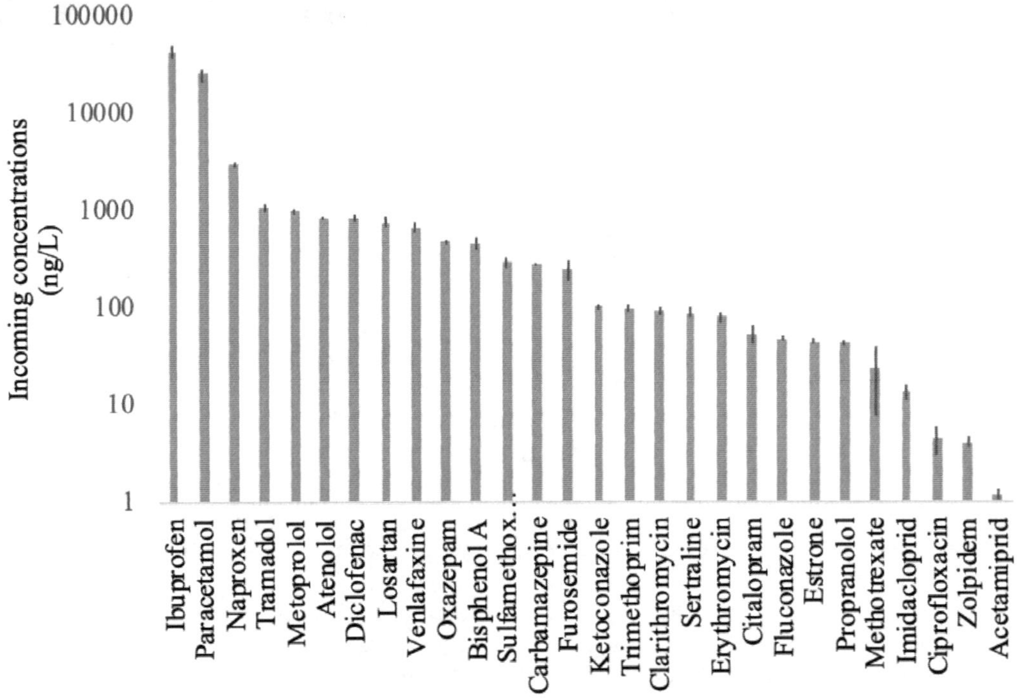

Figure 2. Average influent concentrations of the selected micropollutants to the WWTP. The bars indicate standard deviation.

Negative removal is reported for compounds having higher concentrations in the effluent compared to the influent. Negative removal might be explained by: (i) desorption processes; (ii) transformation of pharmaceutical conjugates; (iii) inaccuracies of the sampling method; (iv) heavy matrix effects, due to the high content of organic matter in the influent wastewater, supressing the MS/MS signal.

4. Conclusion

The activated sludge reactor, being the first reactor in line, contributed to most of the removal from the water phase for the easily degradable compounds. Additional removal of the degradable compounds was observed in the trickling filter and in the MBBR units. The most persistent compounds remained stable through all biological treatment steps.

The limited removal efficiency provided by the carrier biofilm is partly due to already low influent concentrations, due to efficient removal in the activated sludge system but may also be attributed to the short hydraulic times applied in the MBBRs (~ 1.3 h, figure 1). This suggests that the optimization of biological treatment strategies towards OMP removal would be restricted by the retention times applied in the units.

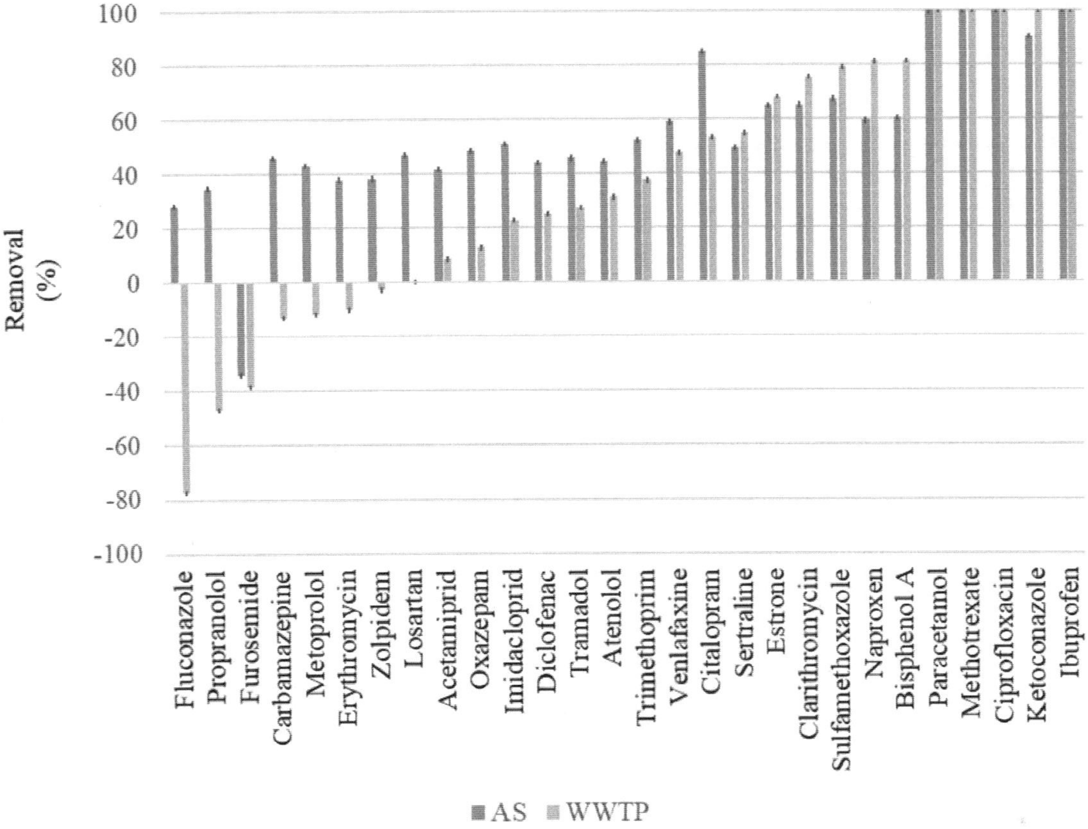

Figure 3. Comparison of average removal efficiencies of the activated sludge unit and of the entire WWTP. The bars indicate standard deviation.

5. References

[1] Falås P, Wick A, Castronovo S, Habermacher J, Ternes TA and Joss A 2016 Tracing the limits of organic micropollutant removal in biological wastewater treatment *Water Research* **95** 240–9

[2] Falås P, Longrée P, La Cour Jansen J, Siegrist H, Hollender J and Joss A 2013 Micropollutant removal by attached and suspended growth in a hybrid biofilm-activated sludge process *Water Research* **47** 4498–506

[3] Svahn O and Björklund E 2019 High flow-rate sample loading in large volume whole water organic trace analysis using positive pressure and finely ground sand as a spe-column in-line filter *Molecules* **24**

[4] Falås P, Andersen H R, Ledin A and la Cour Jansen J 2012 Occurrence and reduction of pharmaceuticals in the water phase at Swedish wastewater treatment plants *Water Sci. Technol.* **66** 783-91

[5] Paxéus N 2004 Removal of selected non-steroidal anti-inflammatory drugs (NSAIDs), gemfibrozil, carbamazepine, β-blockers, trimethoprim and triclosan in conventional wastewater treatment plants in five EU countries and their discharge to the aquatic environment *Water Sci. Technol.* **50** 253–60

[6] Svahn O and Björklund E 2019 Extraction Efficiency of a Commercial Espresso Machine Compared to a Stainless-Steel Column Pressurized Hot Water Extraction (PHWE) System for the Determination of 23 Pharmaceuticals, Antibiotics and Hormones in Sewage Sludge *Appl. Sci.* **9** 1509

IOP Conf. Series: Materials Science and Engineering 1209 (2021) 012017 doi:10.1088/1757-899X/1209/1/012017

Alternatives how to mitigate urban flooding and CSO in urbanized areas used in European cities and towns

M Csóka[1], G Rózsa[1], I Marko[1] and Š Stanko[1]

[1] Department of Sanitary and Environmental Engineering, Faculty of Civil Engineering, Slovak University of Technology in Bratislava, Radlinského 11, 810 05 Bratislava, Slovakia

E-mail: marek.csoka@stuba.sk

Abstract. Urban flooding and combined sewer overflow in city, or town areas represents potential risk in environmental, economic, or social aspects. The goal of this study is to process and evaluate efficiency of individual solutions to reduce occurrence of flooding in urban areas caused by intense rainfall events. The known conventional solutions are stormwater chambers, or storage drains. The new trend in reduction of stormwater drainage into combined sewer system are parts of blue-green infrastructure. Blue-green infrastructure represents environmental urban infrastructure which consists of sensitively selected urban vegetation combined with ingenious hydrological elements of urban city drainage. The study also deals with experimental usage of decentralized real time control, based on a gossip-based algorithm of moveable gates in sewage network. Experiment was proposed for drainage system of the city of Cosenza in Italy. Evaluation will assess application value of individual proposed solutions for the reduction of combined sewer overflow for Slovak republic and its urban cities, or towns.

1. Introduction

Majority of European cities have mostly combined sewer system. Combined sewer systems are designed to collect and transport combination of sanitary wastewater and stormwater throughout single pipe into the wastewater treatment plant. This type of sewer systems was preferred in the past because its simplicity and low costs [1]. However, wastewater treatment plant may not be able treat the increased amount of sewage during rainfalls or snowmelts periods and during these events untreated wastewater is discharged throughout combined sewer overflow (CSO) into the water bodies. If the excess sewage is not released with combined sewer overflow, it would back up into the streets and the basements. CSO are potential threats to environment and to the human health [2].

Extreme rainfall can cause urban flooding, when the urban drainage systems are overloaded with great amount of stormwater runoff. This cause overflowing of stormwater from the manholes into the street surface [3]. The increased frequency of extreme storms can be expected due to the climate change. In combination with growing urbanization where impermeable surfaces as roads, streets, squares, or roofs increase runoff volume, peak, and decrease the time to peak [4]. As an outcome of this is also increased occurrence of urban floods and CSO, thus raise of threat for the environment, human health, and economic assets. Another pollutant is surface runoff and first flush, that is transporting pollution from the streets, or roofs.

Several options were developed to mitigate impact of stormwater in the urban area. Traditional methods as stormwater chambers, or storage drains are accumulating rainwater. But their design is problematic due their size, and they are not contributing to the urban area with other function. New trend in the urban planning are solutions that are not only accumulating stormwater, but they can reduce, or slow down surface runoff, treat stormwater, or reuse it. Concepts like sustainable drainage systems, low impact development, blue-green infrastructure, or rainwater harvesting can lead to urban flood protection against stormwater [5].

2. Blue-green infrastructure

Traditional "grey" urban infrastructure focus on discharging rainwater into pipes and away from cities and towns. This solution is not optional for hydroclimatic issues bounded to urbanisation and increasement of permeable surfaces. It also missing the potential benefits of water as socioeconomic element in urban environment. Blue-green infrastructure (BGI) offers solution for urban areas that are facing climate change challenges. BGI combine hydrological functions (blue) with vegetation systems (green) in urban landscape design. It provides variety of functions such as flood protection, flood reduction, improvement of climate conditions, or social and economic improvements [6].

In table 1 is the list of major functions of BGI.

Table 1. Major functions of Blue-green infrastructure [5].

Function	Principle
Flood and runoff control	Retention and detention, runoff reduction, peak flow mitigation
Stormwater reuse	Reuse for non-drinking water supply, infiltration, groundwater recharge
Social	Creating social places, culture, sport, aesthetic
Landscape design	BGI in relation to dimensional design, and urban design
Water quality	First flush treatment, sedimentation, vegetation, and soil treatment
Biodiversity	Habitat for wildlife, vegetated areas
Collaboration	Collaboration between engineers, landscape planners and stakeholders
Innovation and case studies	Research and innovations, case projects monitoring

The Blue-green infrastructure is designed to copy natural hydrology and ecosystem and it is implemented in combination or individually to manage rainwater on the site it is located. There are several variations of BGI systems, but for simplicity they can be categorized into groups:

- Rain gardens
- Bioswales
- Constructed wetlands
- Retention and detention basins
- Green roofs

The systems such as permeable pavement, rainwater harvesting cisterns, or underground storage tanks are not considered as a BGI, because they do not provide ecosystem function [7].

BGI provide plenty of advantages, but its biggest disadvantages are initial cost, maintenance costs, lack of acknowledgement of government, and usually no design manual, or standards, because every BGI needs to be designed for concrete site with different conditions. Another barrier is already existing grey built-up area that needs to be drastically changed. This solution is proposed in case study project in Olomouc (Czech Republic), where author is proposing implementation of blue-green infrastructure every time the city, or town needs reconstruction of public space. Solution for green roofs structural conditions is to construct only partial green roof on structural stable places of that roof, thus not overloading roof with unbearable weight. Project states that minor element of BGI such as shallow bioswale can improve its area, so even small steps are beneficial [8].

In Europe there are variety of case studies to take lessons from. Denmark – Copenhagen: Lindavang Park (redevelopment of a city park), Sc. Anne Square (redevelopment of a historical place), and

Taasigne Square (enhancement of a residential area). This project mostly focus on flood control, than stormwater utilization [5]. Germany - Postdamer Platz is located in centre of Berlin. The important metro, and train station was redesigned in the early 90s. The main goal of the project was to implement BGI to reduce stormwater runoff to zero [9]. Germany – Hannover-Kroseberg is a residential area with 3000 houses. Project was focused to combine urban life with sustain infrastructure, where BGI is balancing stormwater drainage [6]. Italy – Milan – Bosco Verticale is complex of two towers completed in 2015. Project was designed to be example for sustainable housing with usage of dense vegetation cover. However, water management needs to be improved, because it lacks integrated rainwater management and backflow of irrigation is contributing to the public sewer system, without any other usage [9]. And previously mentioned case study in Olomouc, but this project is not jet constructed.

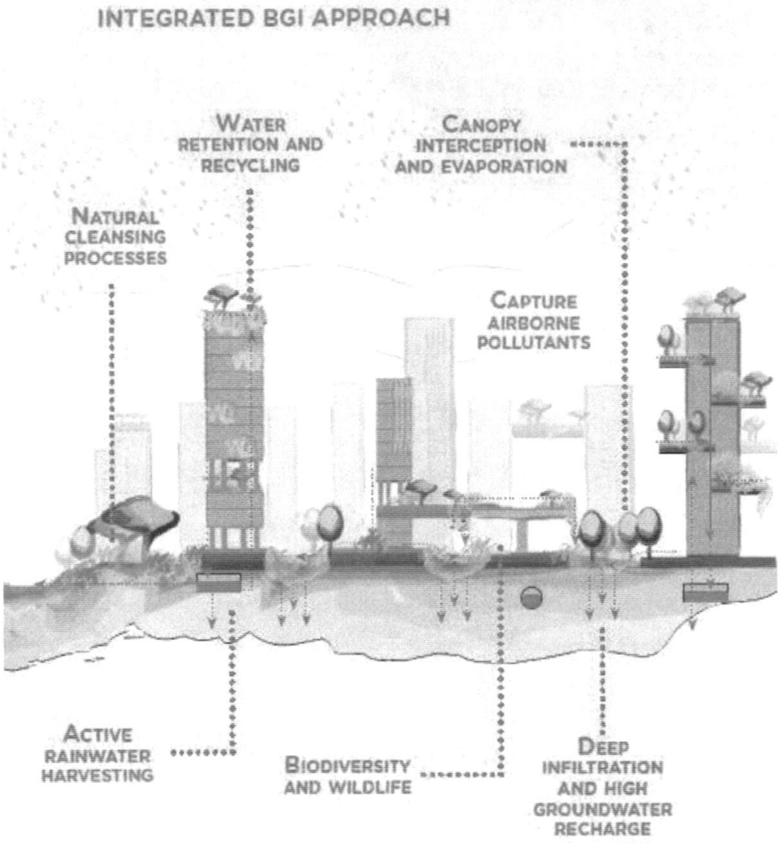

Figure 1. Function of Blue-green infrastructure systems [6].

2.1. Rainwater harvesting

Rainwater harvesting (RWH) in urban areas consist of the accumulation, storage and treatment of stormwater collected from rooftops, courtyards, and other impermeable surfaces onsite. Civil usage is mostly private. Collected water is used for non-drinkable options: toilet flushing, irrigation, or cleaning. The storage tank provides opportunities for rainwater treatment. Sedimentation is mainly used within the storage tank, but it can be enhanced with filters, or post storage disinfection. First-flush diversion can significantly improve quality of water. Storage tanks can be divided into two containers – one for detention and one for retention. The retention can be used for private demands, and detention can help with runoff control [10].

3. Moveable gates with distributed real-time control

The case study of Cosenza city in Italy is using model of combined sewer system with distributed real-time control, with on gossip-based algorithm. The urban drainage system needs to be equipped with sensors – one water level sensor per conduit and flow sensor on the outfall, computational nodes – which can execute the distributed control algorithm, and moveable gates (figure 2) – electronically regulated. The gossip-based algorithm works on communication between nodes. Each node has information only about its closest neighbour node. Nodes provide information about water level measured with the water level sensor and can control moveable gates to increase or decrease water level to provide suitable water level. The simple communication method used in the case study is because nodes compute only with small amount of data, so calculations are fast, and the system is not bound to the "main node" (that can malfunction). In case of node, or gate malfunction the neighbour nodes will adapt to the problem and control od sewer network will continue. The CSO, and urban flood reduction is solved with communication of nodes and moving gates. If one section of sewer network is overloaded with rainwater, then sections that are connected above will reduce flow and sections bellow will increase flow (figure 3) [3].

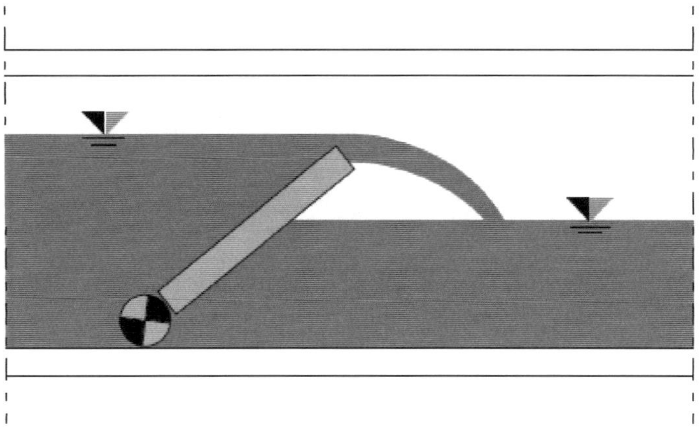

Figure 2. The down-hinged moveable gate [3].

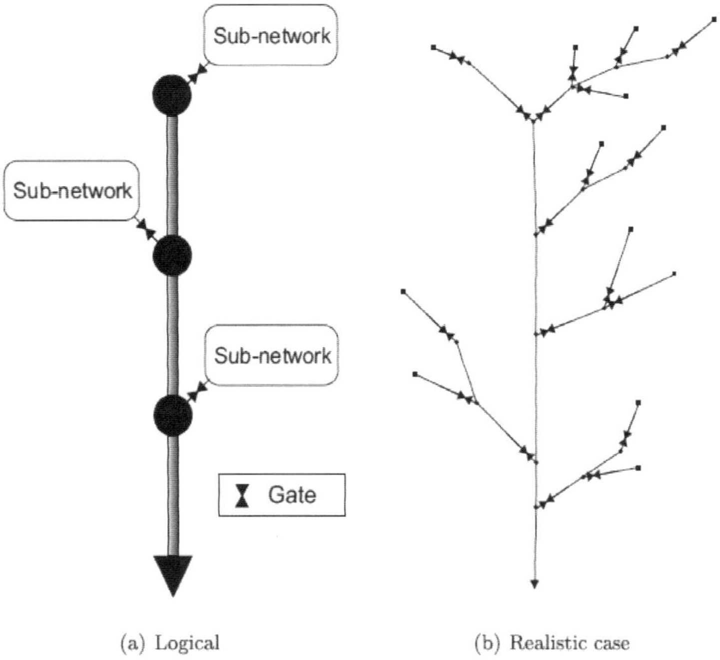

Figure 3. Gates position [3].

4. Potential for Slovak republic

Most of the cities, and bigger towns in Slovakia have built in combined sewer system, which are threatened by increasing occurrence of extreme rainfall events. Nowadays usually new towns, and new parts of growing cities are built with separated sewer system [1]. That alone is reducing CSO and mitigating pollution of environment with untreated raw sewage water, but the untreated surface runoff can be still discharged into the water body.

For Slovak cities can be beneficial to implement elements of BGI during reconstruction, even on small scale, especially in frequently flooded areas. Detention and infiltration can mitigate damage caused by urban flooding, and retention of rainwater can lower temperature the area, or provide social place for people of the neighbourhood. Case studies such as from Ramboll company, or in project for Olomouc can help engineers and architects with first designs and learn from previous mistakes. Pilot projects in Slovakia offers chance to experiment in small scale area, that can provide additional data for future projects.

Rainwater harvesting can assist with water scarcity during summer days, and it is more environmental, and economic solution than to flush toilet, or irrigate crops with drinkable water. Storage tanks can be constructed in towns for private usage, or in new parts of cities, where it can provide rainwater for entire complex, as was in case of Postdamer Platz in Berlin.

Case study in Cosenza needs additional research and case project that can provide additional results. With interested investor maybe one day it can be built within town sewer system.

5. Conclusion

Climate change is causing increased occurrence of intense extreme rainfall, with combination of ubiquitous impermeable surfaces is causing problems in cities and towns. Solutions for stormwater management in Slovakia can be proposed Blue-green infrastructure, that provides sustainable rainwater management, flood protection, and improvement of social, economic, and environmental aspects of urban area. Elements of BGI are based on combination of hydrology (blue) and vegetation (green) systems. Lessons learned from other European countries can provide foundation for pilot project in Slovakia. For Slovak region, as an example with design standards can be case study in Olomouc.

6. References

[1] Urcikán P and Rusnák D 2011 *Wastewater Collection and Treatment, Sewer Networks I. Design of Sewer Systems* (Bratislava: Slovak University of Technology in Bratislava)

[2] Chen J, Liu Y, Gitau M W, Engel B A, Flanagan D C and Harbor J M 2019 *Evaluation of the Effectivnes of Green Infrastructure on Hydrology and Water Quality in a Combined Sewer Overflow Community* (Science of the Total Environment 665) pp 69-79

[3] Garofalo G, Giordano A, Piro P, Spezzano G and Vinici A 2017 *A Distributed Real-Time Approach for Mitigating CSO and Flooding in Urban Drainage Systems* (Journal of Network and Computer Applications 78) pp 30-42

[4] Butler D, Digman Ch, Makropoulos Ch and Davies J W 2018 *Urban Drainage* (London: CRC Press)

[5] Liu L, Fryd O and Zhang S 2019 *Blue-green Infrastructure for Sustainable Urban Stormwater Management – Lessons from Six Municipality – Led Pilot Projects in Beijing and Copenhagen* (Journal Water 11)

[6] Dreiseitl H, Wanschura B 2016 *Strengthening Blue-green Infrastructure in our Cities* (Ramboll)

[7] Liao K-H, Deng S, Tan P Y 2017 *Blue-green Infrastructure: New Frontier for Sustainable Urban Stormwater Management* (Singapore: Springer Nature) pp 203-226

[8] Vítek J, and col 2018 *Rainwater Harvesting – Path to the Blue-Green Infrastructure* (Olomouc)

[9] Well F and Ludwig F 2020 *Blue-green Architecture: A Case Study Analysis Considering Synergetic Effects of Water and Vegetation* (Frontiers of Architectural Research 9) pp 191-202

[10] Campisano A, Butler D, Ward S, and col. 2017 *Urban Rainwater Harvesting Systems: Research, Implementation and Future Perspectives* (Water Research 115) pp 195-209

Acknowledgments
This work was supported by the Scientific Grant Agency of the Ministry of Education, Youth and Sports of the Slovak Republic and the Slovak Academy of Sciences within the project VEGA 1/0574/19 and 1/0574/19, co-funded by the Slovak Research and Development Agency under contract No. APVV-18-0203.

Analysis of CO_2 emissions in municipality Most pri Bratislave

M Garaj[1] **and M Zeleňaková**[1]

[1] Department of Environmental Engineering, Faculty of Civil Engineering, Technical University of Košice
[2] Department of Environmental Engineering, Faculty of Civil Engineering, Technical University of Košice

E-mail: miro@dimatz.com; martina.zelenakova@tuke.sk

Abstract. The most discussed topics in the 21st century at the global level are climate change, carbon neutrality, digitization and globalization. They have impact on the lives of all people, the quality of the environment in which we live, the health of the whole population, the development of cities and urban areas, the development and direction of industry and its new technologies, and last but not least, they shape humanity to which they are giving direction. To meet our 2030 goals, we need to know where we are now from carbon footprint point of view and where CO_2 production comes from around us.

1. Introduction

In the last two decades, we have experienced the 18 warmest years since the beginning of measurements, and extremes of weather such as forest fires, heat and floods are increasingly occurring in Europe and around the world [1]. Europe's forests are ageing and more and more trees are unable to absorb carbon dioxide emissions [2]. All these indicators represent a global challenge, to which we must also respond globally, but in response we can also work individually. In December 2019, EU leaders endorsed the objective of achieving a climate-neutral EU by 2050 [3]. Slovakia must also adapt and honour its commitment to implementing the agreement and setting out a long-term strategy in a socially just and cost-effective way. This background document is a commitment for all of us, and for me personally it is key to further explicitly defining indicators by sector, and by volume of CO_2 production. In terms of the urban zone, seven sectoral producers and their subsectors predominate (Figure 1), with which we work further.

Figure 1. Sectoral breakdown of CO_2 producers taken into account.

2. Study area and data.

2.1. Study area

The benchmark village is called Most pri Bratislave, located in Western Slovakia, only 3 kilometers from Bratislava. The village with an area of 19.01 km^2, is located in the Danubian Lowlands in the north-west part of the Žitný ostrov on the river Little Danube and consists of two catastral areas: Most pri Bratislave and Studené. The village Most is currently experiencing a large construction progress and thanks to the relatively short distance and excellent accessibility to the Capital City, the small provincial village is becoming a satellite town with a current population of 3,716.

The economic structure of the inhabitants of the village and low unemployment rate result in a significant migration of inhabitants to the nearby Capital City, respectively neighbouring towns Senec, Šamorín and Dunajská Streda. Also, the strategic position of the mentioned village on the secondary road number 572, provides conditions for the migration of the population for employment purposes and so the daily frequency of cars is at the level of 6 to 7 thousand cars per day, which is why the transport sector was evaluated as one of the main producers of CO_2.

2.2. Input data

An important key aspect for the processing and correct calculation of the volume of CO_2 production for the benchmark village is to obtain data from publicly available sources, or from the municipality itself, based on which we can take into account several aspects in the recalculation itself. The most important input data to be taken into account is the combination of geographical, demographic and geo-spatial dataset of information. By combining and taking into account these three factors, and by their subsequent analysis, we can move to and reach the most accurate measurable result of production (Figure 2).

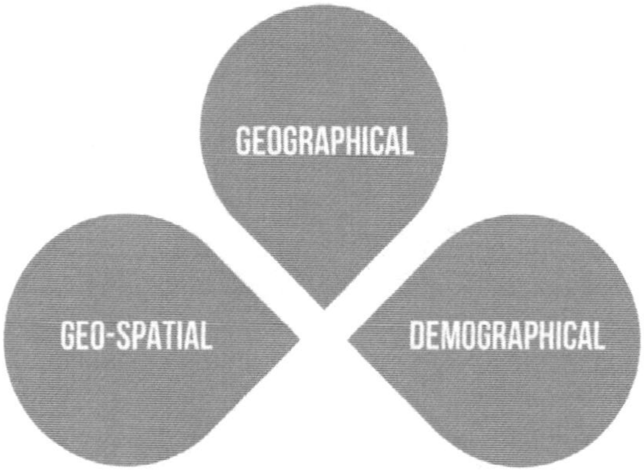

Figure 2. Impacts on selected areas of the village of Most pri Bratislave.

3. Methods and methodology

3.1. Geometric research

Geometric research of spatial urban structure must always take into account geographical and geospatial predispositions, demographic methods of data collection such as UPI (UPI – ground plan) statistics of socio-economic indicators, community development strategy, transport accessibility etc. From the geographical point of view, especially the natural landscape predisposition, morphology, climate, geological, climatic information, and other data provided by the municipality or self-government. In the absence of specific information for further analysis, we have requested these data from other sources or they have been added by us.

Geospatial-mathematical mapping is a method that is based on the quantification of objects using vectors expressed in a coordinate system (x-, y-, and z-axes). Often, the use of simplified geospatial information will help to better classify data information and its subsequent use in the accurate conversion of unit values. This process is the most important for the calculation and integration of the solution as it provides us with absolutely accurate data, which serves as a relevant basis for further processing. Digitization of geospatial information in real scale into digital can also be understood as a "digital twin" – a carrier of data and a large amount of information, which means that it is essentially a virtual model of the real factual situation. Currently, the term Digital Twin means a virtual representation of physical and non-physical objects and entities [4], such as buildings, networks, manufacturing and transportation facilities, but also processes, systems, workers, data, or the entire environment.

3.2. Level of detail (LOD)

Level of detail is the geometric relationship of the x, y, z coordinate system and its complexity of representation in the form of a 3D model polygon mesh. In general, there is a positive correlation, i.e., the lower level of detail of the displayed 3D object is, the faster the display in the software interface is (Figure 3).

LOD 1

Level of detail 1 which is characterized by a lower quality display of a 3D object formed by a small polygon network, without text and materials with basic "x, y, z" information.

LOD 2

Level of detail 2 is a representation of the model in a significantly higher quality, where the tectonics of the facade, division and building openings are visible. LOD 2 is a model formed by a medium polygonal network, the type of material used, a higher accuracy of "x, y, z" information.

LOD 3

It consists of a real depiction of the facade, building openings, technological equipment, building carpentry and all items that correspond to it in real scale so, as from the position of an observer we can perceive the detail of the facade and the architecture itself. Each object carries a texture in real UVW scale, which is standardized by the surface mapping technique using a raster matrix. Level 3 provides high accuracy that is characteristic of the category of BIM models. The model contains GPS coordinates of our territory in the coordinate system S-JTSK (Figure 4).

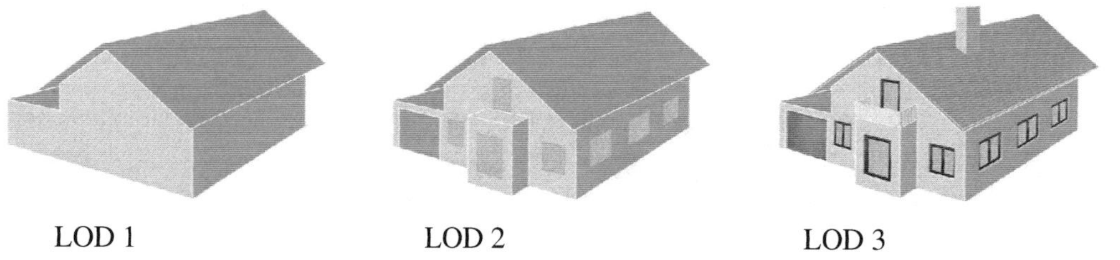

Figure 3. Level of detail

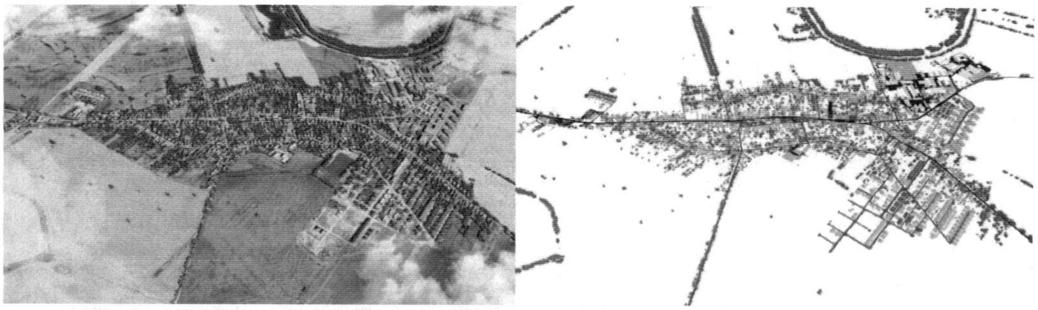

Digital model Most pri Bratislave Digital model LC (Land cover map) LOD1

Figure 4. 3D digital model of village.

Transport as a separate chapter have been analysed by entering the village via secondary road and with one output, we have been able to identify, using the differential protocol, the number of cars leaving the village and the number of cars passing through the village.

We carry out the analysis of **Green areas** on the basis of their diversification into four categories of wood species and their average value of annual wood growth for one vegetation period. Based on measuring the diameter of the trunk at a height of 1.30 meters above ground level and measuring the growth pattern of the crown part, as well as the total height of the tree, we have been able to recalculate the coefficient Y. We work with this coefficient further as with the absolute weight of the tree during the growing season. When multiplying this value by a constant of 120%, which means a multiple of its weight +20% of the root system, - (minus) the point of fiber saturation, which means the moisture content in the wood mass -72.5%, - 50% wood dry matter we get the CO_2 value contained in wood. By simply recalculating the offset of the area in square meters and the total area that the greenery occupies in terms of the map of the land cover, we have been able to identify the quantity of trees for each category and the number of pieces (Figure 5).

year / increment* × Ø² × height × 120% × 72.5% × 50% = CO_2 × 3,67 CO_2 in the atmosphere

* average annual increment per growing season

Coniferous tree / average Volunteer trees / average

Figure 5. 3D tree model.

Modern technologies form a very important part of smart cities, and should contribute to more efficient transport management, greener energy, and well-thought-out waste management, which will increase the sustainability of the local economy. When recalculating the volume of CO_2 production, we have taken into account the demographic factor – the number of the population and multiplied it by the unit value of waste production. According to the Statistical Office of the Slovak Republic, in 2019 the average Slovak produced 435 kg of municipal waste (MW), which was 35% more than ten years ago [5]. The inventory of emissions from the waste sector includes direct greenhouse gas emissions (CH_4, CO_2, N_2O) and indirect emissions (NMVOC), an equally important study in terms of the percentage of CO_2 is [6] a waste management method that also has a significant impact on the carbon footprint.

4. Results

In our work, we have tried to draw attention to the environmental issues that move the world, which to a greater or lesser extent also affect our country, as well as the region that we have focused on in this work. The aim was to calculate and record the most accurate emissions data, especially carbon dioxide in the village of Most pri Bratislave, produced by households as such, transport, other contributors to the production in question, which will subsequently help prepare and implement strategies to reduce greenhouse gases in our territory.

Applying the research methodology, where we have used, among other things, the analysis and synthesis of specialised information, we have been able to recalculate specific results of CO_2 production per person, car, aircraft or production hall of the food industry, based on the formula:

33,200,313 kgCO_2 × 1,419,763 kgCO_2 × 1,078,383 kgCO_2 – 895,451 kgCO_2= <u>34,802,265.00 kgCO_2</u>

*× Transport × Waste - Greenery = CO^2 of the municipality of Most pri Bratislave

*LC – Land cover map

Figure 6. CO2 production.

while we included in the input data various aspects and factors that contribute to the increased result of the emissions concerned in any way whatsoever. In this scientific study, the result of the annual emission production from the cadastral area of Most pri Bratislave, taking into account all aspects and factors, was 34 802 265.00 kgCO^2. In the process of processing the described aspects and factors, I tried to reach the result of this scientific study by pointing to the main producers of CO 2 , but also to highlight the need to address emissions not only from transport as one of the main producers, but also from agriculture, waste management or as already mentioned households as such.

As part of CO 2 reduction, cities and municipalities should focus primarily on reducing emissions from the production of various components in industry and, secondly, on electricity generation, as the energy industry is one of the largest producers of greenhouse gases. The general fact is that countries today have an ecological deficit, which means that their management is environmentally unsustainable from a long-term perspective. In this work, we pointed out several indicators of environmental sustainability, which provide information on how humanity, states, cities, municipalities, households and individuals live on the verge of said sustainability. Despite our findings and efforts to reduce the impact on global warming, this process is irreversible, and we will not feel the effects of its slowdown until many years later. Nevertheless, we believe that the data from our work will help in the preparation of low- carbon to carbon-free strategies for municipalities and cities, which will then act for the benefit of nature for future generations.

5. References

[1] © Európska únia, 2019 2020 Opatrenia EÚ v oblasti zmeny klímy DIO consilium.europa.eu/sk/policies/climate-change
[2] EurActiv/Reuters 2019 Schopnosť európskych lesov absorbovať CO2 sa blíži k limitu DIO euractiv.sk/section/zivotne-prostredie/news/schopnost-europskych-lesov-absorbovat-co2-sa-blizi-k-hranici-021422/
[3] Cabuzel T 2020 European Climate Law DIO ec.europa.eu/clima/policies/eu-climate-action/low_en
[4] Digitálne dvojča: Vedúca technológia inteligentného priemyslu 2019 DIO anasoft.com/emans/sk/home/Novinky-blog/blog/digitalne-dvojca-veduca-technologia-inteligentneho-priemyslu
[5] PROPERTY & ENVIRONMENT s. r. o. 2020 Slováci dosahujú rekordy, ktoré nepotešia DIO odpady-portal.sk/Dokument/105461/slovaci-dosahuju-rekordy-ktore-nepotesia-analyza.aspx
[6] 2020 Nízkouhlíková stratégia rozvoja Slovenskej republiky do roku 2030 s výhľadom do roku 2050 (n.d) DIO minzp.sk/files/oblasti/politika-zmeny-klimy/nus-sr-do-roku-2030-finalna-verzia.pdf

Acknowledgment
This work was supported by the Slovak Research and Development Agency through projects SK-PT 18-0008 "Hydrological risk: from excess to scarcity of water" and SK-PL-18-0033 "Environmental risk assessment in the aspect of adaptation to climate change."

Combined sewer overflow chamber analysis based on CFD modelling.

J Hrudka[1], M Šutúš[1], M Csóka[1], A Raczková[1] and I Škultétyová[1]

[1] Department of Sanitary and Environmental Engineering, Faculty of Civil Engineering, Slovak University of Technology in Bratislava, Radlinského 11, 810 05 Bratislava, Slovakia

E-mail: jaroslav.hrudka@stuba.sk

Abstract. The paper deals with CFD simulation of a real combined sewer overflow chamber using Ansys fluent software. Simulations are created for various structural modifications. Within the results, the hydraulic parameters of the individual are evaluated with a proposal for optimal operation of the given relief chamber.

1. Introduction

Combined sewer overflow chambers are among the most important objects on the sewer network and are therefore subject to high requirements, whether hydraulic, hydrological, hygienic or structural [1]. Their main function, i.e. hydraulic, consists in dividing the inflow at the time of torrential precipitation into a flow continued to the Waste water treatment plant (WWTP) and the flow rate, which from Combined sewer overflow (CSO) is entered into the recipient. Under the rain-free flow rates, CSO drains the entire amount of water into the wastewater treatments and no sedimentation should occur. This principle is mainly solved using a stable and well-situated overflow. If mixed wastewater exceeds the level of the overflow edge, they begin to spill over it and are then routed as shortly as possible to the recipient or rain tank [2]. Another important function of the CSO chambers is related to pollution. Some separator designs can successfully capture larger solids using the edges, but this is not possible with suspended and dissolved substances. Therefore, from the point of view of the protection of the recipient, it is necessary to propose CSO so that the ratio of dilution of sewage and rainwater at the time of overelows is in accordance with the legislative requirements defined in the Ministry of Interior of the Slovak Republic 269/2010 Z. z. The allowed dilution ratios depend on the specific recipient and range from 1:4 to 1:8. The splitting of the flow into two parts therefore allows us to make the stem design behind the CSO chamber more economical, as the design flow rate will be significantly lower. In addition, some CSO or separators also have a cleaning function that reduces pollution of watercourses. Several authors in the study looked at the research of the CSO chambers and their impact on the environment. Their results show a high impact on the environment in the inefficient functioning of chambers [3] [4] [5].

2. CSO chamber Lučenec OK 2D

As a basis for hydraulic assessment using a mathematical model, the existing CSO chamber, which is part of the sewer network of the town of Lučenec. Detailed information about the construction and its associated flow rates was taken from the documents from the water company that operates the drainage network in Lučenec. For the creation of 3D geometry, drawing documentation was provided, prepared from the geodetic focus of the internal dimensions of the CSO chamber. The town of Lučenec is currently sewered by a combined public sewerage system. The sewer network consists of four main sewer collectors A, B, C and D. There are a total of 16 CSO chambers on the sewer network, but currently only 15 are operational. The object of our interest is a CSO chamber marked "OK 2 D" mounted on the collector D and edified into the Krivánsky stream. "OK 2 D" lightening chamber design: CSO we can group in terms of design type with frontal overflow edge, with direct and rounded overflow edge. A pipe with an internal diameter of DN 1400 in longitudinal inclination of 5 ‰ is edging into the chamber. The influent pipe is connected to the throttle by a 450 mm deep trough at the bottom of the pressure chamber. The trough turns to the left, gradually narrowing to the diameter of the throttle DN700. The longitudinal slope of the throttle track is 25.5 ‰ and its length is 3.5 m. The CSO pipe that estuary to the recipient has an internal diameter of DN 800 and its inclination is 26.5 ‰. All pipes are concrete with a circular profile. The overflow edge is rounded and its height from the bottom of the trough is 860 mm. The distance of the overflow edge from the influent pipe is 2.9 m and the distance from the remiding 2.5 m. The corners of the chamber are gradient towards the center. The floor plan internal dimensions of the chamber are 4.2 x 5.4 m and its height from the bottom of the trough is 2.55 m. CSO is constructed from concrete prefabricated slabs 200 mm thick. Access to the chamber is provided by an entrance shaft. All available technical parameters are listed in the following table (table 1).

Table 1. Basic dimensional parameters of the CSO chamber.

Inlet pipe diameter	DN 1400
Throttle pipe diameter	DN 700
Throttle length	3500 mm
Diameter of relief pipe	DN 800
Overflow edge height	860 mm
Elevations of the throttle	179,35 m.a.s.l..
Elevation of top the throttle pipe	180,05 m.a.s.l..
Overflow edge altitude	180,21 m.a.s.l..

3. Hydrotechnical parameters of the CSO chamber „OK 2 D“

Part of the documents from the operator is an assessment of the sewer network of the city of Lučenec, where the actual rainless flow in the CSO chamber is also measured. The measurement was carried out for 72 hours and based on these data the average daily inflow to CSO Q_{24} = 19.31 $l.s^{-1}$. In order to determine the amount of rainwater inedding during the maximum design 15 minutes of rain, a calculation was carried out as part of the assessment. The rain yield was selected from the data of the nearest ombrographic station "Lučenec" with a value of q_{15} = 132,8 $l.s^{-1}.ha^{-1}$. The resulting value of rainwater coming into the discharge chamber is Q_z = 1016,92 $l.s^{-1}$. Currently, the required drain to the VWTP $Q_{ČOV}$ = 115.86 $l.s^{-1}$ and the required dilution ratio based on the valid decision of the District Office of the Environment in Lučenec 1:5. According to the calculation in the assessment, the throttle capacitive flow rate for the pipe profile DN 700 Q_k = 1470 $l.s^{-1}$ implies that the lightening occurs only at a dilution ratio of 1:75. Since the design flow rate based on the sum of sewage and rainwater (Q_d = 1036,21 $l.s^{-1}$) is less than the throttle capacity, there is currently no overflow.

Table 2. Hydrotechnical parameters of the CSO chamber.

Average daily rainless flow Q_{24}	19,31 $l.s^{-1}$
Maximum stormwater flow Q_z	1016,90 $l.s^{-1}$
Design flow Q_d	1036,21 $l.s^{-1}$
Minimum required outflow to the WWTP	115,86 $l.s^{-1}$
Required dilution ratio	1:5
Actual dilution ratio	1:75
Throttle line capacity Q_k	1470 $l.s^{-1}$

4. Boundary condition and CFD model

Ansys FLUENT has been selected to create simulations. For free flow, it is best to use a multi-phase model, in our case VOF. As the primary phase was chosen air, the secondary phase will be water. Primary phase – air constant density 1,225 $kg.m^{-3}$- constant viscosity 1,789 x 10^{-5} $kg.m^{-1}.s^{-1}$ Secondary phase – water constant density 998.2 $kg.m^{-3}$- constant viscosity 0.001003 $kg.m^{-1}.s^{-1}$ It is necessary to define that this is a constant flow and since we are also considering turbulence, we choose a k-ε model for calculation. Gravitational acceleration is considered in the direction of the Y axis and its value is -9.81 $m.s^{-2}$.

Boundary condition:

1) The first marginal condition we have defined all the walls, considering the roughness of concrete 1.5 mm (wall-fluid).
2) Furthermore, it was necessary to define the volume in which air and water (interior-fluid) flow.
3) At the beginning of the inflow pipe, a speed marginal condition (velocity-inlet) was specified, its value together with the cross-sectional water height in the pipe depending on the modelled flow state: • Q_{24} = 19.31 $l.s^{-1}$ = > v = 0.73 $m.s^{-1}$; h = 66 mm • Q_d = 1036.21 $l.s^{-1}$ = > v = 2.28 $m.s^{-1}$; h = 473 mm
4) At the end of the sewer pipe directing to the WWTP, a pressure-outlet condition was specified
5) In the same way, the condition for de-lightening pipes has been defined
6) The pressure-outlet condition has also been defined for the part of the cross-section of the inflow pipe in which there is air. This is due to possible backflow and subsequent sighing in case of higher water level in the pressure chamber.

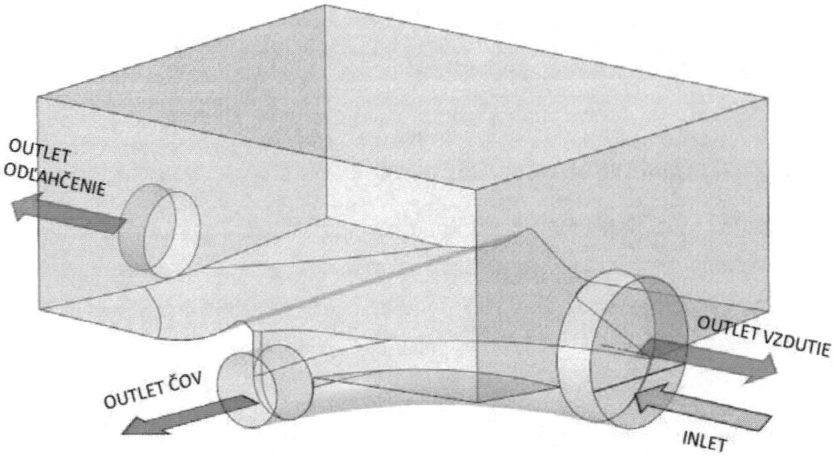

Figure 1. Inlet and outlet boundary condition.

5. Simulations analysis

The first simulation focuses on the current condition of the pressure chamber at the design rain inflow to CSO 19,31 $l.s^{-1}$, corresponding to the cross-sectional speed and height in the influent pipe $v = 0{,}73\ m.s^{-1}$ and h = 66 mm. The marginal runoff condition is specified as a pressure-outlet in accordance with the figure (Figure 1.). CSO is constructed of concrete, therefore the value of the edge condition of the wall roughness was specified as 1.5 mm. We are considering a VOF model where the first phase is modelled as air and the second phase water. Due to the assumption of turbulent flow, we will use the k-ε. The throttle profile is DN 700, which for CSO currently means that overflow does not occur.

Table 3. Boundary condition simulation no 1.

Boundary condition		
Velocity inlet	Velocity "v"	0,73 $m.s^{-1}$
	Water level "h"	66 mm
Pressure-outlet-WWTP		DN700
Pressure-outlet-separated water		DN800
Pressure-outlet- backward flow		-
Wall		0,0015 m
Interior	Faze 1	Air
	Faze 2	water

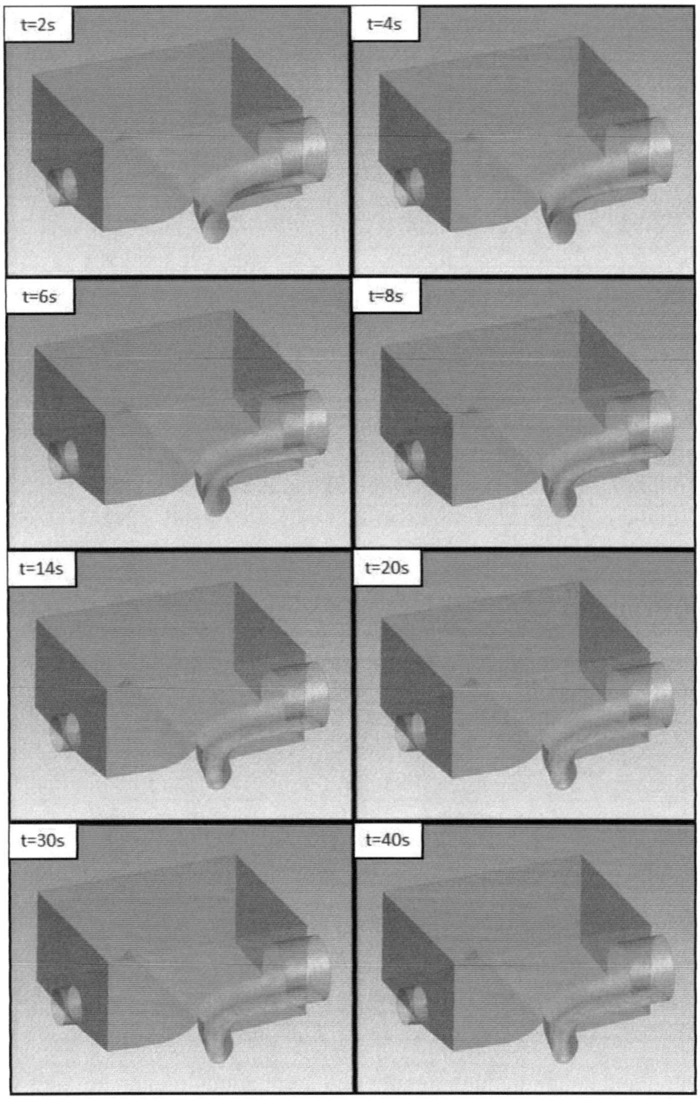

Figure 2. Simulation no 1.

When analyzing the simulation, we may notice that at the time of the rain-free inflow to the pressure chamber, a minimum amount of water flows seamlessly at the bottom of the drain trough into the throttle, pointing to the WWTP. As the drain pipe has sufficient capacity, it is not possible to fill it at this design flow rate. We can conclude that the course of water flow in CSO, created using a mathematical model, is in accordance with the current state of play. The graphical output of the simulation is shown in the figure 2.

Simulation no 2:

The second simulation focuses on the current condition of the pressure chamber at the maximum design inflow to CSO at the time of rain of 1036,21 $l.s^{-1}$, corresponding to the cross-sectional speed and height in the influent pipe $v = 2{,}28$ $m.s^{-1}$ and $h = 473$ mm. The throttle profile is DN700, which for CSO currently means that overflow does not occur. Other peripheral conditions and model settings are consistent with simulation 1.

Table 4. Boundary condition simulation no 2.

Boundary condition

Velocity inlet	Velocity "v"	2,28 m.s^{-1}
	Water level "h"	473 mm
Pressure-outlet-WWTP		DN700
Pressure-outlet-separated water		DN800
Pressure-outlet- backward flow		-
Wall		0,0015 m
Interior	Faze 1	Air
	Faze 2	water

The simulation results in a smooth filling of the CSO chamber up to approximately t = 25 s. There is no overflow, only a minimal amount of water gets behind the overflow edge at the site versus the influent, which is probably due to the greater force of the water flow in this place. As with simulation No. 1, we can conclude that the flow of water in CSO, created using a mathematical model, is in line with the status quo.

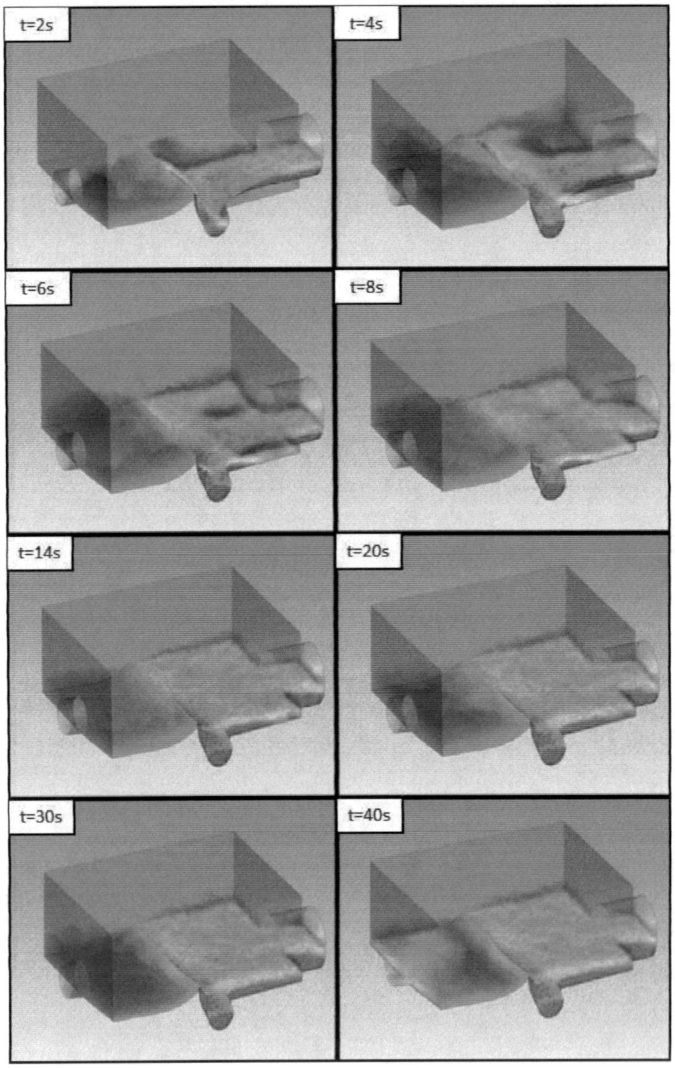

Figure 3. Simulation no 2.

6. Conclusion

When analyzing the results of a variant simulating the current state, it was confirmed that the Combined sewer overflow chamber in the current design is inefficient and there is no misalification, or the overflow amount of water is minimal. The assessment of the simulation aimed at reducing the throttle profile also confirmed the assumption that such modification could make the CSO chamber more efficient. It follows that the CSO chamber in question is inoperative and needs to be reconstructed in order to achieve efficient operation.

7. References

[1] Urikan P and Rusnák D 2004 Stokovanie a čistenie odpadových vôd: Stokovanie I. Navrhovanie stokových sietí (Bratislava: Vydavateľstvo STU v Bratislave)

[2] Harwood R and Saul A J 2001 Modelling the Performance of Combined-Sewer Overflow Chambers *Water and Environment Journal* 300–4

[3] Montserrat A, Bosch L, Kiser M A, Poch M and Corominas 2015 Using data from monitoring combined sewer overflows to assess, improve, and maintain combined sewer systems *Science of The Total Environment* **505** 1053–61

[4] Garofalo G, Giordano A, Piro P, Spezzano G and Vinci A 2017 A distributed real-time approach for mitigating CSO and flooding in urban drainage systems *Journal of Network and Computer Applications* **78** 30–42

[5] Hernes R R, Gragne A S, Abdalla E M H, Braskerud B C, Alfredsen K and Muthanna T M 2020 Assessing the effects of four SUDS scenarios on combined sewer overflows in Oslo, Norway: evaluating the low impact development module of the Mike Urban model *Hydrology Research* **51** 6 1437–54

Acknowledgments

This work was supported by the Scientific Grant Agency of the Ministry of Education, Youth and Sports of the Slovak Republic and the Slovak Academy of Sciences within the project VEGA 1/0727/20 , co-funded by the Slovak Research and Development Agency under contract No. APVV- 18-0203.

The proposals for intensification of a non-capacity wastewater treatment plant

T Kubínová[1] and M Kyncl[1]

[1] Department of Environmental Engineering, Faculty of Mining and Geology, VSB-Technical University of Ostrava, Czech Republic

E-mail: tereza.kubinova@vsb.cz

Abstract. The article deals with the possibilities of intensification of mechanical-biological wastewater treatment plant with intermittent Flexidiblok type activity with denitrification. This wastewater treatment plant is located in a municipality with more than double of the equivalent population than the capacity of the wastewater treatment plant and therefore does not meet current legislative requirements. Proposals for the intensification of the wastewater treatment plant offer the possibility of reconstructing the existing wastewater treatment plant to the required capacity or rebuilding this wastewater treatment plant to a pumping station. Part of the article is an economic-technological comparison of both proposals.

1. Introduction

The production of wastewater by anthropogenic activities is inevitable, which is one of the reasons that wastewater treatment is a necessary element in the treatment and management of water resources in the environment, to improve water reuse and environmental protection. [1, 2] As villages and urban areas grow, which includes the number of people connected to the sewer network, it is necessary to intensify some wastewater treatment plants to meet the wastewater treatment parameters set by legislation. [3] Due to the diverse technical condition, equipment and age, there are several ways to intensify a non-capacity wastewater treatment plant. [4, 5] A thorough comparison of the various designs must be made and the service life parameters of the technical equipment and the structural or investment limits must be taken into account and evaluated [6].

2. Proposals for the intensification of the wastewater treatment plant

2.1. Description of the municipal problem

The proposals for intensification followed the need to connect part of the city district to the sewerage system. In the part of the area with the existing sewerage system, there was already a wastewater treatment plant, which was built at the end of the 20th century and thus corresponded to the technological state and compliance with the requirements of the legislation of the time. [7, 8] The existing mechanical-biological wastewater treatment plant was built in order to remove only carbon pollution (BOD5). Nowadays, the current legislation requires all wastewater treatment plants in the category above 500 PE population equivalent to have a system for the elimination of nitrogen pollution (nitrification). The existing treatment plant is unable to meet these requirements due to the sizing of the biological part. Existing wastewater treatment plant had approximately 710 inhabitants connected to it and the capacity of the plant was only 600 PE [9].

IOP Conf. Series: Materials Science and Engineering 1209 (2021) 012020 doi:10.1088/1757-899X/1209/1/012020

In order to connect the inhabitants via new sewerage system to the treatment plant, it was necessary to propose the intensification of this plant. Another reason was fulfilling the requirements of current legislation valid in Czech Republic, specifically the European Council Directive 91/271 / EEC, the Water Act No. 254/2001 Coll., the Water Supply and Sewerage Act No. 274/2001 Coll. and Government Regulation No. 401/2015 Coll. [8, 9] The specific wastewater production was considered in accordance with the water supply and sewerage development plan of Moravian-Silesian Region (WSSDP MSR) for the urban district in question at 120.0 ls^{-1}. [10] The design of the intensification was based on the design parameters listed in table 1.

Table 1. Main capacity parameters of the system.

Variant	Number of connected inhabitants n [-]	Amount of wastewater Q [m^3year^{-1}]	Design size PE [-]
Actual state	710	31 098	600
Forecasted state	1 400	61 320	1 400

2.2. Description of the existing wastewater treatment solution

The treatment plant has been designed as a conventional mechanical-biological treatment plant with intermittent operation (SBR reactor) of the Flexidiblok type with denitrification, shown on Figure 1. The wastewater treatment plant complex included two SBR reactors, which alternated during operation. The reinforced concrete tank had dimensions of 13.5 m long, 4.8 m wide and 5.2 m deep and was divided by walls into a reactor, a sludge collector, and an emergency storage tank. The capacity of the wastewater treatment plant corresponded to 600 PE according to original project documentation. All operation of the treatment plant was controlled by computer programs [8].

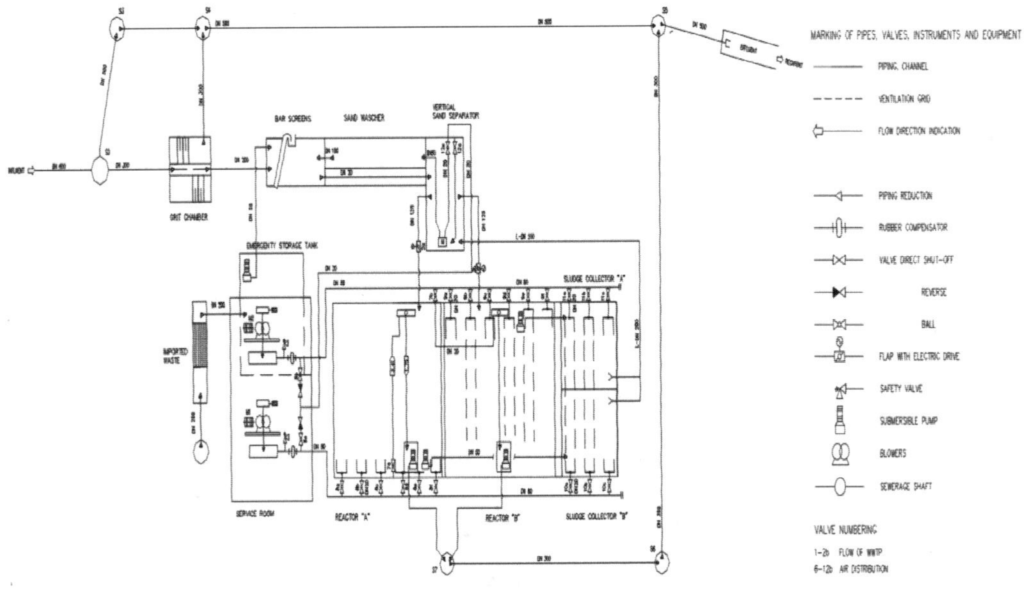

Figure 1. Existing wastewater treatment plant schema.

2.3. Reconstruction of the existing wastewater treatment plant to the required capacity

First plan proposed reconstruction of local wastewater treatment plant to the required capacity. The best option for sewerage would be to design a combined sewerage system. A significant part of the existing sewerage system would be maintained and supplemented with a newly built sewerage system, particularly in the area not yet sewered. Newly build sewer system is designed as a pressure sewer due to the morphology of the terrain. The construction of a pressure sewer is conditioned by the fact that every property that wants to connect to the public network must have its own small pumping station equipped with a positive displacement pump. In the south-western part of development two gravity sewers need to be constructed. The pressure sewer lines from the northern part of the development will be led into these sewers, through which sewage water will be fed to the reconstructed local wastewater treatment plant [7, 9].

It would also be necessary to carry out a complete reconstruction of the existing wastewater treatment plant. The whole mechanical pre-treatment will be newly built, including a grit chamber, a new blowers station and a new activation tank with pre-denitrification and secondary settling tank. Existing activation tanks could be left and used to store excess sludge [7, 9].

According to a capacity forecast of 1 400 PE and the development plan *WSSDP MSR*, the specific water consumption was determined as 120 $lcap^{-1}d^{-1}$. [10] Daily wastewater production was calculated at 168 m^3d^{-1}. The amount of ballast water was determined as 20 % of the daily production of wastewater equal to 33.6 m^3d^{-1}. Due to the large area and terrain's fragmentation, it was necessary to consider a high maximum amount of rainwater equaling to 56 764.8 m^3d^{-1} according to the operating rules [8]. The total daily flow was set at 57 315.8 m^3d^{-1}. The total amount of undissolved substances was determined to be 77 kgd^{-1}. Biological oxygen demand 84 kgd^{-1}, chemical oxygen demand 168 kgd^{-1}, total nitrogen in the inflow 15.4 kgd^{-1} and total phosphorus in the inflow 3.5 kgd^{-1} [11, 12, 13].

Due to the large area of the drainage basin and a large amount of wastewater, a relief chamber needs to be built on the main sewer, which serves for separation of the sewage flow delivered to the treatment plant during rainfall and the flow of discharged sewage without treatment directly into the recipient. [11] The overflow edge of the relief chamber could have a length of 1.5 m and the overflow height is equal to 0.05 m with with a DN 250 outlet to the sump. A dilution ratio of 1:7 was used for the calculation. [8, 11, 14] Hydrotechnical calculations of the relief chamber for a given number of population equivalent are shown in table 2.

Table 2. Hydrotechnical parameters of the relief chamber.

Theoretical flows	Chamber inflow		Chamber discharge	Effluent discharge
	$Q_{W.}$ $[ls^{-1}]$	$Q_{Comp.}$ $[ls^{-1}]$	$Q_{Crit.}$ $[ls^{-1}]$	$Q_{Eff.}$ $[ls^{-1}]$
Flows for 600 PE	2.7	571	21.6	549
Flows for 1 400 PE	6.0	663	48.0	615

The reconstructed treatment plant would consist of the objects shown in Figure 2. The most important technological parameters of selected objects are listed below:

- Storm water basin - The detention time in the tank at an inflow of 47.9 ls^{-1} and an outflow of 5 ls^{-1} will be 0.3 hours. It would take 2.8 hours to drain the entire volume of the basin.
- Primary settling tank with grease trap - The detention time at maximum inflow should be 1 hour. In order to keep this detention time, the volume of the tank has been chosen to be 18 m^3. To meet the hydraulic surface loading condition, the tank area needs to be equal to 3.6 m^2.
- Activation tank - An activation system with pre-denitrification was designed as a low load activation, with a tank volume of 900 m^3 volume and a maximum inflow of 5 ls^{-1}, the volumetric material load according to BOD5 will be 0.24 $kgm^{-3}d^{-1}$, the sludge load according to BOD5 0.07 $kgkg^{-3}d^{-1}$and a detention time of 50 hours.
- Secondary settling tank - Circular settling tank with horizontal flow was designed with volume of the tank chose to be 33 m^3. At this volume and inflow of 5 $ls^{-1}1$, the detention time will be equal to 1.8 hours and at the tank area of 11.3 m^2, the surface hydraulic load will be equal to 1.6 $m^3m^{-2}h^{-1}$. Depths of the individual zones of the settling tank are as follows:
    - sedimentation zone - 1.7 m
    - transition zone - 1.0 m
    - thickening and storage zone - 0.7 m

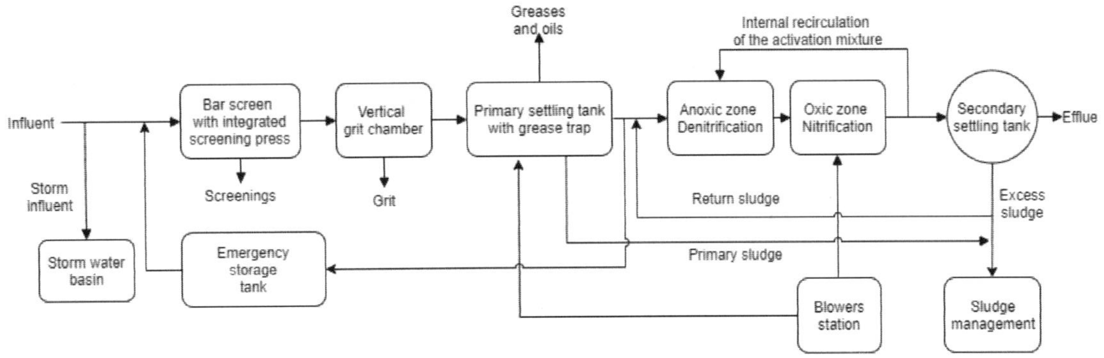

Figure 2. Reconstructed wastewater treatment plant block scheme.

2.4. Rebuilding of the wastewater treatment plant to pumping station

This intensification proposal consists in the abolition of the existing wastewater treatment plant and the rebuilding of this treatment plant to pumping station. All wastewater from the city district will be discharged to the central wastewater treatment plant, which has sufficient capacity, using the newly built sewerage system and pumping station.

The existing sewerage system in the south-eastern area connected to local wastewater treatment plant will be maintained. In the south-western part of the development, one new gravity sewer would be built to carry sewage to existing local wastewater treatment plant. Due to the morphology of the terrain, a new pressure sewerage system will be built in the northern part of the development. The construction of a pressure sewer is conditioned by the fact that every property that wants to connect to the public network must have its own small pumping station equipped with a positive displacement pump. The main branch of the pressure sewer will be led to the central wastewater treatment plant. A relief chamber needs to be constructed on the main sewer. The hydrotechnical calculations of the relief chamber for a given number of equivalent inhabitants are shown in Table 2. [7, 9]

The existing wastewater treatment plant could be used for the construction of the pumping station. The prerequisite here are the construction of the grit chamber, the installation of automatic bar screens with integrated screening press. The current emergency storage tank could be used as pumping station and the activation tanks could be used as a storm water basin with a volume of about 100 m^3. [9]

A forecast of 650 PE was assumed for designing the conversion of the existing wastewater treatment plant at the pumping station. The specific water consumption according to the development plan WSSDP MSR was determined as 120 $lcap^{-1}d^{-1}$. [10] Daily wastewater production was set at 78 m^3d^{-1}. The amount of ballast water was determined as 20 % of the daily production of wastewater equal to 15.6 m^3d^{-1}. Due to the large area and terrain's fragmentation, it was necessary to consider a high maximum amount of rainwater equaling to 56 765 m^3d^{-1} according to the operating rules [8]. The total daily flow was set at 57 085 m^3d^{-1}. The total amount of undissolved substances was determined to be 35.7 kgd^{-1}. Biological oxygen demand 39 kgd^{-1}, chemical oxygen demand 78 kgd^{-1}, total nitrogen in the inflow 7.2 kgd^{-1} and total phosphorus in the inflow 1.6 kgd^{-1}.

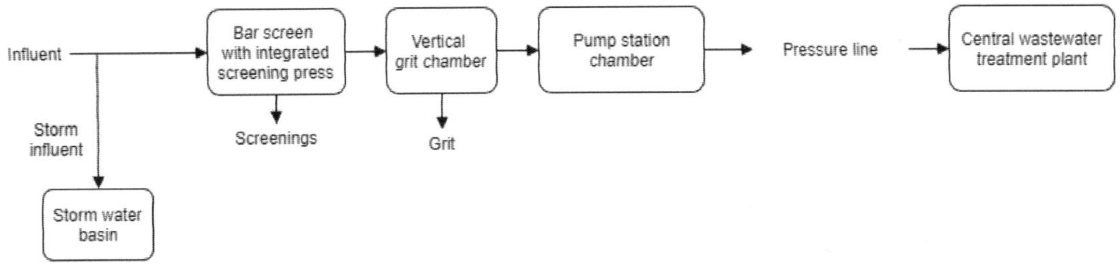

Figure 3. Block scheme for rebuilt wastewater treatment plant to pump station.

The rebuilt treatment plant would consist of the equipment shown in Figure 3. The most important technological parameters of the designed objects are the following:

- Storm water basin - In order to keep the costs associated with wastewater treatment plant intensification as low as possible, it would be possible to use the existing activation tank as a storm water basin with a volume of about 100 m3. The storm water basin will ensure the accumulation of wastewater during the rain before their pumping for mechanical pre-treatment and subsequently to the pumping sump and the central wastewater treatment plant.
- Pump station chamber - The existing emergency tank could be used as pump station chamber. This tank is built as an underground closed storage tank with a volume of 30 m^3. With the maximum rain inflow from the upstream rainwater delay and the size of 5 ls^{-1}, the detention time in the storage space of the pumping station will be equal to 1.6 hours. Minimum or switch-off level will be equal to 0.2 m from the bottom of the sump, therefore the volume of the dead space will be equal to 1.47 m^3. The switch-on level is at a height of 2 m from the bottom of the sump, resulting in working space height equal to 1.8 m from the switch-off level and a volume of 13.23 m^3, the emergency level at a height of 2.2 m from the bottom of the sump and thus 0.2 m from the switch-on level, thus defining a safety margin of 1.47 m^3. The maximum level in this case is at a height of 3.5 m from the bottom of the sump and therefore 1.3 m from the emergency level. These levels limit an emergency volume of 9.6 m^3. This volume should provide 0.53 hours for emergency purposes at a maximum inflow of 18 m^3h^{-1}.
- Pressure line - The discharge from the pumping station here is designed as a discharge line made of polyethylene DN 125 mm with a total length of 795 m.
- Central wastewater treatment plant - From the pumping station, wastewater would be pumped to the central treatment plant. This treatment plant was reconstructed to a capacity of 8.666 PE in 2012 and is ready to accommodate higher material and hydraulic loads, which are related to the gradual connection of the surrounding municipalities and localities to the sewerage system and sewage pumping stations [15].

2.5. Economic comparison of proposed variants

An important element in choosing the type of wastewater treatment plant intensification is the financial complexity of the project. An indicative calculation of investment and operating costs was performed for both proposals. The calculation of investment costs is determined according to the unit prices of the given items, which do not include the surface treatment in the route of the road. [9] The estimated costs for the reconstructed wastewater treatment plant and for rebuilt wastewater treatment plant to pump station are given in Table 3 and Table 4.

Table 3. Estimated costs for reconstructed wastewater treatment plant in thousands of CZK.

	Construction objects	Annual operation costs	1-year costs	10-year costs
Pressure sewerage system	14 722	25	14 747	14 972
Gravity sewerage system	10 125	5	10 130	10 175
Domestic pumping stations	8 685	6 207	14 892	70 755
Local wastewater treatment plant	16 800	11 200	28 000	128 800
Total costs	50 332	17 437	67 769	224 702

Table 4. Estimated costs for rebuilt wastewater treatment plant to pump station in thousands of CZK.

	Construction objects	Annual operation costs	1-year costs	10-year costs
Pressure sewerage system	24 759	48	24 807	25 239
Gravity sewerage system	4 898	3	4 901	4 928
Domestic pumping stations	9 315	3 105	12 420	40 365
Local wastewater treatment plant	1 600	629	2 229	7 890

Central wastewater treatment plant	0	2 533	2 533	25 330
Total costs	40 572	6 318	46 890	103 752

When comparing investment costs, the total value of the initial investment is similar in orders of magnitude. From a more detailed point of view, the costs of the pump conversion design are lower, mainly due to missing need of costly expenses for the reconstruction of the existing treatment plant.

For an overall comparison of the two proposals, it is appropriate to include both investment and operating costs. Therefore, the calculation of the estimated 10-year operating costs in the sum of the initial investment expenses was performed. When comparing the estimated total costs, the first proposal is based on 224 702 thousand CZK and the second proposal for 103 739 thousand CZK. We therefore see a lower estimated financial complexity for rebuilt wastewater treatment plant to pump station proposal.

3. Conclusion

Two methods for the intensification of the wastewater treatment plant and the sewage disposal of the city district have been proposed. The first proposal planned the reconstruction of the local wastewater treatment plant for the required capacity. The second intensification proposal was based on the abolition of the existing wastewater treatment plant and the replacement of this treatment plant with a pumping station. All wastewater from the city district would be discharged to a central wastewater treatment plant with sufficient capacity.

4. References

[1] Samer M 2015 Wastewater treatment engineering (Croatia: InTech)
[2] Russell DL 2006 *Practical wastewater treatment* (Giorgia: John Wiley & Sons)
[3] Makisha N 2016 Restoration and renovation of wastewater pumping stations in case of emergency *Procedia engineering* **165** pp 1087-1091
[4] Liu D H and Lipták B G 2020 *Wastewater treatment* (Boca Raton: CRC Press)
[5] De Chiara M, De Rosa R, Giuliani A, Guadagnuolo S, Leopardi A, Pucci L, and Torregrossa D 2020 A model-based approach for energy optimization of real wastewater pumping station. *Frontiers in Water-Energy-Nexus—Nature-Based Solutions, Advanced Technologies and Best Practices for Environmental Sustainability* pp 525-528
[6] Makisha N and Gogina, E 2017 Scientific approach and practical experience for reconstruction of wastewater treatment plants in Russia *E3S Web of Conferences* **22**
[7] Roupcova K 2011 Dodatek č. 1 ke kanalizačnímu řádu obce Hlučín – Bobrovníky: Aktualizace (Hlučín: Vodovody a kanalizace, s.r.o.)
[8] VaK Hlučín s.r.o. 2018 *ČOV Bobrovníky – provozní řád pro trvalý provoz* (Hlučín: Vodovody a kanalizace, s.r.o.)
[9] Jokl M, Kubík J and Vachovec R 2018 Hlučín – Bobrovníky, Malánky, vyhodnocení variant splaškové kanalizace (Brno: AQUA PROCON s.r.o.)
[10] Malchárková E 2003 *Územní plan města Hlučína* (Strahovice: Česká komora architektů)
[11] Hlavínek P, Mičín J and Prax P 2001 *Příručka stokování a čištění* (Brno: NOEL 2000)
[12] ČSN 75 6401 2006 Čistírny odpadních vod pro vice než 500 ekvivalentních obyvatel (Praha: ÚNMZ)
[13] ČSN 75 6101 2004 Stokové sítě a kanalizační přípojky (Praha: ÚNMZ)
[14] Sommer L 1993 *Malé čistírny odpadních vod* (Praha: Agrospoj)
[15] Koneko spol. s.r.o.2012 ČOV Hlučín - Jasénky: Provozní řád pro trvalý provoz. (Ostrava: KONEKO s.r.o.)

Acknowledgments
This article was written in connection with the grant program - Support for Science and Research in the Moravian-Silesian Region 2020.

IOP Conf. Series: Materials Science and Engineering 1209 (2021) 012021 doi:10.1088/1757-899X/1209/1/012021

The effectiveness of SuDS measures in reducing the rainwater runoff from urbanized areas

I Marko[1], R Wittmanová[1], J Hrudka[1] and Š Stanko[1]

[1] Department of Sanitary and Environmental Engineering, Faculty of Civil Engineering, Slovak University of Technology in Bratislava, Vazovova 5, 812 43 Bratislava, Slovakia

E-mail: ivana.marko@stuba.sk; reka.wittmanova@stuba.sk; jaroslav.hrudka@stuba.sk; stefan.stanko@stuba.sk

Abstract. Over the last few years, climate change is one of the most important phenomena in the 21st century. It is bringing significant changes and negatively affecting the environment. This unwanted phenom causes substantial changes in climatic and hydrological characteristics that are manifested mainly in urban agglomerations. It affects extreme weather changes, the occurrence of natural disasters, water shortages, and other phenomena threatening the quality of the environment. On the other hand, there are measures close to nature based on the sustainable development of urbanized areas. Their main goal is to reduce the volume and rate of runoff and the concentration of undesirable substances in rain runoff from urban areas. Therefore, in the research study, we focus on evaluating published studies that have adopted the SuDS practices. We will compare the effectiveness of SuDS measures in capturing flood peak from surface runoff, pollution loads, and their impact on water quality. The result of this study could help select SuDS measures depending on the need to address the problem in the river basin.

1. Introduction

The fast development of urbanization brings several specific problems associated with changing the city's spatial structure. The increase in the number of paved areas, the development of transport and technical infrastructure, the way of treated rainwater, and other factors significantly affect the quality of life. These changes to the city spatial structure have negatively impacted the hydrological cycle and, subsequently, all processes associated with it (infiltration, precipitation, etc.) [1] [2].

By replacing permeable ecosystems with impermeable paved areas limits the penetration of atmospheric water into the soil horizon, which in turn causes: increased runoff volume, increased flood wave peak, increased urban rainwater contamination, increased likelihood of surface runoff, reduced natural infiltration, etc. Several studies have confirmed the negative impact of this unnatural phenomenon on urban river basins due to the flood risk of urban agglomerations, the burden on sewage systems, and the possible degradation of the quality of aquatic ecosystems [3].

Due to the increasing trend of population migration to cities [4], the percent of urbanized areas has increased several times in recent years. In some cities, impermeable areas occupy more than 60 % of paved surfaces. The problem is that they support the natural infiltration of rainwater and adversely affect the change of the city's microclimate and thus the environment [5]. A serious problem in urban areas is the rainwater runoff, which cannot infiltrate the soil in such unnatural conditions. It thus increases the amount of surface runoff in the basin.

Due to the inappropriate way of dealing with rainwater from urban areas, there was a need to build new modern systems - a central urban drain. In the 1970s, the principle of urban drainage was based on a temporary delay in surface rainwater runoff [9]. Such a disposal method was to regulate the rate of replenishment of surface water reserves to ensure the required quality. This way of rainwater management did not solve the main problem of large cities caused by climate earth. As a result, a completely new area was developed to preserve the natural hydrological regime of the area as much as possible.

The new concept of rainwater management is based on the support of sustainable development of urban areas and the use of primary elements of blue-green infrastructure [7]. It combines nature-friendly measures (wetlands, green roofs, green belts) with technical measures (surface and subsurface infiltration) into one integrated system. The integrated rainwater management system is a complex of facilities and equipment supporting evaporation, infiltration, and slow runoff of rainwater. Their main goal is to eliminate the intensity of surface runoff and thus reduce the risk of damage to the urban environment. The implementation of these measures in urbanized areas would significantly affect the hydrological and climatic conditions of the area, address the crisis of water scarcity, restore the hydrological regime of the area and change the impacts of climate change [8] [9]. This affirmation was confirmed by hundreds of studies that have looked at evaluating the effectiveness of these measures in different rain characteristics and in different spatial areas.

2. Methodology

This paper aimed to obtain an overview of the effectiveness of Sustainable Urban Drainage System (SuDS) objects based on many studies, articles, books, and other documents. Due to the wide application of SuDS measures in a research study, we focus on evaluating published studies that have adopted SuDS procedures. We will compare the effectiveness of SuDS measures in capturing flood peaks from surface runoff, pollution, and their impact on water quality. The study results could help in the selection of SuDS measures depending on the need to address the problem in the river basin. Given that there is a wide range of objects and facilities for rainwater management, we focused on the most commonly used measures.

3. Sustainable Urban Drainage System in urban areas

Civilization has been using water-related activities and rainwater management since ancient times. Many nations (Romans, ancient Greeks) already used tanks to collect rainwater from paved areas at that time [10], other cultures used rainwater collected from paved areas to irrigate fields [11], and underground rainwater tanks were also known [12]. At that time, it was already known that rainwater management consists of three primary stages: capture, accumulate, and use rainwater at the point of impact. Today, the application of SuDS is diverse and includes a number of measures that positively impact the environment.

There is no uniform definition for the term "rainwater management" that is accepted worldwide. It varies from country to country according to industry, original purpose, principle, concept, and technique.

In general, terms associated with the management of rainwater in the urban environment are used to cover a range of methods of collecting and storing surface runoff from paved areas. In foreign literature, it is possible to meet the term Low Impact Development (LID), Stormwater Control Management (SCMs), Sustainable Urban Drainage System (SuDS). In Slovakia is used the term rainwater management.

The new method of rainwater management is operated by different principles and principles than those applicable to traditional - conventional drainage of rainwater from paved areas. It seeks to use fewer resources and energy than the traditional way, using vegetation and thus reducing greenhouse gas emissions. Based on many studies that rainwater management is a way of draining urbanized areas.

The focus of these measurements is not to drain but to capture, infiltrate or temporarily retain rainwater at the point of impact. It is a complex of objects and facilities that imitate the natural hydrological regime of the basin and solve the problem of the city's microclimate. They support evaporation, reduce the air temperature, ensure the replenishment of groundwater reserves, enable the reduction of surface runoff, reduce the concentration of pollutants, create a vegetation environment, affect the reduction of carbon, etc. [13]. These measures support processes such as filtration, evapotranspiration, absorption, biodegradation, percolation, and phytoremediation. With certainty, it can be stated that it is a set of objects and devices. The primary function is to prevent the formation of surface runoff in urban areas and reduce possible environmental damage.

Unlike the traditional way of draining rainwater runoff from urban areas into a sewer or water body (rivers, lakes), an integrated rainwater management system seeks to drain rainwater naturally. The drain of surface runoff into the sewer network is perceived as the last solution when all other variants are exhausted [14][35].

There are several dozen rainwater management measures can be applied and ensure efficient water management while meeting the basic criteria. SuDS objects include green roofs, infiltration systems, filter strips, filter drains, swales, bioretention system, trees, permeable pavement, attenuation storage tanks, detention basins, and other nature friendly. Each of these objects and facilities fulfils basic functions:

- Controlling the necessary water quality
- Managing the quantity of surface runoff
- Creating and maintaining a better place for nature and people [9].

Examples of SuDS measures in Slovakia are shown in figure 1.

Figure 1. Demonstration of revitalization of some parts of Bratislava using SuDS measures [15].

4. The effectiveness of SuDS measures

The effectiveness of the integrated rainwater management system has been monitored based on established criteria, as confirmed by hundreds of studies evaluating the effectiveness of these measures under different rain characteristics and different spatial areas. Due to their impact on the environment, the most frequently monitored parameters were the quality and quantity of surface runoff.

4.1 Effectiveness of SuDS measures in reducing the volume and rate of surface runoff

The effectiveness of the applied SuDS measures on a given river basin in terms of reducing the volume and rate of surface runoff has been assessed by many authors in their research. E.g., Berndtsoon, J.C., (2010) [16], Zhu, H. et al. (2019) [17], as well as others note in their studies that some measures, like vegetation roofs, bio-retention systems, and permeable paved areas, play a significant role in reducing rainwater runoff from urban areas and can capture almost the total volume of rain surface outflow. For example, with bio-retention systems, it is possible to reduce rainwater runoff and maximum flow from 23 to 97 % [18], [19]. By implementing green roofs, it is possible to reduce the volume of runoff in the range from 39 to 99%, depending on the type of roof (intensive extensive) [16], [18]. Reduction of the surface runoff volume is also possible using permeable surfaces, the function of which is a temporary storage of outflow and slow infiltration into the soil. Permeable areas can provide a reduction of runoff volume from 50 to 94% [17], [18], [20].

The effectiveness of these measures depends on several factors, like the weather conditions of the river basin (temperature and humidity), the intensity and duration of the rainfall event, and the type and characteristics of the measure (thickness of filter layers, the material used, infiltration capacity, etc.). During rainy events of lower intensity and short duration was recorded the high effectiveness of these measures. At higher intensities, the effectiveness of SuDS measures was slightly lower. Processes that are able to reduce surface runoff by about half (like infiltration and evapotranspiration) play an important role in reducing the volume of rainwater runoff. By capturing the total volume of the outflow (99%), they ensure complete relief of the sewer network. The percentage of effectiveness of the selected object is shown in the table 1.

Table 1. Percentage expression of captured runoff volume from urban area using SuDS measurs.

SuDS measured	Green goof [16], [18],	Filter strips [21] [22] [24]	Swales [18]	Bioretention systems [18][19]	Permeable pavements [17] [18] [20]
Runoff volume redustion [%]	23 - 99	35 - 95	14 - 98	23 - 97	50 - 94

4.2. Effectiveness of SuDS measures in reducing the concentration of pollutants

Other studies have focused on the effectiveness of SuDS measures in reducing the concentration of pollutants, given that some measures take place in processes like filtration, sedimentation, adsorption, etc. The most frequently monitored parameters were total insoluble matter content (TSS), total nitrogen content (TN), total phosphorus content (TP), and heavy metals, as are Pb, Cu, Zn. Monitoring of these parameters concerned the high risk of environmental degradation of human health [5], [24], [25]. Berndtsson C.J. (2010) [16], Ahiablame, M.L., et al. (2012) [18], Griffin, D.R. (2018) [26] carried out research studies comparing dozens of researches, technical reports, projects, government documents, etc. Among other things, these authors evaluated the effectiveness of removing contaminants from surface runoff in their studies by comparing different measures.

The authors of the research studies note that the removal eficiency of undesirable substances varied in the range from 0 to 99%, while the effectiveness of their removal and capture is conditioned by several important factors: the measure used, characteristics of filter material (thickness, type of material), surface runoff concentration, physicochemical properties of pollutants, equipment maintenance, service life equipment, geometric characteristics of the equipment, vegetation used, geological conditions of the river basin, etc. [16], [17], [18], [19]. Many authors have noticed a significant decrease in some pollutants (mainly nutrients) by up to half using bio-retention systems [19], [27]. Green roofs have proven to be a suitable type of measure to reduce the concentration of heavy metals (up to 93%) [16] [27]. Monitoring the effectiveness of swales was found to reduce nutrients from 14 to 98 % and heavy metal more than 90% [27].

The effectiveness of implemented SuDS facilities and equipment in reducing the content of individual sediments in surface runoff is shown in table 2.

Table 2. Percentage expression of reduction of contaminants occurring in rainwater runoff in urban area.

SuDS measured		Green goof [16], [27],	Filter strips [21] [22] [23]	Swales [18]	Bio-retention systems [18] [27]	Permeable pavements [17] [18]
TSS	%	-	> 80	30 - 97	47 - 99	58 - 94
TP		26-80	43 - 95	24 - 99	29 - 99	10 - 78
TN		-	34 - 74	14 - 61	32 - 99	-
Zn		> 96	24 - 51	68 - 93	62 - 97	73 - 99
Cu		> 96	25 - 65	-	43 - 99	20 - 99
Pb		> 93	-	< 29	31 - 97	74 - 99

5. Conclusion

In recent decades many developed countries, which have experienced rapid development of urban areas and infrastructure, face significant problems associated with inappropriate drainage of surface runoff and subsequent local floods. Given the assumption of possible risks (water crises, degradation of water quality of water bodies, changes in climatic conditions), these countries have developed an integrated and intelligent rainwater management system in urban agglomerations. By implementing these measures, they have significantly increased the quality of life in urban areas, confirmed by the number of studies.

Based on the data given in Table 1 and Table 2, it can be specified that the implementation of SuDS measures can be adapted to be affected by rapid urbanization and climate change. Adaptation of these measures would significantly reduce the concentration of pollutants in the air and on paved areas while reducing the risk of degrading the quality of water bodies and soil, especially by reducing the biodiversity and aesthetic value of the city. [4], [18]

The new rainwater management system in Slovakia is in the development phase, which is not yet one of the leading measures to regulate surface runoff from paved areas. However, some situations would require an integrated approach. At present, it aims to separate and regulate surface water runoff in urban river basins and thus reduce the load on the sewer system and the recipient during extreme rains.

6. References

[1] Gao B, Gao L, Xu D, Zhang M, Qu X and Li Y 2019 A novel method for evaluating the potential release of trace metals associated with rainfall leaching/runoff from urban soils *Science of The Total Environment* **664** 37–44

[2] Li Ch, Liu M, Hu Y, Shi T, Qu X and Todd Walter M 2018 Effects of urbanization on direct runoff characteristics in urban functional zones *Science of the Total Environment* **643** 301–11

[3] Mamoon A A, Jahan S, He X, Joergensen E N and Rahman A 2019 First flush analysis using a rainfall simulator on a micro catchment in an arid climate *Science of The Total Environment* **693**

[4] United Nations 2018 *World urbanization Prospects* [cit. 23. 02. 2021]. Available online https://population.un.org/wup

[5] Angrill S, Petit-Boxi A, Morales-Pizón T, Josa A, Rieradevall J and Gabarrell X 2017 Urban rainwater runoff quantity and quality – A potential endogenous resource in cities? *Journal of Environmental Management* **189** 14–21

[6] Trojan D M, Gulliver S J and Fairbairn J D 2019 Groundwater Impacts from Stormwater Infiltration Practices *Encyclopedia of Water: Science, Technology, and Society* (New York: John Wiley & Sons)

[7] Ghafrani Z, Sposito V and Faggani R 2017 A Comprehensive Review of Blue-Green Infrastructure Concepts *International Journal of Environment and Sustainability* **6** 1 15–36

[8] Bigurra-Alzati A C, Ortiz-Gómez R, Vázquez-Rodríguez A G, López-León D L, Lizárraga Mendiola L 2021 Water Conservation and Green Infrastructure Adaptations to Reduce Water Scarcity for Residential Areas with Semi-Arid Climate: Mineral de la Reforma *Water* **13** 1 45

[9] Woods Ballard B, Wilson S, Udale - Clarke H, Scott T, Ashley R and Kellagher R 2015 *The SuDS Manual* Version 5 (London: CIRIA) p 964

[10] Phoca I and Valavanis P 1999 *Rediscovering ancient Greece: Architecture and city planning* (Athens: Kedros Books) 2nd edition

[11] Crasta F M, Fasso C A, Patta F and Putzu G 1982 Carthaginian–roman cisterns in Sardinia *Proceedings of the International Conference on Rain Water Cistern Systems* (Honolulu: Water Resources Research Center) pp 37–45

[12] Abdulla A F and Al-Shareef W A 2009 Roof rainwater harvesting systems for household water supply in Jordan *Desalination* **243** 1–3 195–207

[13] Brown R R, Keath N and Wong T H F 2009 Urban water management in cities: historical,

current and future regimes *Water Science and Technology* **59** 5 847–55
[14] Vítek J, Stranský D, Kabelková I, Bareš V and Vítek R 2015 Hospodaření s dešťovou vodou v ČRR *Rainwater management in the Czech Republic* (Prague: ZO ČSOP Koniklec) p127
[15] Capital city of SR Bratislava 2021 *Living places, program of renewal of public spaces in Bratislava* Available online https://bratislava.sk/sk/sprava/zive-miesta-program-obnovy-verejnych-priestorov-bratislavy
[16] Berndtsson C J 2010 Green roof performance towards management of runoff water quantity and quality: A review *Ecological Engineering* **36** 4 351–60.
[17] Zhu H, Yu M, Zhu J, Lu H and Cao R 2019 Simulation study on effect of permeable pavement on reducing flood risk of urban runoff *International Journal of Transportation Science and Technology* **8** 4 373–82
[18] Ahiablame M L, Engel A B and Chaubey I 2012 Effectiveness of Low Impact Development Practices: Literature Review and Suggestions for Future Research *Water, Air, & Soil Pollution* **223** 7 4253–73
[19] Zhang L, Ye Z, Shibata, S 2020 Assessment of Rain Garden Effects for the Management of Urban Storm Runoff in Japan *Sustainability* **12** 23 1–17
[20] Myers B, Beecham S, Van Leeuwn J 2011 Water Quality with Storage in Permeable Pavement Basecourse *Water Management* **164** 7 361–72
[21] Ding Y, Dong F, Zhao J, Peng W, Chen Q and Ma B 2020 Non-Point Source Pollution Simulation and Best Management Practices Analysis Based on Control Units in Northern China *International Journal of Environmental Reserch and Public Health* **17** 3
[22] Li R and Kou Y-M 2021 Effects of Shallow Water Table Depth on Vegetative Filter Strips Retarding Transport of Nonpoint Source Pollution in Controlled Flume Experiments *International Journal of Environmental Research* **15** 1
[23] Hou W, Zhou H, Liu L, Zhao G, Yang Z and Yu Y 2017 Performance of Grass Filter Strip in Copper and Zinc Removal in Surface and Subsurface Runoff *MATEC Web of Conferences* **95** 18002
[24] Charters F J, Cochrane T A and O´Sullivan D A 2021 The influence of urban surface type and characteristics on runoff water quality *Science of The Total Environment* **755** 1
[25] Gikas G D and Tsihrintzis A V 2012 Assessment of water quality of first-flush roof runoff and harvested rainwater *Journal of Hydrology* 466–7 115–126
[26] Griffin D R 2018 *Principles of Stormwater Management* (New York: CRC Press) p 350
[27] Davis A P, Shokouhian M, Sharma H and Minami C 2006 Water quality improvement through bioretention media: nitrogen and phosphorus removal *Water Enviroment Research* **78** 284–93
[28] Bäckström M 2002 Sediment transport in grassed swales during simulated runoff events *Water Science and Technology* **45** 41–9

Acknowledgments
This work is supported by „Excelentných tímov mladých výskumníkov STU ", funded by the Rectors office of the STU in Bratislava (from September 2020 to September 2021), the Scientific Gant Agency of the Ministry of Education, Youth and Sports of the Slovak Republic, and the Slovak Academy of Sciences within the project VEGA 1/0574/19, co-funded by the European Regional Development Fund and by the Slovak Research and Development Agency under the contract No. APVV-18-0203.

Assessment of streamflows in the Hornád river basin and analysis of trends

P Nagy[1]

[1] Technical university of Kosice, Faculty of Civil Engineering, Institute of Environmental Engineering, Vysokoskolska 4, 042 00 Kosice, Slovakia

E-mail: patrik.nagy@tuke.sk

Abstract. Climate change is a global phenomenon. The more frequent occurrence of dry periods, which last longer but also extreme rainfall, needs to be reduced for better water management. During the dry season, the quantity and quality of surface and groundwater decreases. Water is important for agriculture, agriculture and ecosystems. This study was focused on the occurrence of trends in daily flows in the Hornád basin at selected hydrological stations for the period 1960-2011. The Mann-Kendall trend test is used to evaluate trends in hydrometeorological time series.

1. Introduction

Climate change is having an impact on the intensity of extreme events, such as extreme rainfall, flash floods and the increasing incidence of drought and the duration of drought [1]. The areas of arid and semi-arid regions are expected to increase and account for more than 50% of the earth's surface [2,3,4]. Changes in the volume and spatial distribution of water will affect the supply of surface and groundwater to industry, agriculture, aquatic ecosystems and the supply of domestic water [5]. The water cycle represents the evaporation, dilution and outflow of water. Climate change has accelerated urbanization and increased demand for water. It is expected that the increasing occurrence of dry periods will lead to fluctuations in groundwater and surface water levels, which may affect water quality and quantity [6]. The growing population and urbanization of the country are putting increasing pressure on water resources.

The subject of this study is the occurrence of trends in Slovakia in the Hornád basin at selected water meter stations for the period 1960-2011.

2. Methods and study area

2.1. Methods

The Mann-Kendall test is a non-parametric test used to evaluate trends in hydrometeorological time series. The Mann-Kendall test is used to determine the trend in daily streamflow on hydrologic stations in basin Hornád. The Mann-Kendall ZMK value is taken into account in order to write increasing or decreasing expressions. Decreasing (increasing) expression is used for time series with a negative (positive) ZMK value. The default ZMK value is given as, [7]

$$Z_{MK} = \begin{Bmatrix} \frac{S-1}{V(S)} \; for\, S > 0 \\ 0 \; for\, S = 0 \\ \frac{S+1}{V(S)} \; for\, S < 0 \end{Bmatrix}$$

$$V(S) = \frac{n(n-1)(2n+5)}{18}$$

$$sgn(x_j - x_k) = \begin{Bmatrix} +1 if\,(x_j - x_k) > 0 \\ 0\; if\; (x_j - x_k) = 0 \\ -1\; if\; (x_j - x_k) > 0 \end{Bmatrix}$$

V (S) is the variance and S is the kendall sum statistic. Differences between each successive value are calculated so as not to show positive (+1), negative (-1) and neutral (0) signs, sgn (...) with eq., Where xj and *xk* are time series values in time cases as, respectively from the given time series {x1, x2,..., xk,..., xj,..., xn} sn observations. Finally, the trend is considered significant if the Z_{MK} is higher than the significance levels [8].

2.2. Study area

The study area is located in eastern Slovakia. The Hornád basin is formed by the rivers Hornád, Torysa and Hnilec. The hydrological station Ždaňa and Kysak are located on the river Hornád, the hydrological station Švedlár is located on the river Hnilec and the hydrological stations Košické Olšany and Prešov are on the river Torysa (Figure 1.) The largest average flow is at the station Ždaňa and the smallest at the station Prešov (Table 1.) Data were provided by the Slovak Hydrometeorological Institute. Figure 1. presents the location of water meter stations in the Hornád basin.

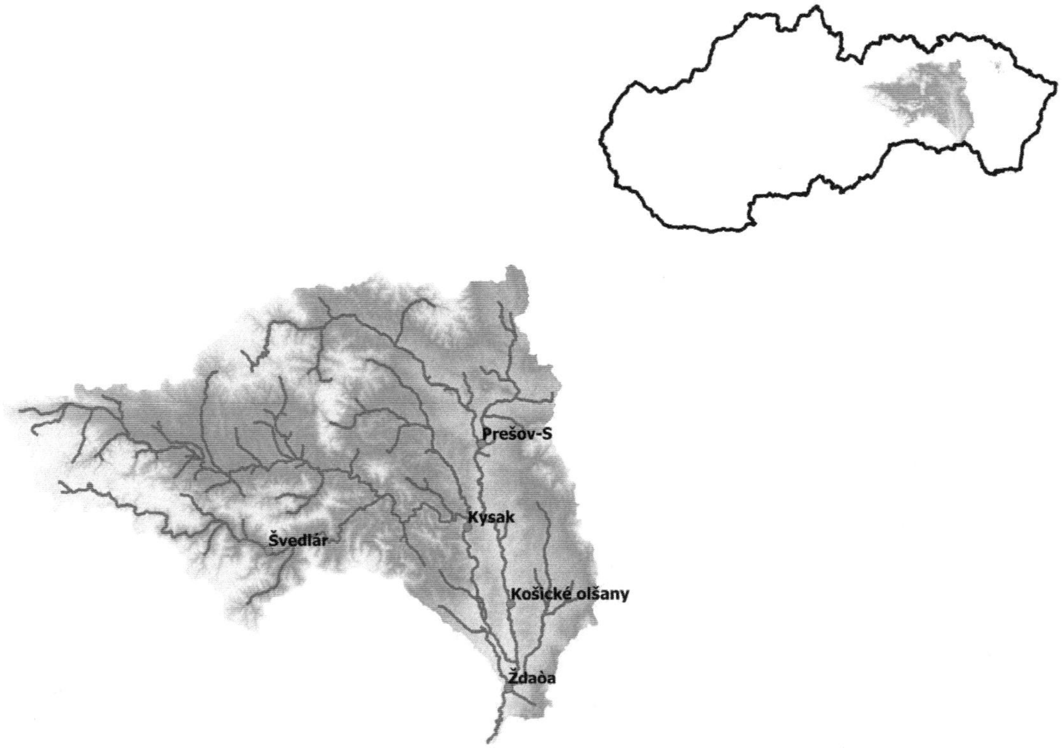

Figure 1. Basin Hornád and hydrologic stations.

Table 1. Hydrologic station in basin Hornád for season 1960-2011

Stations	Average [m3/s]	Max[m3/s]	Min [m3/s]
Ždaňa 8930	29.4	772.25	3.9
Kysak 8690	17.7	454.2	2.7
Švedlár 8540	3.5	110	0.2
Košické Olšany 8870	7.8	292.4	0.5
Prešov 8840	2.0	97.8	0.1

3. Results

In daily flows, the trend occurred at the stations Ždaňa, Kysak and Košické Olšany. The trend did not occur in the hydrological stations Švedlár and Prešov (Table 2.). As the computed p-value is lower than the significance level alpha=0.05, one should reject the null hypothesis H0, and accept the alternative hypothesis H1.

Table 2. Results Mann-Kendall trend test in hydrologic stations Hornád

Hydrological station seasons	Mann-Kendall trend test		Test interpretation
Ždaňa 1960-2011	Kendall's tau	0,052	H0: There is no trend H1: There is a trend **There is a trend**
	S	9341768,000	
	Var(S)	761208253895,333	
	p-value (Two-tailed)	< 0,0001	
	alpha	0,05	
Kysak 1960-2011	Kendall's tau	0,055	H0: There is no trend H1: There is a trend **There is a trend**
	S	9846415,000	
	Var(S)	757487118198,333	
	p-value (Two-tailed)	< 0,0001	
	alpha	0,05	
Švedlár 1960-2011	Kendall's tau	0,004	H0: There is no trend H1: There is a trend **There is no trend**
	S	750990,000	
	Var(S)	761208043005,333	
	p-value (Two-tailed)	0,389	
	alpha	0,05	
Košické Olšany 1960-2011	Kendall's tau	0,056	H0: There is no trend H1: There is a trend **There is a trend**
	S	10021144,000	
	Var(S)	761208490482,667	
	p-value (Two-tailed)	< 0,0001	
	alpha	0,05	
Prešov 1960-2011	Kendall's tau	-0,005	H0: There is no trend H1: There is a trend **There is no trend**
	S	-957993,000	
	Var(S)	725120198037,667	
	p-value (Two-tailed)	0,261	
	alpha	0,05	

The trend occurred only in stations with higher average daily flows and also with higher maximum daily flows, it follows that the water in the lower reaches of the river drains faster from the area and does not stay in the country.

4. Conclusion

In this study, trends were evaluated using a non-parametric Mann-Kendall trend test from given flows at selected water meter stations in the Hornád basin. The results show that the trend occurred in lower hydrological stations where there may be an outflow of surface water from the area. The obtained results confirmed the previous research studies in Slovakia [9,10,11, 12]. This study is the subject of further research to assess the impact of precipitation on water flows in river basins and water retention in the country. In times of climate change, it is necessary to retain water in the country, due to the more frequent occurrence of droughts and periods. In wet periods, when precipitation is more pronounced, it is necessary to retain water in the country so that it is possible to overestimate the water in river basins during periods of drought.

5. References

[1] Zhang Q 2019 Urbanization impacts on greenhouse gas (GHG) emissions of the water infrastructure in China: Trade-offs among sustainable development goals (SDGs). *Journal of Cleaner Production* 474-486

[2] Cunningham C 2017 Climate change and drought in Brazil. *Reduction of vulnerability to disaters: From knowledge to action* pp 361-375

[3] Huang J 2016 Accelerated dryland expansion under climate change *Nature Climate Change* pp 166-171

[4] Rajsekhar D and Gorelick S M 2017 Increasing drought in Jordan: Climate change and cascading Syrian land-use impacts on reducing transboundary flow *Science Advances*

[5] Xu J, Zeng Z 2012 A discrete time optimal control model with uncertainty for dynamic machine allocation problem and its application to manufacturing and construction industries. *Applied Mathematical Modelling* pp 3513-3544

[6] Quante M, Colijn F and Nöhren I 2016 The North Sea Region ClimateChange Assessment (NOSCCA): What happens in the southwest of BACC? *Multiple drivers for Earth system changes in the Baltic Sea region* 180

[7] Mann H B 1945 Nonparametric tests against trend. *Econometrica: Journal of the Econometric Society* pp 245-259

[8] Güçlü Y S 2020 Improved visualization for trend analysis by comparing with classical Mann-Kendall test and ITA *Journal of Hydrology*

[9] Zelenakova M 2014 Trends in minimal stream flows at eastern Slovakia In: *9th International Conference Environmental Engineering (9th ICEE) - Selected Papers*

[10] Zelenakova M 2016 et al Flood occurrence evaluation based on the heights of water level in the rivers In: 3rd International Conference Water Resources and Wetlands pp 153-157

[11] Nagy P 2020 Et al Identification of dry and wet years in eastern Slovakia using indices In: *Advances in Environmental Engineering (AEE 2019)* 444: 012041

[12] Kubiak-Wojcicka K et al 2021 Identification of Extreme Weather Events Using Meteorological and Hydrological Indicators in the Laborec River Catchment Slovakia *Water*

Acknowledgments

This work was supported by the Slovak Research and Development Agency under the Contract no. APVV-20-0281. This work was supported by project of the Ministry of Education of the Slovak Republic VEGA 1/0308/20 Mitigation of hydrological hazards, floods and droughts by exploring extreme hydroclimatic phenomena in river basins.

IOP Conf. Series: Materials Science and Engineering 1209 (2021) 012023 doi:10.1088/1757-899X/1209/1/012023

Single-step and double-step heath recuperation of waste water from thermal baths

A Predajnianska[1] and J Takacs[1]

[1] Slovak University of Technology in Bratislava, Faculty of Civil Engineering, Department of Building Services

E-mail: predajnianska.anna@gmail.com

Abstract. Thermal baths in Slovakia are producers of wastewater, which hides considerable energy potential. The wastewater from the thermal pools has often exceeds the maximum permissible value of the temperature of the wastewater discharged into the water recipient. As a result, there is undesirable damage to the environment, which results in sanctions for the operators of these facilities. Our aim is to present the concept of a single- and double-step heat recuperation system of waste pool water using applications of various types of heat exchangers or heat pumps. The aim of this application is to ensure a suitable temperature to the discharged waste pool water, efficient use of the energy potential of the waste water and saving of primary energy in the form of geothermal water, thus extending the life of the entire system.

1. Introduction

Slovakia´s energy intensity is one of the highest in the Member States of the European Union. High energy prices also force recreational facility operators to look for suitable and environmentally friendly ways to save operating costs. The most appropriate way is to prioritize renewable energy sources. Combinations of renewable energy sources with other technical equipment belong to the highly efficient and operationally reliable way of saving energy, environment and operating costs. The advantage is their ability to respond to individual requirements in the environment in which they are located [3]. Pursuant to Article 9 of the Regulation of the European Parliament and the Council of the EU 2018/1999 on the Management of the Energy Union and Climate Action of 11 December 2018, the Ministry of Economy of the Slovak Republic prepared the Integrated National Energy and Climate Plan for 2021-2030 [1, 2]. This plan updates the current energy policy. Energy policy priorities include, but are not limited to: developing energy infrastructure, reducing energy intensity and promoting the use or renewable energy sources [1].

There are 32 perspective areas in Slovakia with geothermal potential of 215 MW of heat output, provided that the geothermal water is not returned to the geological subsoil. In these areas we can find up to 176 registered geothermal wells, with geothermal water temperature ranging from 30 °C to 130 °C [4, 5]. Geothermal energy in Slovakia is used up to 60 % in recreational facilities. It has also been used in centralized heat supply systems, heat transfer stations, but also in agriculture for heating greenhouses and foilhouses [6, 7]. Fields of geothermal energy utilization and installed heat output is summarized in Table 1. Until 1990, there was not much interest in heat pumps in Slovakia due to low energy prices. However, the situation has changed and they are being applied more and more. The task of these devices is to transform or pump the low-temperature energy of the environment to a higher temperature level by supplying additional energy of higher quality, which is represented by electrical, mechanical or high-temperature thermal energy [8, 9]. The dimensioning of the heat pump depends on

IOP Conf. Series: Materials Science and Engineering 1209 (2021) 012023 doi:10.1088/1757-899X/1209/1/012023

economic investments, but also on the conditioned of the surrounding environment in which it will operate [3].

Table 1. Fields of geothermal energy utilization [6, 7].

Field	Installed heat output (MW)
Recreational facilities	128,3
Systems of centralized heat supply	35,9
Greenhouse and foilhouse heating	34,8
Heat pumps	10,4
Fish farming	5,6
Sum	215,0

2. Recreational facilities

In Slovakia two pool systems are used in thermal baths:

- flow pool system,
- circulation pool system.

Flow pool systems operate on the principle of direct inflow of geothermal water into pools. This system can be used if the geothermal water has a temperature below 50 °C and has a suitable mineralisation. If geothermal well provides a geothermal water with higher temperature than 50 °C, this water cannot be discharged directly into the pool and must be diluted with cold water in the mixing chamber. From this system waste pool water is discharged into recipient without further usage. The disadvantage of flow systems is their high energy consumption and the fact that they require a large yield of the source of geothermal and cold water. Flow based pool system filled with only geothermal water can be seen on figure 1. Flow based pool system filled with mixed geothermal and cold water can be seen on figure 2.

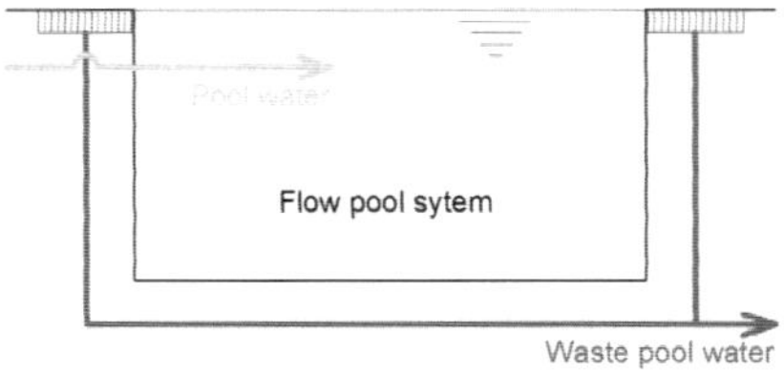

Figure 1. Scheme of the flow pool system filled with only geothermal water [Author].

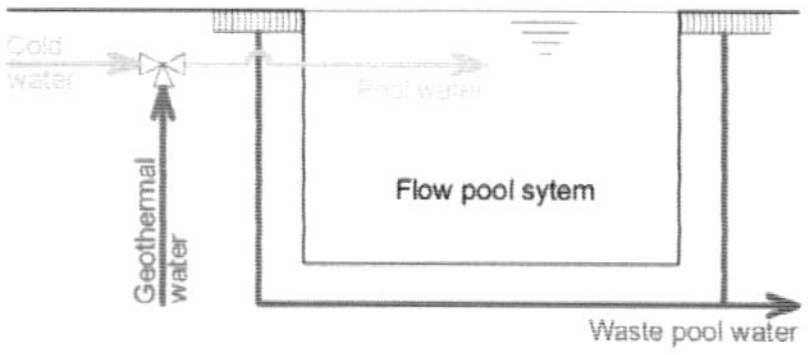

Figure 2. Scheme of the flow pool system filled with mixed geothermal and cold water [Author].

More technically demanding, but still more used are the circulation systems of swimming pools, which include a buffer tank. Geothermal water is mixed with cold water in a buffer tank to reach the required pool water temperature. The mixed water is drained from a buffer tank into a sand filter from where it flows directly into the pool. Pool water flows again into the buffer tank. Only after overfill the buffer tank does the waste pool water flow out of the system, which is discharged into the recipient. Circulation pool system can be seen on figure 3.

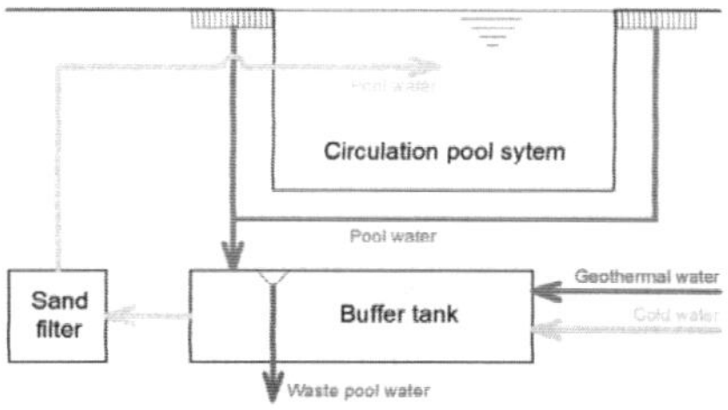

Figure 3. Scheme of circulation pool system filled with mixed geothermal and cold water [Author].

3. Current situation in recreational facilities

Recreational facilities use thermal water to fill their pools. In many cases, the degree of utilization of geothermal energy in thermal baths is less than 50%. Another problem in many recreational facilities is that they discharge waste pool water at too high a temperature. The maximum temperature of waste pool water 26 °C, which can be discharged into the water recipient, is prescribed by Law no. 364/2004 on water and on the amendment of the Law of the Slovak National Council no. 372/1990 Coll. on offenses as amended (Water Law) and Regulation of the Government of the Slovak Republic laying down requirement for achieving good water status no. 269/2010. Exceeding the prescribed temperature may result in a sanction for environmental pollution up to a maximum of 160 000 €. Therefore, it would be appropriate to propose energy-saving measures in recreational facilities, thanks to which we could achieve important aspect. These aspect are reducing the temperature of the waste pool water to the prescribed temperature, increasing the degree of utilization of geothermal energy, extending lifetime of the geothermal energy system and reducing or completely eliminating sanctions for environmental pollution.

4. Energy-saving measures

The proposal of energy-saving measures consists in the reuse of waste energy potential, which has waste pool water leaving the swimming pools. Thanks to these measures, it would be possible to achieve cooling of the waste pool water. In the article, we want to show up possibility of one-step and two-step energy-saving measure by including heat recovery exchanger in first step and by including heat pump as a second step of waste pool water cooling.

4.1. One-step energy-saving measure

This proposal consists in the inclusion of a recuperative heat exchanger in the circuit of waste pool water. The task of the heat exchanger will be to preheat the cold water using heat from the waste pool water. Waste pool water with higher temperature transfer its heat to cold water. The preheated cold water is transported to a mixing chamber, where it is mixed with geothermal water. Since we have heated the cold water due to the waste heat, a lower flow of geothermal water and a higher flow of cold water in the mixing chamber will be needed to reach the same desired pool water temperature. In addition, the rate of geothermal energy use will increase and the lifetime of the geothermal well will be extended [10]. One-step energy-saving measure can be seen on figure 4.

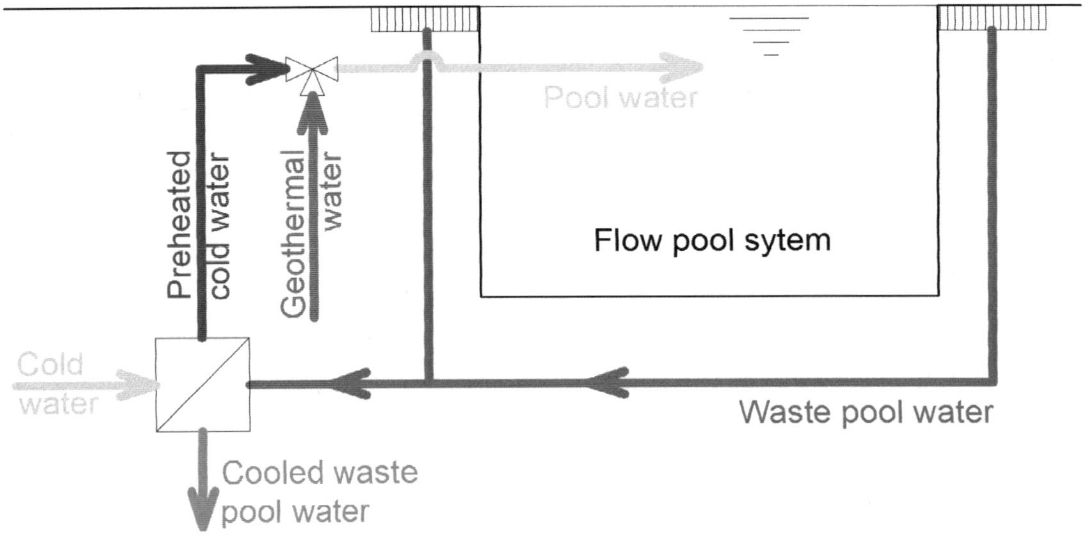

Figure 4. Scheme of one-step energy-saving measure by including heat exchanger into the waste pool water circuit [Author].

4.2. Two-step energy-saving measure

The second stage of cooling waste pool water is the inclusion of a heat pump after the first cooling circuit. After cooling waste pool water in the heat exchanger, it is led to the evaporator of the heat pump, where it is cooled to such a value to meet the requirements for the protection of environment. Heat pumps can cool the waste pool water to lower values, it is only necessary to select a suitable heat pump [10]. Two-step energy-saving measure can be seen on figure 5.

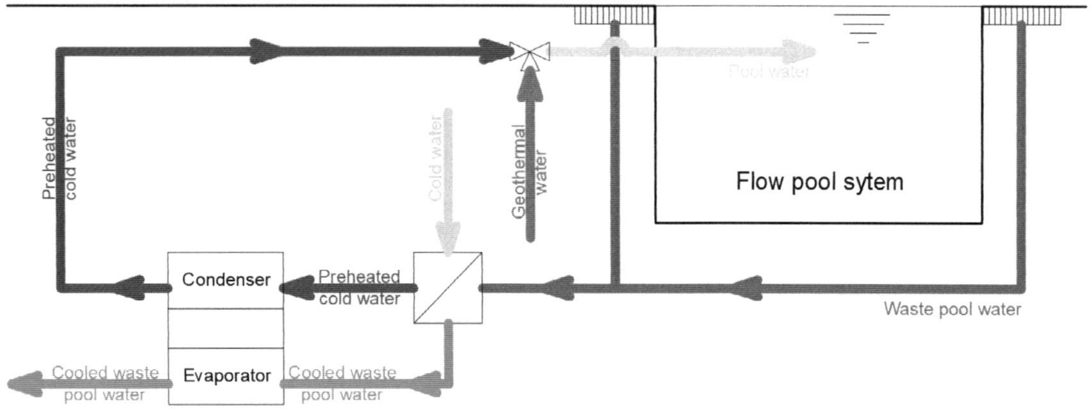

Figure 5. Scheme of two-step energy-saving measure by including heat exchanger and heat pump into the waste pool water circuit [Author].

5. Energy balance calculation

Before recuperation of waste heat proposal we need to know energy balance of the pools in recreational facility. In cooperation with Thermal Corvinus in Veľký Meder, Slovakia we made energy balance calculation of covered recreational flow based pool filled with mixed geothermal and cold water, which area is 295.0 m^2, depth is 1.25 m, volume is 368.0 m^3. Geothermal water with temperature 70 °C and cold water with temperature 15 °C, are mix in mixing chamber to desire pool water temperature 38 °C. Operating time of pool is 69 days and pool was out of order for 7 days. In the calculation we consider the summer operation of the pool. Input data can be seen in Tab.2 [11].

Table 2. Input values for energy balance calculation [11].

Recreational pool	Area	Volume	Depth	Geothermal Water	Cold Water	Pool Water	Service	Out of order
	(m^2)	(m^3)	(m)	(°C)	(°C)	(°C)	(days)	(days)
	295.0	368.0	1.25	70	15	38	69	7

For this calculation we made a simple table with which we are able to calculate energy balance of solved swimming pool. Calculation results such as water flow in process of filling and operation of the pool, geothermal and cold water temperature are shown on Figure 6. Calculation results showed that in process of filling is needed geothermal water flow 5.46 l/s and cold water flow 5.90 l/s to desire pool water flow 11.36 l/s. In process of service of poll is needed geothermal water flow 0.71 l/s and cold water flow 0.77 l/s to desire pool water flow 1.48 l/s. In process of operation there is waste pool water with temperature 38 °C which is discharged into recipient without another usage. Waste pool water is still hot enough to discharge it into recipient without environmental pollution. Waste pool water with this high temperature still has low-energy potential. This is why we need to make increase efficiency proposal [11].

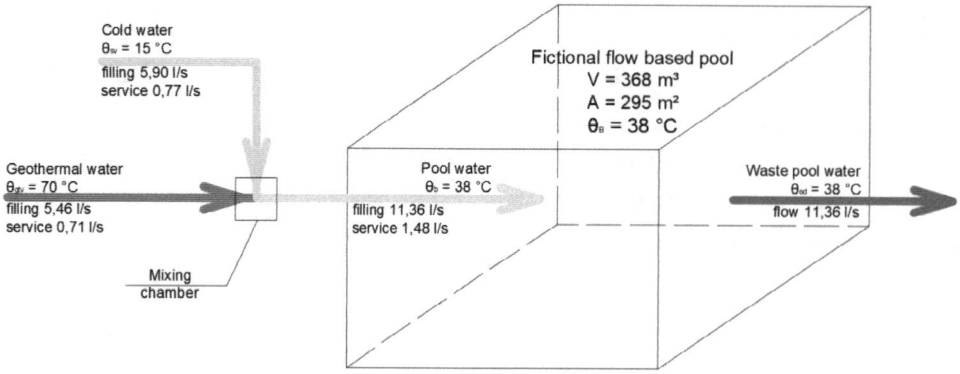

Figure 6. Scheme of recreational flow based pool without recuperation of waste heat [Author].

We summarize energy balance calculation results in graph which can be seen on figure 7. Here we can see graphical display of energy used during filling and operation of solved pool. During filling of this pool we need geothermal water volume of 1 947.47 m^3 mix with cold water volume of 2 102.53 m^3 to desire pool water volume of 4 050.0 m^3. During service of this pool we need geothermal water flow of 12 216.10 m^3 mix with cold water volume of 13 175.91 m^3 to desire pool water volume of 25 392.0 m^3 during the whole operation time of pool which is 69 days per year. Calculation showed, that rate of the usage of geothermal energy is 71.1 % and the rest 28.9 % is considered as waste energy. Waste pool water with temperature of 38 °C is discharged into recipient without another usage. With right increase efficiency proposal we can secure many important aspects which will be described in next part [11].

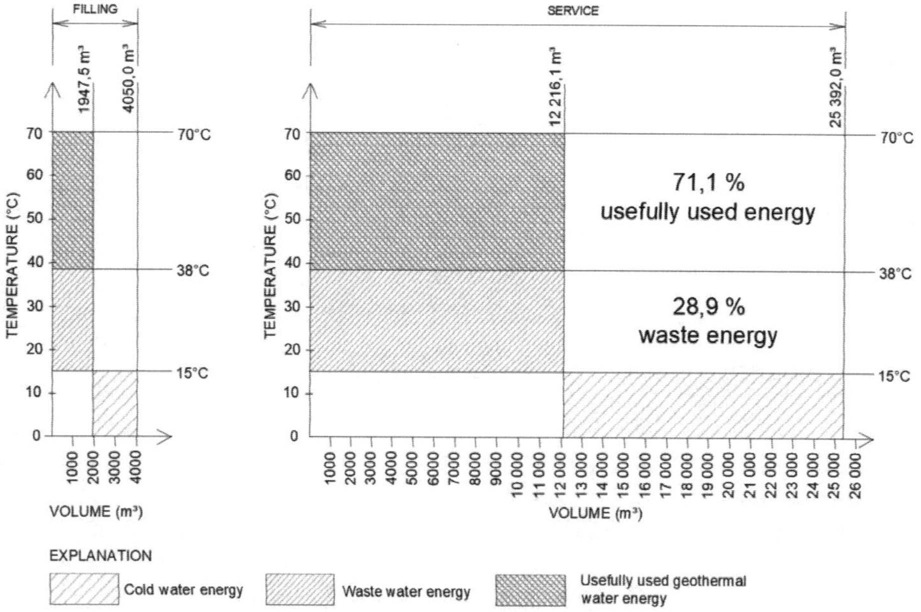

Figure 7. Graphical display of energy used during filling and operation time of pool without recuperation of waste heat [Author].

In this calculation we use only one-step waste pool water cooling, by including recuperation heat exchanger into the circuit of waste pool water. This application can be seen on Figure 8. Waste pool water and cold water are pass through the heat exchanger. Waste pool water transfers its heat to the cold water through heat transfer surface. In this case we increase cold water temperature from 15 °C to 27 °C. Preheated cold water is transported to the mixing chamber. Based on preheated cold water we are able to achieve reduction of geothermal water volume to desire the same pool water flow. This will also increase rate of the geothermal water usage, because we preheat cold water thanks to heat from waste pool water [11].

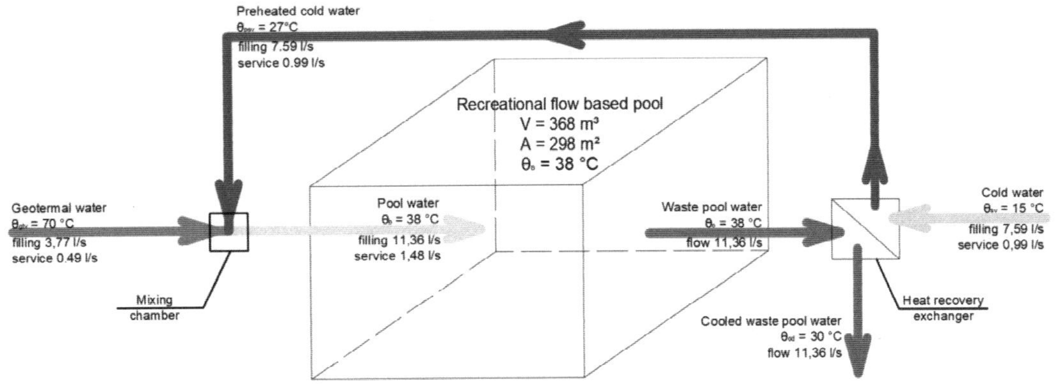

Figure 8. Scheme of recreational flow based pool with recuperation of waste heat [Author].

Energy balance calculation results can be again seen in graphical display of energy used during filling and service of pool with recuperation (figure 9). As we can see, we are not able to secure changes during filling of the pool, but in process of operation of the pool we can decrease geothermal water flow from 0.71 l/s (without recuperation) to 0.49 l/s (with recuperation), decrease geothermal water volume from 1 947.5 m3 to 1 343.53 m3 during filling and decrease geothermal water volume from 12 216.1 m3 to 8 407.29 m3 during operation time of the pool. In this case we was able to decrease waste pool water temperature from 38°C to 30°C, increase rate of the geothermal water usage from 71.1 % to 80.0 % and decrease waste pool water energy from 28.9 % to 20.0 %. By this arrangement we can increase the lifetime of whole open geothermal energy system. Water flow and water volume results of system without recuperation are summarized in Tab. 3 and results for system with recuperation in Tab. 4 [11].

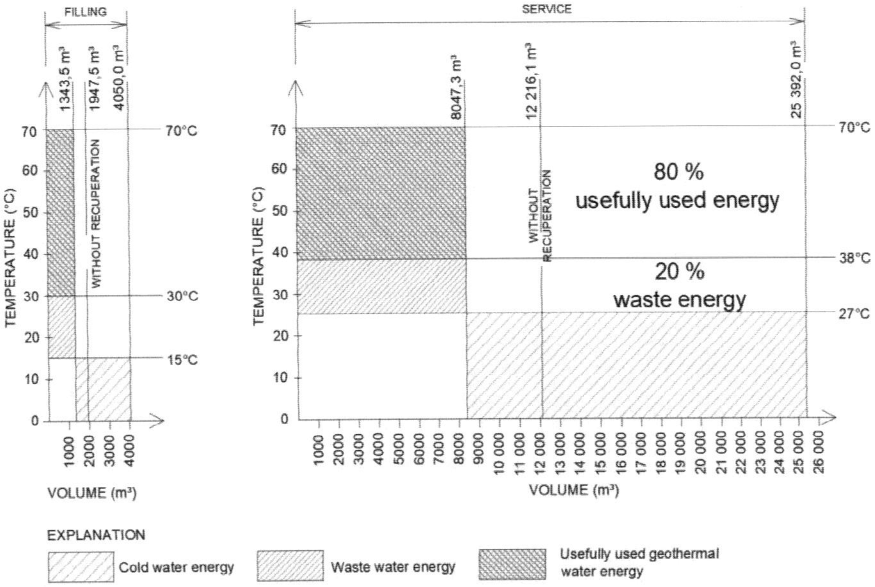

Figure 9. Graphical display of energy used during filling and operation time of pool with recuperation of waste heat [Author].

Table 3. Water flow and volume of system without recuperation [11].

Operating condition	PW flow (l/s)	GTW flow (l/s)	CW flow (l/s)	PW volume (m^3)	GTW volume (m^3)	CW volume (m^3)	Rate of GTW usage (%)
Filling	11.36	5.46	5.90	4 050.0	1 945.47	2 102.53	71.1
Service	1.48	0.71	0.77	25 392.0	12 216.1	13 175.91	

Table 4. Input values for energy balance calculation [11].

Operating condition	PW flow (l/s)	GTW flow (l/s)	CW flow (l/s)	PW volume (m^3)	GTW volume (m^3)	CW volume (m^3)	Rate of GTW usage (%)
Filling	11.36	3.77	7.59	4 050.0	1 343.53	2 704.47	80.0
Service	1.48	0.49	0.99	25 392.0	8 407.3	16 984.7	

6. Conclusion

In this paper, we want to show the possibility of one-step and double-step heat recovery from waste pool water, thanks to which important aspects could be achieved:

- Reduce the temperature of the waste pool water to a maximum temperature of 26 °C by means of the one-step heat recovery system by including heat exchanger into the circuit of waste pool water. The cold water can be preheated from 15 °C to 27 °C to reduce the geothermal water demand. This system is suitable for those geothermal water sources where the yield of the geothermal water is high.
- Reduce the temperature of the waste pool water to temperature of 15 °C by application of one- and double-step heat recovery system by including heat exchanger and heat pump into the circuit of waste pool water. The cold water can be preheated from 15 °C to 32 °C to reduce the geothermal water demand. This system is suitable for those geothermal water sources where the yield of the geothermal water is low. In the winter months when pools operation are limited, the outputs from the heat pump (condenser side) can be used for active thermal protection systems of recreational facility buildings, possibly deicing of roads or sidewalks.
- Increase the energy use of geothermal water and thus ensure the extension of the lifetime of an open geothermal system.
- Reduce or eliminate the sanctions for environmental pollution that must be paid by operators of recreational facilities.
- To contribute to the fulfilment of the commitments made by Slovakia in connection with Directive 31/2010 on Energy Efficiency, which enshrines an increase in the share of the use of renewable energy sources and a reduction in the production of greenhouse gases.

7. References

[1] Integrated National Energy and Climate Plan for 2021-2030. Prepared according to the Regulation of the EP and the Council (EU) no. 2018/1999 on governance of the Energy Union and climate action. Ministry of Economy of the Slovak Republic. Bratislava. October 2019. Available on the Internet: https://www.mhsr.sk/uploads/files/zsrwR58V.pdf

[2] Regulation 2019/1999 of the European Parliament and of the Council of 11 December 2018 on energy governance and climate action. Available online: https://eur-lex.europa.eu/legal-content/SK/TXT/?uri=CELEX:32018R1999

[3] Jandačka J, Holubčík M, Patsch M and Vantúch M 2016 *Modern heat sources for heating. Žilina: EDIS – publishing center ŽU, 264 p.* ISBN 978-80-554-1230-6

[4] Franko O 1995 *Atlas of Geothermal Energy of Slovakia* Dionýz Štúr State Geological Institute in Bratislava p 233

[5] Franko O 1995 *Atlas of Geothermal Energy of Slovakia,* Dionýz Štúr State Geological Institute in Bratislava p 233

[6] Predajnianska A and Takács J 2020 *Utilization of low-temperature energy from thermal baths* TZB Haustechnik, Volume 14 Issue 3 pp 22-25 ISSN 1803-4802

[7] Predajnianska A and Takács J 2020 *What about low-temperature energy from thermal batsh?,* In proceedings of 28th International Scientific and Professional Conference Heating 2020, High Tatras 10 – 14 february 2020 pp 147 – 150 ISBN 978-80-89878-58-1

[8] Mudrá M 2019 *Streamlining the operation of the TP5 heat source for the West housing estate in Brezno [diploma thesis],* Bratislava, Slovak University of Technology in Bratislava, Faculty of Civil Engineering, Department of Building Services p 106

[9] Petráš D, Lulkovičová O, Takács J and Furi B 2009 *Renewable energy sources for low temperature systems,* Bratislava Jaga Group, s. r. o. p 223 ISBN 978-80-8076-075-5

[10] Mudrá M, Predajnianska A and Takács J 2020 *The concept of heat recovery from the wastewater of thermal baths with the application of a heat exchanger and heat pumps,* In proceedings of conference Complex renovation of department houses 2020, Bratislava, 19 – 20 November 2020 pp 44 – 47 ISBN 978-80-973813-1-8

[11] Predajnianska A and Takács J 2020 *Recuperation of waste heat produced by public pools and low-enthalpy energy utilization,* In proceedings of conference E-nova 2020, Internationale Konferenz. Technologie- und Klimawandel. Enerige-Gebäude-Umwelt 26 – 27 November 2020 pp 501 - 507 ISBN 978-3-7011-0460-4

Acknowledgments
This article was supported by the Ministry of Education, Science, Research and Sports of Slovak Republic through a grant VEGA 1/0303/21, VEGA 1/0304/2021 and KEGA no. 005STU-4/2021.

IOP Conf. Series: Materials Science and Engineering 1209 (2021) 012024 doi:10.1088/1757-899X/1209/1/012024

Development of the level of sorting of municipal waste in the selected locality

G Rózsa[1], R Wittmanová[1], J Hrudka[1] and I Škultétyová[1]

[1] Department of Sanitary and Environmental Engineering, Faculty of Civil Engineering, Slovak University of Technology in Bratislava, Vazovova 5, 812 43 Bratislava, Slovakia

E-mail: gergely.rozsa@stuba.sk; reka.wittmanova@stuba.sk; jaroslav.hrudka@stuba.sk; ivona.skultetyova@stuba.sk

Abstract. Waste management is one of the essential areas that everyone in the world should carry about. Waste collection and treatment help to reduce the risk of the spread of various diseases. Municipal waste is waste produced in households by the population. Municipal waste can be sorted into other components, such as plastic, paper, glass, metals, and biodegradable waste. These types of waste are in most urbanized areas collected separately. The sorting level offers accurate information about each urbanization, to what extent its population is involved in separated waste collection. The paper focuses on analyzing the waste collection system in the selected locality and will examine the local waste sorting level development during the last few years.

1. Introduction

Waste management is one of the most critical areas that everyone in the world should address. A separate collection of individual types of waste, such as paper and plastics, is already established everywhere in Europe and almost every country globally. In Slovakia, the obligation of the separate collection has been introduced since 2010. First, five basic categories for separate collection from the population were defined, such as paper, plastics, metals, glass, and mixed municipal waste [1]. Since 2013, the collection of biodegradable waste from the population has also been introduced. The municipality or city must pay the costs of waste collection and disposal or incineration. As the number of separated components of municipal waste increases, waste collection and disposal costs should also decrease. Separation is one step in the waste hierarchy that helps to reduce landfill quantities [2].
Certain types of waste are dangerous to humans and the environment. These are, for example, accumulators, batteries, medicines after the expiry date, which requires special conditions for collection, storage, and disposal. Therefore, it is necessary to have a collection system for other types of waste. The simplest solution is to be close to the living areas of the places, where residents can take sorted waste. Then there is a civic amenity site, where residents can hand over the waste not collected from households.

The efficiency of the civic amenities site lies in the quality of the sorted waste. The conditions must be met for the waste to be stored in the right place or the correct container in the civic amenity site. To comply with all regulations and instructions, it is recommended that there be a person in the civic amenity site who controls all this. By following these conditions, we save nature and reduce waste that would otherwise end up in landfills.

Sorted components of municipal waste can be processed and used in the waste management cycle [2]. The unsorted municipal waste ends up in landfills or is recovered for energy [3]. Slovakia sets a landfill fee for municipalities and cities according to the percentage of the level of sorting of waste from the site.

2. Methodology

Waste management is an activity aimed at preventing and limiting the generation of waste and reducing its danger to the environment by Slovak Republic Act no. 79/2015 Coll. on waste as amended [4]. Waste management is the collection, transport, recovery, sorting, and disposal of waste, including the supervision of these activities and the subsequent care of landfills. It includes the actions of a trader or broker [2,4].

Various systems are used for waste collection. The choice of collection system depends on the amount and type of waste, the mode of transport, and local conditions. According to this method, waste collection systems can be divided into containers and non-container systems [5].

The amount, composition, and type of waste determine the required capacity of the system and its components, especially in the case of separate collection. The structure of the settlement, population density, and construction method is influenced by the determination of collection points, the number and size of collection containers. The area's topography is decisive, for example, for the collection system, the transport options, the size of the collection vehicle used. However, it can also affect the size and type of collection container. The choice of system should be based on financial requirements and the environmental impact of the proposed solution [6].

The definition of municipal waste is not uniform throughout the world. In Slovakia and the Czech Republic, municipal waste is perceived in the sense of common practice only as group 20 of the Waste Catalog, in the narrower sense the so-called mixed municipal waste (20 03 01) [7]. However, in the old E15 countries, municipal solid waste is understood in a slightly broader sense, including so-called commercial waste and other waste from the local community when collected together [8].

2.1. Separate waste collection

Separate waste collection is focused mainly on the types of waste that can be recovered by processing. In Slovakia, municipal waste is sorted in households and collected according to a predetermined method from the municipality. Depending on the location of the collection containers, two basic methods are known. In the first method, the sorted waste is collected in collection containers on their land, from where the sorted waste is collected using collection vehicles. Moreover, the separated waste must be brought to a collection point in the second method, distinguishing between collection points and civic amenity sites [6,9].

3. Calculation of the level of waste sorting in Slovakia

The level of sorting in Slovakia is calculated according to the formula set by Act 329/2018 on fees for waste disposal and the amendment of Act no. 587/2004 Coll. on the Environmental Fund and on the amendment of specific laws as amended. The law specifies the conditions of the calculation. The self-government in calculating the classification of municipal waste in the relevant calendar year is based on data from the records of types and quantities of waste for the previous calendar year. The municipality is responsible for the correct calculation of the level of municipal waste sorting [10].

The following formula is used to calculate the level of municipal waste sorting for the relevant calendar year:

$$LS_{\mathrm{MSW}} = \frac{m_{\mathrm{component}\,1} + m_{\mathrm{component}\,2} + m_{\mathrm{component}\,n}}{m_{MSW}} \tag{1}$$

where the: LSMSW is the value of municipal waste sorting for the previous calendar year expressed as a percentage. The result should be rounded to 2 decimal places [10].

In the numerator of the formula, the values of m are a component. It is the weight of the sorted component of municipal waste collected in the municipality in the previous calendar year. The weight of municipal waste components is expressed in kilograms. The list of sorted components included in the amount of sorted municipal waste is set out in government regulation (see Table 1) [10].

The denominator of the formula contains the values of mMSW, which is the weight of municipal waste generated in the municipality for the previous calendar year. The weight of municipal waste generated is also expressed in kilograms [10].

Table 1. List of separable components of municipal waste that are used to calculate the level of sorting of municipal solid waste [10].

Waste code	Name of the waste
20 01 01	paper and cardboard
20 01 02	glass
20 01 03	multilayer composite materials based on cardboard (cardboard-based composites)
20 01 04	metal packaging
20 01 08	biodegradable kitchen and restaurant waste
20 01 10	clothing
20 01 11	textiles
20 01 21	fluorescent lamps and other wastes containing mercury
20 01 23	discarded equipment containing chlorofluorocarbons
20 01 25	edible oils and fats
20 01 26	oils and fats other than those mentioned in 20 01 25
20 01 33	batteries and accumulators specified in 16 06 01, 16 06 02, or 16 06 03 and unsorted batteries and accumulators containing these batteries
20 01 34	batteries and accumulators other than those mentioned in 20 01 33
20 01 35	discarded electrical and electronic equipment other than those mentioned in 20 01 21 and 20 01 23, containing dangerous parts
20 01 36	discarded electrical and electronic equipment other than those mentioned in 20 01 21, 20 01 23, and 20 01 35
20 01 38	wood other than that mentioned in 20 01 37
20 01 39	plastics
20 01 40	metals
20 01 40 01	copper, bronze, brass
20 01 40 02	aluminum
20 01 40 03	lead
20 01 40 04	zinc
20 01 40 05	iron and steel
20 01 40 06	tin
20 01 40 07	mixed metals
20 02 01	biodegradable waste

The municipality is obliged to publish information on the level of municipal waste sorting for the previous calendar year by 28 February of the relevant calendar year on its website and the municipal's official notice board. Furthermore, at the request of the landfill operator, the Slovak Environmental Inspectorate, the district office, and the Environmental Fund, it is obliged to provide information on the level and method of calculating the level of municipal waste sorting for the relevant calendar year within ten days of their request [10].

Table 2. Amount of municipal waste landfilling fee according to the level of sorting for the given year [11].

	Municipal waste sorting level x [%]	Fee for the year in euro for t		
		2019	2020	2021and following years
1	$x \leq 10$	17	26	33
2	$10 < x \leq 20$	12	24	30
3	$20 < x \leq 30$	10	22	27
4	$30 < x \leq 40$	8	13	22
5	$40 < x \leq 50$	7	12	18
6	$50 < x \leq 60$	7	11	15
	$x > 60$	7	8	11

The amount of the fee for depositing municipal waste in a landfill depends on the level of sorting. The government approved a three-year transition period, during which the amount of the deposit fee was increased by individual percentages each year. In the worst category, where the level of sorting according to formula (1) is less than 10%, the landfill fee has almost doubled in three years. According to the degree of secrecy, the specific fees for individual years are set by Government Decree no. 330/2018. since 2019, municipalities have been paying for the disposal of municipal waste in a landfill according to the level of sorting [10]. At that time, a transitional period began, as individual municipalities already care about classification. In 2021, these fees stabilized, at least for the next few years. According to the level of sorting, the amount of the fee for municipal waste disposal can be seen in table no. 2. It is now up to the municipalities, towns, and villages to ensure the best possible system for collecting sorted municipal waste.

4. Development of the level of sorting in the selected locality

The selected locality is located in the southern part of Slovakia, lies on Rye Island, which is one of the areas with the most significant accumulation of groundwater in Europe. The most crucial groundwater collectors of the solved area are gravelly sediments. At depths of around 2,000 m, hot, highly mineralized waters were verified by drilling. The depth of the groundwater level under the terrain in the area of interest ranges from 3 m to 5 m.

The selected locality belongs to the Association of Municipalities of the Upper Rye Island in Waste Management with its registered office in Šamorín (ZOHŽO) and consists of 31 municipalities and towns. Specifically: Šamorín, Báč, Bellová Ves, Blahová, Blatná na Ostrove, Čakany, Dunajská Lužná, Hamuliakovo, Horná Potôň, Horný Bar, Hubice, Hviezdoslavov, Janíky, Kalinkovo, Kvetoslavov, Kyselica, Legnice, Macov, Mierovo, Miloslavov, Nový Život , Oľdza, Potônske Lúky, Rohovce, Rovinka, Štvrtok na Ostrove, Tomášov, Trnávka, Trstená na Ostrove, Veľká Paka and Zlaté Klasy.

The level of waste sorting in the selected locality has been developing positively over the last three years. Input data, as well as the percentage of sorting, can be found in Table 3. Since 2018, when we have set a formula (1) in Slovakia to calculate the level of sorting, this number in the selected locality has improved by almost 20% percent. During the first year of implementation in the selected locality, the level of waste sorting reached almost 34%. Municipalities, to reduce the landfill fee, began to pay more attention to separate waste collection, which was already seen in the calculation for 2019, where they reached a level of 47.5%. In 2020, they already produced 53.16%, which means that more than half of the waste in the selected locality is already separated from mixed municipal waste.

Table 3. Input data and level of sorting for the last three years.

	Mixed municipal waste [t]	Sorted waste components that can be included in the formula [t]	Level of waste sorting [%]
2018	16382.5	10152.46	33.94
2019	17170.99	20311.4	47.5
2020	17557.49	26516.18	53.16

5. Conclusion

The new bill aimed not to demand as much money as possible from municipalities for fees for depositing municipal waste in landfills. The aim was to motivate municipalities to carry out more excellent sorting, ensure a better municipal waste collection system, and provide the population with better services and opportunities for the collection of sorted municipal waste.

According to Eurostat data from 2016, Slovakia was one of the Member States of the European Union with the lowest share of recycled or composted municipal waste. We managed to separate only about 23 percent of the total amount of waste. Most municipal waste is landfilled in Slovakia. We can state that the new law and the charging system helped to increase the level of classification that results from this study. Municipalities began to pay more attention to collecting sorted components of municipal waste, which we can also see in the numbers. In the selected locality in three years, the level of sorting was achieved from 34% to 53%. This is almost a 20% increase and means that more than half of the waste produced in the selected location will no longer be landfilled but will be reprocessed.

6. References

[1] Slovak Republic Act No. 409/2006 Coll. 409 2006 *ACT of 24 April 2006 on Waste Act and on Amendments to Certain Acts*

[2] Trošanová M, Škutétyová I and Rusnák D 2018 *Pollack Periodica* **13** 2 117–28

[3] Mininni G, Zuccarello R Di B, Lotito V, Spinosa L and Di Pinto AC 1997 *Water Science and Technology* **36** 11 211–18

[4] Slovak Republic Act No. 79/2015 Coll. 79 2015 *ACT of 17 March 2015 on waste and on amendments to certain acts*

[5] Rodrigues S, Martinho G and Pires A 2016 *Journal of Cleaner Production* **113** 374–87

[6] Čermák O 2007 *Waste management* (Bratislava: STU Publishing House) p 106

[7] 2000/532/EC, Commission Decision of 3 May 2000 replacing Decision 94/3/EC establishing a list of wastes pursuant to Article 1(a) of Council Directive 75/442/EEC on waste and Council Decision 94/904/EC establishing a list of hazardous waste pursuant to Article 1(4) of Council Directive 91/689/EEC on hazardous waste, Available online https://eurlex.europa.eu/legal-content/EN/ALL/?uri=celex%3A32000D0532

[8] Hrabčák M 2014 *Čo sú to vlastne komunálne odpady? OdpadyPortal* Available on internet: https://www.odpady-portal.sk/Dokument/102076/co-su-to-vlastne-komunalne-odpady.aspx

[9] Agovino M and Musella G 2020 *Land Use Policy* **91** 104408

[10] Slovak Republic Act No. 329/2018 Coll. 329 2018 *ACT of 17 Oktober 2018 on fees for waste disposal and on the amendment of Act no. 587/2004 Coll. on the Environmental Fund and on the amendment of certain laws as amended*

[11] Slovak Republic Government regulation No, 330/2018 Coll. 330 2018 *Government regulation of 6 November 2018 on fixing the rates of the landfill fees and the details of the redistribution of revenue from the landfill fees*

Acknowledgments
This work was supported by the Scientific Grant Agency of the Ministry of Education, Youth and Sports of the Slovak Republic and the Slovak Academy of Sciences within the project VEGA 1/0727/20, co-funded by the Slovak Research and Development Agency under contract No. APVV-18-0203.

IOP Conf. Series: Materials Science and Engineering 1209 (2021) 012025 doi:10.1088/1757-899X/1209/1/012025

Flood risk assessment methodology

M Šugareková[1]

[1] Department of Environmental Engineering, Faculty of Civil Engineering, Technical University of Košice

E-mail: maria.sugarekova@tuke.sk

Abstract. Paper presents a proposal of the flood risk assessment methodology. The theoretical background is based on the multi-index conceptual models, which consists of three layers – objective layer, index layer and indicators layer. The fulfilment of the layer is based on the Reports on the course and consequences of floods in the territory of the Slovak Republic.

1. Introduction

There are plenty assessments of flood risk based on calculation exposure and vulnerability elements. The elements can contain various factor that cause flood. Flood is one of the most destructive natural hazards worldwide. Flood occurs annually, sometimes with a smaller risk but sometimes with maximum risk with untold damage. There are several flood risk assessment methodologies, most of them includes the base equation to calculate risk – probability x consequence. The relation of risk calculation can be modified according to needs of the study. Flood risk assessment in the Hornád river basin also consist of modified equation, which has been supplemented by probability elements [1]. Also, risk can be assessed by combining the flood susceptibility and its negative consequence. The consequence is divided into human health and financial impact [2]. Each country has its own aspects according to which the flood risk is assessed. Authors Tincu et al. [3] in the study focused on detailed vulnerability aspect analysis to quantify the flood risk in the context of land use planning. Study performed at Bago River Basin [4] is aimed to quantitative assessment of flood hazard and risk quantitively including the effectiveness of dam control for flood protection.

Presented paper aimed to develop a new flood risk assessment methodology based on multi-index conceptual models. The next section describes a brief theoretical background, and an overview to the proposed methodology.

2. Materials and methods

Following chapter is divided into three sub-chapters. Each of them offers a brief view to used background theory of the proposed methodology.

2.1. Background theory

Multi-index conceptual models (MIC model) were developed in the 2020. The model includes two parts: the construction of multi-index system and analysis procedures in ArcGIS. The MIC model contains three layers – object, index, and indicator layer, and the indicator layer includes 13 indicators. The study was performed on the Yangtze River basin, which is the biggest river in China. The most important step in the MIC model is in establishing the filling the indicator layers. The Yangtze River appertains to the object layer. The index layer comprehended hazard, vulnerability, and exposure, which are elements necessary to calculate the flood risk. And then, the 13 indicators are divided into the three index layers.

Authors of the study considered many factors involved in flood. The indicators have been selected by theoretically relevance to the flood occurrence (figure 1) [5].

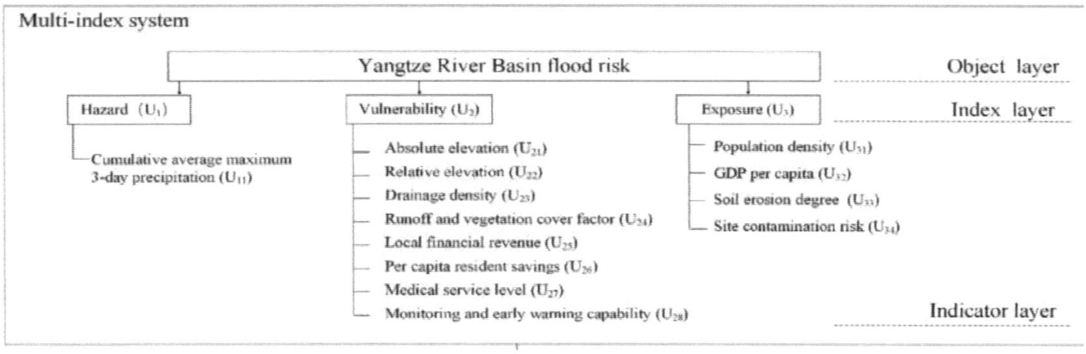

Figure 1. Multi-index conceptual models [5].

Then, the importance of the flood risk indicator was evaluated by analytic hierarchy process. The results of the study are MIC model, and the maps of risk related to each mentioned index.

2.2. *Data collection*

Proposed flood risk assessment methodology is based on online available data due to easy access and accessibility of the necessary data. The website of Ministry of environment of the Slovak republic [6] is publicly available, and so all documents related to the issue covered. Reports on the course and consequences of floods in the territory of the Slovak Republic are prepared annually. The reports contain information about causes and occurrence of floods, meteorological causes of floods, hydrological situation and flood occurrence, evaluation of floods, an overview of expenditures spent on the implementation of flood protection and flood rescue work, monetary compensation and flood damage and summary information about causes and occurrence of floods.

2.3. *Proposed methodology*

Flood risk assessment methodology is based on MIC models [5]. The index layer also contains elements of hazard, vulnerability, and exposure. The difference is in object and indicator layer. In proposed methodology, the object layer should contain the whole study area, i. e. territory of Slovak Republic. The proposed methodology is designed to assess the whole study area because of the input data. The Reports which the proposed methodology is based on contains data related to the individual self-governing (regions). Slovak republic has 8 individual self-governing (regions) (table 3). So, the methodology is designed to assess the whole area at first, then it can be extended and modified to assess smaller study area, for example the individual river catchment.

Table 1 represents proposed indicators according to available data obtain from the Reports on the course and consequences of floods in the territory of the Slovak Republic, and index according to MIC models.

The second difference between MIC models and proposed flood risk assessment methodology is in the indicator layer. While the MIC models uses theoretically relevance to the flood indicators, the proposed methodology is based on available data from Reports on the course and consequences of floods in the territory of the Slovak Republic. To the pre-prepared template were fitted the input data from the mentioned Reports. The main idea of the proposed methodology was to simplify the access to data. An assessor has easy access to input data. Everything needed is available on the online web page of the Ministry of environment of the Slovak republic.

Also, authors of the MIC models used analytic hierarchy process to evaluate the importance of the flood risk indicators, and in the proposed methodology, the weight was assigned according to division of the total amount 100%. In this case, the analytic hierarchy method was not used.

IOP Conf. Series: Materials Science and Engineering 1209 (2021) 012025 doi:10.1088/1757-899X/1209/1/012025

Table 1. Content of index and indicator layers.

HAZARD	
Indicators	Weight
Total precipitation	10
Declaration number of 3rd degree of flood activity	10
Extent of flooded area	10
VULNERABILITY	
Indicators	Weight
Affected habitants	20
Protected landscape areas	10
Condition of flood protection structures	10
EXPOSURE	
Indicators	Weight
Flooded buildings	6
Flooded non-residential buildings	6
Damage utilities	6
Affected animals	6
Another flood damage	6

Proposed content of indicators layer can be found in the above-mentioned Reports [6], except for *Protected landscape areas* and *Condition of flood protection structures*. However, some of the indicators have sub-indicators, and they are divided into smaller groups (table 5).

The main aim of the proposed methodology is to develop a new method to assess the flood risk. A risk calculation will be used to reach the goal. But first data collection must be performed to obtain the required values. The proposed methodology will assess the flood risk in terms of time. Data from 2001 to 2019 are available. Hence, the data from 2001 to 2010 are used as a benchmark to assess the data from 2011 to 2019. Average values are calculated from the data from 2001 to 2010, and the average values serves as a benchmark for further evaluation and assessment. The benchmark is necessary to calculate value of the index. Every index has a weight wi, which is divided into the indicators and sub-indicators. The weight w_i is multiple with a point value of sub-indicator. Most of the sub-indicators are defined in the unit of measures, therefore there must be a reference point (benchmark) circumscribed. The benchmark is obtained through the average's values of individuals indicators. The data are related to individual self-governing regions of Slovak Republic, except of the total precipitation data. These are related to regions – East, Central and West Slovak region.

3. Results

Following subchapters offers a brief overview of the main indexes – hazard, vulnerability, and exposure. In the shown tables there can be seen results calculation. The method of results calculation was easy – there was an arithmetic average method used for all necessary values.

3.1. Hazard

Index layer hazard contains three indicators – total precipitation, declaration number of 3rd degree of flood activity and extent of flooded area. Total precipitation values were obtained from Hydrometeorological reports [6], which describes a current state of weather in the affected year. Unfortunately, data of total precipitation are available from the year 2004, so the benchmark value is calculated from fata from six years (table 2). Also, the data are related to East, Central and West Slovakia region, which are sub-indicators of indicator total precipitation.

Table 2. Total precipitation [mm].

	2004	2005	2006	2007	2008	2009	2010	SUM Σ	AVERAGE
East Slovakia region	926	960	794	913	935	920	1276	6724	960,57
Central Slovakia region	946	1061	857	118	981	977	1415	6355	907,86
West Slovakia region	657	771	659	708	671	756	1036	5258	751,14

Table 2 represents data collection of total precipitation values from the years 2004 to 2010. The final column shows the average values, which serves as a benchmark.

Data of declaration number of 3rd degree of flood activity are available from the year 2001 to 2010, and the data are related to 8 self-governing regions of Slovak Republic (table 3).

Table 3. Declaration number of 3rd degree of flood activity.

	2001	2002	2003	2004	2005	2006	2007	2008	2009	2010	SUM Σ	AVERAGE
Bratislava self-governing		6	1			10			4	31	52	10,4
Trnava self-governing		4			3	12	3		20	65	107	17,83
Trenčín self-governing			1	1	1	4	1	4	16	38	66	8,25
Nitra self-governing		17	3		5	38	2		9	56	130	18,57
Žilina s elf-governing	7		2				13	3	23	181	229	38,16
Banská Bystrica self-governing	1	1	1		1		2		39	101	146	20,86
Prešov self-governing	39		6	6	1	13	20	9	46	438	578	64,22
Košice self-governing	16		12	9	13	28	3	55	13	324	473	52,55

Table 3 shows how many times was 3rd degree of flood activity declared in the years from 2001 to 2010.

Extent of flooded area is indicator which is divided into three sub-indicators – agricultural land, urban land and forest land. The data are also related to 8 self-governing regions of Slovak Republic, so the indicator of extent of flooded area contains 24 averages values. The data are defined in area measure hectare.

3.2. Vulnerability

Similarly, as the hazard index, vulnerability index contains three indicators – affected habitants, protected landscapes areas and conditions of flood protection structures. The affected habitants' indicator is divided into three sub-indicators – homeless habitants, evacuated and rescued habitants, injured habitants, deaden and missing habitants. During the study time period from the year 2001 to 2010, the affected habitants occurred annually. The most devastated flood occurred in the year 2010. The flood is considered as a the biggest in the past 200 years, as evidenced by flood consequences, affected habitants, flooded area and the flood damage. As an example, Table 4 shows the record of affected habitants in the year 2010.

Table 4. Affected habitants in the year 2010.

	Bratislava self-governing	Trnava self-governing	Trenčín self.governing	Nitra self-governing	Žilina self-governing	Banská Bystrica	Prešov self-governing	Košice self-governing
Affected habitants	161	1834	784	866	125	1978	4447	1225
Homeless habitants	4		55		4			12
Evacuated and rescued habitants	249	192	10	124	4	25	177	231
Injured habitants		1				5		
Deaden and missing habitantns	1							

For comparison, in the year 2005, there were 2 affected people in the Banská Bystrica self-governing region.

The value of the protected landscapes areas is calculated as a ratio between area of the protected landscapes area and the self-governing area, also in hectare. The value is then multiple with the weight of indicator. The sub-indicator condition of flood protection structures has not yet been assigned a value.

3.3. Exposure

The exposure index contains the most extensive indicators. The division is shown in table 5.

Table 5. Exposure indicators division.

EXPOSURE	
Flooded buildings	Residential houses
	Family houses
	Other habitable buildings
Flooded non-residential buildings	Industrial buildings and warehouses, tanks, and silos
	Farm buildings and warehouses, stables, and barns
	Cultural heritage
	Hospitals, medical and social facilities
	Other non-habitable buildings
Damage utilities	Railways, cableways, and other tracks
	Motorways, expressways, 1st class roads
	Road's 2nd and 3rd classes, sidewalks
	Local, special-purpose and forest transport communications
	Drainage channel, sewerage, culverts
	Water sources, water treatment plants, WWTPs
	Long-distance oil and gas pipelines
	Local gas distribution
	Long-distance and local water and steam distribution
	Long-distance and local electricity distribution

	Forest warehouses
	Other utilities
Affected animals	
Another flood damage	

The values, which belongs to the exposure indicators according to Reports [6], they aren´t described in the unit measure, so, they don´t need the benchmark. The values will be multiple with the weight of the exposure indicators. Table 5 serves as a view to the content of the indicator layer.

4. Conclusion

Submitted paper presents a brief overview to the proposed flood risk assessment methodology. The introduction is dedicated to the current state of the situation, and in the materials and methods section described the main background of the proposed methodology and the method of data acquisition. In the present time, flood risk assessment methodology is still in progress. The main aim of the methodology is to develop an easy and available concept to assess flood risk in the time period [7,8,9]. The essence of the methodology is in the data availability and austerity.

In the further studies, first, the benchmarks from all sub-indicators will be obtain. Second, the proposed methodology will be extended by scale of the flood risk, based on the available data from the year 2001 to 2010, and according to the mentioned scale, the flood risk occurred in the years 2011 and 2019 will be assess. The final methodology can serve to assess flood risk on the global level (Slovak republic), local level (self-governing region) or regional level (river catchment), according to the data availability.

5. References

[1] Zelenakova, M Preliminary flood risk assessment in the Hornad watershed. In: 5th International Conference on River Basin Management, 2009, pp.15-24

[2] Thai Pham B et al 2021 Flood risk assessment using hybrid artificial intelligence models integrated with multi-criteria decision analysis in Quang Nam Province, Vietnam *J Hydrology* **592** 125815

[3] Țîncu R et al 2020 Quantitative micro-scale flood risk assessment in a section of the Trotuș River, Romania *Land Use Policy* **95** 103881

[4] Shrestha B B 2020 Quantitative assessment of flood risk with evaluation of the effectiveness of dam operation for flood control: A case of the Bago River Basin of Myanmar *International Journal of Disaster Risk Reduction* **50** 101707

[5] Zhang D et al 2020 A GIS-based spatial multi-index model for flood risk assessment in the Yangtze River Basin, China *Environmental Impact Assessment Review* **83** 106397

[6] Ministry of the environment of Slovak Republic
Available online: https://www.minzp.sk/voda/ochrana-pred-povodnami/informacie/

[7] Zelenakova, M et al Flood occurrence evaluation based on the heights of water level in the rivers. In: 3rd International Conference Water Resources and Wetlands, 2016, pp 153-157

[8] Nagy, P et al Identification of dry and wet years in eastern Slovakia using indices. In: *Advances in Environmental Engineering (AEE 2019)*, 2020, 444: 012041

[9] Kubiak-Wojcicka, K et al Identification of Extreme Weather Events Using Meteorological and Hydrological Indicators in the Laborec River Catchment, Slovakia. *Water*, 2021, 10, 1413

Acknowledgments

This work was supported by project of the Ministry of Education of the Slovak Republic VEGA 1/0308/20 Mitigation of hydrological hazards, floods, and droughts by exploring extreme hydroclimatic phenomena in river basins and project HUSKROUA/1702/6.1/0072 Environmental Assessment for Natural Resources Revitalization in Solotvyno to prevent the further pollution of the Upper-Tisza Basin through the preparation of a complex monitoring system.

IOP Conf. Series: Materials Science and Engineering 1209 (2021) 012026 doi:10.1088/1757-899X/1209/1/012026

Hydrodynamic assessment of object on the sewer network in the city of Trnava

M Šutúš[1], G Rózsa[1], M Csóka[1], Š Stanko[1] and I Škultétyová[1]

[1] Department of Sanitary and Environmental Engineering, Faculty of Civil Engineering, Slovak University of Technology in Bratislava, Radlinského 11, 810 05 Bratislava, Slovakia

E-mail: marek.sutus@stuba.sk; gergely.rozsa@stuba.sk; marek.csoka@stuba.sk; stefan.stanko@stuba.sk; ivana.skultetyova@stuba.sk

Abstract. As in most Slovak cities, there is a combined sewer network in the city of Trnava. Therefore, combined sewer overflow (CSO) must be provided on these networks in order to mix rainwater and sewage water. The purpose of these CSO‘s is to reduce the amount of rainwater supplied to the treatment plant and to discharge part of the rainwater to the recipient. However, these waters contain various contaminants, including coarse solids, fine suspended solids, and solutes. Due to this, reduced water quality was recorded in the Trnávka recipient. This reduced quality is also due to the increased number of reliefs as prescribed by Government Regulation n.269/2010. The aim of the work is the analysis and selection of unsuitable relief chambers that do not meet the given limits. Hydrodynamic simulations are performed on these chambers, which demonstrate the need to reduce the amount of rainwater or reconstruction of CSO‘s.

1. Introduction

According to [1] we can observe elongating sections between dry and rainy periods. This change in the redistribution of precipitation over time is due to climate change. The IPCC confirmed an increase in the incidence of meteorological phenomena. Due to these climate changes, the functioning of sewerage systems is affected and thus the number of pluvial floods is increased.

Another problem is the high rate of urbanization due to high number of impermeable surfaces. The volume of water discharged by the sewer network is higher and the concentration time of runoff is shorter. During torrential rains, the network capacity may not be enough [2].

The problem of CSOs is global. The problem with the capacity of the combined sewer system is assessed by researchers in Italy, Norway and Spain by assessing CSOs as well as uniform sewer systems. They developed an assessment procedure using mathematical models [3,4,5].

The sewer network in the city of Trnava is combined, which means that rainwater and sewage water are drained together. In 2015, the total length of the sewer network was more than 111 km. 36 adjacent municipalities are also connected to the network. There are 22 CSOs in Trnava. According to [6], if the amount of CSOs is higher than 10, an expert assessment must be made as they pose an environmental risk.

In the §6, section 4 of [6], “the minimum dilution of municipal wastewater must reach a mixing ratio determined by a value of 1:4. In determining the mixing ratio, the current state of knowledge and economically sustainable possibilities of technical solutions for wastewater disposal are taken into account. When discharging wastewater from combined sewer overflows, the limit values of pollution indicators are not determined."

According to §6, section 5 of [6], if flow time through the sewer network to the CSO is longer or equal to 15 minutes, the maximum limit of overflows is 15 cases per year. If the flow time is less than 15 minutes, number of overflows may not exceed 20 cases per year.

In our case the flow time was calculated to be more than 15 minutes so the limit for the number of overflows is 15.

2. Calculations

To assess the current situation, we first had to determine the values of the mixing ratio and the number of reliefs. The mixing ratio was calculated using (1) and the number of reliefs using (2). Of the total number of CSOs 22, 9 were CSOs unsatisfactory.

$$n = \frac{Q_{odt}}{Q_{24}} - 1 \tag{1}$$

$$m_0 = \left[(1-\tau) * \left(Q_m * \frac{n_0}{Q_d}\right)^{0,83} * \left(1 + c_1 * \frac{log}{p}\right) + \tau\right]^{-3} \tag{2}$$

Table 1 provides a summary of unsatisfactory CSOs with values of mixing ratio and number of overflows.

Table 1. Summary of CSO.

	Mixing ration [-]	Number of overflows [-]
CSO-2	2,7	35,6
CSO-7	6,9	18,3
CSO-11	1,2	51
CSO-12	2,8	63,6
CSO-13	11,8	39,3
CSO-14	4,1	39,9
CSO-17	4,2	20,5
CSO-20	11,5	35
CSO-21	18,6	20,8

3. Simulations

A relatively expensive monitoring system would be needed to assess each CSO. However, we can use the method of computer modeling to determine the phenomena that take place in the CSO. It is effective in assessing the various load conditions that can occur. Simulations were performed on CSO -7 in the Ansys Fluent 2021 R1 program.

3.1 3D model

Dimension of CSO are (h x w x l) 3480 mm, 1850 mm and 2200 mm. The inflow 1 has diameter of DN 400 mm, inflow 2 diameter of DN 700/900 mm, outflow to wastewater treatment plant (WWTP) has diameter of DN 500 mm and overflow sewer has diameter of DN1000/650 mm.

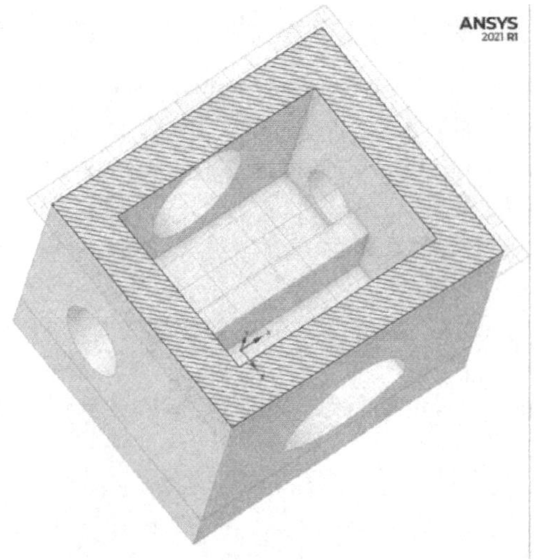

Figure 1. Model of CSO-7.

To meet the limits given by [6], it was economically most advantageous to increase the height of the overflow edge to meet the number of reliefs in CSO-7.

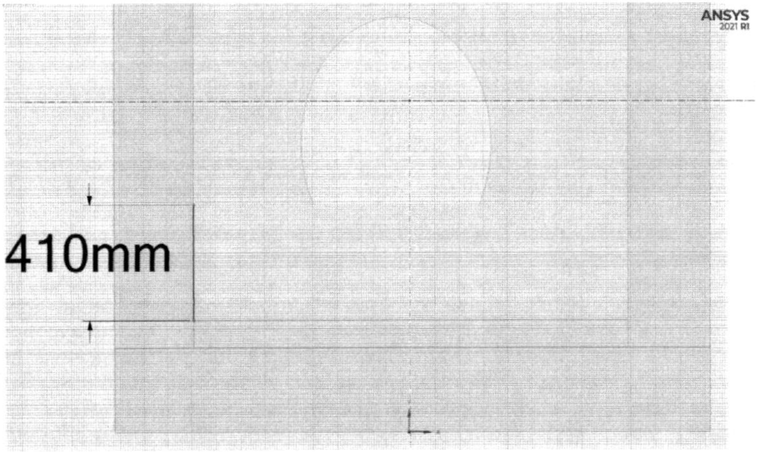

Figure 2. Original overflow edge.

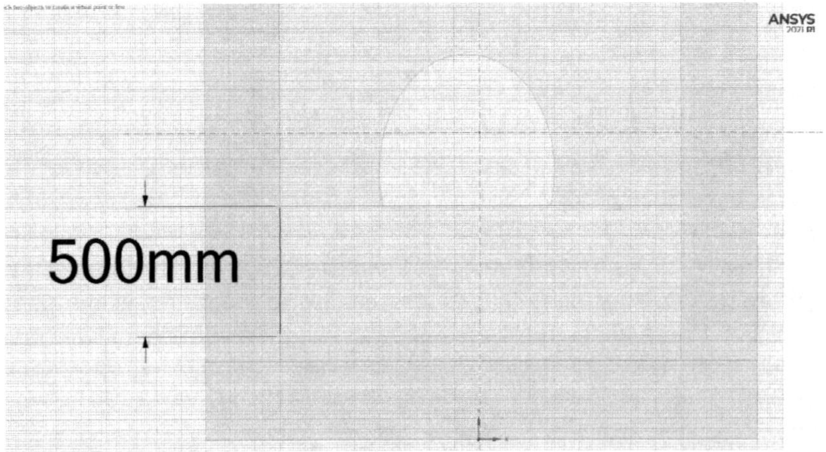

Figure 3. Increased overflow edge.

Calculation used the k-epsilon turbulent model and SIMPLE method. The value of 0,5 m.s^{-1}, which was measured, was chosen as the initial flow velocity. In the next scenario, a flow velocity of 2 m.s^{-1} was chosen, which should correspond to the flow velocity during rains. This value is only approximate as it is very dangerous to be in the CSO during torrential rains.

4. Results

Flow velocity was compared in simulations. Velocity was measured in the points with the same coordinates. The results are summarized in Table 2.

Table 2. Velocity comparison.

	Height of overflow edge											
	410 mm						500 mm					
	Velocity [m.s^{-1}]											
	2,0			0,5			2,0			0,5		
Time [min]	**WW TP**	**Edge**	**Out let**	**WW TP**	**Edge**	**Out let**	**WW TP**	**Edge**	**Out let**	**WW TP**	**Edge**	**Out let**
1	2,21	1,85	2,19	0,51	-	-	2,02	1,74	2,09	0,50	-	-
5	2,18	1,78	2,10	0,48	0,29	0,39	1,95	1,62	1,86	0,48	0,18	0,39
10	2,18	1,75	2,08	0,46	0,28	0,38	1,98	1,59	1,86	0,46	0,19	0,36
15	2,20	1,79	2,10	0,48	0,30	0,39	2,05	1,65	1,88	0,46	0,28	0,38

5. Conclusion

The flow velocity at the outflow to WWTP, the overflow edge and the overflow sewer were compared. The results in Table 2 show the deceleration of the flow at the overflow edge, which means that it is overflowed a smaller amount of water. The increase in the height of the overflow edge resulted in a decrease in the value of the number of reliefs from 18,3 to 14,3 which complies with the legislation. The value of the mixing ratio has increased from 6,9 to 8,4 which is positive in our case.

6. References

[1] Gdina M, Grajciar M, Chocholová L et al. 2017 The Seventh National Communication of the Slovak Republic on Climate Change. Under the United Nations Framework Convention on Climate Change and the Kyoto Protocol (Bratislava)

[2] Van Ootegem L, Elsy V, Van Herck K and Tom C 2015 Multivariate pluvial flood damage models. Environ. *Impact Assess. Rev* **54**

[3] Montserrat A, Bosch L, Kiser M A, Poch M and Corominas L 2015 Using data from monitoring combined sewer overflows to assess, improve, and maintain combined sewer systems *Science of The Total Environment* **505** 1053-1061

[4] Garofalo G, Giordano A, Piro P, Spezzano G and Vinci A A 2017 distributed real-time approach for mitigating CSO and flooding in urban drainage systems *Journal of Network and Computer Applications* **78** 30-42

[5] Hernes R R, Gragne A S, Abdalla E M H, Braskerud B C, Alfredsen K and Muthanna T M 2020 Assessing the effects of four SUDS scenarios on combined sewer overflows in Oslo, Norway: evaluating the low impact development module of the Mike Urban model *Hydrology Research* **51** 6 1437-54
[6] Government Regulation n. 269/2010 *Regulation of the Government of the Slovak Republic laying down requirements for achieving good water status, in Slovak*

Acknowledgments
This work is supported by the Scientific Grant Agency of the Ministry of Education, Youth and Sports of the Slovak Republic, and the Slovak Academy of Sciences within the project No. APVV-18-0203 and No.VEGA 1/0574/19.

IOP Conf. Series: Materials Science and Engineering 1209 (2021) 012027 doi:10.1088/1757-899X/1209/1/012027

Life-cycle assessment of a photovoltaic panel: Assessment of energy intensity of production and environmental impacts

M Vácha[1], J Kodymová[1] and V Lapčík[1]

[1] Faculty of Mining and Geology, VŠB – Technical University Ostrava, 17. listopadu, 708 00 Ostrava, Czech Republic

E-mail: martin.vacha.st@vsb.cz

Abstract. A number of articles have already been published on energy recovery from the sun using solar panels and their environmental impacts. However, in this article, we assess the impact of solar panel technology, and use separately obtained data based on the disassembly of a specific photovoltaic panel into discrete parts. The aim of this article is to list all the environmental impacts of this panel per unit of energy produced and at the same time to focus primarily on deciphering the energy intensity of individual phases of the life cycle of photovoltaic panel production. An analytical method of Life-cycle assessment using the environmental software version SimaPro 9.0.049 with an integrated Ecoinvent 3 database was used to determine the environmental impacts. Throughout the work, we focus on the data obtained, which shows that the process of photovoltaic panel production itself is very energy-intensive, especially in the phase of photovoltaic cell production and solar glass production. In other phases, which is the production of individual parts of the photovoltaic panel, its use, and subsequent recycling, they do not contribute so much to the overall energy balance. In the environmental impact assessment, the most affected aspects were human health, followed by climatic change, resources, and the ecosystem quality came last. In all four of the above categories, the influence of the photovoltaic cell production phase was determined to be dominant.

1. Introduction

The need to change the dependence of developed and developing countries on fossil fuels to generate electricity and replacing them with renewable sources is not a new idea. The effects of fossil fuels overuse are visible all around us and have been demonstrated in many studies [1], [2]. The EU ranks as a world leader in promoting the idea of a transition to renewable energy sources. It promotes its positions both politically (Environmental strategies and actions plans, Law) and through support for environmentally oriented development and research (Funding). Environmental impact assessment is also an integral part of research and development.

By 2038, when electricity generation in coal-fired power plants is to be discontinued in the Czech Republic (decided on 04.12.2020 by the Coal Commission of the Czech Republic), a total of 9,800 MW of installed electricity power should be shut down in the Czech Republic.

As early as 2030, renewables under the most economically advantageous scenario (Bloomberg NEF model) should account for about 12,000 MW of installed electrical power. Roughly half from this capacity should be provided by wind power plants and half by photovoltaics (PV). The output of coal-fired power plants would have already fallen substantially at this time, mainly due to the economic disadvantageousness [3].

It can therefore be stated that in order to meet the above-mentioned requirement for installed power plant capacity of 6,000 MW in the Czech Republic, it will be necessary to increase their installed capacity by about 3,820 MW_e (the current installed capacity of photovoltaic power plants in the Czech Republic is about 2,180 $MW_{e)}$. It is difficult to predict whether or not the above installed electrical power in photovoltaic power plants will be realized in the Czech Republic, but in any case, we should at least try. Of course, these should be mainly PV panels on the roofs of industrial buildings, housing buildings and community energy facilities. The construction of photovoltaic parks is currently undesirable.

According to data from [4] for 2020, solar energy is the third most widely used renewable resource, on global scale, in terms of installed capacities. Hydropower holds the first place, followed by wind power. For 2020, the amount of energy produced from the sun is reported at 844 TWh [4].

In general, photovoltaic (PV) systems connected to the network are the fastest-growing part of the energy market [5]. The solar panel and control unit are the main components of the representative PV system [6]. PV panels are made of various materials which are defined in three generations: silicone-based module (first generation), thin-film module (second generation), and nanomaterials (third generation) [7]; [8].

The "*Communication from the Commission to the Council and the European Parliament*" [9] shows that the Life-Cycle Assessment (LCA) given by ISO 14040:2006 [10] is the best choice for assessing potential environmental impacts. It is an analytical method evaluating the environmental impacts of products, technologies and services with regard to their entire life cycle, i.e., from cradle to grave (sourcing of raw materials, production of material, production of product, use, recycling or ecological disposal). Environmental impacts are assessed on the basis of the effects of energy and material flows [11].

The LCA analysis can be used to develop energy policy, develop, research and make decisions about energy technology costs and their impact on the environment [12]. Other studies assessed the life- cycle of the mono-silicon façade system and its ecological footprint, with a greenhouse gas emission rate of 2.81 years and 60.5 g of $CO_{2\,eq}$/kWh [13] followed by Hadi and Heidari (2021) [14] with their work on environmental and economic performance assessment of the photovoltaic façade consisting of three types of photovoltaic panels CdTe, DSS and Perovskite with the help of LCA, LCC analysis. The most common use of the LCA method is to assess the life-cycle of the product "cradle-to-grave" or "cradle-to-cradle" for environmental impact assessment in all stages of the product life cycle, from extraction of raw material to use [15]. Another example of the use of LCA analysis is the work involved in environmental impact assessment systems using PV panels in combination with solar water heat collectors [16]; [17], etc.

The aim of this work is to examine the environmental impacts during the production, use and recycling of a particular poly-Si panel in the OFF-GRID wiring and its energy performance throughout the process and to determine the energy payback time for a potential user or investor using LCA.

2. Methodology

PV panels are used to generate electricity from solar energy and operate on the principle of photovoltaic effect [18]. For the purpose of the research, we used the solar panel Victron Energy 115 Wp/12V, see the characteristics in table 1.

This panel is connected in the so-called island connection, i.e., with the accumulation of electricity [19]. This panel was used for energy production in the period 1st January 2020 up to 1st January 2021. The measured data were then recalculated to the lifetime of the panel stated by the manufacturer for 25 years with a final value of 1.998347 MWh. The calculation of the final value produced by the PV panel included losses reported by the panel manufacturer, which are set at 10% in the first 10 years of operation and 20% in the period of 25 years of operation, were included in the calculation of the resultant value of the sell-out and not the EM panel. These figures were averaged and in the first 10 years losses were attributed to 1% per year and subsequent 15 years at 0.66% per year to reach the resulting 20% over 25 years of operation. Performance degradation related to Potential Induced Degradation (DIP) [19] was not calculated in this study.

Table 1. Characteristics of 115 Wp Poly-Si PV module.

Item	Description
Dimensions	1015 x 668 x 30 mm
Weight	8.0 kg
Number of cells per module	36
Electrical performance guarantee	10 years 90% + 25 years 80% performance
Maximum power (P_{max})	115 Wp
Tolerance (Tol)	0~+3%

To check the amount of energy obtained, the measured data were compared with the data obtained in Photovoltaic Geographical Information System (PVGIS) [20]. All positional data (place installed, world side of the installation, slope, shielding, spectral effects, temperature, irradiation intensity), data on the technology of the PV panel used, and maintenance was entered into the system. The PVGIS - ERA5 database, which is intended for Europe with the European Center for Medium-Range Weather Forecasts (ECMWF) evaluation algorithm, was used for the calculation. Losses caused by other components and the system (regulator, wiring, batteries) were not included in the calculation. The predicted value of the energy produced was 15.69% higher in the period.

The environmental software version SimaPro 9.0.049 with an integrated Ecoinvent 3 database was used for the environmental impact assessment (LCA) and energy performance assessment of the PV panel (LCI). The SimaPro procedure is based on LCA legislation [10] and divides assessment into four main stages.

2.1. Goal and Scope

The aim of the study is to assess the life cycle of the and determine the energy performance of the LCI component Poly-Si PV panel Victron Energy 115 Wp/12 V from the extraction of raw materials, their processing, transport to the production plant, the actual production of individual components, subsequent transport to the production plant of PV panels, the production process of PV panels, the transport of PV panels to the place of installation, their use and final recycling. Furthermore, the aim is to determine the environmental impacts of the entire chain of events from raw material extraction to the recycling of the PV panel.

The work can also serve as a source of information for potential investors and for mapping the product system and clarifying the possibilities of innovation in the process.

For the calculation of LCA, only materials for the production of the PV panel are counted. Materials for the production of the controller, cabling, the battery are not taken into account in this work.

2.2. System boundary

The boundaries of the system are set from cradle to grave and relate only to materials used for the production of PV panels. Other system components were not taken into account.

2.3. Life cycle inventory

Data for the PV panel for working with SimaPro are given in Table 2 and were obtained in cooperation with REsolar s.r.o., Asekol a.s. dealing with recycling of PV in the Czech Republic and information sheets of the manufacturer of the PV panel Victron Energy. The data below were fed into Simapro software, which makes it easier to examine the modeling and any analysis of demanding life cycles, estimating the environmental impact of the individual components of the PV panel over their lifetime [21]; [22]. For processes with thermal and energy demands, due to ignorance of accurate data, the medium values of the Ecoinvent 3 database were used. Also, for determining transport distances, mean values were introduced (20000 km). Given the size of China, where the panel was made, the exact place of production of the panel is not known.

Table 2. Input material.

Material	
Aluminum	0.75 kg
Solar glass	5.92 kg
Silicone	0.21 kg
EVA	0.60 kg
Tedlar	7 g
Copper/ wire drawing copper	35 g/ 52 g
Silver	12 g
Lead	3.3 g
Plastic	0.405 kg
Tin	5 g
Paper	1.08 g

2.4. Life cycle assessment

The environmental impact assessment was evaluated using LCA analysis SimaPro impact 2002+ V2.15 with a focus on CED (Cumulative Energy Demand) version 1.11 and EPBT (Energy PayBack Time) method. CED is a method estimating the direct and indirect amount of life cycle energy from production to disposal using coinvent version 2.0 data, which is based on higher heating values (HHV). EPBT is an indicator for evaluating the energy production system [23] needed during the life cycle of the LF panel in time. EPBT would be calculated according to formula (1).

$$EPBT = EP_{25}/CED \quad (1)$$

Where EP_{25} (Energy Produced) is the energy produced by the PV panel over 25 years of guaranteed panel life (measured value) and CED is the total amount of energy used to produce, operate and recycle the FV panel (SimaPro). RPUT (Return Per Unit of Time) [year] was calculated using a formula (2).

$$RPUT = SL/EPBT \quad (2)$$

Where SL (Service Life) is a time use after that, which the manufacturer guarantees the functionality and performance of the PV panel.

3. Results and Discussion

According to the displayed results, see Figure 1, it is clear that the greatest burden is directed at human health, followed by climate change, resources, and last but not least, the quality of the ecosystem. In all these cases, was the biggest impact on the negative environmental impact the production of photovoltaic cells (human health 35.00%, ecosystem quality 4.68%, climate change 19.90%, resources 19.01%). The CED method was used to assess the energy performance of the PV panel - see Figure 2, from which it is clear that the energy production needed to produce the panel in full 73.90% still comes from non-renewable fossil fuel sources. The most energy-intensive phase is the production of photovoltaic cells. The total energy needed for the production, use, and recycling of the PV panel was set at 0.650 MWh. When comparing the energy produced with the energy needed for production, use and ecological disposal, we conclude that the panel's return is 8.13 years.

The calculated energy payback time (EPBT) of 8.13 years in a location in the Czech Republic is quite long compared to other studies. For example, V. Frhenakis et al. (2011) [24] indicates the return for

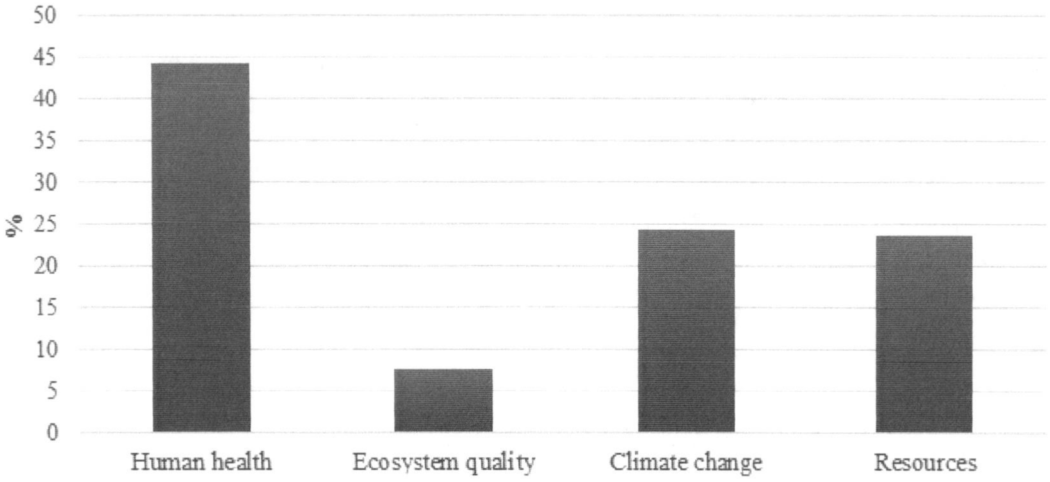

Figure 1. Environmental impact assessment.

Poly-SI solar panel 2.5 years for southern European regions and 3.5 years for areas in the Netherlands. More Studies of Alsema E.A. (2000) [25] and Jungbluth N. (2005) [26] agree on a return of 3 to 6 years on Poly-Si solar panels depending on the efficiency of modules in the Swiss region. With a result for Poly-Si cells, the Palanov N. study (2014) [27] ranks with a 2.5-year return in Denmark. Studies show differences, mainly depending on the amount of solar radiation. But it is not the only reason for

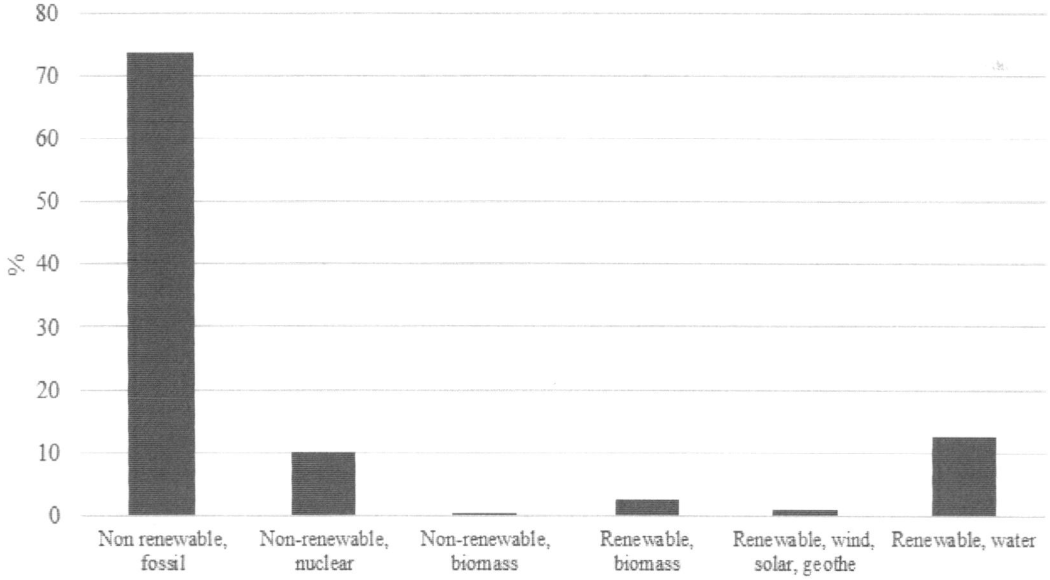

Figure 2. Assessment of the energy performance of the PV panel.

the different outcomes. Another important role in the different efficiency of PV panels is played by the production technology and materials used in the production of PV cells [28]. Among the current technologies that increase efficiency and reduce energy and financial costs are considered passivated emitter and rear cell (PERC) described in more detail in the study by Jia X. et al. (2021) [29] or Lunardi

M. M. et al. (2018) [30] and the use of Perovskite as a stand-alone or complementary material to silicon [31]. Another process that has a significant impact on energy return and environmental protection and has a significant impact only now and in the near future, when the life of the first generations of PV panels to a greater extent, the recycling of photovoltaic modules is described in many studies, e.g. Heath GA et al. (2020) [32] or Dias P. et al. (2021) [33]. We also see a big difference in the overall involvement of the entire solar system. If it is a system on the rooftop [34] or portable charger [35] and also in the chosen LCI methodology in this work, as mentioned above, the Impact2002 + methodology is used. Work Espinosa and Krebs (2014) [36] uses a methodology Recipe 2008 endpoint level, another in his work uses a methodology 2001 IPCC (Intergovernmental Panel on Climate Change) [37] or ILCD (International Reference Life Cycle Data System) which appears in the work of Maranghi et al. (2019) [38].

4. Conclusions

In the current state of knowledge, there is a lot to improve in solar technologies. Solar energy as renewable energy is highly sought, but it has its shortcomings and we need to continue to improve this technology by using new technical processes both in production and in ecological disposal, which only after many years is coming to the forefront and becoming an important part of the life cycle assessment.

Also, the environmental impacts are still significant and it is necessary to focus primarily on the production of photovoltaic cells, which is still the largest burden in all four evaluated categories (Figure 1) and is also the most energy-intensive part of PV panel production.

The study also shows that the predicted data for energy production are significantly overestimated to real measured values. This discrepancy may have arisen from a very different course of weather and the amount of sunshine, especially in the summer months, when excessive precipitation and days without direct sun were recorded over the measured period of time.

The next important step should be to try to extend the life of the entire PV System, which would have a positive contribution on the energy balance of PV panels.

At present, another adverse effect on the production of PV panels resulting from COVID epidemic worldwide is being shown. Due to the fourfold increase in the cost of feedstock for the production of silicon cells after the previous decade of decline, the prices of solar modules are rising sharply, which may result in a slowdown in the use of solar energy (BloombergNEF source).

5. Reference

[1] P. RK, M. LA and (eds.) 2014"Climate Change 2014: Synthesis Report " *Intergovernmental Panel on Climate Change* p 151

[2] Bruckner T, Bashmakov IA and Mulugetta Y 2014 Energy Systems, *Intergovernmental Panel on Climate Change,* vol. chapter **7**, no. 511-97

[3] Gavriněv V 2020 "Konec uhlí může přijít brzy, dle vědců dřív, než si dokážete představit," Seznam

[4] Our World in Data - Modern renewable energy generation by source, Word, [Accessed 15 5 2021Online]. Available: https://ourworldindata.org/renewable-energy

[5] Rüther R and Zilles RJEp 2011 "Making the Case for Grid-Connected Photovoltaics in Brazil," *Energy Policy,* no. vol **39**, pp. 1027-30

[6] Hou G, Sun H, Jiang Z, Pan Z, Wang Y, Zhang X, Zhao Y and Yao QJAE 2016 Life Cycle Assessment of Grid-Connected Photovoltaic Power Generation from Crystalline Silicon Solar Modules in China," Applied Energy, no. vol. 169, pp. 882-90

[7] Liu X, Hoekman S, Robbins C and Ross P 2015 Lifecycle climate impacts and economic performance of commercial-scale solar PV systems: A study of PV systems at Nevada's Desert Research Institute (DRI), *Solar Energy,* no. vol. **119**, pp. 561-72

[8] Allouhi A, Saadani R, Buker MS, Kousksou T, Jamil A and Rahmoune M 2019 Energetic, economic and environmental (3E) analyses and LCOE estimation of three technologies of PV grid-connected systems under different climates, *Solar Energy,* no. vol. **178**, pp. 25-36

[9] Communication from the Commission to the Council and the European Parliament, Brussels: Commission of the Eropean Communities, 2003

[10] ISO 14040:2006 Environmental management Life cycle assessment, Principles and framework, 2006

[11] Kočí V, Posuzování životního cyklu, Chrudim: Ekomonitor, 2009, ISBN: 978-80-86832-42-5

[12] Fthenakis VM and Kim HC 2011.Photovoltaics: Life-cycle analyses, *Solar Energy,* vol. **85**, no. Issue **8**, pp. 1609-28

[13] Perez MJR and Fthenakis V 2011 Lifecycle Assessmant of Façade BIPV in New York, *37th IEEE Photovoltaic Specialists Conference,* pp. 3271-76

[14] Hadi M and Heidari A 2021 Development of an integrated tool based on life cycle assessment, Levelized energy, and life cycle cost analysis to choose sustainable Facede Integrated Photovoltaic Systems, *Journal of Cleaner Production 293,* vol. **126117**

[15] Muralikrishna IV and Manickam V 2017 Chapter Five - – Life Cycle Assessment, Environmental Management, *Science and Engineering for Industry,* pp. 57-75

[16] Mahmud MAP, Huda N, Farjana SH and Lang C 2018 Environmental Impacts of Solar-Photovoltaic and, *Energies,* no. vol. 11(9), p. **2346**

[17] Ardente F, Beccali G, Cellura M and Brano VL 2005 Life cycle assessment of a solar thermal collector, *Renewable Energy,* no. vol. **30**, pp. 1031-54

[18] Haselhuhn R and Maule P 2017 Fotovoltaické systémy, Plzeň: Česká fotovoltaická asiciace, ISBN 978-80-906281-5-1

[19] Haselhuhn R, Fotovoltaika - Budovy jako zdroj proudu, Ostrava: HEL, 2010, ISBN 978-80 86167-33-6 ,pp. 100-3

[20] Tajbar S, Rafiq L, Bibi S and Saidullah M 2020 Photovoltaic geographical information system module for the estimation of solar electricity generation: A comparative study in Khyber Pakhtunkhwa Pakistan, *Journal of Energy Systems,* no. vol. **4**, pp. 12-21

[21] Starostka-Patyk M 2015 New Products Design Decision Making Support by SimaPro Software on the Base of Defective Products Management, *Procedia Computer Science,* no. vol. **65**, pp. 1066-74

[22] Iswara AP, Farahdiba AU, Nadhifatin EN, Pirede F, Andhikaputra G, Muflihah I and Boedisantoso R 2020 A Comparative Study of Life Cycle Impact Assessment using Different Software Programs, *IOP Conference Series:Earth and Environmental Science 506. IOP Publishing **12002***

[23] Parisi ML, Maranghi S, Vesce L, Sinicropi A, Di Arlo A and Basosi R 2020 Prospective life cycle assessment of third-generation photovoltaics at the pre-industrial scale: A long-term scenario approach, *Renewable and Sustainable Energy Reviews,* no. vol. **121**, p. **109703**

[24] Frischknecht R, Kim HC, Frischknecht R, Raugei M and Sinha P and Sticki M 2011 Life Cycle Inventories and Life Cycle Assessments, *International Energy Agency Photovoltaic Power System Programme Task 12,* no. Report **T12-02**:2011, p.5

[25] Alsema EA 2000 Energy Payback Time and CO2 Emissions of PV Systems, *Progress in Photovoltaics Research and Applications* ***8****,* pp. 17-25

[26] Jungbluth N 2005 Life cycle assessment of crystalline photovoltaics in the Swiss ecoinvent database, *Progress in Photovoltaic: Research and Applications **13(8)**,* pp. 429-46

[27] Palanov N 2014 "Life-cycle assessment of Photovaltaic systems : analysis of environmental impact from the production of PV system including solar panels produced by Gaia Solar," pp. 77-78

[28] Haselhuhn R and Maule P 2017 Fotovoltaické systémy, Plzeň: Česká fotovoltaická asociace, ISBN 978-80-906281-5-1

[29] Jia X, Zhou Ch, Tang Y and Wang W 2021 Life cycle assessment on PERC solar modules, *Solar Energy Materials and Solar Cells,* vol. article number **111112**

[30] Lunardi MM, Alvarez-Gaitan JP, Chang NL and Corkish R 2018 Life cycle assessment on PERC solar modules, *Solar Energy Materials and Solar Cells,* no. vol. **187**, pp. 154-59

[31] Lunardi MM, Ho-Baillie AWY, Alvarez-Gaitan JP, Moore S and Corkish R 2017 A life cycle assessment of perovskite/silicon tandem solar cells, *Progress in Photovoltaics* ***25(8)***, pp. 679-95

[32] Heath GA, Silverman TJ, Kempe M, Deceglie M, Ravikumar D, Remo T, Cui H, Sinha P, Sinha P, Libby C and Shaw S et al. 2020 Research and development priorities for silicon photovoltaic module recycling to support a circular economy, *Nature Energy volume* ***5***, pp. 502-10

[33] Dias P, Schmidt L, Lunardi MM, Chang NL, Spier G, Corkish R and Veit H 2021 Comprehensive recycling of silicon photovoltaic modules incorporating organic solvent delamination–technical, environmental and economic analyses, *Resources, Conservation and Recycling,* p. **105241**

[34] Parisi ML, Maranghi S and Basosi R 2014 The evolution of the dye sensitized solar cells from Grätzel prototype to up-scaled solar applications: A life cycle assessment approach, *Renewable and Sustainable Energy Reviews,* no. vol. **39**, pp. 124-38

[35] Tsang MP, Sonnemann GW and Bassani DM 2016 Life-cycle assessment of cradle-to-grave opportunities and environmental impacts of organic photovoltaic solar panels compared to conventional technologies, *Solar Energy Materials and Solar Cells,* vol. **156**, pp. 37-48

[36] Espinosa N and Krebs FC 2014 Life cycle analysis of organic tandem solar cells: When are they warranted?, *Solar Energy Materials and Solar Cells,* no. vol. **12**, part **B**, pp. 692-700

[37] Azzopardi B and Mutale J 2010 Life cycle analysis for future photovoltaic systems using hybrid solar cells, *Renewable and Sustainable Energy Reviews,* no. vol. **13**, Issue **3**, pp. 1130-34

[38] Maranghi S, Parisi ML, Basosi R and Sinicropi A 2019 Environmental Profile of the Manufacturing Process of Perovskite Photovoltaics: Harmonization of Life Cycle Assessment Studies, *Energies* ***12 (19)***, p. 3746

[39] Florides M, Makrides G and Georghiou GE 2021 Electrical and Temperature Behavior of the Forward DC Resistance With Potential Induced Degradation of the Shunting Type in Crystalline Silicon Photovoltaic Cells and Modules, IEEE Journal of Photovoltaics, vol. 11, pp. 16-25

IOP Conf. Series: Materials Science and Engineering 1209 (2021) 012028 doi:10.1088/1757-899X/1209/1/012028

Development of design rainfalls: A review

R Wittmanová[1], I Marko[1], M Šutúš[1] and Š Stanko[1]

[1] Department of Sanitary and Environmental Engineering, Faculty of Civil Engineering, Slovak University of Technology in Bratislava, Radlinského 11, 810 05 Bratislava, Slovakia

E-mail: reka.wittmanova@stuba.sk; ivana.marko@stuba.sk, marek.sutus@stuba.sk, stefan.stanko@stuba.sk

Abstract. Design rainfalls are essential parameters in designing and assessing combined and stormwater sewers and are still used today. They create idealized intensity profiles with statistically determined parameters. Their gradual development is based on the development of computer technology, measuring instruments, and measurement campaigns' progress. However, climate change is causing changes in precipitation events' characteristics, so it is important to focus on reassessing them and adapting to current conditions. The paper summarizes the knowledge about the development of design rains to reveal their potential for the future.

1. Introduction

Design rainfalls are one of the basic parameters in designing and assessing combined and stormwater drainage networks [1], as they represent calculated volumes of stormwater. They are defined as idealized rainfall intensity profiles with a statistically determined periodicity "p" or a return period "T" [2; 3]. The periodicity of the design rainfall expresses the probability of a given rain event during the year [4]. Therefore, the stormwater drainage system and combined sewer network design are based on rainfall and periodicity, which is determined based on the type of site or according to the requirements of the relevant bodies of the state water administration. Recommended design periodicities are given in the standard in table 4 of STN EN 75 6101 [5]. The most important properties of the design rainfalls are the value of the maximum rain intensity and its position within the duration of the storm event, the duration of the rain event, the intensity of precipitation, and the time structure of the rain [3; 6]. In design and assessment, design rainfalls are used as:

- Individual model rains:
  - a) Block rainfall.
  - b) Synthetic rainfalls.
  - c) Individual historical rainfalls.
- Time series of historical rains:
  - a) A series of selected most important historical rainfalls.
  - b) Annual time series of histograms of historical rainfalls.
  - c) Synthetic rainfall time series [2].

The benefits of design rainfalls have been briefly summarized by a study by Balbastre-Soldevil et al. (2019). Design rainfalls guarantee a constant level of quality and operational criteria (operational level) in different urban river basins and simplification of hydrological-hydraulic calculations compared to methods using a series of historical rains. Their advantage is the possibility of design and assessment in areas with no long-term measurement campaign. Thus, there are no relevant inputs to the calculations [3].

The reliability of mathematical simulations is greatly influenced by the limitations of design rainfalls and their weaknesses. It is often assumed that the periodicity of a rain event is equal to the periodicity of the calculated volume of rainwater produced by the storm event. However, this argument is hampered by local conditions. In the case of a territory with a high degree of occupancy and representation of impervious areas, the previous statement is to some extent true. However, in areas with less urbanization, conditions such as soil saturation and runoff losses must also be considered. Another disadvantage is that the design rains are based on IDF curves derived from different types of rain events. The individual elements of the sewer systems are sensitive to the unique characteristics of the design rainfall. The results of some analyses are affected by design rain errors that are not adapted to the problem. An example is identical design rains to determine peak flows in a sparsely populated area network and an area with a high degree of urbanization [7].

2. Evolution of design rainfalls

Design rains are still used in the design of sewer networks. Several methods of creating design rainfalls are available in the literature. Veneziano and Villani (1999) summarized these methods in the following categories:

- determination of the simple geometric shape of the design rainfall from one point of the IDF curve,
- design rainfall derived from IDF curves,
- design rainfalls, the time distribution derived based on actual rain events, not only using information from IDF curves [3; 7].
- simulations from stochastic models [8; 9].

2.1. IDF curves

IDF curves are the basis for methods for determining design rains. "Intensity-duration-frequency curves," i.e., IDF curves, are defined as a mathematical relationship between the intensity, duration, and repetition time of a rain event [10]. Their derivation dates to the early 1930s and is associated with using a rational method of calculating design flows in the design of sewer networks. Gradually, this method spread throughout the world [11; 12]. As a result, IDF curves are widespread in sanitary engineering and hydraulic objects, e.g., sewers, retention tanks, and combined sewer overflow chambers [13]. The IDF curves represent a graphical representation of the probability of the intensity of fallen precipitation over some time [10]. To derive the IDF curves, follow these steps, which have been summarized in several studies:

- Determination of extreme rainfall intensities for a specific period by analysing maximum annual intensities.
- Application of theoretical distribution functions to time series of extreme precipitation intensities for all durations (Gumbel distribution, Pearson distribution type III, GEV distribution - Generalized Extreme Value).
- Calculation of rain intensity for all durations and repetition times based on selected distribution functions.
- Construction of IDF curves according to empirical formulas using regression [12].

The availability of rainfall data dramatically influences the choice of method for determining IDF curves. To derive IDF curves for specific regions in Africa, Awadallah and Awadallah (2013) used satellite data from the Tropical Rainfall Measuring Mission to determine the relationships between maximum daily and monthly intensities [14]. Ombadi et al. (2018) developed methodologies for transforming satellite surface collisions into point collisions, which they then used to generate IDF curves [15]. In their study, Lutz et al. (2020) focus on the introduction of GEV-division into the method of determining IDF curves for Oslo city conditions [16].

Efforts are currently underway to update outdated IDF curves around the world. Given the change in the characteristics of rain events caused by climate change, it is necessary to reassess these curves or consider future changes [17]. Climate change is causing an increase in the intensity of precipitation events, and changes in their periodicity are also noticeable. These facts call for an update of the IDF curves under different climate scenarios [18].

2.2. Design rainfalls

The oldest and best-known design rainfalls derived from IDF curves include the "Chicago Design Storm" or the Keifer and Chu model [19], the block rainfall (referred to in the literature as the Unit Hydrograph method) [11], or the "Alternative Blocks Method" [20]. The design rain designed by Šifald (1973) also has a simple course, the time course of which is divided into three parts. The middle part has a rectangular shape, and the outer parts taper linearly towards the ends [9]. The design rainfalls Yena and Chowa (1980) [21] and Desbordesa (1978) have a triangular course of the hyetogram. They are suitable for the design of sewer networks for small urban catchments [8].

The second group of design rainfalls is those that are derived from actual rain events. A pioneer in this area was researcher Huff (1967), who analysed and processed up to 261 rain events in Illinois. It categorized rain into four groups based on which quarter of the precipitation event occurs the most substantial precipitation [9; 7].

Huff's method was applied in the creation of design rains for the territory of Canada. In his study, Hogg focused on the variability of the time distribution of rain in different parts of the country over 1 to 12 hours. The results were applied in 35 areas of the country [22].

In the United Kingdom, rainfall information for any area was obtained from map data related to the Wallingford Procedure. Together with the Flood Studies Report, which was based on analysis of a wide range of rain events, they were used as the main design tools. Subsequently, the Flood Estimation Handbook was published in 1999, replacing the previous publication [23].

Pilgrim and Cordery proposed the Average Variability Method (AVM) based on a detailed analysis of the structure of a series of selected historical rainfalls. AVM is mainly used in Australia [24; 3]. For the Mediterranean, García-Bartual and Andrés-Doménech (2017) developed a design rain defined by a two-parameter gamma function (G2P). The design rain parameters correspond to the extreme hydrological conditions of the given area. The design rain of G2P is characterized as a sharp increase in rain intensity at the beginning, followed by a gradual decrease in intensity until the end of the rain event [25].

Research on design rains is still ongoing. Balbastre-Soldevila et al. compared 11 known design rainfalls in the conditions of the coastal areas of Spain with an emphasis on the maximum intensity of precipitation, the total cumulative depth of precipitation, and the time course of the rain event [64]. The study by Pan et al. focuses on comparing design rains for an urbanized area of Guangzhou (China) [26].

2.2.1. Model rain of constant intensity - block rain

The simplest design rain used to design sewer networks is block rain. A constant intensity characterizes it throughout the collision event. It is determined by the duration "t," the periodicity "p," and the constant yield "q" or the intensity "i." It defines the most intense part of the rainfall. Due to its course, it does not correspond to natural rains [4]. The calculation of block rain yields was adapted to the conditions in Slovakia based on data on block rains processed by Šamaj and Valovič (1973). The equations for calculating yields were published by Urcikán and Horváth (1979). The calculation method is based on rain gauge data from 68 measurement stations [2].

2.2.2. Synthetic model rains with varying intensity

They belong to the group of model rainfalls, which is close to a natural rain event. Synthetic rains have a variable intensity due to their time distribution. They are more suitable for more detailed calculations, which do not neglect the formation of surface runoff and assimilate non-stationary flow in the inlets. Since the intensity profile is as close as possible to the real precipitation, it has an ascending part, where the intensity of rainfall increases. A central part with maximum intensity and a descending part represents the rain event's gradual end [2]. The most used synthetic rainfalls are:

- Triangular synthetic rainfall: It is the design rain by the triangular time distribution of a rainfall event. It was initially designed for small catchments and sewer networks but had many practical uses [2]. The position of the maximum rain intensity is located in the first third of the time distribution of the rain event. The total depth of rainfall is determined as a multiple of the duration of the rain and the average intensity. American scientists Yen and Show designed this synthetic rain, and their study was published in 1980. The study analysed more than 7,000 precipitation events in three localities [22]. The modification of the triangular design rain was dealt with in Watt et al. (1986), which was developed according to Canadian rain gauge data. The proposed design has 60 minutes, and its parameters were evaluated for 45 rainfall measurement stations in Canada [27].
- Šifald's synthetic rainfall: is the most suitable design rainfall for use in our conditions. It results from the processing of measured data in the area of former Czechoslovakia. The course of this rain and its total volume are divided into three phases. The maximum intensity is in the second phase, located in about the second quarter of the time distribution. The duration of the maximum intensity is 25% of the duration of the entire rain event and represents up to 56% of the total volume of the original block rain. The intensity of rain in the first phase - the first quarter of the duration of the rain has an increasing character and occupies 14% of the volume of block rain. The last phase has a declining character and represents 30% of the volume of block rain [3].
- Keifer and Chu synthetic rainfall: It is a synthetic histogram based on the total rainfall curve of block rainfalls. This hyetogram has the same periodicity as the total hip rain yield curve used to determine it. The study was conducted for the city of Chicago [2]. The model is based on IDF curves, so it is assumed that the amount of precipitation at maximum intensity and persistence t_d is the same as the amount of precipitation at the IDF curve [28]. The method was intended for use in the design and assessment of sewer networks with a duration of design rain of up to 3 hours [22]. The precipitation intensity varies throughout the precipitation, so the method may also be called the instantaneous intensity model. The method also includes equations for determining the time position of the maximum intensity and the redistribution of intensities before and after this value - determination of the ascending and descending curve [28]. In order to be able to use the Keifer and Chu model also in the conditions of Slovakia, these equations were modified using the block rain yield equation. The primary condition for determining these equations is the equality of the rainfall of synthetic rain and the rainfall of the block rain at the same time "t" [4].
- Huff's synthetic rainfall: The Huff (1957) design rainfall determination method was developed to observe extreme precipitation events in the United States' Midwestern region and describe the time distribution of precipitation. It is used to describe the hyetogram and to predict the outflow of river basins [26]. The Huff (1957) design rainfall determination method was developed to observe extreme precipitation events in the United States' Midwestern region and describe the time distribution of precipitation. It is used to describe the hyetogram and to predict the outflow of river basins [26]. The results of the observations processed 261 storm events lasting from 3 to 48 hours [22]. According to the time distribution, Huff classified the maximum intensity of rain, which can be in the first, second, third, or fourth quartile [29]. He presented a mean (50%) hyetogram for each quartile. He further proposed curves that represent a different range of 10-90% [30]. The advantage of this method is its usability in any territory. An excellent example of the versatility of use is the Pan et al. study. (2017), where they compare the course of several design rainfalls in the conditions of the Guangzhou city basin [26].

3. Conclusion

The aim of the article is to summarize the knowledge about the development of design rains to reveal the potential of their use in the future due to climate change. Design rains are one of the basic parameters in designing and assessing combined sewer systems and are still used today. They create idealized intensity profiles with statistically determined parameters. Their gradual development was based on the development of computer technology, measuring instruments, and the progress of measurement campaigns. Climate change causes changes in the characteristics of precipitation events, so it is essential to focus on their reassessment and adaptation to current conditions.

Many studies deal with the adaptation and transformation of design rains, considering the changes caused by global warming and considering different local conditions. Design rains and IDF curves will always be part of sewer networks' calculation and design procedure, so it is essential to update them and adapt them to current conditions.

4. References

[1] Liu Y, Engel B A, Collingsworth P D and Pijanowski B C 2017 Optimal implementation of green infrastructure practices to minimize influences of land use change and climate change on hydrology and water quality: Case study in Spy Run Creek watershed, Indiana *Sci. Total Environ.* **601-2** 1400-1411

[2] Urcikán P and Rusnák D 2004 *Stokovanie a čistenie odpadových vôd. Stokovanie I. Navrhovanie stokových sietí.* (Bratislava: Slovenská technická univerzita v Bratislave) p 324 ISBN 80-227 2136-0

[3] Denault C, Millar R G and Lence B J 2006 Assessment of possible impacts of climate change in an urban catchment *JAWRA* **42** 3 685-97

[4] Peleg N, Blumensaat F, Molnar P, Fatichi S and Burlando P 2017 Partitioning the impact of spatial and climatological rainfall variability in urban drainage modeling *Hydrol. Earth Syst. Sci.* **21** 3 1559-72

[5] STN EN 13508-1 Investigation and assesment of drain and sewer systems outside buildings. Part 1: General Requirements.

[6] Grum M, Jorgensen A T, Johansen R M and Linde J J 2006 The effect of climate change on urban drainage: an evaluation based on regional climate model simulations *Water Sci. Technol.* **54** (6-7) 9-15

[7] Berggren K, Olofsson M, Viklander M, Svensson G and Gustafsson A M 2011 Hydraulic Impacts on Urban Drainage Systems due to Changes in Rainfall Caused by Climatic Change *J. Hydrol. Eng.* **17** 1 92-8

[8] Sztruhár D and Sokáč M 2005 *Hydrológia urbanizovaných území.* (Bratislava: Slovenská technická univerzita v Bratislave)

[9] Kourtis I M and Tsihrintzis V A 2021 Adaptation of urban drainage networks to climate change: A review *Sci. Total Environ.* **771** 145431

[10] Pan C, Wang X, Liu L, Huang H and Wang D 2017 Improvement to the Huff Curve for Design Storms and Urban Flooding Simulations in Guangzhou, China *Water* **9** 6 411

[11] Gong Y, Liang X, Li X, Li J, Fang X and Song, R 2016 Influence of Rainfall Characteristics on Total Suspended Solids in Urban Runoff: A Case Study in Beijing, China *Water* **8** 7 278

[12] Rusnák D 2018 *Významnosť okrajových podmienok hydraulického posudzovania stôk* (Bratislava: SPEKTRUM STU)

[13] Watt W E, Chow K C, Hogg W D and Lathem K W 1986 A 1-h urban design storm for Canada *Canadian Journal of Civil Engineering* **13** 3 293-300

[14] Na W and Yoo C 2018 Evaluation of Rainfall Temporal Distribution Models with Annual Maximum Rainfall Events in Seoul, Korea *Water* **10** 10 1468

[15] Huff F A 1967 Time distribution of rainfall in heavy storms *Water Resour. Res.* **3** 4 1007-19

[16] El-Sayed E A H 2016 Development of synthetic *Ain. Shams. Eng. J.* **9** 4 1949-57

[17] Elsebaie I H 2012 Developing rainfall intensity–duration–frequency relationship for two regions in Saudi Arabia *J. King Saud Univ. Eng. Sci.* **24** 2 131-40

[18] Sun Y, Wendi D, Kim D E and Liong S-Y 2019 Deriving intensity–duration–frequency (IDF) curves using downscaled in situ rainfall assimilated with remote sensing data *Geosci. Lett.* **6** 17

[19] Wang A, Qu N, Chen Y, Li Q and Gu S 2018 A 60-Minute Design Rainstorm for the Urban Area of Yangpu District, Shanghai, China *Water* **10** 3 312

[20] Gong Y, Liang X, Li X, Li J, Fang X and Song R 2016 Influence of Rainfall Characteristics on Total Suspended Solids in Urban Runoff: A Case Study in Beijing, China *Water* **8** 7 278

[21] Rivard G Design 1996 Storm Events for Urban Drainage Based on Historical Rainfall Data: a Conceptual Framework for a Logical Approach *Journal of Water Management Modeling* **4**
[22] Ellouze M, Habib A and Riadh S 2009 A triangular model for the generation of synthetic hyetographs *Hydrol. Sci. J.* **54** 2 287-99
[23] Veneziano, D., Villani, P. 1999 Best linear unbiased design hyetograph *Water Resources Res.* **35** 9 2725-38
[24] Keifer C J and Chu H H 1957 Synthetic storm pattern for drainage design *J. Hydraul. Eng.* **83** 1 25
[25] Adams B J and Howard C D D 1986 Design Storm Pathology *Canadian Water Resources Journal* **11** 3 49-55
[26] Chow V T, Maidment D R and Mays L W 1988 *Applied Hydrology* (New York: McGrow-Hill)
[27] Prodanovic P and Simonovic S P 2004 *Generation of Synthetic Design Storms for the Upper Thames River Basin. Assessment of Water Resources Risk and Vulnerability to Changing Climatic Conditions* ISSN 1913-3200
[28] Kellagher R 1981 *The Wallingford Procedure - for design and analysis of urban storm drainage* (Wallingford: HR Wallingford Ltd)
[29] Pilgrim D H A and Cordery I 1975 Rainfall Temporal Patterns for Design Floods *J. Hydraul. Eng.* **101** 81-95
[30] García-Bartual R A and Andrés-Doménech I A 2017 A two-parameter design storm for Mediterranean convective rainfall *Hydrol. Earth Syst. Sci.* **21** 5 2377-2387

Acknowledgments
This work was supported by the Scientific Grant Agency of the Ministry of Education, Youth and Sports of the Slovak Republic and the Slovak Academy of Sciences within the project VEGA 1/0727/20 , co-funded by the Slovak Research and Development Agency under contract No. APVV-18-0203, and by the grant "Young researchers Grant by STU" under the Monitoring and removal of SARS-CoV-2 virus RNA fragments in wastewater using ferrate title, which is financed from the central sources of the Technical University of Bratislava.

IOP Conf. Series: Materials Science and Engineering 1209 (2021) 012029 doi:10.1088/1757-899X/1209/1/012029

Parasitic Architectural Forms (PAF) S01.E02 "Methodology and Ontology"

T Baroš[1] and D Katunský[1]

[1] Institute of Architectural Engineering, Faculty of Civil Engineering,Technical University of Košice, Vysokoskolska 4, Kosice 042 00, Slovakia

E-mail: tomas.baros@tuke.sk; dusan.katunsky@tuke.sk

Abstract. This paper is to present the chosen methodology for research and thus approach the design process through biomimicry. The described methodology is applied in parasitic architecture research, focusing on developing the experimental design of architectural structure using artificial intelligence and thus approach and support the central hypothesis of the work - the effort to create self-designing and self-growing / developing architecture. Furthermore, the paper outlines the ontology as a model of knowledge, a source of data for the developed software © PAF A.I. and a source of inspiration influencing the final design of the proposed architectural structure. The way in which such an ontological resource is processed for research purposes and how it participates in the final design within the verification and validation of the whole conceptual design is also described. Besides, the paper explains why the ontology in question is essential and what was established as its content.

1. Introduction

Investigation between architecture and biology should require a consideration of models that capture and cultivate the dynamic reciprocity of the less obvious organic systems of architecture and the more obvious living complexities of biological systems [1]. Biomimicry is an alternative solution and approach in the searching for sustainable design in architecture. The inspiration from solutions refined and developed by nature is a driving force in architecture, resulting in great and quality architecture works. Biomimicry is a science that studies nature's models and then emulates these processes, systems, strategies and forms to sustainably solve human problems [2]. Biomimicry being the most advanced in the line of holistic approaches to design. It is important for architectural design to consider social aspects for it will ultimately influence people and behaviour, relationships and influences between entities and its environment [3].

2. Methodology

Dialectics is a philosophical method used to frame research as a debate or discourse, presenting both sides of the argument in order to reason out a conclusion. Dialectical thinking is a process of first presenting, discussing its antithesis and then elaborating a synthesis of these two positions by the way of conclusion. It is a rigorous method of theoretical exploration that addresses more than one position and is honest about the disagreements while being able to say something specific in conclusion [4]. The underlying methodology of biomimicry is to gain an understanding of the fundamental principles of a biological process or adaptation and to subsequently adapt these concepts for bio-inspired design applications to solve specific technical challenges. An analysis can be approached from a variety of angles, focusing on how things or organisms are created, function, behave in nature [5].

Focus is shifting from artefacts to systems, from transactions to relationships, from design as craft to design as thinking, from habits of destruction to awareness of the need for resilience. In order to define quality ideas there are used qualitative descriptors in data presenting [3].

2.1. Methodology of research

Research often begins with a question. While this is a simple idea, it offers a great deal of flexibility to the researcher and expresses the fundamental character of the activity as grounded in a spirit of curiosity [4]. In the beginning, the vision represented by a research question *"What if architecture could be self-designed, self-growing, self-developed, self-healing?"* is framing research focus on issues, whereas defining problematics and theoretical framework is crucial. Recognizing the fields, problems and applications for which local biomimicry design has or can have an impact [3]. These established research ideas are translated into the hypothesis as a conducting scientific problem.

Hypothetical considerations that will be subjected to argumentation in the research process are: Supplementary secondary façade and roof elements capable of improving conditions affecting the properties and functioning performance of the host structure/building by its structure and morphology; Architectural structures capable of being self-designed, self-growing, self-developing and self-healing as innovation and solution to the problematics; Involving a combination of biomimicry, Algorithms Aided Design, Parametricism and A.I. into the design process; The reverse design of an architectural structure (architectural-structural design). When hypotheses are defined, which are then processed into an initial task. Task for research in question is describing, documenting, analyzing and verifying the proposed design method of designing the architectural structure. It generated theses:

Development of experimental design of architectural structures; The creation of a theoretical agenda - manifesto "Selfie Architecture!" - a document describing the researched and developed approach of the experimental designing of an architectural structure with the help of artificial intelligence and biomimicry; The creation of Grasshopper software plugin tool "Parasite"; Architectural-structural, physics, material and production conception of Parasitic Architectural Forms (PAF); The creation of groups of parametric simulations in the Grasshopper software environment as a basis for creating a product software (© PAF A.I.) capable design and assess the architectural structure autonomously.

The most crucial part and essence of biomimicry research - observation. This part of the research process methodologically takes place in parallel on three levels, establishing the biomimicry line; observation of samples; observation of the experimental course of selected processes in laboratory conditions.

The clearly defined role of research further passes into the elaboration of the biomimicry line, using the methodology of the top-down approach, i.e. the approach where abstract concepts are identified first and gradually specialized into specific contents. Understanding the basic steps required for a successful and deep fusion of biological and engineering knowledge. The design inspired biologically is typically thought of as being problem-based, motivated by the effort and desire to generate an improved different solution to a particular technical challenge. The inversion of this approach is to take biological solutions as a starting point and seek out particular technical challenges for which the system is appropriate. Solution-based approaches may encourage the application of biological principles in ways that would not be immediately obvious when viewed from a problem-based perspective [3]. This step of observation in the methodology is accompanied by a detailed historical survey and a biological survey (literature, case studies, experiments, researches). The findings are processed into conclusions - review and the information obtained and example references are further processed and included in the source of knowledge.

When observing samples, after their preparation, purification, description and preparation, occurs the analysis takes place by a selected device where the expected outputs from individual instruments (SEM, TEM, Confocal microscopy) or their combination offer a suitably detailed view of the structure of an observed sample. In the case of selected coral samples, transversal thin-sections will be prepared to obtain the outlines of calicular-sections. The empty calicular cavities will be filled with epoxy resin under vacuum conditions before sectioning. Individual sections will be photographed under a binocular microscope at the required several time magnifications. Subsequently, I will be able to identify the pattern crucial for the design and concept of the simulated structure by using the box-counting fractal analysis method. The fractal description of the identified pattern as an input to the simulation process will be processed using the Flow-based programming method processed by the Algorithms-aided Design tool (Grasshopper®), later forwarded in the same way directly to the simulation of creating a structure. There will be applied fractal, generative and parametric spatial geometry creation methods of the adapted pattern and Jacob's genetic regulatory mechanism in addition to the mentioned approach.

The observation of the experimental course of selected processes focused on growth will take place under laboratory conditions. Depending on the nature of the observed phenomenon, a time-lapse or slow-motion recording will be captured and subsequently, the courses of changes in the observed phenomenon over time, will be analyzed, which will be digitally processed into the required photo frame sequences at clearly defined time intervals. These will be then used as a starting material for the box-counting fractal analysis method on behalf of the use of fractal dimensions to quantify shape changes and decisive changes in the dimensions of observing processes. The detected dependencies and the obtained data in the form of parametric information are incorporated into the simulation, which will be processed using the Flow-based programming method processed by the Algorithms-aided Design tool (Grasshopper®). Recording equipment, in this case, represents the use of two 4K cameras with the possibility of 60fps slow-motion/time-lapse recording (Samsung S21 Galaxy). Processing records into photo frame sequences will be executed by Adobe Premiere Pro software®.

Essential to developing a methodology to collect the relevant data references and create a meaningful data set of references to accomplish the aim of this work [5]. Reference acquisition will occur using the bottom-up approach where most specific concepts are identified first and gradually generalized into abstract concepts. What is considered to be a reference during the creation of an ontology source and how the individual references are designated (in more detail I discuss in chapter "3.2. Ontology Content - References"). The process of collecting references as the initial content of knowledge for design development is subject to processing through the "Graphical Reference Observation System", a custom parametric radar graph. The analytical-observational graph containing the assessed parameters, criteria and properties (thematically divided into 36 radii according to nature/essence/character, five parameters/criteria/properties on each line) serves as a network. For the adept-reference to become a reference that will be further worked on, i.e. to get into the ontology source, it is necessary to capture it in a "network" - a graph. Adept-reference admission and its subsequent processing are conditioned by a minimum of three evaluable parameters of different characteristics (value 1 or 2), thus ensuring the region's creation in the graph by the reference. Such a region plays an essential role for further assessment. Thus, the system prevents the entry of references that are irrelevant or unsuitable for research and concept, thus limiting the number of references and the size of the ontology source and providing suitable content for further processing. Applicated Rating scale (0-2) in references valuation: 0 - not at all or negligibly; 1- partially; 2- completely or significantly. The sum of all achieved points from the five parameters is the resulting point on the graph's radius. Each criterion, parameter or characteristic is assigned a value depending on the content of the reference.

The reference proceeds into the *Ontology classification analysis* process. At this stage, references are assigned a designation according to the "DNA" enrollment of the reference. An algorithmic function evaluates the notation of the reference with output in the form of an incidence matrix. It is a knowledge representation by references to a knowledge-based model.

The next step is an *Ontology source creating* in the form of a *PAF Ontology Map*. Through the design of information and data and their visualization, I am able to find solutions. The involvement of self-learning in an ontological resource will show links to references and relationships that I would not find by conventional methods. A *PAF Ontology Map* is a form of processing a vast amount of data into an acceptable, "digestible" form that we can work with more clearly. It is an interactive topological graph, a stereographic projection on a sphere, which is bijective. The notation of a graph by an incidence matrix represents how the references are connected and combinations of relationships between the individual references.

The re-submission of the reference into the Graphical Reference Observation System with a focus on combinations and various multiple variations proposed by PAF Ontology Map aims to create the most diverse basis for creating the final design and clues to help navigate research and thus furthermore specific survey of the references by a more detailed definition of the field of research. The larger the reference's region contained in the graph, the more suitable the reference is. Simultaneously, the higher the probability of its use in the final concept because choosing the most suitable adepts is based on the most significant region penetration of different combinations of references. Combinations are determined based on an algorithm in the PAF Ontology Map. As this system is open throughout the research (2019-2023), thus the collection is ongoing parallel and independently.

Evaluation and defining highlights of research and conception of design and selecting the most suitable adepts and their combinations leads to conceptual Adaptation & Transformation by Middle-out approach, where most important concepts are identified first and then generalized and specialized into other concepts. The result of the above procedure is the Proposal prototype of an architectural structure represented by two alternatives - reinterpretation and extraction of knowledge into prototype design - entering further into the process of complex simulation of the prototype. Many references entering the system as a source of inspiration are considered an advantage in the design process. The more reference, the more interesting the concept can be. It is a selection of non-traditional references, inspirations in the biomimicry approach, the most suitable candidates for combinations into the concept. Using these tools, we are able to extract exciting influences on the overall concept and thus obtain a conceptual basis that we would not be able to find in any other way.

The output of these three parts of the research observation phase becomes simulations describing, documenting, and analyzing what is observed. This whole is followed by the final processing of partial simulations into a simulation of an architectural structure prototype, again using the Flow-based programming method processed by Algorithms-aided Design tools (Grasshopper® + Rhinoceros® + ArchiCAD®). Such a complex simulation, the thesis, will be submitted to the verification into the verification process as an antithesis. At this stage, the "form-finding simulation" is involved in the process, which works with several other software, through which the correctness of the design of a selected segment of the architectural structure is tested in several cycles and repetitions until the desired results are obtained. At the same time, the assessed segment is adjusted in parallel.

The resulting verified prototype simulation model is analyzed. The overall course of research and development of architectural structures' experimental designing and the acquired knowledge is evaluated and transformed, processed into results. Part of the research and results is the output of the obtained data through an interactive interview on the web platform (www.p-a-f.eu), representing the opinion of the scientific and lay public as well as professionals from practice on the topic. Therefore, the expected final results are the manifesto: Selfie Architecture!; preparation of physical model creation for experimental testing of application of PAF in laboratory conditions; conceptual and computational materials for developing software - product ©PAF A.I..

Informative-analytical survey website for analysis and synthesis of obtained data play an essential role and is also part of the research.

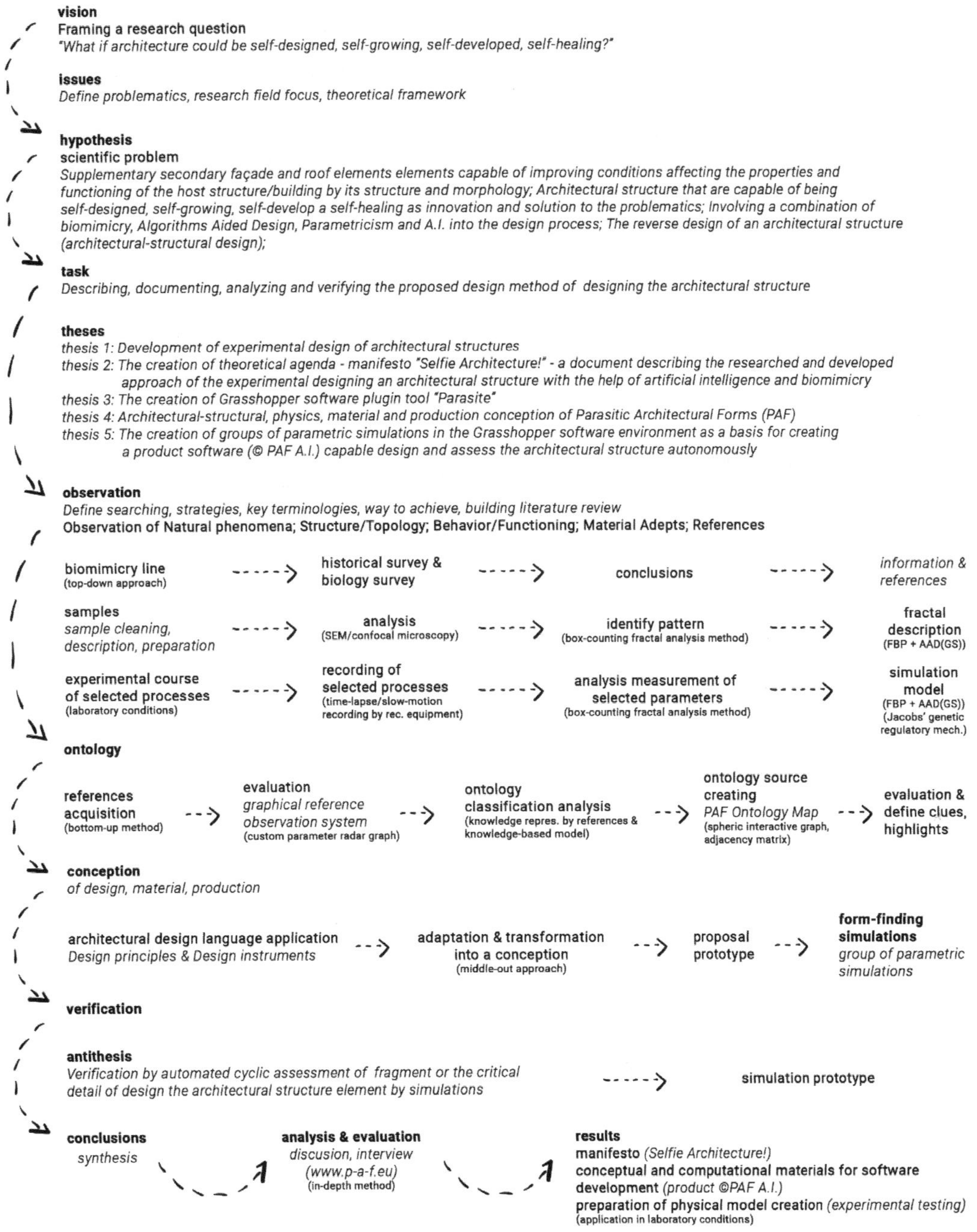

Figure 1. Graphic representation of the research methodology.

3. Ontology

Ontology is the part of philosophy that studies what it means to exist [6]. Also, Ontology is defined as "specification of a conceptualization" and "as a set of definitions of a concept, a representational vocabulary", "and a declarative model of the terms and relationships in a domain" or "the theory of objects" [7].

Ontology is a multimodal representation of architectural design knowledge to linking explicit and implicit types of information. It describes individuals, their properties and characteristics, and relations between them (relationships associated with design factors include environmental, structural, physical and contextual aspects), defines the basic terms, contains classes as concepts of a specific domain and rules, structures information and expresses relations between different concepts and elements [8].

3.1. The essence of Ontology as source for Biomimicry research and A.I.

Practical utilization of the developed ontology is aimed for design reasoning and support, and there are several ways in which information can be explored. The benefit of the combinatory representation is that implicit and explicit types of knowledge are integrated, where visual, numerical and procedural information and design rules are incorporated, thus the explicit meaning is associated with pieces of information that can be used and adapted to suit the needs of a design problem. The current architectural practice could be improved by an interactive comprehensive source of knowledge capable of parametric variation of the model based on the information received from ontology. Conceptual information models are the most expressive for modelling applications and structuring information bases since they rely on abstraction mechanisms inspired by cognitive sciences and artificial intelligence [8]. Such ontology is a valuable source with multidisciplinary content that develops proper models for each unique field to conceptualize the content, specify and represent its knowledge seems essential in this case. Also, provide a wide variety of data and make data gathering and distribution complicated exciting [7].

3.2. Ontology Content - References

Through visualizing information, we can see the patterns and connections that matter and design that information, so it makes more sense or tells a story or allows us to focus only on the crucial information. We need relative figures connected to other data so that we can see a fuller picture, which can lead us to changing our perspective. Visualizing information is a form of knowledge compression; it is a way of squeezing an enormous amount of information and understanding into a small space. Design is about solving problems and providing elegant solutions, and information design is about solving information problems. Visualizing information can give us a rapid solution to those kinds of problems, and even when the information is terrible, the visual can be quite beautiful and often, we can get clarity or the answer to a simple question very quickly [9]. Design knowledge representations by reference are based on cognitive aspects. Design knowledge and know-how were embodied in referents represented in multimodal formats and structured in a source whose objective is to provide a possibility for exploration and knowledge reuse during the early stages of architectural design. [8]. In this research designation, *References* represents the relevant reference, contentually as references in this meaning it could be case studies, construction and buildings references, artworks, installations, phenomena, experiments, inventions, patents, theories, methodologies, design approaches, research projects, design proposal ideas, scientific facts, terms and formulas, precedent situations in design and research, historical references, fauna and flora, data, information technology tools and software. In my work, I am inspired by various influences. There are no limits, no restrictions, for me in this regard. In addition to the usual references, I also work with non-traditional ones such as filmography, scenography, photography and the influences of pop culture. Finally, Čapka's R.U.R. inspired the robotization of the world. The whole work conceived as development, the process of creation. This coding system was created for the purpose of building up PAF Ontology Map as a main information source during research and also as a part of educational source for PAF A.I. The complexity of information is key in the whole design process. A huge amount of information and data affects the overall design. Each reference from the ontology is part of the concept; in a way, it relates to the topic and becomes a partial inspiration for creating the final prototype design. Each reference entering the ontology has been reviewed to the extent of the available literature or information documentation, and a brief review is part of that reference.

4. Conclusion

The absence of a comprehensive description of the biomimicry of research and approach to the creation of an architectural design and overall biomimicry research methodology leads to aiming to bring this main research task - defining, describing, documenting, analysing and verifying the proposed design method of designing the architectural structure with the help of biomimicry and A.I. Development in technology and material science enabled the growth and implementation of ideas of biomimicry. There are visible proofs in contemporary design that using nature as a source of ideas can lead to a higher level of sustainability. Biomimicry definitions indicate that a radical approach to design is inevitable [3]. Implicit knowledge is difficult to communicate and formalise and is usually accumulated through a learning process and practice (i.e. analogues and metaphors to which designers refer during creative work). Machine learning is an exciting field of investigation that can fill the gap of computational resources for the conceptual design phases of architectural projects and address the broader needs of the profession in design analysis and development. Numerous Case studies have proven the great diversity that the algorithm can use to fulfil both aesthetic and functional architectural necessities. For example, data-driven methods that selectively analyse the genotypes could be very interesting for future design method research. Both explicit and implicit types of knowledge are utilised in the design process, and advanced computational applications should offer them to designers and support the transfer between these types [8][10][11].

5. References

[1] Sabin J E and Jones P L 2008 Nonlinear systems biology and design: surface design*Biomimetics Concepts of Nature and Technology* pp 54-65
[2] Rao R 2014 Biomimicry in Architecture *International Journal of Advanced Research in Civil,Structural,Environmental and Infrastructure Engineering and Developing* **1** 3
[3] Zejnilović E and Husukić E 2015 Biomimicry in Architecture *International Journal of Engineering Research and Development* **11** 7 75-84
[4] Lucas R 2016 *Research methods for architecture* (London: Laurence King Publishing Ltd)
[5] Lurie-Luke E 2014 Product and technology innovation: What can biomimicry inspire? *Biotechnology advances* **32** 8 1494-505
[6] Cambridge Dictionary Avialible online https://dictionary.cambridge.org/
[7] Andaroodi E, Andres F, Einifar A, Lebigre P and Kando N 2006 Ontology-based shape grammar schema for classification of caravanserais: a specific corpus of Iranian Safavid and Ghajar open, on-route samples *Journal of Cultural Heritage* **7** 4 312-328
[8] Aksamija A, Iordanova I 2000 Computational Environments with Multimodal Representations of Architectural Design Knowledge *International Journal of Architectural Computing* **8** 4
[9] McCandless D *The beauty of data visualization* Ted-Ed lecturen Avialible online https://www.youtube.com/watch?v=5Zg-C8AAIGg
[10] Pal As I S and Basu P 2018 Artificial intelligence in architecture: Generating conceptual design via deep learning *International Journal of Architectural Computing* **16** 4 pp 306-327
[11] Navarro-Mateu D and Cocho-Bermejo A 2020 Evo-devo strategies for generative architecture: colour-based patterns in polygon meshes *Biomimetics* **5** 2 23

Acknowledgments

The paper presents partial research results of project VEGA 1/0674/18 “Theoretical and experimental analysis of architectural-structural shapes and fragments of building envelope structures designed for harsh weather conditions”.

Biaxial testing of textile material – methodology and instrument assembly procedure

N Freiherrova[1], M Hornakova[1], D Juracka[1], L Stulerova[2] and L Kapolka[2]

[1] VSB – Technical University of Ostrava, 17. listopadu 2172/15, 708 00 Ostrava-Poruba, Czech Republic
[2] Institute of Structural Engineering, Faculty of Civil Engineering, Technical university of Kosice, Vysokoskolska 4, 042 00 Kosice, Slovakia

E-mail: nela.freiherrova.st@vsb.cz

Abstract Membrane structures are becoming popular because of the potential usage in structures with higher aesthetic claims. For the roofing of these structures, different textile materials are used. These special materials offer an alternative to conventional roofing materials and allow, besides its lightweightness, also a possibility to roof a difficult floor plan and big span. When designing such a construction, there are some challenges related to the properties of the specific material. In order to exploit the full potential of textile membrane materials, it is necessary to choose an appropriate material model for numerical modelling, which takes into account its nonlinear behaviour. The two most important material characteristics needed for characterizing the behaviour are Young's modulus of elasticity and Poisson's ratio. Since the material is orthotropic in most cases, it is necessary to test the material in two directions; therefore, these characteristics need to be obtained from the biaxial test. This contribution is focused on the research of the methodology of biaxial tests and test instrument assembly procedure, which will be used for the following testing of textile materials.

1. Introduction

For tensile membrane structures, the technical textile materials are used for roofing of wide-span structures with a difficult floor plan without a need for intermediate supports. The design of membrane structures can offer an efficient solution, which benefits of lightweight of the structure, low cost, and small environmental impact [1, 2]. Today's membrane materials are in most cases made of fabrics woven in 2 perpendicular directions. There are two main types of textile membrane materials: polyester fabrics with a polyvinyl chloride coating (PVC) and glass fibre fabric with a polytetrafluoroethylene coating (PTFE). However, modelling and designing structures of these materials face some challenges as they exhibit nonlinear and anisotropic behaviour. It is essential to take into account the material characteristics (modulus of elasticity E and Poisson's ratio ν) of the used fabric materials and choose an appropriate material model for numerical modelling. However, the models are still simplified because there is not enough measured data in the field of membrane structures and textile materials [3].

The material characteristics are determined based on the biaxial tensile tests performed on cross samples. Since membrane structures are a relatively new topic in the field of civil engineering, there is not a lot of standardized information about the design and testing methodology. That is why the methodology greatly depends on the experiences of individual institutions and design engineers. The standard from Japan [4] is often considered as a default when setting the biaxial testing procedure. With the newly published standard [5], standardized guidelines for biaxial testing procedures and methodologies for deriving material constants are available.

The aim of the contribution is to present the testing methodology for the determination of material characteristics (modulus of elasticity E and Poisson's ratio ν) of the membrane material Flexlight Advanced 902 S2, manufactured by Serge Ferrari. At first, the uniaxial test needs to be done to determine the tensile strength. The results are then compared with the technical report of the material provided by the manufacturer for validation. The obtained value of loading from the uniaxial test will be then used for cyclic loading in the biaxial test. However, due to the absence of a standardized measuring device in our laboratory, that allows testing the sample biaxially, the uniaxial measuring device Zwick/ROELL Z150 needs to be specially modified to ensure the attachment and loading of the cross-shaped sample in the perpendicular direction.

2. Measured textile material

Textile material Flexlight Advanced 902 S2 by Serge Ferrari group was chosen for the experimental testing. Serge Ferrari company designs, manufactures, and distributes innovative, flexible composite materials over the world and is renowned for its lightweight, durable, and recyclable materials.

The tested material is mainly used for roofing outdoor spaces, sport hall roofs, and shading structures. The advantages of the Flexflight Advanced 902 S2 are, according to the supplier, its long-lasting lifetime (over 20 years) and easy maintenance, consistent colour per translucency, high dimensional stability due to Precóinraint technology, which is patented. The declared characteristics are given in Table 1 [6]. Tensile strengths in the warp/weft direction based on the uniaxial test are calculated and compared to the provided data in the following chapter.

Table 1. Mechanical characteristics of Flexlight Advanced 902 S2 [6].

Material characteristic	Value
Tensile strength warp [daN/5cm]	420
Tensile strength weft [daN/5cm]	400
Warp modulus of elasticity E_w [kN/m]	708
Weft modulus of elasticity E_f [kN/m]	774
Poisson's ratio warp-fill ν_{w-f} [-]	0.32
Poisson's ratio fill-warp ν_{f-w} [-]	0.49

3. Testing methodology

3.1. Uniaxial tensile test

As a first step, the uniaxial tensile tests need to be prepared to determine the input values of the loading force during the biaxial tests (see Figure 1, left). For the determination of input values for biaxial tests, it is customary to adhere to Factor 5 [2]. It recommends the operating stress value is determined as 1/5 of the force obtained in uniaxial tests. The initial prestress usually depends on the curvature of the surface and the structure of the weave and is given as 1/5 - 1/10 of the working stress. The initial prestress is therefore around 1/25 of the operating stress. It is not a general rule that the same prestress value is applied to the warp and weft. The results of tensile strength in warp/weft direction are obtained based on the uniaxial tests as well.

According to the standard [8], the material was tested in both directions with 5 samples, which is considered as the minimum number of samples for the test, with dimensions of the measured area 50 x 100 mm (Figure 1, right); the tensile strength R_m [MPa] in both directions was determined as the mean value based on the thickness a_0 [mm], width b_0 [mm] of the specimen and the value of ultimate force F_m [kN] (see Table 2).

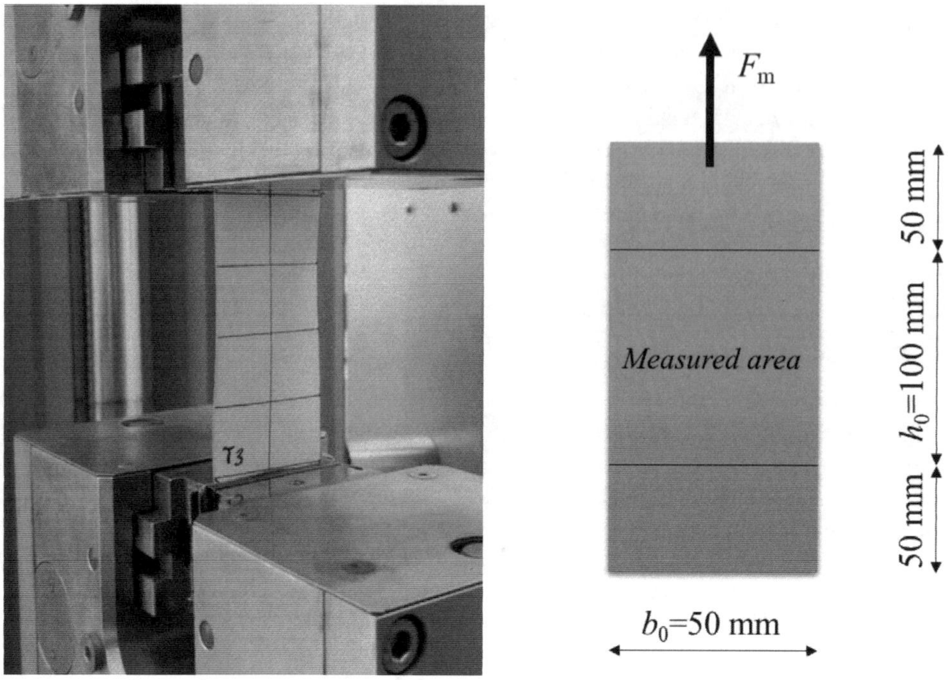

Figure 1. Uniaxial test on Zwick/ROELL Z150 (left), dimensions of the specimen (right).

Table 2. Measured characteristics of Flexlight Advanced 902 S.

#	Specimen ID	Thickness a_0	Width b_0	Ultimate force F_m	Tensile strength	R_m
		[mm]	[mm]	[kN]	[MPa]	[dN/5cm]
			weft direction			
1	BLC/T1	0.767	49.96	3.82	100	383.5
2	BLC/T2	0.765	49.77	3.77	99	378.7
3	BLC/T3	0.767	49.75	3.84	101	387.3
4	BLC/T4	0.767	49.83	3.87	101	387.3
5	BLC/T5	0.768	50.02	3.74	97	372.5
Average value		-	-	**3.808**	**99.6**	**381.9**
			warp direction			
6	BLC/L1	0.774	50.15	4.02	104	402.5
7	BLC/L2	0.778	50.31	4.04	103	400.7
8	BLC/L3	0.776	49.81	3.73	96	372.5
9	BLC/L4	0.777	49.73	3.84	99	384.6
10	BLC/L5	0.776	49.93	3.86	100	388.0
Average value		-	-	**3.898**	**100.4**	**389.6**

Tensile strength R_m was determined based on ultimate force F_m [N], thickness a_0 [mm] and width b_0 [mm] of the sample according to the eq. 1 and then transferred to the required units.

$$R_m[MPa] = \frac{F_m}{a_0 * b_0}, \tag{1}$$

After the conversion related to the size of the sample area (1MPa=10 daN/cm^2, then multiplying by 5 and a_0 to daN/5cm), the final average tensile strength was determined as 390 daN/5cm in the warp direction and 382 daN/5cm in the weft direction. These units are specific for membrane materials (as shown also in Table 1). Comparing with the data provided by the manufacturer (420 daN/5cm in warp, 400 daN/5cm in weft direction), there is 8% difference in warp direction and 5% difference in weft direction, which can be considered as a negligible difference compared to the data provided by manufacturer.

3.2. Biaxial test according to the standard [5]

Since the conventional mechanical test methods based on the uniaxial tensile test are not always suitable for the design of membrane structures, the biaxial test offers a solution for the determination of tensile stiffness properties for numerical modelling, because it is possible to measure along two sides simultaneously. Therefore, the biaxial test equipment shall be capable of simultaneously applying loads to the specimen with a specified load ratio in both directions. Load, strain or displacement shall be measured as well. The strain field may be measured using Digital Image Correlation techniques.

The specimen shall be taken at random from a lot without any signs of damage or dampness. A minimum of three laboratory specimens are required for the biaxial test and should be cut from positions taken at random but at least 1 m from the end of the piece from an area, which is not creased or has visible faults. The specimen shall be of sufficient size to provide a representative sample of the coated fabric as a whole and the geometry should produce a field of homogeneous strain in the centre.

At first, the specimen needs to be mounted using suitable clamps, which secure that the material will not slip in the clamps during the test. Minimum stresses or strains shall be applied to the sample before the test, and it shall be reported. Then, the specimen is loaded in both directions with prescribed magnitudes and ratios. Load ratios are used to define a load cycle and combined into a single test to create a load profile. Load and strains shall be recorded every two seconds for test durations of up to 8 hours, for the test duration beyond 8 hours it shall be recorded in frequencies sufficient to capture the behaviour of the material.

3.3. Instrument assembly procedure for biaxial testing

Currently, there are only a few certificated types of equipment over the world for the biaxial testing of textile membrane materials. This equipment can be found, e.g., in Italy [7], where it was used to provide strain-controlled biaxial tests, or Brussel [3], where it has been proven, that there is a significant impact on the derived material parameters when the used test protocol is changing. The standard [5] gives a general proposal of the schemes of these devices, some of them can be seen in Figure 2.

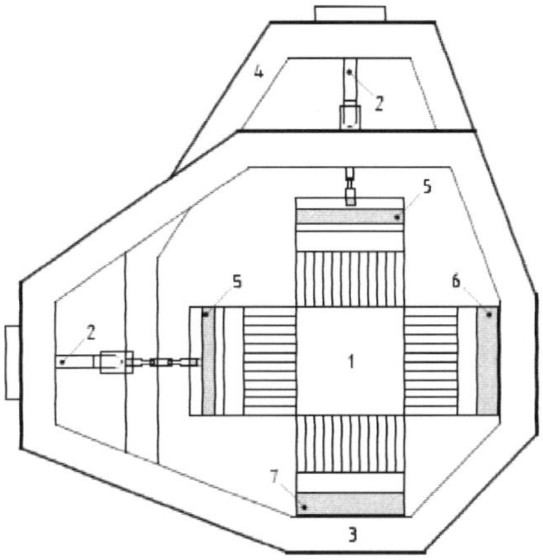

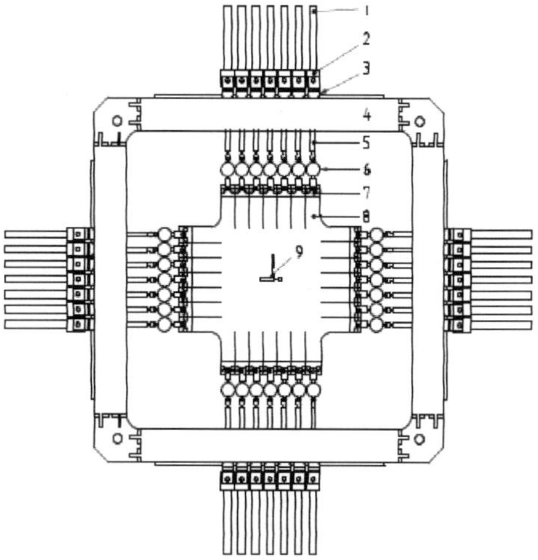

Figure 2. Biaxial test equipment – examples from the standard [5].

IOP Conf. Series: Materials Science and Engineering 1209 (2021) 012030 doi:10.1088/1757-899X/1209/1/012030

Based on the available information related to the equipment for biaxial tests, it is needed to develop an alternative solution – a suitable type of attachment of the cross-sample in the perpendicular direction, which could be used in combination with the device for uniaxial tensile tests Zwick/ROELL Z150, which is available in the laboratory of the Centre for Advanced Innovation Technologies (CAIT) at VSB – Technical University of Ostrava. The scheme of the proposed design of the device can be seen in Figure 3, it consists of a uniaxial testing device and a steel frame, which allows mounting the perpendicular parts of the sample and slightly prestress the sample in the perpendicular direction to the loading. The dimensions of the cross-sample are limited by the testing device to 150 x 150 mm with an extra 50 mm at each side to mount the material. During the biaxial test, stress – strain diagrams will be recorded with the help of the sensors. Digital Image Correlation technique will be used to capture the strain field. To measure the characteristics in both directions, the specimens will be moved from the starting position to the other direction and loaded, which should solve the problem of subsequent loading in the perpendicular direction.

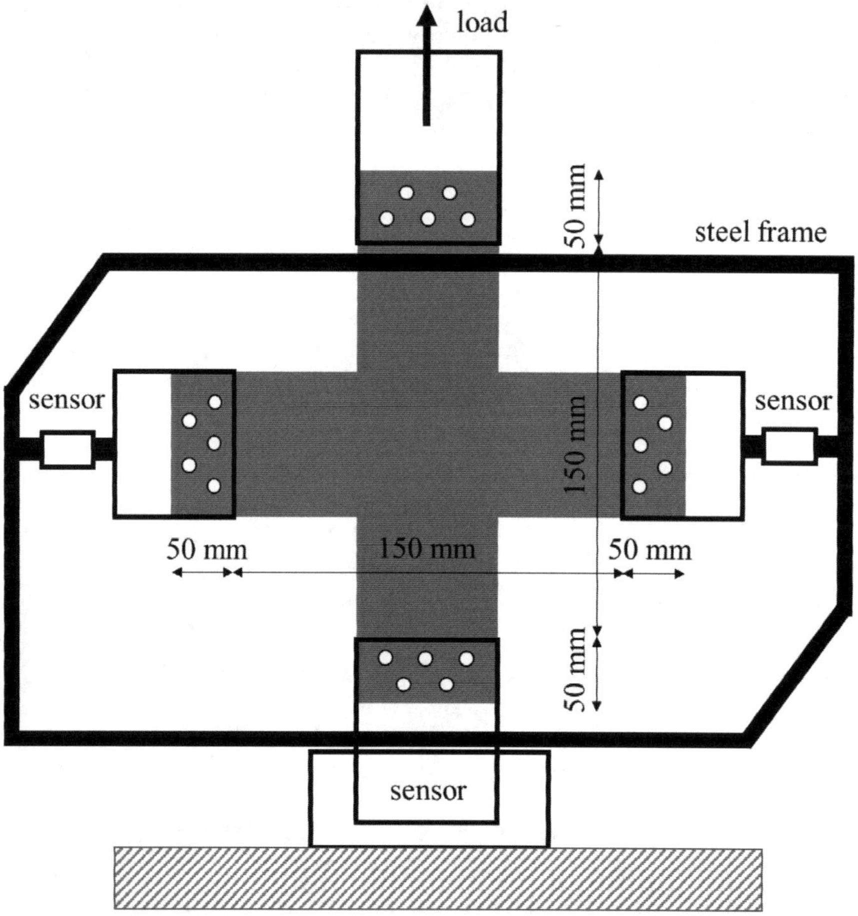

Figure 3. Proposed scheme of the device and dimensions of the specimen for the biaxial test.

After the equipment is prepared, the biaxial tensile test will be performed to determine the modulus of elasticity E and Poisson's ratio v of the selected membrane material. The standard [5] gives 6 options, how to calculate the values of tensile stiffness and Poisson's ratio from biaxial load – strain test data.

4. Conclusion

The aim of this contribution was to study the methodology of biaxial tests and prepare a modification of an instrument assembly procedure for biaxial testing. The textile material selected for testing is Flexlight Advanced 902 S2, which was at first tested in the uniaxial tensile test at Zwick/ROELL Z150, and its tensile strength was determined and compared to the data provided by the manufacturer. It was found that the results are not fundamentally different from the declared values.

However, because of modifying the existing equipment in terms of biaxial test, the methodology in this case needs to be modified regarding the capabilities of the developed device as well. Therefore, the size of the testing sample needs to be adjusted. There is also not an option of subsequent cyclical loading in the perpendicular direction after one cycle in the first direction of the sample is finished. The proposed solution is to start with cyclical loading in one direction and after that, the cross-sample in the equipment will be switched and the second direction will be cyclically loaded as well. For the next cycle, the specimen will be switched back to the starting position and the process will repeat. The results of the performed biaxial tests in CAIT will be compared with the results of the modulus of elasticity E and Poisson's ratio v specified in the data sheets by the manufacturer. These tensile characteristics are key inputs for modelling of the membrane structures; therefore, it is necessary to pay attention to the correct execution of the tests and verification of the obtained data, especially when an alternative device or testing methodology is used.

5. References

[1] Stulerova L, Kapolka L and Kmet S 2020 Numerical nonlinear analysis of a membrane structure subjected to dynamic load *IOP Conference Series Materials Science and Engineering* (Bristol: IOP Publishing) p 867

[2] Forster B and M Mollaert 2004 *European Design Guide for Tensile Surface Structures* (Brussel: TensiNet) pp 1- 354

[3] Craenenbroeck M V, Puystiens S, Hemelrijck D V and Mollaert M 2015 Biaxial testing of fabric materials and deriving their material properties – A quantitative study *International Association for Shell and Spatial Structures (IASS) Symposium (Amsterdam)*

[4] Membrane Structures Association of Japan 1995 *Testing Method for Elastic Constants of Membrane Materials (MSAJ M-02-1995)*

[5] EN 17117 - 1 2018 *Rubber or plastics – coated fabrics – Mechanical test methods under biaxial stress states: Part 1: Tencile stiffness properties* (Brussel: CEN)

[6] Biaxial test based on "MSAJ/M-02-1995 2010 Testing Method for Elastic Constants of Membrane Materials *Certificate of testing. School of Civil Engineering and Geosciences* (Newcastle: Newcastle University)

[7] Beccarelli P, Colsante G, Novadi G, Stimpfle B and Zanelli A 2013 Strain-controlled biaxial tests of coated fabric membranes *Proc. Tensinet Symposium,* ed H. Bogner-Balz, M. Mollaert and E. Pusat

[8] ISO 13934-1 2013 Textiles – *Tensile properties of fabrics – Part 1: Determination of maximum force and elongation at maximum force using the strip method* (Brusel: European Committee for Standardization)

Acknowledgments

This work was supported by the Doctoral grant competition VSB - Technical University of Ostrava, reg. no. CZ.02.2.69/0.0/0.0/19_073/0016945 within the Operational Programme Research, Development and Education, under project DGS/TEAM/2020-010" Analysis of the behavior of membrane fabrics under biaxial loading".

IOP Conf. Series: Materials Science and Engineering 1209 (2021) 012031 doi:10.1088/1757-899X/1209/1/012031

Theoretical Analysis of Structural I-beams Using Various Types of Transparent Wood

V Karľa[1], V Bajzecerová[2] and J Kanócz[3]

[1] Department of Architectural Engineering, Faculty of Civil Engineering, Technical University in Košice, Vysokoškolská 4, 042 00 Košice, Slovakia
[2] Department of Structural Engineering, Faculty of Civil Engineering, Technical University in Košice, Vysokoškolská 4, 042 00 Košice, Slovakia
[3] Department of Architecture, Faculty of Arts, Technical University in Košice, Letná 9 042 00 Košice, Slovakia

E-mail: viktorkarla@gmail.com

Abstract. In this article, a numerical analysis of various types of I-beam with web from transparent wood was performed and compared with standard OSB board and glass web. Properties of transparent plywood which have not yet been produced from new kinds of transparent wood samples were calculated using laminate theory. Results of analysis show, that the transparent wood composites prove to be very promising structural material for future use within the load-bearing structures of buildings. The best properties showed the transparent wood composites infiltrated by cellulose nanofibers (CNF) and polyvinyl alcohol (PVA).

1. Introduction

Transparent wood is relatively new material, which is still only being fabricated in laboratories and its properties are being investigated. However, it does seem to be a very promising material for applications in load bearing structures of buildings as mechanical properties of various transparent wood composites are superior to the original wood and polymer, which these composites were created from. To further prove this idea, we compared I-beam with web from various types of transparent wood to I-beam with web from OSB board and glass.

1.1. Brief history of transparent wood

German scientist Siegfried Fink created the very first sample of transparent wood in 1992 to study three-dimensional wood structures [2]. He was inspired by Spalteholz, who studied coarse body organs in a similar way in 1914 [3]. Although Fink was the first who produced a sample of transparent wood, he did not study the material itself.

Scientist at Swedish KTH [4] and at University of Maryland [5] rediscovered Fink's article between 2015 and 2016 and started to be interested in the transparent wood composite itself. They created transparent wood in a two-step process. First they removed lignin from wood samples and subsequently infiltrated these samples with polymer, which has a matching refractive index to delignified wood. At KTH it was poly methyl methacrylate (PMMA) and in Maryland it was epoxy resin. It was then discovered that such prepared transparent wood has better mechanical properties then the original wood in both studies. Transparent wood also showed interesting optical properties with high optical transmittance and high haze, which might be useful in applications where there is a need for both,

IOP Conf. Series: Materials Science and Engineering 1209 (2021) 012031 doi:10.1088/1757-899X/1209/1/012031

sunlight and privacy in the interior. Researchers suggested that transparent wood could one day replace glass as a more environmentally friendly material.

In order to delignify wood, though, there is necessary to use some aggressive chemicals, thus the statement of environmental friendliness of transparent wood did receive some criticism. At KTH they tried to solve this problem by creating a transparent wood, where lignin was not removed, only altered with use of less aggressive chemicals [6]. It was called lignin retaining transparent wood and showed similar properties to delignified transparent wood samples. It also had some advantages because it was easier to manipulate with lignin retaining samples and transparent wood prepared by this method showed even better mechanical properties. A minor disadvantage was slightly worse optical properties, but those were not any dramatic.

The greatest challenge considering transparent wood composites, however outstanding their mechanical and optical properties might be, is production of thick pieces. At KTH they acetylated delignified samples before infiltration of polymer and managed to produce a 1 cm thick transparent wood sample using this method [7].

Still the production of thick samples is complicated and it requires long time to delignify and infiltrate such pieces. Lamination of smaller blocks into transparent plywood material was also performed at KTH and again showed very promising mechanical properties with lesser anisotropy [8].

A team of researchers from Nanjing Forestry University, Yangzhou University and University of North Texas addressed the problem of producing large transparent wood samples by creation of transparent fibre wood, where they only delignified small fibres of wood and added them into PMMA matrix to create large samples with great optical and reasonable mechanical properties [9].

Researchers from University of Maryland, University of Colorado and Forest Products Laboratory in Madison recently created transparent wood infiltrated by PVA. This kind of transparent wood showed even better mechanical and optical properties than previously created samples [10].

The so far most ecological transparent wood types were created by Korean and Vietnamese researchers, who infiltrated delignified samples by chitosan (CTS) and cellulose nanofibers (CNF). These types of transparent wood had even better mechanical properties than PVA transparent wood, but only very thin samples were created and their optical properties are not among the best so far [11].

2. Researched structure

Although there are various ways transparent wood can be used in load bearing structures [12], in this article, we are focusing on I-beams. For the analytical purposes, their flanges are modelled from massive natural wood with cross section of 85 x 45 mm and their web is modelled from various types of transparent wood, OSB and glass with cross section of 10 x 150 mm. The web is immersed by 20 mm into flanges on its upper and lower side. The I-beam is 4000 mm long (Figure 1).

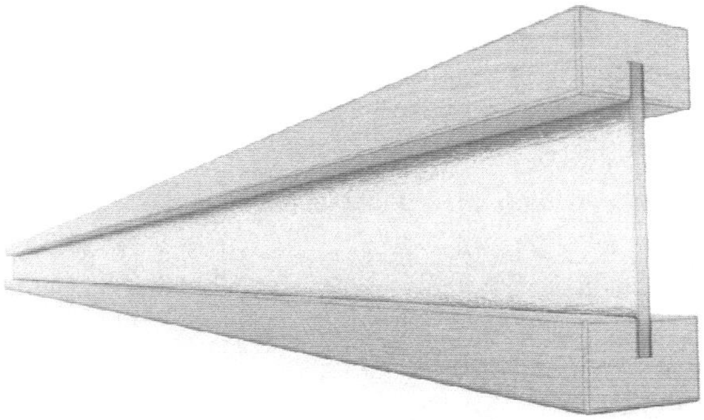

Figure 1. Rendering of the assessed I-beam.

IOP Conf. Series: Materials Science and Engineering 1209 (2021) 012031 doi:10.1088/1757-899X/1209/1/012031

3. Materials for the analysis

Transparent wood composites used as the web material of the I-beam are modelled according to above mentioned types of transparent wood. The first material we chose was transparent veneer infiltrated by PMMA and two types of plywood materials created using this veneer [8]. Then we chose PVA infiltrated transparent wood [10] and all bio-based transparent wood types (with chitosan and cellulose nanofibres) [11]. For them we also used laminate theory to calculate properties of plywood made from these materials [13]. Two types of plywood were considered – CP type (layers are oriented perpendicularly to each other) and Qi type (each layer is rotated by 45° to previous one). The last type of transparent wood we chose was transparent fibre wood [9]. Mechanical properties of materials used are listed in table 1.

Table 1. Mechanical properties of materials used for analysis (each material is 10 mm thick, with plywood, there are 5 layers, each layer 2 mm thick).

Material	Direction	Young's modulus E (GPa)	Shear modulus G (GPa)	Poisson's ratio μ (-)	Tensile strenght f (MPa)
Timber	longitudinal	11.00	0.37	0.48	16.00
PMMA TW veneer [8]	longitudinal	4.30	1.57	0.37	62.50
	transverse	2.40	0.99	0.21	14.60
PMMA TW_CP [8]	longitudinal	4.10	1.60	0.28	50.10
	transverse	3.90	1.54	0.27	44.90
PMMA TW_Qi [8]	longitudinal	3.90	1.57	0.24	45.40
	transverse	3.50	1.43	0.22	42.00
PVA TW veneer [10]	longitudinal	3.85	1.33	0.45	143.00
	transverse	3.10	1.14	0.36	67.00
PVA TW_CP	longitudinal	3.56	1.33	0.41	112.60
	transverse	3.41	1.23	0.39	97.40
PVA TW_Qi	longitudinal	3.63	1.29	0.40	112.60
	transverse	3.47	1.26	0.38	97.40
CTS TW veneer [11]	longitudinal	10.81	4.09	0.32	171.85
	transverse	3.48	1.58	0.10	26.50
CTS TW_CP	longitudinal	7.95	4.09	0.17	113.71
	transverse	6.47	2.84	0.14	84.64
CTS TW_Qi	longitudinal	8.40	3.63	0.10	113.71
	transverse	6.89	3.19	0.08	84.64
CNF TW veneer [11]	longitudinal	15.33	5.51	0.39	258.84
	transverse	4.07	1.85	0.10	31.27
CNF TW_CP	longitudinal	10.93	5.51	0.16	167.81
	transverse	8.66	3.83	0.13	122.30
CNF TW_Qi	longitudinal	11.44	4.89	0.08	167.81
	transverse	9.14	4.31	0.06	122.30
PMMA T fibre W [9]	-	2.20	0.90	0.23	46.80
OSB board	-	3.80	1.08	0.3	9.90
Glass	-	70.00	28.70	0.22	20.00

CP – plywood with veneer layers (0/90/0/-90/0)
Qi – plywood with veneer layers (0/45/90/-45/0)

4. Numerical model of I-beam

Theoretical analysis was performed using ANSYS Workbench 17.2 software as in previous analysis [1]. The parameters of the analysed I-beam are shown in figure 2. The flanges of the I-beams were from solid timber. All materials listed in table 1 were used for the beam web, a total of 15 different models were prepared. The Timber, OSB, and glass were modelled as linear elastic isotropic material, while the transparent wood veneer and plywood were modelled as a linear elastic orthotropic material. Contacts between the web and the flanges were rigid. For the calculation, Newton-Raphson's iterative method with considering large deflection was used. The maximum size of the mesh element was 5.0 mm, the symmetry of the beam was used. The beams were loaded with the uniform area load of 0.012 MPa which is corresponding to 1.0 kN/m of beam load.

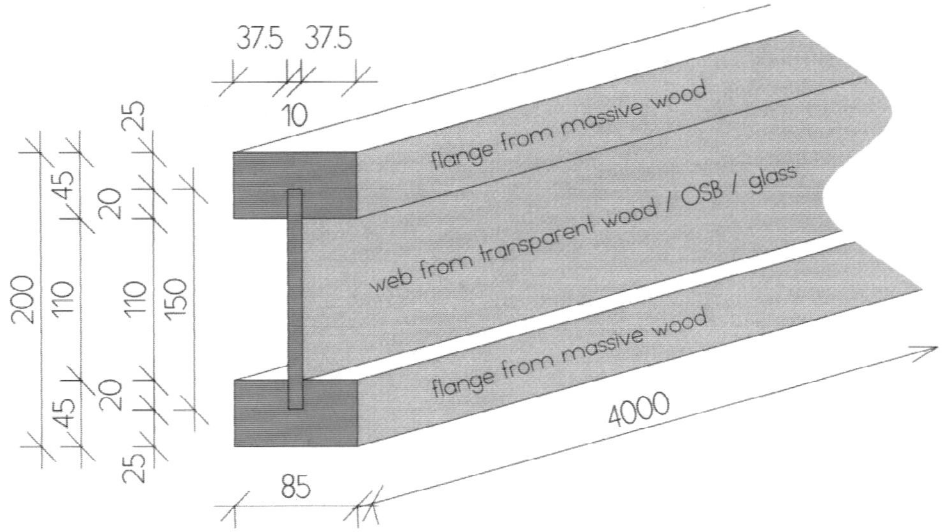

Figure 2. Parameters of I-beam used in numerical analysis (in millimetres).

5. Results of structural analysis of I-beam

The values of deflection in the beams mid-span are compared in figure 3. The smallest deflection was calculated for glass. Considering transparent wood composites, the smallest deflection was calculated for transparent wood veneer infiltrated by chitosan (CTS TW veneer). The greatest deflection was observed for transparent fibre wood (PMMA T fibre W).

Further results of the numerical analysis are shown in table 2. All the materials range in deflection from approx. 6.5 to 7.5 mm, which is somewhere in the middle of the limit deflection set at 13.3 mm. Only glass has a lower deflection of 4.9 mm. However, it has the highest normal stress of all, 15.2 MPa, while its tensile strength is of 20 MPa. Degree of utilization is therefore at 76.2% for glass. All of transparent wood composites show better results in this respect, even better than OSB board (10.7% of the degree of utilization). Their ratio of normal stress and tensile strength ranges from 0.8% to 2.4%. Considering maximum normal shear stress, all analysed materials are relatively equal. Their values are between 1.74 (PMMA TW_Qi) and 2.12 MPa (glass).

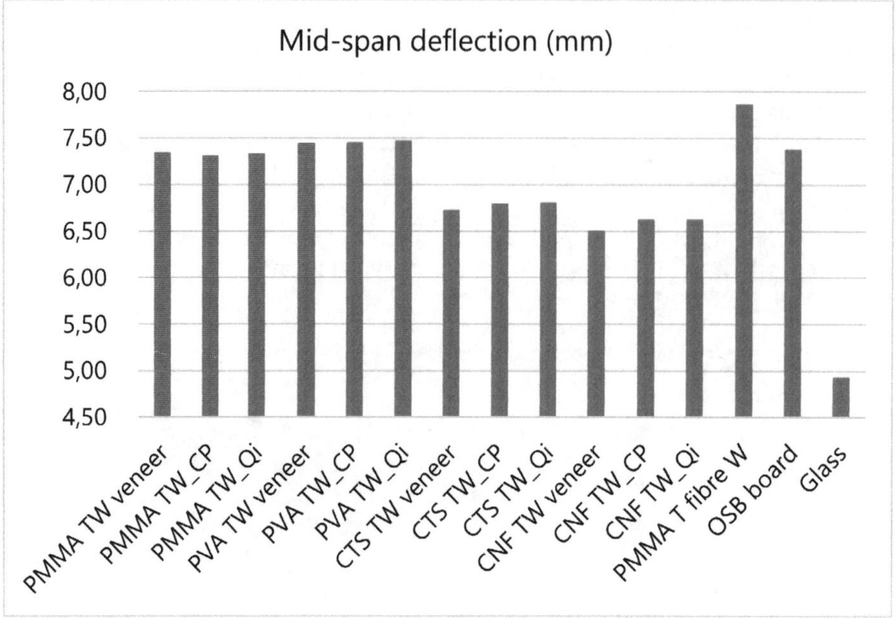

Figure 3. Calculated deflections of I-beams with different web materials.

Table 2. Results of numerical analysis of I-beam with different type of wall.

Material of I-beam web	Deflection (mm)	Max normal stress σ_x (MPa)	Max Tensile strength f_x (MPa)	σ_x / f_x (%)	Max normal stress σ_y (MPa)	Tensile strength f_y (MPa)	σ_y / f_y (%)	Max shear stress τ_{xy} (MPa)
PMMA TW veneer [8]	7.334	1.239	62.5	2.0%	0.887	14.6	6.1%	1.818
PMMA TW_CP [8]	7.303	1.145	50.1	2.3%	1.278	44.9	2.8%	1.755
PMMA TW_Qi [8]	7.327	1.093	45.4	2.4%	1.240	42	3.0%	1.740
PVA TW veneer [10]	7.436	1.126	143	0.8%	1.282	67	1.9%	2.039
PVA TW_CP	7.442	1.016	112.6	0.9%	1.381	97.4	1.4%	1.980
PVA TW_Qi	7.461	1.032	112.6	0.9%	1.461	97.4	1.5%	1.940
CTS TW veneer [10]	6.717	3.060	171.85	1.8%	0.225	26.5	0.8%	2.117
CTS TW_CP	6.789	2.244	113.71	2.0%	0.486	84.64	0.6%	1.940
CTS TW_Qi	6.798	2.389	113.71	2.1%	0.688	84.64	0.8%	1.918
CNF TW veneer [11]	6.501	4.276	258.84	1.7%	0.207	31.27	0.7%	2.259
CNF TW_CP	6.617	3.059	167.81	1.8%	0.476	122.3	0.4%	2.015
CNF TW_Qi	6.617	3.225	167.81	1.9%	0.593	122.3	0.5%	1.988
PMMA T fibre W [9]	7.863	0.607	46.8	1.3%	1.669	46.8	3.6%	1.609
OSB board	7.372	1.059	9.9	10.7 %	1.353	9.9	13.7 %	1.741

Glass	4.924	15.231	20	76.2%	0.698	20	3.5%	2.122

best 2nd best 3rd best worst

The above data prove that transparent wood composites are suitable as the web material of I-beams. Also it is proven that veneer materials are globally a better fit for this purpose than transparent plywood materials. The smallest deflection is recorded for transparent wood veneer impregnated by cellulose nanofibers (CNF TW veneer). The lowest normal stress in x direction was calculated for transparent fibrewood (PMMA T fibre W), while the smallest ratio of normal stress and tensile strength in x direction was again for veneer material, this time impregnated by PVA (PVA TW veneer).

A comparison of the limit loads at deflection of 13.3 mm and the ratio of normal stress and tensile strength of the investigated materials are shown in table 3. Transparent wood composites show the best values in this respect as well.

Table 3. Comparison of the limit load at deflection L/300 recommended for beams of ceiling structure [14] and the corresponding degree of utilization in terms of stress.

Material of I-beam web	Limit load at the deflection of 13.3 mm = L/300 (kN/m)	σ_x / f_x (%)	σ_y / f_y (%)
PMMA TW veneer [8]	1.82	3.6%	11.0%
PMMA TW_CP [8]	1.83	4.2%	5.2%
PMMA TW_Qi [8]	1.82	4.4%	5.4%
PVA TW veneer [10]	1.79	1.4%	3.4%
PVA TW_CP	1.79	1.6%	2.5%
PVA TW_Qi	1.79	1.6%	2.7%
CTS TW veneer [10]	1.98	3.5%	1.7%
CTS TW_CP	1.96	3.9%	1.1%
CTS TW_Qi	1.96	4.1%	1.6%
CNF TW veneer [11]	2.05	3.4%	1.4%
CNF TW_CP	2.01	3.7%	0.8%
CNF TW_Qi	2.02	3.9%	1.0%
PMMA T fibre W [9]	1.70	2.2%	6.0%
OSB board	1.81	19.3%	24.7%
Glass	2.71	206.2%	9.5%

best 2nd best 3rd best worst

6. Conclusion

This analysis of mechanical properties of transparent wood composites proves these to be very promising for future use as load-bearing structures of buildings. Although very thin, these materials are more than able to carry the necessary load while still having a giant reserve of the degree of utilization.

The best mechanical properties were observed for transparent wood composites infiltrated by cellulose nanofibers (CNF) and polyvinyl alcohol (PVA). Very good optical properties of PVA infiltrated transparent wood are an advantage. The main advantage of cellulose nanofibers is at the field of ecology. Compared to other types though, their main disadvantage so far, are their weaker optical properties. Transparent fibrewood also showed some potential. Its main disadvantage is its low modulus, causing it to deflect more at lower stress than other materials. However, it is capable to bear great load and its deflection for the set rules was still quite low.

In general, it can be stated that veneer transparent wood materials are more suitable for such applications (web of the I-beams), but all of the transparent wood materials seem to be a good fit here.

Though, to be certain about it, larger samples of transparent wood would have to be made and tested, whether their mechanical properties are such as it is in relatively small samples which were tested so far.

7. References

[1] Kanócz J, Bajzecerová V and Karľa V 2020 Analysis of mechanical properties of I-beam with web from transparent wood *IOP Conf. Ser.: Mater. Sci. Eng.* **867** 012017
[2] Fink S 2009 Transparent Wood – A New Approach in the Functional Study of Wood Structure *Holzforschung - International Journal of the Biology, Chemistry, Physics and Technology of Wood* **46** 40–408
[3] Spalteholz W 1914 *Über das Durchsichtigmachen von Menschlichen und Tierischen Präparaten und Seine Theoretischen Bedingungen* (Leipzig: S. Hierzel)
[4] Li Y, Fu O, Yu S, Yan M, and Berglund L 2016 Optically Transparent Wood from a Nanoporous Cellulosic Template: Combining Functional and Structural Performance *Biomacromolecules* **17** 1358-1364
[5] Zhu M, Song J, Li T, Gong A, Wang Y, Dai J, Yao Y, Luo W, Henderson D, Hu L 2016 Highly anisotropic, highly transparent wood composites *Advanced Materials* **28** 5181–5187
[6] Li Y, Fu Q, Rojas R, Yan M, Lawoko M and Berglund L 2017 A new perspective on transparent wood: Lignin-retaining transparent wood *ChemSusChem* 10 3445-3451
[7] Li Y, Yang X, Fu O, Rojas R, Yan M and Berglund L 2017 Towards centimeter thick transparent wood through interface manipulation *Journal of Materials Chemistry A.* **6** 1094-1101
[8] Fu Q, Yan M, Jungstedt E, Yang X, Li Y, and Berglund L 2018 Transparent plywood as a load bearing and luminescent biocomposite Composites Science and Technology **164** 296-303
[9] Wang X, Zhan T, Liu Y, Shi J, Pan B, Zhang Y, Cai L, Shi S 2018 Large-Size Transparent Wood for Energy-Saving Building Applications *ChemSusChem* **5** 3855-62
[10] Mi R, Li T, Dalgo D, Chaoji,C, Kuang Y, He S, Zhao X, Xie W, Gan W, Zhu J Y, Srebric, J, Yang R, Hu L 2019 A Clear, Strong, and Thermally Insulated Transparent Wood for Energy Efficient Windows *Advanced Functional Materials* **30** 1907511
[11] Le Van H, Mwongeli R, Panicker P, Agumba D, Pham H, Kim J 2021 All-biobased transparent wood: A new approach and its environmental-friendly packaging application Carbohydrate Polymers **264** 118012
[12] Katunský D, Kanócz J, Karľa V 2018 Structural Elements with Transparent Wood in Architecture *International Review of Applied Sciences and Engineering - IRASE* 101-106
[13] Nettles A T 1994 *Basic Mechanics of Laminated Composite Plates* (Alabama: NASA Marshall Space Flight Center) 35812
[14] STN EN 1995-1-1+A1/NA 2008 Eurocode 5: *Design of Timber Structures. Part 1-1: General Rules and Rules for Buildings* National Annex

Acknowledgments
This contribution is supported by the Scientific Grant Agency and project KEGA 046TUKE-4/2019 supported by the Cultural Grant Agency of the Ministry of Education, Science, Research and Sport of the Slovak Republic and the Slovak Academy of Sciences. This paper was elaborated with financial support project with ITMS code: 26220220182 (TECHNICOM).

Methods for identifying non-value-adding activities in construction processes

D Klosova[1] and M Kozlovská[1]

[1] Department of Technology, Economics and Management in Civil Engineering, Faculty of Civil Engineering, Technical university of Kosice, Vysokoskolska 4, Kosice 042 00, Slovakia

E-mail: daria.klosova@tuke.sk; maria.kozlovska@tuke.sk

Abstract. Construction projects are characterized by low productivity due to many wasteful activities in construction processes. Wasteful or non-value-adding activities result in quality and profit loss in many construction projects. Consequently, construction needs more effective strategies for process innovation at all stages of the construction project life cycle. Analysis of the current state has shown the possibilities of improving construction processes through the implementation of technologies, methods and procedures of lean construction. Such improvements can help increase the efficiency and productivity of construction projects. Therefore, it is necessary to promote more effective construction waste management by identifying the causes of non-value-adding activities during the construction phase, and apply the gained data to eliminate wastes during the planning phase. The paper presents the most appropriate technologies and methods for collecting data on non-value-adding activities.

1. Introduction

Construction projects are characterized by complex coordination of various functions and interconnections of a large number of stakeholders involved in the project [1], which ultimately leads to delays in various construction activities, as well as cost and time overruns [2]. Introducing lean into construction can help increase productivity, reduce inventory, improve product quality, and reduce lead time by eliminating waste or non-value-added activities [3].

Lead time is the time between the initiating of a process and its completing. In construction projects, lead time usually refers to the amount of time that elapses between placing and ordering an item and delivering it, but it can also refer to the time it takes to prepare drawings or complete approval processes. Lead time can be affected by different variables, such as market conditions, holiday periods, strikes, transportation, inclement weather, and so on. Cycle time starts when the actual work on the jobsite begins and ends when the work is completed. In other words, cycle time or process time measures the execution or productivity rate and adds value to the product. Whereas lead time measures all the time invested in a production system, including time that does not add value to the product, such as waiting time or transportation time [4].

The Lean construction philosophy was based on the concepts of lean manufacturing. Its main aim is managing and improving the construction process with profitability and customer needs in mind [3]. Lean methodology brings together stakeholders at all levels to create a multidisciplinary team whose main goal is continual improvement [5]. Lean is a new paradigm in construction planning that approach efficiency rather than schedule, and value rather than cost. Therefore, the best way to achieve value is to integrate lean in early planning stages [3]. An important characteristic of Lean is flexibility because it has been successfully applied in a variety of areas like healthcare, IT, retail, government, transportation, commercial and public sector, supply chain, manufacturing and construction [6].

Lean philosophy has led to a new perspective on what value is. As defined in [7], value-added (VA) activity contributes to the transformation of a product or service to the final result and creates value for which the buyer is willing to pay. Whereas, essential non-value added (ENVA) activity support the performance of value-adding activities but do not add value. Finally, non-value-added (NVA) activity only uses resources but does not create value [2]. The time spent on NVA activity is different in different projects, especially in the construction sector [2][8], where each construction object has its own unique characteristics. In previous study [9], we described one of the most popular classification of non-value adding activities according to principles of Lean thinking, namely: waiting, motion, transportation, defects, overproduction, over-processing, inventory, and making-do. Identifying the incidence of non-value-adding activities during the construction process, can help managers to easily identify the best solutions and ways to apply any new technique for reducing the amount of waste, leading to increased project productivity [10].

The research sector plays a significant role in improving construction management and productivity, but its effectiveness is directly dependent on the quality of the data collected. Data mean evidence and information from observation or experience. Otherwise, the data themselves have no real meaning. The meaning of the data can only be obtained through their adequate interpretation. The efficiency of data collection is crucial for the overall quality of research. Therefore, it is very important to pay due attention to the data collection process to ensure the reliability of the future results [11]. The set of variables during data collection should be clearly conceptualized, defined, measured, and interrelated to avoid conflicting results [12].

Construction site data (personnel, equipment, materials) can provide information for better decision-making process, help to identify weak spots and waste in production process, but collection, analysis and visualization are very energy intensive and almost never undertaken. Furthermore, access to real-time information and subsequent technologies that can collect, process and visualize data without much human effort is limited [13]. Therefore, it is very important to identify existing and potential methods of data collection that can be used for effective implementation of lean principles in construction industry in order to reduce lead time and thereby increase productivity.

This study aims to identify methods and technologies suitable for collecting and processing data on wasteful activity on construction sites.

2. Data collection methods for implementing lean production - case study

Case study analysis was carried out through a literature review to summarize the data collection methods used to support lean implementation. A literature review found evidences of using various data collection methods to waste identification in different industries.

Alwi et al. investigated non-value-adding activities and causes of their occurrence within contracting organizations in Australia. Their research was based on quantitative approach utilizing the results of a questionnaire survey distributed to 45 contracting companies. The response rate was nearly 60%. Results showed that waiting for instruction, poor quality site documentation, weather, unclear drawings, design changes, poor design, slow drawing revision, and unclear specifications were identified as key causes of non-value-adding activities [10].

Hossain et al. investigated rate of non-value adding activities in selected construction processes in Astana, Kazakhstan using observations. Hossain et al. observationally investigated the level of non-value-added activities in selected construction processes in Astana, Kazakhstan. Typical construction processes were chosen as the subject of the research, namely the installation of an aerated concrete block, installation of a floor slab, installation of a bracket, tile cutting, and wall painting [2].

Other study tried [14] to identify the causes of non-value adding activities during the client briefing phase in order to minimize the occurrence of them in design process. For this study, the authors used methods of literature review, questionnaires, and interviews with registered architects, structural designers, project managers, quantity surveyors, and contractors in Cape Town. However, according to the authors, this method did not give positive results, and authors failed to identify and minimize the causes of waste during construction client briefing process.

A study from India [3] used a questionnaire to analyze the views of various project participants on causes of waste in a construction project based on the Lean philosophy. A statistical tool called the Relative Importance Index (RII) was used to correctly interpret the result. RII is a technique used to determine the relative importance of each parameter and factor included in a survey and rank them in order of importance.

In order to monitor the progress of construction, inspectors still need to visit the construction site in person or spend a lot of time to process the data obtained from the construction site using vision and laser technologies. However, these inspection methods lack real-time automation and data exchange. As a consequence, there is a health risk of physical interaction between workers and the inspector, loss of energy, loss of data and waste of time. Khairadeen et al. proposed a real-time construction inspection system called iVR. The system helps visualize interactive indoor inspections on-site and provides numerical data through the integration of Kinect 3D scanning, augmented reality, and visual programming [15].

Proposed system consists of five modules [15]:

- iVR-location finder (finding the best location for a laser scanner on a construction site),
- iVR-scan (capturing point cloud data on indoor on-site activities),
- iVR-prepare (processing and converting 3D scan data into 3D model.),
- iVR-inspect (immersive virtual reality inspection)
- iVR-feedback (visualization of inspection report from the workplace using augmented reality).

The main task of this system is to transfer information from the construction site to the construction office and vice versa by integrating a monitoring device with augmented reality technology into one platform, that is synchronized between the construction office and the construction site in almost real time [15].

Cheng and Teizer presented a real-time resource location data collection and visualization framework designed for construction safety and activity monitoring. This framework is based on streaming data from real-time positioning sensors to a real-time data visualization platform. The tracking technology used is an active radio frequency identification (RFID) technology called Ultra Wideband (UWB). The tagged resources were workers; truck crane cabin, its hook and four outrigger positions; and any temporary vehicle (s) entering the pit. Each tag collected spatio-temporal data and sent the data to a local server. The server then calculated the necessary information such as the position, speed, and direction of each resource. The processed information was then published in the virtual world by linking to the corresponding 3D model with a unique object identifier (ID) [13].

The study conducted by Owad et al. presents a comprehensive lean methodology to identify the root causes of patient flow problems in emergency departments in Saudi Arabia. This research adopts a mixed research strategy, involving observations for investigating the current patient flow, an integrated lean methodology (Lean Six Sigma for process mapping and the A3 problem-solving sheet as visual tools to identify sources of waste), followed by validation on a case study. However, the authors stated that the lean Value Stream Mapping tool could not be used for purpose of this study, because the exact time required at each stage of the treatment process could not be recorded [5].

In another example, a garment manufacturing company applied a stopwatch observation method for the selected product to enhance productivity and increase market competence by eliminating waste, standardizing work, and balancing lines [16].

Logesh and Balaji [17] developed a lean implementation concept and evaluated it at a leading electrical component manufacturing facility in India that produces switches and switch boards. Data for this study was collected through direct observation of the actual production line of electrical switchboards and records on a shift basis. In addition, mind mapping sessions, meetings, and interviews were held among technical assistants, team and managerial levels.

Suryoputo et al. used value stream mapping, Job safety analysis, 10 beats and Rapid Entire Body Assessment to identify and reduce existing waste in the music manufacturing company. In this research data collection was performed directly from by observations of production operator [18].

Another study analyzed the impact of implementing lean principles on the example of Polish cafeterias. Employee interviews and casual observation were used as the data collection method [19].

3. Results and Discussion

The efficiency of data collection is crucial for the overall quality of research. Data can be quantitative or qualitative, but their presence is a necessary part of empirical research. Quantitative data collection is based on surveys, experiments, gathering data using measurement techniques or equipment, structured interviews, or questionnaires, while qualitative data collection is classified as observation, unstructured and semi-structured interviews, projective methods, diaries, verbal protocol, unstructured questionnaires, and documentary inspection. Besides, with the increase in qualitative and quantitative research in the field of construction management, new ways of collecting data have emerged [11].

The identified methods of data collection that can be used to monitor construction productivity are presented below.

- **Surveys** are appropriate method for documenting beliefs, attitudes, perceptions, or knowledge within a predefined sample of people. Surveys questions are usually open-ended in a free-text format. Good open-ended questions should be specific enough to provide consistent responses from respondents, yet broad enough to offer a wide range of responses [14]. The survey can be conducted in person or face-to-face, by phone or the respondent can fill it out on his own. Self-completion of the survey will be the cheapest and fastest method of obtaining data for researchers in conditions of limited time and financial resources Online surveys have many advantages over paper surveys. For example, in web surveys, data is stored electronically, which can help obtain data much faster than a paper survey, and the data is automatically compared as results are obtained. The online survey feature is useful for large-scale research and helps reduce data entry errors. Otherwise, this method is often chosen not because of its suitability for obtaining research data, but because of its low cost and relative speed. The choice of survey tool does not matter, whether it be a website, paper material, a telephone survey or an interview, the main requirement for collecting data is the provision of evidence [11].

- **Interviewing** is an information gathering method in which data is often recorded and transcribed. This method is based on a series of predefined questions or a set of areas of interest that are asked individually to people. Interviews can follow a pre-written script and look like a survey, or they can allow the respondent to express their thoughts more freely. The interview data can be used to create themes, theories, and models. Research questions can often be answered through both surveys and interviews, but interviews tend to provide richer and more detailed data than surveys. However, interviews require more time and resources to conduct and, more importantly, to analyze [20].

- **Focus groups** allow information to be collected in a group discussion process, when the sum of the experiences of a group of people can offer more than an individual experience. Focus groups are a way to capture similarities and differences of opinion. The optimal number of participants in the group to create a rich discussion is from 8 to 10 people [20].

- **Observation** is a way to collect information using the senses: hearing, vision, touch, and smell. Observations are ideal for documenting, understanding, and exploring actions, activities, relationships, culture, or ways of doing things [20]. Observation is an important tool for understanding and obtaining information about a phenomenon [2]. Observation is the process of recognizing and recording people's behaviour, events and objects [16][21]. Direct observation of an existing process can also serve as a tool for collecting data related to a research problem [16].

- **Textual or content analysis** is ideal for documenting the context of certain practices, for investigating changes in official, institutional, or organizational views on a particular topic or area, or for exploring the experiences, views and results of other people work. Documents for analysis may include government reports, newspapers or research articles, organization policies and protocols, recordings, films, letters, photographs, art, checklists, meeting notes or any other textual source [20].

- **Photogrammetry**
  Photogrammetry uses photographs taken with a conventional digital camera as input and generates a three-dimensional cloud of points of the displayed area using computational algorithms. Photos can be collected by an employee or using unmanned platforms. For example, photogrammetry can be used to measure the volume of excavated soil for daily assessment of productivity during earthworks [22].

- **Camera analysis**
  Construction cameras are used for time-lapse and real-time video, as well as high-resolution photography. Construction cameras help document project progress, monitor workers, and allow all stakeholders such as investors, clients, and project managers to see project status remotely [23]. To completely cover a large construction site, it is necessary to install a network of cameras at different locations, elevations, and angles, because a single camera can only cover a limited field of view [24].

- **Audio signals /Acoustic system**
  The idea of using audio signals as a source of information provides an alternative solution for data collection. Audio signals recorded and collected on construction sites reflect the relevant activities that are taking place and can provide useful information for project managers. For example, construction equipment often generates unique sound patterns when performing various tasks, and therefore, by processing this data, we can potentially extract a large amount of information regarding basic activities [24].

- **Wearable technologies**
  There are various wearable technologies that can provide continuous monitoring of a wide range of parameters, mainly used for safety, security, and analysis of processes on a construction site. Wearable technologies are based on various systems such as ultra-wide band (UWB), radio frequency identification (RFID), magnetic field, Bluetooth, radar, ultrasonic, sonar, global positioning system (GPS), laser, video, and static camera. An extensive set of wearable physiological sensors can include an EMG (electromyography) sensor for monitoring muscle activity, an ECG / ECG (electrocardiogram) sensor for monitoring heart activity, an EEG (electroencephalography) sensor for monitoring electrical activity in the brain, a blood pressure sensor, a tilt sensor for controlling the position of the torso, a respiration sensor and motion sensors used to assess the user's activity [25], [26]. Another system used in wearable sensors for calculating heart rate, tracking activity level and physical activity performance is infrared technology. Infrared is electromagnetic energy in which the wavelength is slightly longer than that of red light [26]. One example of a commercially available infrared sensor is a heart rate t-shirt designed to reduce the risk of heart attacks in athletes. The electrodes, printed on the shirt fabric, are fed to a processing unit, which transmits and processes ECG data wirelessly to a mobile phone via Bluetooth. It can detect abnormal heart rhythms that can lead to sudden cardiac arrest [27].

- **Mobile software solutions**
  Mobile devices and applications can also be seen as a source of a lot of information and important data. For instance, mobile time tracking solutions allow employees to simply clock

in and out directly from their mobile phones or from a centralized kiosk with real-time timestamps and GPS tags. Geofencing technology automatically ensures that employees only clock in and out in the workplace. Mobile time tracking apps allows managers to easily find out who is working where and when, view location stamps on a map, manage employee absence and shift requests, receive real-time reports and forms of daily activity, and automate the flow of information [28].

The application of the surveys and observation is evident in many studies. Evidently, surveys are still a common data collection tool [11]. Manual visual observations and routine progress tracking are used to provide feedback on progress and productivity rate, and for planning and monitoring safety [15]. However, traditional construction control methods are time consuming, labor intensive, energy consuming, inefficient and error prone [29]. Therefore, it is necessary to develop more reliable and automated monitoring of construction progress so that it works with lower costs and less labor by reducing the physical interaction between the builder and the inspector [15]. Thus, the development of a method for monitoring construction projects in near real time is an urgent need in the construction industry [19].

4. Conclusion

Data is something that is highly valued in the modern world. This article explores various data collection methods that can be used in the construction industry to identify non-value-added activities at a construction site. Effective methods for identifying non-value-added activities can help construction managers apply the best solutions and techniques to reduce waste in construction processes, which will undoubtedly lead to increased project productivity.

The literature review showed that practices of manually recording and analyzing data is still at the forefront of data collection. These methods are expensive, prone to data entry errors, and result in too small datasets for effective and successful project management. Technologies such as RFID, UWB, laser scanning and augmented reality are also used in the construction industry to collect and analyze data, but to a much lesser extent, which indicates that there is a great potential for development in this direction. Automated construction monitoring is a necessary solution for accurate and continuous performance monitoring on construction sites that can help overcome the limitations of manual labor.

In addition, a combination of different technologies can have a better effect than using one method, especially in construction. Furthermore, it seems necessary to develop / use individual performance monitoring methods for investigation of non-value-added activities for each specific construction process.

5. References

[1] Hampson K D 1997 *Proc. of Innovation Systems and the Construction Industry: An International Workshop* (Montreal: Canadian National Research Council, Institute for Research in Construction)

[2] Hossain A Bissenova A and Kim J R 2019 Investigation of Wasteful Activities Using Lean Methodology: In Perspective of Kazakhstan's Construction Industry Buildings **9** 113

[3] Rajkumar K B and Biswas A P 2016 Identification of Non-Value Adding Activities using Lean Technology in Construction *Int. J. of Engineering Research & Technology (IJERT)* IJERTV5IS03089

[4] Tyagi S Choudhary A Cai X and Yang K 2015 Value stream mapping to reduce the lead-time of a product development process *Int. J. of Production Economics* **160** 202–12.

[5] Al Owad A Samaranayake P Karim A and Ahsan K B 2018 An integrated lean methodology for improving patient flow in an emergency department - case study of a Saudi Arabian hospital *Prod. Planning and Control* **29**(13) 1058–81

[6] Veres C Marian L Moica S and Al-Akel K 2018 Case study concerning 5S method impact in an automotive company *Proc. Manufacturing* **22** 900–5

[7] O'Connor R. and Swain B 2013 *Implementing Lean in Construction: Lean Tools and Techniques an Introduction* (London: CIRIA C730)

[8] Horman M J and Kenley R 2005 Quantifying levels of wasted time in construction with meta analysis *J. Constr. Eng. Manag* **131** 52–61

[9] Kozlovská M Klosova D 2020 Wastes perception in construction projects life cycle *Waste Forum* **4** 265-72

[10] Alwi S Hampson K and Mohamed S 2002 Non Value-Adding Activities in Australian Construction Projects *Proc. of the Int. Conf. on Adv. in Design, Constr. Constr. Manag. And Maint. of Build. Str.* ed Wangsadinata W (Bandung: RIHS) pp 270–80

[11] Carter K and Fortune C J 2004 Issues with Data Collection Methods in Construction Management Research *Proc. of the Twentieth Ann. Conf. of Association of Researchers in Constr. Manag.* pp 939-46

[12] Sapateiro C M Antunes P Johnstone D and Pino J A 2017 Gathering big data for teamwork evaluation with microworlds *Cluster Comput.* **20** 1637–59

[13] Cheng T and Teizer J 2013 Real-time resource location data collection and visualization technology for construction safety and activity monitoring applications *Automation in Constr.* **34** 3–15

[14] Nghona X Crowe J and Ndihokubwayo R 2010 Identification of the Causes of Non- Value Adding Activities during the client briefing process proceedings *Proc. of the fifth built environment conference* (South Africa: ASOCSA) pp 144-161

[15] Khairadeen Ali A Lee O J Lee D and Park C 2021 Remote Indoor Construction Progress Monitoring Using Extended Reality *Sustainability* **13** 2290

[16] Mulugeta L 2021 Productivity improvement through lean manufacturing tools in Ethiopian garment manufacturing company *Materials Today: Proceedings* **37**(2) 1432-36

[17] Logesh B and Balaji M 2021 Experimental Investigations to Deploy Green Manufacturing through Reduction of Waste Using Lean Tools in Electrical Components Manufacturing Company *Int. J. of Precis. Eng. and Manuf.-Green Tech.* **8** 365–374

[18] Suryoputro M R Sari A D Burhanudin R and Sugarindra M 2017 Lean production design using value stream mapping and ergonomics approach for waste elimination on buffing panel upright process *IOP Conference Series: Materials Science and Engineering* **277**(2017)012015

[19] Gładysz B Buczacki A and Haskins C 2020 Lean Management Approach to Reduce Waste in HoReCa Food Services *Resources* **9**(12) 144

[20] Paradis E O'Brien B Nimmon L Bandiera G and Martimianakis M A 2016 Design: Selection of Data Collection Methods *J Grad Med Educ* **8**(2) 263–64

[21] Chisosa D F and Chipambwa W 2018 An exploration of how work study techniques can optimize production in Zimbabwe's clothing industry *J. Textile Apparel Technol. Manage.* **10** (3) 1–11

[22] Bügler M Borrmann A Ogunmakin G Vela P A and Teizer J 2017 Fusion of Photogrammetry and Video Analysis for Productivity Assessment of Earthwork Processes *Comp.-Aided Civil and Infrastructure Eng* **32** 107-123

[23] TrueLook 2017 Improve Worker Productivity with the Watchful Eye of Construction Cameras [online] https://www.truelook.com/blog/improve-worker-productivity-with-the-watchful -eye-of-construction-cameras/

[24] Cheng Ch-F Rashidi A Davenport M A and Anderson D V 2017 Activity analysis of construction equipment using audio signals and support vector machines *Automation in Constr.* **81** 240-53

[25] Jovanov E Milenkovic A Otto C and Groen P C 2005 A wireless body area network of intelligent motion sensors for computer assisted physical rehabilitation *J. Neuroeng. Rehabil.* **2** 6

[26] Awolusi I Marks E and Hallowell M 2018 Wearable technology for personalized construction safety monitoring and trending: Review of applicable devices *Automation in Constr.* **85** 96-106

[27] Seshadri D R Li R T Voos J E Rowbottom J R Alfes C M Zorman Ch A and Drummond C K 2019 Wearable sensors for monitoring the internal and external workload of the athlete. *Npj Digit. Med.* **2** 71

[28] Regan R 2020 7 Best Construction Time Tracking Software Solutions [online] https://connecteam.com/construction-time-tracking-software/

[29] Scott S and Assadi S 1999 A survey of the site records kept by construction supervisors *Constr. Manag. Econ.* **17** 375–82

Acknowledgments
The activity presented in the article is part of the VEGA research grant No. 1/0557/18 "Research and development of technological and product innovations of modern construction methods in the context of the principles of Industry 4.0" and APVV project No.17-0549 "Research of knowledge and virtual technologies for intelligent designing and realization of building projects with an emphasis on their economic sustainability".

Experimental verification referring to the new shape of the perforated steel strip

A Petrík[1]

[1] Department of Steel and Timber Structures, Faculty of Civil Engineering, Slovak University of Technology in Bratislava, Radlinského 11, 810 05 Bratislava, Slovakia

E-mail: adam.petrik@stuba.sk

Abstract. Perforated steel strips are used as reliable shear connectors many years. During this time, many different shapes of the openings or recesses were developed to improve the durability and resistance of the shear connection. Behaviour of these strips were mainly verified by experimental research, but many parameters could not be measured. Numerical analysis provides not only easy and fast way of testing the behaviour of specimens, but also the possibility to improve the geometry of the strips. Modified steel strips can be deeply investigated in simulations and subsequently verified by push-out tests in laboratories. The combination of these two methods can bring the spectrum of the information that can complexly describe the behaviour, durability and also the weak points of the shear connection, especially in case of the new or modified shape.

1. Introduction

Numerical simulations provide effective and economical way of developing new components not only in mechanical engineering, but increasingly also in civil engineering. Initial parameters of the newly developed components or its modification can be easily tested by numerical simulations and afterwards, it can be evaluated by experimental tests. This procedure is focused on complex understanding of behaviour and analysis of important data related to the evaluated component in environment of computer model in order to adjust material, geometrical or other inputs.

This procedure can be also used in analysis of perforated steel strips used in composite connections. Instead of developing the new shape of perforated steel strip, the optimization of geometry of existing one can be done with the aim of increasing the parameters of the composite connection. Afterwards, the computer analysis of final shape must be verified by the experimental tests to confirm the correctness of the numerical model and investigate the behaviour and failure model of test specimens.

2. Developing of the numerical model

To obtain appropriate results of the numerical simulations, correctly developed model is vital with precisely defined geometrical and material parameters capturing the real specifications of the structures. It is advisable to set every property in numerical model equally to the laboratory experiment specimen to get ideally identical results. Due to the size of the model and the subsequent amount of the finite elements and computational time, it is possible to prepare only fragment of the construction, but only if it does not impact on the resistance value or the global behaviour of the model.

In order to analyse the composite connection with the perforated steel strips, in the initial phase of the development of appropriate numerical model, the fragment of the connection with the properly defined boundary conditions appears to be appropriate approach to obtain the correct results. In case of

IOP Conf. Series: Materials Science and Engineering 1209 (2021) 012033 doi:10.1088/1757-899X/1209/1/012033

the composite connection with the perforated steel strips, the push-out tests are principal in determining the resistance value and the overall behaviour of the connection. Because of this fact, the fragment of the real push-out test specimen of composite connection with the precise geometry of MCL perforated steel strip [1] was prepared in 3D software for the purpose of the insertion in FEM software ANSYS [2] to performed numerical simulations. This type of model was also previously used in analysis [3] and proved the good correlation with the experimental tests.

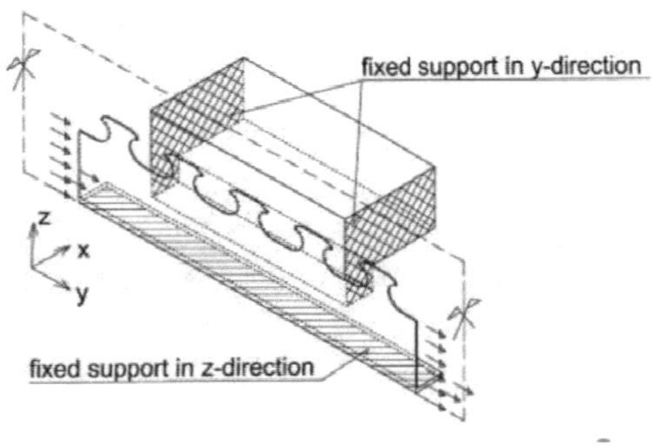

Figure 1. Scheme of boundary conditions of model.

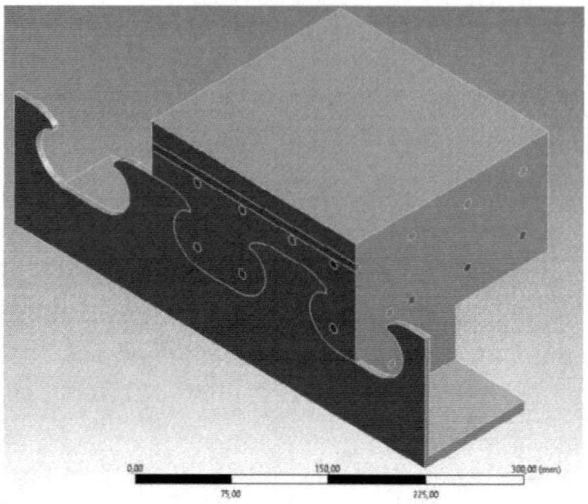

Figure 2. The model of the composite connection in ANSYS software.

One of the main advantages of well-prepared numerical model is the possibility to obtain important parameters that cannot be measured during the experimental tests (stresses or strains in the whole model) or monitor the behaviour of the specific part or component of the connection (deformation of components of composite connections), alternatively to identify the mechanism of failure.

Unlike the analysis [3], the reinforcing steel rebars were added to the model and the material parameters of this model of composite connection were defined elastically and plasticly as follows: steel part by true stress-strain curve for S355 steel; concrete part by Menetrey-William material model [4] of concrete C30/37 and reinforcing steel B 500B by bilinear stress-strain curve.

It can be claimed that the prepared numerical model captures as much as possible the geometry, material parameters and arrangements of the perforated steel strip and reinforcing rebars of the experimental test specimen.

3. Numerical analysis, optimization and developing the new shape

After the developing the numerical model and defining the parameters, the numerical simulations of original MCL perforated steel strips were performed. The results confirmed ductile behaviour of the connection with the appropriate value of the resistance value.

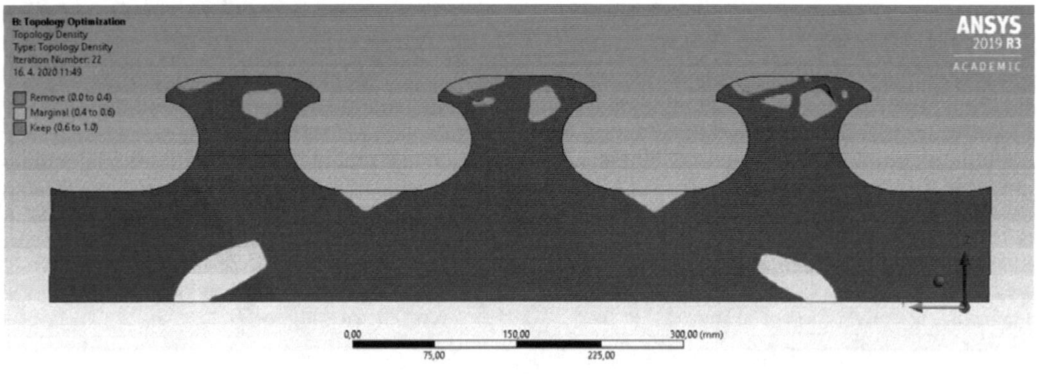

Figure 3. Results of the initial topology optimization of the MCL perforated steel strip (holes represent omitted material after topology optimization).

After the initial analysis, the optimization procedure, consisted of several topology optimizations was performed in ANSYS software. Based on the results of the topology optimization (figure 3), the final shape (figure 4) of modified MCL steel strips was developed to increase the resistance value, at the least possible extent reduce the negative effect on fatigue resistance and deformation capacity of the composite connection by creating the recesses in the triangular area between the shear connectors. To maximize the effect of these recesses to increasing the resistance value, it is advisable to place there an extra bar of reinforcing steel.

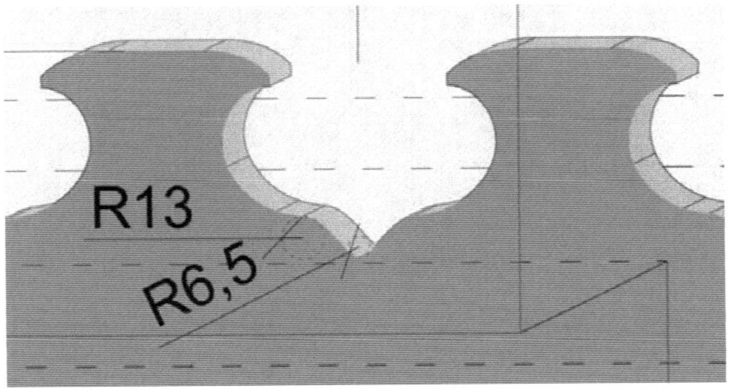

Figure 4. The final shape of modified MCL perforated steel strip with the geometric parameters of the recesses between the connectors.

The assumptions about increasing the load-bearing parameters were confirmed by numerical simulation, that proved almost 20% increase of the resistance value of the composite connection (figure 5) with still appropriate ductile behaviour according to EN 1994-1-1 [5].

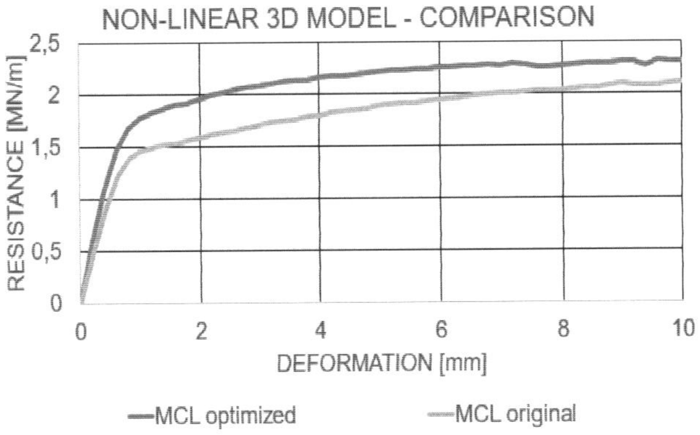

Figure 5. Comparison of the resistance values related to the original and optimized shape of MCL perforated steel strip.

4. Experimental verification of the numerical analysis

Verification of such modifications as created is not only possible through the numerical simulations, but it is necessary to validate them with the experimental tests. Due to the simplifications incorporated in the model (e.g., fragment of the connection), the results from simulations may be inaccurate and can lead to inexact conclusions. Also, testing procedure sometimes includes operations (e.g., repeated loading with a lower force value before full loading), that is usually omitted in numerical simulations.

To verify the parameters gained from numerical simulations and, also, validate the modification of the perforated steel strip, two test specimens for push-out tests of composite connections with perforated steel strips were prepared – one with the original MCL perforated steel strip and the other one with modified MCL perforated steel strip with the geometry gained from topology optimization. Overall dimensions, geometric parameters and arrangement of the components were identical, except the additional bar of reinforcement placed in the recesses between the shear connectors (figure 6 and figure 7). The materials of the test specimens were chosen correspondingly to the parameters in numerical model to provide the comparison of the experimental and numerical test in the end of the research (confirmed by the material tests of test specimens after the experimental tests).

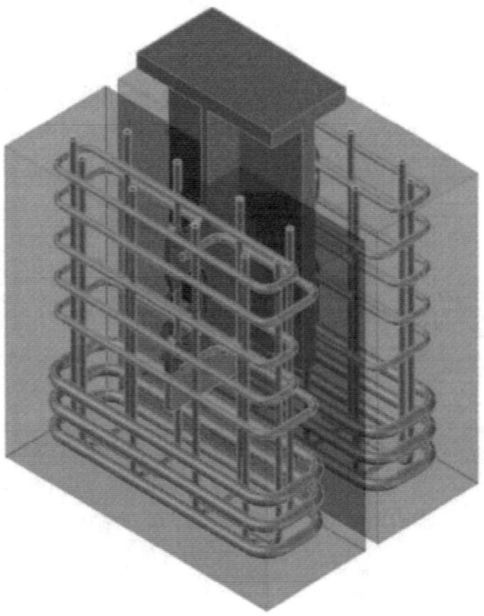

Figure 6. 3D visualization of test specimen of the composite connection with the original MCL perforated steel strip.

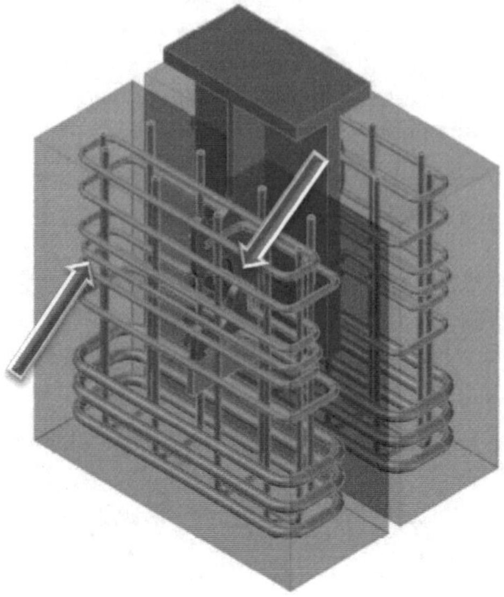

Figure 7. 3D visualization of test specimen of the composite connection with the modified MCL perforated steel strip.

The test procedure follows push-out test recommendations in the norm EN 1994-1-1 [5] – the test specimen is initially loaded by 40% of the expected maximum of the resistance value and then released to 5% of the maximum value. This procedure is repeated for 25 times and afterwards, the test specimen is fully loaded up to its damage. During the whole test, the displacement value and the applied force are measured, and this time, the strain gauges are placed on the perforated steel strips to measure the stress at the beginning of the test before the plastic behaviour. These parameters are important to measure in order to compare them with the results from numerical simulations for the purpose of their validation.

Figure 8. One of the test specimens before the concreting – the perforate steel strips (lubricated surfaces) placed in the formwork with the reinforcing steel.

Figure 9. The test specimen after removal of the formwork before the push-out test.

The test specimen with the original MCL perforated steel strip behaved as expected from previous tests in universities abroad [3] – during the first phase of loading with lower load the displacement was continuously increasing because of the cracks in the concrete area. The maximum value of the applied force was 354,8 kN at displacement of 12,35 mm, which compared to the value from the numerical simulation equalled to 374,4 kN provide good compliance of the numerical simulation to experimental test. The failure of the specimen occurred at the displacement equalled to 31,9 mm. The experiment confirmed very good ductile behaviour of the composite connection.

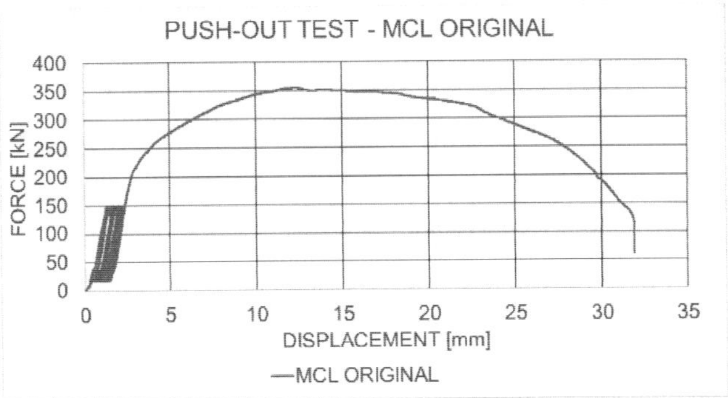

Figure 10. Results of the push-out test related to the composite connection with original MCL shape of the perforated steel strip.

Based on the numerical simulation, the second test specimen should provide higher resistance value compared to the first specimen with original MCL perforated steel strip. However, the behaviour of the test specimen changed and during the first phase of the testing procedure, only minimal increments of the displacement were measured, that could be caused by the increase of the rigidity of the connection by adding extra reinforcement in the recesses between the shear connectors. Only after the increase of the applied force for almost 80% of the maximum applied force (twice the value given in EN 1994-1-1 [5]), the cracks in the concrete occurred, while there was also a sharp increase of the displacement value to expected value after the repeated loading procedure.

The maximum value of the applied force was 423,7 kN at displacement of 10,55 mm, which also compared to the value from the numerical simulation equalled to 418 kN provide good compliance of the numerical simulation to experimental test and confirmed expected increase of the resistance force by the modification of the geometry of the perforated steel strip. The failure of the second specimen occurred at the displacement equalled to 29,4 mm. Interesting fact is that from the displacement value approx. 10 to 18 millimetres, there is a plateau of the resistance value with the difference up to 10 kN. The experiment confirmed the assumption that the small modification of the area between the shear connectors of the MCL perforated steel strip does not adversely affect the ductility of the composite connection.

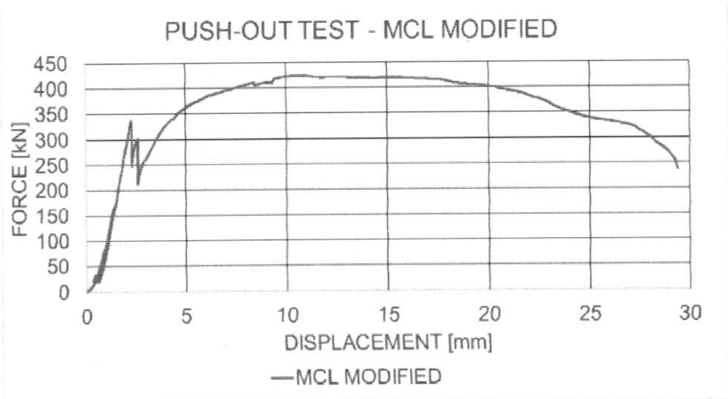

Figure 11. Results of the push-out test related to the composite connection with modified MCL shape of the perforated steel strip.

Despite thorough protection of strain gauges placed on the perforated steel strip, they failed to measure after the initial slip of the steel strip in concrete, that damaged the gauges or their connection with the cables. In the future, different type of the protection should be applied.

Both test specimens' failures were caused by the shear or bending failure of one of the shear connectors in the narrowest area (figure 12 and figure 13). However, global presumed mechanism of failure of the composite connection consisting of combined steel-concrete failure did not occur, presumably because of the omission of the large fraction of aggregate in the concrete.

Figure 12. Cracked shear connector of the original MCL perforated steel strip.

Figure 13. The shear failure of one of the shear connectors in the narrowest area

5. Conclusion
The main aim of this research was to prove the efficiency and correctness of the well-prepared numerical model compared to the experimental tests. Based on the previous research, the existing numerical model was modified to be identical to test specimen for the test in every way. Moreover, by topology optimization, the shape of the original MCL perforated steel strip was modified to increase the load-bearing capacity with the unchanged ductility.

Experimental test in laboratories showed very good compliance of the results between these two approaches. This was vital to have complex view on the connection with the perforated steel strip – there is a possibility to investigate the stresses or strains in every finite element of numerical model and global behaviour and mechanism of failure of test specimens in laboratory tests. If all the strict conditions for compliance with these approaches are met, it is possible to analyse in detail any problems in civil engineering through the numerical simulations and subsequently verified by the experimental tests.

6. References
[1] Deutsches Institut für Bautechnik 2018 *Allgemeine bauaufsichtliche Zulassung, Stahlverbundträger mit Verbunddübelleisten in Klothoiden- und Puzzleform* (Düsseldorf: FOSTA)
[2] Ansys, Inc. *Software ANSYS 2019: User's Manual* Available online https://ansyshelp.ansys.com
[3] Lorenc W, Kozuch M and Seidl G 2013 Zur Grenztragfähigkeit von Verbunddübeln mit Klothoidenform *Stahlbau* **82** pp 196-207
[4] DYNARDO GmbH 2013 *multiPlas - User's Manual - Release 4.1.8 for ANSYS 14.5 - Elasto-plastic material models for ANSYS - General multisurface plasticity* (Weimar: multiPlas)
[5] European Committee for Standardization 2004 *EN 1994-1-1: Eurocode 4: Design of Composite Steel and Concrete Structures. Part 1.1: General Rules and Rules for Building* (Brussels)

Acknowledgments
This paper has been supported by the Scientific Grant Agency of the Ministry of Education, science, research and sport of the Slovak Republic and the Slovak Academy of Sciences – grant VEGA 1/0773/18.

IOP Conf. Series: Materials Science and Engineering 1209 (2021) 012034 doi:10.1088/1757-899X/1209/1/012034

Measurement of the runoff coefficient of extensive greenroof

J Vystrčil[1], O Nespěšný[1], K Šuhajda[1], D Bečkovský[1] and P Selník[1]

[1] Faculty of Civil Engineering, Institute of Building Structures, Brno University of Technology, Veveří 331/95, 602 00 Brno, Czech Republic

E-mail: jan.vystrcil@vutbr.cz

Abstract. Describes the procedure of experimental measurement of the runoff coefficient C, both of individual layers and the entire composition extensive green roofs. Experimental measurements make it possible to determine the reference behaviour of runoff characteristics, namely runoff coefficient C, with emphasis on the simulation of the real behaviour of extensive green roofs. The aim is an elementary description of the structural and physical behaviour of extensive green roofs. For the needs of experimental measurement, the dimensional and shape limits of test specimens are described, the conditions for conditioning of individual specimens, the boundary conditions of execution and individual steps of the experiment. Then is specified the method of evaluation and subsequent verification of measured data. The result of the experimental measurement is the amount of drained water from the tested specimens of the extensive green roof at time t, which shows a nonlinear behaviour. From the set of measured data, it is then possible to predict the behaviour of extensive green roofs in real conditions and to determine the runoff coefficient C of the tested specimens. These data represent reference values for the subsequent design of sub-elements and structures of buildings.

1. Introduction

According to Scharf B, Kraus F, in publication Green roofs and Greenpass [1] The United Nations has identified climate change as the greatest threat to human life. Following these results there is an increasing emphasis on the sustainable development of urbanism, architecture and society as a whole. Frequently addressed topics following climate change are the management of drinking water and rainwater, the elimination of the formation of heat islands in urban areas [2], the impact of greenery on the quality of urban environments [3]. One of the important measures to reduce heat islands, partial retention of rainwater on the construction site and reduce the risk of flash floods in cities during torrential rains is to design a of green roofs on buildings.

One of the parameters of the green roof formation is the runoff coefficient *C*. It is a dimensionless quantity indicating the ratio of the runoff rainwater to the total precipitation for the observed period. This parameter fundamentally affects the following sub-elements of buildings, such as rainwater drainage elements, retention tanks, infiltration objects etc. Furthermore, it can fundamentally affect the price for rainwater wastewater discharged into the public sewerage network.

The currently valid standards in the Czech Republic use only simplified table values to calculate the amount of rainwater drained from the roofs. ČSN EN 12056-3 [4] states for the calculation of the amount of rainwater drained from the roof, the runoff coefficient $C = 1.0$, unless national and local regulations and practices provide otherwise. ČSN 75 6760 [5] states the runoff factor *C* for the same calculation, with a slope of the roof planes of up to 5%, according to the following division 1.:

Table 1. Selected values of runoff coefficient according to ČSN 75 6760 [5].

Type of drained surface type of surface treatment	Tilt of surface <1%	1%-5%	>5%
	Runoff coefficient of drained water C		
Roofs with a permeable top layer of thickness up to 100 mm	0.7	0.7	0.7
Roofs with a permeable top layer of thickness over 100 mm to 250 mm	0.4	0.4	0.5
Roofs with a permeable top layer of thickness up to 250 mm	0.3	0.3	0.3

However, the mentioned standards do not state under which conditions this coefficient was determined.

The German regulation FLL Green roof - Guidelines for the planning, construction and maintenance of Green roofs [6], uses the nomenclature according to DIN EN 12056-3, where it is used for the designation C_s for the runoff coefficient or the runoff reference value. In this case, it is clearly described that this is a peak coefficient. However, the values of the runoff coefficient C_s are given below, according to Tab. 1, also only depending on the layer thickness.

Table 2. Values of runoff coefficient according to FLL Green roof - Guidelines for the planning, construction and maintenance of Green roofs [6].

Roof pitch up to 5°		Roof pitched greater than 5°
For 50 cm layer thickness	$C_s = 0.1$	
For 25 – 50 cm layer thickness	$C_s = 0.2$	
For 15 – 25 cm layer thickness	$C_s = 0.3$	
For 10 – 15 cm layer thickness	$C_s = 0.4$	$C_s = 0.5$
For 6 – 10 cm layer thickness	$C_s = 0.5$	$C_s = 0.6$
For 4 – 6 cm layer thickness	$C_s = 0.6$	$C_s = 0.7$
For 2 – 4 cm layer thickness	$C_s = 0.7$	$C_s = 0.8$

There is determined an exact procedure for determining these runoff coefficients of individual thicknesses of the green roof composition. To obtain the runoff coefficient are first all layers fully water saturated. Furthermore, the specimens are evenly sprinkled with simulated rain at an intensity of 300 l / (s * ha) for 15 minutes. Individual specimens are tested in the unrooted vegetation state. The developed rooting of the vegetation further slows down the outflow of rainwater from the green roof composition. The measurement is therefore performed on the safe side in terms of possible supersaturation of adjacent parts of buildings.

The above-mentioned methods for determining the runoff coefficient *C* of green roof formations were insufficient by the research team in view of the ever-increasing amount of building materials and construction solutions. Furthermore, these current values very often do not provide a sufficiently accurate description of the real behaviour of the green roof composition. However, this is essential for the design of related parts of buildings such as the mentioned sewer networks, storage and retention tanks and infiltration facilities. For these reasons, a methodology for testing and determining the runoff coefficient *C* of individual technical layers and entire green roof composition was determined. That will describe the behaviour of green roof assemblies with a high emphasis on simulation of real behaviour.

2. Methodology

The purpose of the measurement is to determine the reference behaviour of the runoff characteristics, namely the runoff coefficient *C*. The aim is to obtain the basic structural and physical behaviour, which can be further used in common construction practice. Either individual technical layers or the entire composition from the waterproofing layer to the contact surface with the weather are measured. In the case of extensive green roofs, it is a layer of drought-loving plants in the form of a vegetation carpet, in the case of an intensive or semi-intensive roof, it is a surface formed by an intensive substrate.

2.1. Conditioning of test specimens

2.1.1. Conditioning of technical layers. First, the measured components or the composition are sprinkled 27 mm in 15 minutes per 1 m^2. They are then removed and left in bulk for natural drying, the required moisture content of the specimen to be measured being 20 ± 3%.

2.1.2. Substrate conditioning. First, the measured components or the composition are sprayed with 27 mm in 15 minutes per 1 m^2. They are then removed and left in bulk for natural drying, the required moisture content of the specimen to be measured being 25 ± 5%.

2.1.3. Conditioning of vegetation for measurement purposes. For the extensive variant, always using surface vegetation (vegetation carpet according to type) - Before the measurement, the carpet must be stored for at least 48 hours in the installed figure/position. For measuring purposes semi-intensive and intensive green roofs, it is not used vegetation.

2.2. Custom measurement

2.2.1. Slopes for measuring layers. The value of the proven drainage coefficient must correspond to the slope of the designed green roof or differ by a maximum of 3 ° (inclusive) for roofs with a slope of up to 12 °. In other cases, it may differ by a maximum of 6 ° (inclusive). (Measurements are made for flat roofs on a slope of 1 and 5 °. In the case of a sloping roof, measurements are made on a slope of 10 and 20 °).

2.2.2. Test conditions. The reference precipitation is evenly distributed precipitation over time and area with a total volume of 27 mm/m^2 over 15 minutes at constant intensity. The reference precipitation must irrigate the whole area at a temperature of 18 to 22 ° C, at normal atmospheric pressure, a constant relative humidity of 45-60% and no wind up to 0,2 m/s. The permissible measurement deviation is 3%.

2.3. Test specimens

The tested composition is realized on a reference square area of 1.4 to 2.3 m^2 in the entire composition of green roof at the measured slope. The outflow from the outlet line in the length of 1.2 to 1.5 m is continuously monitored by measuring the mass of the drained water in the interval of 5 seconds for 2 hours from the beginning of the sprinkling of the specimen by reference precipitation.

2.4. Boundary conditions for measuring partial layers of green roofs.

Differences for measuring separate layers of a composition. All the above apply with the addition of: Separate technical layers are loaded with a load of 80 kg/m^2 for extensive aggregate roofs of the 4-8 mm fraction, 150 kg/m^2 for semi-intensive and intensive variants of the 4-8 mm fraction.

2.5. Determination of runoff coefficient

The measured data according to the issue showed the need to distinguish between the peak runoff coefficient C_{peak} and the reference runoff coefficient *C*. The reference runoff coefficient *C* is determined by equation (1):

$$C = \frac{(C_{1200}+C_{1500} + C_{1800})}{3} \quad (1)$$

Where C is the reference runoff coefficient, C_{1200}, C_{1500} and C_{1800} are the partial runoff coefficients at 1200, 1500 and 1800 second tracking.

The partial runoff coefficients C_i are determined by equation (2):

$$C_i = \frac{Q_{\text{drained}}}{Q_{\text{rained}}} \tag{2}$$

Where C_i represents the partial runoff coefficient at the observed time t, $Q_{drained}$ represents the value of the effluent from the specimen at the relevant time t and Q_{rained} represents the sum of the simulated precipitation per test specimen at time t.

The peak runoff coefficient C_{peak} is determined by equation (3):

$$C_{\text{peak}} = \frac{Q_{900}}{Q_{\text{rained}}} \tag{3}$$

Where C_{peak} represents the peak runoff coefficient at 900 s from the start of sprinkling the specimen, Q_{drained} represents the value of the effluent from the specimen at 900 s and Q_{rained} represents the total sum of the simulated rain on test specimen.

2.6. Testing equipment

For the purpose of measuring runoff parameters and subsequent determination of runoff coefficient *C*, the research team created a test bench, which allows easy mounting of test specimens, their easy inclination to the appropriate inclination and is equipped with fittings with measuring and reading devices, ensuring simple operation and export of measured data. The test bench is shown at Figure 1.

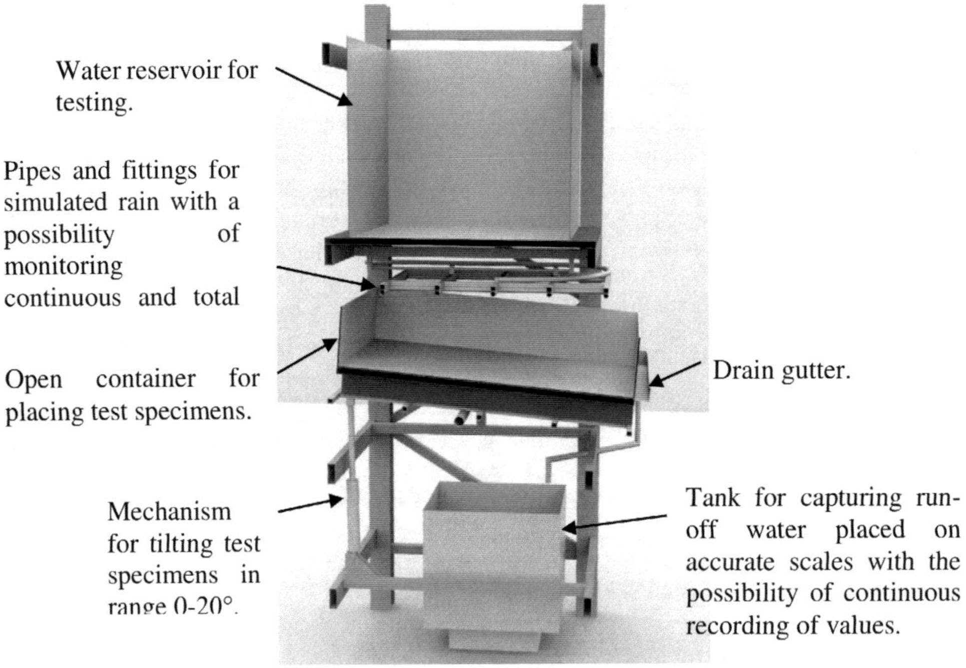

Figure 1. Section of axonometry 3D model test bench.

3. Results of testing

The result of measuring the runoff parameters according to the methodology described above is a graph of the course of runoff of rainwater over time with uniform sprinkling of specimens by simulated rain. Based on these data, the stated runoff coefficients are determined. An example of the measurement output is evident from Figure 2, when a specimen of one composition of an extensive green roof was measured at two different tilts.

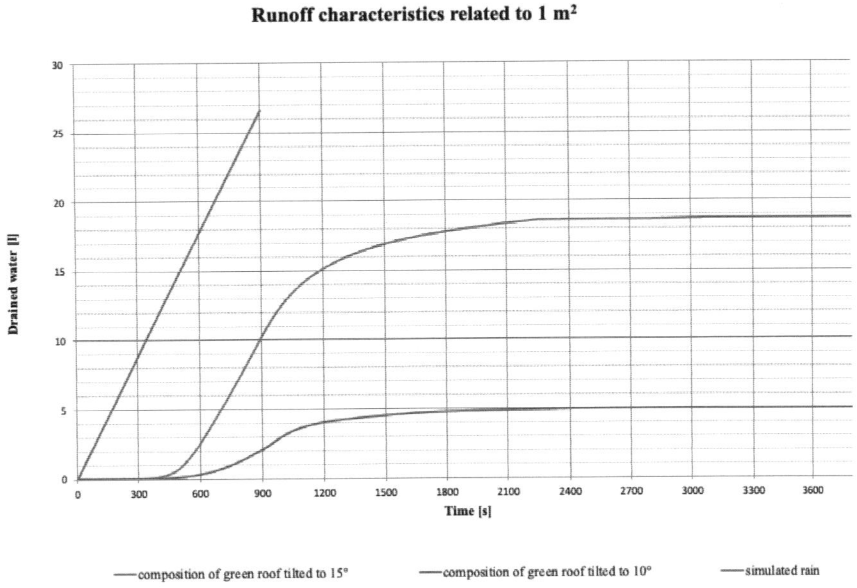

Figure 2. Comparison of runoff characteristics of the specimen green roof cladding in two different tilts.

A comparison of the course of selected materials for hydroaccumulation layers in the same slope is evident from Figure 3.

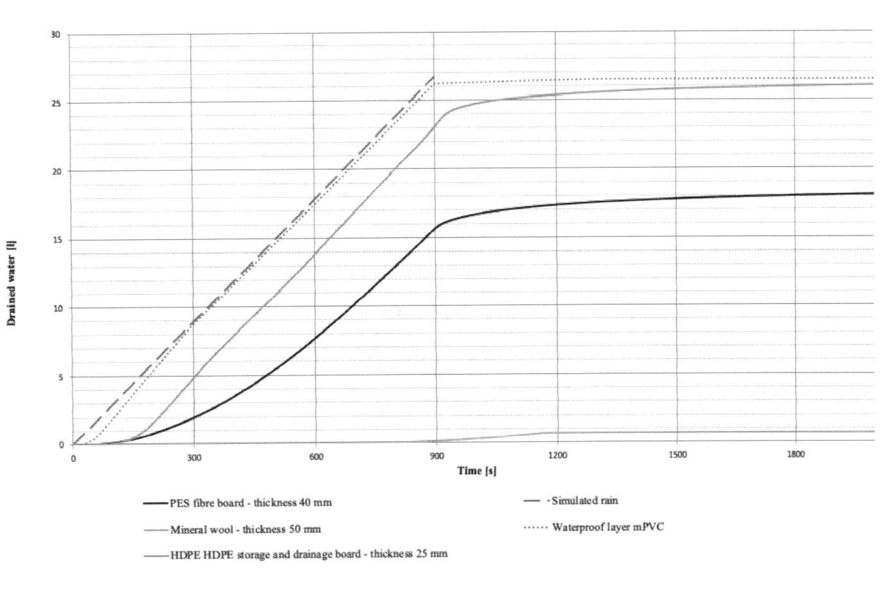

Figure 3. Comparison of runoff characteristics of different materials for water reservoir layer in tilt 3°.

4. Conclusion

The main output of the measurement is the determined reference coefficient of runoff C. Other secondary outputs are the peak runoff coefficient Cpeak and a graph of the course of rainwater runoff during the composition. Thanks to these outputs, it is possible to very quickly compare individual materials or entire compositions of green roofs for a suitable individual design of the green roof structure not only with regard to the type of technical layers, but also according to the conditions of the locality. It also provides more accurate data for the design of related parts of buildings such as retention, accumulation and infiltration facilities, etc. The average runoff coefficient C has a major impact on the calculation of the price for the discharge of rainwater into sewers. The precise determination of the individual compositions of green roofs can therefore have a significant effect on the cost of operating a building during its overall life cycle.

5. References

[1] Scharf B and Kraus F 2019 Green Roofs and Greenpass *Buildings* **9** 205
[2] Filho W L, Icaza L E, Neht A, Klavins M and Morgan E A 2018 Coping with the impacts of urban heat islands. A literature based study on understanding urban heat vulnerability and the need for resilience in cities in a global climate change context *Journal of Cleaner Production* **171** 1140–9
[3] Chun B and Guldmann J-M 2018 Impact of greening on the urban heat island: Seasonal variations and mitigation strategies *Computers, Environment and Urban Systems* **71** 165–76
[4] ČSN EN 12056-3 2001 *Vnitřní kanalizace – Gravitační systémy – Část 3: Odvádění dešťových vod ze střech – Navrhování a výpočet,* (Praha: Český Normalizační Institut)
[5] ČSN EN 75 6760 2014 *Vnitřní kanalizace* (Praha: Úřad pro technickou normalizaci, metrologii a státní zkušebnictví)
[6] The FLL 2018 *Green Roof Guidelines – Guidelines for the planning, Construction and maintenance of green Roofs* (Bonn: Landscape Development and Landscaping Research society e.V. (FLL)) pp 1–15 Available online www.fll.de

Acknowledgments

The paper was produced with support for specific university research at Brno University of Technology, FAST-J-21-7521 (2021) and research and development programme the MPO TRIO FV40357, supported by Czech Ministry of Industry and Trade.

IOP Conf. Series: Materials Science and Engineering 1209 (2021) 012035 doi:10.1088/1757-899X/1209/1/012035

Silicate conductive composites with graphite-based fillers

Š Baránek[1], V Černý[1], G Yakovlev[2] and R Drochytka[1]

[1] Faculty of Civil Engineering, Institute of Technology of Building Materials and Components, Brno University of Technology, Veveri 95, 602 00 Brno, Czech Republic
[2] Department of Geotechnical Engineering and Building Materials, Kalashnikov Izhevsk State Technical University, Studencheskaya Str. 7, 426009 Izhevsk, Russia

E-mail: baranek.s@fce.vutbr.cz; cerny.v@fce.vutbr.cz; gyakov@istu.ru; drochytka.r@fce.vutbr.cz

Abstract. Electroconductive composites are modern materials that are commonly used in many industries such as construction industry and machine-building industry. For example, these materials can be useful as sensors for monitoring changes in constructions, shielding stray currents from electrification networks, shielding electromagnetic radiation in operating rooms, cathodic protection against moisture or overvoltage protection of buildings. The topic of this post is the research of electrically conductive silicate composites with graphite-based fillers. In this research will be tested composites with different ratio and types of graphite and monitor their electroconductive properties like impedance, and physical-mechanical properties like compressive and tensile strength. The post describes basic properties and interactions of silicate electrically conductive composites with graphite fillers. It was found that by adding 10 % wt. graphite into silicate composites, impedance is reduced by 50% and compressive strength by 40%. The flexural tensile strength depends mainly on the roughness of the particles, where the coarser flaky particles transfer the load better and increase the strength while very fine graphites reduce the flexural tensile strength. Furthermore, it has been found that very finely ground synthetic graphites are most suitable for achieving low impedance of composites.

1. Introduction

Composite materials create a wide range of options that can be used to make a variety of materials designed specifically to current market requirements, research, and development. The composites are consisting of two or more substances with different chemical components that differ in physical and mechanical properties.

In most cases, one or more substances form a so-called filler, which represents a discontinuous phase in composite materials. By mixing fillers into the matrix, many physical, chemical, and mechanical properties of the composite material can be modified, such as bulk density, compressive strength, tensile strength, bulk stability, thermal and electrical conductivity, vapor and gas permeability, absorbency, and many other characteristics. Another benefit of fillers can be a reduction in production costs.

Another component of composites is a matrix, which has the main task of bonding all components into one integral material, forming a continuous phase of the material, and acting as a binder. Its function is also to maintain the desired shape and affects the final appearance of the material. The matrix contributes to the mechanical and physical properties of the entire composite material. It also has the task of protecting the fillers or reinforcement from external influences, determines the thermal conductivity, fire resistance of the composite and others. Matrices are most often based on silicate materials such as cement, based on polymer, such as epoxies and polyesters, or based on geopolymer.

Electroconductive composites can be defined as composites with a certain filler content that provide suitable electroconductive properties. With the help of electrically conductive fillers, a stable, perfectly electrically conductive network is created in the material, which can conduct electric current. For solid materials, the electrical conductivity can be divided into surface and internal conductivity. Surface electrical conductivity depends mainly on the moisture of the material, while internal conductivity is related to the structure, amount and properties of conductive fillers used [1].

As electrically conductive fillers, carbon-based fillers such as carbon black, graphite, carbon, carbon nanoplates or graphene, carbon nanotubes, carbon tubes, micro-silica and others are most commonly used, and metal-based fillers such as however, steel wires, fabrics, or powders for these fillers are less effective and are not suitable for silicate composites [1][2][3].

Impedance, analogous to specific electrical resistance (resistivity) is a property of a material when a direct current passes through it, impedance is a "resistance" for alternating current. It is a complex quantity that consists of an imaginary and a real component. For accurate measurement, it is necessary to precisely define the measured quantities (voltage and current).

Once there is a whole, perfectly interconnected electrically conductive structure in the composite material, the impedance of the material itself decreases significantly, this limit is called the percolation threshold and results in the subsequent addition of the material no longer affecting the impedance of the composite [3][4][4].

The properties of silicate composites significantly depending on the type and amount of filler used. With the addition of carbon-based fillers, there is a significant decrease in physical and mechanical properties but also a significant reduction in impedance. These properties are interrelated and depend on the amount of filler. By comparing the different amounts of different types of electrically conductive fillers and their influence on the physically mechanical and electrically conductive properties, no one has study it so far and it is a relatively neglected topic.

In construction, these materials are becoming key elements for monitoring the degradation of building structures, shielding stray currents from electrification networks, shielding electromagnetic radiation in hospital rooms, cathodic protection against moisture, overvoltage protection of buildings, resistance-heated concrete or signal shielding for prison facilities [1].

2. Materials

To verify the effect of the amount of graphite on the electrically conductive and mechanical properties of the composite were designed composites consisting only of cement and graphite were created to eliminate the negative effects of any electrically non-conductive fillers used. Four types of industrially produced graphite were used as conductive filler, representing natural and synthetic graphites, coarse and fine fractions and one with improved electroconductive properties by surface treatment (graphite C). The most important parameters of the graphites used are listed in Table 1.

Table 1. Properties of used electrically conductive fillers.

Graphite	Genesis	Particle type	Specific surface [m²/Kg]	Impedance [Ω]
A	Natural	Flat flakes	1087	0.98
B	Natural	Irregular flakes	12187	1,45
C	Natural	Irregular with adjustment	19879	0.93
D	Synthetic	Irregular	12499	1.19

Portland cement CEM I 42.5 R according to EN 197-1 was used as a binder for the silicate composite. It is a commonly used cement with a designation with a faster increase in strength than ordinary cement of a similar class. The basic properties of CEM I 42.5 R cement according to the manufacturer are summarized in Table 2.

Table 2. Properties of cement.

Type	Density [kg/m^2]	Specific surface [m^2/kg]	Impedance [Ω]
Cement CEM I 42,5 R	3110	391	$7.66 \cdot 10^6$

Graphite was dosed into the composite by weight, in proportions of 4,7,10 and 13% by weight of cement. These values were chosen deliberately on the basis of experience from professional publications [1], where they state that the percolation threshold of graphites is about 10% by weight. Consequently, higher doses of graphite than the percolation threshold would not be effective. Cement and water were also dosed by weight. The water coefficient wc = 0.35 was used. The composition of the mixtures is given in the following table. The properties were verified for each type and dose of graphite on 9 test samples, in total the properties were verified on 144 samples.

Table 3. Composition of the mixture.

Designation of the mixture	Cement (%)	Graphite (A, B, C, D %)
A 4	96	4
A 7	93	7
A 10	90	10
A 13	87	13
B 4	96	4
B 7	93	7
B 10	90	10
B 13	87	13
C 4	96	4
C 7	93	7
C 10	90	10
C 13	87	13
D 4	96	4
D 7	93	7
D 10	90	10
D 13	87	13

3. Methods

Important parameters were determined on all input raw materials, which were the impedance determining the conductivity of the raw materials and the specific surface area, which indicates the grain size and character of the particles. Furthermore, the shape of the grains was specified using an electron microscope.

The specific surface area of the electrically conductive fillers and cements used was determined using BET methods according to ISO 9277: 1995 (E) [6].

The impedance of raw materials and formed composites was determined using a GW Instek LCR-6020 measuring device at a voltage of 1.0 kHz. The impedance of the raw materials like graphite and cement was measured using a mould with electrodes printed on a 3D printer, which is shown in Figure 1. The impedance of the composites was determined on samples of 40x40x160 mm, which is shown in Figure 2, in which copper electrodes are built in. Figure 3 at a distance of 120 mm. The course of measuring the impedance of composites is shown in Figure 4. Impedance measurementFigure 4 the device is connected directly to the electrodes of the composite.

Figure 1. Measuring device GW Instek LCR-6020.

Figure 2. Composite test specimen for impedance measurement.

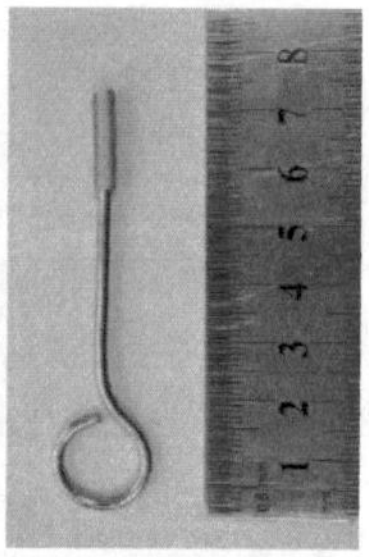

Figure 3. Electrode.

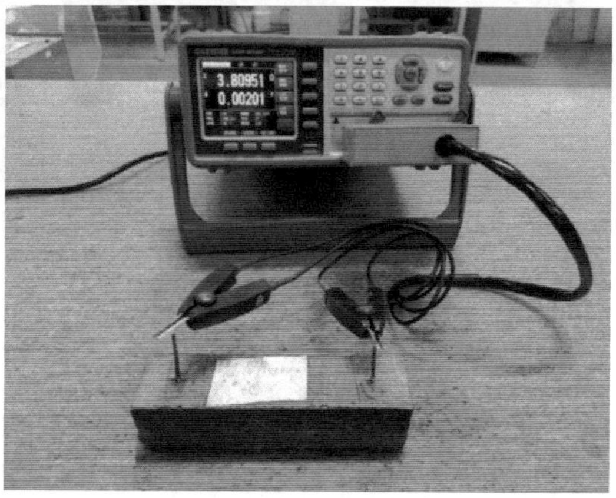

Figure 4. Impedance measurement.

The physical and mechanical properties of the composites were determined after 28 days of maturation in a standard environment on test specimens 40x40x160 mm. The flexural tensile strength was determined on the samples using three-point bending, according to ČSN EN 12390-5. Compressive tensile strength was determined on fractions of samples from the determination of flexural tensile strength in accordance with ČSN EN 12390-3 [7].

4. Results

The results of impedance measurements and determination of compressive strength and flexural tensile strength after 28 days for composites with different amounts and types of graphite are evaluated in Figure 5, Figure 6 and Figure 7.

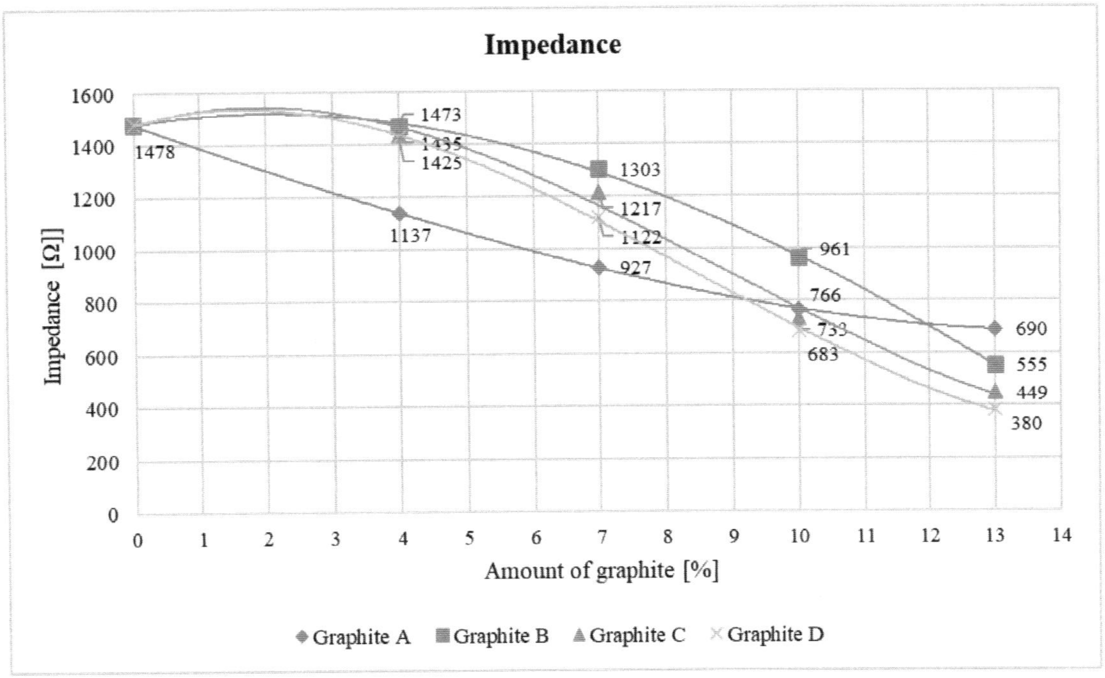

Figure 5. Influence of graphite amount on impedance of composite.

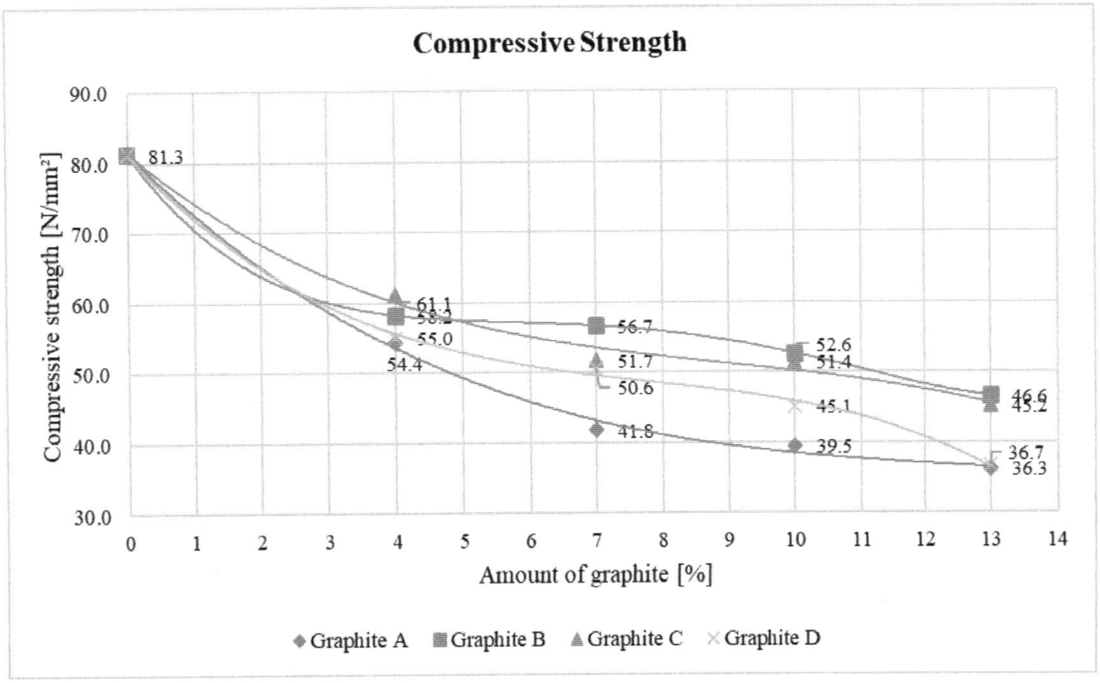

Figure 6. Influence of graphite amount on compressive strength of composites.

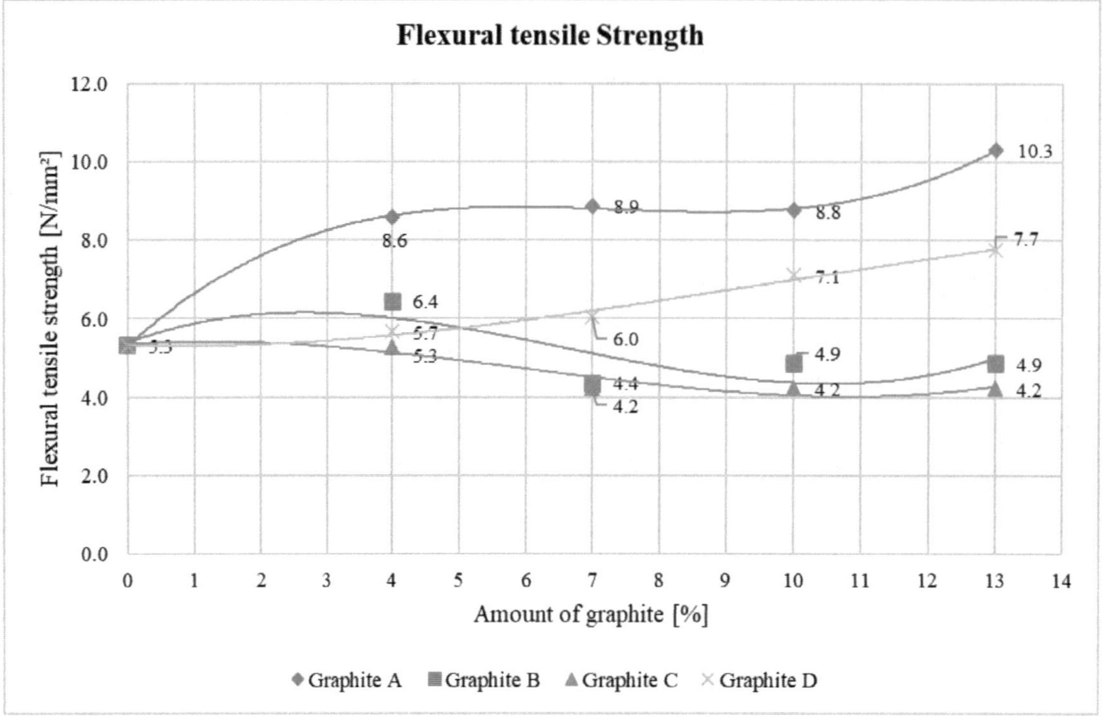

Figure 7. Influence of the amount of graphite on the tensile strength in bending of composites.

With the addition of graphite to the composite, its impedance decreases significantly. The highest decrease was recorded at a dose of 13% by weight, but according to the trend, the percolation threshold, which is given around 10% by weight, probably did not be reached. Impedance was reduced at doses of 4/7/10 and 13% by an average of 8/22/47 and 65%.

The physical and mechanical properties are also affected by the addition of graphite to the composition, the flexural tensile strength has been improved using coarse graphite type (graphite A) and (graphite D) with medium grain size and irregular volume type, this improvement being relatively affected by compaction. In the case of compressive strength, a significant decrease was recorded already with the addition of 4% graphite, when the strength in the first decreased by 30%. At a dose of 7% the compressive strength decreased by 38%, at 10% graphite the strength decreased by 42% and at a dose of 13% graphite the compressive strength decreased by 50%.

From the above results, with the addition of graphite to composite, mainly affects the impedance and compressive strength of the composite, these values decrease from a long-term point in proportion to the increasing dose of filler.

5. Conclusion

The aim of this paper was to monitor the electroconductive and physical-mechanical properties and their interactions in silicate composites with a carbon-based filler. Various doses and types of graphite were tested and at the same time the impedance, compressive strength and flexural tensile strength of composites were monitored.

By introducing suitable types of graphite into the silicate composite, its electrically conductive, physical, and mechanical properties can be significantly influenced. Graphite as a filler is an effective material to reduce impedance. At a dose of 10% graphite the impedance is reduced by 50%, with the compressive strength decreasing by about 40%. The values of these properties decrease symmetrically according to trends with each added percentage of filler. The flexural tensile strength is mainly affected by the grain size of the filler, with coarser types of flake-type graphite having a higher ability to transmit forces in the material, while fine graphite types reduce the flexural tensile strength.

Synthetic graphite and graphites with irregular and very fine particles have been shown to be more effective in achieving lower impedance values. This is due to the larger specific surface area of the particles, and thus the higher probability of contact of irregular grain surfaces in the composite matrix. The different impedance of the graphites themselves as the raw material does not affect the impedance of the entire composite. It can be said that the impedance of the composite depends mainly on the suitability of the size, shape of the particles and the effective amount of the conductive filler used in composite.

6. References

[1] Wu J, Liu J and Yang F 2015 Three-phase composite conductive concrete for pavement deicing *Construction and Building Materials* **75** 129–135 (Amsterdam: North-Holland/American Elsevier)

[2] Terashita K, Tanaka T, Nishimura K and Miyanami K 1993 Continuous kneading for electrically conductive composite materials and evaluation of filler dispersion state *Bulletin of the University of Osaka Prefecture, Series (An Engineering and Natural Sciences* vol 421) pp 51–67

[3] Rhee I, Lee J S, Kim Y A, Kim J H and Kim J H 2016 Electrically conductive cement mortar: Incorporating rice husk-derived high-surface-area graphene: Incorporating rice husk-derived high-surface-area graphene *Construction and Building Materials* **125** 632–42

[4] Xie P, Gu Ping and Beaudoin J J 1996 Electrical percolation phenomena in cement composites containing conductive fibres *Journal of Materials Science* **31** 15 4093–97

[5] Kim G M, Park S M, Ryu G U and Lee H K 2017 Electrical characteristics of hierarchical conductive pathways in cementitious composites incorporating CNT and carbon fiber *Concrete Composites* **82** 165–75

[6] ISO 9277:1995(E) *Determination of the specific surface area of solids by gas adsorption using the BET method*

[7] ČSN EN 12390-5 2020 *Testing hardened concrete – Part 5: Flexural strength of test specimens*

[8] ČSN EN 12390-3 2019 *Testing hardened concrete - Part 3: Compressive strength of test specimens*

Acknowledgments

The reported study was funded by RFBR according to the project № 19-53-26011, and specific research project № FAST-J-21-7491 „Study of the structure of electrically conductive silicate composites containing carbon-based fillers“.

IOP Conf. Series: Materials Science and Engineering 1209 (2021) 012036 doi:10.1088/1757-899X/1209/1/012036

Impact of dampness to changes in the mechanical properties of solid fired bricks

V Bartoň[1], J Šnédar[1] and P Cikrle[1]

[1] Institute of building testing, Faculty of Civil Engineering, Brno University of Technology, Veveří 331/95, 602 00 Brno, Czech Republic

E-mail: 176012@vutbr.cz

Abstract. This work deals with the monitoring of changes in the mechanical properties of solid fired bricks depending on their dampness using non-destructive methods. Decreases of first natural frequencies by the resonance method and increase of passage times of ultrasonic waves depending on increasing dampness are monitored. The elements were firstly fully saturated and then slowly dried so that it was possible to record the values of the first natural frequencies and the passage times of the ultrasonic waves at different dampness. It is not possible to record values in all dampness, so the measured values were interpolated by regression models. A polynomial of the 2nd degree seems to be the most suitable. Dampness corresponding to the minimum natural frequencies and the maximum passage times of the ultrasonic waves were performed on these regression models. This research is the first step in determining the durability criteria for ceramic products, especially solid fired bricks. In the future, durability criteria should help in the reconstruction of historic buildings to assess whether the element that will be exposed to the weather is durable or not. These tests are completely non-destructive, which means that the tested element can be subsequently used in construction.

1. Introduction

The oldest brick buildings were made of bricks fired on sun. They are about 9 thousand years old. They can be found in the Mesopotamia (present Anatolia and Iran). The first evidence about using of solid fired bricks are from Mesopotamia too. They are dated about 3,000 BC, but their production was extremely difficult.

In the Bohemian area the brickwork expanded during the Gothic period. The oldest maintained brick building in the Bohemian area is church in the small town called Plasy. Massive production of fired solid bricks was enabled by invention of Hoffmann kiln in second half of 19th century [1]. Brickmaking industry had been very conservative and constant in producing technologies until the Industrial Revolution. Even though, brickmaking industry has been developing really slow. This industry prospers mostly in the areas, where brick articles are needed. The requirement for building of new brickyards was presence of clay, possibly access to the railway or stream transport right to the brickyard [2].

The solid fired bricks are replaced by more modern products, such as clay bricks THERM or calcium-silicate bricks. These modern types of brick are better than the solid fired bricks in thermal insulation, mechanical properties and in simplification of building process [3].

The solid fired bricks may be overtopped by newer and more modern products, but a huge part of existing buildings is made from them. That is why we are surely going to work with the solid fired brick in reconstructions of historic constructions. During the reconstructions, we can use new solid fired bricks, because it is easy to control their productions and quality. Nowadays, we frequently speak about the reuse of building materials, thus also the solid fired bricks that has been obtained from the demolitions of other objects. The monitoring such bricks is difficult. There are many testing methods, that can be used, but most of them is destructive. That is why, it can be tested scant number of elements. The elements are damaged during these testing and then they cannot be used in construction anymore. Besides, only fundamental properties such as flexural strength or compression strength can be determined by this test [4]. It is not possible to "look" to inner structure of elements and reveal inner imperfections and defects that influence quality of these elements.

The significant element that affects mechanical and physical properties of bricks is dampness or rather absorbability. The frost resistance is directly depended on dampness (absorbability). The frost resistance is ability of the ceramic products to withstand repeated frosting and defrosting. Common causes of low frost-resistant level are imperfections and defects in inner structure, or textural defects, that cannot be find by common tests. There exist corelation between the frost resistance and absorbability. There is an assumption that the elements with lower absorbability can have better resistance against freeze. The frost-resistant level is determined by change of water to ice, where its volume is increased by 10% and the pressure in the elements can by up to 200 MPa [5]. The dampness also affects mechanical properties of bricks such as flexural strength or compression strength. This affects are published widely in writings of other authors. In these writings, there is usually considered only with maximal dampness or absorbability and are monitored the changes in properties of elements in drained state. It is usually considered with linear trend as it is mentioned in [4], where compression strength is reduced linearly with increasing dampness. The linear decrease of mechanical properties in dependency on increasing dampness seems very inaccurate as it is said in [1]. In this publication, there is monitored impact of dampness on changes in modulus of elasticity or compression strength of masonry. It has parabolic trend in both cases.

This research is the first step to imply new criteria of durability of the solid fired bricks by non-destructive methods such as resonance impulse method or ultrasonic impulse method. These new criteria of durability should be helpful during the future reconstructions of historic buildings made of the solid fired bricks. The criteria are going to be related to several factors such as value of natural first frequencies itself and the graph of curve of natural first frequencies (explicitness of vertices and "pure" graph). Last but not least factor is going to be decrease of frequencies between elements that are saturated by water and that are drained. These criteria of durability are going to be able to perform in-situ and specified them by non-destructive methods. A major number of elements can be effectively test by these methods and potentially reuse in buildings or other constructions. These methods are very simple to provide and relatively cheap. Each worker will be able to provide these methods, right after a short teach-in.

2. Materials and Methods

The goal of this research was observation of changes in the mechanical properties in dependence on dampness of the solid fired bricks by non-destructive methods, namely the resonance impulse method and ultrasonic impulse method by the device Pundit PL-200 from Proceq. The properties that were observed by resonance impulse method are the natural first frequencies of longitudinal oscillation– F_L, transverse oscillation – F_F and torsional oscillation – F_T. The passage time of ultrasonic wavs in longitudinal and transverse orientation were observed by the device Pundit PL-200 from Proceq. The sequences of measurements were based on the valid standards [6] and [7] that are determined for testing of concreate. The principles of these standards can be applied on other materials, thus also on the solid fired bricks.

For test, there were picked up 10 test samples (fired solid bricks), that were marked by numbers 1 ~ 10. The samples no. 1 ~ 7 are modern and from different present brickyards located in Czech Republic (Figure 2). The samples no. 8 ~ 10 are historic. The sample no. 8 is from Baroque period, the sample no. 9 is from Renaissance and the sample no. 10 is from Gothic period (Figure 3).

At first, the samples were drained until stable weight, that means when its weight does not change for 24 hours draining in temperature 110 °C about more than 0.2 % from weight gained by previous measurement. Then natural first frequencies and the passage times of ultrasonic waves were determined. After determination of required values from the samples at drained state, the samples were put into the container with drinking water. They were put there in the way they did not touch. Then the water was made to boil and maintained boiling for 4 hours. The samples were kept in the water for next 24 hours. This sequence corresponds to [8].

Then, the samples were drained slowly at the temperature 75 °C. Every hour, the natural first frequencies and the passage times of ultrasonic waves of every sample were determined after they ware cool down. These sequences were chosen so that can be recorded a big amount of values corresponding to different dampness. The samples were drained again until they get stable weight. Their natural first frequencies and the passage times of ultrasonic waves were determined again for control, if the samples were not damaged during the test.

Figure 1. Sample no.2 - modern solid fired brick.

Figure 2. Sample no. 10 - gothic solid fired brick.

3. Results and Discussion

Even thou there is a huge number of measured values, it is not possible to record every dampness and their corresponding natural first frequencies and the passage times of ultrasonic waves. That is why, the measured values were interpolated with appropriated regression model by [9]. Even through doubts, the linear regression model was chosen. The projection matrix *H* was made. Diagonal elements h_{ii} were plot on the graph in dependence on dampness. From the diagonal elements h_{ii} of matrix *H* it can be assumed that the quadratic value is missing (obvious parabolic shape at Figure 3).

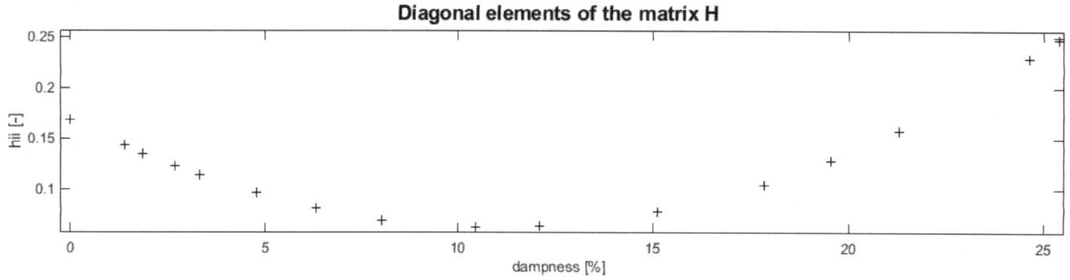

Figure 3. Diagonal elements of the projection matrix "H" depending on dampness – linear regression model.

Afterwards, the degree 3 polynomial was chosen for regression model. The *t-test* of significance of regression parameters were made. The regression parameter β_3, which is the parameter of the degree 3 polynomial, is statistical insignificant according to the test. The degree 2 polynomial is indicated to be the most suitable regression model for natural first frequencies and the passage times of ultrasonic waves in dependence to dampness:

$$y = \beta_0 + \beta_1 \times x + \beta_2 \times x^2 \qquad (1)$$

where:

- y is a vector of observed values (longitudinal, transvers or torsional frequencies, resp. the passage time of ultrasonic waves),
- β_i is a parameter vector,
- x is a vector of dampness.

The display of defects and faults in every single dampness is different mainly in elements with less quality. That is why, there are outlying values in regression models. For detection of these values, the Jackknife resampling of residua with significance level $\alpha = 0.05$ was used. The detected values were eliminated from regression models afterwards. Coefficient of determination R^2 is mostly above the value 0.9 (close to 1.0) that means the regression model was chosen properly.

In the Figure 4 there is dependence of the natural first frequencies and the passage time of ultrasonic waves on the dampness of the elements with average quality (element no. 1). The longitudinal and transverse frequencies together with the passage time of ultrasonic waves have very similar graph of the curve. In the case of the passage time of ultrasonic waves the graph of the curve is concave – more worse properties of the element mean the prolongation of the passage time. In this case the minimal values of the frequencies of the longitudinal and transverse waves occurs when the element has around 60 % of water absorption and the maximal passage time of ultrasonic waves occurs when the element has around 40 % of water absorption. When the elements have higher dampness, the properties are getting better again. In some case, the properties are as good as in drained state.

The torsional frequencies of elements given by the resonance impulse method have different behavior of the curve. In this case, usually there is no improvement in the properties of the elements with higher dampness. In some elements, the torsional frequencies start to increase at some point, but the increase is not so significant as in frequencies of the longitudinal and transverse waves. The minimal frequencies of the torsional frequencies occur when the element has higher dampness. The graph of the curve of torsional frequencies – dampness has mostly only decreasing behavior. The minimal values of the frequencies of the torsional frequencies occurs when the element has fully saturated (100% dampness), besides the exception. The reason may be the impact of defects and imperfections in inner structure, that display differently in torsional oscillation. For the verification of this thesis another research is needed.

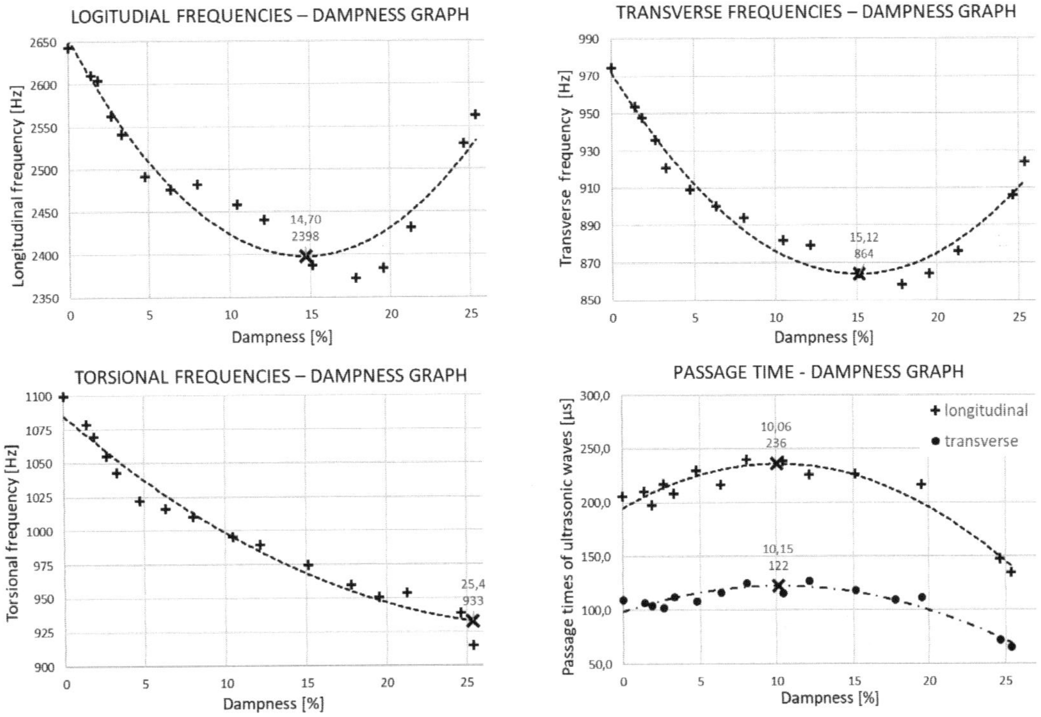

Figure 4. Natural first frequencies and passage times of ultrasonic waves depending on increasing dampness. a) top left - longitudinal frequencies b) top right - transverse frequencies c) bottom left - torsional frequencies d) bottom right - passage times of ultrasonic waves.

The absorbability of the tested elements defers a lot (13 % ~ 30%). To by possible compare values of single elements, the dampness corresponding to minimal natural first frequencies (or maximal passage times of ultrasonic waves) was converted to percentual saturation relative to absorbability of every element. The water absorbability of the elements and percentual saturation in minimal frequencies and maximal passage times are written in Table 1.

Table 1. Water absorbability of elements and percentage saturation at min frequency/max passage times of ultrasonic waves.

element	absorbability	Percentage saturation at minimum frequency F_L	F_F	F_T	at max. passage times of ultrasonic waves longitudinal	transverse
[-]	[%]	[%]	[%]	[%]	[%]	[%]
1	25.4	57.9	59.6	100.0	39.6	40.0
2	13.8	69.4	76.6	100.0	51.1	43.9
3	20.3	83.3	81.4	100.0	25.4	26.7
4	14.0	50.4	51.9	100.0	33.9	34.1
5	17.4	62.5	60.3	100.0	39.7	32.7
6	30.1	100.0	100.0	100.0	44.8	40.0
7	19.9	63.8	60.8	70.5	46.8	41.7
8	22.1	66.1	55.0	70.5	42.7	41.5
9	12.7	56.6	60.9	61.9	33.5	38.9
10	24.2	49.9	52.1	55.8	45.9	44.2

For explicitness, the percentual saturation of the elements with corresponding minimal frequencies were put into the graph (Figure 5). The relations written higher are visible in the graph. The percentual saturation relative to the minimal longitudinal oscillation – F_L and transverse oscillation – F_F gets nearly the same values. The median of the percentual saturation for longitudinal oscillation is 63.2 % and for transverse oscillation is 60.6%. The percentual saturation related to minimal torsional frequencies – F_T occurs when the element is fully saturated with water. That is why the median of the percentual saturation for torsional oscillation is 100%.

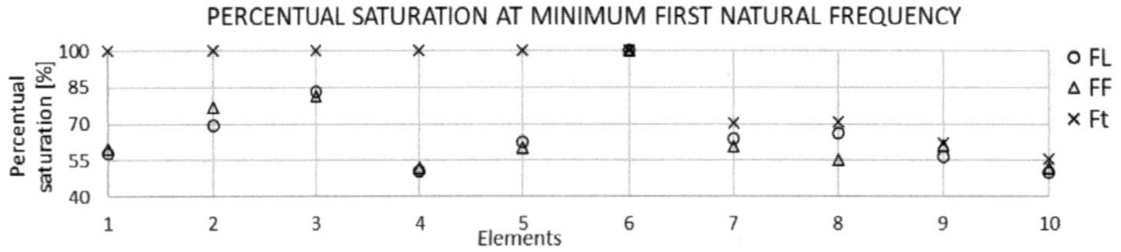

Figure 5. Percentual saturation at minimum frequency of individual elements.

The percentual saturation of the elements with corresponding maximal passage times of ultrasonic waves in longitudinal and transverse direction were put into the graph (Figure 6) by analogy. The maximal passage times occurs when the element has lower percentual saturation then in resonance method (Figure 5). The median of the percentual saturation for longitudinal direction is 41.2 % and for transverse direction is 40.0 %. The graph of the increase of the passage times of ultrasonic waves in dependence to dampness is similar to the graph of the graph given by longitudinal and transverse frequencies measured by resonance impulse method. The lower values of the percentual saturation are caused by different principle of each method.

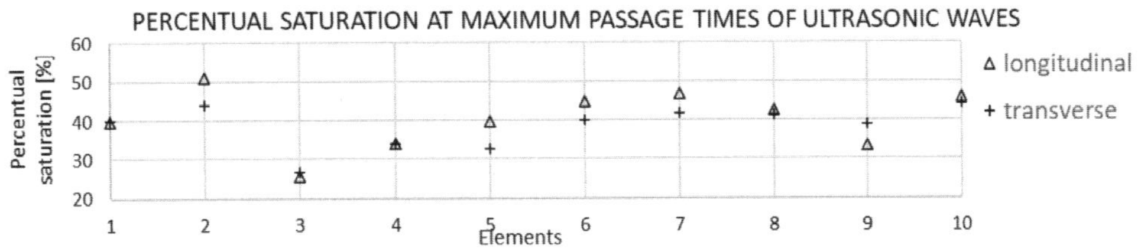

Figure 6. Percentual saturation at maximum passage times of ultrasonic waves.

4. Conclusion

The natural first frequencies and the passage times of ultrasonic waves were monitored in various dampness of the solid fired bricks. The outcome of the research shows that the dependence of monitored properties on dampness has the shape of the degree 2 polynomial. Similar outcomes are mentioned in [1]. The least favourable mechanical properties are around 40 % (the ultrasonic impulse method) and 63 % (resonance method – transverse and longitudinal oscillation) of the percentual saturation of the elements. The impact of the dampness on the mechanical and physical properties is significant factor in qualification of existing constructions, especially in material such as the solid fired bricks. Because it is quite an absorbable material and its properties are more influenced by its dampness, then, for example, the properties of concreated structures [3], [10].

It is nearly impossible to determine the absorbability of single elements and the dampness when the properties of the elements are the most problematic of elements built into the structure. It is necessary to know that the decrease of the properties is not linear. It is important to considered that the decrease of mechanical and physical properties is parabolic. It is necessary to focus on this problem mainly at brickwork constructions that are unheated and permanently unused, e.g. the lowest floors of castles, museums or sacral buildings, where the relative dampness is higher than 60 % and mass dampness of brickwork is higher than 10 % [1]. In these cases, the consideration of linear progress of the decrease of properties depending on dampness of brickwork would be most remarkable mistake.

5. References

[1] Witzany J 1999 Defects and reconstruction of brick buildings (Prague: ČSSI, Technical library of an authorized engineer and technician)
[2] Holub P and Anton O 2020 99 Brno brickyards. Historical development of building materials from fired clay and their production in the city of Brno (Brno: BUT, Vutium)
[3] Marshall D, Worthing D, Health R and Dann N 2013 Understanding Housing Defects: Edition 4 (London: Imprint Estates) pp 1-560
[4] ČSN 730038 2014 Assessment and verification of existing structures - Supplementary guidance. (Prague: ÚNMZ)
[5] Sokolář R 2014 Ceramics: module BJ01_M01. Study aids (Brno: BUT)
[6] ČSN 73 1372 2012 Non-destructive testing of concrete - Testing of concrete by resonance method (Prague: ÚNMZ)
[7] ČSN 73 1371 2011 Non-destructive testing of concrete - Method of ultrasonic pulse testing of concrete (Prague: ÚNMZ)
[8] ČSN 72 2609 2017 Method of tests for masonry units - Specific properties of clay masonry units (Prague: ÚNMZ)
[9] Anděl J 2019 Statistic methods- fifth edition (Prague: Matfyzpress)
[10] Carter C B and Norton M G 2013 Ceramic materials: science and engineering Second edition (New York: Springer)

Acknowledgments

The research was created within the Specific research BUT FAST-J-21-7469 - Prediction of durability of masonry elements based on the identification of their internal defects using non-destructive testing methods.

IOP Conf. Series: Materials Science and Engineering 1209 (2021) 012037 doi:10.1088/1757-899X/1209/1/012037

Study of the properties of chemically resistant repair mortar with the use of secondary raw materials

P Figala[1], R Drochytka[1] and R Hermann[1]

[1] Brno University of technology, Faculty of Civil Engineering, Institute of Technology of Building Materials and Components, Antonínská 548/1, 601 90 Brno, Czech Republic

E-mail: figala.p@fce.vutbr.cz

Abstract. This paper deals with the research of a new silicate-based repair mortar modified with selected secondary raw materials. The aim of this work is to develop a chemically resistant material suitable for use in an extremely aggressive environment of sewers. The monitored parameters include key physical-mechanical characteristics, resistance to sulphate ions and to the attack of aggressive biogenic sulfuric acid. Chemical resistance was tested by simulating the exposure environment in laboratory conditions, according to the methodology of DIN 19573. The obtained results show that by suitable modification of the reference mortar it is possible to maintain the values of physical-mechanical characteristics and improve the chemical resistance of test samples.

1. Introduction

The sewerage system represents a characteristic exposure environment with high requirements for the durability of used building materials. When remediation sewers, the aim is to restore or extend the original service life of the entire building structure. Repair mortars intended primarily for the reprofiling of the damaged concrete surface represent a main material in the remediation of the useful properties of the remediated building structure. The requirements for this type of mortar are, among other things, easy applicability, high strength, resistance to mechanical damage and resistance to chemically aggressive environments.

The service life of a concrete structure is significantly shortened due to chemical degradation. The rapid degradation of cement composites in the environment of sewers is mainly due to the fact that due to their chemical composition and high calcium content they have limited resistance to acid attacks [1]. Conventional concrete sewer elements do not have significantly increased chemical resistance. New concrete elements should be treated with secondary protection in the form of a chemically resistant coating before installation into the structure. During remediation, the application of these coatings is complicated due to the increased humidity.

Problems with corrosion of concrete waste systems have been known since the beginning of the last century [2, 3]. In some cases, the service life of concrete drains is limited to less than 10 years [2]. The high concentration of H_2S (Hydrogen sulfide) in sewers has the greatest effect on reducing the service life of concrete structures. Hydrogen sulfide is formed by the decomposition of organic matter in sewage by anaerobic bacteria found in the sewer system. If there is a sufficient amount of oxygen in the fluid, chemical and biological processes take place that oxidize H_2S to e.g. H_2SO_4 [4].

The process of biocorrosion of concrete can be described in three steps: Neutralization of the concrete surface, which creates a suitable environment for the growth of appropriate bacteria oxidizing sulfur

IOP Conf. Series: Materials Science and Engineering 1209 (2021) 012037 doi:10.1088/1757-899X/1209/1/012037

compounds. This occurs in an environment with a higher concentration of H_2S. Fresh concrete can reach values in the range of pH 11-13, while acid gases reach pH <6. In the next step, the H_2S on the surface of the sewer oxidizes. This can occur not only biologically but also chemically. In both cases, it is an aerobic process in which two molecules of oxygen (O_2) are attached to oxygen-free hydrogen sulfide (H_2S) to form sulfuric acid (H_2SO_4). In the third step, a reaction occurs between H_2SO_4 and the hydration products of the cement [5]. The reaction of sulphates and sulphites on the concrete surface most often takes place by the reaction of portlandite ($CaOH_2$) and other minerals. These are very complex physical and chemical processes [6], they include microscopic processes (diffusion of sulfates and sulfites, formation of new products with lower strengths or expansion character in pores and capillaries and the associated development of crystallization pressure), and macroscopic processes (gradual disintegration of cement composite). [7, 8]. Characteristic minerals formed during this type of cement composite degradation include gypsum ($CaSO_4{\cdot}2H_2O$) and ettringite ($Ca_6Al_2(SO_4)_3(OH)_{12}{\cdot}26(H_2O)$).

The formation of these minerals is accompanied by significant volume changes, as the newly formed compounds contain a large amount of mainly water molecules, which were not contained in the original compounds. For example, the formation of gypsum increases the volume to 124% of the original volume of the mineral and the formation of ettringite even to 227% of the original volume [9], other sources indicate the expansion of ettringite by up to 700% of the original volume [10].

The process of concrete degradation due to the attack of sulphates or sulphites is accelerated by cyclic drying and wetting of the composite [11] and the ambient temperature also has a very significant effect [12].

A many of researches are focused on the possibilities of reducing the degradation of cement composites due to sulphates and sulphites by partial substitution of Portland cement with suitable additives or secondary raw materials. [13, 14] It was found that pozzolanic additives (high temperature fly ash, silica fume, metakaolin, etc.) have a positive effect on the structure of cement composite, pore distribution or their size. However, the results of other studies show [15] that, despite this positive effect of the addition of secondary raw materials, a negative effect on sulphate corrosion resistance may increase with increasing binder substitution rate. The aim is therefore to find the optimal degree of substitution of the cement binder with suitable secondary raw materials.

Due to the expensive remediation work, materials and technologies with sufficient durability but they must also be affordable. Thus, cement-based materials are still a suitable alternative, making it one of the most widely used variants of remediation systems. However, suitable materials must be selected for the design of such remediation materials. There is currently no comprehensive standard in Europe that provides information on materials suitable for use in these exposure conditions. However, there are several national standards that address the design of exposure concentration levels to define exposure classes related to aggressive chemical environments. [4] An example is the German standard DIN 19573 [16], which includes the design of exposure environments for chemically aggressive solutions of Na_2SO_4 and H_2SO_4 with defined concentrations, as well as requirements for tested materials based on cement binder.

2. Experimental details

2.1. Materials

The results presented in this work focus on the design and study of selected characteristics of raw material variants of cement-based repair mortar for the remediation of concrete sewer structures. This development is part of the cooperation between Brno University of Technology, the Faculty of Civil Engineering and a renowned manufacturer of building materials. The mentioned project is still ongoing, that is why only approximate mixtures are shown in Table 1. For the same reason, Table 1 only summarizes the composition of the mixture of admixtures, which includes, for example, plasticizers, crystalline admixtures, polypropylene fibers, and other admixtures. The mixture of the repair mortar is designed to function in the intended remediation system as a high-quality remediation mortar for manual application on small areas or for the repair of more complex profiles.

The aim of the project is to modify the existing mortar to meet the requirements for repair mortars according to DIN 19753 [16]. The reference mass (RM-ref) is a commercially available Portland cement-based repair mortar which does not have any increased chemical resistance. The reference mass was modified by partial substitution of the binder with selected secondary raw materials (high temperature fly ash (A), amphibolite filler (B) and Thermal power plant slag (C)) in the amount of 10 to 30%. Individual modified raw material variants whose results are presented in this work are marked RM-10(20, 30) -A(B, C). For example, mixture RM-20-A is a repair mortar with 20% binder substitution by fly ash. In other respects, the composition of the raw material variants is similar to that of the reference mass.

The fly ash used meets the requirements for fly ash for construction purposes in accordance with Czech standard ČSN 72 2071. Amphibolite filler (further: filler) is a secondary raw material from the treatment of aggregates in the quarry. This secondary raw material was used in the form supplied by the producer. Thermal power plant slag, or coal slag (further: slag) was delivered in a fraction of 0–8 mm. Pre–treatment of slag for research purposes began by adjusting it to a 0–4 mm fraction using a jaw crusher. In the next step, the slag was adjusted to predetermined specific surface area 450 m^2/kg (± 50 m^2/kg). This value of the specific surface area corresponds to the value measured for the substituted cement. The grinding drum for discontinuous grinding was selected for grinding.

Table 1. Approximate composition of test recipes of repair mortar mixtures.

Raw material	The average proportion of raw materials in the developed mixture (wt.%)	
Cement CEM I – 42.5 R	20.3	30.6
Mixture of admixtures (wt.% of m_c) [a]	0.84	1.84
Secondary raw material [b]	2.9	10.2
Finely ground limestone	3.0	6.0
Silica sand 0.18 – 1.0	34.0	37.0
Silica sand 0.6 – 1.0	17.0	21.0
Silica sand 1.0 – 4.0	3.0	6.0

[a] Mixture of plasticizers, crystalline admixtures, other admixtures and polypropylene fibers.

[b] Substitution of cement in the rate of 10-30%, depending on the type of secondary raw material: A - fly ash, B - amphibolite filler, C – coal slag.

2.2. Fabrication of test specimens

For the correct interaction of all components contained in the repair mortar, all input raw materials were homogenized using a container homogenizer. To ensure perfect mixing of all raw materials, the minimum homogenization time was set at 30 minutes.

The production of the fresh mixture was carried out in accordance with standard EN 196 -1: 2016. The mixing time was 270 ± 5 seconds. After mixing, the consistency of the fresh mortar was determined in according with standard EN 1015-3. The water / cement ratio (w) was chosen so that the consistency of the individual fresh mortars corresponds to the selected optimal consistency of 137.5 ± 5 mm. The influence of the additive of individual secondary raw materials on the consumption of mixing water was negligible and the value of "w" was almost identical for all tested mixtures.

All fabrication was performed in a laboratory mixer with forced circulation at the set speed in accordance with the standard ČSN EN 196-1 and in laboratory conditions (23 ± 2 °C, 55 ± 5% relative humidity). The setting of the mixer was chosen for perfect mixing of the dry mixture with the mixing water and activation of the powder admixtures by intensive mixing. All fabricated test specimens were cured for 24 h at more than 90% humidity and temperature of 20 ± 1 °C, before being demolded and after that were cured in water until the measurements were made.

2.3. Test methods

2.3.1. Physical-mechanical characteristics. As mentioned above, the consistency of fresh mortars was determined according to EN 1015-3. The compressive strength was measured in accordance with EN 12190:1998. Mortar specimens were fabricated with dimensions of 40 mm × 40 mm × 160 mm and the compressive strength of the repair mortars was tested at an age of 28 days. The water absorption of the test specimens was subsequently determined on fractions of these test specimens. The water absorption was measured in accordance with Czech standard ČSN 72 2448.

2.3.2. Test for sulphate resistance. The test specimens (10 × 40 × 160 mm) were exposed to an aggressive sulphate solution with a concentration of 29.8 g / l (44 g / l Na_2SO_4) for 91 days. To measure the length changes, measuring pegs were attached to the surface of the test specimens. These measuring pegs and the entire measurement process were in accordance with EN 12617-4.

The reference specimens were left throughout the exposure only in a saturated solution of $Ca(OH)_2$. Initial length between measuring pegs were measured and recorded at an age of 60 ± 3 days. Length changes of test specimens were recorded at regular intervals, and standard DIN 19573 define boundary value for length change ($\Delta\varepsilon$ <0.8 mm / m), which is determined by comparing test specimens exposed to both environments.

2.3.3. Test for resistance to biogenic sulphuric acid attack. A solution of sulfuric acid (H_2SO_4) with a concentration of pH = 0 ($c(H_2SO_4)$ =1.0 mol / l) was used to determine of the chemical resistance of repair mortars to biogenic sulphuric acid attack. The test specimens (40 × 40 × 80 mm) were exposed to a chemically aggressive solution for 14 days.

The test for resistance to biogenic sulfuric acid attack was started at the age of the test specimens of 60 ± 3 days. The pH value of the bath is checked and kept constantly topped up by titration on a daily basis. At the end of storage in acid, any loose material is cleaned off the test samples with a soft brass wire brush under running water. Prisms of h = 40 mm (± 1 mm) are sawn from the middle of each of these prisms. The cut surfaces are ground square and parallel. Reference prisms are prepared from the reference specimens in the same way. After storage for 2 days at 21°C (± 2°) and at 60% (± 10%) relative humidity, compressive strength is measured using the compression test as per EN 196 -1:2016.

The relative residual compressive strength of the corroded mortar test samples is used as the measure of the level of corrosion. Relative residual compressive strength is the residual compressive strength of corroded mortar specimens compared to the compressive strength of specimens after the equivalent length of time stored under water.

The effective depth of corrosion $X_{f,D}$, related to compressive strength, is obtained from the test load applied the corroded prisms at the end of testing and the test load on the reference prisms that have meanwhile been stored in water.

3. Evaluation of Experimental Testing

The results of the determination of the compressive strength and water absorption of 28-day-old test specimens are given in table 2.

Table 2. Selected physical–mechanical properties of individual developed mixtures.

Developed mixtures	Water absorption (%)	Compressive Strength (N/mm^2)
RM-ref	7.1	76
RM-10-A	7.8	76
RM-10-B	8.1	57
RM-10-C	7.5	71
RM-20-A	8.4	73
RM-20-B	8.6	52
RM-20-C	8.1	67
RM-30-A	9.1	61
RM-30-B	9.5	40
RM-30-C	8.8	58

The compressive strength of the test specimens of the modified mixtures decreased with increasing degree of cement substitution. However, even with the substitution of 30% of the cement with fly ash, the compressive strength decreased by only 20% after 28 days of maturation, and with the 30% coal slag admixture, the compressive strength decreased by 24%. In the case of 20% cement substitution, there was only a negligible decrease in compressive strength in most raw material variants. The water absorption has also achieved good results in the case of mixtures with a medium degree of binder substitution. In view of the above, a 20% cement substitution rate was chosen as optimal for chemical resistance testing.

The course of deformation of the test specimens exposed to Na_2SO_4 solution for 91 days is represented graphically in figure 1.

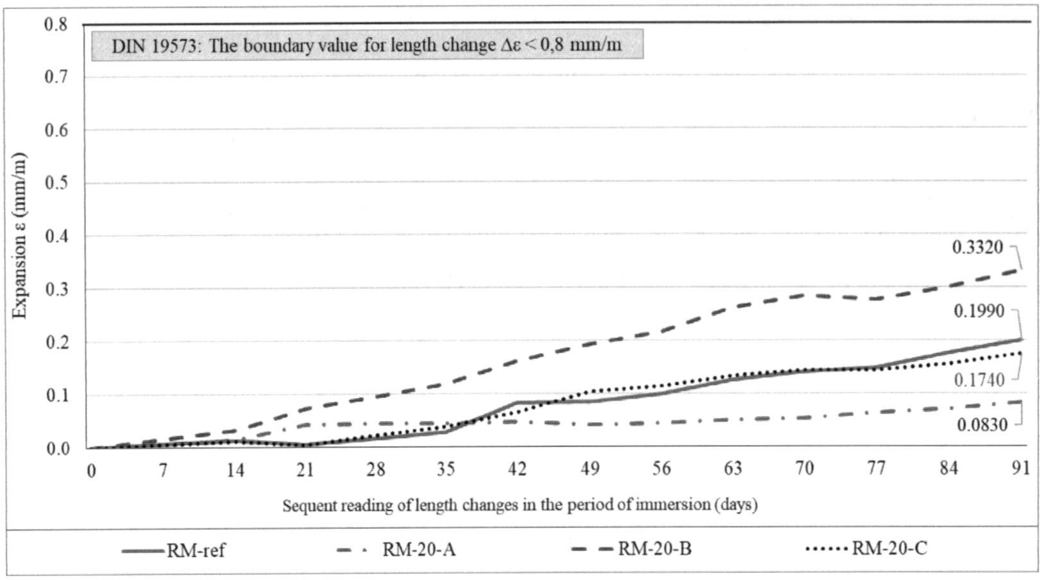

Figure 1. Graphical representation of the course of deformation of test specimens exposed to sulphates in time.

The results of the determination of the sulphate resistance of the test samples of the treated mixtures show that the resistance to external sulphate attack was the highest in the mixture with 20% substitution of cement by fly ash. Compared to the reference mixture, RM-20-A mortar showed a 58% reduction in the length change of the test specimens after 91 days of exposure. In the case of the RM-20-C mixture (mass with coal slag) there was a reduction in length changes of 12% compared to the reference mixture.

The next Figure 2 graphically presents the results of determining the relative residual compressive strength of corroded test specimens exposed to an aggressive solution of sulfuric acid at a concentration of pH = 0 ($c(H_2SO_4)$ =1.0mol / l) for 14 days. Standard DIN 19573 defines that the minimum relative residual strength of chemically resistant repair mortars must be > 55% for this exposure solution.

It is clear from the above results that the resistance of the modified mixtures to biogenic sulphuric acid attack does not compare with the previous results of resistance to attack by the sulphate solution.

When compared with the reference mortar, the highest relative residual compressive strength was determined for RM-20-B, which is a repair mortar with the addition of amphibolite filler (5% increase in strength). The lowest measured value was determined for the mixture with the addition of fly ash (8% decrease in strength).

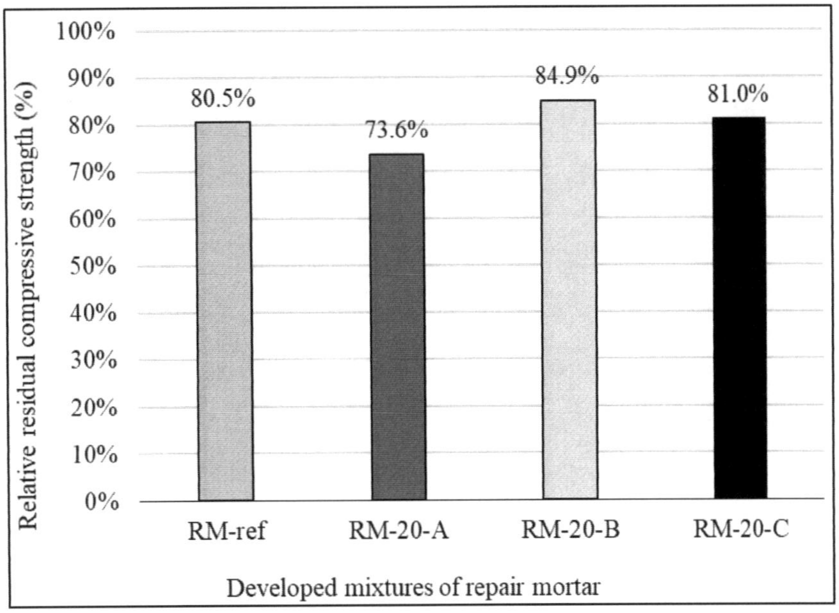

Figure 2. Graphical representation of the relative residual compressive strength of the corroded mortar test samples after exposed to a chemically aggressive solution of sulfuric acid (H_2SO_4) with a concentration of pH = 0 (c(H_2SO_4) =1.0mol / l) for 14 days.

4. Conclusion

The aim of this work was to verify the properties of new mixtures for chemically resistant repair mortar modified by the additive of selected secondary raw materials. Based on DIN 19573, several specific apparatuses for testing the chemical resistance of test specimens have been assembled. These devices have been optimized for research purposes.

Based on previous research, a set of repair mortar mixtures modified by partial substitution of the cement binder with potentially suitable secondary raw materials was compiled. The properties of the individual mixtures were assessed on the basis of experimental tests. The result of this work is an optimal mixture for a new chemically resistant repair mortar.

The results of this work show that it is possible to successfully increase the chemical resistance of the developed repair mortar by partial substitution of the cement binder with suitable secondary raw materials. The effect of individual secondary raw materials differed depending on the observed physical-mechanical characteristics or type of chemically aggressive environment.

From the results it can be deduced that the most balanced results were achieved in the case of the mixture RM-20-C. This is a mixture with 20% substitution of cement by coal slag. This mortar was characterized by low water absorption, great compressive strength and also it achieved an improvement in chemical resistance compared to the reference mortar, in both tested exposure environments.

The next phase of research and development of the new mortar will focus on a detailed study of the influence of individual secondary raw materials on the properties of the repair mortar with emphasis on the resistance to sulphate attack and biogenic sulfuric acid attack. Among other things, the influence of secondary raw materials on the structure of the cement composite and the formation of corrosion products will be evaluated.

The existing set of laboratory analyses will be extended by a complex of physicochemical analyses for a more detailed characterisation of the evolving corrosion processes.

5. References

[1] Park J S 2007 *Necessity of rehabilitation and current state of domestic sewage pipe network* J Korean Soc Civ Eng **55** pp 135–43

[2] Olmsted F H and Hamlin H 1900 *Converting portions of the Los Angeles outfall sewer into a septic tank* Engineering News pp 317-318

[3] Parker C D 1947 *Species of sulphur bacteria Associated with the Corrosion of Concrete* Nature 159 (4039) pp 439-440

[4] Grandclerc A, Guéguen-Minerbe M, Nour I Dangla P and Chaussadent T 2017 *Impact of cement composition on the adsorption of hydrogen sulphide and its subsequent oxidation onto cementitious material surfaces* Construction and Building Materials **152** pp 576-586

[5] Yuan H, Dangla P, Chatellier P and Chaussadent T 2015 *Degradation modeling of concrete submitted to biogenic acid attack* Cement and Concrete Research **70** pp 29-38

[6] Monteiro P J M and Kurtis K E 2003 Time to failure for concrete exposed to severe sulfate attack. Cement and Concrete Research **33***(7)* pp 987-993

[7] Yin G-J, Zuo X-B,Li X-N and Zou Y-X 2020 An integrated macro-microscopic model for concrete deterioration under external sulfate attack Engineering Fracture Mechanics p 240

[8] Ledererova J et al 2009 *Biocorrosive effects on building structures* (Biokorozní vlivy na stavební díla) first ed. Prague: Silikátový svaz

[9] Parande A K, Ramsamy P L, Ethirajan S, Rao C R K and Palanisamy N 2006 *Deterioration of reinforced concrete in sewer environments* Proceedings of the Institution of Civil Engineers - Municipal Engineer **159***(1)* pp 11-20

[10] Chang H B and Choi Y C 2020 *Accelerated performance evaluation of repair mortars for concrete sewer pipes subjected to sulfuric acid attack* Journal of Materials Research and Technology **9***(6)* pp 13635-13645

[11] Xie F, Li J, Zhao G, Zhou P and Zheng H 2020 *Experimental study on performance of cast-in-situ recycled aggregate concrete under different sulfate attack exposures* Construction and Building Materials p 253

[12] Hossack A and Thomas M D A 2015 *The effect of temperature on the rate of sulfate attack of Portland cement blended mortars in Na_2SO_4 solution* Cement and Concrete Research **73** pp 136-142

[13] Al-Akhras N M 2006 *Durability of metakaolin concrete to sulfate attack* Cement and Concrete Research **36***(9)* pp 1727-1734

[14] Hou P et al 2019 *Comparison study on the sulfate attack resistivity of cement-based materials modified with nanoSiO_2 and normal SCMs: Pore structure and phase composition* Construction and Building Materials p 228

[15] Liu Z Hu W Hou L and Deng D 2018 *Effect of carbonation on physical sulfate attack on concrete by Na_2SO_4* Construction and Building Materials **193** pp 211-220

[16] DIN 19753 2017 Draft on mortars for construction and rehabilitation sewer system La Plaine Saint-Denis Cedex

Acknowledgments

This paper has been worked out with financial support from the state budget through the Ministry of Industry and Trade of the Czech Republic within the project FV20149, “Unified system for remediation of chemically attacked and exerted building constructions” and from the budget under the project FAST-J-21-7312 “Study of the properties of a chemically resistant silicate repair mortar based on secondary raw materials”.

Study of polymer-based adhesive mortar with higher durability

R Hermann[1], P Figala[1], L Mészárosová[1] and R Drochytka[1]

[1] Faculty of Civil Engineering, Brno University of Technology, Veveří 331/65, Brno, Czech Republic

E-mail: Hermann.r@fce.vutbr.cz; Figala.p@fce.vutbr.cz; meszarosova.l@fce.vutbr.cz; Drochytka.r@fce.vutbr.cz

Abstract. This paper deals with the study and development of polymer-based adhesive with high filling ratio of secondary raw materials and waste materials. The goal of this paper is to develop adhesive mortar with the highest filling rate of secondary raw materials and waste materials as possible while preserving very high physical-mechanical properties, including flexural and compressive strength, pull-off bond strength and abrasion resistance. High-temperature fly ash, waste slag and waste packaging glass are used in this paper as fillers. The resulting mortar shows high physical-mechanical properties, including high abrasive resistance and very high bonding strength to a large variety of building materials including concrete, steel, glass, and tiles.

1. Introduction

Today, the construction industry is increasingly facing problems related to the ecological and economic demands of the production of building materials, which is also reflected in the demands on the construction of new structures. The materials used in the implementation of new buildings, but also in the implementation of remediation works are increasingly demanding in terms of strength characteristics, but also the consumption of material. Today, buildings are no longer built according to historical principles. Careful calculation is performed for each construction to ensure optimal usability of the structure while ensuring minimum material requirements. There are also economic demands attached to this. Thus, the construction industry is facing an increasing demand for high-quality materials, which, however, would be economically available and have a low environmental impact. One way to achieve this is to use secondary and waste raw materials as an available and inexpensive filler in polymeric materials, which are otherwise very economically and environmentally demanding.
During 2021, the price of polymer binders rose by up to 100% in some parts of the world. For this reason, this issue is becoming increasingly relevant. Secondary and waste raw materials that can be used for this purpose include, for example, high-temperature fly ash, waste glass, or waste slag from a thermal power plant which are used in this paper. The chemical composition of used materials is shown in Table 2 and sieve analysis is shown in figure 2.

2. Materials

2.1. Epoxy resin

For the purposes of this work, an epoxy resin based on a combination of bisphenol A and F was used. Amine-based hardeners were used for curing. The mixing ratios of the resin and the hardener are as follows: 62 parts of resin and 40 parts of hardener (by weight). The binder was then mixed with a filler component consisting of high temperature fly ash, waste glass and waste slag from a thermal power plant. The grain size of the fillers was adjusted by crushing and grinding to the required grain size of 0 - 500 μm. The epoxy resin used had the following properties:

Table 1. Epoxy resin properties.

Properties	Value
Viscosity at 25°C [mPas]	1000-1400
Density at 20°C [$g \cdot cm^{-3}$]	1.07
Pot life (40°C after) [min]	20
Exothermic temperature [°C]	170

2.2. High-temperature fly ash (HFA)

High-temperature fly ash is formed during the burning of pulverized coal in thermal power plants. Combustion takes place at temperatures of 1200–1700°C. The fly ash used for this work comes from electrostatic precipitators, which capture the fly ash from the combustion flue gas. The fly ash meets the properties of the ČSN EN 450 standard, which prescribes the suitability of fly ash for use in concrete and mortar. This means that the fly ash contains at least 45% amorphous SiO and at the same time the maximum amount of free CaO, MgO and alkalis is monitored. In addition to concrete and mortar, fly ash from high-temperature combustion can also be used as a microfiller in the production of polymer composites. The specific weight of the used fly ash reaches the range 2400-2500 $kg \cdot m^{-3}$. Due to the high content of SiO_2 and relative inertness to a number of chemical compounds, it is a suitable filler for the developed material [1].

2.3. Waste glass (WG)

Waste packaging glass is a type of glass that is created during the sorting of glass packaging and can no longer be recycled. This is due, for example, to its color inhomogeneity or the impossibility of perfect sorting on automatic lines, which prevents its further use to produce glass containers. However, it is still a glass phase, which is produced by a glass stem containing approximately 50% silica sand, 10% Na_2CO_3 (soda), 12% $CaCO_3$ (limestone), 18% crushed shards and about 4% other, additional components. Glass, which can no longer be used, is a significant environmental problem. Glass generally has a very high hardness, according to the Mohs scale about grade 6. At the same time, glass is very resistant to most chemical compounds, except hydrofluoric acid [2].

2.4. Waste slag from thermal power plant (WS)

When burning coal, flammable shale, or other solid fuels in thermal power plants on grids and in furnaces, solid fuel residues are formed, which are partially melted. The resulting material is contaminated with organic residues of the original fuels. Its further use can therefore cause certain problems in building materials. Waste slag from a thermal power plant contains, for example, residues of unburned coal, free CaO and MgO, or several soluble salts. Slag is landfilled in warehouses near thermal power plants. The quality of the slag increases in proportion to the landfill time. The longer the time, the higher the quality of the slag, as the residual soluble salts are washed away by the rain. However, if the repository is not maintained, grass and trees grow in the repository, which introduces other organic residues into the slag, which are not suitable for use in building materials [2].

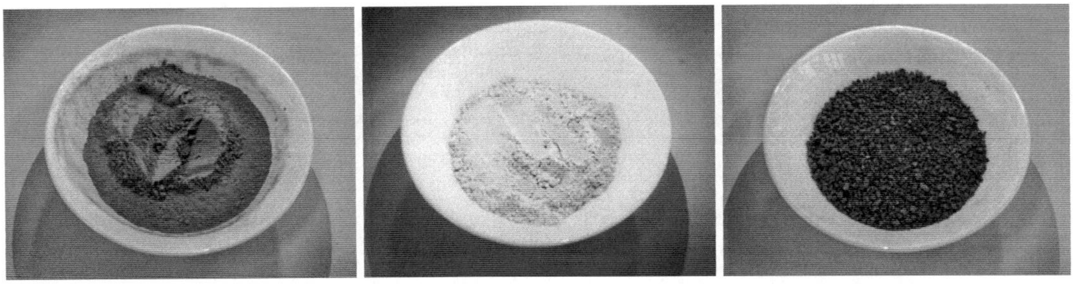

Figure 1. Sieve analysis of used waste and secondary raw materials.

The chemical composition of the secondary and waste raw materials used is shown in the following table 2.

Table 2. Chemical composition of raw materials [wt%].

	Cl	P	SiO_2	SO_3	P_2O_5	Al_2O_3	CaO	Fe_2O_3	K_2O	MgO	MnO	Na_2O
HFA	<0.40	<0.05	55.6	0.4	<0.1	26.4	1.5	5.9	1.9	1.05	0.06	0.29
WG	-	-	71.0	-	-	1.7	10.4	0.7	0.3	2.3	-	13.0
WS	-	-	51.0	1.1	0.34	22.4	4.3	0.9	3.3	1.7	-	1.77

For the purposes of this work, pre-treatment of secondary and waste raw materials was performed using a jaw crusher and a laboratory ball mill. The grain size of the filler was adjusted to a fraction of 0–500 μm and the grain size was decomposed according to the following sieve analysis curve (figure 2).

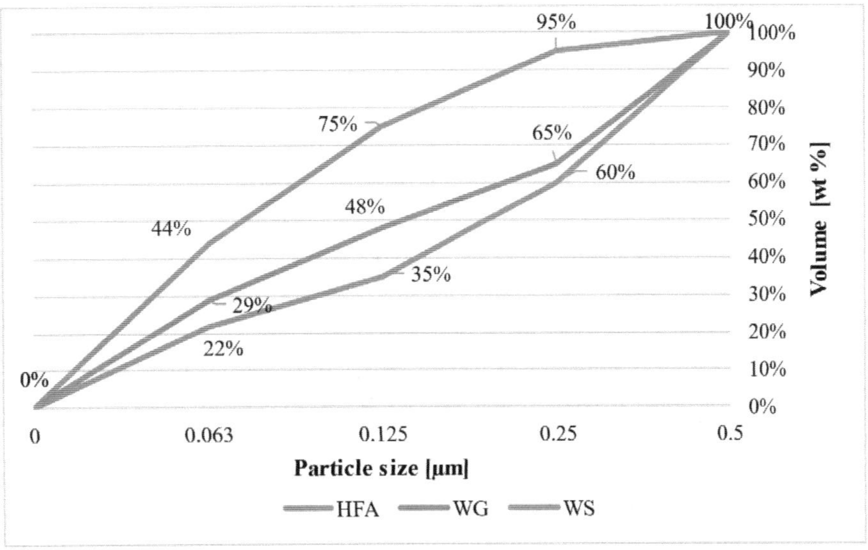

Figure 2. Sieve analysis of used waste and secondary raw materials.

The above sieve analysis curves have been adjusted to approximate ideal granularity curves. Only the filler consisting of high-temperature fly ash achieves a finer curve, as this raw material achieves a very high fineness without pre-treatment. For the purposes of this research, the highest possible level of filling with individual types of fillers was carried out, while maintaining the workability in the masonry way, with a spatula. In general, fillers whose specific surface area is higher, or whose absorbency is higher, can be dosed in a significantly smaller dose than in the case of non-absorbent fillers. The following recipes describe the recipes used.

Table 3. Composite formulations.

Filling rate	WS	WS	WS, HFA	HFA	HFA, WG	WG	WG
Epoxy resin [wt%]	30.4	28.0	25.5	23.1	18.2	18.2	15.8
Hardener [wt%]	19.6	18.0	16.5	14.9	11.8	11.8	10.2
Filler [wt%]	50.0	54.0	58.0	62.0	66.0	70.0	74.0

3. Methods

Epoxy resin and hardener were mixed to prepare the samples. The ingredients were mixed thoroughly with a low-speed stirrer (300-400 rpm) for 2 minutes. A slow-running stirrer was used to prevent bubbles from entering the mass. The entire contents of the filler were slowly mixed into the liquid component and mixed for the necessary time to ensure perfect homogeneity. This time was longer than two minutes. After mixing, the mixtures were placed by masonry or casting in test moulds measuring 20x20x100 mm to determine flexural tensile strength and compressive strength, then into test moulds measuring 70x70x50 mm to determine wear resistance and the materials were applied to concrete and steel base for determining pull-of bond strength.

3.1. Apparent density

After curing of the 20x20x100 mm samples, the apparent density was determined according to ČSN EN ISO 845 standard by weighing the test samples and measuring its dimensions [3].

3.2. Compressive and flexural strength

Polymer composite was applied and compacted into silicon moulds of dimensions 20x20x100 mm. The samples were used for the determination of compressive strength and three-point flexural strength. Support distance for three-point flexural strength was 80 mm. For the determination of compressive strength, the 20x20 mm loading plates were used to distribute force evenly on the test sample sides. Three test samples of each filling rate and each filler were prepared. All test samples were cured in a laboratory environment (21°C and 55% relative humidity) for 7 days. Samples are shown in figure 3.

Figure 3. Samples for the flexural and compressive strength testing.

3.3. Pull-off bond strength

The pull-off bond strength was determined on test samples applied to concrete tiles and steel sheets. The strength was determined according to the ČSN EN 1542:1999 standard by using the Dyna pull-off tester with hand crank. All materials were applied to dry and clean surfaces. Materials were applied in thickness of approximately 3mm. The pull-off bond strength was tested after 7 days of curing [4].

3.4. Wear resistance

The wear resistance was determined on samples of dimensions 70x70x50 mm according to the ČSN EN 13892-3:2015. The wear resistance was tested on Böhm disk. Three test samples of each filling rate and type of filler were used. The wear resistance was tested after 7 days of curing [5].

4. Results

For each filler, the maximum filling rate was verified while maintaining workability in a masonry manner. This led to a different maximum filling for different types of fillers. Compressive and flexural strength was determined by a modified method according to ČSN EN 13892-2 used by Hermann et al. [6, 7].

4.1. Apparent and bulk density

The apparent density (D) of epoxy resin was 1.05 g·cm^{-3} in a laboratory environment (21°C). The bulk density of used fillers is shown in table 4.

Table 4. Bulk density of used fillers.

Filler type	Bulk density [g·cm^{-3}]
High-temperature fly ash (HFA)	2.41
Waste glass (WG)	2.56
Waste slag (WS)	2.61

The apparent density (D) of test samples is shown as blue line in figure 4.

4.2. Flexural and compressive strength

The results of flexural (σ_t) and compressive (σ_c) strength of test samples are shown in figure 4. The flexural and compressive strength was determined according to modified testing method stated in ČSN EN 13892-2. The flexural strength was tested, and the compressive strength was tested on the halves of samples, which resulted from the test of flexural strength.

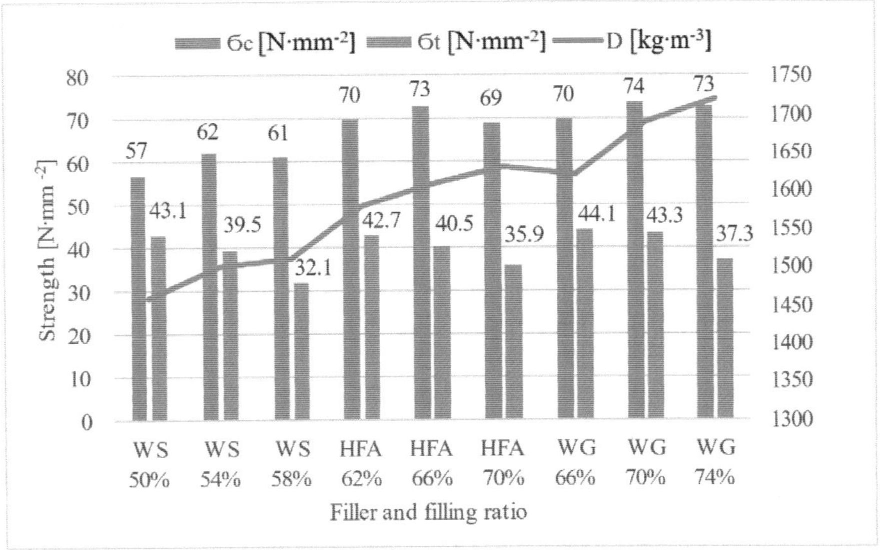

Figure 4. Flexural and compressive strengths and apparent density of test samples.

From the above figure it can be determined that with the increase of filling ratio, the flexural strength tends to decrease, while the compressive strength increases. This is due to different behaviour of epoxy resin and fillers. The epoxy resin has very high flexural strength while relatively low compressive strength, which is increased by the filler. The apparent density increases with the increase of filling ratio. When the maximum filling ratio is achieved, there are several bubbles introduced into the mass of samples and the apparent density does not increase anymore.

4.3. Pull-off bond strength

The pull-off bond strength of all tested samples was greater than tensile strength of tested concrete tiles, which was between 3.3 N·mm^{-2} and 3.6 N·mm^{-2}. The bond strength to steel sheets of all test samples was greater than the maximum tensile force of DYNA pull-off tester, which is 16 kN. The pull-off bond strength was tested on a circle area of 50mm diameter. Which, according to the 16 kN maximum cap of pull-off tester is approx. 8.1 N·mm^{-2}. The samples after testing are shown in figure 5.

Figure 5. Samples after the pull-off bond strength test.

4.4. Wear resistance

The wear resistance was tested on samples with dimensions 70x70x50 mm on laboratory Böhme disc according to according to the ČSN EN 13892-3:2015. The results are shown in following figure 6.

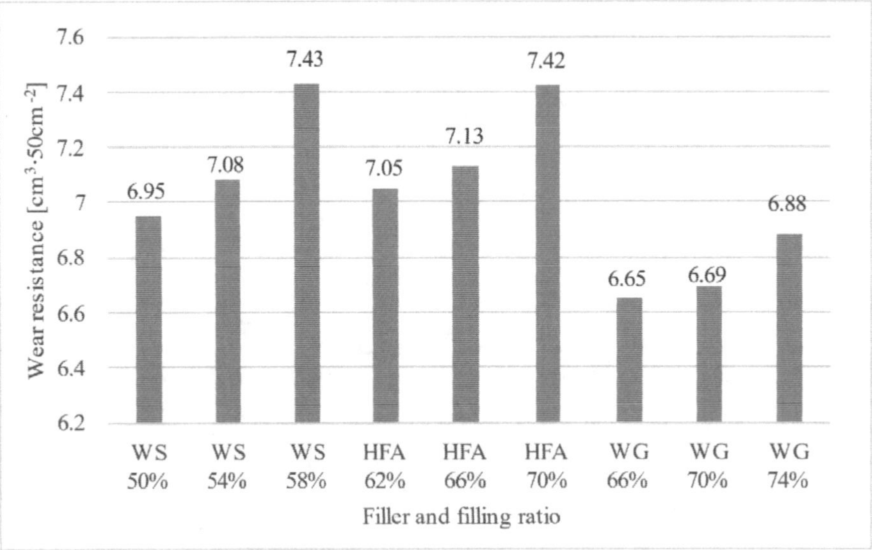

Figure 6. Wear resistance of test samples.

In the above figure, the wear resistance of samples is shown. The wear resistance decreases (the volume of abrased material is higher) with the increase of filling ratio. Which implies, that the bearer of the wear resistance is the epoxy resin itself and when the filling ratio is increased, the wear resistance decreases.

5. Summary

In this paper the possibility of using waste and secondary raw materials as filler in epoxy polymer composite was studied. Maximum filling ratio of each filler was determined by a workability test. The requirement was that the polymer composite would be still applicable in a masonry way by spatula. The A/F epoxy resin in combination with amine-based hardener was thoroughly mixed and then filled by waste and secondary raw fillers with a grain size of 0-500 μm (figure 2.). For purposes of this paper, the waste slag from nearby power plant, high-temperature fly ash and waste packaging glass were used. It was found that all the selected materials can be used as fillers in epoxy-based mortar/adhesive up to a point. The lowest filling ratio was achieved with waste slag. That is probably due to a large number of impurities and organic residues which exhibit high absorption; therefore, the epoxy resin was absorbed into the material and could no longer function as a binder between grains of filler. The high-temperature fly ash has very low absorption and the filling ratio of this material was higher. The material also exhibits very high specific surface area; therefore, it needs a lot of binder to wet all of grains. The high-temperature fly ash had also the highest amount of very fine grains. The highest filling ratio was observed when waste packaging glass was used. The same behaviour can be observed in cementitious materials. This is probably due to shape of grains of this material and very low absorption. The flexural and compressive strengths were determined. With increase of amount of filler in the composite, the flexural strength tends to decrease, this is due to the fact, that the epoxy resin is the carrier of flexural properties. With the increase of filler in the material, the compressive strength increased, this is due to the fact, that the filler in combination with resin is the carrier of compressive strengths. It was determined that the maximum filling ratio, at which the physical-mechanical properties do not decrease too severely, was 54% for the waste slag, 66% for the high-temperature fly ash and 70% for the waste packaging glass. The pull-off bond strength was tested. The results show that all the test samples shown very good adhesive properties to both concrete and steel sheets. The pull-off bond strength to concrete was greater than the tensile strength of used concrete and was above 3.3 $N{\cdot}mm^{-2}$ and the adhesion to steel sheets was greater than 8.1 $N{\cdot}mm^{-2}$. The wear resistance was also tested, which confirms the results of flexural and compressive strengths. The lower values of physical-mechanical properties of the highest tested filling ratios are due to insufficient workability and inability to compact the samples, which resulted in porous samples with lower overall properties.

6. References

[1] Ahmaruzzaman M 2010 A review on the utilization of fly ash *Prog. Energy Combust. Sci.* **36** 3 327–63

[2] Choudhary J, Kumar B and Gupta A 2018 Application of waste materials as fillers in bituminous mixes *Waste Management* **78** 417–25

[3] ČSN EN ISO 845 2010 *Cellular plastics and rubbers - Determination of apparent density* (Prague: Český normalizační institute)

[4] ČSN EN 1542 1999 *Products and systems for the protection and repair of concrete structures – Test methods - Measurement of bond strength by pull-off* (Prague: Český normalizační institute)

[5] ČSN EN 13892–3 2015 *Methods of test for screed materials - Part 3: Determination of wear resistance - Böhme* (Prague: Český normalizační institute)

[6] Hermann R, Hodul J and Keprdová Š 2020 Effect of Specimen Geometry on Selected Physical and Mechanical Properties of Epoxy-Based Building Materials *Key Engineering Materials* **868** 24–31

[7] ČSN EN 13892–2 2003 *Methods of test for screed materials - Part 2: Determination of flexural and compressive strength* (Prague: Český normalizační institute)

Acknowledgements

This paper was supported by the FAST-S-21-7228 project "Analysis of the influence of the size and morphology of microfiller particles from primary and secondary raw materials on the resulting properties of composite materials".

IOP Conf. Series: Materials Science and Engineering 1209 (2021) 012039 doi:10.1088/1757-899X/1209/1/012039

Overview of the presence of predominant microorganisms on the interface of cementitious materials and soil from selected locations

M Hološová[1], A Eštoková[1] and A Luptáková[2]

[1] Technical University of Košice, Faculty of Civil Engineering, Institute of Environmental Engineering, Vysokoškolská 4, 042 00 Kosice, Slovakia
[2] Slovak Academic of Science, Institute of Geotechnics, Watsonova 45, 040 01 Košice, Slovakia

E-mail: miriama.holosova@tuke.sk, adriana.estokova@tuke.sk, luptakal@saske.sk

Abstract. The presented work focuses on the predominant microorganisms on the interface of cementitious materials and soils from different localities. Soils contain microorganisms that can affect the underground concrete structure positively or negatively. Different types of microorganisms are adapted to the environmental conditions depending on the quality and quantity of elements in the soil. Methodology of the work was based on screening 5 selected localities in the Košice city from which soil was taken in the nearness and in the contact with the concrete structure. These samples were further analysed. In this paper the presence of bacteria in the soil around the construction is confirmed.

1. Introduction

Soil will create an aggressive environment for constructions if it contains microorganisms that can destroy the material by their action. Biocorrosion is described as any change in the properties of a technical product dependent on the vital activity of the organism, and the causal organism of these changes can be microorganisms as well as higher animals or humans [1].

Foundation of buildings is associated with the choice of concrete components used according to the standards appropriate to the environment, load, or technological process. As the foundations of structures or subsurface structures are in direct contact with the soil, the functional capability, durability and consequently the service life of such constructions may be impaired not only due to physical or chemical interaction of soil substrate with the object, but also corrosive influence of microbiological activities. Bacteria and fungi are the main microorganisms responsible for biodeterioration [2]. According to [3] the most prominent types of bacteria are sulphur and nitrifying bacteria. The best-known representative of sulphur bacteria is the Acidithiobacillus genus. Especially species Acidithiobacillus thiooxidans can gradually oxidize sulphur from the present H_2S to H_2SO_4, which results in lowering the pH of the concrete to 2-3 [1]. The pH of the concrete should be in the range of 12.5-13.5 [3].

Efforts to minimize the biocorrosion degradation of concrete can be led according to [3] in two basic ways, which can be combined: 1. Creation of concrete from components with higher resistance to the effects of aggressive environments; 2. Application of biocides directly to the concrete mass or surface application of biocides. To be able to implement the proposed methods, it is necessary to know the groups of predominant representations of microorganisms and microorganisms that are most involved in biocorrosion. Table 1 presents main compounds contained of sulphur in the soil.

Table 1. Predominant compounds containing sulphur in the soil [3].

Anhydrite	$CaSO_4$	Present in evaporic rocks
Barite	$BaSO_4$	Common venous mineral in rocks
Celestine	$SrSO_4$	Rarely in some claystones
Epsomite	$MgSO_4.7H_2O$	Present in evaporic rocks
Gypsum	$CaSO_4.2H_2O$	Common in soils and rocks
Jarosite	$KFe_3(OH)_6(SO_4)_2$	Pyrite weathering product
Marcasite	FeS_2	Particles in chalk and limestone
Mirabilite (Glauber's salt)	$NaSO_4.10H_2O)$	Present in evaporic rocks
Pyrite	FeS_2	Common in soils and rocks
Pyrrhotine	FeS_2	Rarely found in soils and rocks
Organic sulphur		Common in peat

Sulphates present in soils or wastewaters rich in the content of sulphur compounds at concentrations of about 20-40 $mg.l^{-1}$ are anaerobically reduced by sulphur-reducing bacteria to H_2S. Sulphide dissolved in H_2O is adsorbed on the surface of concrete walls, where it is aerobically oxidized by sulphur bacteria to the form as a biogenic sulphuric acid. This causes undesirable corrosive changes due to lowering the pH of concrete with the formation of ettringite, which is known for its high crystallization pressures with subsequent disruption of the concrete structure - increases porosity, absorbency, and the formation of unstable zones [3].

Other type of microorganisms such as bacilli can influence concrete construction in positive or negative way. More information offers Table 2.

Table 2. Influence of selected gram-positive bacteria on concrete structures [6].

Researchers	Findings	Type of Bacteria	Concentration of bacteria added
Nosouhian et al. (2015)	Carbonate producing bacteria to improve concrete characteristics	Sporosarcina pasteurii and Bacillus subtilis	BS: 6 x 10^9 cells/mL SP: 9 x 10^7 cells/mL
Guadalupe et al. (2014)	Bio-based via bacteria-mediated calcium carbonate production improved mechanical properties and bonding strength of concrete	Bacillus cohni	1.2 x 10^7 bacterial spores/gr +15% calcium lactate
Irwan et al. (2014)	Survive in the alkaline environment with anaerobic conditions. Improve concrete properties	The Ureolytic bacteriaand Sulphate Reduction Bacteria	3% UB 0.1-0.5 mol/L & 5% SRB 0.1-0.5 mol/L
Wang et al. (2011)	Polyurethane has potential to be used as bacterial carrier for self-healing of concrete cracks	Bacillus sphaericus	10^9 cells/mL

Jonkers et al. (2011)	Self-healing capacity of concrete to seal the crack	Bacillus cohni	2.4 x 10^8 spores/cm3 cement stone
Tittelboom et al. (2009)	Repair crack in concrete	Bacillus sphaericus	20 g/L BS+20 g/L urea
Ghosh et al. (2009)	Strength of bacteria modified mortar increases	Bacillus hewanella	10^4 -10^6 BS cells/ml
Muynck et al. (2008)	Improves the durability of cement materials	Bacillus sphaericus	10^3 spores/g mortar
Ramakrishnan et al. (2005)	Improvement of concrete durability	Bacillus pasteruii	1 x 10^8 cells/ml

Bacillus of the family Bacillaceae is a genus containing gram-positive aerobic and sporulating rods and cocci. The spores of Bacillus strains are thermoresistant. They are common in nature, especially in soil, and most species are (27) non-pathogenic. Some of the species of the genus Bacillus are used as sources of antibiotics (Bacillus subtilis - bacitracin, B. polymyxa - polymyxin). The type of strain is B. subtilis. Pathogenicity to humans and animals is manifested mainly by B. anthracis and B. cereus. B. anthracis forms massive gram-positive mallets with straight ends, an endospore in the middle and a capsule made of pathological material. B. cereus occurs in the human environment, in soil, dust, accidentally contaminates materials for microbiological examination. B. subtilis has all the characteristics of the genus and, with special metabolic possibilities, forms endospores, the lifespan of which is estimated up to thousands of years. It is the subject of intensive study in scientific circles and is used for the preparation of recombinant proteins [7]. Division of G+ and G- provides Figure 1 and Figure 2.

The objective of the work was to investigate the soil and interface composition and to found if the microorganism present in the soil samples would also be identified on the surface of the constructions.

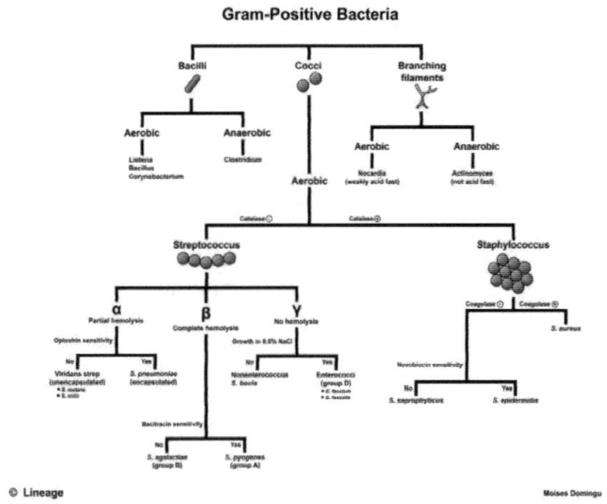

Figure 1. Gram-Positive Bacteria division [4].

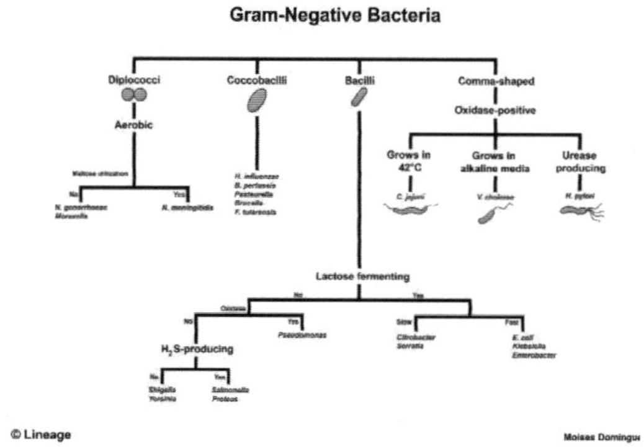

Figure 2. Gram-Negative Bacteria division [5].

2. Methodology

The methodology of sampling the soil near to the constructions and at the interface with the constructions was preceded by a part of the work with the methodology focused on sampling of cement composites (plaster systems on the surface of prefabricated apartment buildings or concrete systems of other buildings) from various urban areas across Košice, where biocorrosion was present according to visual observation. In this part of the work, samples were taken from 14 different localities, in the case of five of them it was possible to take samples from the soil, which was in contact with concrete foundation constructions.

Table provides summary of sampling with information such as location, type of construction or distance of the sampling point from the constructure.

Table 3. Sampling points information such as location, type of construction or distance of the sampling point from the construction.

N° of sample	Location	Street	Type of construction	Type of the sampling point
1	Ťahanovce	Americká trieda	concrete upperbridge	Soil near to the construction
2	Ťahanovce	Americká trieda	concrete upperbridge	Interface
3	Sever	Boženy Nemcovej	Stairs of a panel apartment building	Soil near to the construction
4	Sever	Boženy Nemcovej	Stairs of a panel apartment building	Interface
5	Dargovských hrdinov	Maršala Koneva	Elevated concrete foundation	Soil near to the construction
6	Dargovských hrdinov	Maršala Koneva	Elevated concrete foundation	Interface
7	Juh	Košťova	Garage	Soil near to the construction
8	Juh	Košťova	Garage	Interface
9	Západ	Ružínska	Garage	Soil near to the construction
10	Západ	Ružínska	Garage	Interface

2.1. Meteorological conditions during sampling
The air temperature according to the Slovak Hydrometeorological Institute was 25°C, 17% cloudiness, air pressure – 1017 hPa, wind speed 9 m/s from the North.

2.1.1. Air quality

Table 4. Hour concentrations of pollutants [μg/m^3].

PM_{10}	31	good
$PM_{2,5}$	21	good
O_3	99	increased
SO_2	3	very good
CO	523	very good
NO_2	16	very good
NO_x	25	good

2.2. Sampling
The soil near to the constructions was excavated by hoe tool with forks to the depth of 10 cm. The soil was subsequently removed from this area in the first step. In the second step, the sampling was concentrated on the soil in close contact with the structure, thus it was possible to obtain the soil from the interface of the construction and the soil. Samples were taken with a clean spoon into plastic bags and then stored in the refrigerator for further processing.

2.3. Preparation of a microscopic specimen

Samples from the bag were weighed on pre-weights with the bag. To extract the samples, a portion of the soil sample was poured with a sterile 3 ml spoon into 10 ml of sterile water, and the suspension in the beakers was shuffled.

After sedimentation of the soil samples in the beakers, a drop of the solution was applied from the phase interface using a Pasteur pipette to a sterile slide. After drying and flame fixation followed the gram staining.

2.4. Gram staining

Gram staining is an empirical method that serves to divide bacterial species into two large groups - gram-negative and gram-positive. The division is based on the chemical and physical properties of their cell wall. The cell walls of G + bacteria have a much thicker layer of peptidoglycan than the cell walls of G- bacteria. G-cells also have several lipids in the outer cell wall.

2.4.1. Hucker's modification – procedure

1) Preparation of an extract of the test culture or a mixture of G + and G- bacteria in the medium of the slide subsequently spread the sides of the control G + and G- cultures. After drying and flame fixation, the preparation is overlaid for 20 seconds with a crystal violet solution over filter paper.
   a) a. In this step, both cell types (G + and G-) are stained with crystal violet dye, whereby both stain purple.
2) The dye is decanted, and the preparation can be rinsed very briefly with running water (about 20 s).
3) Pour the preparation over about 20 sec by Lugol's solution.
   a) a. In this point, the preparation is inoculated with iodine in a solution of potassium iodide, which forms a complex with the dye in both cells.
4) The Lugol's solution is decanted in the same direction as before or can be rinsed in the same direction as before.
5) Stain the slide 20-30 sec with ethanol or acetone. The decolorizing agent is dropped onto an inclined slide so that the dye is rinsed in the same direction as the crystal violet and Lugol's solution.
   a) In this point, the complex is leached only from G-bacteria (the outer lipid layer of the cell membrane dissolves), which discolours.
6) Rinse the specimen with running water again in the same direction.
7) Staining is done by pouring the preparation to 30 - 60 sec with diluted carbolfuchsin or safranin.
8) Rinse the slide again with running water and dry it between filter papers.
   a) G + cells remain purple after Gram stain, G- cells are red (or pink).

2.4.1.1. Analysis of results of Gram staining

Gram-positive bacteria are stained with crystal violet dark purple to blue-black. Gram-negative bacteria are red or pink because they are stained with ethanol (or acetone) and stained with the contrast dye carbolfuchsin (or safranin).

Optical microscope was used to evaluate the Gram's method and to detect the presence of G + and G- bacteria by Olympus BX51 Fluorescence Microscope Cutaway Diagram which is polarizing microscope with performance in polarized light. It is a combination of UIS infinity-corrected optics and Olympus original optical-design [8]. Software support of microscope device was used by the Stream Start program.

3. Results

Table 5 provides simple classification of number of bacteria in the samples. The sign "+" is an indication to express the number of bacteria in the sample. More the pluses, more of the bacteria are present in the sample. Following figures 3 – 8 presents examples of representatives for individual classification. Upper image is taken by cell phone camera with 12Mpx resolution. Lower image is taken by Stream Start software which reflects images of the sample through the optical microscope. For better orientation both performances are used. The left side of the figures represents 50x zooming and right side of the figure represents 100x zooming of the microscope.

IOP Conf. Series: Materials Science and Engineering 1209 (2021) 012039 doi:10.1088/1757-899X/1209/1/012039

Table 5. Classification of number of bacteria in the samples.

N° of sample	Number of bacteria in the sample
1	++
2	+++
3	++
4	+
5	+++
6	++
7	+
8	++
9	+
10	+++

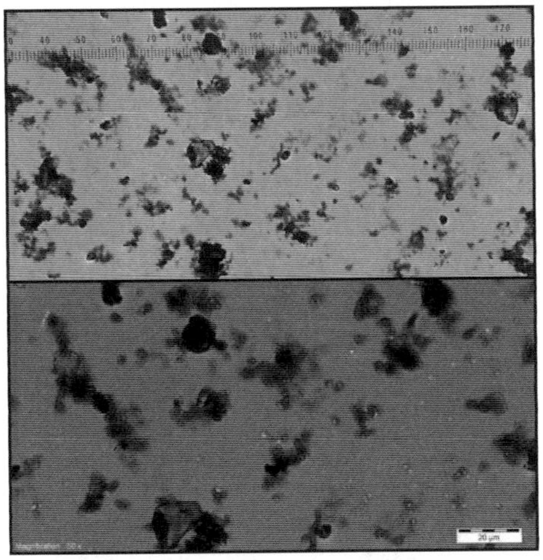

Figure 3. Sample 1, 50x zoom (++).

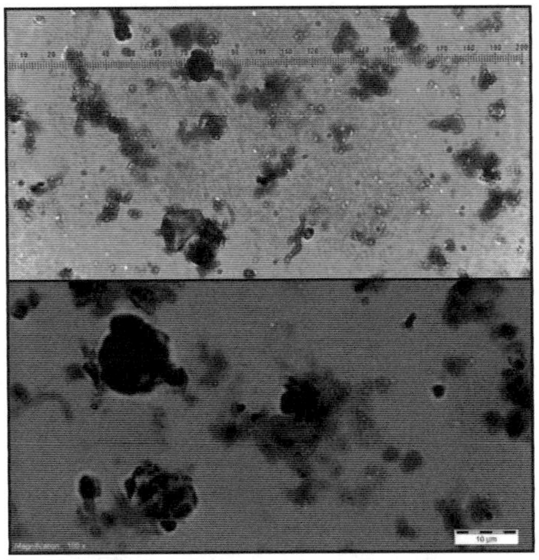

Figure 4. Sample 1, 100x zoom (++).

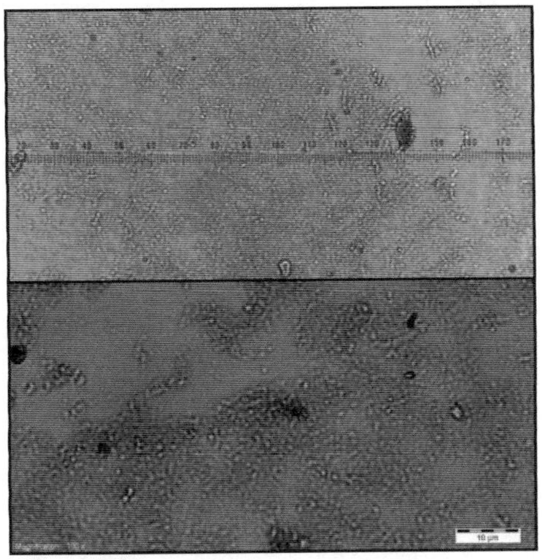

Figure 5. Sample 2, 100x zoom (+++).

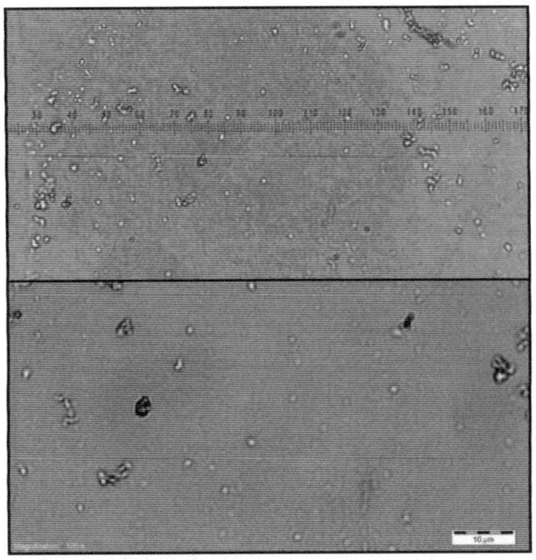

Figure 6. Sample 4, 100x zoom (+).

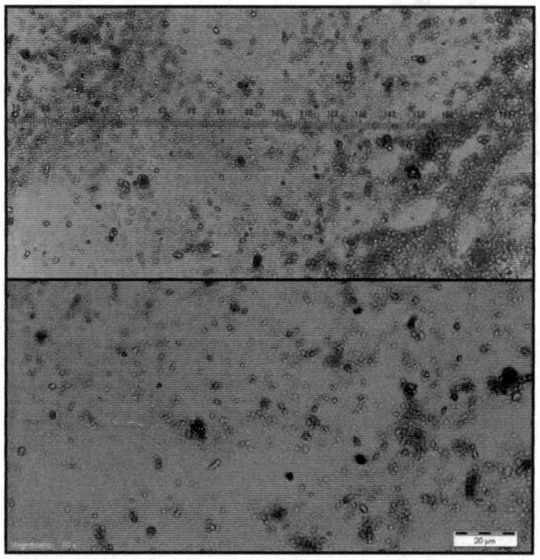

Figure 7. Sample 5 (+++).

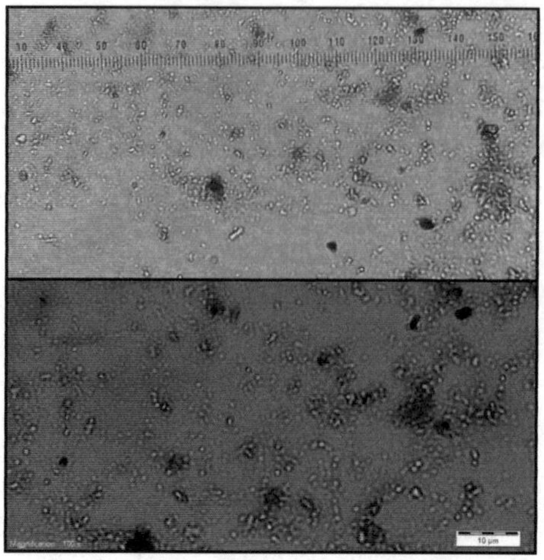

Figure 8. Sample 5 (+++).

4. Conclusion

By visual observation it is possible to declare that all the samples, including those from the construction surface, contain G+ and G- bacteria in different shapes, such as dots for cocci or stick and even spirilla for bacilli, despite to the graphic performance. The screening of 5 selected locations from Košice proved presents of microorganisms by performing the Gram staining. There is no correlation between number of bacteria in soil on the interface or near to the construction. Also, the type of construction has no reflect to the numbers of bacteria.

5. References

[1] Wasserbauer R 2000 Biological degradation of buildings 1 Praha ABF (in Czech)
[2] Bertron A 2014 Understanding interactions between cementitious materials and microorganisms: a key to sustainable and safe concrete structures in various contexts *Materials and Structures (Electronic Materials* vol 47) ed J Provis (Springer Science+Business Media)
[3] Ledererova J et al. 2009 Biocorrosive influences on construction works vol 1 (Silicate Association for the Research Institute of Building Materials, a.s.) chapter 7 pp 141-162 (in Czech)
[4] Available on: https://step1.medbullets.com/microbiology/104192/gram-positive-bacteria
[5] Available on: https://www.grepmed.com/images/702/identification-differential-gramnegative-microbiology-diagnosis
[6] Irwan, J M and Teddy T 2017 An overview of bacterial concrete on concrete durability in aggressive environment *Pertanika J Sci Technol* vol 25 (Universiti Putra Malaysia) pp 259-264
[7] Smolinska M 2017 Ecology and taxonomy of microorganisms *(Electronic Materials* https://fns.uniba.sk/uploads/media/Ekologia_a_taxonomia_mikroorganizmov.pdf) (in Slovak)
[8] Available on: https://www.olympus-lifescience.com/en/microscope-resource/primer/techniques /fluorescence/ bx51fluorescence/

Acknowledgments
This work has been supported by VEGA 2/0142/19.

The use of construction and rubber recycled material for the production of cement composites with thermal insulation function

J Charvát[1], J Svoboda[1], K Máčalová[1], L Chlebková[1] and R Papesch[1]

[1] Department of Environmental Engineering, Faculty of Mining and Geology, VSB - Technical university of Ostrava, 17. listopadu 2172/15 708 00 Ostrava-Poruba Czech Republic

E-mail: jakub.charvat@vsb.cz

Abstract. This article presents the results of the possible use of construction waste and rubber granulate as a filler substitute for the production of cement composites with thermal insulation properties. It describes the individual components of the materials necessary for the designing of recipes and their properties. The article presents the thermal insulation properties and strength characteristics of the designed recipes of non-traditional cement composite after 7, 14, 28 and 90 days.

1. Introduction

Halls, houses and other buildings are demolished all over the world every day. These demolitions create a lot of waste material, which can be further processed, but it is often stored in landfills or it is illegally stored. These procedures create burden for the environment both from the point of view of landfilling of this material as well as from the point of view of the processes of obtaining new material. The volume of construction demolition waste that is produced in Europe each year reaches 46% of the total volume of waste. This percentage corresponds to approximately 820 million tonnes of construction demolition waste. In Great Britain, for example, 50% of waste in landfills is construction demolition waste. In the United States, it represents one third of landfill waste, in Canada, it is 35% and in Australia, it is 20-30% [1–3]. In the Czech Republic, the annual production of construction demolition waste amounted to 18.7 million tonnes between 2014 and 2016 [4]. The European Union wanted to encourage the reuse of this type of material by charging storage in landfills. However, some companies started to process this waste illegally, e.g. as material for various sub-bases or as filling for sacrificial formwork [2,5]. At the same time, there are many studies dealing with the reuse of demolition and other types of waste materials [6–14].

2. Materials and Methods

2.1. Cement

CEM III/A 32.5 N cement, originating from the manufacturer Považská cementáreň a.s. Ladce, was used for the production of the test mixtures. The production and properties of the cement adhere to ČSN EN 197-1: 2012[15]. The properties of the used cement CEM III/A 32.5 N are presented in Table 1.

Table 1. Properties of CEM III/A 32.5 N cement.

Cement CEM III/A 32.5N				
		Measuring unit	Required values EN 197-1	Achieved values in PCLA, a.s., Ladce
Setting start:		minute	min. 60	229.0 (±10.0)
Compressive strength:	2 days	$N \cdot mm^{-2}$	min. 10	21.8 (±0.9)
	28 days	$N \cdot mm^{-2}$	min. 42.5 - max. 62.5	47.4 (±0.7)
Specific surface:		$m^2 \cdot kg^{-1}$	not prescribed	339.0 (±7.0)
Content Cr (VI):		ppm	max. 2	0.0 (±0.0)

2.2. Mixing water

Water from the water supply system was used according to the standard of ČSN EN 1008 – Mixing water in concrete - Specifications for sampling, testing and assessment of water suitability [16]. A cementing compound, which serves as binder in concrete, was formed after the water had been mixed with the cement.

2.3. Filler properties

All components of the filler were subjected to a sieve analysis which was performed according to ČSN EN 933-2- Determination of grain size - Sieve analysis [17].

Construction recyclate

Construction recyclate fr. 0/4 mm was chosen as the first type of filler. Construction rycyclate was obtained from demolition waste of a family house using treatment processes. The recyclate contained burnt bricks, ceramics and mortar. The representation of these components is shown in Figure 1. Figure 1 clearly shows that the construction recyclate consisted of 29% of bricks, 37% of mortar and 34% of ceramics. The grain composition of the construction recyclate is presented graphically in Figure 2. Mean grain size d_{50} = 0.78 mm.

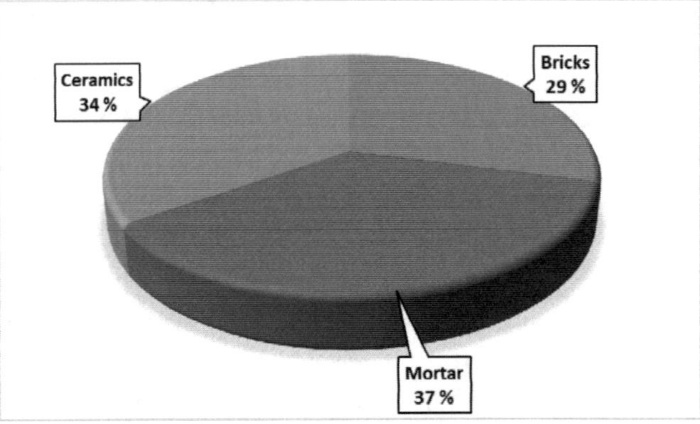

Figure 1. Proportion of individual components of construction mixed recyclate.

Rubber recyclate

Recycled rubber originates from tire recycling and was supplied by Bonus CB s.r.o. company in two fractions designated as dust (fraction 0/1 mm) and fine crushed material (fraction 1/3 mm). The results of the sieve analysis of the recycled rubber are graphically shown in Figure 2. The mean grain size of the fraction 0/1 mm d_{50} = 0.68 mm, for the fraction 1/3 d_{50} = 2.25 mm.

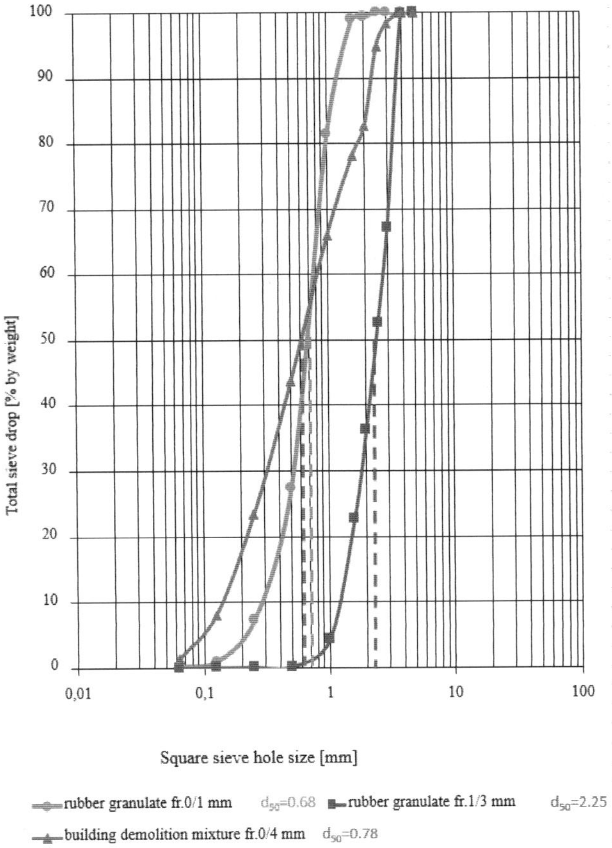

Figure 2. Results of determination of the grain size curve: grain size curve for fr. 0/4 mm mixture; a sample of rubber granulate with a grain size of 0/1 mm; sample of rubber granulate with a grain size of 1/3 mm; sample of mixed construction recyclate with a grain size of 0/4 mm.

2.4. Designing experimental recipes

6 concrete mixture recipes were designed for the production of experimental test specimens from concrete based on recycled rubber and construction recyclate. They are presented in Table 2.

Table 2. Composition of experimental mixtures.

Mixture	Cement CEM III/A 32.5N [g]	Mixing water [g]	Mixed construction recyclate [g]	Rubber granulate fr. 0/1 mm and 1/3mm in ratio 50:50 [mL]	Rubber granulate fr. 0/1 mm and 1/3 mm in ratio 40:60 [mL]
R1 + 10%	569	285	181	990	0
R1 + 20%	569	285	362	880	0
R1 + 30%	569	285	543	770	0
R2 + 10%	569	285	181	0	990
R2 + 20%	569	285	362	0	880
R2 + 30%	569	285	543	0	770

2.5. Production of the test specimens

ČSN EN 206+A1 standard served as a template for the production of the test specimens. Rubber granulate and construction recyclate were used as the filler. Blast furnace cement CEM III 32.5 N was used as the binder. Water from the water supply system was used as the mixing water. The mixture was prepared in a laboratory cement paste mixer Betonsystem BS MI-CM5AX and was placed in moulds.

The test specimens were in the shape of beams measuring 40x40x160 mm. The cement mixture was placed in moulds in two layers each, and after each layer, the mould was placed on a ZSC-40 jolting table and compacted with 60 strokes. Excess mixture was removed after the compaction and the surface was levelled with a trowel. To maintain the cement hydrating process and to prevent it from drying out, the moulds were covered with foil. 24 hours after the mixture was placed into the moulds, the specimens were demoulded and placed in a water bath for 3, 7, 14, 28 and 90 days. The water in the water bath had a temperature of 20 ± 1 °C.

2.6. Determination of the thermal conductivity coefficient

ISOMET 2114 (Applied Precision Ltd., Bratislava, Slovakia) instrument was used to measure the coefficient of thermal conductivity of the examined samples at a specific surface temperature. The measurement was performed using the attached measuring probes with a measuring range of 0.3–2.0 $W \cdot m^{-1} \cdot K^{-1}$ on the prepared specimens measuring 140 × 160 × 40 mm. After removing the formwork, the specimens were stored in a humid environment for 28 days and then dried to a constant weight. The measuring probe was placed on the surface of the prepared test specimen and, after the temperature values of the probe and the specimen was stabilized, the measurement of the thermal technical properties was initiated [18].

3. Result and Discussion

The results of thermal conductivity measurements of the prepared composites are graphically presented in Figures 3 and 4.

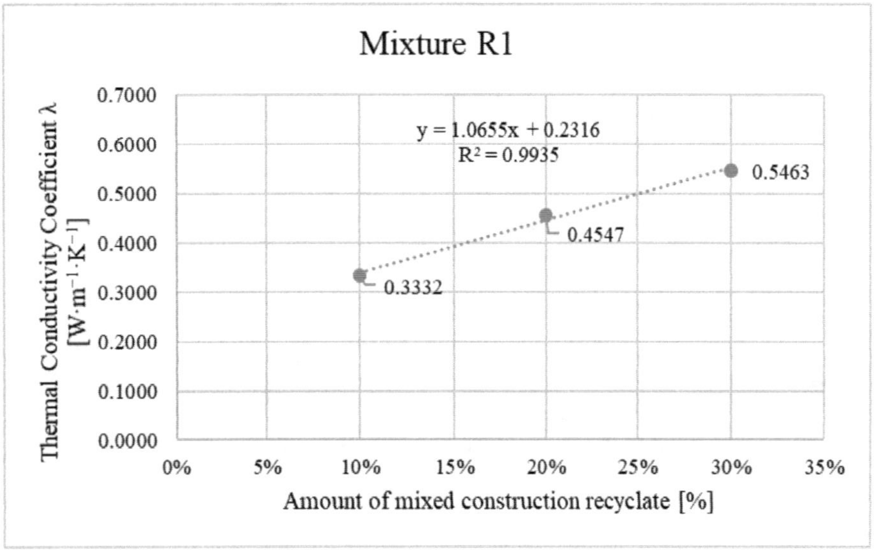

Figure 3 – thermal conductivity coefficient of composites according to recipe R1.

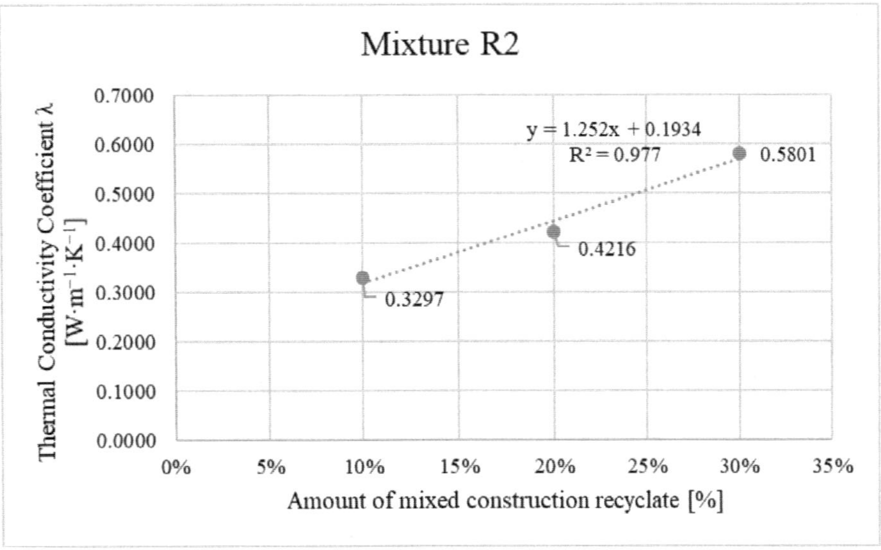

Figure 4 – thermal conductivity coefficient of composites according to recipe R2.

Figures 3 and 4 clearly show a linear increase in the coefficient of thermal conductivity depending on the increasing replacement of recycled rubber with construction recyclate. In the case of the R1 mixture, the thermal conductivity coefficient increased by 36.46% when the replacement of recycled rubber with construction recyclate increased by 10%. With an increase of 20%, the coefficient of thermal conductivity increased by 66.93%. The R2 mixture showed the same increasing trend. When the share of construction recyclate in the mixture increased by 10%, the coefficient of thermal conductivity increased by 27.87%. With further increase of the replacement of recycled rubber with construction recyclate by 10%, the coefficient of thermal conductivity increased by 75.95%.

3.1. Determination of the strength of the test specimens

The determination of the strength characteristics was performed according to ČSN EN 196-1 and the compressive and flexural strengths of individual mixtures are presented graphically in Figures 5 - 8.

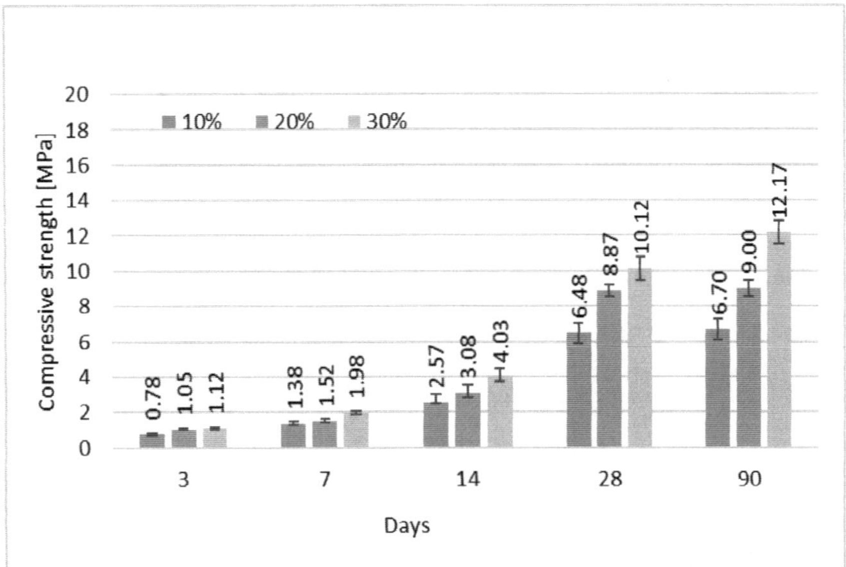

Figure 5. Overview of compressive strengths of R1 cement composites with 10%, 20%, and 30% substitutions.

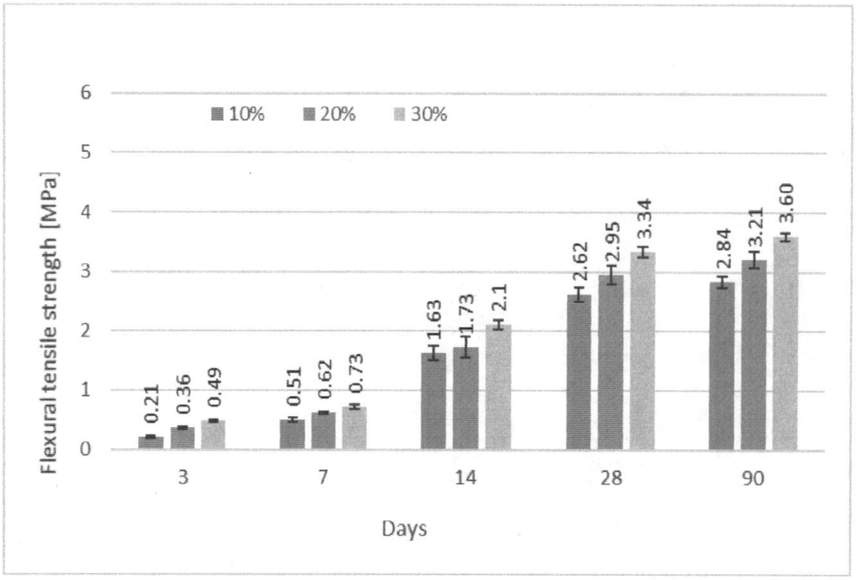

Figure 6. Overview of flexural strengths of cement composites of recipe R1 with 10%, 20%, and 30% replacement.

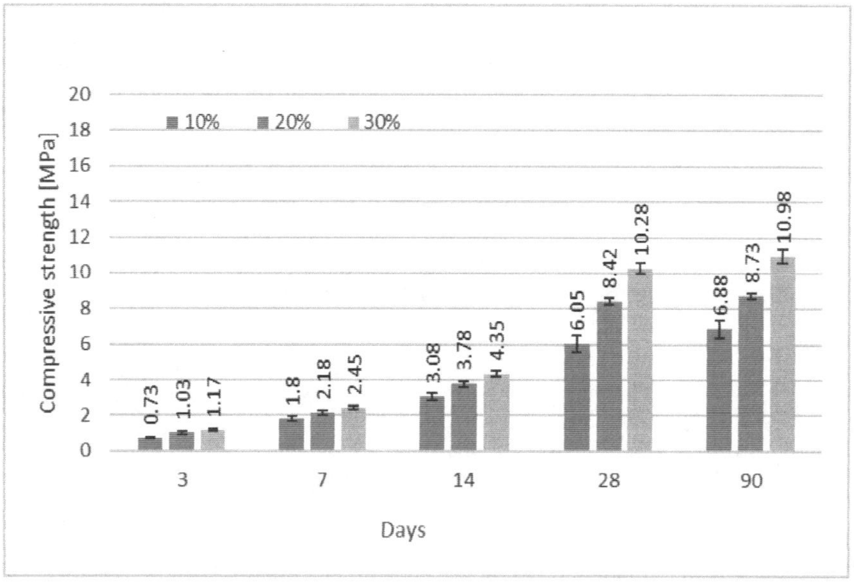

Figure 7. Overview of compressive strengths of R2 cement composites with 10%, 20%, and 30% substitutions.

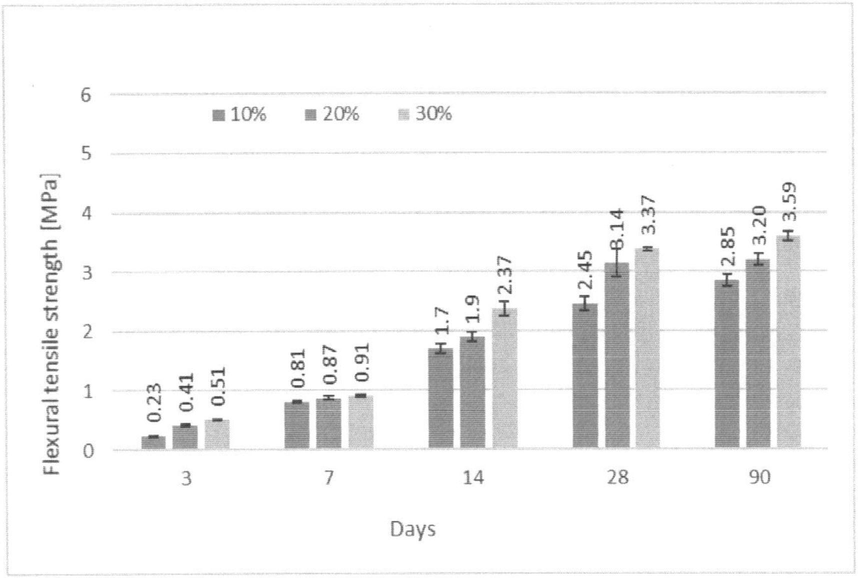

Figure 8. Overview of flexural strengths of cement composites of recipe R2 with 10%, 20%, and 30% replacement.

The results of the strength characteristics of the prepared cement composites according to recipes R1 and R2 show an increasing trend with the increasing replacement of recycled rubber with construction recyclate. As expected, the increasing age of the specimens goes hand in hand with the increasing final compressive strength and flexural strength. Figures 5 - 8 clearly present that the highest increase in strength was recorded for all test specimens between 14 and 28 days of age. The values of flexural strengths of the test specimens after 28 and 90 days are almost identical, from which it can be assumed that the hydration process of the binder after 28 days is already app. 90% completed.

4. Conclusion

The article presents the results of an experimental research dealing with a new type of cement composite based on recycled tires with a fraction of 0/1 mm and a fraction of 1/3 mm and demolition recyclate with a fraction of 0/4 mm. The research results show:

According to the performed strength tests, we can draw the following conclusions:

- It is possible to replace natural aggregate fraction 0/4 mm with recycled rubber fraction 0/1 mm and 1/3 mm in combination with demolition recyclate.
- The increasing share of the replacement of recycled rubber with construction demolition recyclate is accompanied by the increasing compressive and flexural strengths.
- The highest increase in strength always occurred between 14 and 28 days of aging.
- The values of flexural and compressive strengths of the test specimens after 28 and 90 days are almost identical, from which it can be assumed that the hydration process of the binder after 28 days is already about 90% completed.
- If the compressive strengths of R1 and R2 recipes are compared, the R1 recipe achieves better results in case of 3, 28 and 90 days of aging. The R2 recipe achieves better results in case of the aging period of 7 and 14 days.
- If the flexural strengths of R1 and R2 recipes are compared, recipe R1 achieves better results only after 90 days. The R2 recipe achieves better strength results after 3, 7, 14 and 28 days.

5. References

[1] Gálvez-Martos J-L, Styles D, Schoenberger H and Zeschmar-Lahl B 2018 Construction and demolition waste best management practice in Europe Res *Con. and Rec.* **136** 166–78
[2] Chen J, Hua C and Liu C 2019 Considerations for better construction and demolition waste management: Identifying the decision behaviors of contractors and government departments through a game theory decision-making model *J. Clean. Prod.* **212** 190–9
[3] Esin T and Cosgun N 2007 A study conducted to reduce construction waste generation in Turkey B*uild. and Env.* **42** 1667–74
[4] *Třídění Odpadu* Available online https://www.trideniodpadu.cz/
[5] Lu W 2019 Big data analytics to identify illegal construction waste dumping: A Hong Kong study *Res., Con. and Rec.* **141** 264–72
[6] Junak J 2016 Utilization of Crushed Glass Waste in Concrete Samples Prepared with Coal Fly *Ash Sol. St. Phen.* **244** 102–7
[7] Hospodarova V, Stevulova N, Briancin J and Kostelanska K 2018 Investigation of Waste Paper Cellulosic Fibers Utilization into Cement Based Building Materials *Build.* **8** 43
[8] Stevulova N, Estokova A, Holub M, Singovszka E and Csach K 2020 Characterization of Demolition Construction Waste Containing Asbestos, and the Release of Fibrous Dust Particles *App. Sci.* **10** 4048
[9] Smolakova M, Estokova A, Vaclavik V, Soucek K and Zajicova V 2018 Determination of durability of mortar with slag exposed to bacterial environment *IOP Conf. Ser.: Mater. Sci. Eng.* **385** 012052
[10] Ondova M, Stevulova N and Estokova A 2012 The Study of the Properties of Fly Ash Based Concrete Composites with Various Chemical Admixtures *Proc. Eng.* **42** 1863–72
[11] Dvorský T, Václavík V, Šimíček V and Aleš B 2015 Research of the Use of Waste Rigid Polyurethane Foam in the Segment of Lightweight Concretes *Inz. Miner.* **2** 51–6
[12] Klus L, Václavík V, Dvorský T, Svoboda J and Papesch R 2017 The Utilization of Waste Water from a Concrete Plant in the Production of Cement Composites *Build.* **7** 120
[13] Hospodarova V, Stevulova N, Vaclavik V and Dvorsky T 2017 Influence of Cellulosic Fibres on the Physical Properties of Fibre Cement Composites *IOP Conf. Ser.: Mater. Sci. Eng.* **251** 012015
[14] Stevulova N, Hospodarova V, Estokova A, Singovszka E, Holub M, Demcak S, Briancin J, Geffert A, Kacik F, Vaclavik V and Dvorsky T 2019 Characterization of Manmade and Recycled Cellulosic Fibers for Their Application in Building Materials *J. of Ren. Mat.* **7** 1121 45
[15] ČSN EN 197-1 (722101) 2001 *Cement - Část 1: Složení, specifikace a kritéria shody cementů pro obecné použití* (Praha: Český normalizační institut)
[16] ČSN EN 1008 (732028) 2003 *Záměsová voda do betonu - Specifikace pro odběr vzorků, zkoušení a posouzení vhodnosti vody, včetně vody získané při recyklaci v betonárně, jako záměsové vody do betonu* (Praha: Český normalizační institut)
[17] ČSN EN 933-2 (721184) 1997 *Zkoušení geometrických vlastností kameniva - Část 2: Stanovení zrnitosti - Zkušební síta, jmenovité velikosti otvorů* (Praha: Český normalizační institut)
[18] Svoboda J, Dvorský T, Václavík V, Charvát J, Máčalová K, Heviánková S and Janurová E 2021 Sound-Absorbing and Thermal-Insulating Properties of Cement Composite Based on Recycled Rubber from Waste Tires *App. Sci.* **11** 2725

Acknowledgments

This article was partially supported by:
- Grant of SGS No. SP2021/21, Faculty of Mining and Geology, VSB – Technical University of Ostrava, Czech Republic.
- Moravian-Silesian Region – grant number RRC/10/2018 and RRC/02/2020
- Považská cementáreň, a.s.,, Ul. Janka Kráľa,, 018 63 Ladce, SlovakiaThis research was funded

IOP Conf. Series: Materials Science and Engineering 1209 (2021) 012041 doi:10.1088/1757-899X/1209/1/012041

Analysis of the behaviour of 3D samples printed by FFF/FDM technologies under bending stress with a focus on infill

D Juračka[1] and M Kawulok[1]

[1] Department of Structural Mechanics, VSB - Technical University of Ostrava, Ludvika Podeste 1875/17, 708 00 Ostrava - Poruba, Czech Republic

E-mail: david.juracka@vsb.cz

Abstract. The focus of this paper is to explore the possibilities of optimizing 3D printed elements produced by FFF/FDM technology based on bending tests according to ČSN EN ISO 178 (64 0607) with variations in the settings of the printing itself. The principle of FFF/FDM is the printing of a continuous fiber made of thermoplastic material, which is applied by machine to the previously printed fiber. There are many combinations of possible print settings, and one of them is the geometry of the inner density with variable density. Their resulting maximum values of the achieved load were then compared with the weight (amount of material used) and printing time. The result is a comparison to achieve economical printing with the greatest possible load capacity.

1. Introduction

3D printing technology is relatively new (1980 - first patent attempt [10]; author Hideo Kodama), yet very widespread in many fields. It is used mainly in the creation of prototypes. It is used to print otherwise difficult-to-manufacture elements and is even used in the fashion and design industries [1]. The printing of habitable buildings from concrete mixtures is also beginning to occupy an interesting place for practical use [2]. In these cases, it already has a great influence on the way in which the printing is performed, so that the resulting shape and its fillings have satisfactory mechanical properties with respect to weight.

The aim of this project is to explore the possibilities of optimizing 3D printed elements produced by FFF (Fused Filament Fabrication) or otherwise called FDM (Fused Deposition Modeling) based on bending tests according to the standard [11] with variations in the settings of the print itself.

The principle of FFF / FDM is the printing of a continuous fiber made of thermoplastic material, which is applied by machine to the previously printed fiber. The cross-section of the support element cannot be considered homogeneous, because of gaps between the individual fibers. Due to the printed fibers, the resulting elements have an anisotropic character - whether the stress is conducted along the fibers, across or at a certain angle [3-5] - Figure 1. The load-bearing capacity of the fibers themselves is further affected by material properties, temperature of the print head and heat bed, layer thickness, fiber width and print speed.

Another important aspect is the filling of the printed element. Here, it depends on the type of geometry pattern used and the percentage fill inside the element. Some works have already dealt with this topic [6-9]. However, in this case, only the filling itself without external perimeters is tested.

IOP Conf. Series: Materials Science and Engineering 1209 (2021) 012041 doi:10.1088/1757-899X/1209/1/012041

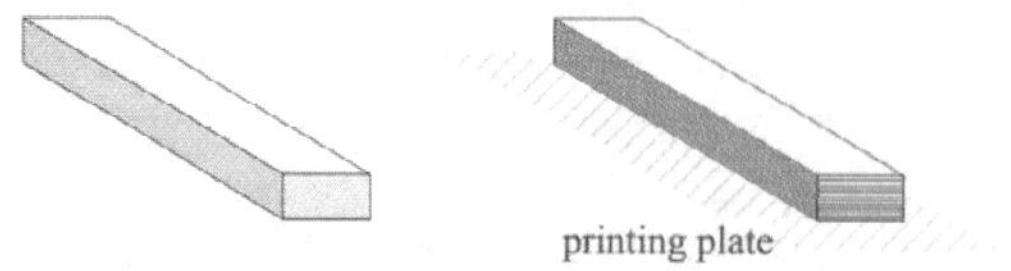

Figure 1. Main difference between perfect and local sample using FFF / FDM.

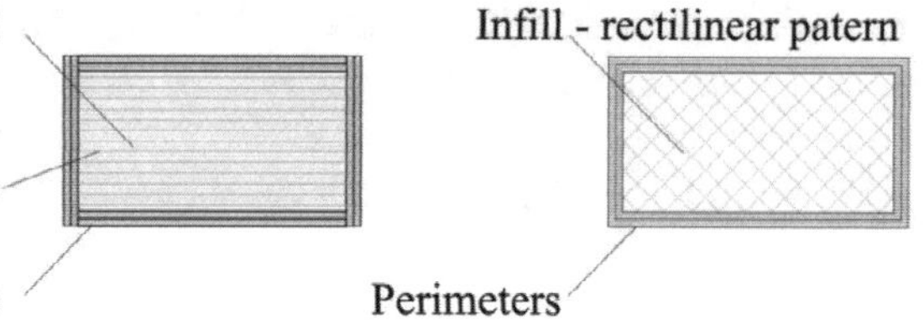

Figure 2. Test specimens at different fill densities - rectilinear pattern.

This paper examines the dependence of the maximum load on the geometry of the inner fill pattern and its density at three-point bending. Subsequently, the parameters of the material used, and the print length are compared.

2. Sample preparation and testing

Three types of pattern geometry (rectilinear, gyroid, honeycomb) and 3 fill densities (25%, 50%, 75%) were selected for testing - Figures 3 - 5. The original Prusa I3 MK3S + 3D printer with PrusaSlicer software was used for their production and material Prusament PLA. When preparing the print parameters, the outer perimeter and walls were completely omitted so that only the inner filling itself was tested. Another main parameter, which is the same for each sample, is a nozzle width of 0.4 mm with a printed fiber height of 0.2 mm. After creating virtual patterns in PrusaSlicer and setting their properties, a gcode containing information for the printer was exported. It then lays the molten thermoplastic material PLA on the given trajectories and then layer by layer produces these samples.

A standard [11] was used for the testing procedure. The dimensions of the samples are given by the standard - 80x10x4 mm. These dimensions were changed in accordance with the standard to 80x20x20mm to compensate for the sensitivity of the measuring device - mechanical press FP 10. A total of 9 sets of samples are created (3 patterns x 3 fill densities) and 5 samples are tested in each set to get average values.

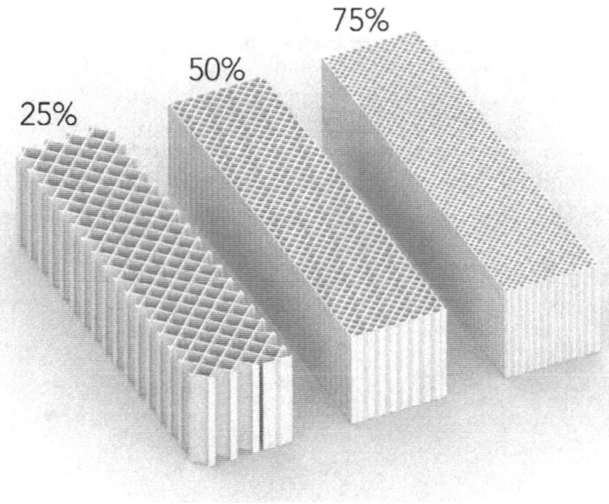

Figure 3. Test specimens at different fill densities - rectilinear pattern.

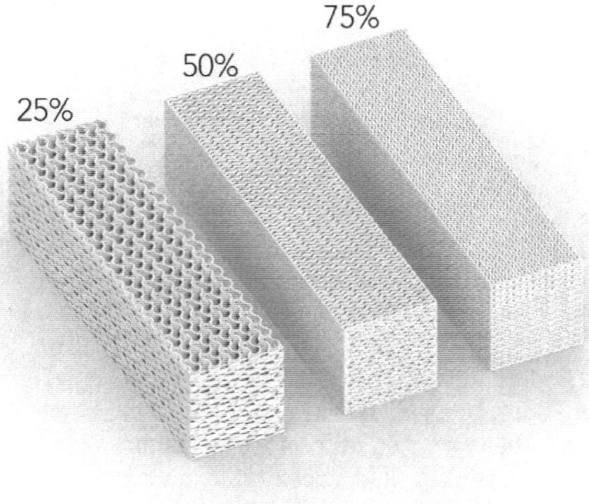

Figure 4. Test specimens at different fill densities – gyroid pattern.

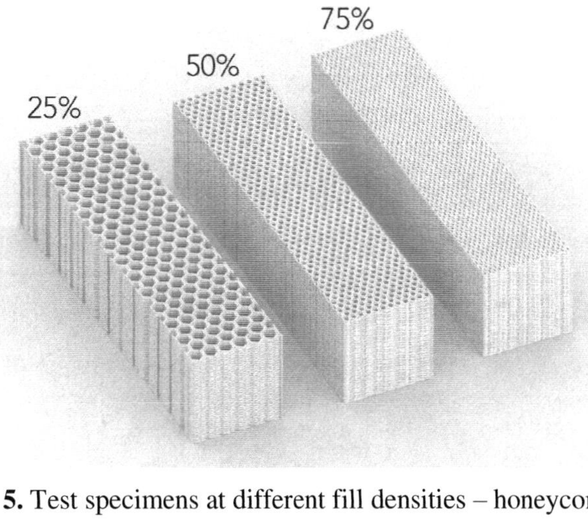

Figure 5. Test specimens at different fill densities – honeycomb pattern.

All samples were then weighed and the dimensions checked. If a printing error occurred, such as a torn fiber or local sintering, the damaged sample was replaced. The temperature of the samples was measured before the measurement itself.

A mechanical press FP 10 was used for the measurement, where the sample was placed according to the standard - Figure 6. The test aims to achieve the maximum force for the bending sample and from this value the maximum bending stress will be calculated. The individual procedures during the measurement and the individual distances and dimensions are given by the standard.

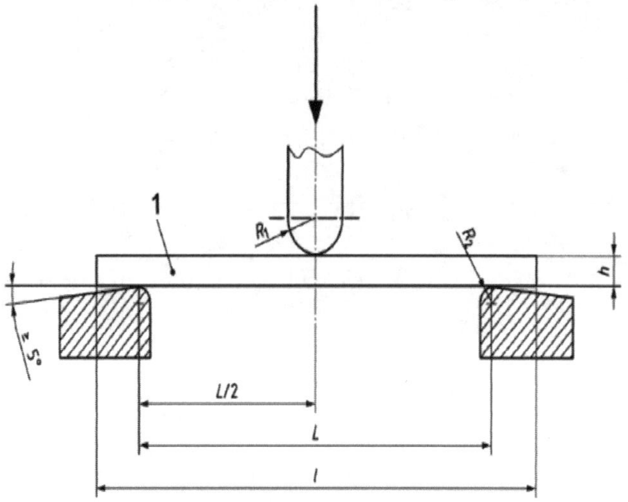

Figure 6. Position of the specimen at the beginning of the test [11].

3. Measurement results and their comparison

The tested samples had an average temperature of 23.7 °C before the test. This was followed by the test itself, where the sample was placed in a test rig according to the standard, and bending testing of the samples began. During this time, the values of the loading force *F* were recorded. As soon as the specimen reached its maximum load capacity and the values began to decrease due to the failure and deformation of the specimen, the test was terminated. The resulting values were then recalculated to determine the maximum bending stress. All averaged values are written in Table 1.

The formula from the standard [11] for calculating the normal stress under bending stress has the form:

$$\sigma_f = \frac{3FL}{2bh^2}, \tag{1}$$

where σ_f is the normal bending stress [MPa], F is the loading force [N], L is the span [mm], b is the width of the rectangular cross-section [mm] and h is the thickness of the rectangular cross-section [mm].

Table 1. Results - after averaging the sample values.

Pattern	Infill [%]	Weight [g]	Maximum force to bending *F* [N]	Maximum load capacity σ_{fm} [MPa]
Rectilinear	25	9.6	180	2.16
	50	19.2	743	8.91
	75	28.0	1950	23.40
Gyroid	25	9.6	603	7.24
	50	18.2	1701	20.41
	75	25.9	2955	35.46
Honeycomb	25	12.2	1023	12.28
	50	22.0	2584	31.01
	75	30.4	3676	44.11

The resulting calculated maximum bending stress values converted to the graph in Figure 7 clearly show that it depends very much on the geometry of the pattern used. From the three selected patterns, it can be clearly seen that in terms of maximum bending stress, the honeycomb pattern reached the highest values, which had on average about 2 times higher load capacity than the weakest rectilinear pattern. At the same time, on the same graph, a sharper increase in carrying capacity can be seen in the rectilinear pattern with an increase in its density, while in the more complex one the increase is flattened. In some cases, the sample was broken when lower load values were applied than in others. This phenomenon was common in samples with low fill density. It can be explained by the fact that at this density the drawn fiber has fewer adjacent joints in the horizontal plane, and is therefore more prone to damage, especially if there is an indistinguishable printing imperfection at some point.

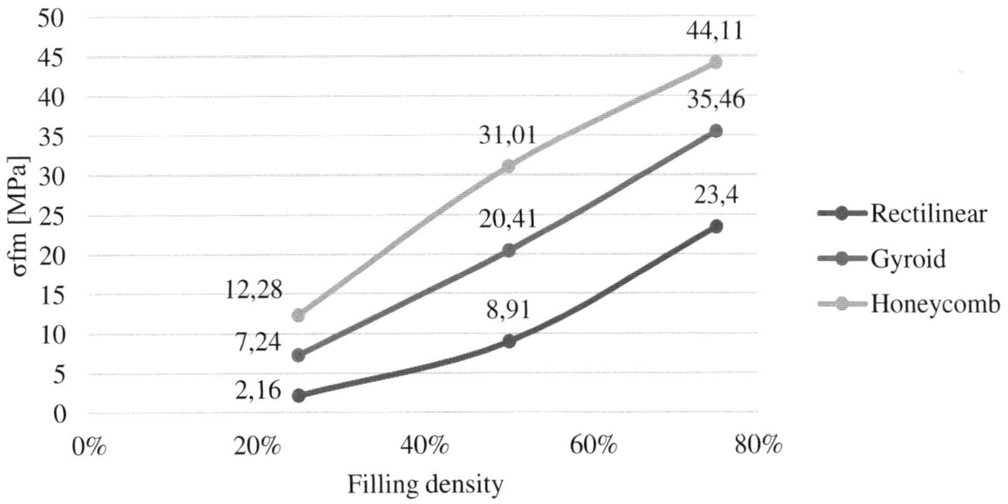

Figure 7. Maximum load capacity in 3-point bending vs filling density (with curves of probable course of values).

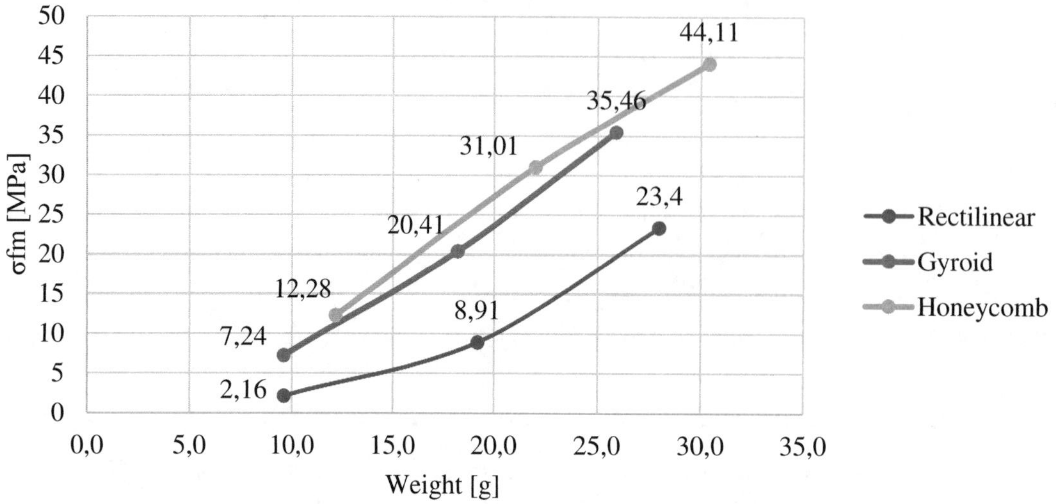

Figure 8. Maximum load capacity in 3-point bending vs weight (with curves of probable course of values).

Another possible comparison of results is the economic side. Figure 8 compares the maximum values of flexural stress with the weight of the sample or material consumed. Although the patterns have the same fill density percentage, their weight varies. It is smaller for simple patterns and larger for more complex ones. However, with the gyroid pattern, it is clear, that even with a slightly lower weight it achieves significantly greater load capacity than a simple pattern.

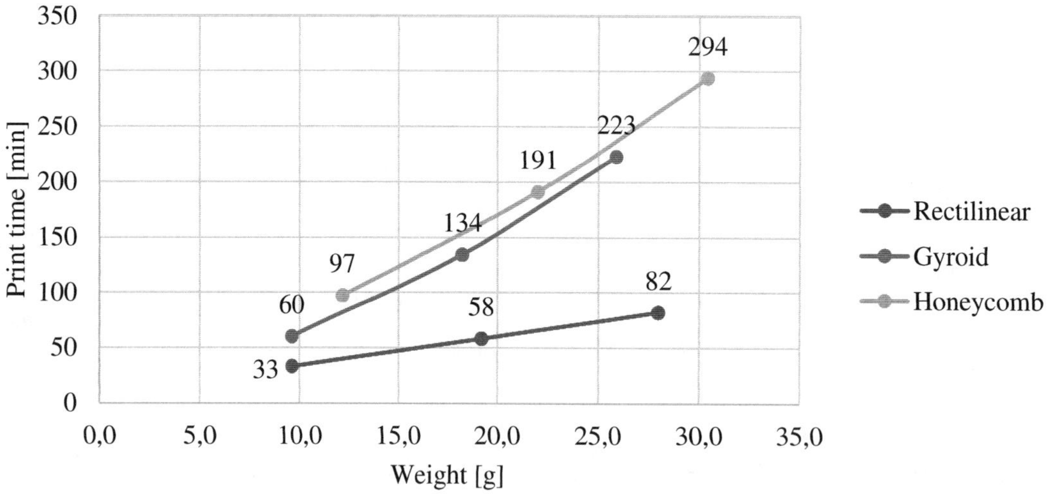

Figure 9. Print time vs weight (with curves of probable course of values).

The last comparison in Figure 9 shows the relationship between the weight (material used) and the printing time of one sample. From this point of view, it can be clearly seen that with more complex geometry there is a sharp increase in the need for machine time for its production, up to 3 times to obtain 2 times more load capacity. The increase in complexity acquires a slightly exponential course with increasing density, while a simple rectilinear pattern has a clear linear course.

4. Conclusion

The results show that the geometry of the inner filling pattern significantly affects the load capacity. As the density of each pattern increases, so does its bending resistance. At the same time, it is shown that simple patterns (rectilinear) have a faster increase in load capacity at a higher density than more complex ones (honeycomb), in which, on the contrary, this increase slows down. Another finding is the non-uniformity of the density (even during the increase) of the pattern with the weight. Here, the gyroid pattern came out best from the ratio of weight to bending resistance.

Another aspect is the economic side. The compared weight-to-machine time parameters for sample production showed that a simple rectilinear pattern requires up to 3x longer time for 75% density than a complex honeycomb pattern to obtain approximately 2x greater load capacity.

There are many parameters that can be further tested - fiber thickness, high or low temperature, other types of pattern geometry, the effect of external perimeters, and many more. It is important to examine these aspects so that they can then be properly combined to have the greatest desired effect on the element.

5. References

[1] Končić J and Ščapec J 2018 3D print additive technology as a form of textile material substitute in clothing design – interdisciplinary approach in designing corsets and fashion accessories *Industria Textila* **69** 3 190–96

[2] Guangchao Ji, Ding T, Xiao J, Du S, Li J and Duan Z 2019 A 3D Printed Ready-Mixed Concrete Power Distribution Substation: Materials and Construction Technology *Materials* **12** 9

[3] Ravindrababu, Suraj, Yunus Govdeli, Zhuo Wei Wong and Erdal Kayacan 2018 Evaluation of the influence of build and print orientations of unmanned aerial vehicle parts fabricated using fused deposition modeling process *Journal of Manufacturing Processes* **34** 659–66

[4] Wu W, Geng P, Li G, Zhao D, H Zhang and Zhao J 2015 Influence of Layer Thickness and Raster Angle on the Mechanical Properties of 3D-Printed PEEK and a Comparative Mechanical Study between PEEK and ABS *Materials* **8** 9 5834–46

[5] Dizon J R C, Espera A H, Chen Q and Advincula R C 2018 Mechanical characterization of 3D printed polymers *Additive Manufacturing* **20** 44-67

[6] Dudescu C and Racz L 2017 Effects of Raster Orientation, Infill Rate and Infill Pattern on the Mechanical Properties of 3D Printed Materials *ACTA Universitatis Cibiniensis* **69** 1 23–30

[7] Khan S F, Zakaria H, Chong Y L, Saad M A M and Basaruddin K 2018 Effect of infill on tensile and flexural strength of 3D printed PLA parts *IOP Conference Series: Materials Science and Engineering* **429**

[8] Rismalia M, Hidajat S C, Permana I G R, Hadisujoto B, Muslimin M and Triawan F 2019 Infill pattern and density effects on the tensile properties of 3D printed PLA material *Journal of Physics: Conference Series* **1402**

[9] Rebenaque A G and González-Requena I 2019 Study of bending test of specimens obtained through fdm processes of additive manufacturing *Procedia Manufacturing* **41** 859–66

[10] Kodama H *Stereoscopic figure drawing device* Japan JPS56144478A

[11] ISO 178:2019 2019 *Plastics - Determination of flexural properties* (Brusel: European Committee for Standardization)

Acknowledgments

The financial supports of the grant program financed by Ministry of Education, Youth and Sports of the Czech Republic through VSB – TU Ostrava (SGS SP2021/83) is highly acknowledged.

IOP Conf. Series: Materials Science and Engineering 1209 (2021) 012042 doi:10.1088/1757-899X/1209/1/012042

Mechanical fracture parameters of concrete drill-core specimens supported by a slenderness ratio study

D Lisztwan[1, 3], I Kumpova[2, 3], P Danek[1], P Frantík[2] and Z Kersner[2]

[1] Brno University of Technology, Faculty of Civil Engineering, Institute of Building Testing, Veveri 331/95, 602 00 Brno, Czech Republic
[2] Brno University of Technology, Faculty of Civil Engineering, Institute of Structural Mechanics, Veveri 331/95, 602 00 Brno, Czech Republic
[3] The Institute of Theoretical and Applied Mechanics of the Czech Academy of Sciences, Prosecka 809/76, 190 00 Prague 9, Czech Republic

E-mail: Dominik.Lisztwan@vutbr.cz; Ivana.Kumpova@vutbr.cz

Abstract. The detailed analytical and experimental investigation of the fracture behaviour of quasi-brittle materials is an endeavour which has been ongoing worldwide for many years. Such materials are usually characterized in terms of their mechanical fracture parameters, which are determined based on the evaluation of quasi-static fracture experiments. One of the most commonly used building materials with a quasi-brittle response is concrete, which is most often based on a cement matrix. It is sometimes also necessary to characterize concrete included in existing structures. In this case, test specimens are obtained by core drilling, and the investigation is conducted with the requirement to maximize the number of parameters obtained while minimizing the number of test specimens drilled from the structure. This paper focuses on the mechanical fracture parameters of core-drilled specimens taken from a selected concrete structure. Tests were performed on cylindrical specimens with a chevron-notched stress concentrator in the three-point bending configuration in order to determine modulus of elasticity, fracture toughness and fracture energy. Subsequently, theoretical compressive strength was estimated and tests for the determination of compressive strength values were performed focusing on dependence on the slenderness ratio, i.e. the relationship between the compressive strength and the length to diameter ratio of the cylindrical specimens. In relation to the obtained mechanical fracture parameters, selected specimens were analysed and three-dimensionally characterized via high-resolution X-ray computed tomography.

1. Introduction

The detailed analytical and experimental investigation of the fracture behaviour of quasi-brittle materials is an endeavour which has been ongoing worldwide for many years [1–3]. Materials with a quasi-brittle response are most often characterized in terms of their basic mechanical fracture parameters, which are usually determined based on the evaluation of quasi-static fracture experiments. One of the most commonly used building materials with a quasi-brittle response is concrete, which is most often based on a cement matrix. Concrete test specimens are prepared from the developed concrete mixtures in appropriate mould sand tested at the required stage of maturity, or – as in the case of the presented work – can be obtained from existing structures for the purposes of construction research or the evaluation and specification of load tests to assess their current state. The success of a fracture experiment and the accuracy of its results depends on several factors: the chosen method of loading (which must be adequate for the needs of the experiment), the determined range of the applied load, and also the selection of suitable measuring methods and equipment along with their most advantageous location on the structure. As regards the computational support for loading experiments, i.e. measuring the deformation of structures, determining the degree of their degradation or estimating their load-bearing capacity, it is necessary to have accurate data on as many input mechanical fracture parameters as possible. At the same time, it is necessary to maximize the amount of obtained parameter values while minimizing the number of test specimens taken from the structure.

The diagnostic analysis and evaluation of existing concrete structures is a very complex discipline. The successful evaluation of the current condition of a structure requires knowledge of the physical and mechanical parameters of the materials from which it is made. In the case of a new structure, the properties of the materials used are defined using specimens made during the construction process itself. When determining the material characteristics of existing structures, one of the few methods that can be used for obtaining specimens is core drilling. The core-drilling method used to collect specimens from concrete structures uses special diamond drill bits, which vary in size (diameter, length). The right choice of drill core size depends on several factors: the size of the structure, the maximum aggregate size, and the importance of preventing damage to the tested structure. The smallest core diameter that can be drilled from a structure is around 25 mm (micro cores). Nevertheless, cores as small as that are used only in special cases. Cores of diameters from 50 to 100 mm are most commonly used, though it is the case that the smaller the core diameter, the less damaging it is to the tested structure [4].

Existing codes and requirements specify the dimensions of specimens that can be used to evaluate the compressive strength of concrete f_c. An important factor which influences the subsequent compressive strength of concrete is the slenderness (*L*/*D* ratio) of the tested specimens [5]. Most often, standard specimens (made in a mould) have an *L*/*D* ratio of 1.0 (cube strength) or 2.0 (cylinder strength). In this case, some codes allow the compressive strength of the drill-core specimens with an *L*/*D* ratio in this range to be transformed into cube or cylinder strength using a correction factor [6].

Structural concrete can be also characterized via parameters obtained via the evaluation of fracture tests, and this paper presents the results of the evaluation of three-point bending fracture tests conducted on core-based concrete specimens with a chevron-type notch. The data obtained from fracture tests in the form of load vs. displacement and load vs. crack mouth opening displacement diagrams are analysed to obtain values for selected mechanical fracture parameters: modulus of elasticity E, fracture toughness K_{Ic}, and fracture energy G_F, which are determined based on the linear elastic fracture mechanics approach and Work-of-Fracture method. Within the study of the fracture behaviour of different materials, in recent years attention has been given to the use of X-ray imaging methods for the two-dimensional and three-dimensional characterization of internal structure and its alteration caused by various types of loading [7], of which the most modern approach is time-lapse X-ray computed tomography (4D-XCT) performed during fracture experiments [8, 9]. This method allows for the observation of changes in material structure caused by the loading process in three-dimensional models, and also supplements the determination of fracture parameters with data on the geometric and morphometric parameters of the damage that arises, which can be observed not only on the surface of the test specimen but also throughout its entire volume.

The article focuses on the options for obtaining as much information as possible about the material of existing structures while minimizing the size and number of samples taken from them and thus the damage caused to them as well. Tests were performed on cylindrical specimens with a chevron notch stress concentrator in the three-point bending configuration in order to determine modulus of elasticity, fracture toughness, and fracture energy. Further attention was paid to the determination of compressive strength values and their dependence on the slenderness of the core-drilled specimens in relation to the standardized *L*/*D* ratio [6, 10]. Selected specimens were tomographically monitored during the loading process, providing additional information on ongoing morphometric changes and damage propagation.

2. Materials and methods

2.1. Specimen sampling and preparation
For the experimental part, a collection of 27 specimens with a diameter of 50 mm was taken from the same cement-based concrete structure with the use of the diamond core drill method. For the standard compressive strength test, six groups of three specimens were prepared, each group differing in the length to diameter ratio (*L/D* ratio) of the specimens, which was set to 0.5, 1.0, 1.5, 2.0, 2.5, and 3.0.

For the determination of mechanical fracture parameters using the three-point bending test, a seventh group of three specimens were prepared with a length of 150 mm and equipped with a chevron-type notch. After the test, the ends of these specimens were used for complementary compressive strength tests as well.

The eighth and last group contained six specimens, each with a different *L/D* ratio conforming to the six ratios stated above, which underwent 4D-XCT scanning. These specimens were tomographically scanned twice – first in their initial state before loading, and then after applying the load. The numbers of samples in the groups subjected to different test methods, their labelling and nominal dimensions are listed in Table 1, while photographs of the specimens and their loading configurations are shown in figure 1.

Table 1. Groups of specimens and their nominal dimensions.

Group	Spec. (–)	Diameter (mm)	Length (mm)	*L/D* ratio (–)	Method
CN_05_1–3	3	50	25	0.5	Compressive strength
CN_10_1–3	3		50	1.0	Compressive strength
CN_15_1–3	3		75	1.5	Compressive strength
CN_20_1–3	3		100	2.0	Compressive strength
CN_25_1–3	3		125	2.5	Compressive strength
CN_30_1–3	3		150	3.0	Compressive strength
CN_L1–3	3		150	3.0	Three-point bending
CN_05–30_4X	6		diff.	diff.	X-ray tomography

2.2. Determination of mechanical fracture parameters
In this paper, the evaluated fracture experiment is a three-point bending test applied to cylindrical concrete specimens with a chevron-type notch. Load vs. midspan deflection diagrams (*F–d* diagrams) and load vs. crack mouth opening displacement diagrams (*F–CMOD* diagrams) were recorded during the specimen loading. The fracture tests were carried out using an electromechanical testing machine with a measuring range of 0–250 kN. The displacement increment was required to be constant – the speed of the induced displacement of the upper support was equal to 0.02 mm/min. A photograph of the test setup is shown in figure 1.

The above-mentioned recorded diagrams were used together with information about specimen dimensions, the cross-section and initial notch as input data for CheF academic software [11]. The values obtained for selected mechanical fracture parameters – modulus of elasticity E, fracture toughness K_{Ic}, toughness G_{Ic}, and fracture energy G_{F} – were determined based on the linear elastic fracture mechanics approach [1] and Work-of-Fracture method [12].

The first, almost linear part of the *F–CMOD* diagram (namely a selected point from this part with values F_i and corresponding $CMOD_i$) was used to estimate the modulus of elasticity E with the help of geometrical factor g_0 [13].

Figure 1. Specimens subjected to compressive and fracture tests. A) Photograph of the specimens. B) Illustration of the three-point bending fracture test setup with a chevron notched core-based concrete specimen. C) Illustrations of compressive tests.

$$E = g_0 \cdot \frac{F_i}{CMOD_i} \cdot \frac{1}{D},\ g_0 = 20.8 - 19.4 \cdot \frac{a_0}{D} + 142.3 \cdot \left(\frac{a_0}{D}\right)^2, \quad (1)$$

where D is specimen diameter, a_0 is the depth of the initial notch.

The maximum load $F_{\max}$ was used for fracture toughness K_{Ic} assessment with the help of geometrical factor $A_{\min}$ [14] and span length S:

$$K_{\mathrm{Ic}} = A_{\min} \cdot \frac{F_{\max}}{D^{1.5}},\ A_{\min} = \frac{S}{D} \cdot \left[1.835 + 7.15 \cdot \frac{a_0}{D} + 9.85 \cdot \left(\frac{a_0}{D}\right)^2\right] \quad (2)$$

The toughness G_{Ic} was then calculated as follows:

$$G_{\mathrm{Ic}} = \frac{(K_{\mathrm{Ic}})^2}{E}. \quad (3)$$

The work of fracture value W_{F} was assessed from the complete F–d diagrams [11]:

$$W_{\mathrm{F}} = \int F(d)\mathrm{d}d, \quad (4)$$

and then the specific fracture energy G_F value was calculated:

$$G_F = \frac{W_F}{A_{lig}}, \tag{5}$$

where A_{lig} is the area of a ligament.

The following compressive tests were performed on the same electromechanical testing machine. The increment of the displacement was 0.05 mm/min. During the tests, the actual force and displacement of the upper girder were recorded directly from the testing machine's sensors. For the purpose of eliminating the effect of the testing machine's stiffness, the displacement of the loading plates was measured externally with the use of a pair of LVDT transducers. Three specimens from each *L/D* ratio group were used for the compressive test, which was applied until each specimen failed. The fourth specimen of each group was first scanned using X-ray computed tomography and then subjected to compressive testing. However, in these tests the specimens were not tested until failure occurred in order to maintain their integrity for the X-ray scan after the test. Thus, the maximum force applied to each scanned specimen was calculated as 90 % of the minimum force applied at the failure of the three specimens of each *L/D* ratio group tested previously. After a second X-ray scan, these specimens were loaded to failure. The last step of the experimental part of the investigation was to determine strength parameters via compressive tests on the adjusted ends of the specimens used earlier in the three-point bending tests. In the case of the specification of the relationship between the compressive strength and slenderness of concrete, these adjusted ends were set up in two more groups with *L/D* ratios of approximately 0.6 and 0.7, respectively.

The compressive strength was primarily determined from the performed compressive tests using the equation below.

$$f_c = \frac{F_{max}}{A_c}, \tag{6}$$

where F_{max} is the maximum force at the failure of the specimens, and A_c is the initial cross-section area of the specimens' loading surfaces.

2.3. X-ray computed tomography and the TORATOM device

Due to their ability to pass through opaque materials, X-rays are widely used to provide images of the internal structure of objects. Radiation is emitted from a generator (X-ray tube), passes through the examined object, where attenuation occurs proportional to the material characteristics of the object, and hits a detector, where 2D image information is generated. If a number (typically hundreds to thousands) of such images is acquired at different angles of rotation of the examined object relative to the generator-detector system, modern computational algorithms (e.g. filtered back projection) can reconstruct a three-dimensional model of this object, which can be handled further virtually without the need to manipulate the original specimen. Thus, industrial X-ray computed tomography (XCT) is considered to be a non-destructive method and is suitable, inter alia, for observing changes caused by various physical and chemical factors.

The advanced TORATOM (Twinned Orthogonal Adjustable Tomograph) device [15] depicted in figure 2 combines two X-ray tube–detector pairs in an orthogonal arrangement with a shared high-precision rotary tomographic table. This arrangement allows the use of imaging pairs individually but also simultaneously (dual-source or dual-energy XCT) during the data acquisition process for tomographic reconstructions. Projection magnification can be changed from about 1.2× to 100×. The device is designed modularly and allows, among other things, the quick exchange of various types of detectors available at the workplace. Concerning the type of detector used and its native pixel size, it is possible to change the resolution of CT reconstructions, i.e. the size of one spatial point of the model (voxel), from about 0.18 mm to micrometric. Very high stable resolution is achieved thanks to the use of a fully motorized CNC-controlled positioning system mounted on an actively damped anti-vibration platform.

Six specimens with different *L/D* ratios were tomographically investigated in their initial state and after loading was applied up to 90 % of the previously determined compressive strength value. These conditions ensured that irreversible changes at the micrometric level were reflected in the structure of the material, but they stopped short of causing the specimen to fail. One specimen with an *L/D* ratio of 1.5 presented within this paper was additionally supplemented by tomography after the completion of the compressive strength test, e.g. after the maximum force was reached.

The specimens were scanned using one X-ray tube–detector pair. A Microfocus reflective-type X-ray tube (XWT-240-SE, X-Ray WorX, Germany) operating at an accelerating voltage of 230 kV and a target power of 48.3 W was used to generate the radiation. The projections were acquired by a flat panel detector (Dexela 1512NDT, Perkin Elmer, USA) with a GOS scintillator, a matrix of 1536 × 1944 pixels, a 74.8 μm pixel size and an active area of 114.9 × 145.4 mm.

Figure 2. TORATOM tomography system with a concrete specimen: 1) Detector holder, 2) Detectors of different types, 3) X-ray tubes, 4) High-precision tomographic rotary stage, 5) Object under investigation, 6) Active damped anti-vibration table with CNC positioning system.

By setting the distance between the X-ray tube and detector to 359.6 mm and the distance between the X-ray tube and the specimen axis to 240.5 mm, a magnification of ~1.5× was obtained, leading to a voxel (one spatial point) size of ~50 μm in the resulting 3D volumetric models. However, due to the requirement for high resolution, the size of the active area of the detector was a limiting factor and for samples with an *L/D* ratio higher than 1.5 it was necessary to perform an XCT scan of each half of the sample separately and then virtually bind their models together. Thus, a total of 19 XCT scans were performed. For each scan, 1800 projections were taken as an average of 4 images with an exposure time of 230 ms.

Data treatment was performed using VG Studio Max (Volume Graphics, Germany) and the free open-source software ImageJ. After the reconstruction of virtual 3D models using logarithmization and a filtered back-projection algorithm, the time-lapse models were mapped to each other by rigid transformations. To be able to monitor the changes, differential models between the initial and final state were calculated. From this difference, it was possible to segment the cracks, observe their pattern and determine their geometric parameters (including the fractal dimension) by applying the box-counting algorithm embedded in the BoneJ plugin in ImageJ free open-source software [16]. Furthermore, the size and direction of displacements were calculated based on local non-linear transformations using a digital volume correlation module embedded in VG Studio Max.

3. Approximation of the strength factor

Approximation of the strength factor [17] was carried out during the derivation of an updated simplified analytical model of the compressive strength of a concrete specimen similar to that found in [18].

3.1. Derivation of the root of approximation

The assumption is that the main crack will only slightly deviate from a plane. Such a plane will have an angle α according to a plane perpendicular to the cylinder axis. The strength factor c can then, by taking a simplified model into account, be approximated by a relationship:

$$c_{\mathrm{root}}(L/D) = \frac{1}{\sin\alpha} = \frac{\sqrt{D^2+L^2}}{L} = \frac{\sqrt{1+(L/D)^2}}{L/D} = \sqrt{\frac{1+(L/D)^2}{(L/D)^2}}. \tag{7}$$

3.2. Generalization of the approximation

The relationship (7) can be parametrized by adding a variable exponent p, which helps with the empirical data fit:

$$c_{\mathrm{p}}(L/D) = \sqrt{\frac{1+(L/D)^p}{(L/D)^p}}, \tag{8}$$

for which the value of $p = 3$ will be used here.

However, concrete specimens tend to have a slightly lower steepness, which the following update takes into account:

$$c(L/D) = q\left(\sqrt{\frac{1+(L/D)^p}{(L/D)^p}} - 1\right) + 1, \tag{9}$$

where $q = 0.7$ is the steepness parameter (which is reduced by 30 % according to (8)).

The measured compressive strength is then given by the expression:

$$f_{\mathrm{c}}(L/D) = c \cdot f_{\mathrm{ca}}, \tag{10}$$

where f_{ca} is the asymptotic value of compressive strength, which is assumed to be independent of the length L of the specimen.

4. Results

4.1. Fracture parameters obtained from three-point bending tests

The following table illustrates the results gained from the aforementioned three-point bending tests. The fundamental geometry values of the specimens along with their mechanical fracture parameters such as modulus of elasticity, fracture toughness, fracture energy, etc. are shown in table 2: measured values, mean values, standard deviations (StDev), and coefficients of variation (CoV).

Table 2. Cylinder bending specimens with a chevron notch: input and outputs.

	Parameter	Symbol	Specimen CN_L1	CN_L2	CN_L3	Mean	StDev	Unit	CoV (%)
Inputs	Diameter of specimen	D	49.73	49.67	49.72	49.70	0.03	mm	0.1
	Notch tip depth	a_0	3.50	3.47	3.93	3.63	0.26	mm	7.1
	Notch depth	h_0	9.88	9.36	10.01	9.80	0.35	mm	3.5
	Span of specimen	S	140.0	140.0	140.0	140.0	0.0	mm	0.0
Outputs	Modulus of elasticity	E	21.7	22.2	15.1	19.7	3.94	GPa	20.0
	Critical stress intensity factor	K_{Ic}	0.755	0.679	0.859	0.764	0.090	$MPa \cdot m^{1/2}$	11.8
	Critical energy release	G_{Ic}	26.3	20.8	48.7	31.9	14.8	$N \cdot m^{-1}$	46.3
	Fracture energy	G_F	103.2	128.2	167.4	132.9	32.39	$J \cdot m^{-2}$	24.4
	Fracture energy ratio	G_F/G_{Ic}	3.93	6.16	3.43	4.50	1.45	–	32.2

4.2. *Fracture parameters and their relation to the slenderness ratio*

In table 3, values obtained from compressive strength tests on specimens with different length to diameter ratios are shown. Besides the values from the tests for each specimen, the results include mean compressive strength and CoV values for each group of specimens.

Table 3. Results of the compressive tests.

Specimen	Diameter	Length	Length to diameter ratio	Density	Maximal force	Compressive strength	
	D	L	L/D	D_c	F_{max}	f_c	
	(mm)	(mm)	(–)	$(kg \cdot m^{-3})$	(N)	$(N \cdot mm^{-2})$	
CN_05_1	49.73	25.55	0.51	2269	162840	**83.8**	$f_{c,mean}$
CN_05_2	49.67	25.69	0.52	2284	169650	**87.6**	83.1
CN_05_3	49.73	25.68	0.52	2282	152960	**78.8**	CoV (%)
CN_05_4X	49.70	25.53	0.51	2293	159726	**82.3**	4.4
CN_10_1	49.77	50.50	1.01	2315	93370	**48.0**	$f_{c,mean}$
CN_10_2	49.77	50.03	1.01	2277	94820	**48.7**	45.8
CN_10_3	49.79	50.42	1.01	2295	85430	**43.9**	CoV (%)
CN_10_4X	49.74	50.54	1.02	2278	82922	**42.7**	6.5
CN_15_1	49.72	75.19	1.51	2294	73330	**37.8**	$f_{c,mean}$
CN_15_2	49.77	74.85	1.50	2302	72890	**37.5**	39.1
CN_15_3	49.76	75.40	1.52	2291	80180	**41.2**	CoV (%)
CN_15_4X	49.80	75.34	1.51	2293	77950	**40.0**	4.6

Table 4. Results of the compressive tests (cont.).

Specimen	Diameter	Length	Length to diameter ratio	Density	Maximal force	Compressive strength	
	D	L	L/D	D_c	F_{max}	f_c	
	(mm)	(mm)	(–)	(kg·m^{-3})	(N)	(N·mm^{-2})	
CN_20_1	49.81	100.82	2.02	2283	71620	**36.8**	$f_{c,mean}$
CN_20_2	49.72	99.34	2.00	2279	76520	**39.4**	37.3
CN_20_3	49.64	99.73	2.01	2300	74400	**38.4**	CoV (%)
CN_20_4X	49.74	99.86	2.01	2285	67135	**34.6**	5.7
CN_25_1	49.65	125.75	2.53	2256	69510	**35.9**	$f_{c,mean}$
CN_25_2	49.63	124.30	2.50	2266	76730	**39.7**	36.9
CN_25_3	49.74	123.32	2.48	2289	70010	**36.0**	CoV (%)
CN_25_4X	49.73	124.47	2.50	2296	70156	**36.1**	5.0
CN_30_1	49.71	149.65	3.01	2317	68430	**35.3**	$f_{c,mean}$
CN_30_2	49.68	149.95	3.02	2303	65540	**33.8**	36.0
CN_30_3	49.76	149.99	3.01	2287	73200	**37.6**	CoV (%)
CN_30_4X	49.75	149.68	3.01	2273	72467	**37.3**	5.0

Additionally, compressive strength values obtained from the adjusted ends of specimens previously used for the three-point bending test are shown in table 4.

Table 5. Results of the compressive tests performed on the adjusted ends of specimens after the three-point bending test.

Specimen	Diameter	Length	Length to diameter ratio	Density	Maximal force	Compressive strength	
	D	L	L/D	D_c	F_{max}	f_c	
	(mm)	(mm)	(–)	(kg·m^{-3})	(N)	(N·mm^{-2})	
CN_L1a	49.76	62.98	1.27	2264	84150	**43.3**	$f_{c,mean}$
CN_L1b	49.72	63.10	1.27	2257	62290	**32.1**	36.8
CN_L2a	49.75	63.35	1.27	2273	78000	**40.1**	
CN_L2b	49.74	54.43	1.09	2634	71580	**36.8**	CoV (%)
CN_L3a	49.76	63.30	1.27	2286	67500	**34.7**	11.4
CN_L3b	49.74	63.43	1.28	2282	65930	**33.9**	

Figure 3 shows a graphical illustration of the gained values for compressive strength and L/D ratio for each tested group of specimens, including the adjusted specimens after the three-point bending test. These values are supplemented by the cube compressive strength value of the same concrete tested in the form of drilled core specimens. Besides these results, data from the experiment were used for the calculation of the asymptotic value $f_{ca} = 34.95$ N·mm^{-2} of the approximation function (10) by the least squares method. A comparison of the values of the approximation function and correction factors from EN 12390-3 [6] and ASTM C42/C42M [10] is shown in figure 4.

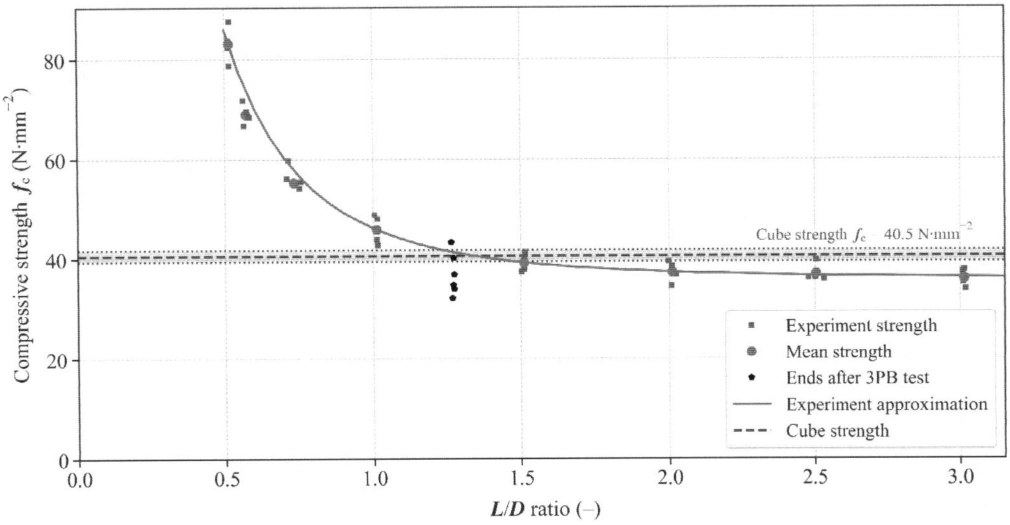

Figure 3. Results obtained from compressive tests. The plot shows the relationship between the compressive strength and the *L*/*D* ratio. In addition, the cube strength and approximation function of the experiment are shown.

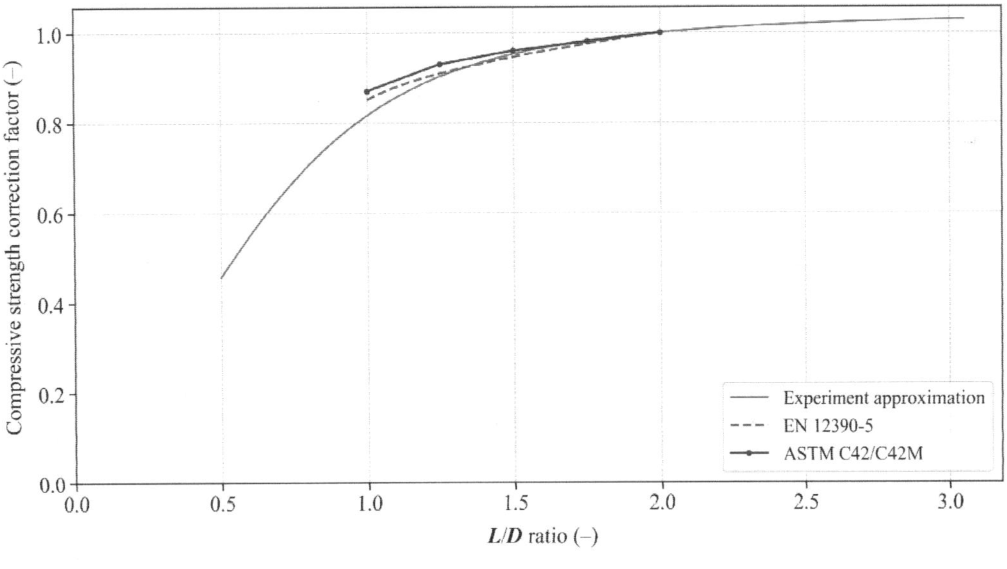

Figure 4. Approximated function and functions from existing codes and recommendations for the compression strength correction factor with relation to the *L*/*D* ratio of concrete specimens.

4.3. Tomographic observation and volume analyses

From the qualitative assessment of the computed tomography results in figure 5, it is obvious that the initiation of cracks within the volume at 90 % of the minimum force at failure is difficult to observe and differences in comparison with the initial state are not clearly visible until the maximum force is reached and damage propagates. Although the given resolution plays a major role in the detectability and visibility of the expected microcracks, these are additionally hidden in the distinctive structure of the material. Thus, differential volumes were used to monitor the cracks' pattern, their segmentation and the determination of geometric parameters. These volumes, represented by an example slice in figure 6, were calculated as the difference of virtual models rigidly transformed to geometrically match as much as possible. Based on digital volume correlation displacement fields, e.g. directions and magnitudes were calculated.

Figure 5. Tomographic slices in the middle of the specimen length oriented parallel to the base. From the left: specimen in its initial state, specimen at 90 % of the minimum force at the failure measured at the same *L/D* ratio, specimen after failure.

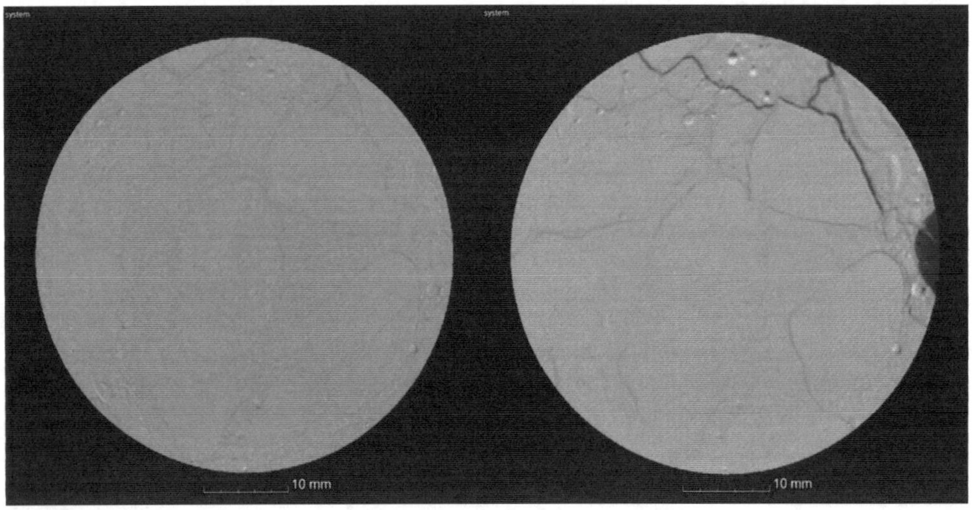

Figure 6. Differential tomographic slices from the middle of the specimen length oriented parallel to the base. Left: the difference between the specimen in its initial state and the specimen at 90 % of the minimum force at failure; right: the difference between the specimen in its initial state and the specimen after failure.

Different 3D representations of the obtained results are shown in figure 7. In the first column, there is a section of the specimen representing the changes to the initial state. The colour coding corresponds to the intensity values normalized to their standard deviation from –0.5 to 1.5. In the second column, there is a section of the specimen representing the displacement magnitude fields. In the third column displacement lines representing the directions of the displacement are provided. The colour coding here corresponds to a displacement from 0 to 1.5 mm. In the fourth column, a red crack pattern is visualised within the grey specimen body.

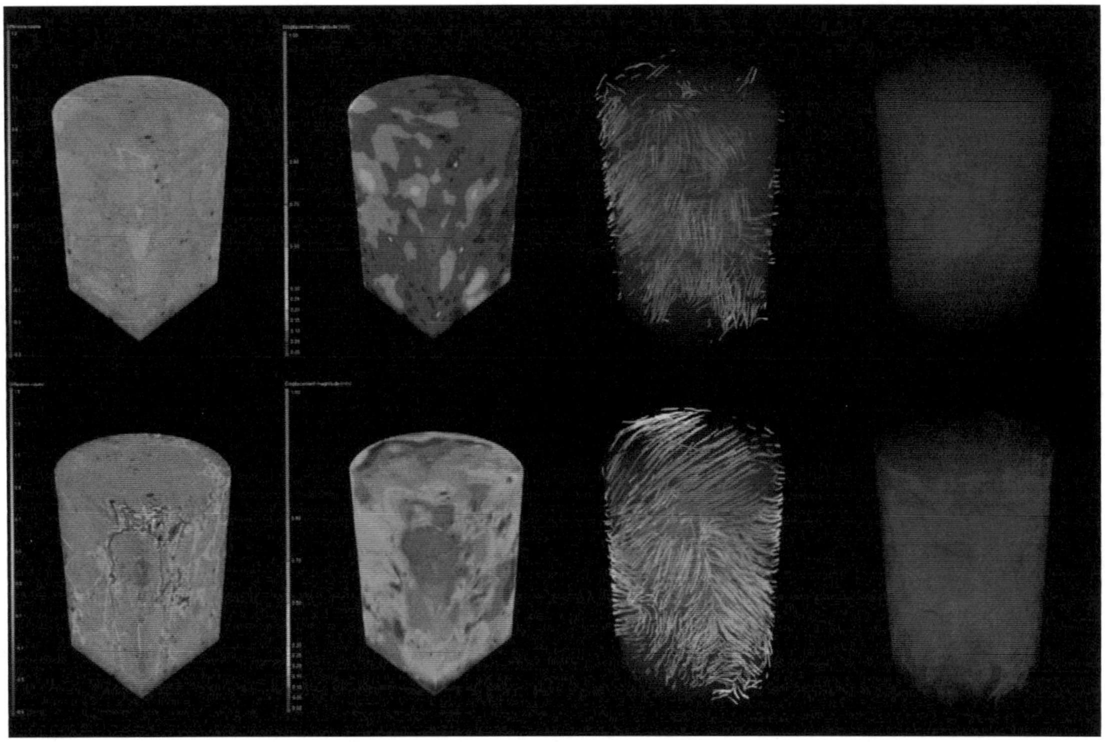

Figure 7. 3D representations of the CT results. Top row: specimen at 90 % of the minimum force at failure; bottom row: specimen after failure. From the left: volume changes compared to the initial state, displacement magnitude, displacement lines, 3D crack pattern.

Table 6. Values obtained based on computed tomography results.

Specimen CN_15_4X state / Volume analyses and characteristics	Initial state	90 % of the minimum force at failure	After failure
Material volume (mm^3)	123169.29	123038.88	119527.45
Void volume, incl. pores (mm^3)	1606.15	1684.90	7710.62
Crack volume (mm^3)	–	939.46	6351.74
Void volume fraction, incl. pores (%)	1.29	1.35	6.06
Crack volume fraction (%)	–	0.75	4.99
Crack volume surface (mm^2)	–	15591.26	67467.10
Displacement magnitude max (μm)	–	1300.00	2733.30
Displacement magnitude mean (μm)	–	77.60	365.30
Displacement magnitude median (μm)	–	73.50	205.54
Crack fractal dimension D (–)	–	2.1129	2.2756
Coefficient of determination R^2 (%)	–	99.9978	99.9931

From the values determined based on the volumetric data summarized in table 6, additional information can be deduced. At 90 % of the minimum force at failure when relatively high displacement values are already being recorded, the material loss and void volume increase is minimal, which indicates the thinning of the material in the interfacial transition zone and the initiation of microcracks among the present pores. However, after failure the material loss is significant and the crack coincides with the present pores. In both cases, the values for the content of the fracture surface are relatively high and the cracks are richly branched, as is also evidenced by the fractal dimension value.

5. Discussion

The influence of the length to diameter (*L*/*D*) ratio on the compressive strength of concrete is not negligible. In order to evaluate tested concrete and make further use of compressive strength values it is necessary to know the cube or cylindrical compressive strength which is used by default. This means that the used samples taken from the structure must have specific dimensions, e.g. length to diameter ratio. In some cases, it is difficult to adhere to the required dimensions and so it is desirable to know the relationship between compressive strength and the *L*/*D* ratio. The performed compressive strength tests show the effect of slenderness ratio on the resulting compressive strength of concrete specimens. Specimens with a slenderness ratio lower than 1.0 show considerable higher compressive strength than specimens with a higher *L*/*D* ratio.

Some codes and requirements contain values for a correction factor that allows the recalculation of the compressive strength value obtained for concrete specimens according to the ratio of their slenderness to cube strength (*L*/*D* = 1.0) or cylindrical strength (*L*/*D* = 2.0) [6, 10]. However most of these values from codes and requirements are specified only in the range from 1.0 to 2.0 due to the above-mentioned need to recalculate only specimens in this range. The supplementation of the code correction factors with the correction factor gained from the approximation shown in figure 4 can help to correct the compressive strength values obtained for specimens of dimensions other than those which are recommended by the standards. This can reduce the core-drilled specimens needed, both in terms of dimensions and numbers.

The specimens made from halves left over after the three-point bending test reached lower compressive strength values than expected. This difference could be caused by multiple saw-cutting (notch cutting, adjusting the specimens for the compressive strength test) that caused damage to the microstructure of the concrete.

The complementary performance of tomographic observation and volume analyses allows the observation of changes in material structure caused by the loading process in three-dimensional models and supplements the determination of fracture parameters with additional knowledge of the geometric and morphometric parameters of the arising damage. The tomographic virtual model can also serve as an input or comparison for modelling and predicting the fracture behaviour of the studied material.

6. Conclusion

The mechanical fracture parameters of concrete drill-core specimens and the response of their internal structure to load with regard to the slenderness ratio have been studied in this paper. Eight groups of concrete core-drilled specimens with various *L*/*D* ratios were set up and the influence of the *L*/*D* ratio on the compressive strength was specified. The results of compressive tests were used for the creation of an approximation function which specifies the influence of the specimen's slenderness on its compressive strength. Three-point bending tests were performed to determine the fracture characteristics of the used concrete, and selected specimens were analysed and three-dimensionally characterized via high-resolution X-ray computed tomography.

In the case of diagnostics and concrete structure evaluation, the amount of taken samples is crucial, and it is desirable to obtain as many values for material characteristics as possible from a single specimen. One of the options is to use the same sample of concrete for more tests, such as in this paper – specimens were subjected to three-point bending tests, and then their adjusted halves that remained after such testing underwent compressive strength tests. The paper also describes the combination of compressive strength tests and X-ray computed tomography.

It has been shown that specimens with a lower *L/D* ratio reached higher compressive strength values than specimens with a higher *L/D* ratio – the advanced approximation derived from an analytical model of the compressive strength of a concrete specimen perfectly fitted into the whole range of slenderness ratios. In future research the influence of the method of cutting and preparing the samples on the mechanical properties of the concrete could be determined, as well as the more detailed relationship between the slenderness of specimens and the compressive strength of concrete with an *L/D* ratio ranging from 0.5 to 2.0.

7. References

[1] Karihaloo B L 1995 Fracture mechanics and structural concrete (New York: Longman)
[2] Shah S P, Swartz S E and Ouyang Ch 1995 Fracture mechanics of structural concrete: applications of fracture mechanics to concrete, rock, and other quasi-brittle materials (New York: Wiley)
[3] Bažant Z P and Planas J 1998 Fracture and Size Effect in Concrete and other Quasibrittle Materials (Boca Raton: CRC Press)
[4] European Committee for Standardization EN 13791 2019 Assessment of In-situ Compressive Strength in Structures and Precast Concrete Components
[5] Panizza M, Natali M, Garbin E, Ducman V and Tamburini S 2020 Optimization and mechanical-physical characterization of geopolymers with Construction and Demolition Waste (CDW) aggregates for construction products *Construction and Building Materials* **264**
[6] European Committee for Standardization EN 12390-3 2020 Testing Hardened Concrete – Part 3: Compressive strength of test specimens
[7] Trawinski W, Bobinski J and Tejchman J 2016 Two-dimensional simulations of concrete fracture at an aggregate level with cohesive elements based on X-ray μCT images *Engineering Fracture Mechanics* **168** pp 204–26
[8] Skarzynski L and Tejchman J 2016 Experimental Investigations of Fracture Process in Concrete by Means of X-ray Micro-Computed Tomography *Strain* **52** pp 26–45
[9] Kumpova I, Vopalensky M, Fila T, Kytyr D, Vavrik D, Pichotka M, Jakubek J, Kersner Z, Klon J, Seitl S and Sobek J 2018 On-the-Fly Fast X-Ray Tomography Using a CdTe Pixelated Detector – Application in Mechanical Testing *IEEE Transactions on Nuclear Science* **65** 2870–76
[10] ASTM International 2012 ASTM C42 / C42M-12 Standard Test Method for Obtaining and Testing Drilled Cores and Sawed Beams of Concrete ASTM International West Conshohocken PA
[11] Svobodová B, Šimonová H and Keršner Z 2018 Calculation of Mechanical Fracture Parameters of Chevron Notched Core Based Concrete Specimen: Software CheF. In Solid State Phenomena: *24th Concrete Days 2017*. Solid State Phenomena. Switzerland: Trans Tech Publications pp 185–188
[12] RILEM TC-50 FMC (Recommendation) 1985 Determination of the fracture energy of mortar and concrete by means of three-point bend test on notched beams, *Materials & Structures* **18** pp 285–290
[13] Backers T 2004 Fracture toughness determination and micromechanics of rock under mode I and mode II loading, dissertation thesis, University of Potsdam
[14] Ouchterlony F 1988 Suggested methods for determining the fracture toughness of rock, Int. J. *Rock Mech. Min. Sci. & Geomech. Abstr.* **25** pp 71–96
[15] Fila T and Vavrik D 2016 A multi-axial apparatus for carrying out X-ray measurements, particularly computed tomography. European patent No. 14002662.6
[16] Doube M, Kłosowski M M, Arganda-Carreras I et al 2010 Bonej: Free and extensible bone image analysis in ImageJ *Bone* **47**(6) pp 1076–79

IOP Conf. Series: Materials Science and Engineering 1209 (2021) 012042 doi:10.1088/1757-899X/1209/1/012042

[17] Frantík P 2021 Approximation of compressive strength of concrete specimen, published online at http://www.kitnarf.cz/publications/2021/2021.06.www_fc_approx/2021.06.www_fc_approx.pdf
[18] Frantík P 2005 Simple model of compressive fracture (in Czech) Proceedings of seminars *Problémy lomové mechaniky* V. Brno

Acknowledgement
The research was carried out as part of the Brno University of Technology specific university research project No. FAST-J-21-7473, with the support of the Czech Science Foundation under project 19-09491S (MUFRAS) and a project for the long-term conceptual development of research organizations, RVO 68378297.

IOP Conf. Series: Materials Science and Engineering 1209 (2021) 012043 doi:10.1088/1757-899X/1209/1/012043

Self-sensing performance of metakaolin geopolymer with carbon nanotubes subjected to repeated compressive loading

C Mizerová[1], P Rovnaník[1], I Kusák[1] and P Schmid[1]

[1] Brno University of Technology, Faculty of Civil Engineering, Veveří 331/95, 602 00 Brno, Czech Republic

E-mail: mizerova.c@fce.vutbr.cz

Abstract. Alkaline activated binders showing enhanced piezoresistive properties have recently attracted increased interest in research of their application in smart self-sensing components. This study is focused on metakaolin geopolymer mortar doped with 0.05 and 0.10% carbon nanotubes, a conductive filler that effectively increases electrical conductivity without considerable deterioration of mechanical properties. Self-sensing performance of composites incorporated with electrodes and attached strain gauge was tested during different regimes of compressive loading cycles with continuous monitoring of strain and resistivity. Although the differences in sensitivity and repeatability were observed, all samples including the reference material have shown good response to applied loading.

1. Introduction

Geopolymers are advancing inorganic binders prepared by alkaline activation of natural or manufactured source materials with prevailing aluminosilicate content [1]. Resulting geopolymeric binders feature highly coordinated microstructure and superior material properties that draw increased research interest in last decades. Apart from superior mechanical performance that have already proven themselves in many structural applications, geopolymers can exhibit exceptional durability, thermal resistance, and lower environmental impact in comparison with traditional cementitious binders [2]. Moreover, the certain material characteristics, different from the ones of OPC open new possibilities for advanced multifunctional geopolymeric composites [3].

The introduction of conductive fillers represents a common approach towards concrete with additional functionalities because enhanced electrical properties can provide a potential for designing elements with self-sensing, self-heating or electromagnetic shielding abilities [4]. Initial attempts in development of self-sensing concrete referring to a material able to detect alterations of its own structural health dates back to 90s. Structural health monitoring in self-sensing structures is performed via measuring the electrical resistivity that can reflect load induced mechanical strain and possible defects but also the external ambience conditions. Such solution of structural element being continuously self-monitored can serve as an alternative to currently used strain gauges that are expensive and ineffective in use [5]. In order to increase the concrete conductivity, various metal or carbon admixtures are used to establish functional conductive network within the composite and improve its self-sensing performance. The most common conductive fillers include carbon fibres, carbon nanotubes, graphite powder, carbon black, steel fibres or graphene [5][6].

Unlike poor conductivity of plain OPC, geopolymers and other alkaline activated materials, especially metakaolin and slag-based binders, exhibit higher conductivity ascribed to increased availability of mobile hydrated alkali cations in the pores of the geopolymeric structure [7] and the presence of Fe [8]. Piezoresistive behaviour of geopolymers described by Lamuta et al. [9] was later confirmed in experimental studies that revealed their applicability for the purpose of self-sensing measurements even without the use of conductive filler [8]. Nevertheless, the conductivity of geopolymers with/without conductive fillers vary greatly and the application of such admixtures remains an effective means to achieve convenient response in self-sensing tests and prevent conductivity loss at completely dry matrix [10].

Carbon nanotubes (CNTs) are single- or multi-walled tubular structures formed by rolling a sheet of graphene (carbon atoms in honeycomb lattice) into a tube with high aspect ratio. CNTs exhibit superlative mechanical properties (flexural strength, Young´s modulus) and excellent electrical behaviour [11]. The key issue of their applicability in self-sensing materials is the uniform distribution within the matrix, well dispersed CNTs result in significantly enhanced electrical performance in spite of very low concentrations usually not exceeding 1% wt. [12]. Alkaline solution used for geopolymer synthesis serves as a surfactant and facilitates the dispersion of CNTs. Besides, the presence of inert nanofillers in geopolymers (CNTs and graphene) contributes to refined pore microstructure of the binder hence improving its durability while acting as a microfiller and nucleation seeds for initial geopolymerization reactions [13]. Higher CNTs dosage is associated with inevitable agglomeration that can increase binder porosity and decrease mechanical performance, surpassing the threshold has rather impact on mechanical than electrical properties. Regarding the contribution to piezoresistive effect, CNTs enhance electrical conductivity and considerably improve the sensitivity in testing of self-sensing performance indicated by gauge factor (GF). Geopolymers doped with CNTs are characterized by much higher GF than cementitious binders incorporated with any type of conductive filler [12].

This paper deals with self-sensing performance of metakaolin geopolymers doped with CNTs subjected to repeated compressive loading. Experimental results are discussed in relation to strain and fractional change in resistivity during loading cycles. Fractional change in resistivity ((R□R0)/R0) is used to report sensitivity and repeatability in self-sensing performance. Moreover, it is an indicator of plastic/elastic deformation development during loading cycles.

2. Experimental part

2.1. Materials

The major source material for geopolymer preparation was metakaolin Mefisto K05 (České lupkové závody, a. s.) with predominant aluminosilicate content (55.01% SiO_2 and 40.94% Al_2O_3). Alkaline activator solution combined sodium silicate and 30% NaOH solution, standardized quartz sand with a maximum grain size of 2.5 mm was used as fine aggregate. Agglomerate powder of carbon nanotubes Graphistrength® C100 was supplied by Arkema, the solid concentrates were dispersed in hot water to prepare 2% CNTs solution. CNTs concentration in mortars was 0.05 and 0.10% by weight.

2.2. Specimen preparation

Metakaolin mortars were prepared in a standard planetary mixer according to following steps. At first, waterglass, NaOH solution, 2% CNTs solution and water were mixed to form a homogeneous suspension. Then metakaolin was added and mixed properly until well combined into a smooth paste. Finally, all fractions of fine aggregate were added. Compositions of all tested mortars are given in table 1.

Fresh mortars were cast into cubic moulds (100 × 100 × 100 mm) and incorporated with four copper mesh electrodes with dimensions 80 × 120 mm. Regular 20 mm distance between electrodes was assured by waterproof plywood board. Specimen were then covered with a plastic foil and left to set under laboratory conditions. Demoulded samples were wrapped in PE foil and stored in laboratory till the age of testing.

Table 1. Mix design proportions of geopolymer mortars.

	REF	CNT0.05	CNT0.10
Metakaolin (g)	900	900	900
Waterglass (g)	748	748	748
30% NaOH (g)	83	83	83
Sand (g)	2700	2700	2700
2% CNT solution (g)	-	22.5	45
Water (g)	103	80.5	58

Fresh mortars were cast into cubic moulds (100 × 100 × 100 mm) and incorporated with four copper mesh electrodes with dimensions 80 × 120 mm. Regular 20 mm distance between electrodes was assured by waterproof plywood board. Specimen were then covered with a plastic foil and left to set under laboratory conditions. Demoulded samples were wrapped in PE foil and stored in laboratory till the age of testing.

2.3. Testing methods
Compressive loading cycles were performed in a multipurpose electromechanic testing machine LabTest ® 6.250 (Labortech) allowing to achieve maximum 250 kN loading force. The repeated loading was carried out perpendicularly to electrodes configurated in a standard serial arrangement. We defined two different courses of linear loading: a) 10 cycles with constant amplitude (10 kN, 300 N/s); b) 30 loading cycles with increasing and decreasing amplitude (7–35 kN, 500 N/s). The applied load took place within the elastic region of the material and did not exceed 25% of the ultimate failure load.

Electrical resistance was measured under AC using a sinusoidal signal generator Agilent 33220A and two multimeters Agilent 34410A. The generated signal was brought to outer electrodes while the inner ones were used to measure the output voltage. The electrical resistance of the measured specimen was calculated from the electric current flowing through the exact resistance R and the voltage that was measured separately on the test sample using Ohm's law, taking into account the internal resistance of the voltmeter. The input frequency and voltage values were 1 kHz and 5 V, respectively. In order to monitor the longitudinal strain, strain gauges were attached to the sides of each specimen.

3. Results and discussion

3.1. Linear compressive loading with constant amplitude (10 cycles)
The first phase of self-sensing performance testing consisted in 10 linear compressive loading cycles with constant 10 kN amplitude. Longitudinal strain measured by the strain gauge in μm/m is shown in Figure 1 (a). We can see that the deformation rates of REF and CNT0.05 samples was comparable around 60 μm/m while the longitudinal strain of CNT0.10 was more than twice higher. Maximal strain of reference sample was slightly decreasing with each loading cycle while the highest deformation of CNT-doped samples was consistent.

Piezoresistive behaviour was observed at all samples, as the changes of loading force are reflected by the change in fractional resistivity in Figure 1 (b). Response of CNT samples were very similar, fractional change in resistivity was growing during the whole experimental course. Increasing resistivity refers to permanent structural defects formed within the binder microstructure causing the interruptions of conductive pathways. Fractional resistivity of reference sample was gradually decreasing but fluctuating in almost constant range, thus featuring better repeatability. Results achieved in self-sensing experiments of alkaline activated slag doped with carbon fibers [13] lead to similar outcome with the formation of microcracks and no improvement of self-sensing performance. On the other hand, 10% graphite [14] or 0.5% carbon black [15] improved the repeatability and sensitivity of fly ash geopolymer specimens during loading with constant amplitude, carbon black was a suitable admixture with high effectivity in low concentration and no harmful impacts on the mechanical performance of geopolymer matrix.

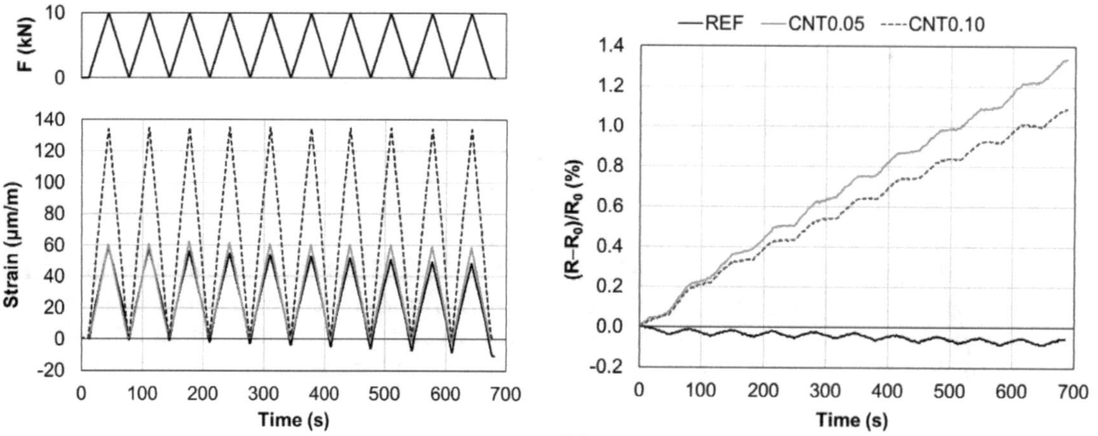

(a) **(b)**

Figure 1. Specimens response during linear compressive loading with constant amplitude: (a) longitudinal strain, (b) fractional change in resistivity (self-sensing performance).

3.2. Linear compressive loading with variable amplitude (30 cycles)

Linear compressive loading with variable amplitude was performed in 30 loading cycles with an amplitude 7–35 kN. During this loading regime, CNT0.10 sample again reached the highest strain values, strain of CNT0.05 was comparable with REF (Figure 2). There were minor alterations in strain peaks with successive loading cycles.

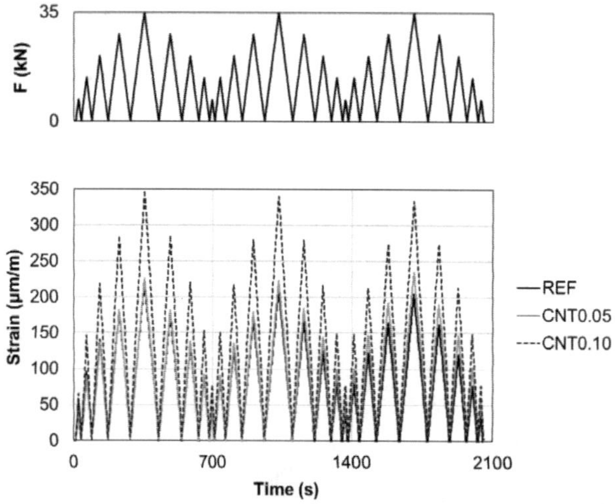

Figure 2. Longitudinal strain during linear compressive loading with variable amplitude.

Self-sensing performance of geopolymers is shown in Figure 3 that displays the fractional change in resistivity response during the loading course. Samples with CNT content were again characterized by resistivity increasing with time, i.e. the irreversible plastic deformation. Similarly, to results in the previous chapter focused on loading force with constant amplitude, CNT0.05 suffered higher resistivity increment than CNT0.10. Increased strain levels of samples with CNT observed in this study are in a contrary of published papers reviewed in **Chyba! Nenašiel sa žiaden zdroj odkazov.** that mostly reported improved elastic modulus of CNT-doped mortars.

Fractional change in resistivity of all samples is displayed in figure 3. Response of REF sample (figure 3 (a)) corresponded well with the repeated loading cycles. There were observed minor changes of total resistivity, the sensitivity was around 0.10–0.15% at maximum loading force. The possible benefit of CNT modification to enhanced sensitivity in self-sensing tests was not established due to continuous resistivity growth (figure 3 (b), (c)), sample CNT0.05 exhibited the less favourable self-sensing performance. On the other hand, REF sample featured very good reproducibility and repeatability.

Application of CNT in this study did not lead to anticipated contribution to self-sensing performance under compressive loading, superior properties were observed in case of geopolymers with graphite [14] or carbon black [15] admixture. Even though the fractional change in resistivity exhibited higher peaks in corresponding loading cycles, these benefits were outbalanced by increasing resistivity.

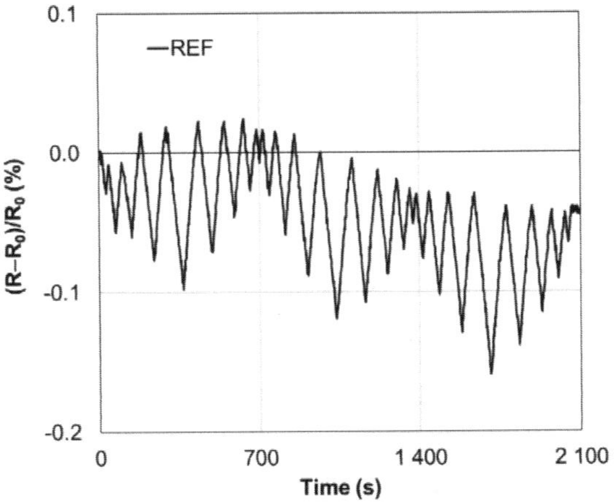

(a)

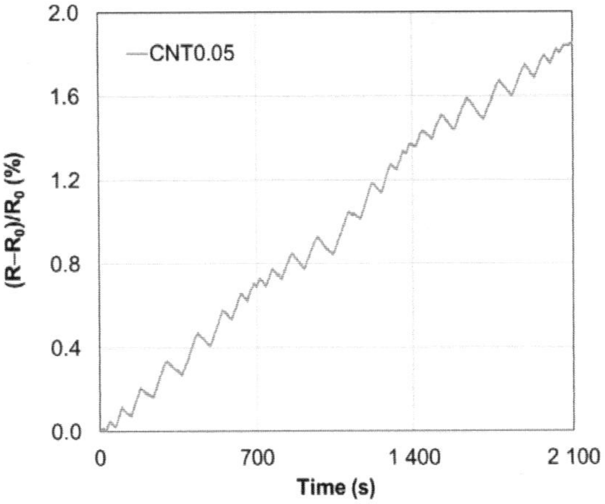

(b)

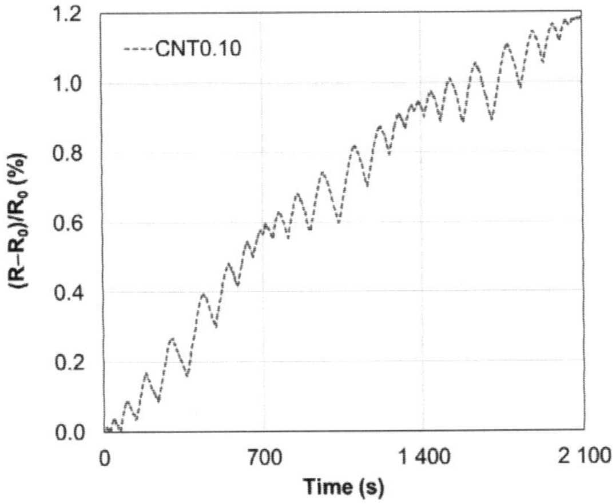

(c)

Figure 3. Self-sensing properties during linear compressive loading with variable amplitude: (a) reference material (REF), (b) sample CNT0.05, (c) sample CNT0.10.

4. Conclusions

The objective of this study was to determine the self-sensing performance of metakaolin geopolymer mortars doped with CNT. Reference material and specimens with 0.05 and 0.10% CNT, respectively, were each subjected to two series of compressive loading cycles with constant/variable amplitude. Measurements of corresponding change in resistivity response have shown that all samples exhibit good self-sensing performance but CNT-doped specimens suffered increasing resistivity related to permanent deformations. The performance of reference material was thus better in repeatability. Research of the mechanical and electrical properties influencing the self-sensing performance will be further studied in following experiments, ex. during loading till fracture.

5. References

[1] Provis J L and Bernal S A 2014 Geopolymers and related alkali-activated materials *Annual Review of Mat. Res.* **44** pp 299-327
[2] Matheu P S, Ellis K and Varela B 2015 Comparing the environmental impacts of alkali activated mortar and traditional portland cement mortar using life cycle assessment. *IOP Conference Series: Mat. Sci. and Eng.* **96** 012080
[3] Tang Z, Li W, Hu Y, Zhou J L and Tam V W 2019 Review on designs and properties of multifunctional alkali-activated materials (AAMs) *Constr. Build. Mat.* **200** pp 474-489
[4] Han B, Zhang L and Ou J 2017 Smart and multifunctional concrete toward sustainable infrastructures (Singapore: Springer) pp 369-377
[5] Han B, Yu X. and Ou J 2014 *Self-sensing concrete in smart structures* (Butterworth-Heinemann)
[6] Chung D D L 2012 Carbon materials for structural self-sensing, electromagnetic shielding and thermal interfacing *Carbon* **50(9)** pp 3342-3353
[7] Hanjitsuwan S, Chindaprasirt P and Pimraksa K 2011 Electrical conductivity and dielectric property of fly ash geopolymer pastes *Int. Jour. of Miner. Metall. Mat.* **18(1)** pp 94-99
[8] Rovnaník P, Kusák I, Bayer P, Schmid P and Fiala L 2019 Comparison of electrical and self-sensing properties of Portland cement and alkali-activated slag mortars *Cem. Concr. Res.* **118** pp 84-91
[9] Lamuta C, Candamano S, Crea F and Pagnotta L 2016 Direct piezoelectric effect in geopolymeric mortars *Mat. & Design* **107** pp 57-64
[10] Vlachakis C, Perry M and Biondi L 2020 Self-sensing alkali-activated materials: a review. *Minerals* **10(10)** p 885
[11] Dai H 2002 Carbon nanotubes: synthesis, integration, and properties *Accounts of Chem. Res.* **35(12)** pp 1035-1044
[12] Su Z, Hou W and Sun Z 2020 Recent advances in carbon nanotube-geopolymer composite *Constr. Build. Mat.* **252** 118940
[13] Rovnaník P, Kusák I, Topolář L and Schmid P 2019 Sensing properties of slag-based geopolymer composite with carbon fibers under compressive loading
[14] Rovnaník P, Kusák I and Schmid P 2021 Self-sensing properties of fly ash geopolymer composite with graphite filler *AIP Conf. Proceed.* **2322(1)** 020016 AIP Publishing LLC
[15] Mizerová C, Kusák I, Topolář L, Schmid P and Rovnaník P 2021 Self-Sensing Properties of Fly Ash Geopolymer Doped with Carbon Black under Compression. *Mat.* **14(16)** 4350
[16] Xie T and Fang C 2019 Nanomaterials applied in modifications of geopolymer composites: a review. *Austral. Jour. of Civil Eng.* **17(1)** pp 32-49

Acknowledgements
This outcome has been achieved with the financial support of the Czech Science Foundation under project No. 19-11516S and BUT junior research project FAST-J-21-7478.

IOP Conf. Series: Materials Science and Engineering 1209 (2021) 012044 doi:10.1088/1757-899X/1209/1/012044

Characterization of materials used for earth architecture in Eastern Croatia

A Perić[1]

[1] Teaching and research assistant, Department of Materials and Structures, Faculty of Civil Engineering and Architecture Osijek, Vladimira Preloga 3, 31 000 Osijek, Croatia

E-mail: aperic@gfos.hr

Abstract. Even though earthen architecture is recognized as cultural heritage in Croatia, most of existing buildings are in deteriorated and heavily damaged state. Until today, there has been no conducted research on earthen architecture in Croatia, nor was any preservation act proclaimed. Therefore, the purpose of this research is to learn more about behaviour of traditional earthen architecture to highlight the importance of sustainable building. For that purpose, earthen houses in Eastern Croatia, mainly Slavonia and Baranja area, were studied and in-situ samples were collected. In order to learn about characteristics and composition of mixtures, five samples were tested. Results gathered within this research will be the first step towards creation of national standard for designing rammed earth structures. In scope of future research, authors are planning to use verified recipe in making of walls that will be tested for seismic resistance.

1. Introduction

Earth architecture was very popular in 19h and 20th century in Eastern Croatia [1]–[3] as well as in most parts of Europe[4]–[7] and even world [8]–[11]. However, with implementing baked bricks and later concrete into civil engineering practice, earth construction faded away. Existing earth constructions in Croatia are recognized as cultural heritage but most of them are in heavily damaged state. To this day, no research has been conducted on earthen architecture in Croatia, nor was any preservation act nor building standard proclaimed.

Earth buildings can be built using different techniques, such as adobe, rammed earth, cob and more. Authors conducted field research and found numerous rammed earth constructions and their high occurrence in Eastern Croatia was also confirmed by Lončar-Vicković and Stober [1] and Živković [2]. For that reason, scope of this research are constructions built using rammed earth technique. Traditionally, rammed earth wall is built by pouring earth mixture between formwork and ramming it using wooden rammer [1]. However, today wooden rammers are usually replaced by pneumatic rammers [12]. After ramming, thickness of a layer usually decreases from about 15 cm to about 10 cm [12]. Process is repeated until desired height of the wall is reached. Layers are distinguished on face of the finished wall, as shown in Figure 1.

IOP Conf. Series: Materials Science and Engineering 1209 (2021) 012044 doi:10.1088/1757-899X/1209/1/012044

Figure 1. Visible layers on face of the RE wall (author's archive).

The aim of this research is to learn more about the behaviour of earthen material in construction. For that purpose, authors conducted field research and visited still existing rammed earth houses and outbuildings in Eastern Croatia (Figure 2). Several buildings have been measured in detail, examined, and photographed, and material samples were collected from walls of houses for characterization.

Figure 2. Rammed earth houses in Eastern Croatia (author's archive).

Collected material was tested in laboratory at Faculty of Civil Engineering and Architecture Osijek. Presented results were gathered from preliminary tests and simplest experiments conducted on five different samples taken from observed houses. Along results from Eastern Croatia, in following paragraph, a digest of results from Europe is presented as well.

2. Material used in earth architecture in Europe

Majority of studies dealing with behaviour of rammed earth structures in Europe to date originate from France and Portugal. Difference in results of material properties are inevitable due to differences in local soil. For that reason, material properties presented in other studies from Europe are good for understanding referent values but should not be the goal.

Earth mixtures used for rammed earth structures usually comprise fine and coarse aggregates [13] in different ratios, depending on location. Optimum moisture content is usually obtained using Proctor compaction test [14], [15]. Moisture content in time of building the wall is crucial and it should be kept at about 10 %, according to Bui, Morel et al [12]. However, during and after curing, moisture decreases. Curing time usually lasts for 28 days, following norms for concrete structures [13], [14], [16], and during that time, moisture content decreases on average to 2 %. However, according to Schroeder [17], rammed earth needs more time to dry completely, so the author recommends 90 days of curing.

Based on 40 reviewed studies published over the last 25 years [10], [13], [23]–[32], [14], [33]–[42], [16], [43]–[51], [17]–[22], dry density varies from 1.53 g/cm^3 to 2,19 g/cm^3, and compressive strength varies even more, that is from 0.21 MPa to 5.89 MPa. Compressive strength is usually determined using uniaxial compression test, using norms for concrete.

3. Material characteristics of samples from Eastern Croatia

Material presented in this study was collected from five rammed earth walls in houses located in Eastern Croatia. During field observation, the authors have found several earthen houses, usually heavily deteriorated and abandoned. With owner's permission, large chunks of material were collected and stored in closed plastic bags in laboratory until testing.

3.1. Moisture content

Immediately after collecting material, moisture content of in-situ material was tested according to BS 1377: Part 2 : 1990 (Figure 3). Moisture (Table 1) ranged from 2.43 to 6,65 %, with average value of 4.38 %. Differences in moisture content between the five specimens can be explained by taking material from different locations in Eastern Croatia area. Furthermore, the specimens were taken from different heights on the wall, varying from 0.3 m to about 1.5 m from the ground. It should also be noted that some walls were exposed to direct sunlight, while others were shaded by vegetation.

Since compressive strength testing was performed approximately one month after collection of samples, moisture content decreased for about 20 %, as seen in Table 1. In time of compressive strength testing, moisture content ranged from 1.95 to 5.59 %, with average value 3.48 %. In Table XY, w_0 refers to moisture determined on the same day when the samples were collected from the field, while w_f refers to moisture determined on the day when the compressive strength was determined.

Figure 3. Samples for determining moisture content.

Table 1. Moisture content of earthen material.

Specimen	w_0 (%)	w_f (%)	Difference (w_f/ w_0) (%)
A2	2.43	1.95	80
B4	6.65	5.59	84
B5	6.48	4.80	74
A7	2.77	2.26	82
A8	3.57	2.82	79

3.2. Dry density

Small blocks of earth material were used to determine the dry density of earth material, according to BS 1377: Part 2: 1990. Dry density (Table 2) was determined one month after collecting samples. Values ranged from 1.56 to 2.09 g/cm^3, with average value of 1.75 g/cm^3.

Table 2. Dry density of earthen material tested.

Specimen	ρ (g/cm^3)
A2	1,.56
B4	1.82
B5	2.09
A7	1.60
A8	1.67

3.3. Compressive strength

Big chunks of earthen material were cut into cubes with side length of approximately 4 cm, using circular saw (Figure 4). From every sample, between two and six cubes were cut, depending on size and shape of the chunk. Compressive test (Figure 5) was conducted using Shimatzu AG-X plus 50 kN testing machine, with stress control 0.01 MPa/s. Values ranged from 1.22 to 3.32 MPa, with average value of 2 MPa.

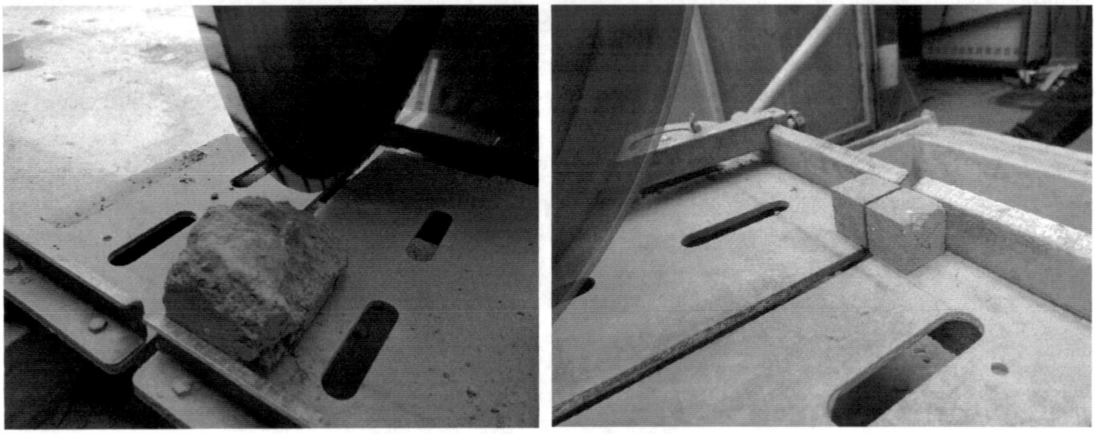

Figure 4. Cutting earth chunks into cubes for testing compressive strength.

Figure 5. Testing compressive strength.

Table 3. Compressive strength.

Specimen	Number of specimens tested	f_c (MPa)
A2	2	1.37
B4	2	2.22
B5	3	1.22
A7	6	3.32
A8	5	1.88

4. Conclusion

Earthen construction in Eastern Croatia have been forgotten and left to decay since no preservation act nor normative have yet been proclaimed. Aim of this research is to learn more about the behaviour of earth architecture from Eastern Croatia and raise awareness of their importance and benefits over conventional construction methods. For that purpose, rammed earth constructions in Eastern Croatia have been observed and material from rammed earthen walls were collected. Preliminary tests were conducted to gather information on material.

From five different samples collected, moisture content was determined of in-situ and material stored in laboratory conditions for one month. Moisture content decreased after a month for 20 % on average. After a month, dry density and compressive strength was determined. Average dry density was 1.75 g/cm^3 and compression strength 2 MPa.

Besides results from preliminary tests, a short review of material properties presented in studies from Europe has been written. It should be noted that the rammed earth properties are highly dependent on granulometry of earth material used for construction. Therefore, it is not recommended to directly compare physical and mechanical properties presented in studies from different parts of Europe but to use that data merely for reference.

Further investigation on material characteristics from Eastern Croatia is planned to determine more characteristics of earth mixtures.

5. References

[1] Lončar-Vicković S and Stober D 2011 Tradicijska kuća Slavonije i Baranje *Ministarstvo turzima Republike Hrvatske; Sveučilište Josipa Jurja Strossmayera u Osijeku, Građevinski fakultet Osijek* pp 1–116

[2] Živković Z 2013 Hrvatsko tradicijsko graditeljstvo *Ministarstvo kulture, Uprava za zaštitu kulturne baštine*

[3] Perić A 2020 Potresno ponašanje zidova tradicijskih kuća od nabijene zemlje *Josip Juraj Strossmayer Osijek*

[4] Gomes M I, Gonçalves T D and Faria P 2014 Unstabilized rammed earth: Characterization of material collected from old constructions in south portugal and comparison to normative requirements *Int. J. Archit. Herit.* **8** 2 185–212

[5] Bui Q B, Bui T T, Limam A and Morel J-C 2015 Discrete element modeling of rammed earth wall *Rammed Earth Construction - Proceedings of the 1st International Conference on Rammed Earth Construction, ICREC 2015* pp 57–61

[6] Serrano S, Rincón L, González B, Navarro A, Bosch M and Cabeza L F 2017 Rammed earth walls in Mediterranean climate: Material characterization and thermal behaviour *Int. J. Low Carbon Technol.* **12** 3 281–8

[7] Jaquin P A, Augarde C E, Gallipoli D and Toll D G 2009 The strength of unstabilised rammed earth materials *Geotechnique* **59** 5 487–90

[8] Ciancio D and Gibbings J 2012 Experimental investigation on the compressive strength of cored and molded cement-stabilized rammed earth samples *Constr. Build. Mater.* **28** 1 294 304

[9] Ciancio D, Beckett C T S and Carraro J A H 2014 Optimum lime content identification for lime-stabilised rammed earth *Constr. Build. Mater.* **53** 59–65

[10] Shrestha K C, Aoki T, Miyamoto M, Wangmo P and Pema 2020 In-plane shear resistance between the rammed earth blocks with simple interventions: Experimentation and finite

element study *Buildings* **10** 3
[11] Lin H, Zheng S, Lourenço S D N and Jaquin P 2017 Characterization of coarse soils derived from igneous rocks for rammed earth *Eng. Geol.* **228** 137–45
[12] Bui Q B, Morel J C, Hans S and Meunier N 2009 Compression behaviour of non-industrial materials in civil engineering by three scale experiments: The case of rammed earth *Mater. Struct. Constr.* **42** 8 1101–16
[13] Liu Q and Tong L 2017 Engineering properties of unstabilized rammed earth with different clay contents *J. Wuhan Univ. Technol. Mater. Sci. Ed.* **32** 4 914–20
[14] Araki H, Koseki J and Sato T 2016 Tensile strength of compacted rammed earth materials," *Soils Found.* **56** 2 189–204
[15] Ávila F, Puertas E and Gallego R 2021 Characterization of the mechanical and physical properties of unstabilized rammed earth: A review *Constr. Build. Mater.* **270**
[16] Hall M and Djerbib Y 2004 Rammed earth sample production: Context, recommendations and consistency *Constr. Build. Mater.***18** 4 281–86
[17] Schroeder H 2011 Moisture transfer and change in strength during the construction of earthen buildings *Inf. la Constr.* **63** 523 107–16
[18] Arrigoni A, Pelosato R, Dotelli G,. Beckett C T S and Ciancio D 2017 Weathering's beneficial effect on waste-stabilised rammed earth: a chemical and microstructural investigation *Constr. Build. Mater.* **140** 157–66
[19] Bui Q B, Morel J C, Hans S and Walker P 2014 Effect of moisture content on the mechanical characteristics of rammed earth *Constr. Build. Mater.* **54** 163–9
[20] Bui Q B, Bui T T and A 2016 Limam Assessing the seismic performance of rammed earth walls by using discrete elements *Cogent Eng.* **3** 1 1–12
[21] Bui Q B, Limam A and Bui T T 2018 Dynamic discrete element modelling for seismic assessment of rammed earth walls *Eng. Struct.* **175** 690–9
[22] Bui T T, Bui Q B, Limam A and Maximilien S 2014 Failure of rammed earth walls: From observations to quantifications *Constr. Build. Mater.* **51** 295–302,
[23] Dong X, Soebarto V and Griffith M 2015 Design optimization of insulated cavity rammed earth walls for houses in Australia *Energy Build.* **86** 852–63
[24] El-Nabouch R, Bui Q B, Plé O and Perrotin P 2017 Assessing the in-plane seismic performance of rammed earth walls by using horizontal loading tests *Eng. Struct.* **145** 153 61
[25] El-Nabouch R, Bui Q B, Plé O and Perrotin P 2018 Characterizing the shear parameters of rammed earth material by using a full-scale direct shear box *Constr. Build. Mater.* **171** 414 20
[26] El-Nabouch R, Bui Q B, Perrotin P and Plé O 2018 Shear parameters of rammed earth material: Results from different approaches *Adv. Mater. Sci. Eng.* 1-9
[27] Gomes M I, Lopes M and De Brito J 2011 Seismic resistance of earth construction in Portugal *Eng. Struct.* **33** 3 932–41
[28] Kosarimovahhed M and Toufigh V 2020 Sustainable usage of waste materials as stabilizer in rammed earth structures *J. Clean. Prod.* **277**
[29] Koutous A and Hilali E 2021 Reinforcing rammed earth with plant fibers: A case study *Case Stud. Constr. Mater.* **14**
[30] Lilley D M and Robinson J 1995 Ultimate strength of rammed earth walls with openings *Proc. Inst. Civ. Eng. Struct. Build.* **110** 3 278–87
[31] Lombillo I, Villegas L, Fodde E and Thomas C 2014 In situ mechanical investigation of rammed earth: Calibration of minor destructive testing *Constr. Build. Mater.* **51** 451–60
[32] Maniatidis V and Walker P 2008 Structural capacity of rammed earth in compression *J. Mater. Civ. Eng.* **20** 3 230–38
[33] Miccoli L, Oliveira D V, Silva R A, Müller U and Schueremans L 2015 Static behaviour of rammed earth: experimental testing and finite element modelling *Mater. Struct. Constr.* **48** 10 3443–56
[34] Miccoli L, Drougkas A and Müller U 2016 In-plane behaviour of rammed earth under cyclic loading: Experimental testing and finite element modelling *Eng. Struct.* **125** 144–52

[35] Miccoli L, Müller U and Fontana P 2014 Mechanical behaviour of earthen materials: A comparison between earth block masonry, rammed earth and cob *Constr. Build. Mater.* **61** 327–39
[36] Miccoli L, Müller U and Pospíšil S 2017 Rammed earth walls strengthened with polyester fabric strips: Experimental analysis under in-plane cyclic loading *Constr. Build. Mater.* **149** 29–36
[37] Mužíková B, Plaček Otcovská T and Padevět P 2018 Modulus of Elascity of Unfired Rammed Earth *Acta Polytech. CTU Proc.* **15** 63–8
[38] Nabouch R et al. 2016 Seismic Assessment of Rammed Earth Walls Using Pushover Tests *Procedia Eng.* **145** 1185–92
[39] Nowamooz H and Chazallon C 2011 Finite element modelling of a rammed earth wall *Constr. Build. Mater.* **25** 4 2112–21
[40] Pakand M and Toufigh V 2017 A multi-criteria study on rammed earth for low carbon buildings using a novel ANP-GA approach *Energy Build.* **150** 466–76
[41] Porter H, Blake J, Dhami N K and Mukherjee A 2018 Rammed earth blocks with improved multifunctional performance *Cem. Concr. Compos.* **92** 36–46
[42] Readle D, Coghlan S, Smith J C, Corbin A and Augarde C E 2016 Fibre reinforcement in earthen construction materials *Proc. Inst. Civ. Eng. Constr. Mater.* **169** 5 252–60
[43] Silva R A, Oliveira D V, Miranda T, Cristelo N, Escobar M C and Soares E 2013 Rammed earth construction with granitic residual soils: The case study of northern Portugal *Constr. Build. Mater* **47** 181–91
[44] Silva R A, Oliveira D V, Schueremans L, Lourenço P B and Miranda T 2014 Modelling the structural behaviour of rammed earth components *Civil-Comp Proc.* **106**
[45] Silva R A, Mendes N, Oliveira D V, Romanazzi A, Domínguez-Martínez O and Miranda T 2017 Evaluating the seismic behaviour of rammed earth buildings from Portugal: From simple tools to advanced approaches *Eng. Struct.* **157** 144–56
[46] Toufigh V and Kianfar E 2019 The effects of stabilizers on the thermal and the mechanical properties of rammed earth at various humidities and their environmental impacts *Constr. Build. Mater.* **200** 616–29
[47] Tripura D D and Singh K D 2015 Characteristic Properties of Cement-Stabilized Rammed Earth Blocks *J. Mater. Civ. Eng.* **27** 7
[48] Wang Y, Wang M, Liu K, Pan W and Yang X 2017 Shaking table tests on seismic retrofitting of rammed-earth structures *Bull. Earthq. Eng.* **15** 3 1037–55
[49] Wangmo P, Shrestha K C, Miyamoto M and Aoki T 2018 Assessment of out-of-plane behavior of rammed earth walls by pull-down tests *Int. J. Archit. Herit.* 1–15
[50] Yamin L E, Phillips C A, Reyes J C and Ruiz D M 2004 Seismic behaviour and rehabilitation alternatives for adobe and rammed earth buildings *13th World Conference on Earthquake Engineering (Vancouver)* 1–10
[51] Zhou T and Liu B 2019 Experimental study on the shaking table tests of a modern inner reinforced rammed earth structure *Constr. Build. Mater.* **203** 567–78

Acknowledgment
This research was supported by Croatian Science Foundation under the project UIP-2020-02-7363 Rammed earth for modelling and standardization in seismically active areas - RE-forMS.

Preparation of triclinic alite with short soakings in a small experimental high-temperature furnace

S Ravaszová[1]

[1] Faculty of Civil Engineering, Brno University of Technology, Veveri 331/95, 602 00 Brno, Czech Republic

E-mail: ravaszova.s@fce.vutbr.cz

Abstract. The article deals with a laboratory preparation of triclinic modification of clinker mineral tricalcium silicate. A substantial part of the article is devoted to the technology and technique of firing a sample of tricalcium silicate, which would in the future allow the study of the development of the crystal lattice structure of this clinker mineral at short isothermal durations, in the order of minutes. As part of the research, a small high-temperature experimental furnace was designed and constructed. Based on the results, we can express the suitability and applicability of this furnace for the study of the formation of triclinic tricalcium silicate at short soakings.

1. Introduction

The current main problem of the industrial cement productions is the variability of their mineralogical composition. This depends on the presence of impurities from the raw materials or fuel and the firing process. The process of synthesis of clinker minerals becomes a decisive factor that determines their subsequent behaviour in industrially produced cements. To characterize and possibly influence the variability of the mineralogical composition of clinker, it is necessary to study the influence of various polymorphs of clinker minerals on the properties of industrially produced clinkers, especially from a chemical and structural point of view [1-3].

In general, more than twenty minerals can be identified in clinker, but only four of them are considered important, namely tricalcium silicate (C_3S, alite), dicalcium silicate (C_2S, belite), tricalcium aluminate (C_3A) and tetra calcium aluminate ferrite (C_4AF). Due to their characteristic properties, each of these minerals influences the specific properties of individual cements. Except for C_4AF, all major clinker minerals show strong polymorphism [4-5].

Most of the standard methods used for laboratory preparation of clinker minerals are relatively time consuming. The composition of the raw material mixture and the setting of the combustion parameters are among the main factors that determine the final composition of the clinker.

The article deals with the study of the course of triclinic alite formation during short soakings. The aim is to assess the suitability of the proposed small experimental high-temperature furnace and its applicability for this purpose.

2. Materials and methods

Two raw materials in chemical purity p.a. were used to prepare triclinic alite, namely calcium carbonate (p.a. 99.0%, Penta) and silica (99.9%, Lachner). The raw materials were dosed in an amount corresponding to the molar ratio for alite in the percentages of 73.6% CaO and 26.3% SiO_2.

The raw material mixture was homogenized by wet milling in the presence of water in a PULVERISETTE 6 planetary mill at a milling speed of 500 rpm for 30 minutes. A 500 ml grinding capsule with 25 pieces of agate grinding balls with a diameter of 20 mm was used.

The wet milling slurry was then dried in a Binder C 170 laboratory oven at 105 °C for 24 hours. The material thus prepared was used to produce tablets for the alite firing itself.

2.1. Preparation of tablets

Samples for firing were prepared from powdered raw material mixture, which was compressed into a cylindrical hole, from which, after compression, cylindrical tablets with a hollow centre were made. This hollow centre serves for hanging the sample on the weld of a thermocouple used to insert the tablet into the furnace.

A custom-made compression mold was used to prepare the tablets, see figure 1a. The mold with a shaped piston was made of hardened steel and ground with high precision to prevent it from seizing up and losing weight. This allows for accurate weighing. By pushing the material to the stop, a constant bulk density of the molding was maintained.

The bulk density of the tablet mentioned in literature [7] is 1.8 g/cm^3 based on which the weight to produce the tablet has been optimized. The production of the tablet itself took place by means of a compression roller using oil and graphite, which was then placed in a hand press, where compression was performed to a constant volume of the tablet.

Figure 1. a) tablet mold; b) made tablet for firing.

2.2. Firing

The firing took place in an electric resistance furnace with a manipulator, which is suitable for firing substances in which reactions take place in the solid phase or with only a partial proportion of the melt. The firing cycle is fully automated, which guarantees high reproducibility of heat treatment and precise adherence to firing parameters.

The sample was placed in a well in the oven cap when the selected oven temperature was reached and stabilized. Then the soaking time of the sample in the furnace was set on the control unit panel. The ejection of the thermocouple weld with a small sample hole occurs after one turn of the crank.

The thermocouple lifted the sample and the cap opened. The closure of the cap occurs after the return of the thermocouple with the sample to the centre of the heating zone of the furnace. After the firing time has elapsed, the shutter opened, and the sample is ejected on the thermocouple to the upper position. After ejection, the cap is closed and the thermocouple leaves the sample in the cap well, where air flows from the fan.

The tablets prepared by an optimized method were subjected to firing in an experimental furnace at a temperature of 1500 °C and with soaking times calculated from the insertion of the sample into the oven at 5, 10, 15 and 20 minutes, with addition of 60 s to warm up the sample.

The XRD analysis was performed with a multifunctional diffractometer XRD Empyrean PANanalytical with Cu anode; Kα was used as the radiation source with para meters of 5° to 90° with a scan-step of 0.01°. The ICSD database (released 2012) was used to qualitatively analyse the diffraction patterns. HighScore plus software was used to identify the individual phases.

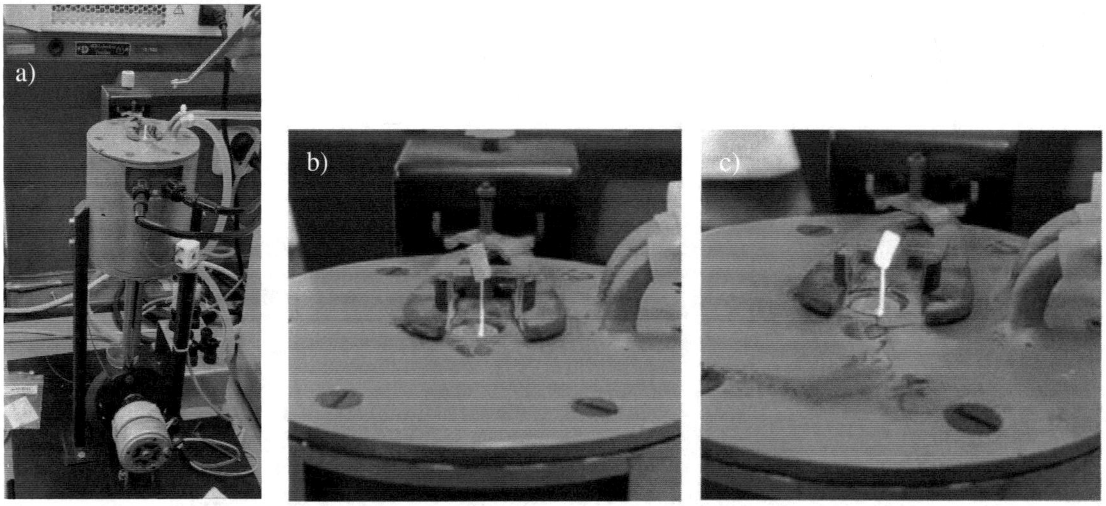

Figure 2. a) high-temperature furnace with manipulator; b) tablet before firing; c) tablet after firing.

3. Results and discussion

The article deals with the study of the progress of alite formation during short soaking times. The aim of this article is to assess the suitability of the proposed small experimental high-temperature furnace and its applicability for this purpose.

3.1. Preparation of tablets

For the first sample preparation, the value of bulk density was taken from the literature, namely 1.8 g/cm^3. The weight of the batch was calculated to be 58 mg. The pouring into the press roller took place at once and the sample was compacted. The pressed tablet could not be removed without damage. Compaction occurred only in the upper part the lower part disintegrated during handling.

Subsequently, the optimization took place in steps. The weight of the batch and bulk density were increased. The individual weights for the samples are given in table 1.

For sample no. 2, the weighing was optimized, the pressing rod and the inner part of the roller were rubbed with oil to prevent the sample from sticking and so that there was no problem during removal. The filling process was optimized and divided into 3rd phases; each poured layer was compacted with a pressing rod. After filling and pressing, the sample was not strong enough, the lower part disintegrated during gentle handling and the batch was glued to the walls of the pressing device.

After further optimization, sample no. 3, the weight and the method of sample separation were increased. The press roller and rod were coated with a combination of oil and hard graphite. The batch was poured in the same way as previous.

Finally, the whole procedure was optimized as follows. The amount of sample was weighed to 73 mg. The material was ground, without additional compaction, and soft graphite was applied to the press. This sample was sufficiently strong and did not disintegrate during handling. With such a procedure, it was possible to repeatedly form 30 pellets without breaking them, see figure 1.

Table 1. Density and weight of individual samples when optimizing tablets for firing.

Sample no.	Weight [mg]	Bulk density [g/cm3]	Result
1	58	1.878	x
2	65	2.034	x
3	70	2.191	x
4	75	2.347	x
5	73	2.285	✓

Weighing 73 mg ($CaCO_3$, SiO_2) for an assumed density of 2.285 g/cm^3. Separating agent soft graphite in combination with oil. Filling process continuously, slowly with simultaneous cellaring, pressing to constant volume to the stop.

3.2. Firing

The prepared tablets were fired in an experimental furnace at a temperature of 1500 °C and soaking times of 5, 10, 15 and 20 minutes. The proposed tablet, prepared by an optimized method, has been shown to be reliable and works very well for firing in a small experimental high-temperature furnace. The manipulator in the oven repeatedly picks it up, inserts it into the furnace and pushes it back out. The tablet does not disintegrate during insertion, firing and removal from the furnace and does not tend to explode or crack.

By XRD analysis using the Rietveld quantitative method, in addition to the main mineral alite, mineral residues such as free lime and belite were also identified in the fired samples, see Fig. 3.

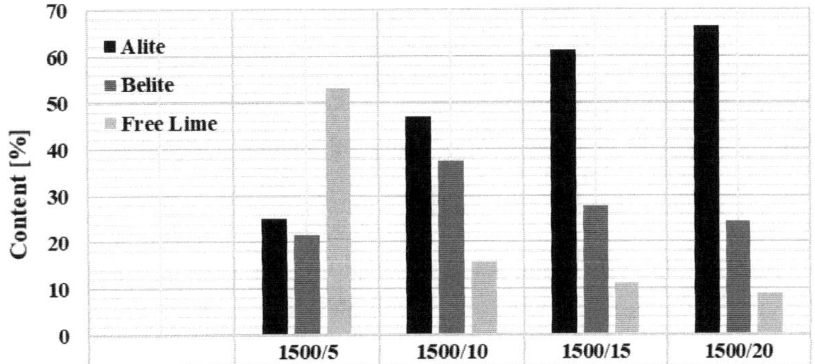

	1500/5	1500/10	1500/15	1500/20
Alite	25.29	46.92	61.32	66.43
Belite	21.59	37.53	27.63	24.4
Free Lime	53.12	15.55	11.04	9.17

Temperature/soaking

Figure 3. Graphical representation of the mineral content during firing.

IOP Conf. Series: Materials Science and Engineering 1209 (2021) 012045 doi:10.1088/1757-899X/1209/1/012045

Based on the results, we can observe an increase in the alite mineral content with increasing firing time. After 5 minutes the alite content in the formula is 25.29%, while after 20 minutes its content increases to 66.43%. With increasing firing time and the related increasing alite content, the free lime content gradually decreases, from 53.12% after 5 minutes to 9.17% after 20 minutes of firing. This is due to the saturation of the mineral belite and its gradual conversion to alite. This is illustrated by the decreasing content of belite, which is gradually transformed into alite, which is confirmed by the results of optical phase analysis, see figure 4.

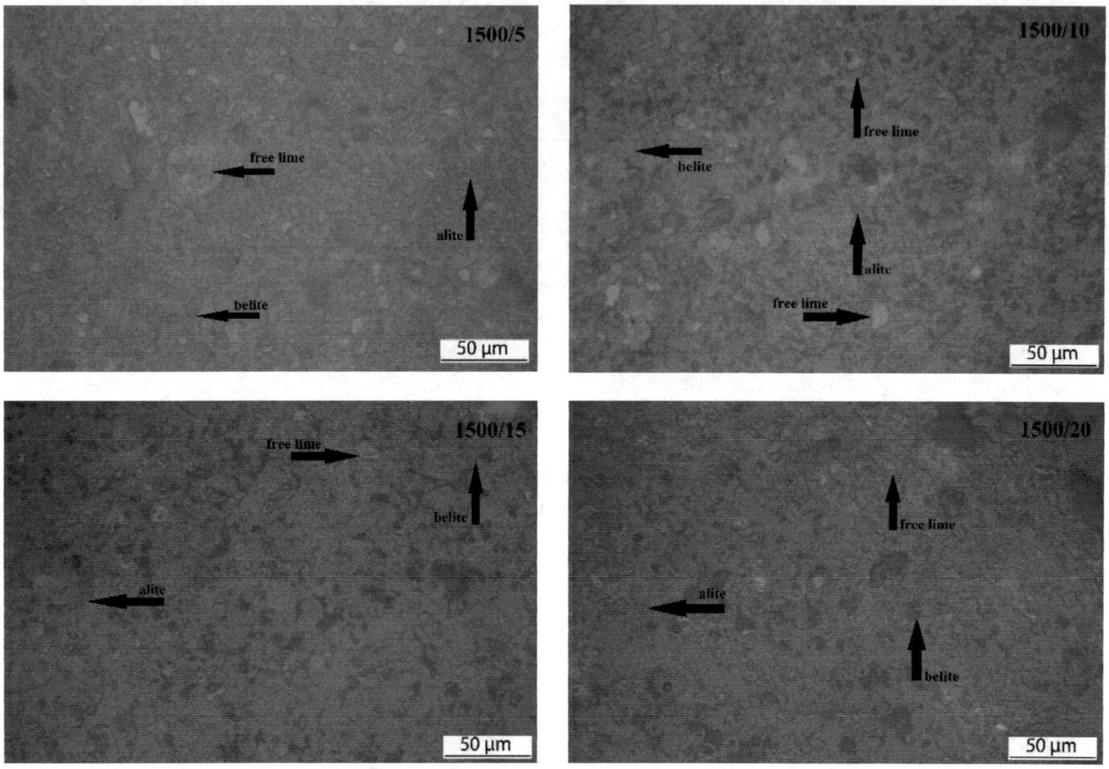

Figure 4. Optical phase analysis.

After only 5 minutes of firing, we can observe small nuclei of alite in the images from the optical phase analysis. As the alite crystals grow, the belite content disappears and the free lime content gradually decreases.

4. Conclusion

The article deals with laboratory preparation of triclinic modification of clinker mineral tricalcium silicate. A substantial part of the article is devoted to the technology and technique of firing tricalcium silicate at short soaking times. A furnace that requires special sample preparation was selected and designed. Therefore, a tablet preparation procedure was designed and optimized, making it possible to repeatedly prepare identical tablets with precise parameters for this type of oven. The resulting tablet weight was 73 mg at a density of 2.285 g/cm^3, which was compressed to a constant volume using an oil and graphite compression roller and a hand press. The firing of the tablet in a furnace with a manipulator using short firing times took place under identical conditions and the obtained results show that all processes take place very quickly. The used furnace with a manipulator is therefore suitable and usable for the study of the course of triclinic alite formation at short soaking times.

5. References

[1] Jeffery J W 1954 The tricalcium silicate phase *Proc. of the 3rd ISCC (London)* p 30–48

[2] Kondo E R and Choi S 1968 Mechanism and kinetics of Portland cement formation for an example of the solid-state reaction in the presence of a liquid phase *Proc. 5th ISCC (London)*
[3] Taylor H F W 1997 *Cement chemistry* 2nd ed (London: Thomas Telford Publishing) p 459
[4] Hewlett P 2004 Lea´s Chemistry of Cement and Concrete 4th ed (Amsterdam: Elsevier) p 1088
[5] Stanek T and Sulovsky P 2002 The influence of the alite polymorphism on the strength of the Portland cement *Cement and Concrete Research* **32** 7 p 1169-75
[6] Li X, Ouzia A and Scrivener K 2018 Laboratory synthesis of C_3S on the kilogram scale *Cement and Concrete research* **108** p 201-7
[7] Chromý S and Zavadil M 1980 *Laboratorní a výpočetní technika: Pec s manipulátorem pro izotermní výpaly laboratorních vzorků* p 62 - 65

Acknowledgments
We acknowledge the financial support to project ″Study of crystallization kinetics of various polymorphs of tricalcium silicate″ (FASt-J-21-7309).

Influence of fly ash addition on physical-mechanical properties of clay bricks – literature review

I Vrdoljak[1]

[1]Faculty of Civil engineering and Architecture, Vladimira Preloga 3, Josip Juraj Strossmayer university, 31000 Osijek

E-mail: ivrdoljak15@gfos.hr

Abstract. During the past years, due to the increase in the number of people and the increasing degree of industrialization, the amount of by-products produced, the so-called waste materials are becoming larger. As this problem grows with age, attempts are being made to find answers to the questions of how to use waste material purposefully in order to reduce the problem of disposing of large quantities of this type of material. One way to reduce the accumulation of waste material is to use it in construction materials. This literature review paper presents the research of the area dealing with the influence of fly ash application on properties of clay brick. The application of fly ash at various percentages in relation to the mass of clay and the influence of fly ash at different firing temperatures are studied.

1. Introduction

Coal combustion in thermal power plants produces two by-products: bottom ash and fly ash. Fly ash is a waste material which is separated from waste gases by filter devices. Nowadays, the proper disposal of fly ash and bottom ash is becoming increasingly difficult due to the extremely high generation of these waste materials. Inappropriate disposal of fly ash is a major problem for human and environmental safety due to heavy metals that fly ash contains. For that reason, more and more research is emerging that seeks to discover how waste materials can be reused. One of the ways is the use of fly ash as an admixture in building materials. In this chapter, the research of the application of fly ash in clay brick and the properties of such composites will be studied. Compared to cement based materials, the impact of fly ash on clay building materials has been much less investigated. According to Yao et al. [1], the percentage of fly ash utilization in clay materials is only 0.26% compared to the total use of this waste material in Europe, while the use of fly ash in cement based materials is 19.63%. In other parts of the world (China, India, USA) the percentage of fly ash utilization in clay materials is similar to that in Europe, while the use in cement based materials is slightly higher than Europe's. As an admixture to some building material, fly ash appeared as early as 1940s [2], [3]. The particle size of fly ash varies between 0.5 μm and 150 μm, while the highest concentration of fly ash is found in particle sizes up to 50 μm [4]. The assumption is that the granulometric composition of fly ash is more important when adding it to a clay based material than to a cement based one. For achieving better results in clay based materials, it is recommended to use equal sizes of clay grain and fly ash grain. A large disparity in the grain size of fly ash and clay would result in reduced homogeneity between the two materials. Uneven grain distribution in clay based materials was reported in Leiva et al. [5] where the grain size of fly ash was much finer than that of the clay. The medium size of fly ash grain size was D50: 20 μm, while the medium size of clay grain size was D50: 86 μm. According to Levia et al. due to reduced homogeneity, larger differences in grain size should be avoided [5]. The process of firing at high temperatures leads to sintering (softening of the particles) which then react with each other and create new chemical compounds that change the properties of the material in relation to the original raw material. This literature review will be presented with the aim of investigating potentially applicable fly ash in construction material as a possible additive that enhances protection against electromagnetic radiation. Before examining the possibility of electromagnetic protection, it is necessary to check the impact of the application of fly ash on other properties of concrete.

2. Analysis of physical-mechanical properties of concrete with fly ash addition

2.1. Density

According to the European standard EN 77-1: 2011 [6] bricks are divided into those with high and low density. Bricks that are considered to have high density must have a density higher than 1000kg/m^3, while low density bricks have a density lower than or equal to 1000kg/m^3. Standard solid brick has an approximate density of 2.0 g/cm^3. Analysing table 1, it can be concluded that in some cases there is an increase, and in some there is a decrease in density, but the density of the composite has never dropped below 1.0 g/cm3. From this it can be concluded that all tested samples can be classified in the category of high density bricks. With the increased proportion of fly ash in clay bricks, fired at lower temperatures (<1000 °C), reduced density of the composite occurs. This can be explained with the lower specific weight of fly ash compared to the clay. Therefore, if fly ash is added as a substitute for clay in a certain percentage, it leads to a decrease in density due to less specific weight of fly ash compared to clay [5]. For example, in a study by Leiva et al. [5], the specific weight of clay bricks fired at 800°C was 1.70 g/cm3, while the specific weight of bricks made with the addition of fly ash was 1.52 g/cm^3. A similar thing happens with Abbas et al. [7] where the addition of 25% fly ash reduced the density from 1.35 g/cm2 to 1.20g/cm^2. Analysing the density column in table 1, it can be seen that in some cases an increase amount of fly ash resulted in increase in density. One of the reasons for the increase is density is the firing temperature. If fly ash bricks are fired at higher temperatures (≥1000 °C), an increase of fly ash increases the density. Leiva et al. [5], Lin et al. [8] and Sultana et al. report an increase in density due to increase in fly ash content when firing at higher temperatures. In a study by Leiva et al. [5], this phenomenon is explained on the basis of conclusion that at 1000 °C, in addition to the vitrification of the brick, there is also the vitrification of the fly ash, thus achieving a more compact and denser material. It should be noted that in the research of Lingling et al. [9], Türkel, Aksin [10] and Quesada et al. [11] despite higher firing temperatures, the increase in density due to the addition of a larger amount of fly ash did not occur. No explanation was found in the reviewed literature as to why there was no increase in density at higher firing temperatures. It can be assumed that the vitrification process has not yet occurred in these cases. If the density of composites is analysed for different firing temperatures with equal propositions of fly ash, the density of bricks with equal composition will be bigger at higher firing temperatures. This phenomenon occurs in all analysed studies.

Table 1. Density of bricks with different amounts of fly ash additive.

Study	Fly ash addition [%]	Firing temperature [°C]	Density [g/cm3]	Source
Abbas et al.	0-25	800	1.35-1.20	[1]
Lingling et al.	50-80	1000	1.61-1.35	[2]
		1050	1.72-1.43	
Leiva et al.	0-80	800	1.70-1.52	[3]
		900	1.72-1.68	
		1000	1.78-2.10	
Quesada et al.	0-50	1000	1.84-1.79	[4]
Lin	0-40	800	1.68-1.78	[5]
		900	1.80-1.84	

		1000	1.86-1.94	
Queralt et al.	40	900	2.0	[6]
		1000	2.30	
		1100	2.40	
Pereira et al	15	1050	1.34	[7]
		1075	1.37	
	20	1050	1.28	
		1075	1.29	
Kayali	100	1000	1.45	[8]
Singh et al.	10-50	-	1.73-1.54	[9]
Türkel, Aksin	0-30	1000	1.65-1.25	[10]
Sultana et al.	0-15	1050	1.70-1.74	[11]

2.2. Compressive strength

Compressive strength of a brick with admixture depends on several factors: the type of clay, type of admixtures, the type of kiln, the firing temperature, heating program etc. The change in compressive strength by increasing the proportion of fly ash is closely related to the density and porosity of the material. By analysing the relationship between compressive strength and density in table 1 and table 2, it can be seen that as the density of the material increases, so does the compressive strength. It can be assumed that similar factors depend on how the density will change with adding fly ash, as well as how the compressive strength will change by altering the proportion of fly ash. In a study by Leiva et al. [5] at higher firing temperatures (1000 °C), the addition of fly ash to the brick increases the density. Therefore, at higher firing temperatures, the addition of fly ash also causes the increase of compressive strength. The author explains that the increase in compressive strength, as a result of the addition of fly ash at higher firing temperatures, is a result of the higher degree of sintering of composites due to fly ash [6]. The sintering process is a process in which fine-grained materials are melted into a solid material at high temperatures, figure 1.

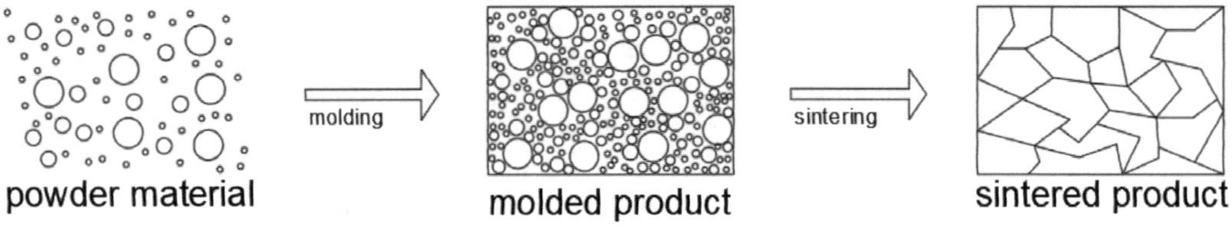

Figure 1. Mass compaction (sintering) [17].

In a study by Karaman et al. [18] the compressive strength behaviour of bricks at different firing temperatures was analysed. It can be seen, when firing at temperatures above 950 °C there is a much larger increase in compressive strength compared to lower temperatures. For example, at firing temperatures of 800 °C, 825 °C and 850 °C, the compressive strengths were 11.18 MPa, 11.57 MPa and 12.75 MPa, while the compressive strength at 950 °C, 975 °C and 1000 °C was 20.01 MPa, 25.99 MPa and 30.32 MPa. The author explains how bigger growth occurs due to vitrification [18]. As with density, in studies by Lingling et al. [9] and Türkel, Aksin [10], the high firing temperature, the compressive strength decreases by increasing the proportion of fly ash. The authors do not provide an explanation regarding the drop in compressive strength due to addition of fly ash. According to the standard EN 771-2: 2011 [6] the prescribed minimum compressive strength of bricks is 5 MPa. Accordingly, all tested samples of the analysed research would meet the minimum regulations, except Pawar, Gerud [19] (poor quality of brick without admixture) and Leiva et al. [5] (very high proportion of fly ash at lower firing temperature (800 °C)). The behaviour of composite bricks when using fly ash with a higher proportion of pulverized powder was investigated in Lingling et al. (figure 2) [9]. Lingling et al. [9] report that additional pulverising of fly ash results in a significant increase in compressive strength. The composite with clay and original fly ash in a ratio of 40:60 has a compressive strength of 39.6 MPa, while the composite with pulverized fly ash and clay in the same ratio reached a strength of 85.9 MPa. At a higher ratio of 20:80 there is a smaller increase in compressive strength compared to less pronounced ratios, but there is still an increase. The composite with original fly ash had a compressive strength of 25.4 MPa, while the one with pulverised fly ash had 37.1 MPa.

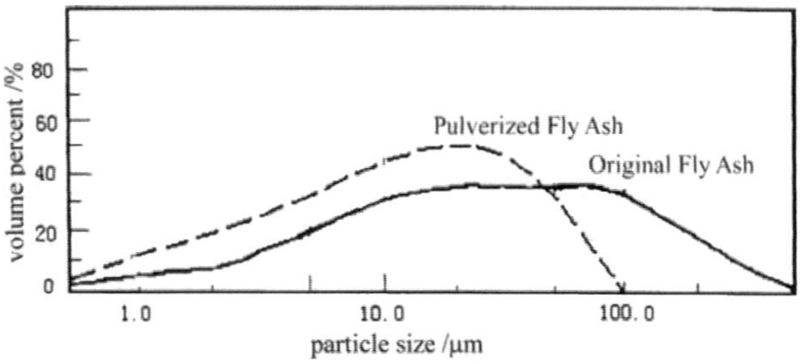

Figure 2. View of the arrangement of original and pulverised particles [9].

Table 2. Compressive strength of bricks with different amounts of fly ash additive.

Study	Fly ash addition [%]	Firing temperature [°C]	Compressive strength [MPa]	Source
Abbas et al.	0-25	800	23.0-7.0	[1]
Pawar, Garud	5-50	1000	4.4-0.8	[19]
Lingling et al.	50-80	1000 1050	50.0-14.7 98.5-24.4	[2]
Leiva et al.	0-80	800 900 1000	9.0-1.0 13.0-7.0 14.0-46.0	[3]
Quesada et al.	0-50	1000	60.0-29.0	[4]
Lin	0-40	800 900 1000	10.0-23.0 22.0-31.0 62.0-63.0	[5]
Kumar et al.	100	-	12.59	[20]
Kayali	100	1000	43.0	[8]
Türkel, Aksin	0-30	1000	35.9-7.0	[10]

2.3. Flexural strength

Flexural strength has been much less studied compared to compressive strength. The studies that have studied the flexural strength are: Abbas et al. [7], Obada Kayali [14], Türkel et al. [10]. In a study by Abbas et al. [7], the flexural strength of the brick without admixture was 6.1 MPa. By adding an increasing amount of fly ash, the flexural strength decreased approximately linearly. For example, with a fly ash content of 15 %, the flexural strength was 4.2 MPa, while with a 25% fly ash content, the flexural strength was 3.0 MPa. The reduction in flexural strength relative to fly ash-free brick is more than 50 %. The reason for the drop in flexural strength by adding fly ash is attributed by the author to a too low firing temperature of 800 °C. The movement of the flexural strength results coincides with the movement of compressive strength and density. In the Türkel, Aksin study [10] also a decrease in flexural strength due to the addition of fly ash occurs. Bricks without fly ash had a flexural strength of 4.4 MPa, while bricks with 30% fly ash had a flexural strength of only 0.4 MPa. The authors do not explain the reason for the significant decline. In the Kayali study [14], bricks with 100% fly ash were analysed. The result shows that the flexural strength at a firing temperature of 1000 °C is 10.3 MPa, which is significantly higher than the flexural strength from Abbas et al. [7] and Türkel et al. [10].

2.4. Water absorption

The absorption property is closely related to the density and strength of the material. Increasing the density of the material reduces the amount of pores, which reduces the ability of the material to absorb water. The sintering process contributes to the reduction of pores. Sintering thickens the porous body at high temperatures, which reduces the porosity and increases the mechanical strength. The influence of grain growth during the sintering process can be seen in figure 3 [17]. Sintering occurs at temperatures from 430 °C to 850 °C [21]. For example, in the Lin study [22], the absorption of bricks without the addition of fly ash at 800 °C was 20.0 %, while by increasing the firing temperature to 900 °C it was 19.0 %. Analysing the research from Table 3, it can be noticed that at lower firing temperatures, due to the increase in the proportion of fly ash, there is an increase in water absorption. According to Abbas et al. [7] this is attributed to the high absorption capacity of fly ash. Leiva et al. [5] when firing samples at 800 °C reported an absorption of samples without fly ash of 10.50 %, while the addition of fly ash of 80.0 % increased the absorption to 18.40 %. However, when firing at higher temperatures, silicate materials and oxygen melt, creating a more compact and dense material that further reduces porosity (vitrification). Vitrification occurs at temperatures above 950 °C. The results of Leiva et al. show that at higher temperatures (1000 °C) the addition of fly ash reduces absorption. Thus, the absorption of the sample without fly ash was 8.80 %, while with the addition of 80.0 % fly ash the sample was 6.0 %. But despite the high firing temperatures (1000 °C), Lingling et al. [9] and Türkel et al. [10] recorded an increase in absorption due to an increase in the proportion of fly ash in the brick. The authors do not explain why this phenomenon occurs.

Table 3. Water absorption of bricks with different amounts of fly ash additive.

Study	Fly ash addition [%]	Firing temperature [°C]	Water absorption [g/cm3]	Source
Abbas et al.	0-25	800	11.0-24.0	[1]
Pawar, Garud	5-50	1000	21.67-26.51	[19]
Lingling et al.	50-80	1000	22.18-31.26	[2]
		1050	17.62-27.86	
Leiva et al.	0-80	800	10.50-18.40	[3]
		900	10.00-16.90	
		1000	8.80-6.00	
Kumar, Hooda	100	1000	9.77	[23]
Lin	0-40	800	20.0-17.0	[8]
		900	19.0-18.0	
		1000	14.0-16.5	
Queralt et al.	40	900	9.0	[6]
		1000	3.20	
		1100	1.50	
Pereira et al.	15	1050	32.10	[7]
		1075	29.80	

	20	1050	37.10	
		1075	35.60	
Kumar et al.	100	-	7.63	[20]
Kayali	100	1000	10.0	[14]
Türkel, Aksin	0-30	1000	21.0-38.0	[10]
Sultana et al.	0-15	1050	7.60-10.28	[16]

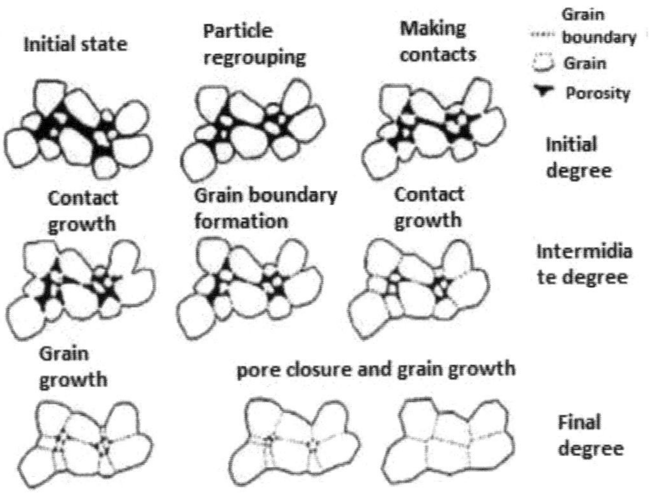

Figure 3. Grain growth during sintering process [12].

2.5. Mass loss

Clay and fly ash lose part of their mass when fired at high temperatures. Some of the reasons for the loss of mass during firing are: loss of water from the clay mixture, combustion of organic matter of plant origin that takes place at temperatures of 200 °C [21], up to 900 °C, burns all carbon except graphite, loss of CO_2 from carbonate at temperatures up to 500 °C [24] etc. The results of weight loss during firing are shown in table 4. From the presented results it can be concluded that clay on average loses a larger amount of mass compared to fly ash. This is expected because the clay contains a larger amount of organic matter and the fly ash has previously been exposed to high temperatures, so it is expected that reheating at high temperatures will not cause major changes in mass.

Table 4. The results of weight loss during firing.

Study	Clay [%]	Fly ash [%]	Source
Lingling et al.	5.66	2.82	[2]
Quesada et al.	12.51	4.74	[4]
Lin	6.80	0.03	[5]
Queralt et al.	N.S.[a]	2.70	[6]
Türkel, Aksin	7.84	0.91	[10]

[a] N.S. – not specified

2.6. Efflorescence

The phenomenon of brick efflorescence is considered to be the appearance of white spots or streaks on the outer surface of the wall. Due to the capillary rise of water to the surface of the brick, it evaporates. This water often contains salts, which crystallize after evaporation and leave a white mark. Some of the reasons why this can happen are: the use of poor quality bricks, a large amount of salt in the water to prepare the bricks, sand consisting of a high concentration of salt and the like. There is little research examining efflorescence. Investigations of efflorescence properties were conducted by Abbas et al. [7], Pawar, Garud [19], Kumar, Hooda [23] and Kumar et al. [20]. According to Abbas et al. [7] efflorescence is reduced by increasing the amount of fly ash. This explains the phenomenon due to the possibility of fly ash to bind salts and lime. Pawar, Garud [19] state that the addition of fly ash in the amount of up to 50% results in only a slight efflorescence process. Kumar, Hooda [23] and Kumar et al. [20] analysed the occurrence of efflorescence on bricks with 100% fly ash. According to Kumar, Hooda [23], if white or grey sediment occurs on 10% of the brick surface, it is categorized as mild efflorescence, if white or grey sediment occurs on about 50% of the surface, efflorescence is considered moderate. And if a grey or white precipitate appears on an area greater than 50%, efflorescence is categorized as very severe. Both studies describe the appearance of grey sediment in the amount of less than 10%.

Figure 4. Brick efflorescence [25].

3. Conclusion

After reviewing the available research, it can be concluded that fly ash as an additive in a certain amount can improve the properties of bricks. Composites fired at higher temperatures will always have higher strengths than those fired at lower temperatures due to the higher degree of vitrification. Despite the fact that fly ash has a lower specific weight compared to clay, after adding fly ash to clay brick and exposing composite to firing process at higher temperatures it can result higher strengths and higher density. The reason for this phenomenon lies in vitrification of clay and partial vitrification of fly ash, which results in more compact and denser material. In certain studies, despite the high firing temperatures, there was no increase in strength. The authors do not give an explanation regarding this phenomenon, but one of the assumptions may be that for some reason in these cases the composite vitrification has not yet occurred. Further research on this topic is suggested to explain the differences in the results.

4. References

[1] Yao Z T et al. 2015 *A comprehensive review on the applications of coal fly ash* (Hangzhou: Earth Science Rev.) pp 105–21
[2] Kluge R W, Sparks M M and Tuma E C 1949 Lightweight Aggregate Concrete *ACI Journal, Proceedings* **9** 45 625–44
[3] Davis R E, Carlson R W, Kelly J W and Davis H E 1937 Properties Of Cements And Concretes Containing Fly Ash *ACI Journal, Proceedings* **33** 5 pp 577–612
[4] Dondi M, Marsigli M and Fabbri B 1997 Recycling of Industrial and urban wastes in brick production: A review *Tile brick Int.* **13** 218–25
[5] Leiva C et al. 2016 Characteristics of fired bricks with co-combustion fly ashes *J. Build. Eng.* **5** 114–8
[6] Group of authors 2011 *EN 771-2: BSI Standards Publication Specification for masonry units Part 1: Clay masonry units* (BSI Stand) pp 1–66
[7] Abbas S, Saleem M A, Kazmi S M S and Munir M J 2017 Production of sustainable clay bricks using waste fly ash: Mechanical and durability properties *J. Build. Eng.* **14** 7–14
[8] Lin K L 2006 Feasibility study of using brick made from municipal solid waste incinerator fly ash slag *J. Hazard. Mater* **137** 31810–6
[9] Lingling X, Wei G, Tao W and Nanru Y 2005 Study on fired bricks with replacing clay by fly ash in high volume ratio *Constr. Build. Mater* **19** 243–7
[10] Türkel S and Aksin E 2012 A comparative study on the use of fly ash and phosphogypsum in the brick production *Proc. Eng. Sci.* **37** 5 595–607
[11] Eliche-Quesada D, Sandalio-Pérez J A, Martínez-Martínez S, Pérez-Villarejo L and Sánchez-Soto P J 2018 Investigation of use of coal fly ash in eco-friendly construction materials: fired clay bricks and silica-calcareous non fired bricks Ceram. Int. **44** 4 4400–12
[12] Queralt I, Querol X, López-Soler A and Plana F 1997 *Use of coal fly ash for ceramics: A case study for a large Spanish power station* Fuel **76** 8 787–91
[13] Fernández-Pereira C, De La Casa J A, Gómez-Barea A, Arroyo F, Leiva C and Luna Y 2011 *Application of biomass gasification fly ash for brick manufacturing* Fuel **90** 220–32
[14] Kayali O 2005 *High Performance Bricks from Fly Ash* (Canberra: World Coal Ash)
[15] Mann H S, Brar G S, Mann K S and Mudahar G S 2016 Experimental Investigation of Clay Fly Ash Bricks for Gamma-Ray Shielding *Nucl. Eng. Technol.* **48** 5 1230–6
[16] Sultana M S, Hossain M I, Rahman M A and Khan M H 2014 Influence of Rice Husk Ash and Fly Ash on Properties of Red Clay J. Sci. Res. **6** 63 421–30
[17] Marković I 2009 *Utjecaj temperature sinteriranja na tvrdoću aluminij oksidne keramike* (Zagreb)
[18] Karaman S, Gunal H and Ersahin S 2006 Assesment of clay bricks compressive strength using quantitative values of colour components *Constr. Build. Mater.* **20** 5 348–54
[19] Pawar A S and Garud D B 2014 Engineering Properties of Clay Bricks With Use of Fly Ash *Int. J. Res. Eng. Technol* 75–80
[20] Kumar R, Patyal V, Lallotra B and Kumar Ashish D 2014 Study of Properties of Light Weight Fly Ash Brick *Int. J. Eng. Res. Appl.* 49–53
[21] Miletić I 2007 *Struktura keramičkoga materijala* (Sisak)
[22] Lin K L 2005 Effects of SiO2 and Al2O3 on the hydration characteristics of MSWI fly ash slag *J. Ind. Eng. Chem.* **11** 6 834–40
[23] Kumar R and Hooda N 2014 An Experimental Study on Properties of Fly Ash Bricks *International Journal Of Research In Aeronautical And Mechanical Engineering* **5** 4 56–67
[24] Schulte E E and Hopkins B G 1996 Estimation of Soil Organic Matter by Weight Loss-On Ignition *Soil Organic Matter: Analysis and Interpretation* **46** pp 21–31
[25] Janssen H, Chwast J and Elsen J 2020 Gypsum efflorescence on clay brick masonry: Analysis of potential efflorescence origins J. Build. Phys. **44** 1 37–66

IOP Conf. Series: Materials Science and Engineering 1209 (2021) 012047 doi:10.1088/1757-899X/1209/1/012047

Methods of calculation of cold-formed structures

E Hodovanets[1], M Al Ali[1] and V Kvocak[1]

[1] Institute of Structural Engineering, Faculty of Civil Engineering, Technical university of Kosice, Vysokoskolska 4, 042 00 Kosice, Slovakia

Email: elvira.hodovanets@tuke.sk; mohamad.al.ali@tuke.sk; vincent.kvocak@tuke.sk

Abstract. Cold-formed structures are a separate type of metal structures and require more complex mathematical models in comparison with massive profiles. The development of main theories was described. A new type of buckling is also taken into account - warping, which is characteristic only of cold-formed bars.

1. Introduction

The roots of the theory of bars and bar systems go back to the 17th and 18th centuries and is closely connected with the development of mathematics in general, especially with such branches of it as differential and integral calculus. The foundations for its formation were laid by G. Galilei, Marriott, Hooke. The main tasks of the theory of bars are to study the strength, stability, rigidity and carrying capacity of bars and bars systems. Starting with J. Bernoulli and L. Euler, mathematical modeling plays an important role in solving these problems, which includes, first, the derivation of the basic equations taking into account the physical and mechanical properties of the materials from which they are made; second, analytical and numerical solution methods for studying their behavior (deformation properties, strength, etc.) when exposed to external mechanical factors (load), temperature, fields of various nature [1].

Soviet scientists made a great contribution to the development of one-dimensional problems of the theory of elasticity - the theory of torsion and bending of prismatic bars, the foundations of which were laid by Saint-Venant in 1855. Solutions were found for a variety of problems on the torsion of bars of various sections and solutions expressed in the Fourier series. The works of S.P. Timoshenko became the basis of the modern theory of constrained torsion of thin-walled bars. He was the first to note the inadequacy of Saint-Venant's torsion theory for thin-walled open-profile members. S.P. Timoshenko showed, using the example of the problem of constrained torsion of a cantilever I-beam, that to accurately find the value of the torsion angle, one should take into account not only free torsion stresses but also bending stresses in the flanges of the I-beam, that is, take into account the effect of deplanation of the section. He also solved the problem of bending-torsional forms of buckling of a bending I-beam [1].

In some cases, the generalized theory of calculation of thin-walled bars under constrained torsion, proposed by Vlasov, is very effective. He proposed a theory of open profile thin-walled bars based on the hypotheses according to which an open profile member is considered as a shell with a non-deformable profile in the transverse plane and the shear deformation in the median plane of the profile is zero. He obtained general differential equations for the deformation of a thin-walled bar under load and also revealed that Saint-Venant's principle of rapid damping along the length of the bar of local stresses from a balanced external load practically does not work. The methods of V.Z. Vlasov became widespread and formed the basis for the calculation of many engineering structures. However, it should be noted that Vlasov considered only open-profile members. The theory of calculation of thin-walled bars of a closed profile was developed by A.A. Umansky. Considering that the effects of constrained torsion in bars with a closed cross-section are less than in bars with an open contour, Umansky introduced a new function for the warping measure, which does not coincide with the derivative of the torsion angle, as was adopted by Vlasov.

The history of the development of the theory of thin-walled bars can be conditionally divided into three stages:

- the first stage (1905 - 1940) - the first theoretical and experimental, devoted mainly to deviations from the law of flat sections when bending beams.
- the second stage (1940 - 1970) is the development of a complete theory of thin-walled bars with an in-depth study of continuous thin-walled beams, calculations of flat and three-dimensional frames, and stability issues.
- the third stage (1970 - up to the present) is the emergence and active development of computing facilities and, in particular, the finite element method [1].

2. Main theories and where they start

The specified groups of methods include the following:

- classical method of structural mechanics;
- method based on S. Timoshenko's theory of bearing capacity;
- method based on the technical theory of thin-walled rods by V. Vlasov;
- method based on semi-shear theory of V. Slivker;
- method proposed in Eurocode-3;
- Finite Element Method;
- empirical method (testing of samples).

2.1. Coulomb torsion theory

The theory of Sh. O. Coulomb, who was one of the first to conduct a series of experiments on torsion and establish some regularities about their deformation, is based on the premise that the cross-section is non-deformable and that there are no displacements along the longitudinal axis of the bar. Based on this, he suggested that when the member is twisted, its sections rotate like rigid platforms in their plane, sliding over each other. For a real bar, it coincides with the low-frequency dispersion of the first torsional wave only under the condition that the cross-section of the bar is circular or annular. It was with such bars that Coulomb conducted his research. For members with a more complex cross-section, the variance is underestimated; therefore, a similar equation proposed by Saint-Venant should be used [2].

2.2. Torsion theory of Saint-Venant

One of the most significant contributions to the development of the theory of torsion of bars was made by B. Saint-Venant. He was the first to obtain an exact solution to the problem of torsion of prismatic bars of an arbitrary profile based on a mathematical formulation. Having constructed his theory, he solved some specific problems for bars with elliptical and rectangular sections, carried out a large series of calculations to determine the torsional stiffness and the distribution of shear stresses, which is necessary to study durability of bars. And although the solutions he obtained are valid only for "free" ("pure") torsion of the bars, modern approximate theories of torsion vibrations use them to some extent, and all calculations related to the torsion of prismatic bars are performed according to the formulas obtained by Saint-Venant. Studying the operation of a prismatic member under the action of torques, statically applied to the ends of the member, equal in magnitude and opposite in direction, Saint-Venant assumed that the torsion of the bar consists of two components:

- torsion along the Coulomb (no curvature of the cross-section);
- longitudinal cross-section displacements uniform along the entire length.

The uniformity of longitudinal displacements (warps) of cross-sections is a distinctive feature of Saint Venant's theory of free torsion [2].

2.3. Theory of constrained torsion
It is customary to call torsion constrained if the warping along the length of the bar is variable. A striking example given by many authors is a bar of arbitrary (non-circular) cross-section, restrained by one end. A torque or a pair of forces is applied to the free end of the bar (Figure 1.). In this case, the warping in the fixed part of the section will be equal to zero, at the free end is nonzero. Such inhomogeneous warping leads to the appearance of normal longitudinal stresses, which in turn cause the appearance of additional shear stresses in the cross-sections with a nonzero torque. Obviously, in constrained torsion, the relationship between torque and twist angle is more complex than in free torsion [2].

The engineering theory of constrained torsion is based on the following hypotheses:

- the cross-section of the bar in its plane is not distorted during deformation;
- the warping of the bar is proportional to the relative angle of torsion.

Thus, the main difference between this theory and Saint-Venant's theory is that the torsion angle is an arbitrary function of the longitudinal coordinate and time. If, as an example, we consider a bar of I-section, along the edges of the shelves of which four identical forces are applied, forming pairs, then these pairs applied to the shelves will cause bending in different directions (Figure 2.), which will be is equivalent to torsion. In other words, the moment created by two pairs of forces will be a bimoment.

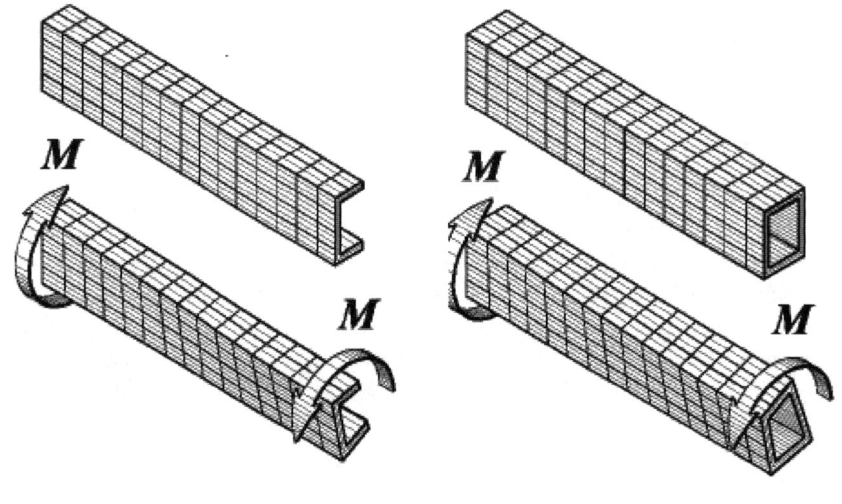

Figure 1. Illustration of closed and open thin-walled constrained bars.

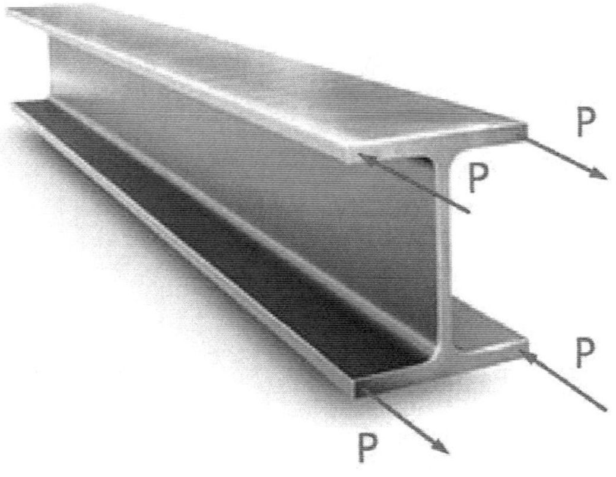

Figure 2. Illustration of a system of longitudinal forces causing the deplanation of a cross section.

2.4. Vlasov's theory
The difference between Vlasov's theory of constrained torsion and Timoshenko's theory is in the expression of kinetic energy. In addition to the kinetic energy of rotation, the kinetic energy of warping is taken into account. In this case, the expression for the potential energy according to the Vlasov theory remains identical to the Timoshenko theory [3].

2.5. The theory of V.I. Slivker
The above theories do not take into account the shear strain energy. Consider the theory proposed by Slivker. Within the framework of this theory, called semi-shear, the torsion angle function and the warping measure function are independent. Then the torsional vibrations of a thin-walled bar will be described by a system of two differential equations. One of the possible ways of constructing approximate theories of calculation is based on dividing stresses and strains into major and minor ones, which are further excluded from the resolving equations. In his refining theory for calculating thin-walled bars, Slivker took into account the effect of shear deformations, including in the expression for the deformation energy of a thin-walled bar that part of it that is caused by the work of shear stresses. When calculating thin-walled bars in the developed semi-shear theory, Slivker proposes to represent tangential stresses as the sum of two terms: tangential bending stresses caused by transverse forces and, and shear torsional stresses caused by a moment of constrained torsion. This theory has some advantages: firstly, it allows one to partially take into account shear deformations and at the same time turns out to be much simpler than the shear theory, and secondly, it is universal for calculating both thin-walled bars of an open profile (based on the theory of Vlasov) and a closed profile (based on the theory of Umansky) due to the similarity of the corresponding differential equations of torsion and functionals of deformation energies [4].

2.6. Calculation provisions of Eurocode-3
The influence of the loss of local stability and imperfections in shape should be taken into account when determining the strength and stiffness of cold-formed elements and decks. The design of compressed members with the edge or intermediate stiffeners should be based on the assumption that the stiffener operates as compressed with continuous partial pinching, like a spring, the stiffness of which depends on the boundary conditions and the bending stiffness of adjacent flat members. The calculation method for local buckling and shape imperfection is described in detail in Eurocode 3. According to Eurocode 3, the effect of local stability loss and shape imperfections should be taken into account when determining the strength and stiffness of cold-formed elements and decks. The effect of the loss of local stability can be taken into account using the geometric characteristics of the effective section, calculated on the basis of the effective widths. When calculating the load capacity taking into account local

buckling, the yield strength fy is taken as fyb for the case of determining the effective width of compressed members according to EN 1993-1-5 [5].

If you do not use the simplified method according to clause 5.5.3, part 1-3 of Eurocode 3, where elastic buckling stresses are determined by linear calculation, the following algorithm can be proposed:

- calculate the elastic buckling stresses for the corresponding buckling waveforms of the full-length element;
- calculate the effective (reduced) width following 5.5.2 parts 1-3 of Eurocode-3 for the parts of the cross-section that have lost local stability at minimum stress.
- calculate the reduced thickness (see 5.5.3.1, parts 1-3 of Eurocode-3) for edge and intermediate stiffeners or other cross-sectional parts subject to form imperfections under minimum stress;
- calculate the resistance to buckling following clause 6.2 of Part 1-3 of Eurocode-3 for bending, torsional forms, or out of the plane (depending on the buckling form) for a full-length element with an effective cross-section [6].

3. Comparison of the main analytical methods

Timoshenko's method differs from Saint-Venant's theory in fact, that the angle of rotation is taken as a function of the longitudinal coordinate x from time t. The angles of rotation, measured experimentally by Timoshenko, were in good agreement with the theoretical ones calculated according to his formulas. Another theory - Vlasov, which differs from Timoshenko in that the term for kinetic energy takes into account not only the torsional energy but also deplanation. This theory is widely used in the field of determining the load-bearing capacity of thin-walled bars of open cross-section, but it has its disadvantages because it assumes a rigid cross-section and does not take into account the local loss of stability. In Slivker's half-shear theory, the angle of rotation and the degree of deplanation are independent functions in the description of kinetic energy. It is suitable for bars of open and closed profiles, which allows the use of a single design model in combined constructions of open and closed profiles. In addition, the accuracy of the calculation is improved by considering part of the shear deformation. The main difference between Umansky's theory is the hypothesis that the longitudinal shifts of points due to deplanation are proportional (with a certain coefficient) to the shift of points in free torsion. This assumption makes it possible to obtain a system of equations describing the torsion of a thin-walled bar of closed cross-section [7].

Table 1. Comparison of theories in mathematical form.

Theory	Shearless theory according to Vlasov	Semi - shear theory according to Slivker	Semi - shear theory according to Umansky
		Basic equations	
Equilibrium equations	$-H' - B_\omega'' = m_x + m_B'$ $-B_\omega' + M_\omega = m_B$	$-H' - M_\omega' = m_x$ $-B_\omega' + M_\omega = m_B$	$-H' - M_\omega' = m_x$ $-B_\omega' + M_\omega = m_B$
Geometric equations	$\chi_x = \theta'$; $\chi_B = -\theta''$; $\gamma_\omega = 0$	$\chi_x = \theta'$; $\chi_B = -\beta'$; $\gamma_\omega = \theta' - \beta$	$\chi_x = \theta'$; $\chi_B = -\beta'$; $\gamma_\omega = \theta' - \beta$
Physical equations	$H = GI_d\theta'$ $B_\omega = -EI_\omega\theta''$	$M_\omega = \frac{GI_d}{\psi - 1}(\theta' - \beta)$ $H = GI_d\theta'$ $B_\omega = -EI_\omega\beta'$	$M_\omega = -EI_\omega\beta'' - m_B$ $H = GI_d\theta'$ $B_\omega = -EI_\omega\beta'$
		Mainly unknown	
Movement	θ – angle of rotation θ' – deplanation rate	θ – angle of rotation β – deplanation rate	θ – angle of rotation β – deplanation rate
Deformation		χ_x – deformation at the corner χ_B – relative deformation γ_ω – shear deformation	

Integral characteristics of normal and tangential stresses	B_ω – bimoment H – moment of free torsion M_ω – moment of constrained torsion
Other characteristics	m_B – bimoment distributed along the entire length m_x – bending moment distributed along the entire length EI_ω – constrained torsion stiffness GI_d – free torsional stiffness ψ – profile shape coefficient

4. Conclusion

First theories connected to thin walled bars and bars systems came to life in 17th century. Physics of thin walled bars with slender cross section differed from standard bars. Thin walled bars are also affected by torsion which is caused by two moments applied with opposite directions. This torsion is called bimoment and it causes deplanation. There are three main theoretical theses considering thin walled bars. Theory for open cross-section profile, closed cross-section profile and general profile. Idea for future is investigated a theory, which works well with all cross-section profiles and doesn't have limitations and disadvantages of previous theories.

5. References

[1] Lampsi B 2018 *Nonlinear wave dynamics and strength of thin-walled bars affected by cross section deplanation* (Nizny Novgorod: Lobachevsky State University of Nizhny Novgorod National Research University)
[2] Rybakov V 2011 *Fundamentals of structural mechanics of steel thin-walled structures. Textbook* (Petrohrad: Publishing house of Polytechnic University)
[3] Vlasov V 1959 *Thin-walled elastic bars* (Erusalem: Israel Program for Scientific Translations) p 508
[4] Slivker V 2005 *Structural mechanics. Variation methods* (Moscow: Publishing House of the Association of Construction Universities) p 736
[5] EN 1993-1-5 2006 Eurocode 3: *Design of steel structures - Part 1-5: General rules – Plated structural elements* (Brussels)
[6] EN 1993-1-3 2006 *Eurocode 3: Design of steel structures - Part 1-3: General rules Supplementary rules for cold-formed members and sheeting* (Brussels)
[7] Hodovanets E and Al Ali M 2020 Vývoj teórie výpočtu tenkostenných prútov *Sborník UIS 2020 SvF TUKE pp 31-6*

Acknowledgments

This work was supported by The Slovak Scientific Grant Agency under the contract No. VEGA 1/0172/20 Stress and strain analysis of load-bearing elements made of steel, glass and composite materials.

Numerical analysis of asymmetrical using of actuators to membrane structure

L Kapolka[1] and L Stulerova[1]

[1] Department of Structural Engineering, Faculty of Civil Engineering, Technical university of Kosice, Slovakia

E-mail: lukas.kapolka@tuke.sk

Abstract. This paper is focused on the numerical analysis of an adaptive membrane structure, the model of which was created on the basis of experimental device in the lab. It was researched out the effect of asymmetric use of actuators on the membrane structure without structural loads. Results of the article describe the behaviour of numerical model. The results gained from numerical model will be compared in the future with the results measured out from the experimental device.

1. Introduction

As a structural category, fabric tensile structures are a special form of lightweight surface structures, which include shells, folded plates, grid-domes and cable nets. In each of these, the continuous spatially curved surface is a critical and integral structural element. In tensile structures, these surface elements, consisting of structural fabric and high strength cables, can carry load in tension only [1].

The primary advantage of tensile members over compression members is that they can be as thin and as light as their tensile strength permits. Consequently, the weight of tensile structures is almost negligible. A further advantage of thin, lightweight tensile components is that they are easy to ship and erect. Cables can be tens of meters long, requiring no splices or internal connections. They can be raised and connected to their end supports by cranes, winches or helicopters, requiring no scaffolds. In fact, the erection time for a fabric structure is much shorter than that for a conventional structure [2].

To make a tensile surface structure work requires a minimum of four support points, one more than is needed for a rigid structural system. The most basic form, therefore, is a four-point structure. One of the four points has to be outside the plane defined by the other three to achieve the double curved surface, which gives the structure its stability and its capacity to carry load [3].

Because of the lack of structural weight, there need to be elements, which resist upward loads from wind suction in addition to the elements, which carry downward loads. In order to generate the structural surface grid, which satisfies all these requirements, there have to be support points at the high points of the surface, others at the low points, and still others located around all sides of the periphery [4].

The pattern of surface stresses which is required for the stability and load carrying capacity of the structure results in horizontal forces at the anchors in addition to the customary vertical forces. This is the price to be paid for the advantages of a tensile structure. The skill and efficiency with which these horizontal forces are anchored or balanced has a large impact on the economy of the structural system [5].

The choice of these support points defines the shape of the structure. Their geometry, combined with the stress pattern assigned to the surface, lead to the form of the structural surface. The final shape is determined with the help of a form finding computer program, which is based on placing all the tensile forces in all the elements in equilibrium. For one given configuration of supports and one internal stress pattern, there is only one equilibrium shape. To change this shape, anchor points or internal stresses have to be changed [5], [6], [7].

This paper focuses on the effect of asymmetric use of actuators on the membrane structure without structural loads. Results of the article describe the behaviour of numerical model.

2. Description of the numerical model

The numerical model was closely described few times in the couple of papers, e.g. [8], [9], [10]. It was based on the parameters of an experimental device located in a laboratory of Technical University of Košice.

In brief, the numerical model of the hyperbolic paraboloid was created in the Dlubal RFEM computing environment, which contains an add-on module for form-finding of the surface and the line elements, on the specified boundary conditions.

It consist of a rigid steel frame, 3.34 x 3.34 x 2.50 m, are dimensions of the frame axes. The steel frame is composed of different cross-sections of the members and metal plate stiffeners. Two anchor rods and two actuators are placed in the four corners of the frame to form the basic shape of a hyperbolic paraboloid. The edge ropes and the membrane surface form the hyperbolic paraboloid. The size of membrane surface floor plan is 2.00 x 2.00 m. The elevation of membrane surface is 1.00 m with saddle in the middle. The initial sag of edge ropes was set up at 0.20 m. The initial stress in the surface was set up at 4.00 kN/m width. The material characteristics of the technical fabric, Ferrari Serge Precontraint 502, were taken from the manufacturer. The material characteristics for the steel and stainless-steel components, were taken as the conventional norm values. Specifications of numerical model elements are shown in the figures 1, 2.

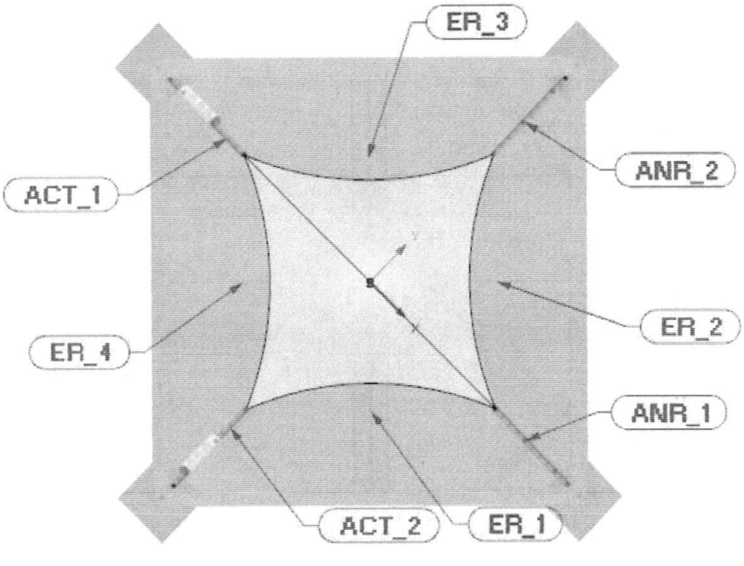

Figure 1. Specification of the numerical model elements – floor plan view.

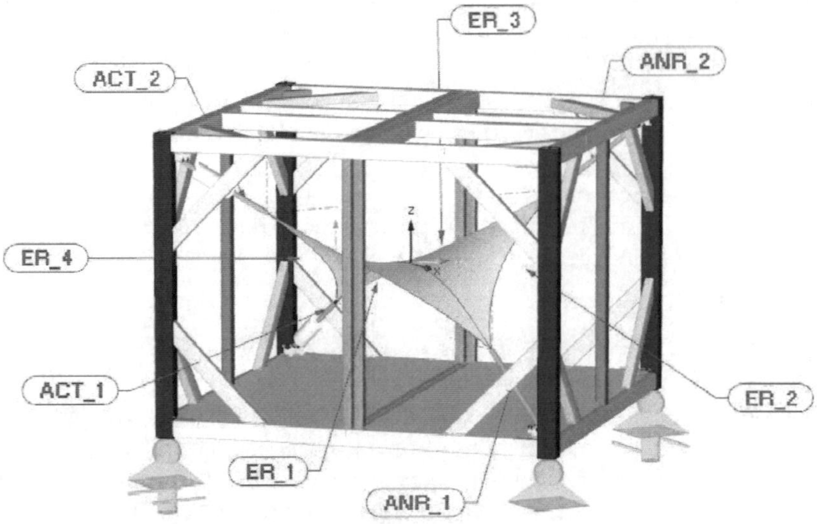

Figure 2. Specification of the numerical model elements – axonometric view – perspective.

3. Numerical analysis

The numerical analysis is based on finite element method (FEM), while the form-finding process is based on the method of update reference strategy (URS) of the finite element mesh [5], which is the mathematical homotopic method to solve inverse problem.

FEM analysis was set up. It used a large deformation analysis with Newton-Raphson method to solve nonlinear equations. In general, the target length of surface finite elements triangular mesh was set up at value 0.050 m. Number of divisions for cables/nonlinearity elements was set up at value 20 per member. Number of divisions for struts elements was set up at value 10 per member. The settings of global calculation parameters was set up: number of iterations – 100, number of load increments – 20.

URS analysis was set up. Tolerance for form-finding convergence criteria was set up at value 0.50. Speed of convergence was set up at value 0.40 to slower the convergence and made higher calculation stability. Maximum number of iteration for form-finding process was 200.

In this paper, load cases of structural loading were not done. The changing of absolute length of the actuator was used to simulate the movement of the actuator. Several combinations of changing length on the top and the bottom actuator were generated separately, to record the behaviour of the structural model.

4. Results of the numerical analysis

To monitor the data, there were made two sections in the numerical model, sections A-A and B-B. The sections were arranged orthogonally to each other in floor plan of membrane structure. Sections started and ended in actuator and anchor rod.

Section A-A monitored the bottom actuator, the bottom anchor rod, also the data of membrane surface in which the section gone through. It means the section A-A was focused on the data of the membrane surface from the zero height to the height of membrane surface saddle.

Section B-B monitored the top actuator, the top anchor rod, also the data of membrane surface in which the section gone through. Section B-B was focused on the data of the membrane surface from the height of membrane surface saddle to the top points of surface.

This paper is focused on points of the numerical model where actuators and anchor rods were placed, also it monitors the point in the middle of the membrane surface. The graphs shown in the figures 3, 5 are researching the dependence of global, nodal deformation in Z direction on the variation of the actuators, separately for each other for every section. The graphs shown in the figures 4, 6 are researching the dependence of the normal (axial) forces of the actuators/anchor rods on the variation of the actuators, separately for each other for every section.

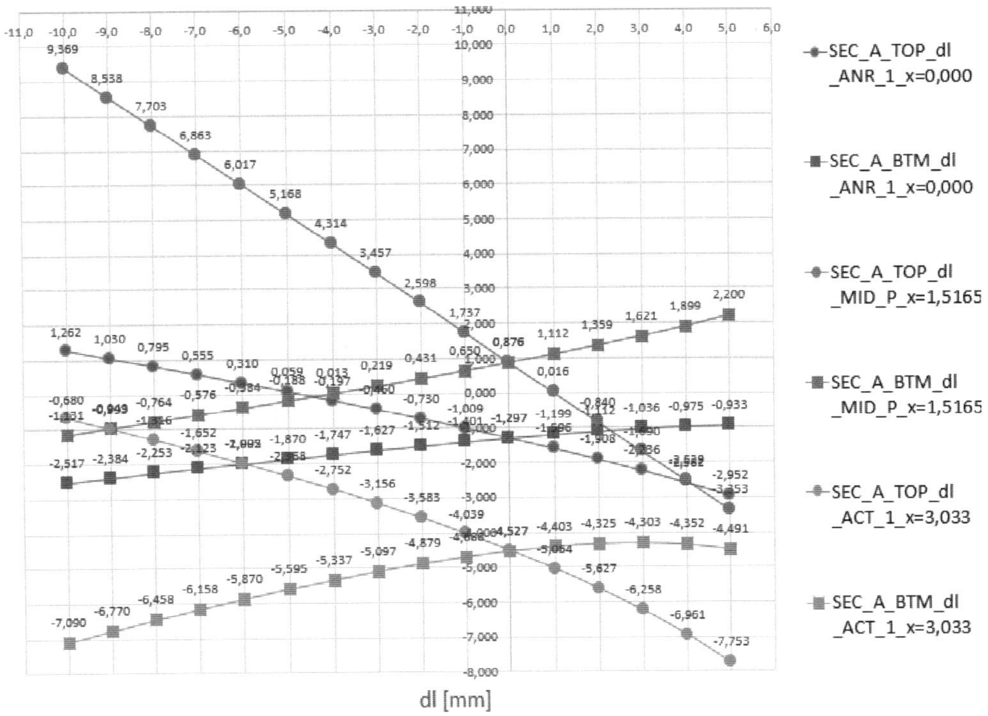

Figure 3. The dependence of the global, nodal deformation in the Z direction on the variation of the actuators – SECTION A-A.

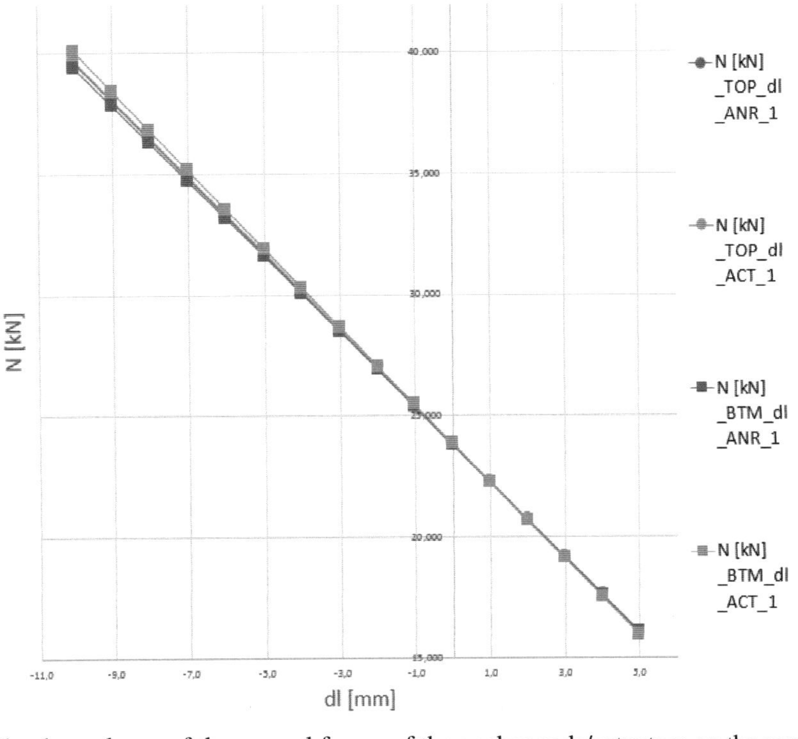

Figure 4. The dependence of the normal forces of the anchor rods/actuators on the variation of the actuators – SECTION A-A.

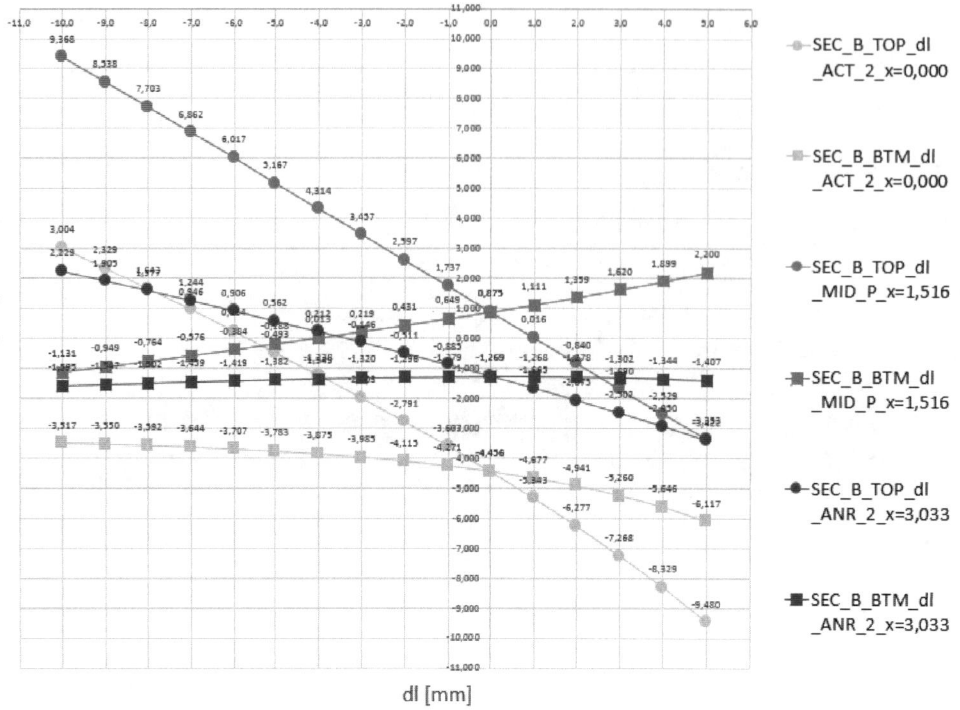

Figure 5. The dependence of the global, nodal deformation in the Z direction on the variation of the actuators – SECTION B-B.

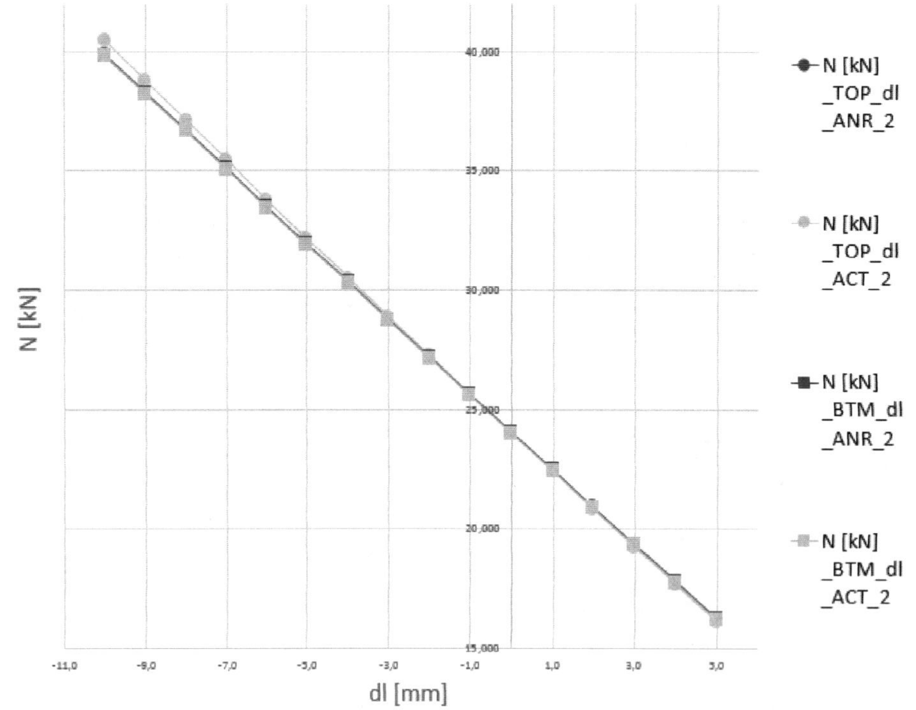

Figure 6. The dependence of the normal forces of the anchor rods/actuators on the variation of the actuators – SECTION B-B.

5. Conclusion

In this paper the effect of asymmetric use of actuators on the membrane structure without structural loads was analysed.

From the shown results, which can be seen on graphs 4, 6, the normal forces in actuators and anchor rods are almost equal in every single step of the variation of actuators, and it does not matter which one of the actuators was varying. This proves the numerical model is in the equilibrium stage in every single position of actuators but as it is shown in graphs 3, 5, there can be a very big difference between dependences on global, nodal deformation on the variation of the actuators. While shortening of top actuator generates a large upward movement of monitored points of structure, shortening of lower member has more of a stabilizing character for construction at the same movement level of actuators.

The results of the article describe the behaviour of numerical model. These results will be compared with a results gained from the experimental device in the laboratory.

6. References

[1] Kmeť S 2012 Modern cable and membrane constructions *Building materials* **8** 48–52
[2] Foster and Mollaert Marijke 2003 *European Design Guide For Tensile Surface Structures* (Germany: Tensinet) p 354
[3] Wakefield D S 1999 Engineering analysis of tension structures: theory and practice *Engineering Structures* **21** 680-90
[4] Horst Berger 1999 Form and function of tensile structures for permanent buildings *Engineering Structures* **21** 669–79
[5] Wüchner R and Bletzinger K U 2005 Stress-adapted numerical form finding of pre-stressed surfaces by the updated reference strategy *Int. J. Numer. Meth. Eng.* **64** 143–66
[6] Henrysson, E. : Conceptual design and analysis of membrane structures. Sweden: p.104. 2012. ISSN 1652-8557
[7] Del Grosso A E and Basso P 2010 Adaptive building skin structures *Smart Materials and Structures* **19** 12
[8] Kapolka L 2019 Analysis of an experimental tension structure *IOP Conf. Ser.: Mater. Sci. Eng.* **566** 012014
[9] Kapolka L, Štulerová L and Kmeť S 2019 Adaptive membrane structure – experimental device *Assembly of steel constructions workers* **19** 79–86
[10] Kapolka L 2020 Comparison of results of an adaptive membrane structure with a numerical model *IOP Conf. Ser.: Mater. Sci. Eng.* **867** 012015

Acknowledgments

This work was supported by The Slovak Research and Development Agency under the contract No.APVV-15-0777 and also by The Slovak Scientific Grant Agency under the contract No.VEGA 1/0129/20.

Crack behaviour at the interface of a surface layer applied on a steel substrate by laser cladding

L Malíková[1,2], P Doubek[2,3], P Miarka[1,2] and S Seitl[1,2]

[1] Institute of Physics of Materials Czech Academy of Sciences, v. v. i., Žižkova 513/22, 616 00 Brno, Czech Republic
[2] Institute of Structural Mechanics, Faculty of Civil Engineering, Brno University of Technology, Veveří 331/95, 602 00 Brno, Czech Republic
[3] OMNI-X CZ s.r.o., Šámalova 60a, 615 00 Brno, Czech Republic

E-mail: malikova.l@fce.vutbr.cz

Abstract. An influence of the bi-material interface between a steel substrate and a thin protective layer applied through laser cladding was investigated. A range of elastic properties and thicknesses of the layer were considered to cover the behaviour of a short crack in the selected materials such as bronze, nickel or cobalt alloys. The special case of the crack terminating directly at the interface was investigated, which is connected to the necessity of application of generalized approaches of linear elastic fracture mechanics. The results contribute to better understanding of fracture response of selected materials and to a more reliable decision on choosing a proper material of the protective layer.

1. Introduction

In technical practice, structural elements which are formed by combining layers of different metallic materials are used, see e.g. [1, 2]. These combinations of materials are created based on the required function either during the design of the structural component or during its renovation and repair [3].

There is a relatively large number of methods of applying surface layers. One of the modern, widely applicable methods is laser cladding technology, a review on this technology can be found in [4]. The principle of laser cladding is that the metal powder or wire is fed to a laser beam, where this material is melted together with the base material of the part and a deposition layer is formed on the surface of the part. A metallurgical bond is formed between the deposition layer and the base material, guaranteeing excellent adhesion between the cladding layer and the part [5]. With the right choice of material combinations, laser cladding technology can replace some older technologies effectively as they are often problematic from an ecological point of view.

In this work, fracture behaviour of a crack terminating at the bi-material interface between the base steel material and a laser-cladded layer of selected materials is investigated.

2. Generalized fracture mechanics

The goal of the paper was to estimate the value of the critical stress that is needed for penetration of the crack terminating directly at the interface to the material of substrate, see [6–9]. It is well known that a crack with its tip at a bi-material interface represents a general stress concentrator, which means that the stress singularity is greater or lower than 0.5 in dependence on the mutual elastic mismatch of both materials. The stress tensor components near the crack tip can be expressed via the following equation:

$$\sigma_{ij} = \frac{H_I}{\sqrt{2\pi}} r^{-p} f_{ij}(p, \alpha, \beta) \tag{1}$$

The equation (1) shows the dependence between the stress s_{ij}, generalized stress intensity factor H_I, radial distance measured from the crack tip r, stress singularity exponent p and known function f_{ij} that is dependent besides other things on the bi-material parameters a and b, see [10]. These constants depend on the elastic properties (Young's modulus E and Poisson's ratio n) of the materials on both sides of the interface as well as on the stress/strain conditions. For the case of the plane strain state, their values can be calculated as (note that index 1 represents the surface layer and index 2 denotes the material of the substrate):

$$\alpha = \frac{\frac{E_1}{E_2}\frac{1+\nu_1}{1+\nu_2} - 1}{4(1-\nu_1)} \quad , \quad \beta = \frac{E_1}{E_2} \cdot \frac{1-\nu_2^2}{1-\nu_1^2} \tag{2}$$

The stress singularity exponent needs to be calculated as $p = 1 - l$, where l is the eigenvalue and for a crack perpendicular to the interface can be determined from the characteristic equation, see [10] for more details:

$$\lambda^2(-4\alpha^2 + 4\alpha\beta) + 2\alpha^2 - 2\alpha\beta + 2\alpha - \beta + 1 + (-2\alpha^2 + 2\alpha\beta - 2\alpha + 2\beta)\cos(\lambda\pi) = 0 \tag{3}$$

The generalized stress intensity factor (GSIF) can be calculated via direct method when the numerical solution of the problem is found. Then, its value can be determined for example from the development of the opening stress ahead of the crack tip. Using equation (1) enables the extrapolation of the dependence $H_I = H_I(r)$ to the location $r = 0$ where the final value of the GSIF can be found.

Finally, when the critical load leading to crack propagation through the interface shall be calculated, a critical value of the GSIF needs to be estimated by means of selected fracture criteria, see the following subsections.

2.1. *Mean tangential stress value criterion*

According to this criterion, the stability condition for a crack is related to the average stress calculated across a distance d ahead of the crack tip, more details can be found for instance in [11]. Mathematically written:

$$H_{IC} = K_{IC} \frac{2d^{\lambda-1/2}}{2-\lambda+g_R} \tag{4}$$

The meaning of the symbols is as follows: H_{IC} represents the critical value of the GSIF, K_{IC} represents the fracture toughness of the material, d is the distance ahead of the crack tip, where the criterion is applied, l is the eigenvalue and g_R is a known function of the parameters a and b as they are defined in equation (2), see e.g. [12].

2.2. *Generalized strain energy density factor criterion*

A similar equation between the critical value of the GSIF and fracture toughness can be derived also within the idea of the generalized strain energy density criterion, where the minimum value of the strain energy density factor is considered:

$$H_{IC} = K_{IC} \cdot d^{p-1/2} \left(\frac{1-2\nu}{(1-p)^2[4(1-2\nu)+(g_R-p)^2]} \right)^{1/2} \tag{5}$$

The meaning of the symbols is as described in the previous sections. It is worthy to note that the elastic and fracture mechanical properties in equations (4) and (5) represent the properties of the material behind the interface.

3. Geometry and material properties considered in the numerical model

A geometry of the numerical model was suggested with regard to the real samples, when a thin surface layer of the thickness between 1 and 3 mm is cladded on a cylindrical steel substrate with the diameter about 82 mm, see figure 1.

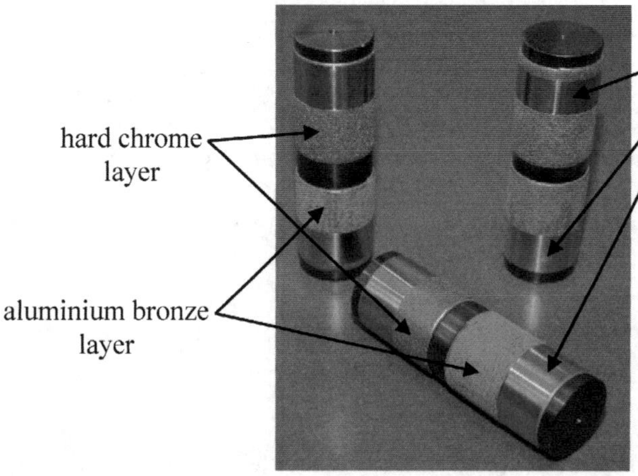

Figure 1. Real samples with laser-cladded surface layers made of hard chrome and aluminium bronze applied on steel cylinders.

Particularly, a simplified bi-material cracked bar under pure tension was modelled according to the schema in figure 2.

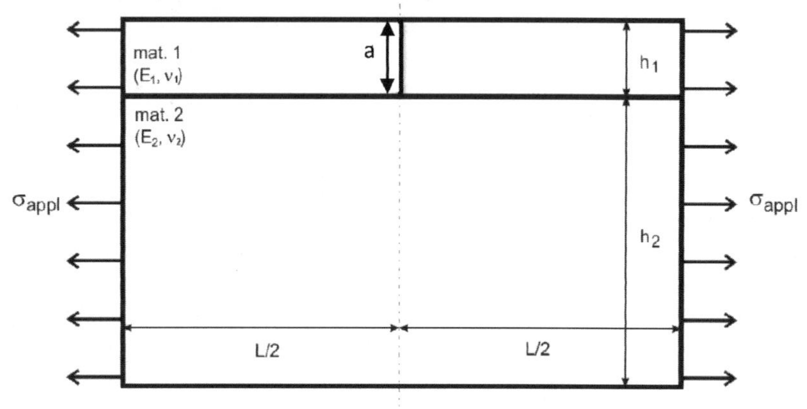

Figure 2. Schema of the bi-material cracked bar under pure tension.

The material properties considered within the parametrical study as well as the dimensions of the numerical model can be found in table 1.

Table 1. Material properties, dimensions and loading applied in the numerical model of the cracked bar under pure tension, see figure 2.

Quantity	Value
h_1	1, 2 and 3 mm
h_2	40 mm
a	h_1
L	$6(h_1+h_2)$
E_1	100 ÷ 300 GPa
E_2	200 GPa
$n_1 = n_2$	0.3
s_{appl}	800 MPa

The values of the Young's modulus of the material of the cladded layer were considered to be 100, 150, 200, 250 and 300 GPa. In the model, quadrilateral 8-node elements 183 were applied and the mesh near the crack tip was refined to emphasize the crack-tip singularity (although its value is different from 0.5 and it holds that $p = 0.43 \div 0.55$ for the studied cases). For calculation of the GSIF via the direct method, always the nodes at the distance of 0.2 to 1 mm from the crack tip were utilized. Values of the stress singularity exponent and GSIF are introduced in table 2.

Table 2. Values of the various Young's modulus of the surface layer and corresponding stress singularity exponents and GSIF values for s_{appl} = 800 MPa.

E_1 [GPa]	100	150	200	250	300
E_1/E_2 [-]	0.50	0.75	1.00	1.25	1.50
p [-]	0.43389	0.47133	0.50000	0.52324	0.54279
H_I [MPa·m^p] for h_1 = 1 mm	62.494	55.157	50.340	46.811	44.045
H_I [MPa·m^p] for h_1 = 2 mm	84.963	77.456	71.914	67.474	63.754
H_I [MPa·m^p] for h_1 = 3 mm	104.570	96.266	89.603	84.006	79.186

The critical distances that appear in the generalized fracture criteria were considered to be $d = r = 1$ mm in agreement with recommendations published for instance in [13–15].

4. Results and discussion

In figures 3 and 4, the dependences of the critical stress on the elastic mismatch (Young's moduli ratio) can be found for the DK_{Ith} and DK_{IC} value of 9 and 60 MPa·m$^{1/2}$ considered as the material parameters of the steel substrate, respectively. The values of the critical tensile stress are calculated from a simple relation:

$$\sigma_c = \sigma_{appl} \cdot \frac{H_{IC}}{H_I} \tag{6}$$

The plots are printed for the three values of the thickness of the surface layer and both generalized fracture criteria are applied.

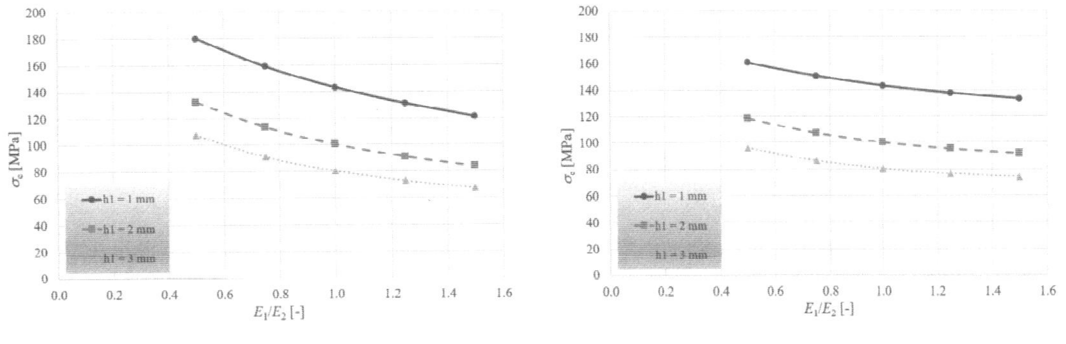

(a) (b)

Figure 3. Dependence of the values of the critical stress (when DK_{Ith} = 9 MPa·m$^{1/2}$) on the elastic mismatch for various thicknesses of the surface layer calculated via (a) mean tangential stress value criterion and (b) generalized strain energy density factor criterion.

IOP Conf. Series: Materials Science and Engineering 1209 (2021) 012049 doi:10.1088/1757-899X/1209/1/012049

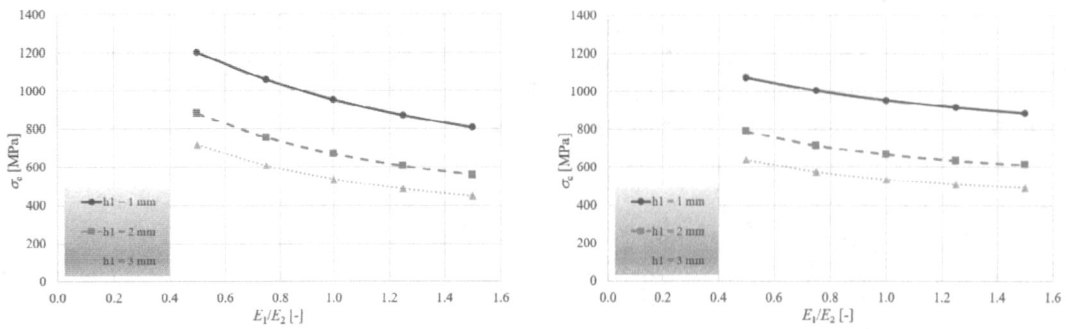

(a) (b)

Figure 4. Dependence of the values of the critical stress (when DK_{IC} = 60 MPa·m$^{1/2}$) on the elastic mismatch for various thicknesses of the surface layer calculated via (a) mean tangential stress value criterion and (b) generalized strain energy density factor criterion.

The results plotted in figures 3 and 4 show how the critical stress necessary for crack propagation through the cladded surface layer/steel substrate interface depends on the elastic mismatch between both layer as well as on the thickness of the surface layer. The following conclusions can be stated:

- The critical stress necessary for beginning of the stable fatigue long crack growth (corresponding to DK_{Ith} = 9 MPa·m$^{1/2}$) is between the values of 60 and 200 MPa based on the ratio of the elastic moduli, surface layer thickness and fracture criterion applied.
- The values of s_c necessary for unstable crack growth (corresponding to DK_{IC} = 60 MPa·m$^{1/2}$) are much higher, between 400 and 1200 MPa.
- The mean tangential stress value criterion gives slightly higher values of the s_c values and it is therefore less conservative than the generalized strain energy density fracture criterion.
- For thinner surface layers, it is necessary to apply higher tensile load for stable/unstable crack propagation – this is of course connected to the total crack length ($a = h_1$).
- The more compliant material of the surface layer, the higher critical stress is necessary for the crack to be able to propagate through the interface to the steel substrate.

Considering the dependences discussed above, rather compliant materials of the cladded layer shall be recommended, such as hard chrome (with Young's modulus of ca 104 GPa), aluminium bronze (E ~ 113 GPa) or copper beryllium alloys (elastic modulus about 130 GPa). On the other hand, choice of different cobalt alloys (that are rather tough with E over 200 GPa) as the surface layer could decrease the fracture resistance of the structure assuming the presence of a crack with its tip at the bi-material interface.

5. Conclusions

A numerical study has been performed in order to assess the fracture behaviour of a crack terminating at the bi-material interface between a surface layer cladded on a steel substrate. Various elastic properties as well as various thicknesses of the surface layer were considered and finite element calculation in combination with generalized linear elastic fracture mechanics was performed. The obtained results show that laser cladding of materials with lower elastic modulus (such as hard chrome, aluminium bronze and/or copper beryllium alloys) can improve fracture response of the bi-material structure when a short crack with its tip at the interface is presented. Both generalized fracture criteria applied brings higher values of the critical stress that is necessary for stable/unstable fatigue crack growth to the steel substrate when a more compliant material is cladded on the surface of the specimen.

6. References

[1] Bhat S, Adarsha H, Ravinarayan V and Koushik V P 2019 *Proc. Struct. Integr.* **17** 21
[2] Khodadad Motarjemi A, Koçak M and Ventzke V 2002 *Int. J. Pressure Vessels Piping* **79** 3 181
[3] Li M, Han B, Song L and He Q 2020 *Applied Surface Science* **503** 144226
[4] Zhu L, Xue P, Lan Q, Meng G, Ren Y, Yang Z, Xu P and Liu Z 2021 *Optics & Laser Technology* **138** 106915
[5] Laser cladding – Laser Therm s.r.o. *Homepage – Laser Therm s.r.o.* [28.6.2021] Available online https://www.lasertherm.cz/eng/technologies/laser-technologies/laser-cladding
[6] Klusak J, Krepl O and Profant T 2019 *Theor. Appl. Fract. Mech.* **104** 102341
[7] Krepl O and Klusák J 2017 *Theor. Appl. Fract. Mech.* **90** 85
[8] Krepl O and Klusák J 2018 *Proc. Struct. Integr.* **13** 1279
[9] Nahlik L, Stegnerova K and Hutar P 2018 *Theor. Appl. Fract. Mech.* **93** 247
[10] Lin K Y and Mar J W 1976 *Int. J. Fract.* **12** 4 521
[11] Nahlik L, Knesl Z and Klusak J 2008 *Engng. Mech.* **15** 2 99
[12] Knesl Z, Knapek A and Bednar K 1998 Evaluation of the critical stress in bonded materials with a crack perpendicular to the interface *Surface modification technologies XI* ed T S Sudarshan, M Jeandin and K H Khor (London: The Institute of Materials)
[13] Seweryn A and Lukaszewicz A 2002 *Engng. Fract. Mech.* **69** 13 1487
[14] Sih G C and Ho J W 1991 *Theor. Appl. Fract. Mech.* **16** 3 179
[15] Susmel L and Taylor D 2008 *Engng. Fract. Mech.* **75** 3 534

Acknowledgments

Financial support from the Faculty of Civil Engineering, Brno University of Technology (project No. FAST-S-21-7338) is gratefully acknowledged.

Evaluation of loading methods for determination of material and elastic constants of cement fiber board

O Nespesny[1], J Pencik[1], J Vystrcil[1] and D Beckovsky[1]

[1] Faculty of Civil Engineering, Institute of Building Structures, Brno University of Technology, Veveří 331/95, 602 00 Brno, Czech Republic

E-mail: ondrej.nespesny@vut.cz

Abstract. Cement fiber boards (CFB) are special group that are used in a wide range of structural civil engineering. For the correct design of a cement fibre board structure, it is important to define their material and elastic constants, which are usually determined by destructive tests. The paper deals with a definition of a suitable method of loading for the determination of basic materials and elastic constants of cement fiber boards reinforced with organic fibers loaded in the mid-plane. The publication compares and evaluates load tests by three-point and four-point bend.

1. Introduction

Composite boards based on cement and organic fibers (referred to as cement fiber boards) are today an environmental, universal and durable building material. It serves as a substitute for natural wood and wood products, such as plywood or oriented strand boards (OSB). The properties of cement fiber boards as a building material use a variety of uses in a wide range of building structures. Those that include cladding of ventilated facades (Figure 1) for the reconstruction of new buildings, internal and external ceilings, cladding of internal and external structures, cable bridges or fire protection structures.

Figure 1. Ventilated facade, application of cement fiber boards 0

Currently, the Hatschek method is one of the most widely used processes for the production of cement fiber boards 0. This production process consists in laying individual thin layers (called monolayers) on top of each other until the required thickness is reached, this layer is then pressed. In this way of production, the reinforcing fibers are divided in a board-plane, they are oriented in the direction of production.

In the case of normal use, cement fiber boards are loaded and tested (Figure 2) primarily in the y-z (y-z) direction, and y-x (x-y). In the case of innovative use, for example in the form of stairs (Figure 3), the slabs are stressed primarily in the x-z (z-x) direction, the methodology for experimental testing in this direction is not defined.

Figure 2. Scheme of loading directions with regard to the direction of production of cement fiber boards produced by the Hatschek method

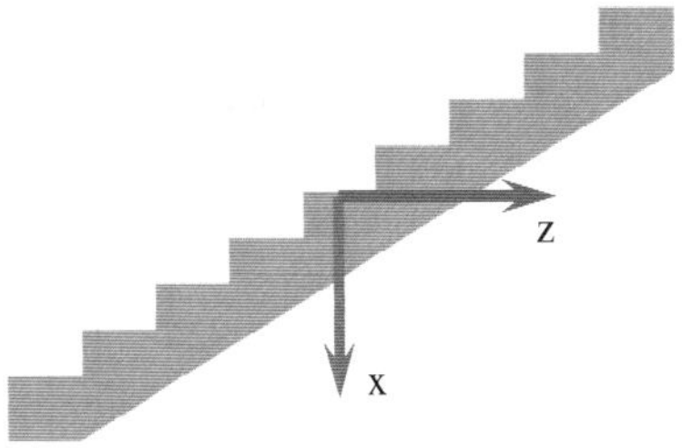

Figure 3. Stairs, the main supporting element of the stair staircase, marking of the primary directions of loading

2. Calculation methods

Determination of material and elastic constants for flat cement fiber boards (direction y-z and y-x) is governed by *ČSN EN 12467 + A2* 0. Here it is recommended to choose test specimens with dimensions of 250 × 250 mm, regardless of the thickness of the board. It is recommended to perform the load by three-point bending (index $_3$) with a constant path increment. Subsequently, the values of modulus of elasticity (*MOE*) and tensile strength in bending (*MOR*) are determined according to equations (1) and (2).

$$MOE_3 = \frac{(F_2 - F_1) \cdot l_s^3}{4 \cdot b \cdot e^3 (f_2 - f_1)} \tag{1}$$

$$MOR_3 = \frac{3 \cdot F \cdot l_s}{2 \cdot b \cdot e^2} \tag{2}$$

where *MOE* is modulus of elasticity in [N/mm^2]; F_1 and F_2 are forces in two points located in the linear part of the loading displacement diagram graph expressing dependence between vertical displacement and applied load in [N]; l_s is a distance between supports, in [mm]; b and e is the width and thickness of specimens, in [mm]; f_1 and f_2 are vertical displacements corresponding with the selected load, in [mm]; F is failure force, in [N].

Ranachowski and Schabowicz describes in 0 as a suitable method of loading (for direction y-z and y-x) by four-point bending (index 4), where the values of modulus of elasticity (*MOE*) and tensile strength in bending (*MOR*) are determined by relations (3) and (4).

$$MOE_4 = \frac{23 \cdot (F_2 - F_1) \cdot l_s^3}{108 \cdot (w_2 - w_1) \cdot b \cdot h^3} \tag{3}$$

$$MOR_4 = \frac{F_{max} \cdot l_s}{b \cdot h^2} \tag{4}$$

where F_{max} is failure force in [N]; h is height, respectively thickness, in [mm]; F_1 is force, in [N], corresponding to 10 % of failure force (F_{max}); F_2 is force, in [N], corresponding to 40 % of failure force (F_{max}); w_1 is vertical displacement of testing specimens corresponding to F_1, in [mm]; w_2 is vertical displacement of testing specimens corresponding to F_2, in [mm].

Equations (1) to (4) are based on European standards and are in principle applicable to flat plates, ie. for the basic directions y-z and y-x (Figure 2).

In the case of defining material and elastic constants in the z-x direction, the procedures from *ČSN EN 408* 0 can be applied analogously, similarly to *Pěnčík* in 0 for wooden beam specimens. The position of the loading forces and the size of the test specimens are described in Figure 5. The modulus of elasticity (MOE_4) in this case can be determined according to equation (5)

$$MOE_4 = \frac{l^3 \cdot (F_2 - F_1)}{bh^3 \cdot (w_2 - w_1)} \left[\left(\frac{3a}{4l} \right) - \left(\frac{a}{l} \right)^3 \right] \tag{5}$$

where a is the distance between the load and the nearest bend test support in [mm] and l the bending span in [mm].

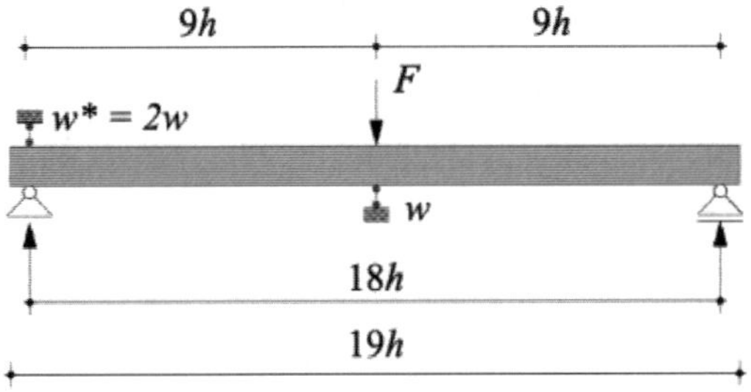

Figure 4. Description of tests under three-point bend loading.

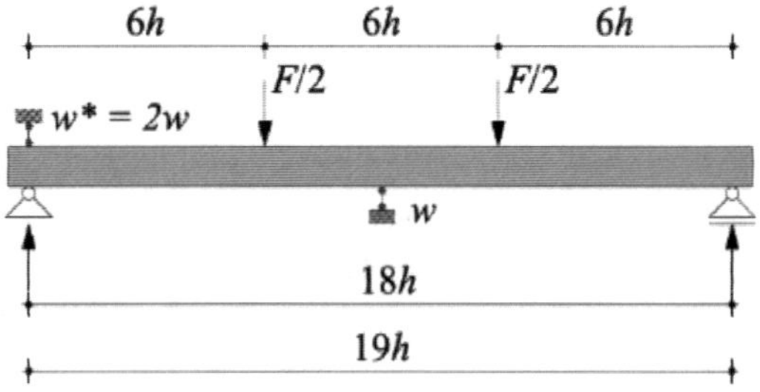

Figure 5. Description of tests under four-point bend loading.

To create a complex material model, ie. in the directions y-x, y-z, x-z (x-y, z-y, z-z) it is appropriate in all cases to choose the size of test specimens defined according to *ČSN EN 408* 0 (Figure 4) and (Figure 5) on the basis of the search. To determine the modulus of elasticity (MOE_3) and tensile strength in bending (MOR_3) under three-point bending loading, it is appropriate to apply equations (1) and (2) on the basis of research. To determine the modulus of elasticity (MOE_4) and tensile strength in bending (MOR_4) under loading by four-point bending, it is appropriate to apply equation (4) and (5) on the basis of research.

3. Comparison of three-point and four-point bend test

The essential difference between the three-point and four-point bend test in the course of internal forces. During the three-point bend test, one point is exposed to the maximum shear force (V) and at the same time to the maximum bending moment (M), the course of internal forces is shown in (figure 6). In the case of four-point bend load, the bending moment is evenly distributed between the forces and thus allows a more accurate determination of the modulus of elasticity during the bending test (Figure 7).

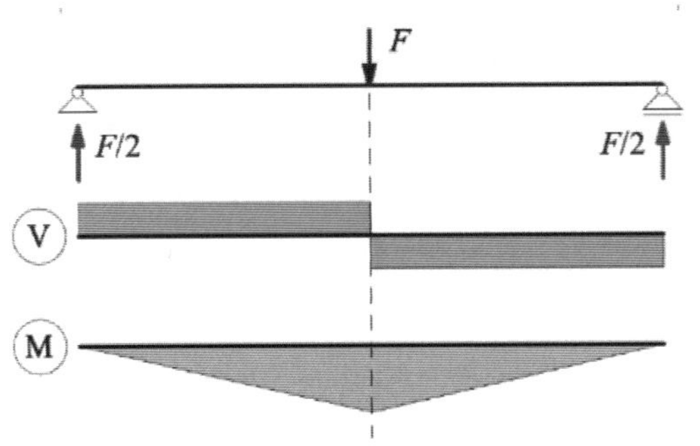

Figure 6. Diagram of the course of shear forces (V) and bending moment (M) under loading by three-point bending.

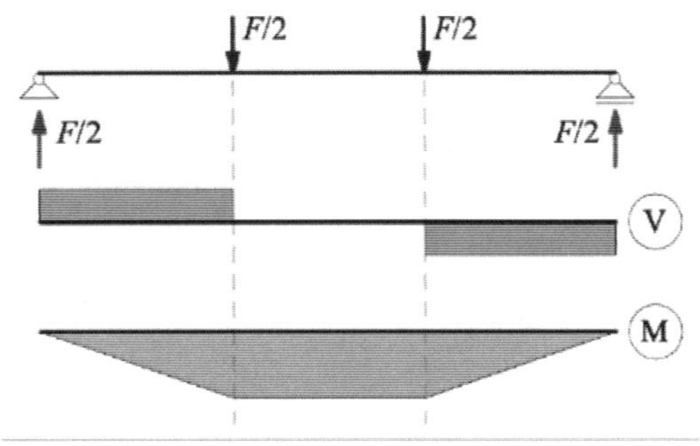

Figure 7. Diagram of the course of shear forces (V) and bending moment (M) under loading by four-point bending.

The four-point bend test is suitable for brittle materials with low shear stress. The magnitude of the flexural tensile strength depends on the toughness and severity of the defects of the test specimens. In the case of cement fiber boards produced by the Hatschek method, delamination and cracking very often occur due to the natural maturation of the material. Exposure of one point to maximum stress (Figure 8) will reduce the measured flexural strength because there is an increased likelihood that cracks will reach a critical length at the applied load. Four-point bend test (Figure 9) is suitable for determining the strength of joints, the location of the joint area is repeatable.

Figure 8. The real test arrangement in three-point bend load.

Figure 9. The real test arrangement in four-point bend load.

4. References

[1] CEMVIN s.r.o., product catalog, available from: https://www.cemvin.cz/upload/katalog.pdf

[2] Ardanuy M, Claramunt J and Toledo Filho R D 2015 Cellulosic fiber reinforced cement-based composites: A review of recent research *Construction and Building Materials* **79** pp 115-128

[3] Delvasto S, Toro E F, Perdomo F and de Gutiérrez R M 2010 An appropriate vacuum technology for manufacture of corrugated fique fiber reinforced cementitious sheets *Construction and Building Materials* **24** pp 187-192

[4] Anon. 2020 *ČSN EN 12467+A2: Fibre-cement flat sheets - Product specification and test methods* (Office for Technical Standardization, Metrology and State)

[5] Ranachowski Z and Schabowicz K 2018 *The Fabrication, Testing and Application of Fibre Cement Boards* (Lady Stephenson Library, Newcastle upon Tyne, NE6 2PA, UK: Cambridge Scholars Publishing)

[6] Anon. 2004 *ČSN EN 408: Timber structures - Structural timber and glued laminated timber - Determination of some physical and mechanical properties* (Office for Technical Standardization, Metrology and State)

[7] Pěnčík J 2015 Tests of Wooden Specimens from Scots Pine (Pinus sylvestris) with the Help of Anisotropic Plasticity Material Model *Drvna industrija* **66** pp 27-33

Acknowledgments

The paper was produced with support for specific university research at Brno University of Technology, FAST-J-21-7449 (2021) and the Technological Agency of the Czech Republic within programme EPSILON, TH04020263 (2019-2021).

IOP Conf. Series: Materials Science and Engineering 1209 (2021) 012051 doi:10.1088/1757-899X/1209/1/012051

Computational modelling of mechanical response of glulam beams due to moisture inducted delamination

J Pełczyński[1] and P A Król[1]

[1] Faculty of Civil Engineering, Warsaw University of Technology, Al. Armii Ludowej 16, 00-637 Warsaw, Poland

E-mail: j.pelczynski@il.pw.edu.pl

Abstract. Glued laminated timber beams are nowadays used as load-bearing beams of large-span structures that operate in various humidity conditions. Typical application areas are aqua parks with high humidity as well as market halls with low humidity. It is related to the possibility of the occurrence of cracks typical for the drying of wood, even with such controlled conditions of production technology as glued laminated timber. Cracks visible on the used girders raise doubts as to the safe operation of the structures. The subject of this paper is the computational simulation and the evaluation of the influence of beam delamination on the mechanical response of the structure. The attention was established on a typical two-span beam of constant height with a slight slope to the horizontal. The numerical analysis was carried out for three variants of the location of potential delamination of different scope. The beams were modeled as a problem of the linear theory of elasticity in a plane stress state with orthotropic material properties. The calculations were made in the Abaqus software environment. The results obtained in the paper allow to determine the areas in which the presence of delamination or cracks should be considered dangerous from the point of view of the safety of operation. Computational analysis is helpful in assessing the safety of structures where cracks appear. Theoretical considerations are supplemented by an example from engineering practice.

1. Introduction

Cracks in wood, taking into account glued laminated timber, are one of the typical defects of this construction material [1]. Cracks may be caused by drying of the wood, but also by exceeding the relatively small allowable stresses in the direction transverse to the wood grain due to mechanical effects [2,3]. The latter may be the result of a design error or faulty operation of the facility. In any case, the appearance of cracks raises doubts as to the failure-free use of the object as it may have a significant impact on the stiffness of structural elements [4,5]. Design standards allow a certain range of cracks. But some questions arise. What will happen to the structure when deep or double-sided cracks appear, covering the entire width of the cross-section? Does the girder lose its initial load-bearing capacity and if so, to what extent? Probably the structure requires repair related to the delamination, but can it be safely operated until repairs? The key question is whether the presence of delamination is acceptable from the point of view of the safety of the building object.

The drying process of wood is difficult to be precisely described mathematically [6]. In the literature, attempts are made to computationally model the influence of moisture changes on the wood cracking process (e.g. [7]), along with experimental validation (e.g. [8]), also in the field of glued laminated timber [1,9]. Such modeling is subject to a high level of uncertainty due to the heterogeneity of the wood, twist of the fibers, knots and other material defects. The few numerical attempts are of a rather general nature and do not lead to design conclusions.

The finite element method allows to relatively easily model the structure in which the delamination occurred, if one only determine the place where it appears deterministically. This remark applies to both single scratches and large-scale delamination in the indicated areas. For the purpose of this study, delamination is understood as a double-sided crack of a glued laminated beam. Modeling is carried out on the example of a real simply supported beam with a slight slope.

The aim of the present study is an attempt to answer the above questions, in particular the analysis and assessment of the impact of individual cracks on the mechanical properties of the structure, as well as the assessment of the safety of the beam operation with full delamination of a large extent occurring in one of the three selected areas of the beam. Modeling was performed in the Abaqus system environment.

2. Computational model

Theoretical analysis of complex tasks of wooden structures within the framework of the theory of elasticity is possible only with the use of the Finite Element Method [10]. As it has been shown in [11], 3D modeling is not always necessary. To which the connections analyzed in the study belong, 2D models are sufficient with the use of orthotropic material [12]. 2D models are also correct in modeling the reinforcements of glued laminated timber [13-14].

The available literature and the authors' own experience justify the selection of the 2D model as sufficient to analyze delamination of beams considered in the present paper. The structure described in the paper is modelled using a 2D linear theory of elasticity in plane stress state. Material properties are considered as an orthotropy with the parameters taken from the design standard (EN 14080:2013) for timber class GL28h: E_1=12600 MPa, E_2=300 MPa, ν_{12}=0.35, G_{12}=650 MPa. The calculations are performed using the Abaqus FEM software [15]. Two dimensional 8-node biquadratic solid plane stress elements (CPS8R) with reduced integration are used [15]. Delamination is modelled with the use of double, not connected nodes in selected areas of beams.

3. Numerical simulation of delamination

The numerical analysis is carried out on the example of a girder with a span of 25 m, height 1.5 m and width 0.22 m, constructed of 30 laminates of GL28h wood, slightly inclined from the horizontal (Figs. 1 and 2). The girder is supported as shown in Figure 1, with the end supports being fixed, and there is a dilatation over the internal wall that may close under load.

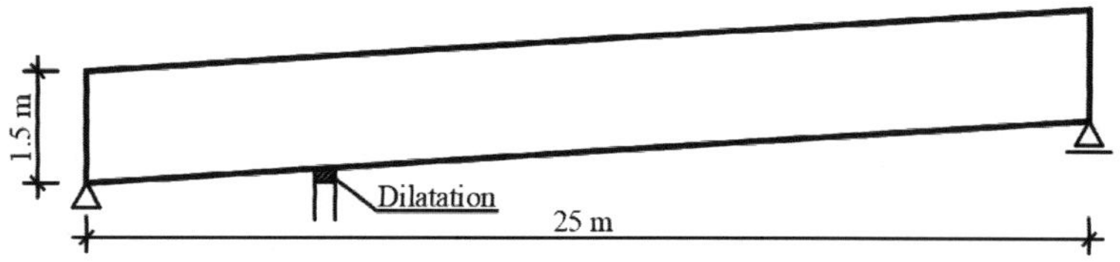

Figure 1. Glulam beam under consideration.

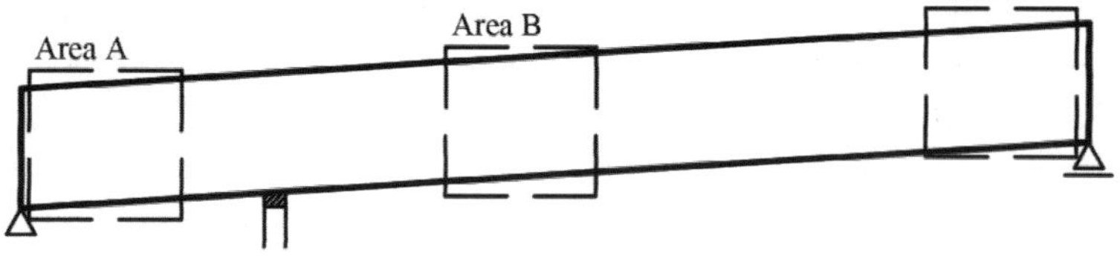

Figure 2. Areas of full delamination for computational modelling.

At first, single 4 m long delaminations located in different areas and places of the beam were considered. In this case, numerical analysis does not indicate significant neither qualitative nor quantitative differences in terms of deflections and normal stresses. This means that a single delamination does not compromise the safety of the beam. The second group of tasks is the delamination of every second layer of glulam in the areas A, B or C (Figure 2). Delamination in area A implies rest the beam on the intermediate support and does not endanger the safe use of the structure. Selected results for A and B delamination are shown in Figs. 3 and 4, in comparison of the non-cracked beam.

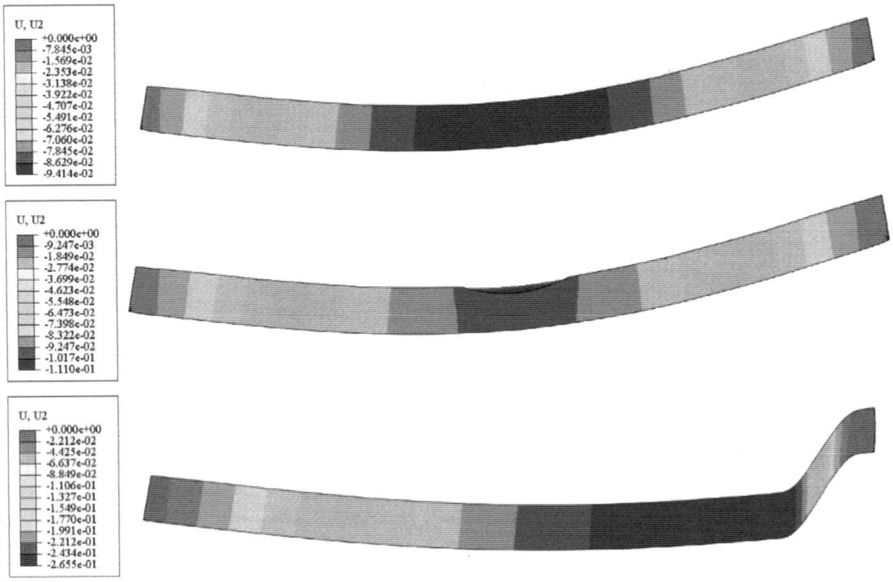

Figure 3. Deformation of delaminated beams [m]: no delamination, delamination in the Area B, delamination in the area C.

Deflections due to delamination in the central part of the beam are not significantly greater than in the original beam. Deflections with delamination in the C region increase significantly, which is a beam's failure condition.

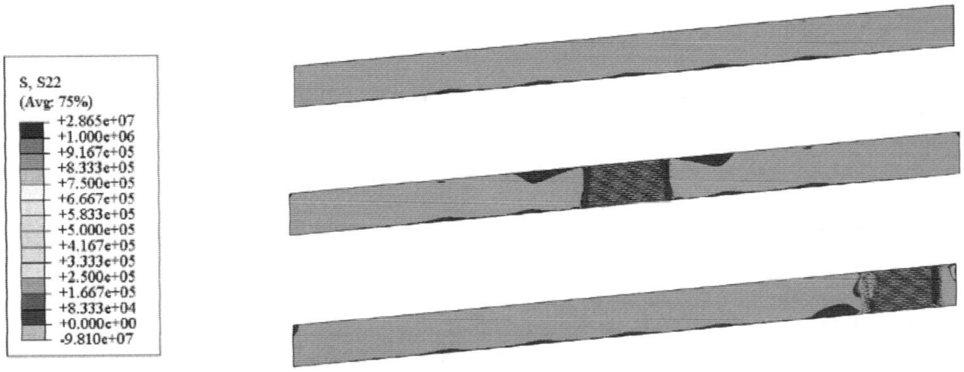

Figure 4. Perpendicular to grain stress in delaminated beams [Pa]: no delamination, delamination in the Area B, delamination in the area C.

The normal stresses along the grain and the shear stress do not change significantly after delamination. Figure 4 shows the areas of the delamination stress concentration, in the safe range.

IOP Conf. Series: Materials Science and Engineering 1209 (2021) 012051 doi:10.1088/1757-899X/1209/1/012051

4. Conclusions

The conclusions of the numerical analysis are as follows:

- A single delamination with a length of about 4 m, in any of the zones A, B, C, does not cause the loss of the beam load-bearing capacity and does not threaten the safety of the facility.
- The impact of delamination in every second layer of wood, with a length of approx. 4 m, on the load-bearing capacity of the structure depends on the zone in which such delamination occurs:
  - Full delamination in zone A will rest the girder on the intermediate wall.
  - Full delamination in zone B does not significantly reduce the load capacity of the beam,
  - Full delamination in zone C leads to a significant reduction of the beam load capacity.

Delamination zones A and B are not dangerous. Delamination in the area C means the beam disaster.

5. References

[1] Thelandersson S and Larsen HJ 2003 *Timber Engineering* (England: John Wiley & Sons, Chichester)

[2] Danielsson H 2010 Design and perpendicular to grain tensile stress in double-tapered glulam beams *Lund University Report* (Lund: Wallin and Dalholm Digital AB)

[3] Al Sabouni-Zawadzka A, Gilewski W and Pełczyński J 2017 Some considerations on perpendicular to grain stress in double-tapered glulam beams *XXVI R-S-P Seminar 20177, Theoretical Foundations of Civil Engineering MATEC (Web of Conferences* vol 117)

[4] Magniere N, Franke S and Franke B 2014 Investigation on elements presenting cracks in timber structures. World Conference on Timber Engineering *WCTE 2014 (Quebec City)*

[5] Mergny E, Mateo R, Esteban M, Descamps T and Latteur P 2016 Influence of cracks on the stiffness of timber structural elements *World Conference on Timber Engineering, WCTE 2016 (Vienna)*

[6] Ylien Y and Jumppanen P 1967 Theory of the shrinkage of wood *Wood Science and Technology* **1** 241–52

[7] Chassagne P, Bou-Said E, Jullien JF and Galimard P 2006 Three dimensional creep model for wood under variable humidity – numerical analysis at different material scales *Mechanics of Time-Dependent Materials* **9** 203–23

[8] Dubois F and Petit C 2005 Creep in wood under variable climate conditions: numerical modelling and experimental validation *Mech. of Time-Dependent Materials* **9** 173–202

[9] Yahyaei-Moayyed M and Teheri F 2011 Creep response of glued-laminated beam reinforced with pre-stressed sub-laminated composite *Construction of Building Materials* **25** 2495–506

[10] Zienkiewicz OC and Taylor RL 2000 *The finite element method. The basis* vol 1 (New Jersey: Butterworth-Heinemann)

[11] Al Sabouni-Zawadzka A, Gilewski W, Pełczyński J and Skowrońska M 2019 3D finite element stress analysis of reinforced double-tapered glulam beams, XXVIII R-P-S Seminar 9–13 September (Žilina) *IOP Conf. Ser.: Mater. Sci. Eng.* **661** 012068 (Bristol: IOP Publishing Ltd)

[12] Gilewski W and Pełczyński J 2017 The influence of orthotropy level for perpendicular to grain stresses in glulam double tapered beams *23rd Internat. Conference Engineering Mechanics 2017 (Svratka)*

[13] Al Sabouni-Zawadzka A, Gilewski W and Pełczyński J 2019 2D computational modelling of the influence of transverse reinforcement on perpendicular to grain stress in double tapered glulam beams. Internat *Conference on Computational Methods in Wood Mechanics – from Material Properties to Timber Structure. CompWood 2019 (Vaxjo)*

[14] Al Sabouni-Zawadzka A, Gilewski W and Pełczyński J 2020 Reduction of perpendicular to grain stress in double-tapered beams with transverse reinforcement *Archives of Civil Engineering* **66** 2 267–83

[15] ABAQUS Documentation 2017

Modal analysis of a membrane structure with actuators

L Stulerova[1], L Kapolka[1] and S Kmet[1]

[1] Technical university of Kosice, Faculty of Civil Engineering, Vysokoskolska 4, 042 00 Kosice

E-mail: lenka.stulerova@tuke.sk; lukas.kapolka@tuke.sk

Abstract. The aim of this paper is the analysis of the modal vibration of the membrane structure. Membranes are defined as structures of the lightweight architecture and they are currently very popular. They have a long history and development, in which they have reached a stage where we can complement them with action elements, also called actuators. These elements can change their length and thus affect a stress state of a membrane, which allows more efficient use. In addition to a static analysis, it is necessary to subject structures to a dynamic analysis, in this case we deal with the natural vibration. This modal analysis deals with the first 5 mode shapes and their dependence on the change in the length of the actuators. This initial calculation will be followed by a forced vibration analysis in the future.

1. Introduction

As alternative to the traditional materials and structures we also know structures of the lightweight architecture. They are very popular last decades although they are not new in construction practice. Typically, it represents cable nets or membranes. These constructions were widely used as temporary dwellings – tents or roofing. They are very light, movable, fast to set and easy to store. Along their positives they had lack of stability under climatic loads. Strong wind or heavy snow caused ruptures and tearing. This is reason why we prestress these structure now. Membranes and cable nets are then stable, safer, transfer load very well and they can serve also for longer time. So original structures are quite different from the form we know them today. They had to undergo to extensive research and evolution from physical models to the sophisticated mathematical methods.

Shin et al. [1] investigated geometrically nonlinear dynamic characteristics out of the plane of an axially moving membrane. Zhang and Shan [2] studied the equilibrium initial shape and free vibration of a prestressed membrane structure using the relaxation method. Pan and Gu [3] discussed the effects of changes in prestress, magnitude, elastic ratio, density, relative deflection and self-weight on square prestressed membranes with respect to structural nonlinearities and the equivalent frequency of the system's natural vibration. Zheng and Liu [4] studied the geometric nonlinear free vibration of orthotropic membranes using the direct integration method and the Lindstedt-Poincare method.

However, development continues unstoppably and scientists evolved these structures even more. To make them more intelligent, they were supplemented with actuator. These elements can change stress state of whole structure with shortening and elongation of themselves and create the adaptive structures. Miura and Furuya [5] described concept of adaptive structures for future space applications. Wang et al. [6] focus on neural networks and their algorithm.

One structure representing adaptive structures can be found in University of Stuttgart [7]. It is a wooden shell structure with over 100 m^2 and it is supported with 4 points where three of them can generate movements and adapt to changes within miliseconds.

IOP Conf. Series: Materials Science and Engineering 1209 (2021) 012052 doi:10.1088/1757-899X/1209/1/012052

2. Numerical model

This article follows up on the analysis of the modal vibration of the membrane structure [8]. It is a structure in the shape of a hyperbolic paraboloid with a floor plan of 2 × 2 m, with a total height of 1 m. Because it is an orthotropic material - technical fabric, the main characteristics differ in two directions perpendicular to each other. The modulus of elasticity in the main (warp) direction has the value E_x equals to 1 057 MPa (value of the Young's modulus between the upper vertices), in the weft direction is the modulus of elasticity E_y equals to 612 MPa (value of the Young's modulus between the lower vertices) [9].

Membrane structures need to be prestressed and stabilized to withstand external loads. In this case, the membrane element is stabilized along its 4 edges by stainless steel ropes with a diameter of 8 mm, with a Young's modulus E equal to 130 GPa. The prestress is considered to be 4 kN/m^2 in both directions of the membrane surface and the prestress of each rope is determined by an absolute sag of 0.2 m.

The membrane structure is anchored by 2 hinged anchor rods and 2 action elements. Because of these actuators it is the adaptive structure. In the numerical model round bars of the cross-sections D46 and D145 were used as the masses of these anchoring elements, which is important mostly in the case of the modal analysis. Because the action elements (actuators) have the ability to change their length, in this case it was considered to be a symmetrical shortening up to -10 mm and a symmetrical elongation of +5 mm, in steps of 1 mm (figure 1).

The analysis was performed by FE software Dlubal RFEM. The surface elements are meshed by triangular finite elements and cables are defined to be divided into 10 parts per element.

The last step before starting natural vibration analysis is the form-finding process. This process is very important because it determines initial equilibrium stress-strain state therefore the shape and geometry configuration. In the mentioned software Dlubal the Update Reference Strategy [10] as form-finding tool is programmed.

Figure 1. The example of shortening the actuator (shortening of -10 mm).

3. Modal analysis of adaptive structure

The modal analysis or natural vibration analysis was performed in the above mentioned Dlubal RFEM software, specifically in the RF-DYNAM Pro add-on module. The results of this analysis are natural shapes, natural frequencies, natural periods and angular frequencies (table 1, table 2). The mode shapes are modified by $\{u_j\}$ T [M] $\{u_j\}$ equals 1 and 10 mode shapes are considered.

For this calculation it is necessary to consider the following: 1 mass state representing self-weight and 15 natural vibration cases (elongation and shortening of both action elements). In these cases, it is absolutely necessary to take into account the change of stiffness from initial form-finding. The settings of calculation are considered as follows: the consistent type of the mass matrix and the Lanczos method for solving eigenvalue problem [11].

Table 1 and table 2 show the first 10 mode shapes in detail. Figure 3 and figure 4 show the dependence of the change in natural frequency on the mode shape for all 10 shortening shapes and 5 elongation shapes.

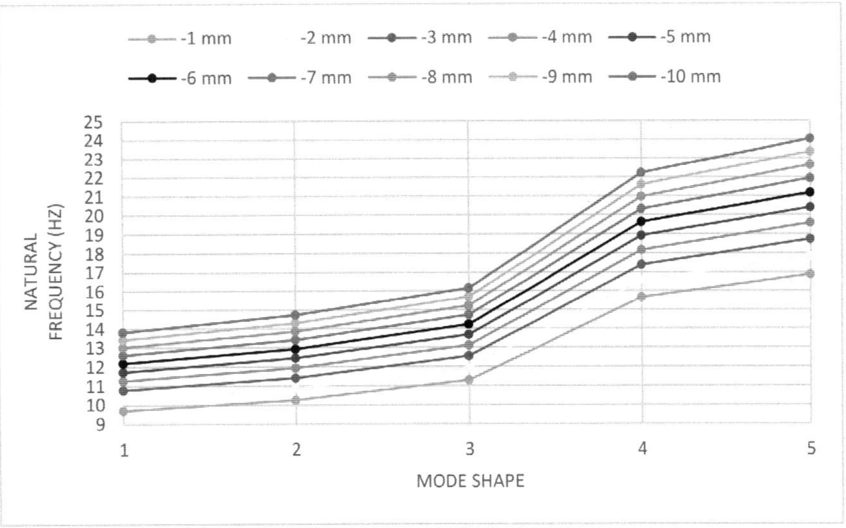

Figure 2. Dependence of natural frequency on natural shapes (shortening of actuators).

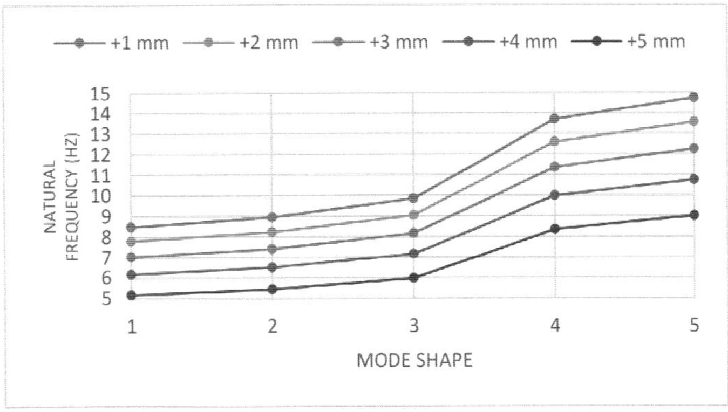

Figure 3. Dependence of natural frequency on natural shapes (elongation of actuators).

Table 1. Important characteristics of natural vibration (shortening of actuators).

	Mode Shape	Angular Frequency ω [rad/s]	Natural Freq. f [Hz]	Natural Period T [s]		Mode Shape	Angular Frequency ω [rad/s]	Natural Freq. f [Hz]	Natural Period T [s]
-1 mm	1	60.925	9.697	0.103	-6 mm	1	76.567	12.186	0.082
	2	64.419	10.253	0.098		2	81.314	12.942	0.077
	3	70.906	11.285	0.089		3	89.335	14.218	0.070
	4	98.397	15.660	0.064		4	123.324	19.628	0.051
	5	105.961	16.864	0.059		5	133.094	21.183	0.047
-2 mm	1	64.396	10.249	0.098	-7 mm	1	79.283	12.618	0.079
	2	68.147	10.846	0.092		2	84.273	13.413	0.075
	3	74.986	11.934	0.084		3	92.545	14.729	0.068
	4	103.939	16.542	0.060		4	127.640	20.315	0.049
	5	111.970	17.821	0.056		5	137.820	21.935	0.046

-3 mm	1	67.670	10.770	0.093	-8 mm	1	81.896	13.034	0.077
	2	71.674	11.407	0.088		2	87.128	13.867	0.072
	3	78.840	12.548	0.080		3	95.637	15.221	0.066
	4	109.162	17.374	0.058		4	131.789	20.975	0.048
	5	117.645	18.724	0.053		5	142.370	22.659	0.044
-4 mm	1	70.776	11.264	0.089	-9 mm	1	84.417	13.435	0.074
	2	75.031	11.942	0.084		2	89.888	14.306	0.070
	3	82.501	13.130	0.076		3	98.621	15.696	0.064
	4	114.111	18.161	0.055		4	135.788	21.611	0.046
	5	123.034	19.581	0.051		5	146.761	23.358	0.043
-5 mm	1	73.736	11.735	0.085	-10 mm	1	86.853	13.823	0.072
	2	78.238	12.452	0.080		2	92.561	14.732	0.068
	3	85.992	13.686	0.073		3	101.506	16.155	0.062
	4	118.823	18.911	0.053		4	139.649	22.226	0.045
	5	128.174	20.399	0.049		5	151.008	24.034	0.042

Table 2. Important characteristics of natural vibration (elongation of actuators).

	Mode Shape	Angular Frequency ω [rad/s]	Natural Freq. f [Hz]	Natural Period T [s]
+1 mm	1	53.238	8.473	0.118
	2	56.200	8.945	0.112
	3	61.886	9.849	0.102
	4	86.095	13.702	0.073
	5	92.671	14.749	0.068
+2 mm	1	48.902	7.783	0.128
	2	51.586	8.210	0.122
	3	56.808	9.041	0.111
	4	79.140	12.596	0.079
	5	85.184	13.557	0.074
+3 mm	1	44.109	7.020	0.142
	2	46.501	7.401	0.135
	3	51.204	8.149	0.123
	4	71.441	11.370	0.088
	5	76.915	12.241	0.082
+4 mm	1	38.688	6.157	0.162
	2	40.768	6.488	0.154
	3	44.876	7.142	0.140
	4	62.716	9.982	0.100
	5	67.567	10.754	0.093
+5 mm	1	32.323	5.144	0.194
	2	34.056	5.420	0.184
	3	37.457	5.962	0.168
	4	52.451	8.348	0.120
	5	56.588	9.006	0.111

4. Conclusion

From the above mentioned results it is possible to state following conclusions: as the shortening of the action elements increases from -1 mm to -10 mm, the natural frequency increases with the same mode shapes. In the case of increasing the elongation of the actuators from +1 mm to +5 mm, on the contrary, the values of the natural frequencies decrease with the same natural mode. This is due to the fact that as the membrane stress state increases during shortening, the structure is more stiff and therefore less prone to the effects of dynamic wind (frequency up to 5 Hz). In the case of elongation of the actuators, on the other hand, the prestress is lower and the membrane structure slacks. The structure becomes less stable to resist static and dynamic effects of the load. From the figures of the dependence of the change in frequency when the action elements are shortened, it can be seen that with the increase of the mode shapes, the natural frequency increases, same with the case with the elongation of the action elements. These changes represent the importance of modal analysis that needs to be considered in the dynamic analysis. In the next phase of the research, it is necessary to complete the current state of analysis with the forced vibration of the membrane structure without and with the use of action elements. The overall dynamic analysis is ultimately necessary compared with the results obtained on a real experimental device and thus verified numerical models with real models.

5. References

[1] Shin C, Chung J T and Kim W 2005 Dynamic characteristics of the out-of-plane vibration for an axially moving membrane *J. Sound Vib.* **286** 1019-31
[2] Zhang H and Shan J 2006 Initial form finding and free vibration properties study of membrane *Xi'an Int. Conf. Architecture and Technology, Proceedings-Architecture in Harmony* 316-20
[3] Pan J and Gu M 2007 Geometric nonlinear effect to square tensioned membrane's free vibration *J. of Tongji Univ. Nat. Sc.* **35** 1450-54
[4] Zheng Z L, Liu C J, He X T and Chen S L 2009 Free vibration analysis of rectangular orthotropic membranes in large defection *Math. Prob. Eng.* **2009** 1-9
[5] Miura K and Furuya H 1988 Adaptive structure concept for future space applications *AIAA J.* **26** 995-1002
[6] Wang J, Belatreche A, Maguire L and McGinnity T M 2014 An online supervised learning method for spiking neural networks with adaptive atructure*Elsevier* **144** 526-36
[7] University of Stuttgart 2012 Adaptive structures: Revolution for sustainable construction *Un. Of Stuttgart* **22**
[8] Stulerova L, Kapolka L and Kmeť S 2020 Numerical nonlinear analysis of a membrane structure subjected to dynamic load IOP Conf. Ser.: Mat. Sci. and Engng. **867**
[9] Inova 2017 Dokumentácia k skúšobnému zariadeniu – 17 120 časť 1 – Plachtostroj
[10] Wüchner R, Bletzinger K U 2005 Stress-adapted numerical form finding of pre-stressed surfaces By the updated reference strategy *Int. J. Numer. Meth. Engng.* **64** 143–166
[11] Dlubal 2020 Manual RF-DYNAM Pro

Acknowledgments

This work was supported by The Slovak Research and Development Agency under the contract No. APVV-15-0777 and also by The Slovak Scientific Grant Agency under the contract No. VEGA 1/0129/20.

IOP Conf. Series: Materials Science and Engineering 1209 (2021) 012053 doi:10.1088/1757-899X/1209/1/012053

Evaluation of dynamic behaviour of an experimentally tested model on a shaking table

D Taušan[1], A Brandis[1] and I Kraus[1]

[1] University of Josip Juraj Strossmayer in Osijek, Faculty of Civil Engineering and Architecture Osijek, Vladimira Preloga 3, 31 000 Osijek, Croatia

E-mail: tausan385@gmail.com

Abstract. The usual structural analysis assumes that buildings are fixed to the ground. This is not always the case, especially for structures with shallow foundations on soft soils. It was learned from the literature review that neglecting soil compliance in the structural analysis may result in significant structural damage during an earthquake event. The seismic performance can greatly differ between buildings rigidly fixed to the ground and buildings for which soil compliance is considered. Therefore, the main goal of the experiment conducted in the laboratory of the Faculty of Civil Engineering in Rijeka was to observe the dynamic behaviour of a structure on soft soil. The model was experimentally tested using a shaking table. The foundation soil was modelled using local river sand. For parametric analysis, a numerical model was done using the computer software SAP2000. The model was calibrated using experimentally obtained results. A comparison between the experimental and numerical models is presented in the paper.

1. Introduction

Dynamic analysis is one of the fields that still requires extensive study for it to be understood completely. Any experiment conducted in the field brings scientists closer to a better understanding of buildings' response to earthquakes and other shake-inducing forces.

Usual structural analysis implies that buildings are fixed to the ground and that the forces are transferred undisturbed between soil and the building.

The assumption that buildings are fixed to the ground is true only to some extent, for example, a building can be taken as fixed to the ground if the foundations are deep enough and if the soil is sturdy enough. In such cases, deviations are small enough to be ignored which results in easier calculations.

In the example, which is the scope of this paper, where the building has a shallow foundation and is constructed on soft soil, deviations would be too great to calculate the building as fixed to the ground. A similar problem was studied in a paper by Sekhar Chandra Dutta, Koushik Bhattacharya, and Rana Roy [1].

There are multiple reasons why a building cannot be considered fixed to the ground if it has shallow foundations on soft soil, and one of them is the unpredictable nature of the soil during the strong shaking forces.

2. The experiment model

For the research purpose shown in this paper, a steel model comprising two rigid blocks and two columns was used in the experiment, as shown in figure 1, which was later modelled in computer software SAP2000.

Figure 1. The model used in the experiment.

The physical model used in the experiment was made to represent the real buildings, most of which are four-story highs with a story height of 3 meters, as shown in a paper by Taranath, Bungale S. [2]. The only material used in a model was steel, where each section was made from different grades of steel as shown in table 1. Geometry of the model components is given in table 2. In table 2, *l* is length of the element, while *b* and *h* are cross-section dimensions. Detailed geometry of the model is provided in figure 2, where the elevation and vertical cross-section are shown.

Table 1. Materials used in the model.

Section	Material
mass block	S355
column	S275
foundation	S235

Table 2. Dimensions of each section used in the model.

Section	Cross section	Dimension *l* / *b* / *h* [mm]
mass block	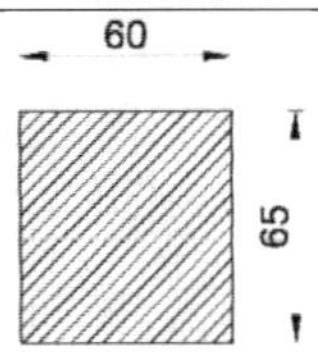	230 / 60 / 65
column	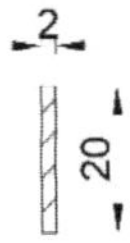	162.5 / 20 / 2
foundation	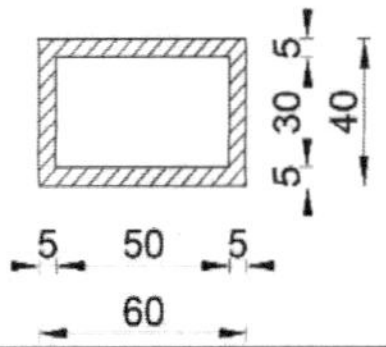	250 / 60 / 40 (wall thickness: 5 mm)

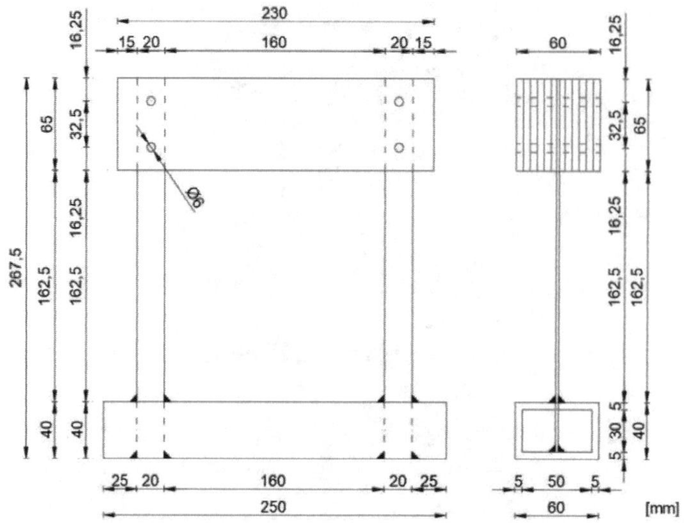

Figure 2. Dimensions of the model used in the experiment.

Only the first two modes were of interest for this experiment, and values of their periods are given in table 3. The model was tested for its dynamic properties in two configurations. Firstly, the model was as fixed into the rigid laboratory floor, and then the same model was tested laid on a layer of sand.

Table 3. Modes and periods of vibrations.

Model	Mode	Period of vibration [s]
fixed to the ground	1st	0.46
	2nd	0.10
on soft soil	1st	0.68
	2nd	0.18

The model used in the experiment, with dimensions shown in table 2 and figure 2, was placed in a plexiglass crate filled with local river sand which was then placed on a shaking table where the shake test was conducted.

The plexiglass crate had an external dimension of 280 mm/470 mm/515 mm (figure 3). The thickness of plexiglass used was 10 mm.

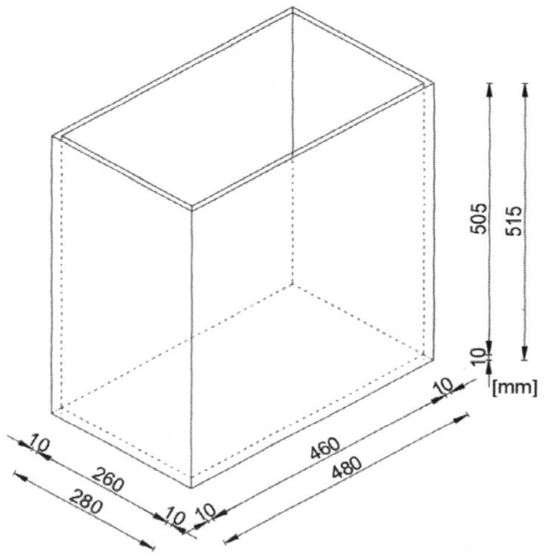

Figure 3. Plexiglass crate used in the experiment.

IOP Conf. Series: Materials Science and Engineering 1209 (2021) 012053 doi:10.1088/1757-899X/1209/1/012053

The plexiglass crate was filled with local river sand of which density was also experimentally determined. The density of the local river sand was determined both for the compact state as well as the loosely state.

Local river sand used in the experiment had uniform granulometric composition which prevented it from being overly compacted during the compacting phase.

The density of the local river sand was acquired through two different methods. Firstly, by weighing the sand and measuring its volume, and secondly by using the electric scale.

The two abovementioned methods yielded four different averaged values. Two for each state of the local river sand, loose and compact as shown in table 4.

Table 4. Materials used in the model.

State	Experiment value [g/cm^3]	E. scale value [g/cm^3]
loose	1.4273	1.5460
compact	1.6234	1.6050

For the experiment, sand in a loose state was used, and the true density of sand was: 1.5460 g/cm^3 Electric scale gave more precise results, and thus its value was used both in the experiment as well as the numerical model. If a lower value of density were to be used, for example, the value from weighing the sand, damping from it would be too large and wouldn't represent the real experiment.

3. The numerical model

As mentioned previously, the numerical model was made based on the experiment model. It was modelled and analysed in a computer software SAP2000, and its behaviour was assessed.

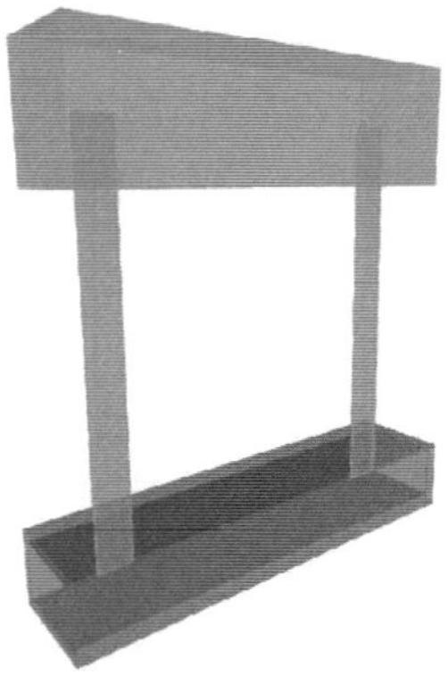

Figure 4. The Numerical model used prepared in the computer program SAP2000.

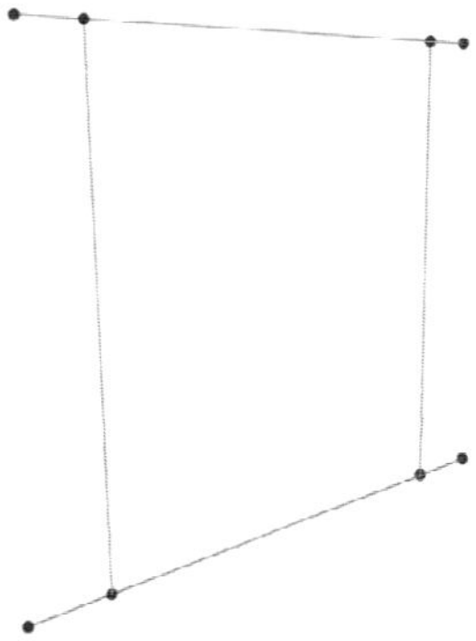

Figure 5. Model created in a computer software SAP2000.

As shown in figure 5 whole numerical model was modelled using frame elements that followed the experiment's model central axis.

Two numerical models were made, one to represent the fixed configuration, and the second one to represent model with a shallow foundation placed on soft soil.

Same as for the experimental model only the first two periods were of interest in both numerical models which are shown in table 5.

Table 5. Modes.

Model	Mode	Period of vibration [s]
fixed to the ground	1st	0.48
	2nd	0.19
with springs and dampers	1st	0.70
	2nd	0.21

Soft soil was modelled using springs with appropriate stiffnesses. Spring stiffnesses were calculated using expressions found in a work published by NIST (National Institute of Standards and Technology) [3]. Values of spring stiffness and damping calculated using input data from table 6 are shown in table 7 and table 8 respectively.

Table 6. Input data used for calculation of spring stiffness and dashpot coefficients.

Data	Symbol	Unit	Value
density	ρ	kg/m^3	1500
S waves velocity	v_s	m/s	135
Shear modulus[a]	G	N/m^2	28200
Footing width	B	m	0.03
Footing length	L	m	0.04
Poisson coefficient	ν		0.3
Foundation depth	D	m	0.04

[a] calculated using the expression: $G = V_s \cdot \rho^2/1000$

The final values used for spring stiffness were divided by the total number of springs used. In this case, a set of 4 springs were used, and thus a total stiffness was divided by 4. In table 7 k_x, k_y and k_z represent dynamic stiffness in the x-axis, y-axis, and z-axis respectively, while k_{xx} and k_{yy} represent dynamic stiffness about the x-axis and y-axis.

Table 7. Total stiffness and stiffness used for single spring.

Symbol	Unit	Value
k_x	N/m^2	16970.55
k_y	N/m^2	19164.72
k_z	N/m^2	17863.54
k_{xx}	N/m^2	45.51
k_{yy}	N/m^2	292.48
$k_{x,used}$[a]	N/m^2	4242.63
$k_{y,used}$[a]	N/m^2	4791.18
$k_{z,used}$[a]	N/m^2	4465.88
$k_{xx,used}$[a]	N/m^2	11.37
$k_{yy,used}$[a]	N/m^2	73.12

[a] calculated using the expression: k/n
where n represents the number of springs

Dampers were used in parallel with springs to connect the model with the rigid base and to simulate the real soil compliance. The values of dampers were also calculated using expressions found in the work published by NIST (National Institute of Standards and Technology) [3].

The values used in a model are shown in table 8. In table 8 c_x, c_y and c_z represent dashpot coefficient for x-axis, y-axis, and z-axis respectively, while c_{xx} and c_{yy} represent dashpot coefficient about x-axis and y-axis.

Table 8. Total dashpots and dashpots used for the damper.

Symbol	Unit	Value
c_x	N/m^2	21.44
c_y	N/m^2	26.25
c_z	N/m^2	23.48
c_{xx}	N/m^2	0.01
c_{yy}	N/m^2	0.00
$c_{x,used}$ [a]	N/m^2	5.36
$c_{y,used}$ [a]	N/m^2	6.56
$c_{z,used}$ [a]	N/m^2	5.87
$c_{xx,used}$ [a]	N/m^2	0.00
$c_{yy,used}$ [a]	N/m^2	0.00

[a] calculated using the expression: c/n
where n represents the number of dampers

4. Dynamic loads

Models described earlier in this paper were subjected to two different dynamic loads shown in figure 6 and figure 7.

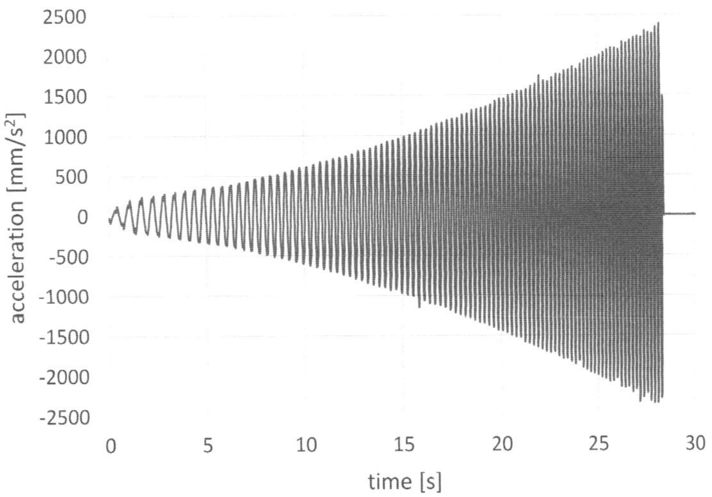

Figure 6. Sinusoidal load.

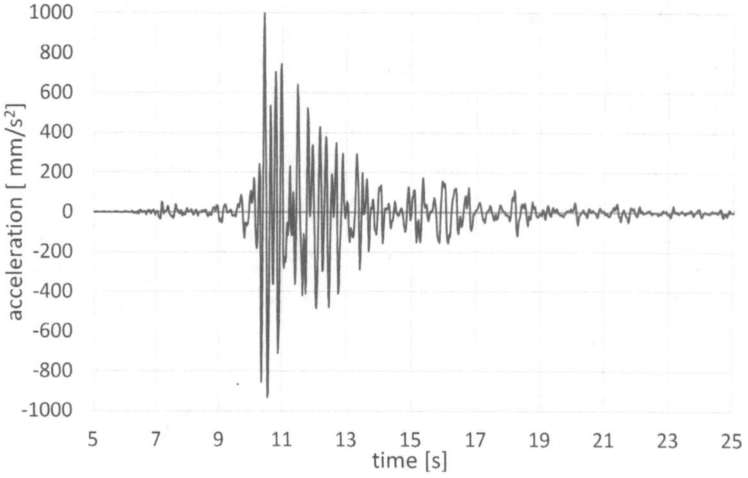

Figure 7. Kobe 1995 earthquake with scaled amplitude.

One of the loads was a regular dynamic load with a sinusoidal shape, while the other was a real earthquake with an amplitude scaled down as it would be otherwise too strong for the models of this size. The sinusoidal load was used to assess the dynamic properties of the model. On the other hand, the scaled Kobe 1995 earthquake record was used to assess the seismic performance of the model.

5. Results

Both numerical and experimental models demonstrated similar behavior for both configurations observed. Figures 9 – 12 show that models founded on soil have larger periods of vibration but also higher values of horizontal displacement when compared to the fixed-base counterpart. Results obtained experimentally and numerically are graphically represented and compared in later chapters. Displacement was acquired from joint 1 for both the experiment and numerical models, as shown in figure 8.

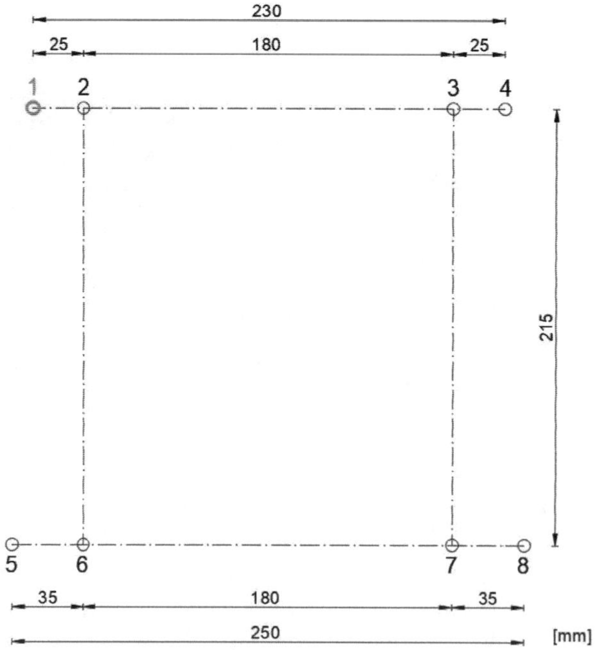

Figure 8. Position of joints in the model.

5.1. Experiment model

The difference between a model fixed to the ground and a model with shallow foundations on soft soil is shown through the displacement of the joint number 1.

Comparison of displacement time-histories for both input motions used in the experiment illustrated large differences between the two configurations of the models as shown in figure 9 and figure 10. Namely, the model placed on sand exploits larger displacements than the fixed-base model.

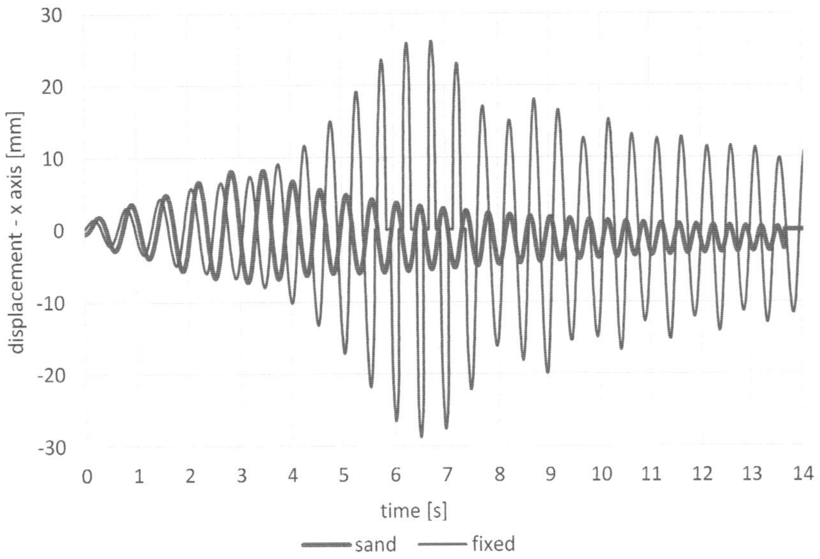

Figure 9. Displacement time histories obtained for sinusoidal acceleration curve of model fixed to the ground and model on soft soil with shallow foundations.

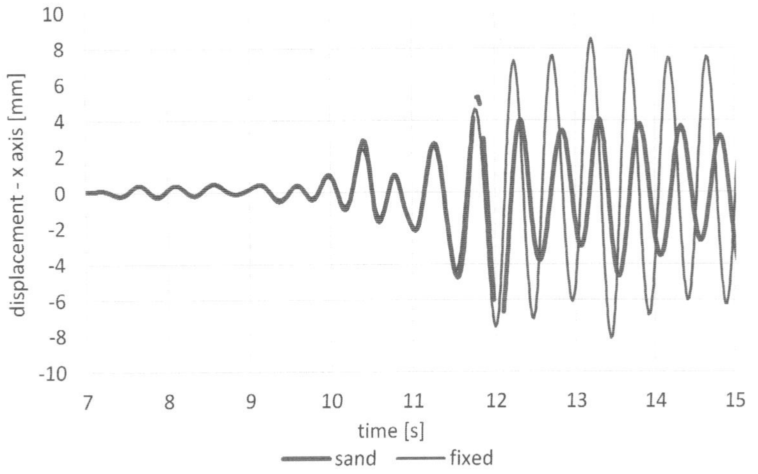

Figure 10. Displacement time histories obtained for Kobe earthquake acceleration curve of model fixed to the ground and model on soft soil with shallow foundations.

5.2. Numerical model

Comparing the results obtained from the numerical model fixed to the ground and numerical model on springs also illustrates large differences in behavior, as shown in figure 11 and figure 12. As for the model experimentally tested, the numerical model placed on sand also exploits larger displacements than the fixed-base model. It is presumed that the match between numerical and experimental models is better for the first few seconds of the simulated earthquake as during that time the soil was still in its original and pseudo-elastic state. After the soil suffered significant deformations in the experiment, the numerical model poorly predicts the dynamic behavior of the system.

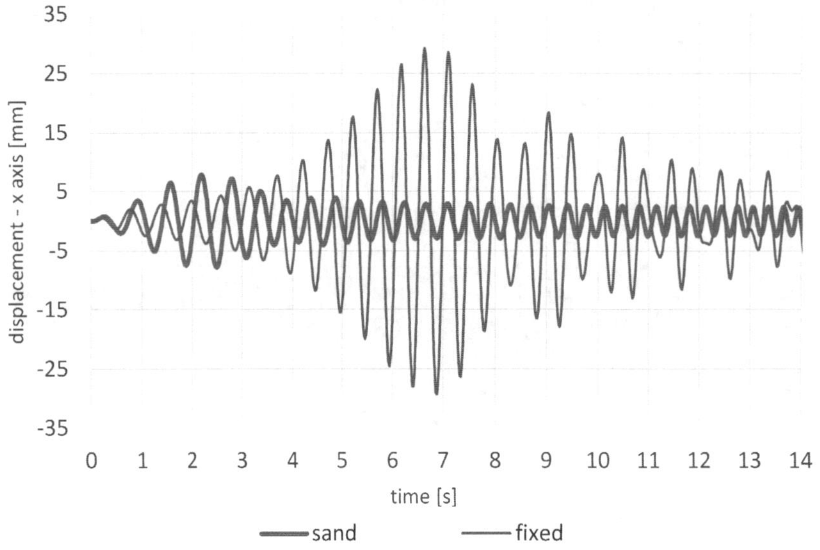

Figure 11. Displacement time histories obtained for acceleration curve between the model fixed to the ground and the model on the spring.

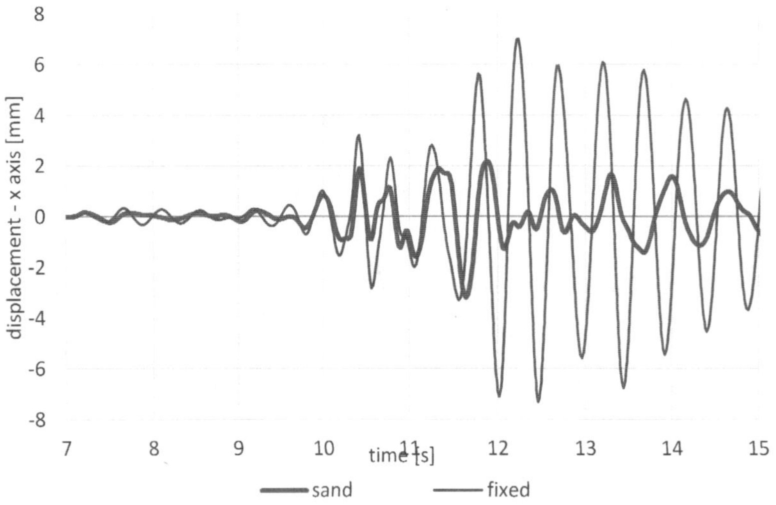

Figure 12. Displacement time histories obtained for earthquake acceleration curve of model fixed to the ground and model on the springs.

6. Comparison of the experiment and the numerical model

While creating the numerical model the goal was to get a response as close as possible to the experiment model while using simple numerical tools and finite elements available.

Numerical model fixed to the ground yielded much better matching with the experiment as the process of making a fixed-base model is well established and chances for errors are minimized. Unlike the model fixed to the ground, a numerical model connected to springs and dampers illustrated greater discrepancies from the experiment model. Thus, it can be concluded that more details should be included in the numerical model representing the soil-structure system. Moreover, non-linearity representing real material (i.e. soil) behaviour should be included in the model representing soil-structure system.

In figures 13-16 are compared displacements from the experiment model and numerical model for both dynamic loads, as well as both states of the model (fixed to the ground and with foundations on soft soil). A better match is obtained for low-intensity sinusoidal input motion than for the simulated Kobe 1995 motion.

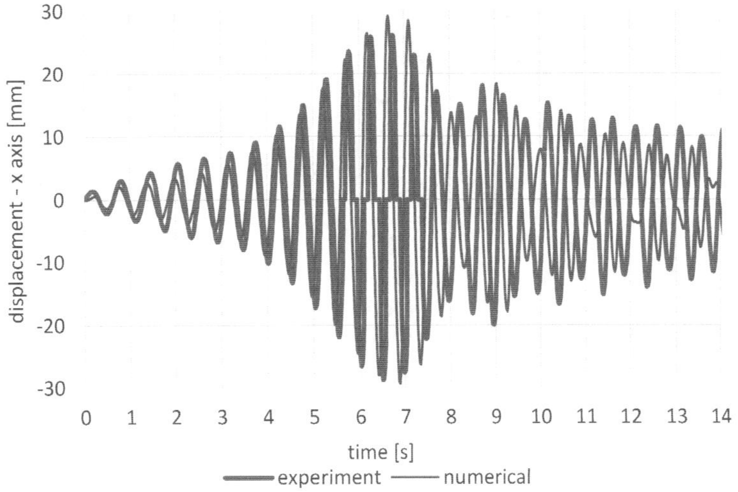

Figure 13. Comparison between displacement at joint 1 in the experiment and numerical models fixed to the ground for sinusoidal acceleration curve.

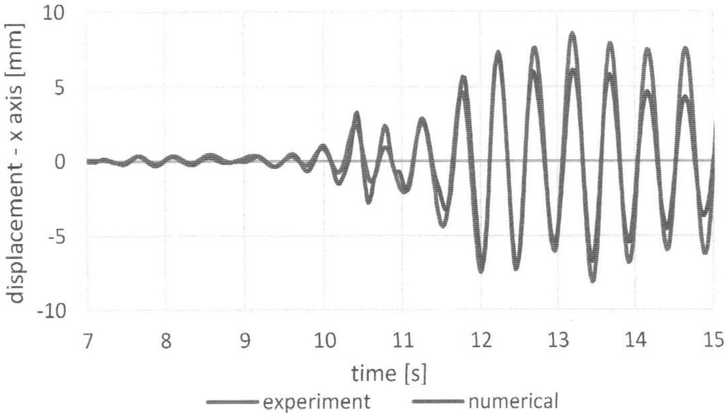

Figure 14. Comparison between displacement at joint 1 in the experiment and numerical models fixed to the ground for Kobe earthquake acceleration curve.

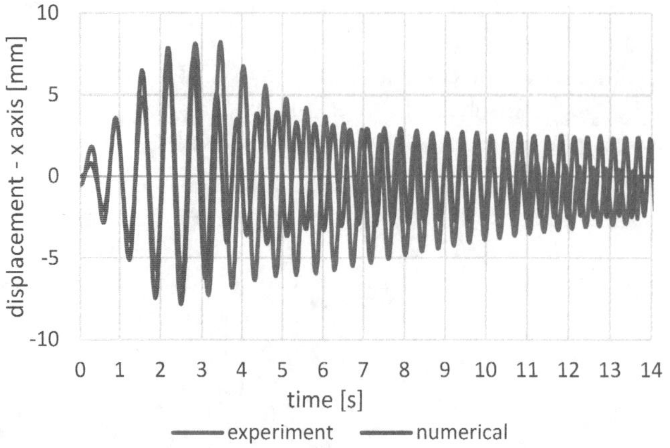

Figure 15. Comparison between displacement at joint 1 in the experiment and numerical models on soft soil for sinusoidal acceleration curve.

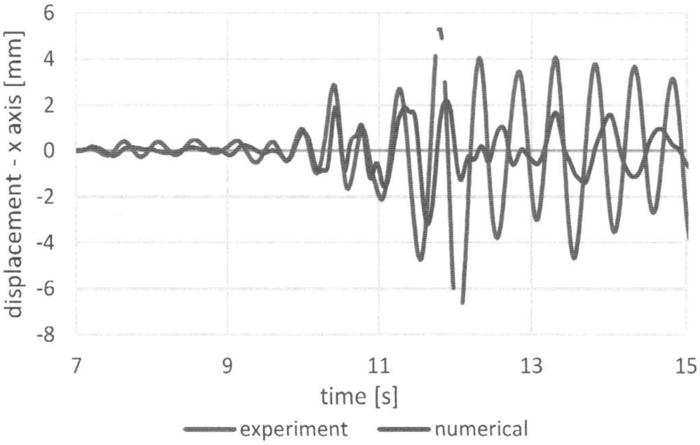

Figure 16. Comparison between displacement at joint 1 in the experiment and numerical models on soft soil for Kobe earthquake acceleration curve.

7. Conclusion

The main goal of this paper was to assess the true behavior of the building with shallow foundations on soft soil exposed to dynamic stresses. Results obtained experimentally as well as by using computer software SAP2000 were compared.

Comparison of displacement time-histories illustrated large differences between the two configurations of the models for both experiment and numerical models. The model placed on sand exploited larger displacements than the fixed-base model. The match between numerical and experimental models is better for the first few seconds of the simulated earthquake as during that time the soil was still in its original and pseudo-elastic state. However, after those first few seconds model poorly predicts the dynamic behavior of the system.

Unlike the model fixed to the ground where the process of making a model is well established and errors minimized, a numerical model connected to springs and dampers is not well defined and thus illustrated greater discrepancies from the experiment model.

This paper is just another example of why including soil compliance while modeling buildings on soft soils is needed. Extensive research is still needed to learn about the real behavior of buildings with shallow foundations founded on soft soil and models which are used to predict the real behavior.

8. References

[1] Sekhar Ch D, Koushik B and Rana R 2004 Response of low-rise buildings under seismic ground excitation incorporating soil-structure interaction

[2] Bungale S. Taranath 2016 *Earthquake Effects on Buildings*, (*Tall Building Design,* Boca Raton: CRC Press)

[3] National Institute of Standards and Technology 2012, *Soil-Structure Interaction for Building Structures*, U.S. Department of Commerce

Acknowledgments

This work was carried out within the framework of the project SYNERGY and supported by Grant No. IZIP_A_02 by the Faculty of Civil Engineering and Architecture Osijek.

IOP Conf. Series: Materials Science and Engineering 1209 (2021) 012054 doi:10.1088/1757-899X/1209/1/012054

Comparison of the level of required prestressing of road bridges according to the historical development of technical standards in Slovakia

J Baran[1] and V Borzovič[1]

[1] Department of Concrete Structures and Bridges, Faculty of Civil Engineering, Slovak University of Technology in Bratislava, Radlinského 11, 810 05 Bratislava, Slovakia

E-mail: jaroslav.baran@stuba.sk; viktor.borzovic@stuba.sk

Abstract. The paper deals with the level of required prestressing designed for box-girder bridges. Mainly focuses on the development of standards used in the design of prestressed concrete road bridges in the Slovak Republic. The standards for bridge design have been amended several times. A parametric study was performed on a model post-tensioned concrete bridge with a box-girder cross-section. The bridge has three-span with a main span of 50 m and end spans of 40 m. The study compares the normal stresses caused by traffic load along the length of the bridge according to various standards and technical regulations. The differences in design principles were investigated while the amount and geometry of the longitudinal prestressing of the bridge were changed as required by corresponding standards. The numerical model is considered as a beam element with neglecting of the torsional effects of the load. The parametric study points out the differences in the normal stresses along the bridge according to different design regulations and standards. The aim of the study is to compare level of prestressing according to the historical development of prestressed bridges

1. Introduction

In practice, we increasingly encounter the need to assess existing structures with respect to changes in technical standards or changes in the use of these structures. With the increasing intensity of traffic comes the demand to increase the load-bearing capacity of bridges. This phenomenon leads to the development of new traffic load models, which we use in their evaluation. In recent years, various reconstructions of bridges have been carried out on the territory of Slovakia, which were built according to old standards. An example is the reconstruction of one of the oldest concrete bridges in Krásno nad Kysucou, built in 1892 [1]. Reconstruction by additional external prestressing of the segment bridge that crosses the Ružín dam from 1967 [2].

This work is focused on the prestressing of concrete bridges in Slovakia. In the case of prestressed bridges, the effect of a gradual reduction of the prestressing force due to prestressing losses or corrosion is often investigated. This effect was investigated experimentally on concrete beams with naturally corroded prestress [3]. Another investigated aspect influencing the load-bearing capacity of bridges is the ratio of reinforcement by longitudinal and transverse reinforcement. The influence of shear reinforcement on the load-bearing capacity of bridges at constant and dynamic loads was investigated in the experiment of simple beams [4].

2. Case of study

The objective of the work is the concrete three-span prestressed bridge with a box-girder cross-section. The bridge is supported by bearings with one fix, which allows free horizontal movements of the bridge. The length of the superstructure is 131 m. The end spans have a length of 40 m and the main span has a length of 50 m. On the bridge there is a road with two-way traffic with a width of 9.5 m. The carriage way is made of a 90 mm thick asphalt mixture. The cross-section of the bridge deck and accessories are presented in figure 1. The total width of the bridge deck, including the kerbs, is 13.6 m. The cross section of the bridge is designed as box-girder with a height of about 2.5 m and a width of the box-girder of about 7.0 m, shown in Figure 1. The lower slab of the chamber increases its thickness from 250 mm to 500 mm at the supports. The wall thickness is 500 mm. The upper slab has a variable thickness. The minimum thickness of the top slab is 250 mm, and maximum in the connection with walls 500 mm. The extension of the cantilever is 2.7 m.

The superstructure of the bridge is made of concrete C35/45 and the reinforcement is made of reinforcement steel grade B 500B according to Eurocode 2. The prestressing of a bridge is composed by 19-strands post-tensioned tendons with a strand diameter of 15.7 mm. The 16 polygonal tendons are running along the bridge, 12 of tendons have geometry A and the rest 4 tendons have geometry B shown in Figure 2.

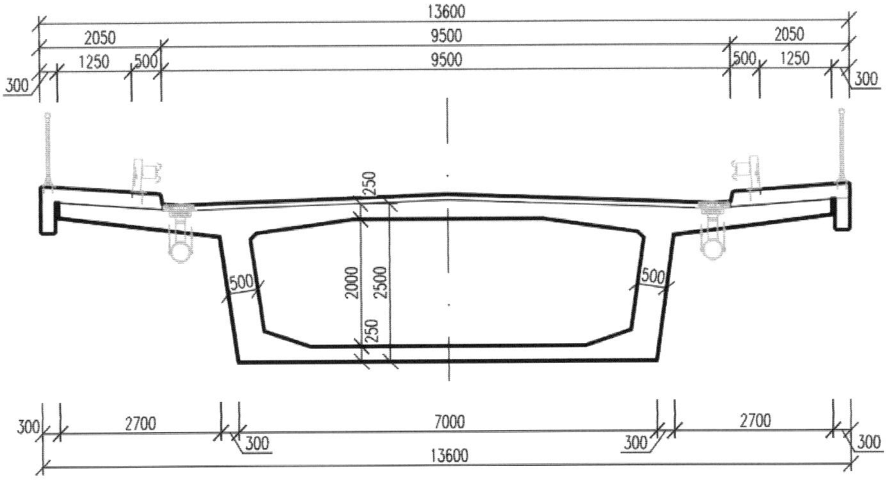

Figure 1. Cross section of box-girder bridge with accessories.

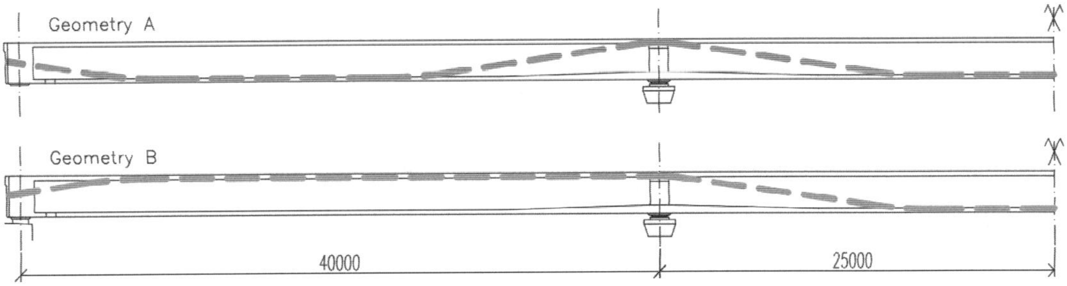

Figure 2. Geometry of prestressing tendons.

3. Applied standards

The loads were applied to the bridge model using load models according to the historical development of standards. There have been 5 important changes in the design standards of bridges in Slovakia, since 1968.

The standard ČSN 73 6203 became valid in 1968. This standard assessed structures according to the theory of allowable stresses. The load model contains two loads: uniformly distributed load 5.4 kN/m2 over entire width and a three-axle vehicle with a weight of 600 kN. The value of pedestrian load was 4.0 kN/m2.

The traffic load patterns of ČSN 73 6203 were updated with amendment A in 1975. The Amendment contains two load models. Load Model 1 involves a uniformly distributed load of 4.0 kN/m2 over the entire width and a three-axle vehicle with a weight of 600 kN. Load model 2 represented a uniformly distributed load of 4.0 kN/m^2 over the entire width with an additional load of 4.0 kN/m2 in a 3.0 m wide lane. The pedestrian load was assumed as area load 4.0 kN/m2.

After the dividing of Czechoslovakia, in 1994 the standard STN 73 6203 became valid in Slovakia. There are three load models prescribed traffic loads in this standard. The pedestrian load was assumed at 4 kN/m2. The load model 1 represented six three-axle vehicles. The total weight of the vehicle was assumed 320 kN. On the rest of the area was applied area load of 2.5 kN/m2. The load model 2 represented an area load of 9.0 kN/m2 on a lane width of 3.0 m supplemented by an area load of 3.5 kN/m2 on a lane for the rest of the width. Load model 3 represented one four-axle vehicle. The total weight of the vehicle was assumed 800 kN.

The next change in design standards of bridges came in 2010 with transferring to common European standards – Eurocodes. The Eurocodes contains one main load model representing traffic load. The load is prescribed as a tandem system (TS) and uniformly distributed load (UDL). National annex adjusted individual loads with adjustment factors α.

The standard named STN EN 1991-2 (2010) became valid in 2010. The Load model dividing the road into four traffic lanes. The first 3 m wide traffic lane with area load of 0.9 x 9.0 kN/m2 and two-axle vehicle with an axle weight of 0.9 x 300 kN. The second 3 m wide traffic lane with area load of 2.5 kN/m2 and two-axle vehicle with an axle weight of 0.9 x 200 kN. The third 3 m wide traffic lane with area load of 2.5 kN/m2 and two-axle vehicle with an axle weight of 0.9 x 100 kN. The fourth traffic lane rests on a residual area with an area load of 2.5 kN/m2.

The standard named STN EN 1991-2 (2020) affected the adjustment factors α in 2020. It changed the values of the α factor from 0.9 to 1.0 for UDL and TS. This change is due to increasing the intensity of traffic.

α is a reduction factor. They are considered depending on the expected traffic or the class of the proposed road. The values of the α factor may correspond to the transport classes in the National Annex. If they were equal to 1, heavy industrial transport was expected in the design. For normal traffic situations such as highways and 1st class roads were allowed to reduce the α factor by 10 - 20% according to STN EN 1991-2 (2010). I applied this rule and used an α equal to 0.9.

4. Description of analysis

For the example model of the bridge, the prestress was designed according to STN EN 1991-2 (2020) to eliminate decompression in the cross-section of the bridge under frequent load combination with the load model according to STN EN 1991-2 (2020). The equation (1) was used to quantify the normal stresses. The same procedure has been used for each standard. Subsequently, the value of the required prestressing force to eliminate decompression for each standard under relevant combination was calculated and compared with the prestressing force according to STN EN 1991-2 (2020). The ratio of prestressing forces to STN EN 1991-2 (2020) is given in percentage in the Figure 8.

When using STN EN 1991-2 (2020) and STN EN 1991-2 (2010), frequent combination of loads was used to assess decompression. When using ČSN 73 6203, ČSN 73 6203 amendment A and STN 73 6203, the main combination was used to assess decompression.

The analytical models of the bridge were created in SOFISTIK analysis software separately for each standard. The torsional effect from loads and the wind loading were neglected.

$$\frac{\mathrm{N}}{\mathrm{A}} + \frac{\mathrm{M}}{\mathrm{W}} = \sigma\ [MPa] \tag{1}$$

N - Normal force from combination according to used standard
M - Bending moment from combination according to used standard

IOP Conf. Series: Materials Science and Engineering 1209 (2021) 012054 doi:10.1088/1757-899X/1209/1/012054

A - Cross section area
W - Section modulus

5. Results and comparisons

The following graphs show the influence of traffic model of individual standards on the magnitude of the normal stresses in cross section along the half length of the bridge. The case of ČSN 73 6203 is shown in Figure 3, the case of ČSN 73 6203 amendment A is shown in Figure 4, the case of STN 73 6203 is shown in Figure 5, In case of STN EN 1991-2 (2010) shown in Figure 6 and the case of STN EN 1991-2 (2020) is shown in Figure 7. For better representation of the results only half of the structure is displayed by the axis of symmetry.

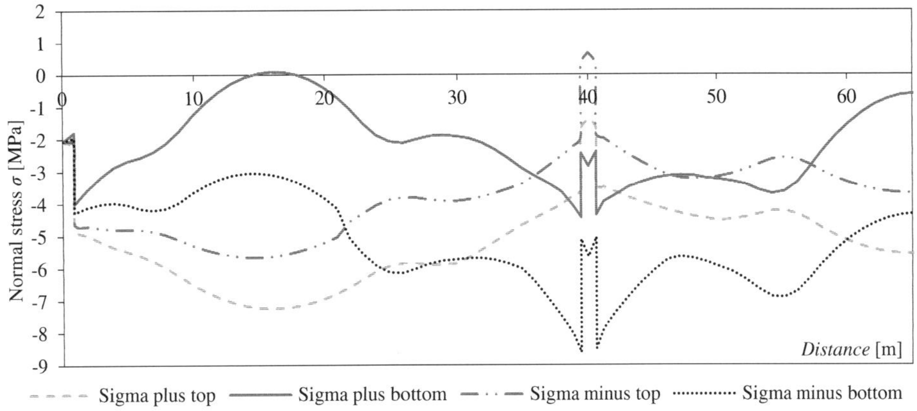

Figure 3. Normal stresses according to ČSN 73 6203 (1968).

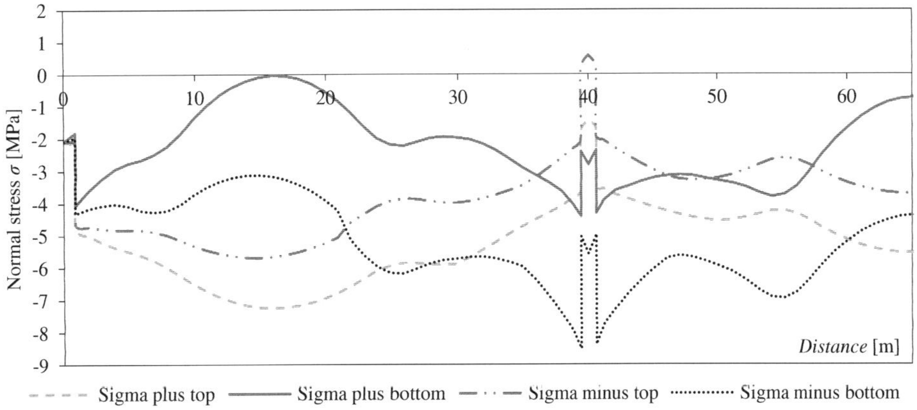

Figure 4. Normal stresses according to ČSN 73 6203 amendment A (1975).

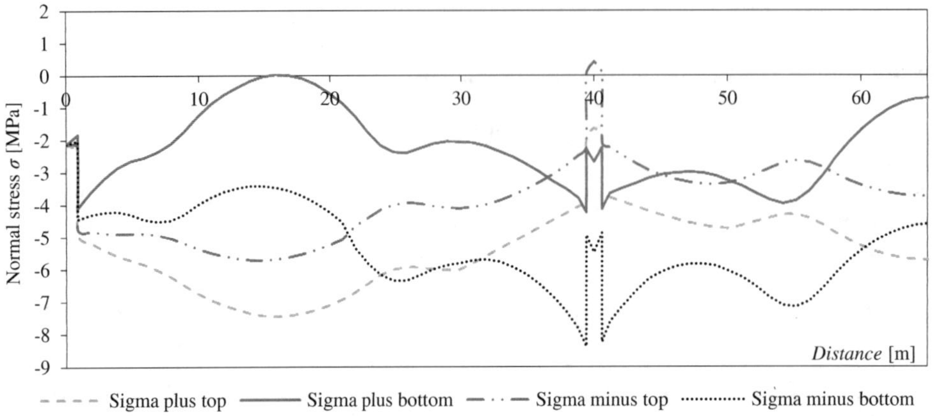

Figure 5. Normal stresses according to STN 73 6203 (1994).

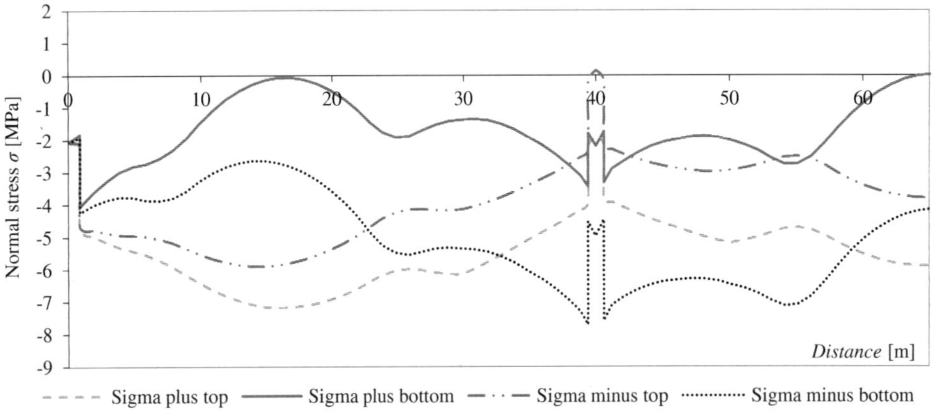

Figure 6. Normal stresses according to STN EN 1991-2 (2010).

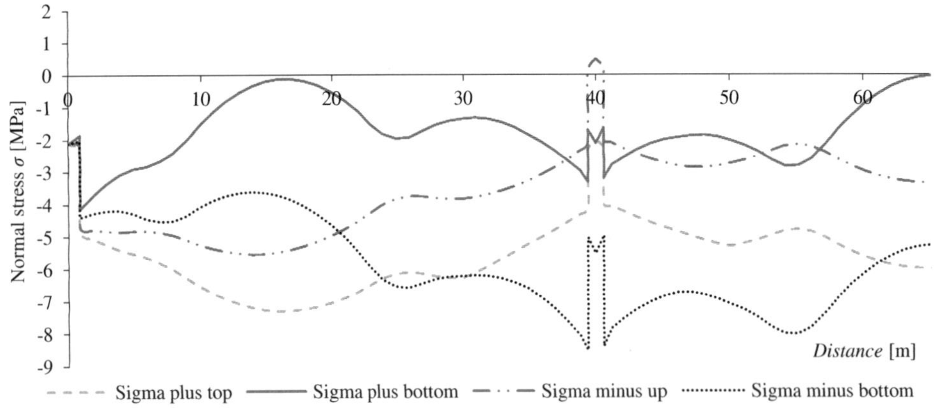

Figure 7. Normal stresses according to STN EN 1991-2 (2020).

Legend of graph lines:

Sigma plus top – Normal stresses from positive values of bending moments on top surface

Sigma plus bottom – Normal stresses from positive values of bending moments on bottom surface
Sigma minus top – Normal stresses from negative values of bending moments on top surface
Sigma minus bottom – Normal stresses from negative values of bending moments on bottom surface

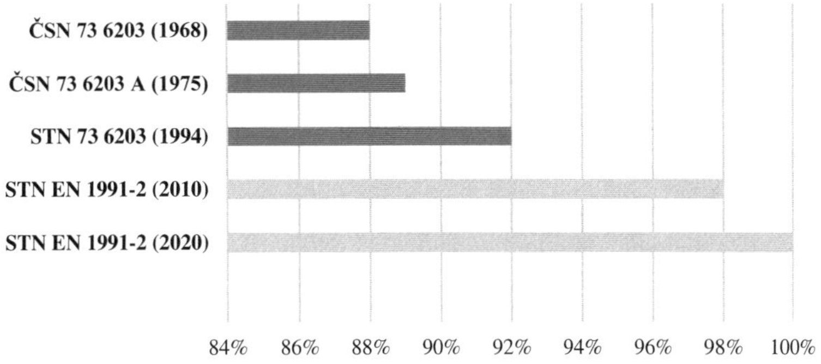

Figure 8. Comparison of amount of required prestressing.

6. Results and comparisons

The presented analysis shows that with the increasing demand for the load-bearing capacity of bridges, the amount of prestressing required to eliminate decompression in cross-section also increases in direct proportion. Differences in prestressing correspond to a gradually increasing traffic load. The biggest difference in required prestressing occurred during the transition from STN 73 6203 to STN EN 1991-2 (2010). It is caused by the introduction of a reduction fp0.1k which represents a decrease of 10% average value of the strength of the prestressing tendon. The maximum difference between the oldest standard ČSN 73 6203 and the newest standard STN EN 1991-2 (2020) reach 12% of prestressing. A comparison of each standard is shown in Figure 8. Another aspect influencing the amount of prestressing is the combination of loads for which decompression is evaluated. The frequent combination for assessing decompression according to Eurocodes has a different combination scheme and also different coefficients compared to older standards using the Main combination. An interesting aspect is also the location of the critical cross-section, which is evident when comparing the stress curves in the cross-section with respect to individual standards. The graphs show the change in the position of the critical section depending on older standards (Figure 3,4,5) and Eurocodes (Figure 6,7). When older standards are used, the position of the critical section is located in end spans and if Eurocodes are used, the critical section appears in the main span.

7. References

[1] Strieška M, Koteš P, and Bujňáková P 2016 The real static behaviour of historic bridge in town Krásno nad Kysucou in Slovakia *MATEC Web of Conferences* **86** 02002

[2] Moravcik M, Kubik P, Kopcak J and Kanda V 2019 Failures and rehabilitation of Slovak first generation segmental bridge erected by balanced cantilever method *FIB 2018 – Proc. for the 2018 fib Congress: Better, Smarter, Stronger* pp 2284-93

[3] Belletti B, Rodríguez J, Andrade C, Franceschini L, Sánchez Montero J and Vecchi F 2020 Experimental tests on shear capacity of naturally corroded prestressed beams *Structural Concrete* **21** 5 pp 1777-93

[4] Hillebrand M, Teworte F and Hegger J 2020 Shear fatigue of prestressed I-beams with shear reinforcement *FIB 2020 – Structural concrete 2020* p 1-15

[5] Moravcik M, Bujnakova P and Bahleda F 2020 Failure and damage of a first-generation precast prestressed bridge in Slovakia *Structural Concrete* **21** 6 pp 2353-62

[6] Bujnakova P and Strieška M 2017 Development of precast concrete bridges during the last 50 years in Slovakia *TRANSCOM 2017: International scientific conference on sustainable, modern and safe transport* p 1-5

Acknowledgments
This work was supported by the Scientific Grant Agency VEGA under the contract No. VEGA 1/0645/20. This work was supported by the Slovak Research and Development Agency under the contract No. APVV-17-0204.

Prepared experimental study for durability verification of FRP reinforced concrete members

S Blaho[1] and K Gajdošová[1]

[1] Department of Concrete Structures and Bridges, Faculty of Civil Engineering, Slovak University of Technology in Bratislava, Radlinskeho 11, 810 05 Bratislava, Slovakia

E-mail: stanislav.blaho@stuba.sk

Abstract. Major advantage of fibre reinforced polymers (FRPs) is their high strength and low weight to strength ratio. These are also the main reasons for a choice for this material in the process of design of reinforced concrete structures. Since there is no corrosion of FRP, this reinforcement could be strongly recommended for concrete reinforcement in aggressive environment. Till today there is no sufficient knowledge of long-term behaviour of FRP-reinforced concrete structures. Design codes give low utilization capacity of FRP materials and are not supposed to be correct according to the real behaviour in a few experiments of last decades. Reduction factors limit the mechanical properties in the range from 0.95 for CFRP to 0.5 for GFRP. In the paper there is presented a prepared and today realized long-term experimental study based on four point bending test on simply supported concrete beams reinforced with GFRP reinforcement.

1. Introduction

In the last period, progressive composite materials for reinforcing concrete structures were recognized. Well known are glass fiber reinforced polymers (GFRPs) and carbon fiber reinforced polymers (CFRPs). For their properties, mainly resistance for corrosion, high strength-to-weight ratio, and in case of GFRP relatively low cost. For short-term behaviour of members reinforced with GFRP reinforcement there is a large database of publications. In comparison with this database, investigation of long-term properties is insufficient.

Arockiasamy et al1., Hall and Ghali2, Vijay3, Abdalla4, Bischoff5, Mia`s et al.6, Fergania et al.7 have made different investigations in the field of short- ang long-term properties of FRPs composites. It is needed to do more investigations and determination of calculation of elements reinforced with FRP reinforcement. Presented paper consists of the conclusion of experimental measurements of short-term deflections and preparation of long-term experimental study.

2. Experimental investigation

Three simply supported concrete beams with the height of 77 mm in average and the width of 150 mm,reinforced with two GFRP bars (f_{fu} = 1000 MPa) with the diameter of 8 mm (figure 1), were tested under fourpoint bending test. The cover of the reinforcement was designed to be 20 mm.

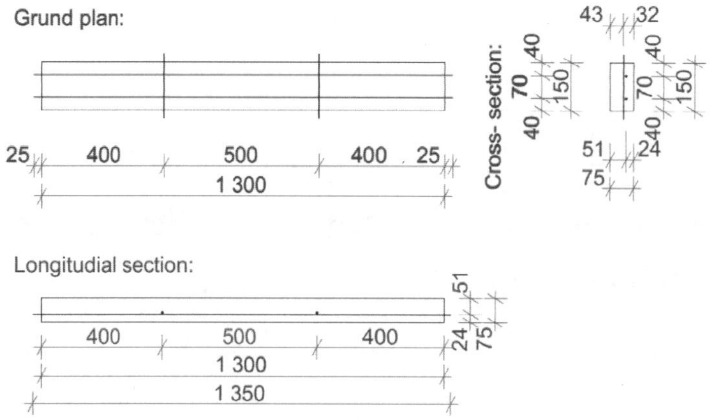

Figure 1. Dimension of the concrete beams.

3. Materials

Before testing the material properties of the concrete were investigated. GFRP bars with helically wrapped surface treatment were used for this study. The properties of the GFRP bars provided by the manufacturer (Table 1) were not verified because good agreement with the results of previous experimental testing carried out at our department during a former investigation had already been achieved.

Table 1. Properties of FRP bars.

Properties	Value from the manufacturer	Value from the previous testing
Tensile strength (MPa)	1000	1070
Modulus of elasticity (GPa)	52	52

The same concrete mix was used for all the specimens. During the concreting of the specimens, cubic(150 x 150 x 150 mm), cylindrical (150 x 300 mm) andprismatic specimens (100 x 100 x 300 mm) were also prepared and tested according to EN 12390 standard for testing hardened concrete just before the beams were tested (figure 2). The results are summarized in Table 2.

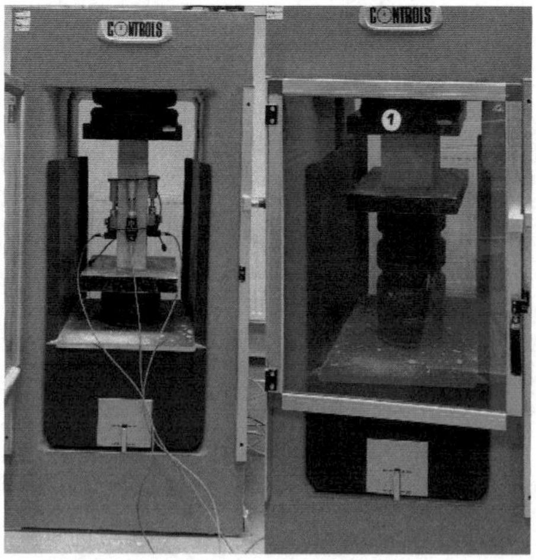

Figure 2. Concrete specimens for testing of material properties.

Table 2. Material properties of the concrete.

Properties	Value
Cube strength (MPa)	32.26
Cylindrical strength (MPa)	24.29
Static modulus of elasticity (GPa)	30.72

4. Experimental testing

All three beams for short-term study (exp01) were tested in same way by a four-point bending test that increased the force in 0.5 kN steps. The span between supports was 1.2 m and forces were acting in the distance of 10 cm from the middle of beam on each side. (figure 3). During the loading force and deflection were monitored on each beam. The main goal of the investigation was to examine the relationship between the deflection at mid-span and the acting force. The parameters obtained during the testing were recorded by computer software. The test was stopped after the cross-section failure. The results are shown in Section 5.

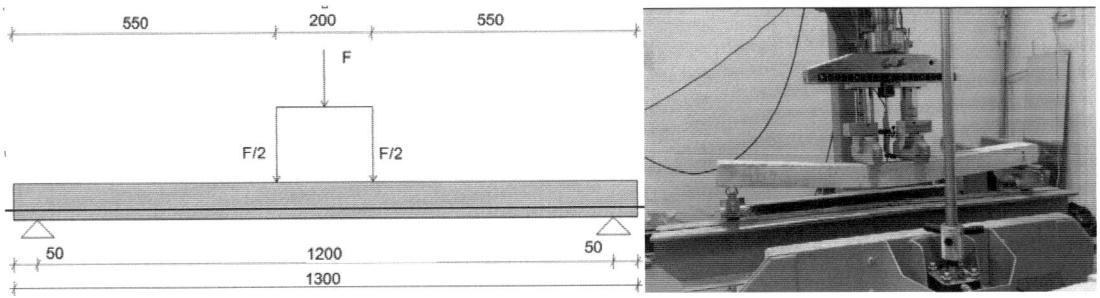

Figure 3. Test set- up.

Contemporary, the next set of experimental study (exp02) was prepared. There were made eight sets of three beams loaded on 25%, 50% and 75% of the bearing capacity obtained from short-term testing. The loading will last for 1 year, 3 years and 10 years for chosen beams. The distances between supports and acting forces are the same as at the short-term tests. On four of these sets the instantaneous deflection was measured immediately after loading and contemporary, dial gauges were installed for long-term measurement of deflection (figure 4).

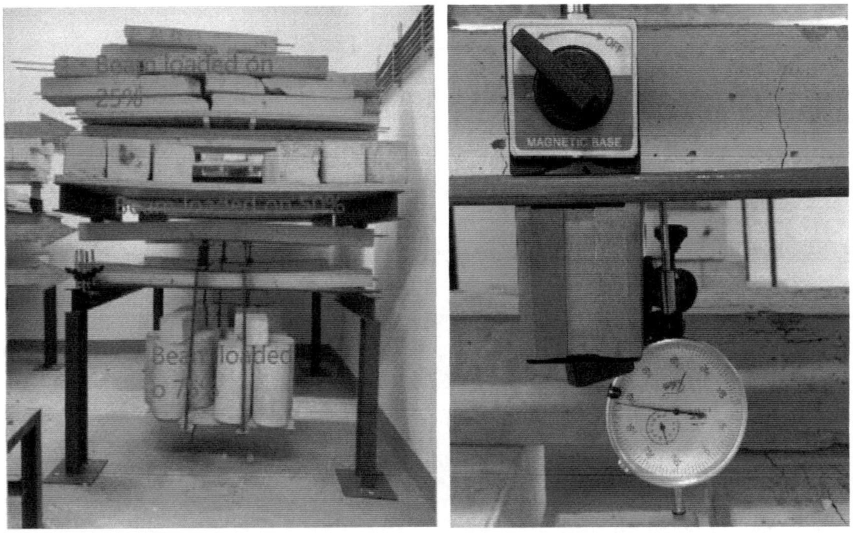

Figure 4. Set of three beams – long-term loading.

IOP Conf. Series: Materials Science and Engineering 1209 (2021) 012055 doi:10.1088/1757-899X/1209/1/012055

5. Experimental results

Results from short-term tests (exp01) of maximum forces and related deflections are presented in Table 3. and in figure 5. Figure 5 shows the load deflection curves measured by linear variable differential transformer (LVDT) for tested beams in comparison with results of immediate deflections from the sets of beams after loading to the chosen level. It is necessary to mention that the timeline of applying the load was considerably different. While in the first case, in power press, the time until reaching the load-bearing capacity was of 15 minutes, in the second case, the time needed to apply all necessary load for one set was circa eight hours.

Table 3. Measured values of the maximal force and deflection during short-term tests.

Beam no.	Maximum force (kN)	Deflection (mm)
N1	10.9	46
N2	8.72	41.5
N3	10.8	48.7
Average	9.6	45.4

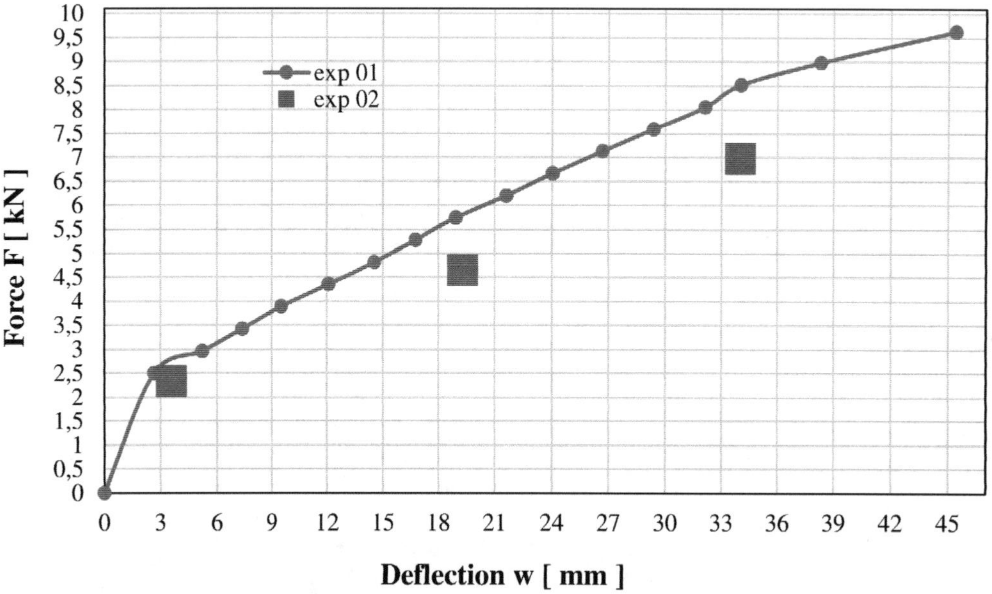

Figure 5. Comparison of result.

6. Conclusion

Four-point tests declare non ductile behaviour of beams. From the results of short-term tests it was possible to set the levels of load for 25, 50 and 75% of achieved bending resistance for long-term loading. These levels were set so that the maximum applied load overcomes the maximum allowed capacity given in foreign design codes for long-term usage of GFRP reinforcement. In comparison of short-term deflections there is a visible difference because of different time duration of applying the load. But the deflection trend is the same. The increase in deflection is continuously monitored on the sets of beams for long-term loading. After one year the first set of beams under long-term loading at different levels of designed resistance will be unloaded and subjected to four-point bending test up to failure.

7. References

[1] Arockiasamy M, Chidambaram S, Amer A, et al. 2000 Time-dependent deformations of concrete beams reinforced with CFRP bars *Compos Part B: Eng.* **31** 577–92

[2] Hall T and Ghali A 2000 Long-term deflection prediction of concrete members reinforced with glass fiber reinforced polymer bars *Can. J. Civ. Eng.* **27** 890–898

[3] Vijay P V 1999 Aging and design of concrete members reinforced with GFRP bars *Graduate Theses, Dissertations, and Problem Reports* 3138

[4] Abdalla H A 2002 Evaluation of deflection in concrete members reinforced with fiber reinforced polymer (FRP) bars *Compos. Struct.* **56** 63–71

[5] Bischoff P H 2005 Reevaluation of deflection prediction for concrete beams reinforced with steel and fiber reinforced polymer bars *J. Struc.t Eng.* **131** 752–767

[6] Mia` s C, Torres L, Turon A, et al. 2010 A simplified method to obtain time-dependent curvatures and deflections of concrete members reinforced with FRP bars *Compos. Struct.* **92** 1833–1838

[7] Fergani H, Di Benedetti M, Mia` s Oller C, et al. 2018 Longterm performance of GFRP bars in concrete elements under sustained load and environmental actions *Compos. Struct.* **190** 20–31

Acknowledgments

This work was supported by The Slovak Research and Development Agency under the contract No. APVV-17-0204 and by the Scientific Grant Agency VEGA under the contract No. VEGA 1/0645/20.

Strengthening flat slabs without shear reinforcement against punching shear by concrete overlay

D Čereš[1] and K Gajdošová[1]

[1] Faculty of Civil Engineering, Slovak University of Technology in Bratislava, Radlinskeho 11, 810 05 Bratislava, Slovakia

E-mail: daniel.ceres@stuba.sk; katarina.gajdosova@stuba.sk

Abstract. Research in this paper presents a theoretical study of increasing in punching shear capacity of the strengthened flat slab by concrete overlay. The parametric study is based on comparison of three different relevant standards design models and presents results how Eurocode 2 (EN 1992-1-1), Model Code 2010 and draft of second generation of Eurocode 2 (prEN 1992-1-1) take into account strengthening by concrete overlay. A reference specimen is represented by a fragment of a flat slab supported by circular column. Influence of concrete toppings depends on thickness and also on reinforcement ratio. In Eurocode 2 and new generation of Eurocode 2 the increase of punching shear resistance of the slab with concrete topping can be taken into account only by reinforcement ratio and thickness of the slab considering the perfect connection and bond between the original slab and new layer of concrete overlay. Model Code 2010 is based on Critical shear crack theory and the reinforcement ratio in concrete topping was considered in equation of moment of resistance and punching shear resistance is calculated by considering the rotation and deformation of the slab. Estimation of results by parametric study are compared by non-linear model from Atena software.

1. Introduction

Flat slabs are a part of a commonly used structural system in multi-story buildings because of the flexibility in architectural design. Disadvantage of flat slab system is suffering from brittle punching-shear failure around columns and also partial walls. My research is focused on enhancing shear resistance of flat slabs. The shear resistance of the reinforced concrete slab without shear reinforcement with concrete overlay was calculated according to three different design models. The first design model considered in this study was Eurocode 2 [1], which is the current valid code for the design of reinforced concrete structures in Europe. The second calculation model was Model Code 2010 [2] [3], which is based on the Critical Shear Crack Theory (Muttoni and Ruiz, 2008) [4] and takes into account rotations. The last design model used was the latest version of the new generation of Eurocode 2[5], which is based on the Model Code 2010. Parametric study was compared with non-linear model created in commercial software Atena which considers non-linear behaviour of materials.

2. Parametric study

Parametric study was carried out on the fragment of flat slab with a thickness of 200 mm and an average effective depth of 154 mm. Next calculation in a parametric study consist of 3 thicknesses of an overlay - 50mm, 100mm and 150mm and each overlay was calculated with three different reinforcement diameters $\phi_{o,1} = 8mm$, $\phi_{o,2} = 12mm$ and $\phi_{o,3} = 16mm$ in two orthogonal directions x and y in a grid of 150mm. Verification consist of total 9 calculations of shear resistance, with different thicknesses of concrete overlay and different reinforcement ratio in each thickness and one calculation for an original fragment of thickness of 200 mm. The specimen was supported by circular column with a diameter of 250mm.

In calculation the characteristic cylinder compressive strength was considered as f_{ck} = 30MPa (concrete C30/37). The maximum aggregate size $d_{g,max}$ was taken 16 mm and nominal cover layer (c_{nom}) was 20mm for diameter of reinforcement 8mm, 25mm for diameter of reinforcement 12mm and 30mm for diameter of reinforcement 16mm. Nominal cover layer differ for each diameter of reinforcement according to minimal cover layer which takes into account cohesion ($c_{min,b}$).

The effective depth of a slab strengthened by concrete topping was determined as the geometric mean value of the effective depth of the original 200mm thick flat slab and the effective depth of a slab strengthened by concrete topping taking into account the reinforcement area.

The reinforcement ratio of strengthened slab was theoretically estimated as the sum of the reinforcement ratio of a slab 200 mm thick reinforced with ϕ16/125mm and the reinforcement ratio of strengthened slab including only upper reinforcement in concrete topping.

Major diference between each design code is a control perimeter u, in which calculations are realised and in Model Code 2010 punching shear resistance also depends on parameter k_ψ, which includes rotations of the flat slab.

Tables show comparison of 4 results of resistance for each design code and also non-linear analysis.

Table 1. Results for concrete overlays reinforced with ϕ8mm.

Concrete overlay thickness [m]	h_d [m]	ϕ_{oi}/n [mm/-]	Eurocode 2 $V_{Rd,c,EC2}$ [kN]	New gen. of EC2 $V_{Rd,c,nEC2}$ [kN]	Model Code 2010 $V_{Rd,c,MC10}$ [kN]	Eurocode 2 $V_{Rd,max}$ [kN]	Non-linear (3D Atena) $V_{Rd,c}$ [kN]
0	0.20	-	475	487	412	766	494
0.05	0.25	8/150	564	563	456	824	553

Table 2. Results for concrete overlays reinforced with ϕ12mm.

Concrete overlay thickness [m]	h_d [m]	ϕ_{oi}/n [mm/-]	Eurocode 2 $V_{Rd,c,EC2}$ [kN]	New gen. of EC2 $V_{Rd,c,nEC2}$ [kN]	Model Code 2010 $V_{Rd,c,MC10}$ [kN]	Eurocode 2 $V_{Rd,max}$ [kN]	Non-linear (3D Atena) $V_{Rd,c}$ [kN]
0	0.20	-	475	487	412	766	494
0.10	0.30	12/150	734	698	557	939	739

Table 3. Results for concrete overlays reinforced with ϕ16mm.

Concrete overlay thickness [m]	h_d [m]	ϕ_{oi}/n [mm/-]	Eurocode 2 $V_{Rd,c,EC2}$ [kN]	New gen. of EC2 $V_{Rd,c,nEC2}$ [kN]	Model Code 2010 $V_{Rd,c,MC10}$ [kN]	Eurocode 2 $V_{Rd,max}$ [kN]	Non-linear (3D Atena) $V_{Rd,c}$ [kN]
0	0.20	-	475	487	412	766	494
0.15	0.35	16/150	988	908	733	1105	1042

3. Non-linear analysis

More accurate results than calculations in valid codes are given mostly by non-linear analysis using computer software. Therefore, the assumptions were verified according to the standards in software Atena. Uncertainties are connected with used material models, particularly of a concrete. Concrete fracture energy G_f depends on the concrete strength, maximum aggregate size $d_{g,max}$ and water/cement ratio. Calculation according to CEB/FIP Model Code 1990 can be determined by equation (1).

$$G_f = G_{f,MC1990} = G_{f0}(f_c/10)^{0.7} = 30 * (30/10)^{0.7} = 64.73 \text{ N/m} \quad (1)$$

where $G_{f0} = 30$ N/m for $d_{g,max} = 16$ mm.
Cylinder compressive strength was chosen as f_c=30MPa and concrete tensile strength according to Atena was calculated as $f_{ct} = 0.267*f_c^{2/3}$. Shear factor expresses how many times the normal stiffness is higher than shear stiffness. For punching failure analyses according to Červenka et al. (2018) the recommended value is of 20. The shear factor was assumed 20 and Poisson's ratio □□= 0.2.

Another uncertainty is the number of elements meshed through the depth of the slab. Kadlec and Cervenka (2015) carried out a parametric study with a different number of brick elements through the slab depth by the Atena software and they found that at least four, better five elements provide reliable results. In my research there were used 4 elements through the depth of the original slab. The square mesh refinement with the element size of 50 mm was used. Models with concrete overlay also used brick mesh with a size of 50mm. I tis important to note, that the key is a right choose of mesh. Chosen mesh for this experiment provided similar or satisfying result in comparison with parametrical study.

Analysed specimen was modelled as 1/4 of the calculated fragment – 1.25m x 1.25m (Figure 1). Modeling consist of 4 models with different thicknesses – 200mm (original slab), 250mm, 300mm, 350mm. Model of original flat slab includes reinforcement ϕ16/125mm. Between reinforcement and concrete perfect bond is considered. Other three models were reinforced as follow: concrete overlay of 50mm was reinforced by ϕ8mm in a grid of 150mm, 100mm overlay was reinforced by ϕ12mm in a grid of 150mm and last overlay with 150mm in thickness was reinforced by ϕ16mm in a grid of 150mm. In parametric study perfect bond between original flat slab and concrete topping was assumed. In Atena software the connection between these two surfaces was predicted as perfect bond. It is clear that in fact it would be problematic or even impossible for us to achieve perfect connection between these two surfaces, comparison between non-linear model and valid design models shows us that even if we predicted perfect connection, the results are different.

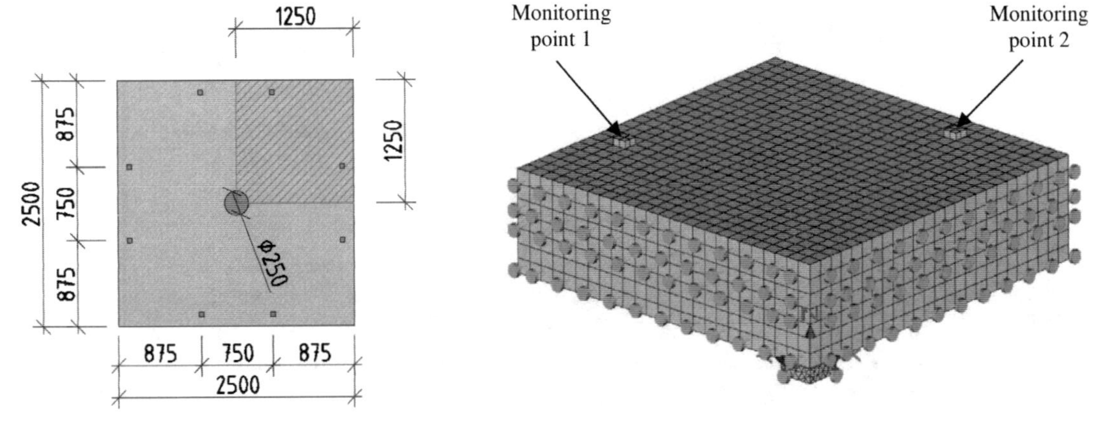

Figure 1. Scheme of analysed specimen and model analysed in Atena software

4. Results of non-linear analysis

The resistance of the flat slab in non-linear model was determined according to failure of the flat slab. Failure of the flat slab is signalized by a sharp drop in force at the same deformation as can be seen from the following diagrams (Figure 2-Figure 5). Load was applied by deformation on the column stub, iterated step by step by Newton-Raphson method in 80 iterations. The diagrams are represented by 2 force components that were measured at the 2 monitoring points of the examined part of the flat slab. The resulting force at which the failure occurs is the geometric mean of these two forces. The first diagram (Figure 1) shows the original slab and rise of the forces is almost linear. The force at failure of the original flat slab is significant. At these user interface settings in the Atena software, the failure force almost coincides with the parametric study, specifically for the Eurocode 2 [1] and the latest version of the new generation of Eurocode 2[5]. Figures 3-4 represent flat slab with concrete overlay and failure of the slab is indistinct.

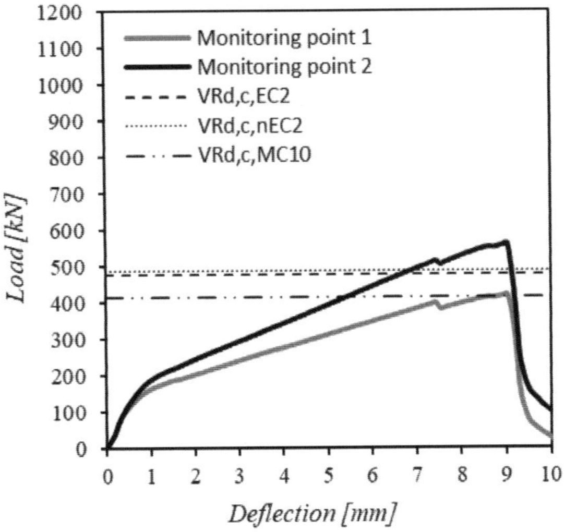

Figure 2. Specimen of original flat slab.

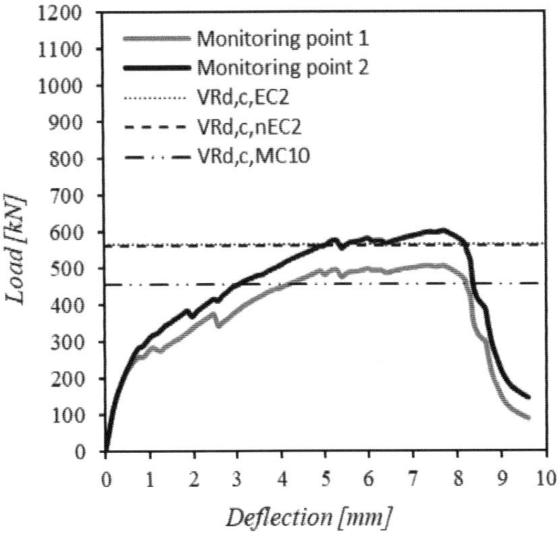

Figure 3. Specimen with 50mm concrete overlay and reinforcement *ϕ8/150.*

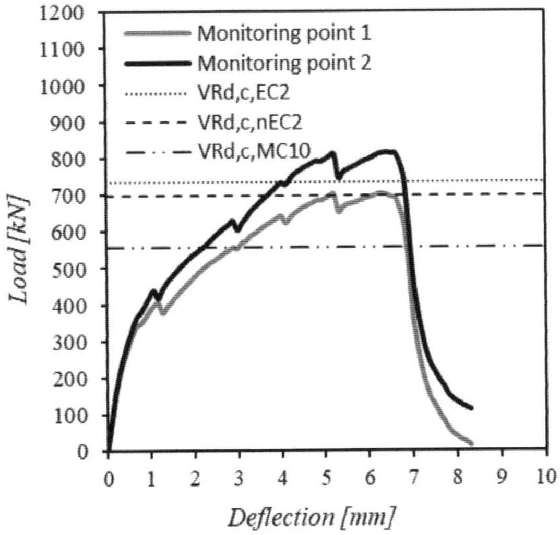

Figure 4. Specimen with 50mm concrete overlay and reinforcement *ϕ12/150.*

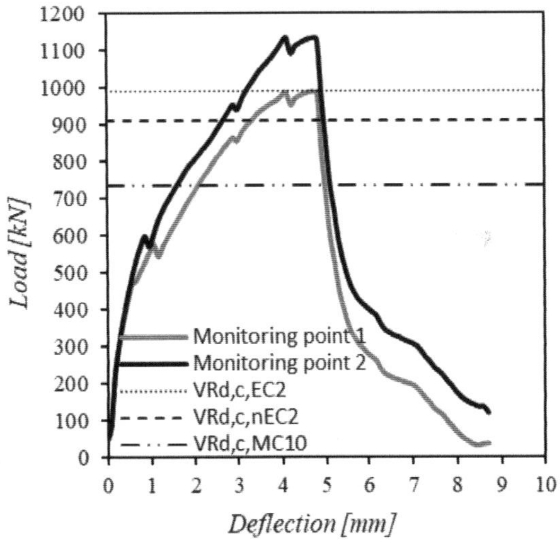

Figure 5. Specimen with 150mm concrete overlay and reinforcement *ϕ16/150.*

5. Conclusions

Figures 2-5 are represented by two force components at the 2 monitoring points. First monitoring point represents the resistance in x-direction and second in y-direction. Difference in monitoring points depends on reinforcement location - effective depth of reinforcement in x,y direction.

The resulting force at which the failure occurs is the geometric mean of these two forces and this value is decisive for comparison with design codes.

Considering the perfect connection between the original slab and the overlay, results from the non-linear model at the specific user settings of the Atena software are similar as the results of the parametric study, specifically for the Eurocode 2 [1] and the latest version of the new generation of Eurocode 2[5]. The Model Code 2010 in each case underestimates the results. Interesting to observe is how the results from the Eurocodes differ with increasing thickness of the concrete overlay comparing to non-linear model and also to see a substantial difference on the safe side with thicker concrete topping.

Setting the correct mesh size has a large effect on the results from the non-linear model. In future research, it would be appropriate to incorporate how these results differ with different mesh sizes, especially mesh of concrete topping. The non-linear model should always be calibrated by experiment. Another uncertainty in the model and the experiment will be the correct application of the concrete topping so that we achieve an almost perfect connection of these two layers.

6. References

[1] Eurocode 2, Design of Concrete Structures - Part 1-1: *General Rules and Rules for Buildings*

[2] Fédération Internationale du Béton (fib) 2010 Model Code, final draft, *Fédération internationale du béton, Bulletin* 65 vol 1 (Lausanne) p 350

[3] Fédération Internationale du Béton (fib) 2010 Model Code, final draft *Federation internationale du béton, Bulletin* 66 vol 2 (Lausanne) p 370

[4] Muttoni A, Fernández Ruiz M and Simões J T 2017 The theoretical principles of the critical shear crack theory for punching shear failures and derivation of consistent closed-form design expressions *Structural concrete* pp 1–17

[5] Final Version of PT1-draft prEN 1992-1-1 ver. 2021-01 work document N 1358 of CEN/TC 250/SC 2 p 383

[6] Cervenka V, Jendele L and Cervenka J 2018 *Atena Program Documentation, Part 1: Theory* (Prague: Červenka Consulting)

[7] Kadlec L and Cervenka V 2015 Uncertainty of Numerical Models for Punching *Fib Symposium Proceedings: Concrete, Innovation and Design (Copenhagen)* pp 1–13

Acknowledgments

This work was supported by the Scientific Grant Agency VEGA under the contract No. VEGA 1/0254/19.

Theoretical research of dimensional optimization and choice of material strength in steel-concrete composite beams

K Hrabovska[1] and J Brecka[1]

[1] Institute of Metal and Timber Structures, Faculty of Civil Engineering, Brno University of Technology, Veveří 95, Brno, 602 00, Czech Republic

E-mail: hrabovska.k@fce.vutbr.cz

Abstract. The aim of this study is to find out how the change of individual parameters will affect the flexural strength of steel-concrete composite beams. The project was focused on the choice of strength of materials and the choice of dimension, specifically the height of the concrete slab and the size of the steel profile. The research aim is to reveal which parameters have dominant influence on the flexural strength and thus facilitate the optimization of the design in practice.

1. Introduction

Steel-concrete composite structures are becoming increasingly popular in the construction industry almost all over the world. Their use is mainly applied to the solution of multi-storey buildings. The composite structures use favourable properties of both materials - steel and concrete, which give us very good strength properties. Concrete has a high compressive strength, therefore it transmits the compressive load and conversely steel has a high tensile strength, thus it transmits the tensile load. Thanks to the combination of both materials an economical and effective cross-section is found. There are many of different combinations of material strengths and dimensions of the individual components. For this reason, our research is focused on the theoretical optimization and the choice of material according to its strength and dimensions of individual parts of composite beams.

The main aim of this research is to find out how the change of individual parameters will affect the flexural strength of steel-concrete composite beams. Specifically, we observe the changes in compressive strength of concrete, the height of concrete slab, yield strength of steel IPE profile and change of size of IPE profile (its area and height). From the relation for plastic calculation of flexural strength, a general mathematic relationship was formulated. This relationship shows the change of flexural strength with regards to the original value, when the values of any of the input parameters are changed. This relationship was further verified using a calculations processor and then a parametric study comparing the obtained flexural strength was performed. All calculations were performed with respect to Eurocodes [1], [2], [3].

2. Present state

Steel-concrete composite structures are designed according to the valid Eurocode 4. This standard considered the use only strength of concrete from C20/25 to C50/60 and steel up to strength class S420. Therefore, nowadays a large part of researchers focus on the use of higher strength materials and their use in practice, e. g. [4]. Another large part of research on composite steel-concrete beams is concentrated on the development and testing of new coupling elements, such as various types of perforated shear connectors or slip-released connectors.

V. Peetha, V. Senthikumar, K. Kalaviani and S. Navaneetha from the Department of Civil Engineering of the Bannari Amann Institute of Technology in Sathyamangalm in India performed a research comparing deformation and stress in composite beams [5]. They compared the stresses and deformations on a T-beam formed by a concrete slab of thickness 80, 100 and 120 mm and three different steel profiles ISMB 150, ISMB 200 and ISMB 250 (ISMB = Indian Standard Medium Beam). The beam was considered as simply supported with stable length of 1500 mm and a slab width 500 mm. The results were obtained using computation software, specifically modelling using the finite elements method in program ANSYS. Their research dealt only with the optimization of the dimensions of partial parts beam and did not include the influence of different strength of concrete or steel used.

The aim of this research is to extend the previously mentioned calculations by considering the changes in the strength of both used materials (steel and concrete), and to find out which parameter has a dominant effect on flexural strength of the composite beam.

3. Methods

The assumption of maintaining the flatness of the cross-section and the validity of the working diagrams apply when calculating the flexural strength of steel-concrete composite beams in the plastic area. The plastic neutral axis is located in most cases in concrete and the following calculations have been performed for this variant.

Consider a composite beam which is created by effective coupling of a concrete slab of width b_{eff}, height h_c and characteristic compressive strength f_{ck} and a steel profile of the IPE series of cross-section A_a, height h_a which is made of steel with a nominal yield strength f_y. A load that causes a simple bend will be considered on the upper surface of the concrete slab. Using the idealization of the stress course, we are able to draw a stress on beam for steel profile and concrete in compression (Figure 1). The force N_a acts on steel IPE profile in its center of gravity. An equally large oppositely oriented force N_c acts on concrete in the center of gravity of the pressed part of the concrete.

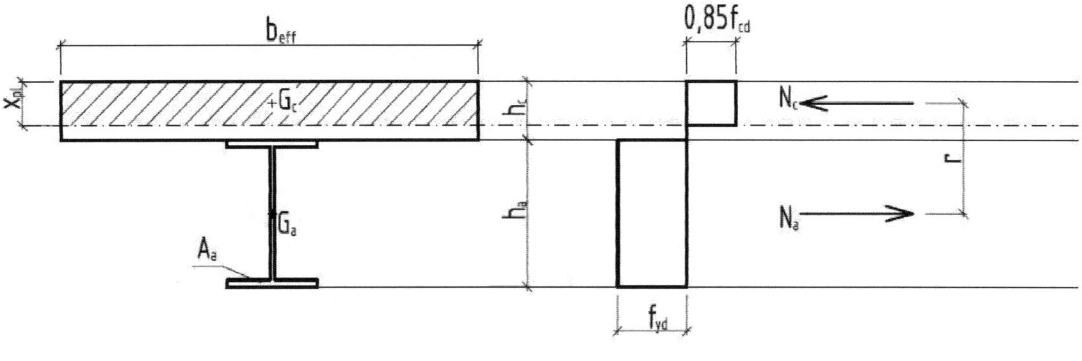

Figure 1. Stress acting on the composite beam as the plastic axis is located in the concrete.

The value of the force N_a acting on the IPE profile is calculated according to the relation:

$$N_a = f_{\mathrm{yd}} \cdot A_a \tag{1}$$

where A_a is the total area of the steel cross-section and f_{yd} is the yield strength of the steel used.

The value of the force N_c acting on the press part of the concrete slab is obtained from the relation:

$$N_c = 0.85 \cdot f_{\mathrm{cd}} \cdot x_{\mathrm{pl}} \cdot b_{\mathrm{eff}} \tag{2}$$

where f_{cd} is the design compressive strength of the concrete used, b_{eff} is the effective width of concrete slab and x_{pl} is the position of the neutral axis from the upper surface of concrete slab.

From the equality of the two forces N_a and N_c the position of the neutral axis can be calculated

$$x_{\text{pl}} = \frac{f_{\text{yd}} \cdot A_a}{0.85 \cdot f_{\text{cd}} \cdot b_{\text{eff}}} \tag{3}$$

The total load-bearing capacity of a steel-concrete composite beam can be calculated from the relation

$$M_{\text{pl,Rd}} = N_a \cdot r \tag{4}$$

where N_a is the force acting on the steel profile and r denotes the arm between the force N_a and N_c, is the distance of the center of gravity of the IPE profile and the center of gravity of the pressed part of the concrete.

The size of the arm r can be calculated as

$$r = h_c + \frac{h_a}{2} - \frac{x_{\text{pl}}}{2} \tag{5}$$

where h_c is the height of the concrete slab, h_a is the height of the IPE profile and x_{pl} is the distance of the neutral axis from the upper surface of the concrete slab.

If we substitute equation (3) into equation (5) and also equation (1) into equation (4) we obtain the relation for the calculation of the flexural strength of the composite beam

$$M_{\text{pl,Rd}} = A_a \cdot f_{\text{yd}} \cdot \left(h_c + \frac{h_a}{2} - \frac{A_a \cdot f_{\text{yd}}}{2 \cdot 0.85 \cdot b_{\text{eff}} \cdot f_{\text{cd}}} \right) \tag{6}$$

Equation (6) served as the default for deriving relations for the new flexural strength $M_{pl,Rd,N}$, expressed as the sum of the original flexural strength and the increment caused by the percentage increase of one of the four examined parameters.

When the height of the concrete slab h_c changes, we get relation for the new flexural strength after the modification by the derivation

$$M_{\text{pl,Rd},N} = M_{\text{pl,Rd}} + (x-1) \cdot A_a \cdot f_{\text{yd}} \cdot h_c \tag{7}$$

where parameter $x = h_{c,N} / h_c$, where $h_{c,N}$ is the new height of the concrete slab.

When the characteristic compressive strength of the concrete f_{ck} changes, we obtain a relationship for the new flexural strength after modifications

$$M_{\text{pl,Rd},N} = M_{\text{pl,Rd}} - \left(\frac{1}{x} - 1\right) \cdot \frac{A_a^2 \cdot f_{\text{yd}}^2}{2 \cdot 0.85 \cdot b_{\text{eff}} \cdot f_{\text{cd}}} \tag{8}$$

where the parameter $x = f_{ck,N} / f_{ck}$, where $f_{ck,N}$ is the characteristic compressive strength of the newly used concrete.

When the yield strength f_y of steel profile changes, we get the relationship for new flexural strength

$$M_{\text{pl,Rd},N} = M_{\text{pl,Rd}} + (x-1) \cdot A_a \cdot f_{\text{yd}} \cdot h_c + (x-1) \cdot A_a \cdot f_{\text{yd}} \cdot \frac{h_a}{2} - (x^2-1) \cdot \frac{A_a^2 \cdot f_{\text{yd}}^2}{2 \cdot 0.85 \cdot b_{\text{eff}} \cdot f_{\text{cd}}} \tag{9}$$

where $x = f_{yd,N} / f_{y,d}$, where $f_{yd,N}$ is the yield strength of the newly used steel.

When the steel profile of the IPE type is changed, both the total cross-section area A_a and the height of the profile h_a will change. We derive the relationship for the new flexural strength

$$\begin{aligned} M_{\text{pl,Rd},N} = M_{\text{pl,Rd}} &+ (x-1) \cdot A_a \cdot f_{\text{yd}} \cdot h_c + (x-1) \cdot A_a \cdot f_{\text{yd}} \cdot \frac{h_a}{2} - (x^2-1) \cdot \frac{A_a^2 \cdot f_{\text{yd}}^2}{2 \cdot 0.85 \cdot b_{\text{eff}} \cdot f_{\text{cd}}} \\ &+ (x-1) \cdot (y-1) \cdot A_a \cdot f_{\text{yd}} \cdot \frac{h_a}{2} + (y-1) \cdot A_a \cdot f_{\text{yd}} \cdot \frac{h_a}{2} \end{aligned} \tag{10}$$

where parameter $x = A_{a,N} / A_a$ and parameter $y = h_{a,N} / h_a$, where $A_{a,N}$ is cross-section area of new IPE profile and h_a is its height.

These derived equations were further verified using a spreadsheet program on specific cases.

4. Results

A simply supported steel-concrete composite beam with the length of 8 m was chosen as a basic pattern for the parametric study. The cross-section consisted of a concrete slab with an effective width b_{eff} = 2000 mm and height h_c = 80 mm made of concrete C20/25 and an IPE 200 made of steel S235.

Changing the IPE profile will cause a change of both the cross-section area of steel profile and its height and thus the position of center of gravity of the profile. Therefore, it was necessary to find the dependence between these two parameters x and y, so that the flexural strength can still be compared depending on only one parameter x, which expresses the proportional increase of one of the four monitored quantities. The dependence of the cross-section area on the section height was monitored on specific profiles of the IPE series from size 200 to size 360 mm. Their values were plotted on the graph shown in Figure 2 and approximated using the trend line so that we obtained the equation $y = - 1.106x^5 + 8.968x^4 -28.77x^3 + 45.26x^2 - 34.32x + 10.93$, which determines the approximate dependence between parameters x and y.

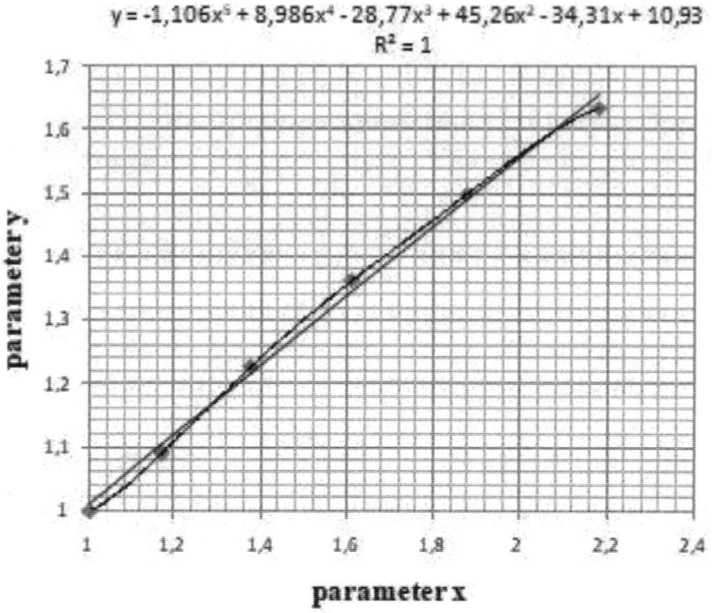

Figure 2. Interdependence of the change in the ration of cross-section area (parameter x) and height (parameter y) of the IPE profile.

Using a spreadsheet program, a parametric study was performed, in which the parameter x was inserted into the relationship (7), (8), (9) and (10), so that it monitored the development of the flexural strength with a growth of the monitored quantities by tens of percents from the value of the basic beam to doubling the parameter. The values of flexural strength were calculated in MPa and it was also determined by how many percent the flexural strength is higher compared with the basic steel-composite beam.

Table. 1 The value of flexural strength in MPa depending on the increase of the parameter x indicating the ration of new value of the monitored quantities to the original value.

x	$M_{pl,Rd,N}$ for change h_c	$M_{pl,Rd,N}$ for change f_{ck}	$M_{pl,Rd,N}$ for change f_y	$M_{pl,Rd,N}$ for change IPE
1.0	135.432	135.432	135.432	135.432
1.1	141.705	136.665	147.483	150.890
1.2	147.979	137.693	159.262	169.523
1.3	154.252	138.562	170.770	189.512
1.4	160.526	139.308	182.008	210.147
1.5	166.800	139.953	193.974	231.039
1.6	173.073	140.519	203.668	252.123
1.7	179.347	141.017	214.091	273.591

1.8	185.620	141.461	224.243	295.756
1.9	191.894	141.857	234.124	318.853
2.0	198.167	142.214	243.733	342.761

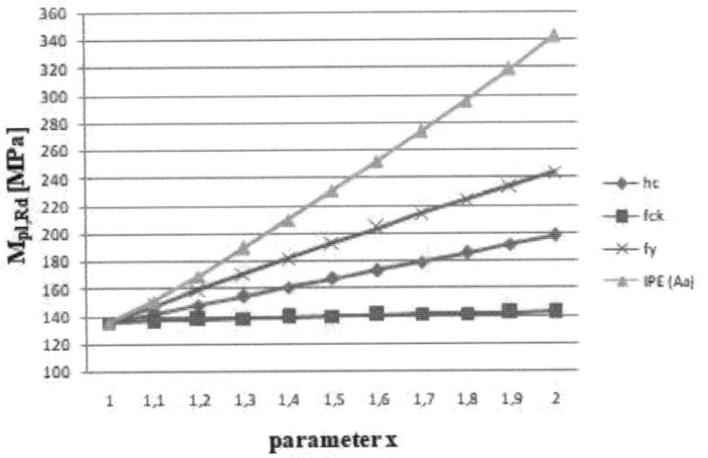

Figure 3. Graph of flexural strength depending on the parameter *x*.

Table. 2 Percentage increase of the flexural strength compared to the basic beam depending on the increase of parameter *x*, indicating the ratio of the new value of the monitored quantities to the original value.

x	$M_{pl,Rd,N}$ for change h_c	$M_{pl,Rd,N}$ for change f_{ck}	$M_{pl,Rd,N}$ for change f_y	$M_{pl,Rd,N}$ for change IPE
1.0	0.00	0.00	0.00	0.00
1.1	4.63	0.91	8.89	11.41
1.2	9.26	1.67	17.60	25.17
1.3	13.90	2.31	26.09	39.93
1.4	18.53	2.86	34.39	55.17
1.5	23.16	3.34	42.49	70.59
1.6	27.79	3.76	50.38	86.16
1.7	32.43	4.12	58.08	102.01
1.8	37.06	4.45	65.58	118.38
1.9	41.69	4.74	72.87	135.43
2.0	46.32	5.09	79.97	153.088

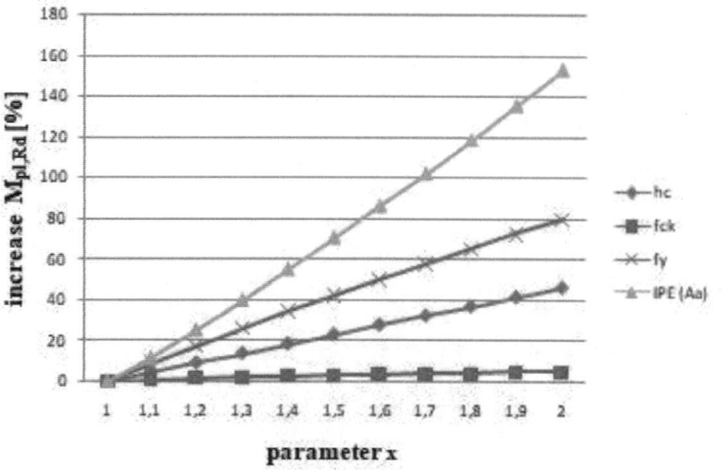

Figure 4. Graph showing how many percent the flexural strength increase depending on the parameter x.

5. Discussion

From equations (7), (8), (9) and (10) it is clear which members of the calculation affect the change of some of the observed quantities. E.g. by comparing relation (7) for changing the height of concrete slab and relation (9) for changing the strength of steel, it is clear that changing of the yield strength will increase the value of the original flexural strength more than changing the dimension of the slab. From the equations, we are able to find out that the dominant influence for the calculation has a change in the steel profile, where there is a change in the cross-section area and its height and thus the position of the center of gravity. On the contrary, the smallest effect is caused by the change of the compressive strength of concrete, where the parameter x appears in equation (8) in the denominator of the fraction and thus there is only a very gradual increase in flexural strength.

The values presented in table 1 and shown in the graph in Figure 3 show the effect of increasing the monitored variables by tens percent from the basic pattern. While a change in the compressive strength of concrete will cause only a negligible increase in flexural strength, by changing the height of the concrete slab we will achieve a significant improvement. The second best evaluated parameter was the change in steel strength and the largest increase is attained by changing the IPE steel profile.

From the values presented in table 2 and shown in the graph in figure 4, it can be seen by how many percent the flexural strength increases compared to the base pattern. When using higher strength concrete, we are talking about an increase in units of percent. When the thickness of concrete slab or the IPE profile material changes, the flexural strength increases by tens of percent. The change in steel causing almost double increase compared to the change in the dimension of the concrete slab and with increasing value of the parameter x this difference gradually decreases. And when changing the steel profile, we can achieve double flexural strength even with a 70% increase in cross-section area. The choice of a larger cross-section has the greatest influence on the flexural strength and increases with increasing parameter x.

6. Conclusion

The results of the study proofed that the dominant influence on the flexural strength has a steel profile, which forms the composite beam. The primary influence is the size of the selected profile and the secondary choice of the strength class of the steel. On the other hand, the height of the concrete slab has a significantly smallest effect on the flexural strength and compressive strength on concrete, almost negligible. This research focused only on flexural strength and was not considered in combination with other types of strain. Equally, the serviceability limit state was not taken into account, where a change in the dimension of either the concrete slab or the steel profile will cause an increase in the clear dimension of the beam, an increase in its weight and the rigidity of the beam.

7. References

[1] ČSN EN 1994-1-1 2006 Eurocode 4: Design of composite steel and concrete structures – Part 1-1: General rules and rules for buildings, Praha, Český normalizační institute

[2] ČSN EN 1992-1-1 2006 Eurocode 2: Design of concrete structures – Part 1-1: General rules and rules for buildings, Praha, Český normalizační institute,

[3] ČSN EN 1993-1-1 2006 Eurocode 3: Design of steel structures – Part 1-1: General rules and rules for buildings, Praha, Český normalizační institute

[4] Uy B 2018 Applications behaviour and construction of high performance steel in steel-concrete composite structures *12th Int. Conf. on Advances in Steel-Concrete Composite Strucrtures (Valencia)*

[5] Preetha V, Senthilkumar V, Kalaivani K and Navaneetha S 2019 Děformation Studies on Steel-Concrete Composite T Beam Section *Int. Conf. on Materials, Manufacturing and Machining (Tamilnadu)* vol 2128, ed A. Megalingam, M. Kumar and R. Shanmuga Prakash (College Park: AIP Conference Proceedings)

Acknowledgments

The article was prepared within the solution of the project of Specific University Research at the Faculty of Civil Engineering of the Brno University of Technology No. FAST-J-20-6430 and No. FAST-J-21-7474.

IOP Conf. Series: Materials Science and Engineering 1209 (2021) 012058 doi:10.1088/1757-899X/1209/1/012058

Effect of the number and arrangement of inclusions on the strength of concrete specimens

M Kovac[1] and S Kusnir[1]

[1] Institute of Structural Engineering, Faculty of Civil Engineering, Technical university of Kosice, Vysokoskolska 4, 042 00 Kosice, Slovakia

E-mail: maros.kovac@tuke.sk; stefan.kusnir@tuke.sk

Abstract. In concrete structures, we identify deformation fields of nodes with applied concentrated stress, which are accompanied by a change in the density and shape of force trajectories. These changes are caused by the presence of large grains of aggregate and/or reinforcing bars (inclusions) in the concrete, clumps of grains of normal size, or the presence of a cavern around the grain of the aggregate. These internal imperfections are among the important parameters that affect the failure process of concrete test specimens, and thus the initiation and development of cracks in concrete and can have a negative impact on the strength of concrete elements. This paper deals with the influence of the number of inclusions in their symmetrical arrangement on a series of test specimens of three different sizes, which are subjected to uniaxial compression. Attention is paid to the selection of a suitable standardization parameter, which will consider the different quality of concrete test specimens

1. Introduction

Concrete is characterized by a variety of its composition - heterogeneity. Heterogeneity affects the mechanical properties of concrete and depends on the ratio of components in the concrete, the size and shape of the aggregate grain used, as well as the size and number of pores and cavities. The heterogeneous nature of concrete, in which large aggregate grains predominate, is believed to control the process of crack propagation in the concrete. In the case of normal concrete, large aggregate grains act as the most pronounced stress concentrators. The mutual arrangement of the inclusions in the concrete matrix directs the course of the force flow, while the weak connection between the inclusions and the matrix leads to the initiation of a crack at their interface. The interfaces of aggregate and cement matrix grains are therefore, on the one hand, obstacles to crack propagation, but on the other hand, they are also a source of their initiation [1].

It has been found that a change in the curvature of the pressure trajectories will cause a local transverse tension and a change in the density of the trajectory will cause a local increase in pressure, i.e., compression. It was further found that the shear stress develops at the peak and below the inclusion. In these areas, which are approximately conical in shape, triaxial stress is created, which causes the matrix not to damage in these areas [2].

The aim of this experimental research is to investigate whether the different number of inclusions in the test specimen affects the achieved nominal strength of the specimens in their symmetrical arrangement with respect to the direction of the load. The resulting nominal strength of test specimens may not only be affected by the number of inclusions, but also by the type of inclusion material and the quality of the inclusion and cement matrix interface - position, number, and size of the interfaces. These input variables can influence the achieved nominal strength of the test specimens and the course of the curve σ - ε. Previous research has shown that different inclusion material has a significant effect on the test specimen failure process [3]. Furthermore, it was found that the mutual distance of inclusions affected the failure process of the test specimens, while the compressive resistance of the specimens decreased with increasing number of inclusions [4].

2. Experimental research

2.1. Material and specimens

The article partially presents the results of extensive research, which consisted of 6 series of experimental tests. Each series consisted of 18 test specimens, where the number of inclusions varied from 1 to 5, with each series containing one specimen without inclusions. To prepare the specimens, dry concrete mix class C16/20 was used. The test specimens were cast into the steel framework where the inclusions were previously glued in the desired position. At the same time, specimens for the determination of the concrete matrix properties were concreted. The strength characteristics of the concrete were determined on the 7th day after concreting. The prismatic compressive strength of the concrete ($f_{cm,pr}$), the flexural tensile strength ($f_{cf,m}$) and the modulus of elasticity of the concrete (E_{cm}) were investigated on specimens measuring 400×100×100 mm (h/d = 4). On specimens with dimensions of 100×100×100 mm the value of concrete splitting tensile strength ($f_{ct,sp,m}$) and cube compressive strength of concrete ($f_{cm,cu}$) were investigated. An overview of the material properties of concrete belonging to specific specimens is given in table 1.

Table 1. Material characteristics of concrete.

Specimen	$f_{cm,pr}$ (MPa)	$f_{cm,cu}$ (MPa)	$f_{cf,m}$ (MPa)	$f_{ct,sp,m}$ (MPa)	E_{cm} (GPa)	ρ_{matrix} (kN/m^3)
SQ-7-200-0-1	17.96	19.31	4.57	2.48	19.72	22.550
SQ-7-200-40-S-2	17.74	18.55	4.49	2.40	18.08	22.346
SQ-7-200-40-S-3	17.96	19.31	4.57	2.48	19.72	23.347
SQ-7-200-40-S-6	17.96	19.31	4.57	2.48	19.72	22.331
SQ-7-200-40-S-7	17.74	18.55	4.49	2.40	18.08	22.630
SQ-7-200-40-S-9	17.74	18.55	4.49	2.40	18.08	22.823
SQ-7-200-40-S-12	18.74	21.43	3.95	3.28	17.75	22.966
SQ-7-200-40-S-15	18.74	21.43	3.95	3.28	17.75	23.027
SQ-7-200-40-S-18	18.34	20.14	4.01	3.13	18.58	22.783
SQ-7-150-0-1	19.02	20.33	4.04	3.34	17.41	22.276
SQ-7-150-30-S-2	18.12	20.27	4.34	3.04	16.51	22.517
SQ-7-150-30-S-3	18.12	20.27	4.34	3.04	16.51	22.113
SQ-7-150-30-S-6	18.12	20.27	4.34	3.04	16.51	22.205
SQ-7-150-30-S-7	15.09	16.73	3.80	2.69	16.76	22.072
SQ-7-150-30-S-9	15.09	16.73	3.80	2.69	16.76	22.137
SQ-7-150-30-S-12	15.09	16.73	3.80	2.69	16.76	22.419
SQ-7-150-30-S-15	18.02	21.27	4.26	3.05	19.17	22.088
SQ-7-150-30-S-18	18.02	21.27	4.26	3.05	19.17	22.565
SQ-7-100-0-1	18.34	20.14	4.01	3.13	18.58	22.880
SQ-7-100-20-S-2	18.34	20.14	4.01	3.13	18.58	22.170
SQ-7-100-20-S-3	18.34	20.14	4.01	3.13	18.58	22.542
SQ-7-100-20-S-6	18.34	20.14	4.01	3.13	18.58	22.416
SQ-7-100-20-S-7	19.02	20.33	4.04	3.34	17.41	22.309
SQ-7-100-20-S-9	19.02	20.33	4.04	3.34	17.41	22.966
SQ-7-100-20-S-12	19.02	20.33	4.04	3.34	17.41	22.984
SQ-7-100-20-S-15	11.39	12.53	2.49	1.83	14.60	22.392
SQ-7-100-20-S-18	11.39	12.53	2.49	1.83	14.60	21.982

In wider experimental research, two types of interfaces were used - a double coherence of steel inclusions in the shape of rollers and a concrete matrix was created. This paper focuses on the presentation of selected results on test specimens, where the interface of type (*S*) - steel inclusion/concrete matrix was used in a symmetrical arrangement of inclusions with respect to the direction of loading. The height of the steel inclusions was identical to the depth of the test specimens, that is 50 mm. The ratio of the side of the test specimen to the diameter of the inclusion (ratio h/ϕ) was constant with a value of 5. The diameter of the inclusions was 40, 30 and 20 mm. The inclusions strength class was S235. Uniaxial compression tests were performed on test specimens of three different sizes, where the slenderness (ratio h/d) of the test specimens varied:

- 200×200×50 mm, $h/d = 4$, $\phi = 40$ mm.
- 150×150×50 mm, $h/d = 3$, $\phi = 30$ mm.
- 100×100×50 mm, $h/d = 2$, $\phi = 20$ mm.

2.2. Arrangement of the test

Tests were performed by using a loading device consisting of rigid steel plates, threaded rods, and a hydraulic jack (figure 5). The use of rigid steel plates created high frictional stress on the contact surface of the loading device and the specimen. Such a test arrangement prevents transverse deformations on the bases of steel plates, which results in the formation of confined zones. These areas are characterized by the action of triaxial compressive stress, and the concrete matrix is not broken in them - the area without cracks [2].

Test specimens were subjected to uniaxial compression. The tests were controlled by force, the load step was about 9.32 kN. During the test, the vertical deformation of the specimen and the exact load values in the loading steps were recorded. The vertical deformation was read using two indicator gauges with an accuracy of 0.001 mm. Figure 6 shows the test specimen SQ-7-200-40-S-18 after the test from its front and back sides. The front of the specimen was coated with lime and then, the chaotically arranged black dots were applied for non-contact measurement of deformation by photogrammetry GOM Aramis system. The use of the method of photogrammetry in compression tests made it possible to obtain information on the initiation, propagation, and development of cracks in the specimens. From previous research, it was found that crack initiation occurred at the interface of the inclusion and the concrete matrix. Then, the increase in load caused further development of cracks at the interface, with their subsequent propagation into the matrix [5].

Figure 1. Arrangement of the compression test.

(a)

(b)

Figure 2. Specimen after the test (a) front side, (b) back side.

2.3. Normalized nominal strength

The different strength characteristics of the concrete of the analysed test specimens were considered in the investigation by standardization of the investigated quantities by means of a standardization parameter. The nominal parameter must be a quantity or several quantities that most significantly reflect the different quality and properties of the concrete. The investigation test specimens are stressed by uniaxial compression. Considering the different qualities of concrete by normalizing the prismatic/cube compressive strength of concrete is the first prerequisite for choosing the optimal parameter. Under the action of uniaxial compression, the formation of cracks (shear and compression) in the samples is a phenomenon that can be explained by the formation of cracks in indirect tension. It is the flexural and splitting tensile strength characteristics of concrete that could be solved by a suitable combination of different material properties of the test specimens with a varying number of inclusions. Finally, one of the important properties is the bulk density of the concrete matrix, which expresses the quality of compaction of the concrete mixture, which affects the achieved strength. Table 2 shows the percentage differences in the compaction rate of the concrete matrix on the specimens of the three different series.

Table 2. Percentage difference of the compaction rate of the concrete matrix.

Series	$\Delta\rho_{matrix}$ (%)
200×200×50	4.55
150×150×50	2.23
100×100×50	4.56

The selection of a suitable combination of quantities into the nominal parameter can be based on the variability of the distribution of the measured values of all specimens by means of the coefficient of variation (table 3).

Table 3. Values of the coefficient of variation.

Nominal parameter	$f_{cm,pr}$	$f_{cm,cu}$	$f_{cf,m}$	$f_{ct,sp,m}$	E_{cm}	ρ_{matrix}
Coefficient of Variation	12.98	13.33	13.86	16.50	7.86	1.55

From the calculations, the most suitable combination is the prismatic compressive strength of concrete ($f_{cm,pr}$), and the flexural tensile strength of concrete ($f_{cf,m}$). This combination was expressed by the geometric mean of the two values and considering the respective compaction rate of each test specimen. The normalization of the quantities was achieved by the ratio of the contact compressive stress in the respective load step and the nominal parameter (1).

$$\sigma/\left(f_{cm,pr}\cdot f_{cf,m}\right)^{1/2}\cdot\rho_{matrix} \tag{1}$$

2.4. Result and discussion

The results of the analysis are presented in three graphs as a relationship between the stress to nominal strength ratio and the displacement to maximum displacement ratio (u/u_{max}). This relationship allowed a clearer comparison of achieved stresses, since different levels of peak stress of the test specimens were observed. Figure 3 shows the results of test specimens measuring 200×200×50 mm, where the number of inclusions varied from 1 to 5 due to their symmetrical arrangement in the direction of the load. The effect of the number of inclusions in the specimens was compared with the specimen without inclusions. An analysis of the stress peak was also performed on a series of specimens measuring 150×150×50 mm (figure 4), and finally on specimens measuring 100×100×50 mm (figure 5).

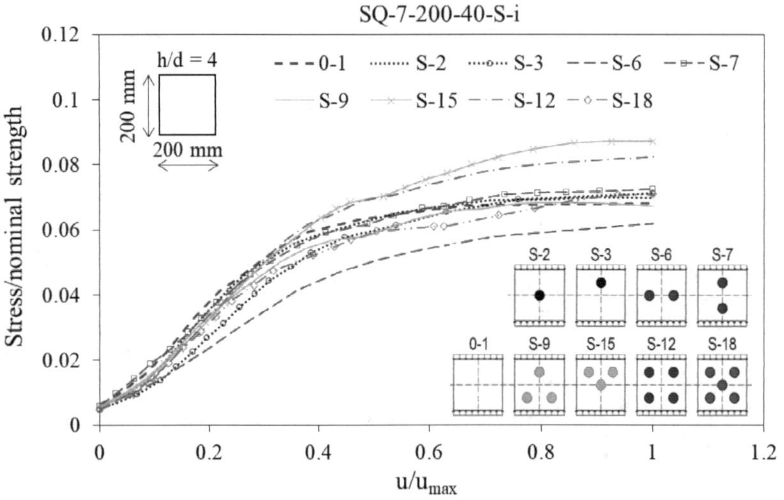

Figure 3. Stress-nominal concrete strength ratio vs displacement-maximum displacement ratio for specimen's size 200×200×50 mm.

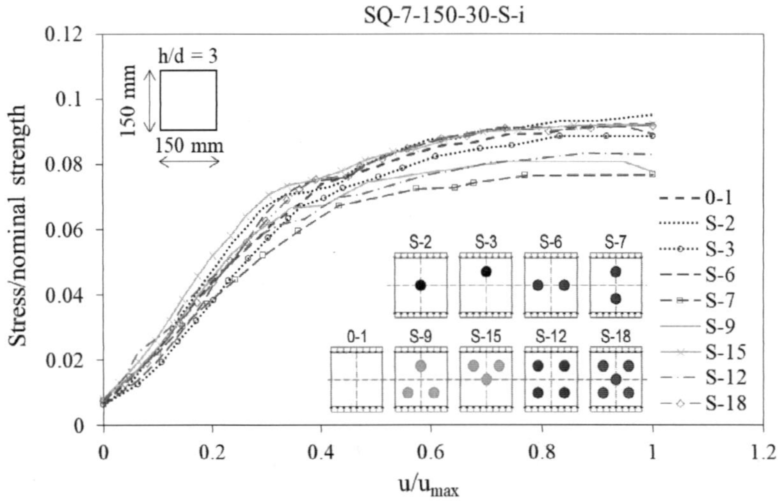

Figure 4. Stress-nominal concrete strength ratio vs displacement-maximum displacement ratio for specimen's size 150×150×50 mm.

The results of the tests showed that the range of the achieved stress peak values decreased as the slenderness of the test specimens decreased. While in the case of specimens measuring 200×200×50 mm (h/d = 4) the percentage difference in peak stress of the whole series was 40.4 %, in the case of the specimens measuring 150×150×50 mm (h/d = 3) and 100×100×50 mm (h/d = 2) the difference was 24.1 % and 12.3 %, respectively. Figure 5 shows that a significant deviation of the nominal strength from the average value was achieved only by the specimen with 5 inclusions (SQ-7-100-20-S-18).

When analysing the whole set of specimens, it is possible to observe an almost linear elastic behaviour of the ascending part of the curve in the initial stages of loading. Its partial curvature and non-linear character arose due to the presence of inclusions in the concrete matrix, where initial cracks formed at the interface. A further increase in compressive stress increased the number and length of cracks, which was accompanied by a greater curvature of the ascending part until the peak stress and unstable crack growth were reached.

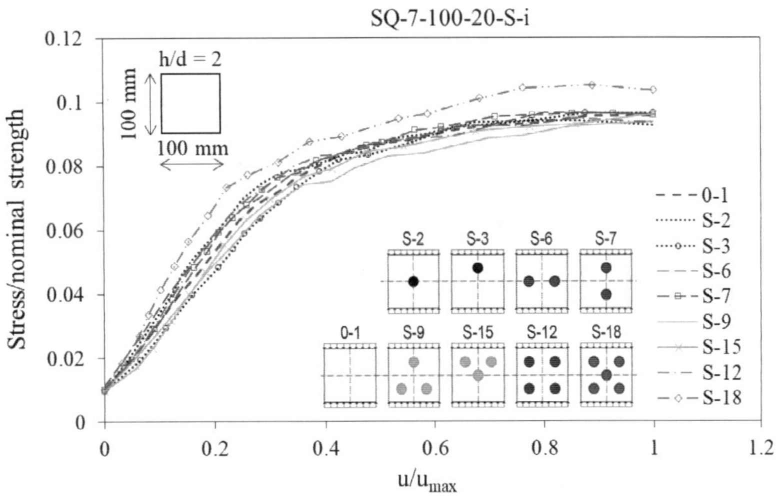

Figure 5. Stress-nominal concrete strength ratio vs displacement-maximum displacement ratio for specimen's size 100×100×50 mm.

3. Conclusion

Based on a series of experimental tests performed, it can be stated that the influence of the number of symmetrically arranged inclusions depended on the size of the test specimens with different slenderness. The effect of the different number of inclusions was found to be most pronounced on the 200×200×50 mm specimens, with this range gradually decreasing as the specimens' size was reduced to 150×150×50 mm and 100×100×50 mm. This phenomenon can be explained by the relative change in the volume of inclusions with respect to the changing size of the test specimen. On the other hand, changing the size of the inclusions will affect the interface area. The influence of a weak interface (number, size) can thus negatively affect the achieved strength of the test specimens due to the presence of a solid inclusion in a relatively soft matrix. Furthermore, it was found that there was no correlation between the measured stress levels between the specimen without inclusions and the specimens with varying numbers of inclusions. This can be solved by repeating the experimental tests and quantifying the results, or by comparing the results of the experiment with a numerical model.

4. References

[1] Vonk R A 1992 *Softening of concrete loaded in compression* PhD Thesis (Netherlands: Eindhoven University of Technology)

[2] Van Mier J G M 1998 Failure of concrete under uniaxial compression: An overview *3rd Int. Conf. on Fracture Mechanics of Concrete and Concrete Structures: Proc. FraMCoS* (Gifu, Japan) ed H Mihashi and K Rokugo (Freiburg: Aedificatio) pp 1169-82

[3] Vervuurt A and Van Mier J G M 1995 Interface fracture in cement-based materials *2nd Int. Conf. on Fracture Mechanics of Concrete and Concrete Structures: Proc. FraMCoS* (Gifu, Japan) ed H Mihashi and K Rokugo (Freiburg: Aedificatio) pp 295-304

[4] Choi S 1998 Nondestructive evaluation of cement-based materials with computer vision *3rd Int. Conf. on Fracture Mechanics of Concrete and Concrete Structures: Proc. FraMCoS* (Gifu, Japan) ed H Mihashi and K Rokugo (Freiburg: Aedificatio) pp 17-24

[5] Kovac M et al 2020 Compressive strength of concrete specimens with inclusions *IOP Conf. Ser.: Mater. Sci. Eng.* ***867*** *012022*

Acknowledgments
This work was supported by The Slovak Scientific Grant Agency under the contract No. VEGA 1/0363/21.

Influence of time intervals of microwave drying of masonry on the final result of drying

V Kvapilová[1]

[1] Faculty of Civil Engineering, Brno University of Technology, Veveří 331/95, 602 00 Brno-střed

E-mail: kvapilova.projekce@gmail.com

Abstract. In the area of drying damp masonry, we already know the discussed method of microwave drying. This method is indestructive, which is why it is considered to be used for listed heritage buildings. Even though there was much research done already on this topic, we are still missing answers for some questions regarding particular methodological procedures. This contribution is focused on the impact of time intervals of microwave drying on the final result of masonry drying. The experiment was done on historical masonry of CPP and on ceramic masonry. Two laboratory samples were made out of each type of masonry, to which we applied microwave drying in different time intervals. Difference was in both time of drying and time of pauses between them. The aim was to set an ideal interval of drying.

1. Introduction

The principle of the microwave drying is the ability to oscillate molecules of water in the masonry material and subsequent heat generation, which is created by the movement and friction of the molecules, causes their conversion to water steam. Such transformed water steam is subsequently just vented out from the construction. This process enables water molecules to get rid of their energy potential, which binds them to the capillaries in the construction material and disables their evaporation. Water temperature is higher than the construction material temperature itself and it was found that the warming up rises from the middle of the material, or better said, it is highest nearest the device antenna. Warmed up area is then spreading in a way that already removed dampness enables spreading of radiation exposure. The above shows clear coherence between time intervals of masonry drying and dampness evaporation from the construction itself. In case of drying intervals being too long and pauses being tooshort there is a risk of overheating the masonry, which is undesirable and can devalue the result. In the same way drying intervals too short in combination with pauses being too long can lead to insufficient final result of microwave drying.

The aim of the experiment is to set an ideal time of drying and pauses between individual drying cycles, so that the maximum efficiency of microwave drying of masonry is reached. This experiment was done on masonry samples of historical CPP and on masonry of modern trim tiles of Therm type [1], [2], [3], [5].

2. Literary overview/description of current state

There were several relevant researche and experiments made on the topic of microwave drying of masonry constructions, many of these were carried out at the Faculty of Civil Engineering in Brno. Specific connection between the drying interval and the final result of microwave drying was not yet exactly determined.

3. Methodology

Tests were done on two samples of historical masonry of burnt clay brick and on two samples of modern masonry of Therm type trim tiles, the masonry was done on cementlime mortar P 2,5. Masonry sample dimensions were 1000 x 300 x 500 (l, w, h - mm). Masonry samples were made in two plastic soaking tubs, the same type of masonry samples were put together in one tub. Such created samples were left in laboratory conditions with air temperature 23°C and relative humidity 55% for a week. After one week since making the masonry, drilling of injection holes was done in an axial distance of 100 mm with 24 mm diameter drill. The drill hole must go through at least one bed joint. Individual drill holes are situated horizontally and in height of 220 mm on the historical masonry, that is between the third and fourth layer of bricks. For the ceramic tiles masonry drill holes were situated in height of 270 mm, that is in second trim tile above first bed joint. Afterwards the holes were cleaned using compressed air.

Into such prepared soaking tubs with individual samples the water was pumped into height approximately 150 mm. After one month when water rise was observed into wall height, dampness measuring was done and microwave drying followed.

Microwave drying was done using a device with a rod antenna, always roughly in the middle of the wall. Two rod antennas with 600 W output were used for the experiment. The antenna composition enables even spreading of microwave energy to full length of drill hole, which also ensures constant heat rise of antenna surrounding. This way the pressure direction of energy growth is created, which positively influences water movement behind the heated area. Water is thus pushed behind the dried masonry area by pressure of water steam, which is getting out of the construction mainly from those drill holes with antennas.

Length of drying interval for first two samples of each masonry type was 5 cycles of 14 minutes each, that is 12 minutes of drying and 2 minutes pause. Length of drying interval for second two samples of each masonry was 5 cycles of 20 minutes each, that is 17 minutes of drying and 3 minutes pause. Dampness measuring was done using gravimetric method at the beginning of the experiment, as well as during the individual drying cycles. During the experiment we also observed external appearance of the wall, dampness maps and possible occurrence of salts on the wall surface.

The result of the experiment will be establishing volume of dampness in the masonry depending on different time intervals of drying individual samples. Also, mutual comparison of final dampness of those samples and overall evaluation of ideal drying cycle time [4], [6], [7].

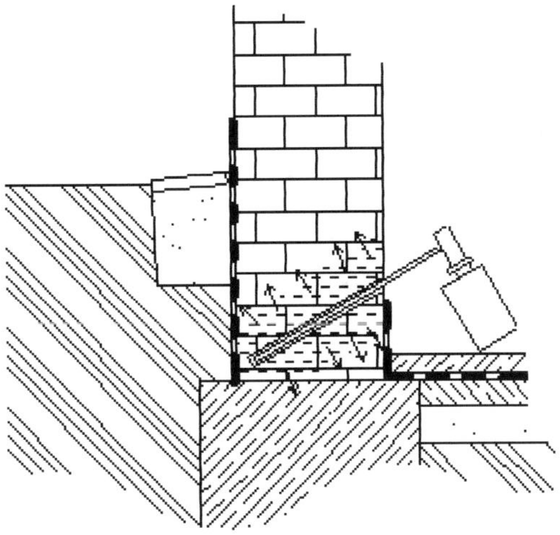

Figure 1. Scheme of masonry drying by rod antenna.

IOP Conf. Series: Materials Science and Engineering 1209 (2021) 012059 doi:10.1088/1757-899X/1209/1/012059

4. Results

During the microwave drying of individual samples, we mainly observed a decrease in their weight in time of drying and from those data we subsequently determined the moisture content after each drying cycle. In the below charts (Fig1 and Fig2) are values of moisture content of individual samples in timeof MW drying.

To explain further it needs to be mentioned, that samples labeled CP20 - full burnt clay brick, and T20 - trim tiles of Therm type were being dried in 20 minutes cycles and samples labeled CP14 - full burnt clay brick, and T14 - trim tiles of Therm type were being dried in 14 minutes cycles.

Table 1. Values of moisture content duringdrying, drying cycle 20 minutes.

	Moisture content [%]					
Sample labelling	Initial	After 20 min.	After 40 min.	After 60 min.	After 80 min.	After 100 min.
CP20	13,60	12,81	12,02	11,14	10,28	9,44
T20	21,79	21,06	20,29	19,48	18,62	17,71

Table 2. Values of moisture content during drying, drying cycle 14 minutes.

	Moisture content [%]					
Sample Labelling	Initial	fter 14min.	fter 28min.	fter 42min.	fter 56min.	fter 70min.
CP14	14,92	13,91	11,52	10,35	9,17	8,69
T14	19.96	19,01	17,87	16,90	16,02	15,37

Based on the above tables and the data they display we can see that the initial dampness was differentfor individual samples, despite the same materials used - bricks of the same origin, trim tiles of Therm type of the same origin and the same mortar, we need to evaluate the real decrease in weight, or better yet, dampness loss. This value can be seen in the following charts.

Table 3. Resulting moisture loss of individual brick samples.

	Overall moisture loss [%]
CP20	30,59 %
CP14	41,76 %

Table 4. Resulting moisture loss of individual trim tiles of Therm type samples.

	Overall moisture loss [%]
T20	18,72 %
T14	22,99 %

5. Discussion

Despite the initial expectations, that there will be higher moisture loss for samples CP20 and T20, the experiment has proved that overall decrease in weight was higher for a drying cycle of 14 minutes, thatis 12 minutes of drying and 2 minutes pause. Difference in moisture loss for brick samples was more than 10%.

Before a final conclusion can be made, we need to discuss the question of masonry homogeneity. Samples made of historical bricks are of the same origin and put together with the same mortar, however, based on their age and original placement we can assume that each sample reached slightly different attributes, which could also cause different initial dampness of individual samples.

If we overlook possible irregularities caused by dissimilarities of the brick material, we can concludethat it is better to use shorter drying cycles, even despite shorter pauses in between the individual dryingcycles.

Furthermore, it showed that moisture loss was higher for samples of historical brickwork, and that was more than two times higher compared to trim tiles of Therm type. The reason for such difference isarguable and we can only presume, that it is the structure of modern trim tiles - hollow walling.

6. Conclusion

From the experiment we can conclude that better results are reached with shorter drying cycles, despitethe shorter pauses in between the individual drying cycles.

The experiment can contribute to other important research works to follow, so that ideal results of microwave drying can be reached, however, it is recommended to broaden those measurements either with more samples or try the experiment with a microwave device with attachable funnel head and on other types of material.

7. References

[1] Pavlát J 2014 *Moisture of Buildings Expert and Investor-engineering office* (Praha) Available online http://www.pavlat-znalec.cz/nktere-vybrane-otazky-z-problematiky-stavebni-technologie/96-vlhkost-staveb

[2] Novotný M 2011 *Problems of utilization of microwave radiation in building constructions: thesis of lecture on professor appointment procedure in the field of Civil Engineering* (Brno: Vutium)

[3] Solař J 2013 *Moisture removal: remediation of damp masonry* (Praha: Grada Profi & hobby)

[4] Jiroušek J Z and Šuhajda K 2013 Masonry Drying, Sterilization of biotic factors using microwave technology *XXII international scientific conference of forensic engineering (Brno)* p 9

[5] Maxwell J C 2019 *Archive of the history of mathematics MacTutor (Skotsko)* Available online https://www-history.mcs.st-andrews.ac.uk/Biographies/Maxwell.html

[6] Alexa T 2010 *Humidity removal in MW connecting mortar by heating* Dissertation (Brno: Faculty of civil engineering department of building structures)

[7] Sobotka J 2015 *Disposal of biotic pests' trough EMW radiation* Dissertation (Brno: Faculty of civil engineering department of building structures)

IOP Conf. Series: Materials Science and Engineering 1209 (2021) 012060 doi:10.1088/1757-899X/1209/1/012060

Application of critical shear crack theory on punching of flat slabs

S Sarvaicova[1] **and V Borzovic**[1]

[1] Department of Concrete Structures and Bridges, Slovak University of Technology in Bratislava, Faculty of Civil Engineering, Radlinského 11, 810 05 Bratislava, Slovakia

E-mail: simona.sarvaicova@stuba.sk

Abstract. This article deals with the punching capacity of a flat slab fragment supported by an internal atypically elongated column. Based on the results of this analysis and the application of Critical Shear Crack Theory, the reliability of two design models was determined. The CSCT model is a mechanical model where the shear force transferred by concrete in shear crack can be determined by accounting for the roughness and opening of a critical shear crack. The crack width is proportional to the slab rotation, which was obtained from a nonlinear program Atena and from experimental test and shear capacity was obtained by integrating the shear strength along the control perimeter. The aim of this analysis was to compare the application of CSCT in non-linear analysis and experimental test to point out the significant difference between obtained results, which shows the importance of experimental tests realization. Non-linear analyses provided unsafe results. Contrary the currently used EC2 model provided safe results when reduction of the control perimeter was applied. The best results were obtained in a combination of the CSCT model with measured rotations of the slab specimen.

1. Introduction

Structural failure of concrete flat slabs caused by achieving their shear resistance is one of the most dangerous types of failure due to its fragile character. The first models to determine the cause of this phenomenon have evolved since 1960s. They were based on various principles, for instance models based on the failure mechanism (Kinnunen and Nylander) [1], on the theory of plasticity (Braestrup-Nielsen) [2] or models based on bending capacity (Moe) [3].

Currently, there are many approaches and models developed for the assessment of punching shear resistance of flat slabs. One of them is Critical Shear Crack Theory proposed by Muttoni and Schwartz [4] in 1991, which later also became the model applied in Model Code 2010 [5]. Like the model developed by Kinnunen and Nylander, the CSCT relates punching capacity to slab rotation. Muttoni and Schwartz [4] assumed that the width of this critical shear crack is proportional to the product of the slab rotation ψ, and the effective slab depth d, as shown in Figure 1.

The subject of this analysis is a fragment of a flat slab supported by a rectangular column with dimensions 0.95 m x 0.15 m. The punching shear resistance of slabs supported by a rectangular column is lower than the shear resistance of slabs supported by a square column with approximately the same length of control perimeter. This phenomenon is due to the fact that the formulas for the evaluation of the punching shear resistance include empirical factors that take into account only the geometry of the column, but do not take into account the deformed shape of the slab.

2. Critical shear crack theory (CSCT)

The CSCT takes into account many factors in assessment of the flat slab shear resistance, which do not appear in currently used Eurocode 2 (EC2) model [6]. This theory assumes that a slab-column connection punching capacity decreases due to increasing slab rotation ψ and associated opening of the critical shear crack. The amount of shear stress which can be transferred across the critical shear crack is assumed to be dependent on the crack roughness, which is estimated as a function of the maximum aggregate size d_g and by the concrete tensile strength. The relation between shear resistance and slab rotation is expressed by formula (1).

$$V_{Rd,c}=k_{\psi}\frac{\sqrt{f_{ck}}}{\gamma_C}b_0 d_v \quad (1)$$

$$k_{\psi}=\frac{0.75}{1+15*\frac{\psi*d}{d_g+16}} \quad (2)$$

where:

k_{ψ} parameter related to a slab rotation evaluated by equation (2),

γ_C partial safety factor [-],for the analysis γ_C was taken equal 1

f_{ck} characteristic concrete cylinder strength [MPa],

b_0 control perimeter length at distance $d_v/2$ from the column face, $b_0 = 2(c_1 + c_2) + 2\pi(d/2)$ [mm],

d_v effective shear depth of the slab [mm],

ψ the slab rotation [-],

d effective depth, which is an average value of effective depths in both orthogonal directions [mm],

d_g maximum aggregate size [mm].

To apply the CSCT, the load-rotation response of a flat slab must be estimated. While evaluating experimental test results, the rotations can be directly measured or can be calculated based on measured deflections by assuming a conical deformation pattern of the slab beyond the column region (Figure 1). In designing structures, the load-rotation relationship can be estimated using nonlinear finite element analysis or simplified design equations.

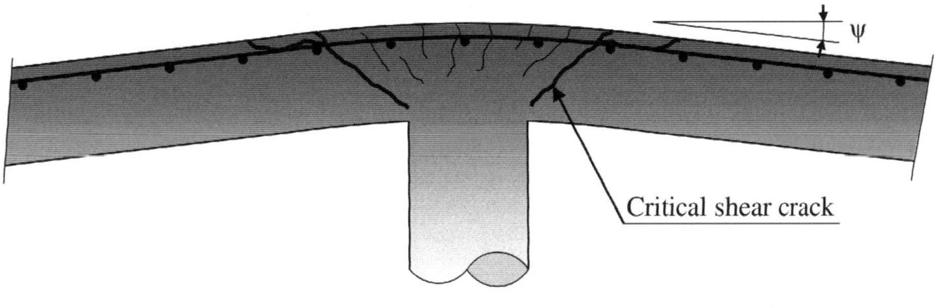

Figure 1. Developing critical shear crack and related slab rotation.

3. Eurocode 2 (EC2) model

The shear resistance in flat slabs without shear reinforcement is ensured by several aspects. It mainly depends on the aggregate interlock in shear crack, compressed concrete area under neutral axis and longitudinal reinforcement. These facts are considered in the empiric formula which determines the punching shear resistance of a slab according to EC2.

$$V_{Rd,c} = \frac{C_{Rk,c}}{\gamma_C} k\left(100\rho_l f_{ck}\right)^{1/3} u_l d \qquad (3)$$

where:

$C_{Rk,c}$ - empirical factor [MPa];
γ_C - partial safety factor, $\gamma_C = 1.5$ [-]; for the analysis γ_C was taken equal 1
k - the size factor, $k = 1 + (200[\text{mm}]/d)^{0.5} \leq 2.0$ [-];
ρ_l - reinforcement ratio, $\rho = (\rho_x\rho_y)^{0.5}$ [-];
f_{ck} - characteristic concrete compressive cylinder strength [MPa];
u_l - shear-resisting basic control perimeter at distance of $2d$ from the face of a column [mm];
d - effective depth [mm].

4. Experimental specimen

The subject of this analysis is a fragment of a flat slab supported by an internal column with significantly rectangular cross-section. The specimen has dimensions 2.5 m x 2.5 m and is 0.2 m thick. The measured concrete cylinder strength was 22 MPa and maximum aggregate size $d_{g.\,max} = 16$ mm was used. The slab fragment was reinforced with steel reinforcement of B500B with yield strength f_{ym} = 550 MPa: top face reinforced with 16 mm diameter bars spaced equally by 100 mm in both orthogonal directions and 10 mm bars spaced equally by 200 mm on the bottom face. The concrete of column reached 89 MPa of cylinder strength. It was reinforced with 12 steel bars with diameter $\varnothing = 12$ mm spaced evenly along the column perimeter.

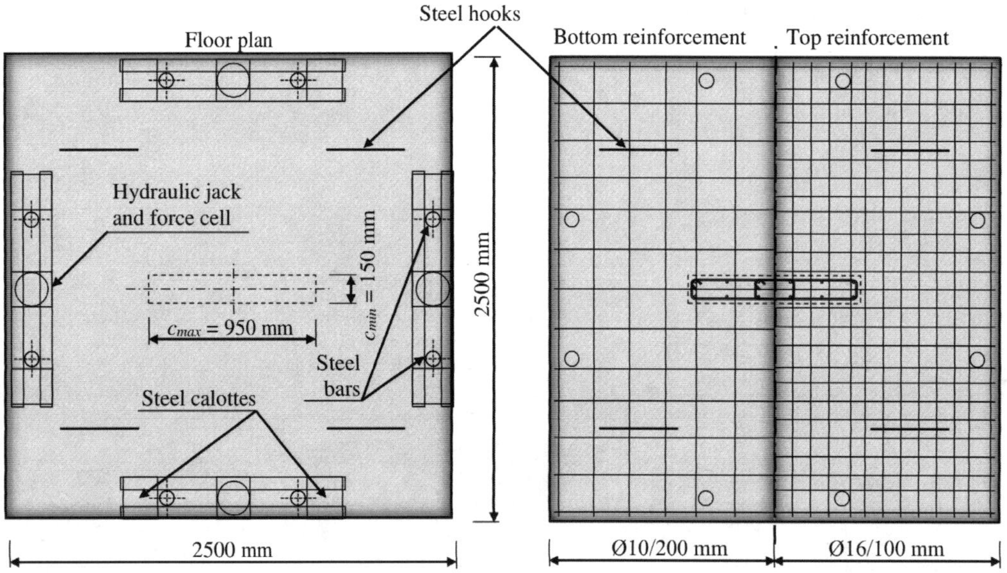

Figure 2. Experimental specimen.

5. Non-linear analysis

The non-linear analysis (NLA) was performed in program Atena, which is based on finite element method [7]. The conditions of specimen symmetry were used in the modelling and therefore only a quarter of the whole specimen was modelled, which significantly shortened the evaluation time.

A cubic mesh of finite elements with side of $a = 40$ mm was used in the model. The specimen was loaded evenly around its perimeter. The load was applied to the specimen through the steel calottes in the form of concentric forces according to the Figure 2.

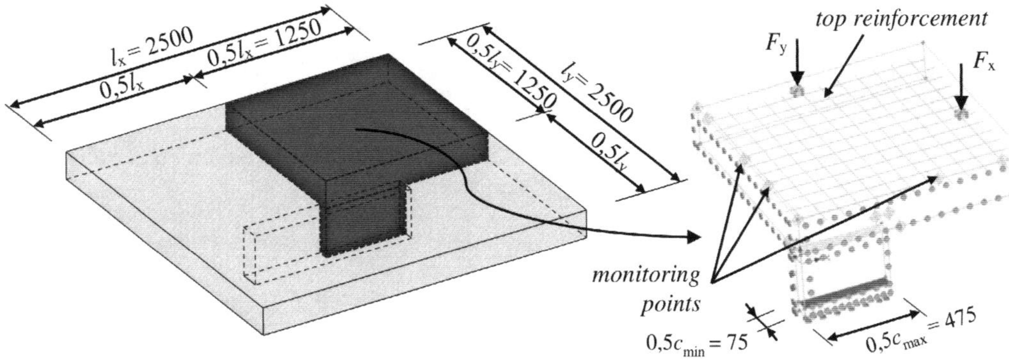

Figure 3. Model of specimen fragment in program Atena.

6. Results

The slab rotations were evaluated based on the resultant nodes' deformations of the modelled specimen fragment. Nodes' deformations were estimated at a distance d from the support face. The rectangular column was divided into several segments according to Figure 4. The rotation in the selected segment was evaluated as the average value of the individual nodes' rotations of a cubic mesh with a side length of a = 40 mm. The shear resistance was determined on individual segments separately. The resultant shear resistance was obtained by integrating the individual values of $v_{Rd.c}$ along the length of the basic control perimeter. The results from this analysis are shown in Table 1.

Table 1. Shear resistance obtained by applying CSCT considering full length of control perimeter.

Segment	$b_{0.i}$ (m)	Ψ_i (mrad)	$k_{\psi i}$ (-)	$v_{Rd.c}$ (kNm^{-1})	$V_{Rd.c}$ (kN)
1	0.95	4.768	0.549	421	400
2	0.95	7.442	0.477	365	347
3	0.52	9.715	0.429	330	169
4	0.30	14.048	0.361	276	82
				$\sum$=	998

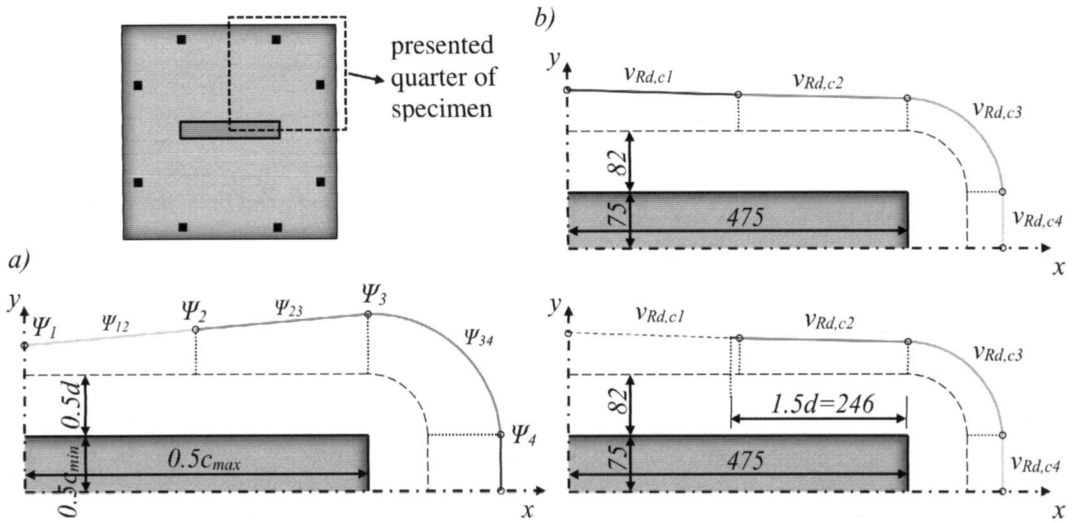

Figure 4. The slab rotations and associated shear resistance of the fragments.

As Figure 4a shows, due to the concentration of shear forces in the column corners, the largest rotations at a distance of 0.5*d* from the support face occur just in the vicinity of the shorter cross-section dimension of the column c_{min}. As a consequence of this phenomenon, the shear resistance $v_{Rd,c}$ in this region reaches the lowest value.

Table 2. Shear resistance obtained by applying CSCT considering reduced length of control perimeter.

Segment	$b_{0.i}$ (m)	Ψ_i (mrad)	$k_{\psi i}$ (-)	$v_{Rd.c}$ (kNm^{-1})	$V_{Rd.c}$ (kN)
1	0.34	4.768	0.549	416	141
2	0.95	7.442	0.477	365	347
3	0.52	9.715	0.429	325	168
4	0.30	14.048	0.361	294	88
				∑=	744

The results of the experimental test showed that the shear resistance of slab-column connection reaches 688 kN. Based on the formula according to the design model EC2, it was concluded that the necessity to reduce the length of the control perimeter is needed. Compared to CSCT, EC2 shows a higher safety of almost 18 %. If the recommendation to reduce the length of the control perimeter to a distance of 1.5*d* from the corners of the support was applied when evaluating according to CSCT, the shear resistance would decrease significantly (Table 2), as the part of the perimeter with the largest share of resistance is excluded.

In cooperation with the Department of Geodesy, Svf STU, a method of measurement using convergent photogrammetry was implemented in the experimental test. The principle of this method consists in the selective measurement of the position of the points signalized by special code marks. Based on the measured deformations of individual points, it was possible to evaluate slab rotations in segments in the same way as in numerical analysis. The results from experimental test are presented in Table 3.

Table 3. Shear resistance obtained by applying CSCT to the experimental test.

Segment	$b_{0.i}$ (m)	Ψ_i (mrad)	$k_{\psi i}$ (-)	$v_{Rd.c}$ (kNm^{-1})	$V_{Rd.c}$ (kN)
1	0.95	8.525	0.453	348	330
2	0.95	8.356	0.457	350	333
3	0.52	8.951	0.444	341	176
4	0.30	9.687	0.430	330	99
				∑=	938

The design model EC2 does not take into account the deformed form of the slab in the evaluation of shear resistance. Therefore, its calculation differs significantly from CSCT method, which causes a significant difference in the results. The reliability of these two models is presented in the Table 4 in consideration of the experimentally measured shear resistance.

Table 4. The shear resistance according to the presented models and evaluation of their reliability.

b_0	$V_{Rd,c}$			V_{NLA}	V_{test}	$V_{test}/V_{Rd,c}$			V_{NLA}/V_{test}
	EC2	**CSCT$_{NLA}$**	**CSCT$_{test}$**			**EC2**	**CSCT$_{NLA}$**	**CSCT$_{test}$**	
(m)	**(kN)**	**(kN)**	**(kN)**	**(kN)**	**(kN)**	**(-)**	**(-)**	**(-)**	**(-)**
full	794	998	938	825	688	0.87	0.69	0.73	0.83
reduced	623	774				1.10	0.89		

The difference between shear resistance obtained by applying CSCT theory from non-linear analysis and the experimental test reaches 10 %. This phenomenon is caused by a significant difference in the actual slab rotations measured in the experimental test. Especially in the first segment the slab rotations reach twice as high values than in the non-linear analysis. The significant difference in the results of shear resistance indicates the importance of experimental tests.

7. Conclusions

The subject of this analysis was the application of CSCT method to evaluate the shear resistance of the flat slab and to compare the results with the design model EC2. Based on this analysis, the following conclusions were derived:

1. The largest rotations near the corners of the support prove the concentration of shear forces in this area in the case of a slab supported by a rectangular column.
2. The segments in the middle of the longer cross-sectional dimension c_{max} have the biggest share in the shear resistance, which are eliminated from the total shear resistance when reducing the length of the control perimeter.
3. It was concluded that a reduction of the length of the control perimeter is actually needed in both design models due to experimentally determined shear resistance.

8. References

[1] Kinnunen S and Nylander H 1960 *Punching of concrete slabs without shear reinforcement* Transaction No. 158 (Stockholm: Royal Institute of Technology) p 112

[2] Braestrup M W and Nielsen M P 1976 *Axisymmetric Punching of Plain and Reinforced Concrete* Report No. R75 (Denmark: Technical University of Denmark) p 43

[3] Moe J 1961 *Shearing Strength of Reinforced Concrete Slabs and Footings under Concentrated Loads* Bulletin D47 (Chicago: Portland Cement Association) p 135

[4] Muttoni A and Schwartz J 1991 Behaviour of Beams and Punching in Slabs without Shear Reinforcement *IABSE Colloquium* **62** 703-8

[5] Fédération Internationale du Béton (fib) 2012 *Model Code 2010 - Final draft vol 1 Fédération Internationale du Béton* Bulletin 65 (Lausanne: Ernst & Sohn) vol 2

[6] EN1992-1-1 2004 *Design of Concrete Structures Part 1-1 General Rules and Rules for Buildings*

[7] Červenka V, Jendele L and Červenka J 2018 ATENA Program Documentation Part 1 Theory (Prague: Červenka Consulting)

Acknowledgments

This work was supported by the Slovak Research and Development Agency under the contract No. APVV-15-0658 and Scientific Grant Agency of the Ministry of Education, science, research and sport of the Slovak Republic and the Slovak Academy of Sciences No 1/0254/19.

IOP Conf. Series: Materials Science and Engineering 1209 (2021) 012061 doi:10.1088/1757-899X/1209/1/012061

Comparison of the effect of GFRP and steel reinforcement in concrete members subjected to compressive stress

Ž Šenšelová[1] and V Borzovič[1]

[1] Faculty of Civil Engineering, Slovak University of Technology in Bratislava, Radlinského 11, 810 05 Bratislava, Slovak Republic

E-mail: zaneta.senselova@stuba.sk

Abstract. Composite materials became more popular and commercially available as reinforcement for concrete elements. Fibre Reinforced Polymer (FRP) bar is an excellent thermal and electrical insulator with high tensile strength and low weight. These assumptions make them a possible substitution for steel reinforcement. Moreover, GFRP is not responsible to corrosion for that are suitable for structures with high humidity and unfavorable environment. GFRP is easier to handle due to its low weight. Also, it has electromagnetic neutrality. But it has some disadvantages. It has a low modulus of elasticity and sensitivity to elevated temperatures. Another drawback and uncertainty with designing is the impact of an alkaline environment, which decreases the long-term strength of GFRP bars. This paper describes a pre-experiment study of concrete elements resistance. The analysis is performed for a cross-section of 200x150 mm for a short concrete column with steel and GFRP reinforcement. The study compares P-M diagrams for steel reinforcement and GFRP reinforcement with different reinforcement ratios. Other characteristics such as tensile strength and modulus of elasticity must be considered to design the GFRP reinforced concrete element. The study also considers the contribution of GFRP reinforcement in compression. The analysis has shown, the shape of interaction diagrams of steel and GFRP reinforcement are significantly different.

1. Introduction

The basic component of FRP reinforcement is long fibers. The common fibers for polymer reinforcement are glass, aramid, basalt, and carbon. All FRP reinforcements behave linearly elastically up to failure under tensile stress. The most common composite reinforcement is with glass fibers (Glass Fibre Reinforced Polymers - GFRP). At a constant high level of load, glass fibers are prone to creep and subsequent sudden rupture. GFRP reinforcements are generally excellent thermal and electrical insulators; they have a high tensile strength of 800 – 1200 MPa. In contrast to their positive properties, they have a low modulus of elasticity of 35-60 GPa, are highly sensitive to humidity and pH in the environment, where long-term exposure to an alkaline environment in combination with long-term stress leads to a significant decrease in tensile strength up to 20 – 40 % compared to short-term strength [1, 2]. For concrete element reinforced by GFRP reinforcement is a typical brittle failure caused by an absence of the plastic branch in a working diagram of GFRP reinforcement. Failure can come suddenly and can occur in two ways: crushing the concrete or tensile rupture of the reinforcement. The failure is not signaled by accompanying phenomena such as a significant increase in deflection and crack width.

The paper deals with the resistance of elements subjected to a combination of axial force and bending moment. The calculation is based on the procedure of Fib bulletin 40: FRP reinforcement in RC structures [1]. The study was performed on column elements, an example of the stress state of a compression column or pile, for column element was used a cross-section of 200x150 mm reinforced by GFRP and steel reinforcement. The parametric analysis will be used as preparation for future experiment programs. Due to parametric analysis was used short-term strength of GFRP reinforcement. The target is to point out the differences between FRP and steel reinforcement with different reinforcement ratios and analyze the impact of GFRP reinforcement for cross-sectional resistance.

Figure 1. An example of GFRP reinforcement with different surface treatment (wrapping, sandblasting).

2. Compressive strength of GFRP reinforcement

In the procedure of Fib bulletin 40 is stated that the compressive strength of GFRP reinforcement can be neglected. Many authors do not consider the compressive strength of GFRP reinforcements. Despite this fact, some authors consider this contribution; even it is not significant. Benmokrane even states in his study that the compressive strength of GFRP reinforcement is only about 40 – 60 % of their tensile strength [3]. However, Tobbi notes that by neglecting the compressive strength of the GFRP reinforcement, er underestimates the maximum axial stress of the element [4]. On this basis, he derived a relationship for resistance in centric pressure, which also includes the pressure contribution, based on the elastic theory and properties of FRP reinforcement:

$$P_0 = 0.85 f_c (A_c - A_f) + \varepsilon_0 E_f A_f \quad (1)$$

Where f_c is the compressive strength of the concrete; A_c is the cross-sectional area of the concrete; A_f is the cross-sectional area of the longitudinal GFRP reinforcement; ε_0 is the ultimate deformation of concrete; E_f is the modulus of elasticity of the GFRP reinforcement.

There are currently no test protocols available to determine the compression behavior of GFRP reinforcement. In article [5,] AlAjarmeh developed a new test method for determining and describing the pressure properties of GFRP reinforcements based on the ASTM D695 test procedure. The GFRP reinforcement is placed into steel pipes on both ends before being filled with cement grout. Rods of four different diameters and three different lengths were used in the testing. According to compression strength testing, the smaller diameter reinforcements had the same compressive strength as their tensile strength. The tensile strength of the larger diameter reinforcements was only 75% and 65% of the tensile strength of the smaller diameter reinforcements, respectively. The length-to-diameter ratio of a part also influences compression behavior. 4 or 8 are ideal ratios.

3. Column element with recycled aggregate concrete

Innovative concrete production technologies must be found for concrete to become a sustainable material. One such solution is recycled aggregate. Load testing on samples is required to establish the material properties of recycled concrete.

A similar study was performed by Raza [4] on column elements from recyclate aggregate concrete reinforced by GFRP (GFRP recyclate aggregate concrete – GRAC) and steel reinforcement (steel recyclate aggregate concrete – SRAC). The experiment study found that compressive resistance of column element from GRAC was low, about 7.79% than compressive resistance of column element from SRAC. In terms of axial compressive resistance and failure behavior, recycled aggregate concrete performed well in GRAC and SRAC columns. Recycled aggregate concrete can be characterized by a lower value of the modulus of elasticity Ec. It allows with a higher strain of concrete used more compressive effect of GFRP reinforcement.

4. P-M diagram

The parametric study used concrete of class C25/30 with material characteristics according to STN EN 1992-1-1 [6] listed in table 1, and steel reinforcement of class B500B and GFRP reinforcement with material characteristics listed in table 2.

Table 1. Material characteristics of concrete.

	Char. value of concrete compressive strength f_{ck} [MPa]	The partial factor of reliability γ_c	The design value of concrete compressive strength f_{cd} [MPa]	Mean value of modulus of elasticity E_C[GPa]
C25/30	25	1.5	16.67	31

Table 2. Material characteristics of reinforcement.

	Char. yield strength reinforcement f_{yk} [MPa]	The partial factor of reliability γ_s	The design yield strength reinforcement f_{yd} [MPa]	Mean value of modulus of elasticity E_S [GPa]
GFRP	1000	1.25	800	50
B500B	500	1.15	434.8	200

Figure 2 shows the working diagram used in the computation.

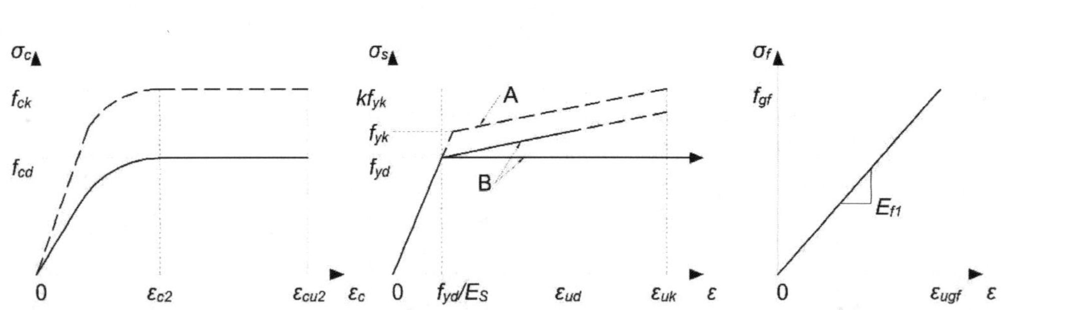

Parabolic-linear stress-strain diagram

Stress-strain diagram for reinforcing steel, A – idealized diagram, B – design diagram

Stress-strain diagram for GFRP

Figure 2. σ-ε diagrams.

The study was performed on column elements with cross-section 200x150mm with two different reinforcement ratios. The cross-section was reinforced symmetrically at the shorter edge with GFRP and steel reinforcement, allowing better eccentric load. The first case of the element is reinforced by 3 bars with a diameter of 10 mm on both sides, as shown in Figure 3a. The second case of the element is reinforced by 3 bars with a diameter of 16 mm on both sides, as shown in Figure 3b. Longitudinal bars are GFRP or steel reinforcement; stirrups are steel reinforcement.

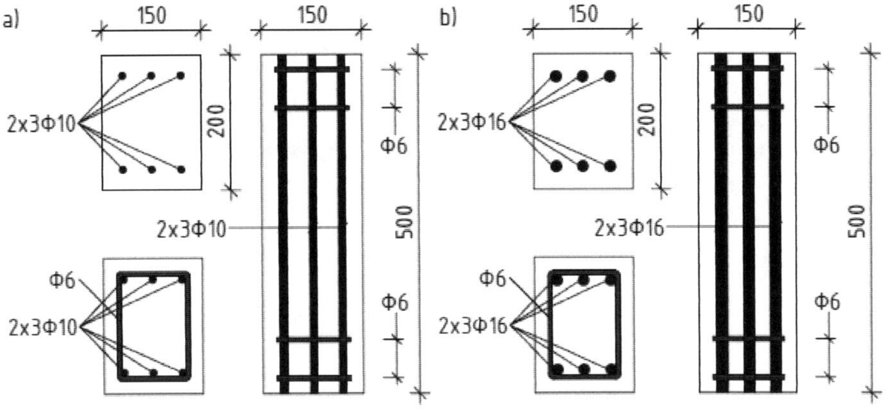

Figure 3. Elements reinforcement schemes.

The first step was to compare cross-sections reinforced by GFRP bars without the contribution of compression pressure and compression pressure contribution. Reinforcement by six bars with a diameter of 10 mm represents a 1.57% reinforcement ratio, and a diameter of 16 mm represents a 4% reinforcement ratio. It can be stated that the pressure contribution is low due to it is limited by the relative limiting deformation of concrete in compression. On the contrary, the maximum bending resistance of the element, including the contribution in compression, decreased.

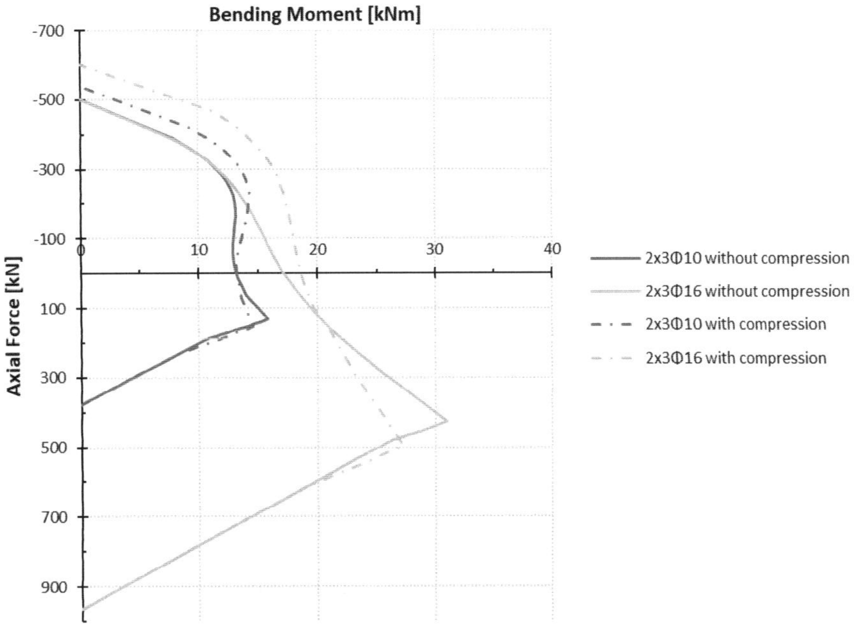

Figure 4. Interaction diagram of element – compare GFRP reinforcement with compression strength and without compression strength.

IOP Conf. Series: Materials Science and Engineering 1209 (2021) 012061 doi:10.1088/1757-899X/1209/1/012061

As shown in Figure 4, compression resistance increases with the counting of contribution strength of GFRP reinforcement. At reinforcement by bars, a diameter of 10 mm compression resistance increase by 7.8%, and reinforcement by bars a diameter of 16 mm increase by 20%. In contrast, theoretical bending resistance decrease by 20% at the element with bars of diameter 10mm, and by 32.8% at the element with bars of diameter 16 mm.

Although the contribution I compression is not significant, we would underestimate the overall resistance to compression and bend by neglecting it.

The analysis deals with the comparison of the resistance of an element reinforced with steel and GFRP reinforcement. Figure 5 shows P-M diagrams for both reinforcement ratios. The calculation was made for the design values of materials.

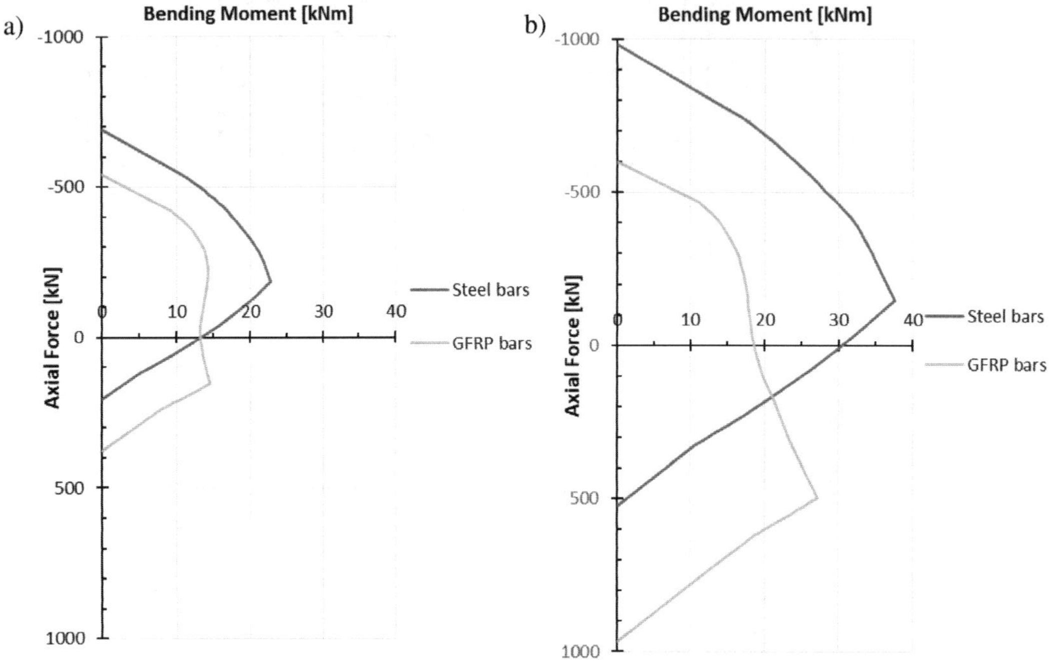

Figure 5. Interaction diagrams – comparison of elements reinforced by GFRP and steel reinforcement
a) element reinforced by bars with a diameter of 10 mm
b) element reinforced by bars with a diameter of 16mm.

As shown in figure 5, the tensile resistance of elements reinforced by GFRP bars is significantly higher than the tensile resistance of elements reinforced by steel bars. On the contrary, compression resistance and bending resistance of elements reinforced by GFRP bars are lower than those reinforced by steel bars.

5. Conclusion

Replace steel reinforcement with composite reinforcement looks at first glance simple. However, the impact of individual elements is markedly different. Resistances of a designed element differ significantly depend on reinforcement ratios, the shape of the cross-section, or serviceable life.

The contribution of GFRP reinforcement compression slightly increases the element's compression resistance and decreases theoretical bending resistance. The slenderness of the element can cause it. The recycled aggregate concrete may be caused by increased overall resistance of the element.

The presented analysis is a preparation for an experimental study, which will be performed on short columns 500 mm long with a cross-section of 200x150 mm. One group will be with reference concrete C25/30, and the other group will be with concrete C25/30 with recycled concrete as a full substitute for aggregates.

6. References

[1] Technical report 2007 *Fib Bulletin 40: FRP reinforcement in RC structures* (Stuttgart: Sprint Digital Druck) pp 12–39 pp 61–67

[2] Benko V, Bilčík J, Gažovičová N, Girgle F, Hollý I and Štěpánek P 2015 *Manuál na navrhovanie GFRP výstuže do betónových konštrukcií* pp 11–50

[3] Benmokrane B, Chaallal O and Masmoudi R 1995 Glass fiber reinforced plastic (GFRP) rebars for concrete structures *Construction and Building Materials* **9** 6 353–64

[4] Tobbi H, Farghaly A S and Benmokrane B 2014 Behaviour of concentrically loaded fiber-reinforced polymer reinforced concrete columns with varying reinforcement types and ratios *ACI Struct. J.* **111** 2 375–86

[5] Al Ajarmeh O S, Manalo A C, Benmokrane B, Vijay P V, Ferdous W and Mendis P 2019 Novel testing and characterization of GFRP bars in compression *Construction and Building Materials* **225** 1112–26

[6] STN EN 1992–1–1 + A1 2015 *Navrhovanie betónových konštrukcií. Časť 1: Všeobecné pravidlá a pravidlá pre budovy*

[7] Raza A and U Rafique 2020 Efficiency of GFRP and hoops in recycled aggregate concrete columns: Experimental and numerical study *Composite Structures* **255** 112986

Acknowledgments

This work was supported by the Scientific Grant Agency VEGA under the contract No. VEGA 1/0645/20. This work was supported by the Slovak Research and Development Agency under the contract No. APVV-17-0204.

IOP Conf. Series: Materials Science and Engineering 1209 (2021) 012062 doi:10.1088/1757-899X/1209/1/012062

The influence of signal noise to measurements at bridge structure

R Štecák[1]

[1] Department of Steel and Timber Structures, Faculty of Civil Engineering, Slovak University of Technology in Bratislava, Radlinskeho 2766/11, 81005 Bratislava, Slovakia

E-mail: robert.stecak@stuba.sk

Abstract. Bridge weight-in-motion (BWIM) system is a method, that provides to identify axle weights. It is a non-destructive method, which allows not only to identify the axle weight, but it can show current shape of the structure, so it has a great potential. There are various methods to do measurements for this system. Mostly, accelerometers or strain gauge are used. Signal noise has significant effects to the results. It could be resonance of the bridge, wind, defect at the support system, defect at the roadway, etc. It is necessary to filter all this effects, to get clear data. There are many ways to do the filtering. Digital filters allow it. Sometimes, this type of filtering could remove important data about the crossing of the vehicle. It could generate inaccuracy of the whole system and create major errors to identified vehicles. It is necessary to find the optimal way, to keep important data and remove all dynamic noise. This paper will investigate the previously mentioned problems. Measurements will be accomplished on a small-scale model of the bridge. Vehicle will be crossing over the bridge, while the bridge will be awakened to the first vibration shape and other frequencies, that will have a great impact to the measurements.

1. Introduction

Measurements of the vehicle weights are quite important. Nowadays, the transport is often overloaded, the intensity is higher. It has significant influence on the shape of the roads and bridges. This paper will focus on the bridges. BWIM system is used to identify axle weight of the vehicles. It is usable to road and train bridges. It is the non-destructive method, so it has no effect to the support system of the bridges. It is much accessible than static scales, which requires stopping the vehicle.

There are many methods to do the measurements. Mostly, the strain gauges or accelerometers are used. With this type of sensors several problems appear. The most influential is dynamic noise. It could happen with wind, defect at the support system, defect of the vehicle, which happens mostly in train bridges. There are several methods to eliminate this influence. Digital filters allow to exclude the noise from the measurements. Digital filter allows to cut some frequencies off and measurements will not include data, which were generated with these frequencies. The most essential element is to find the right frequencies. Using of the wrong values could cut such data off which were evoked by the vehicle and measurement will not be applicable.

2. Bridge weight in motion system

The BWIM system is the method, which allows to detect axle weights without any intervention to the support system of the bridge. This method uses Moses algorithm [2], which is based on the fact that a moving load along the bridge will causes stresses in proportion to the product of the value of the influence line and the axle load magnitude. Moses uses fact that each individual girder stress is related to the moment from the relationship:

$$\sigma_j = \frac{M_j}{W_j} \tag{1}$$

where:
j= 1... G (number of girders)
σ_j = the stress in the jth girder
M_j = the bending moment in the jth girder
W_j = the section modulus
The bending moment can be expressed in terms of strain as:

$$M_j = W_j * \sigma_j = E * W_j * \varepsilon_j \tag{2}$$

where:
E = the modulus of elasticity of the bridge material
ε_j = the strain in the jth girder
Assuming each individual bending moments, the global bending moment is:

$$M = \sum_j^G M_j = \sum_j^G E * W_j * \varepsilon_j = EW \sum \varepsilon_j \tag{3}$$

This algorithm could be used with various kinds of sensors. The most used are strain gauges. With this type of sensors, there is quite a great influence of the noise in the measurement. It is necessary to clear the data.

3. Dynamic noise

The dynamic noise could be caused by many factors. The wind, defect at the bridge construction, defect of the axles, workers at the construction, etc. All these aspects have considerable impact to the results. Each of these aspects could evoke harmonic oscillation and create resonance of the construction. In this situation, vehicle passage is considerably distorted. Accuracy of the measurement is low, and it could cause the wrong identification of the axle weight and then the identification of the vehicle.

3.1. Vehicle damage

In train bridges, the train could cause the enormous influence on the results. It is due to its own length and high possibly of defect of the axle. It is necessary to filter out all this defects in the record.
As is shown in figure [4], there is a derailment of the train. Of course, there is low probability of this scene, but it can show the effect to the measurement. The whole record is not usable.

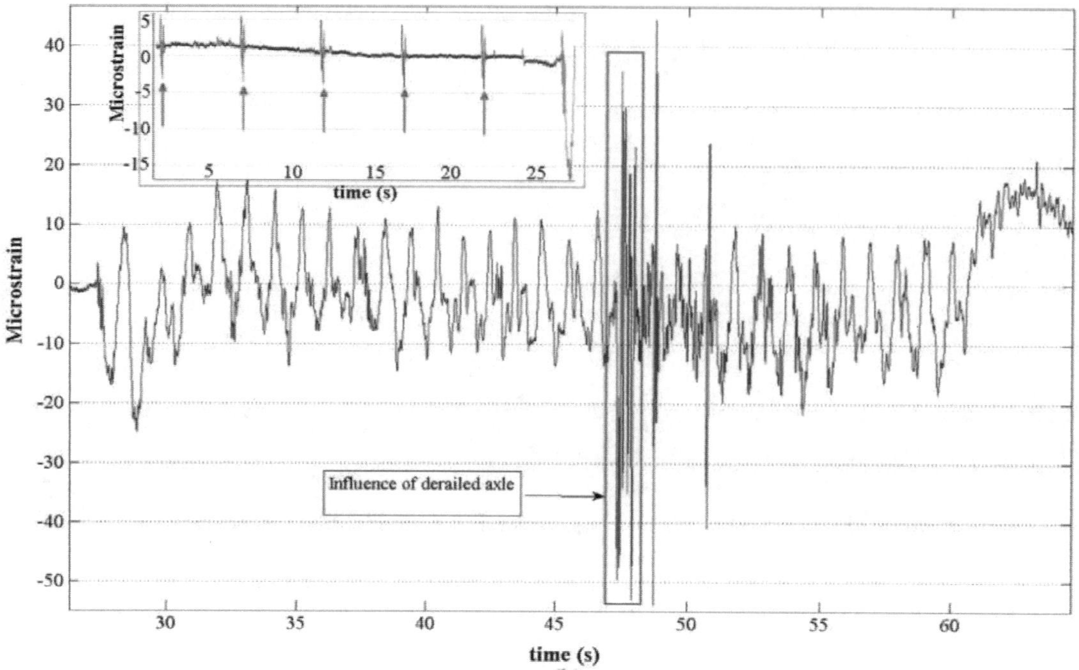

(b)

Figure 1. Strain time series associated to the derailed train in the cross-girder [4].

3.2. Complication at the dilatation

Another factor could be some issue at the bridge. A part of the support system, dilatation or part of the roadway could be damaged. E.g., dilation could cause defects about the identification of the axles. For comparison, there is a passing of the vehicle over the bridge with exceptional dilatation and erroneous dilatation. The strain gauge was located in the middle of the span, at cross-beams. For measurements was used simple quarter bridge, without temperature-sensing (dummy) gauge. There were no change in temperature, because all measurements were done in laboratory.

As is possible to see in figure 4, the influence of dilatation is huge. In the part *a)* of the figure, it is easy to identify the two axles of the vehicle. In the part *b)* it looks like three axles. The erroneous dilatation caused impossibility to identify axles. Another problem is with passing over the dilatation. The strain gauge recorded the moment when vehicle was on the dilatation. Then, the whole measurement is inaccurate. The strain does not return to the original value of the strain when vehicle was over. This phenomenon is caused by the jump of the vehicle over the dilatation. It has quite enough power, to detune the strain gauge and it cannot return to original value.

This kind of complication is impassable to recover. No digital filter will help. It is quite complicated to detect the frequency when vehicle is passing over the dilatation. It is caused by the fact, that passing over dilatation takes not enough time and Fast Fourier Transformation (FFT) is not feasible. The fact, that axles of the vehicle are not detectable is not possible to solve. It is required to correct existing problem with part of the construction.

Figure 2. The truss bridge used for measurement with span of 2.5 meters.

a) b)

Figure 3. Experimental model with dilatation. On the left, a) exceptional dilatation, on the right, b) erroneous dilatation.

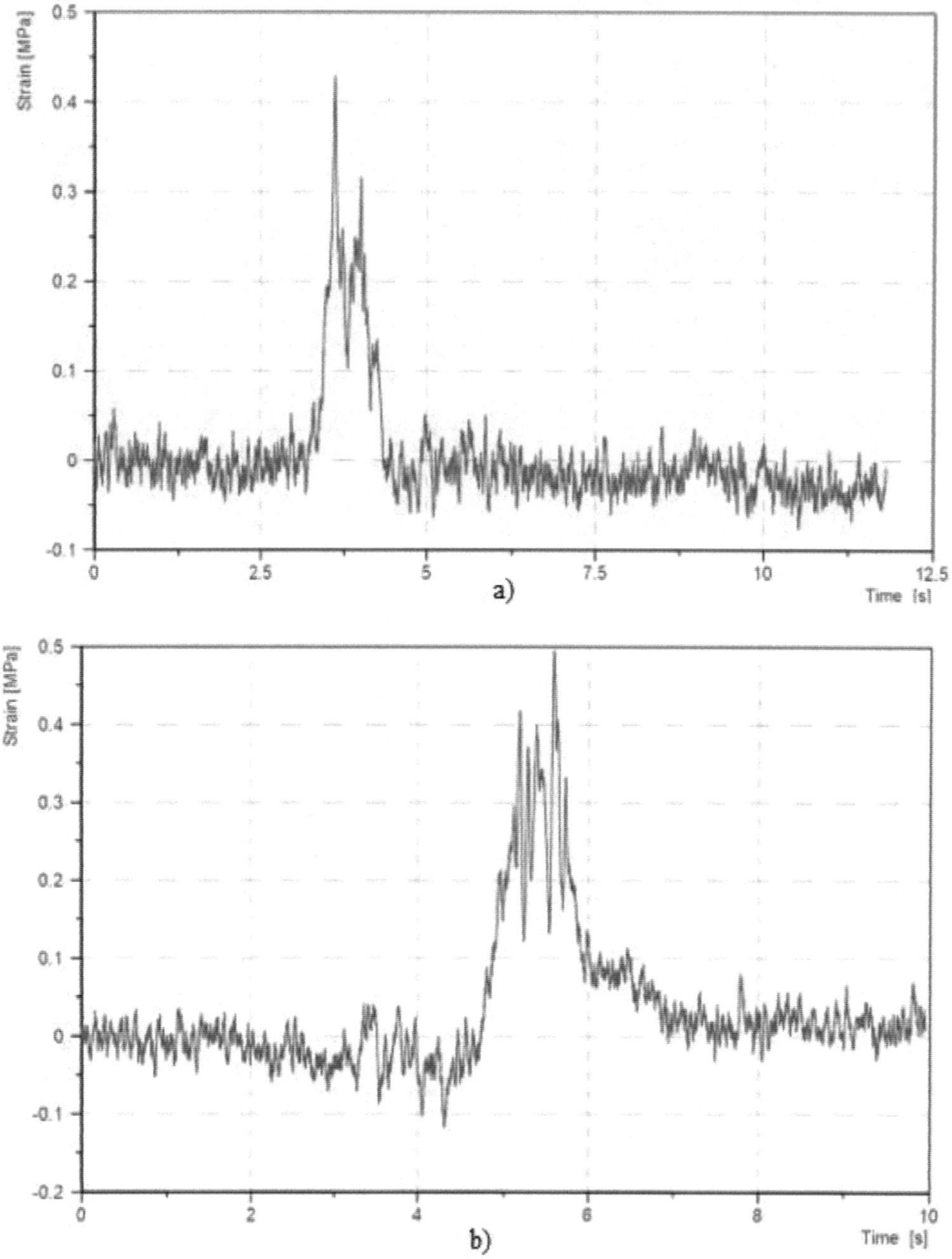

Figure 4. a) Passing of the vehicle over the exceptional dilatation b) Passing of the vehicle over the erroneous dilatation.

3.3. Excitation of the structure

The significant factor, which has a great influence on the measurement is the excitation of the structure. It does not have to be first vibration mode, but any other mode. These shapes could be generated by wind, vehicle, workers at the bridge, etc.

In contrast with previous factor, this one could be fixed in post processing. The significant element is to detect the frequency of the excitation. It is possible in the software, but there are always differences between computational model and reality. The most suitable way is to measure frequencies at the superstructure by accelerometers. This method allows to place many sensors at different members of construction. Then it is possible to detect excitation frequencies by Fast Fourier Transformation (FFT) from the record.

As it is possible to see in figure 5, the FFT shows many frequencies. The most significant frequency is about 14 Hz. It does not mean, that it is the first vibration mode. It only has the highest influence. Lower frequencies were generated by pedestrians. This is exactly the case, where it is necessary to filter out the signal. Lower frequencies could generate disruptive values to the record.

The digital filters must be used, to filter out the values. Generally, Bessel, Butterworth or Chebyshev filters are used. Each type of filter has different algorithm [5]. In this paper Butterworth filter was always used. There are many opportunities with digital filters. Signal could by modified by many ways. In this case, if it is looking like disturbing frequencies, it is possible to use bandstop and not to allow to release part of the record, where the disturbing frequencies are located.

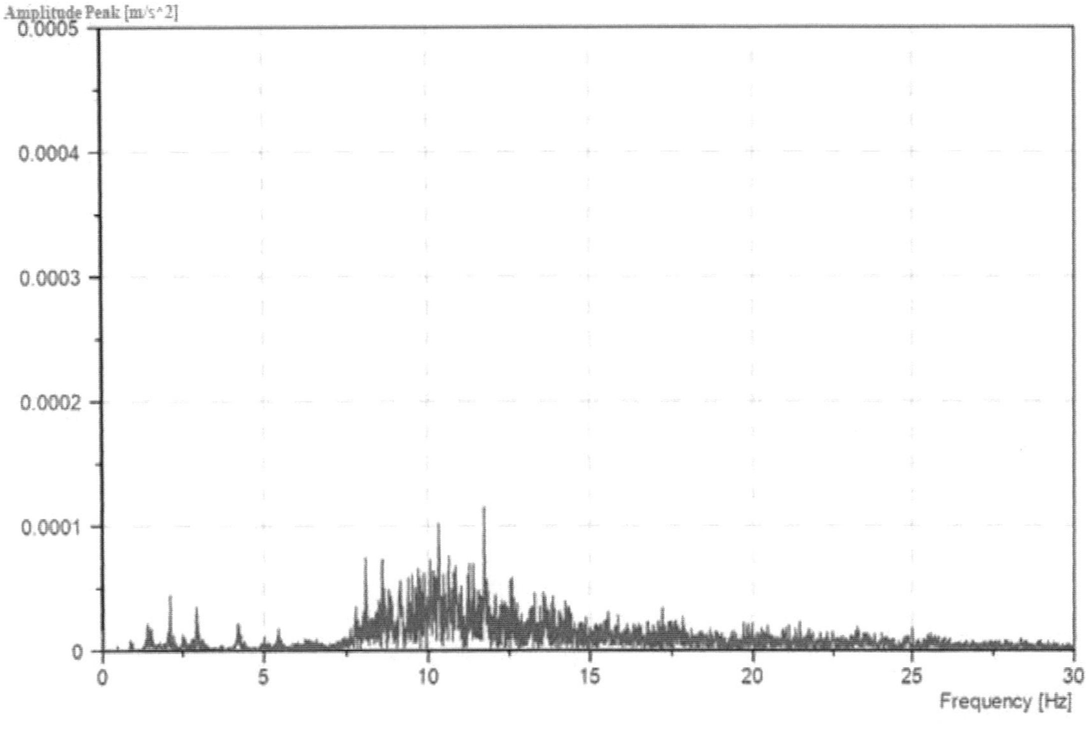

Figure 5. Fast Fourier Transformation with important frequencies of steel console of road bridge.

4. Data processing

In this paper, the attention to the data processing of measurements will be paid, while construction is awaked by exciter to the specific vibration mode. The truss bridge with 2.5 meters span was used. The superstructure was awakened to the most intensive vibration. It is shown in figure 6, frequency 34.5 Hz is the most significant.

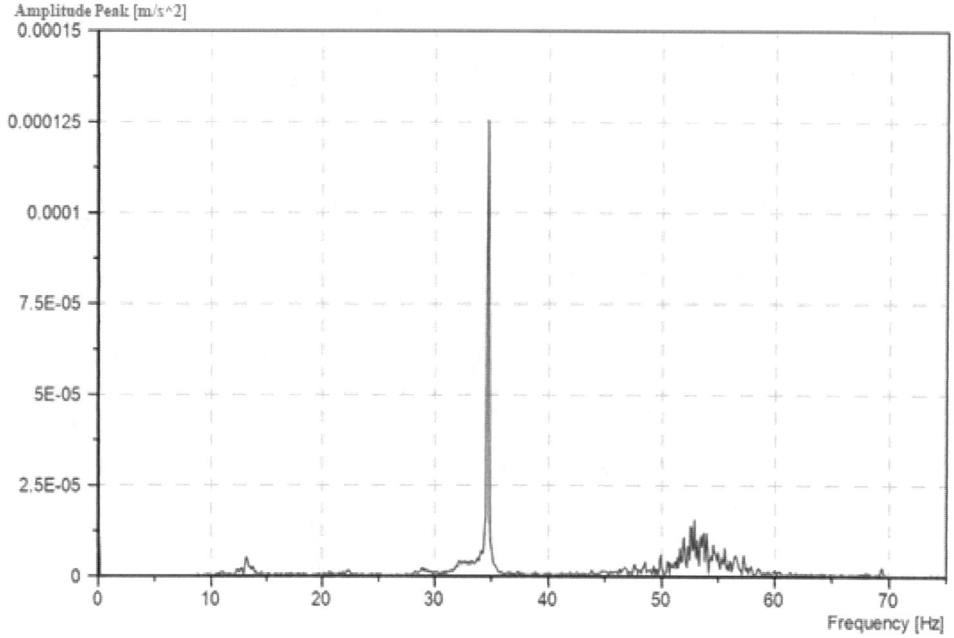

Figure 6. The frequencies generated by the exciter.

During the measurement, the vehicle run over the bridge four times. The exciter was set, to generate the most significant vibration of the construction. After the measurement, the data was filtered by Butterworth filter. It was set the band stop for the frequencies, which were recorded by accelerometer. The band stop frequencies were: 13-14, 34-35, 52-54 Hz. These frequencies from the record are very similar to the model from the software RFEM. The diversity between the measured and calculated frequencies is acceptable. Then, the bridge is of older construction, therefor not all components are in good shape.

As it is possible to see in figure 7, these frequencies have no influence on the measurement. It has the simple reason. They are too high to have influence on the passing of the vehicle. Generally, walking people, passing of the vehicle is quite below frequencies.

Table 1. Frequencies from the software.

Mode no.	Angle frequency *v* [rad/s]	Natural frequency *f* [Hz]	Natural period *T* [s]
1	95.44	15.19	0.066
2	103.291	16.439	0.061
3	169.906	27.041	0.037
4	274.385	43.67	0.023
5	279.972	44.559	0.022
6	358.036	56.983	0.018
7	362.904	57.758	0.017
8	371.434	59.116	0.017
9	557.613	88.747	0.011
10	561.4	89.35	0.011

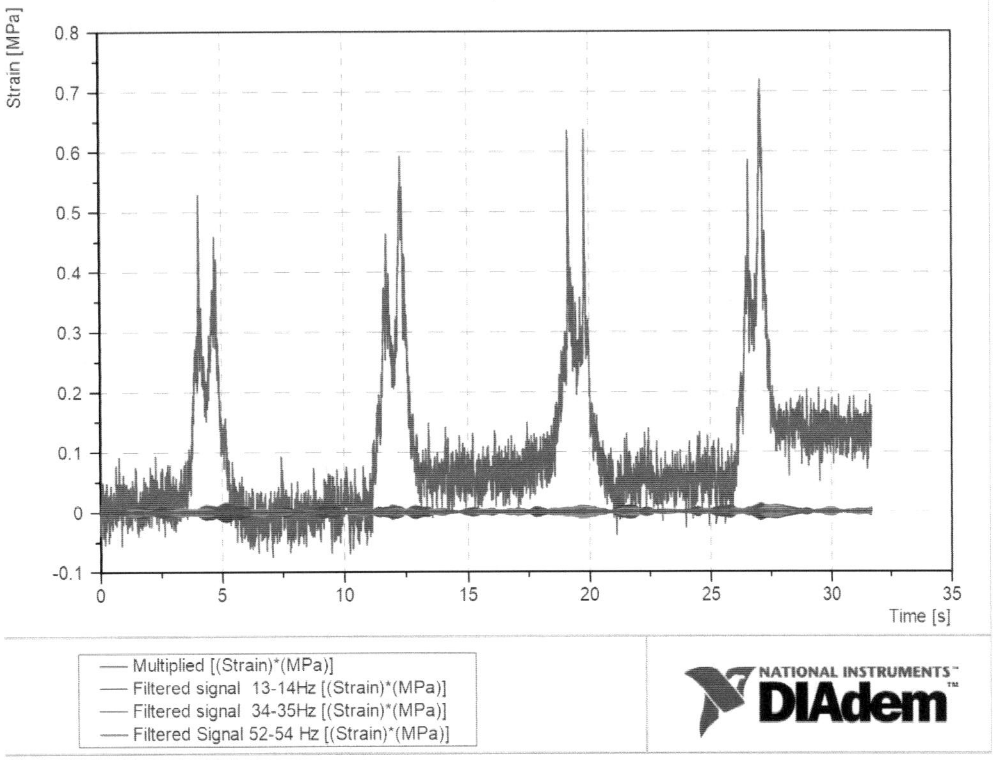

Figure 7. Original data and filtered signal with selected frequencies.

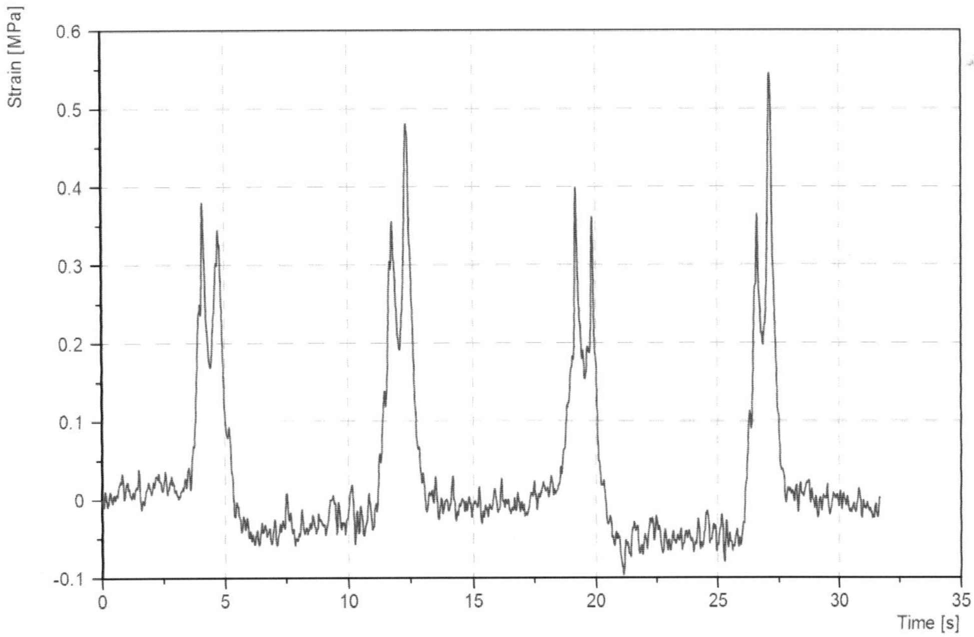

Figure 8. Filtered signal with band pass of the 0.01-4 Hz.

Figure 8 shows, that data with frequencies under 4 Hz include the vehicle pass. If awakening of construction generated values about this range, then it should have significant influence on the results. The high frequencies of the awakening were caused by the small scale of the bridge. This bridge has span of 2.5 meters. The bridge with about 100 meters or more will have resonant frequencies much below. Of course, the exciter must be much more powerful.

5. Conclusion

This paper showed the significance of signal noise at bridge structures. Problems with construction, or dilatation are challenging for the correction in post processing. For instance, the dilatation is kind of issue, that is not possible to be filtered out. This kind of issues could be observed at visual inspection. More complicated obstacle is dynamic noise, which is generated by wind, traffic, etc. The generated frequencies must be identified. As it was shown, traffic generally generates frequencies below 5 Hz. If the construction is rigid and has vibration mode higher than 5 Hz, then this dynamic noise is almost insignificant. At less rigid construction or larger span, the custom shapes are much lower. Then, the vibration awakening of the superstructure of the construction has significant influence on the measurement, and it is crucial to do correct post processing. Digital filters allow to edit data at any range of frequencies. Identification of the problematic frequencies is the most critical step. If this complication is identified correctly, there is almost no problem with identification of the vehicle axles.

6. References

[1] Quilligan M 2003 Bridge Weigh-in-Motion Development of a 2-D Multi-Vehicle Algorithm Trita-BKN Bulletin vol 69 (Stockholm: Byggvetenskap) p 144

[2] Moses F 1979 Weigh-in-Motion System using Instrumented *Trans Engin J of ASCE* **105** 3 233 249

[3] Žnidarič A, Kalin J, Kreslin Mb, Favai P and Kolakowski P 2016 Railway bridge Weigh-in Motion system *Trans Research Procedia* **14** 4010–19

[4] Marques F, Moutinho C, Wei Hua Hu, Cunha Á and Caetano E 2016 Weigh-in-motion implementation in an old metallic railway bridge *Engineering Structures* **123** 15–29

[5] Diniz S R P, da Silva A B E, and Netto L S 2010 *Digital Signal Processing System Analysis and Design* vol 2 (New York: Cambridge University Press) pp 1–913

Acknowledgement

This paper has been supported by the Scientific Grant Agency of the Ministry of Education, science, research and sport of the Slovak Republic and the Slovak Academy of Sciences – grant VEGA 1/0773/18.

Construction types and typologies of existing masonry arch bridges in Croatia

D Trajber[1], D Penava[1], F Rinaudo[2], V Sarhosis[3] and L Abrahamczyk[4]

[1] Faculty of Civil Engineering and Architecture Osijek, The Josip Juraj Strossmayer University of Osijek, Vladimira Preloga 3, 31000 Osijek, Croatia
[2] Department of architecture and Design, Politecnico di Torino, Viale Pier Andrea Mattioli, 39, 10125 Torino, Italy
[3] School of Civil Engineering, Faculty of Engineering and Physical Sciences, University of Leeds, Leeds LS2 9JT, UK
[4] Chair of Advanced Structures, Fakulty of Civil Engineering, Bauhaus-Universität Weimar, Marienstraße 13D, 99423 Weimar, Germany

Email: dtrajber@gfos.hr

Abstract. Historical masonry arch bridges still form an important part of Croatian transportation network. There are approximately 680 masonry arch bridges and culverts currently being used for railways and roadways. Many of these bridges are relatively old (more than hundred years in most cases) but still in usage. Increasing vehicle load and speeds as well as deterioration due to anthropogenic and environmental influence have highlighted the need for reliable assessment of their service condition and regular maintenance. The aim of this study is to provide a review of existing masonry arch bridges in Croatia. Firstly, a historical review of bridges is given showing the time period in which they were built, indicating the materials and design principles used for their construction. Next, bridge typologies are presented as well as their detailed analysis of geometric characteristics for brickwork bridges. Finally, a short review of damages and their impact on serviceability of bridges is given. This review presents masonry arch bridges in Croatia and the need for reliable method of assessing their service condition in order to provide proper maintenance, repairing and retrofitting.

1. Introduction

Masonry arch bridges consists of massive abutments, piers, arch barrel, spandrel and wing walls, and soil fill. However, their structural behaviour is quite complex, due to the materials used for their construction, and numerous interactions between bridge elements. The art of building masonry arch bridges was refined by Roman builders, who used masonry arch for building bridges, aqueducts, sewage systems, etc. Roman bridges were mostly short-spanned bridges with circular arches. After the decline of the Roman Empire knowledge of building masonry bridges declined with it, only to be reintroduced during the 11th and 12th centuries. Since then, a wide variety of masonry arch bridges have been built all over Europe.

Bridges were mostly designed empirically, based on past experiences and by taking measures from existing bridges that proved to be reliable. Later, manuals and technical rules based on simple geometrical relations for determining the dimensions of several bridge components and the safety of construction have been issued. These historical manuals were written over a long period of time, dating from the 15th to early 20th century. During the 18th century application of statics to masonry arches was initiated which offered a more precise approach to bridge design. These methods were essentially based on graphic statics and were used for larger more important bridges, while empirical methods remained popular for smaller bridges. A detailed review of empirical rules can be found in [1]–[3].

There are approximately 300,000 masonry arch bridges in Europe [2] on railway and national road systems networks. Papers e.g. [3]–[5] are just a few of the papers published that present review of masonry bridges in some European countries. Many of these bridges were built over 100 years ago, and during their lifespan have been exposed to environmental hazards, extreme natural events, and ever-increasing traffic loads. These effects' caused material deterioration, damage development and deformations. This decreases the load-bearing capacities of bridges and makes their safety questionable. Therefore, it is important to have reliable methods for assessing the load-bearing capacities of bridges. During the last few decades, important advances have been made concerning the mechanics of masonry arch bridges, such as the development of methods for computation of load-bearing capacities and assessment techniques based on non-destructive testing methods. [6] gives a review of experimental investigations and assessment methods for masonry bridges.

This paper deals with a statistical analysis of masonry arch bridges in Croatia. Therefore, data about bridges have been collected from official organisations (Croatian railways, Croatian roads, County road administration). Additionally, any State archives and Conservation institutes that could have any information about bridges have been contacted too. More detailed data about brickwork bridges that couldn't be obtained through official channels (more detailed geometry and damage on bridges) was gathered by visual in-situ inspection. From the data collected the most common bridge typologies in Croatia were identified, as well as, referenced brickwork bridge based on geometrical survey. After the presentation of the results of statistical analysis a review of damages on brickwork bridges is presented and their impact on serviceability.

2. Masonry arch bridges in Croatia

Masonry arch bridges still form an important part of transportation lifelines in Croatia. There are total of 178 bridges in Croatia, 85% are located on roadways and 15% on railways.

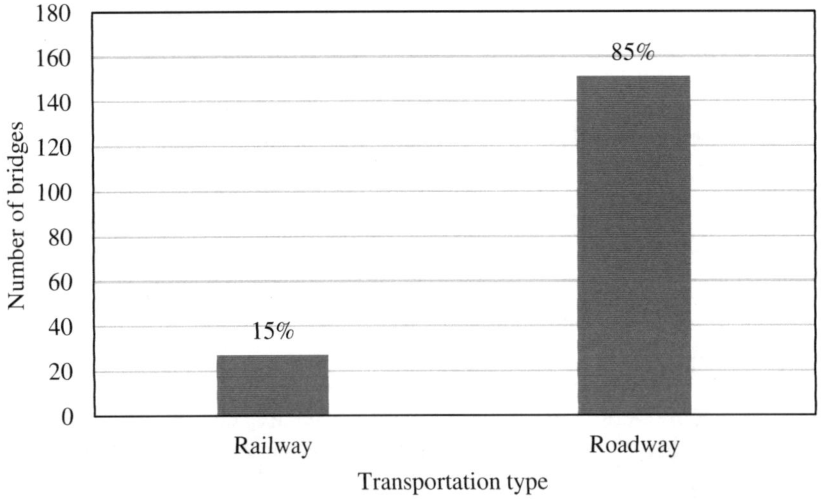

Figure 1. Bridges by transportation type.

Majority of Croatian masonry bridge stock was built during the 19th and 20th century out of locally available materials, natural stone in coastal and mountain regions; and clay bricks in the continental region. Figure 2. presents distribution of bridges by year of construction. Two of the oldest bridges still in the use today, marked as outliners on Figure2. and are shown on Figure 3.

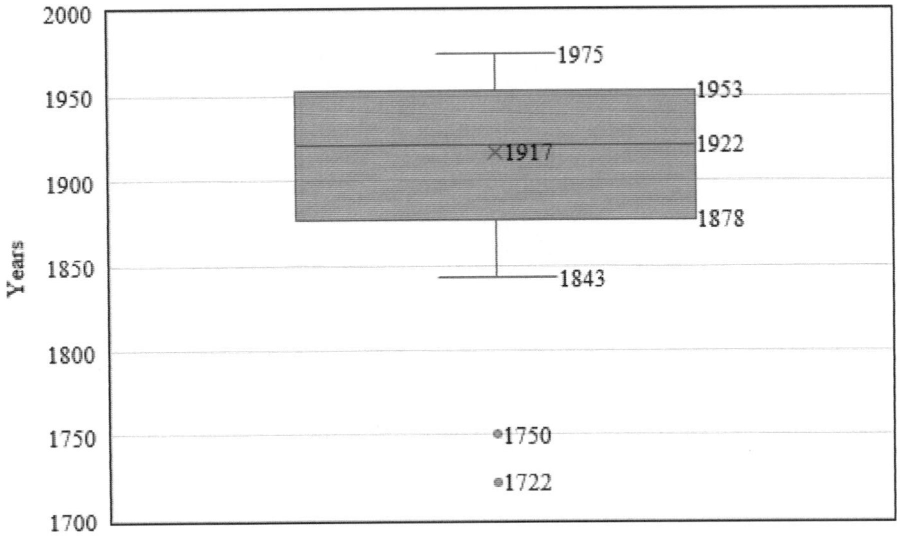

Figure 2. Year of construction of bridges.

a) Old bridge Valpovo

b) Stone bridge on river Dobar (Image taken from Google)

Figure 3. Old bridge Valpovo (a) and Stone bridge on river Dobra (b).

3. Typology of masonry arch bridges

In this section classification by typology of masonry arch bridges in Croatia is presented. So, a modified classification considering materials, arch geometry, arch thickness, number of spans, and the type of the front or spandrel wall according to [5] was used (see Figure 4.). Classification was modified by removing irregular and adding segmental arches to arch geometry, as to authors knowledge there are no bridges with irregular arches in Croatia. This method was chosen because it facilitates collecting and sorting of data, and offers a simple and unambiguous classification.

Material	→	**Brick**
		Natural stone
		Mixed
Arch geometry	→	**Circular arch**
		Parabolic arch
		Elliptical arch
		Segmental arch
		Skewed arch
Arch thickness	→	**Constant**
		Variable
Number of spans	→	**Single**
		Multi
Superstructure	→	**Closed spandrel walls**
		Open spandrel walls

Figure 4. Classification of masonry arch bridges.

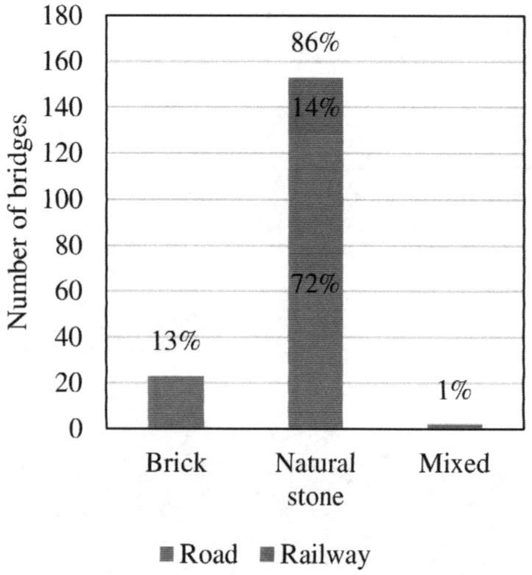

Figure 5. Distribution of bridges by material.

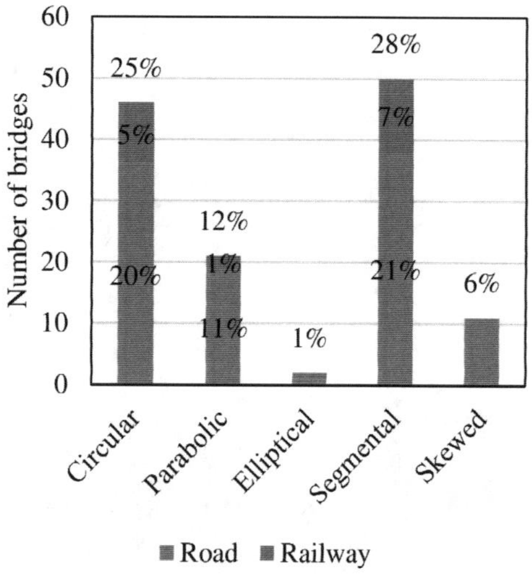

Figure 6. Distribution of bridges by arch geometry.

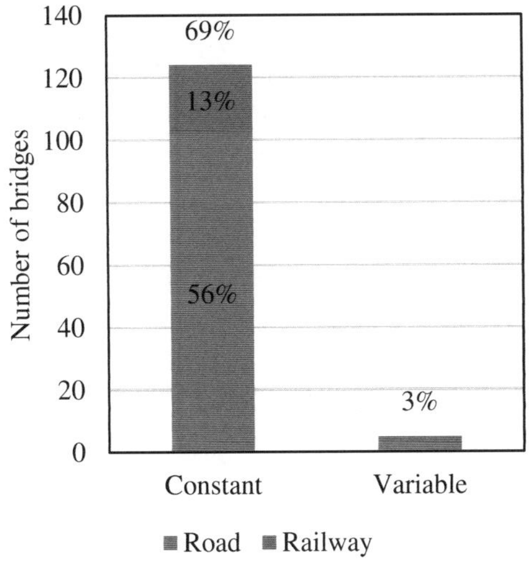

Figure 7. Distribution of bridges by arch thickness.

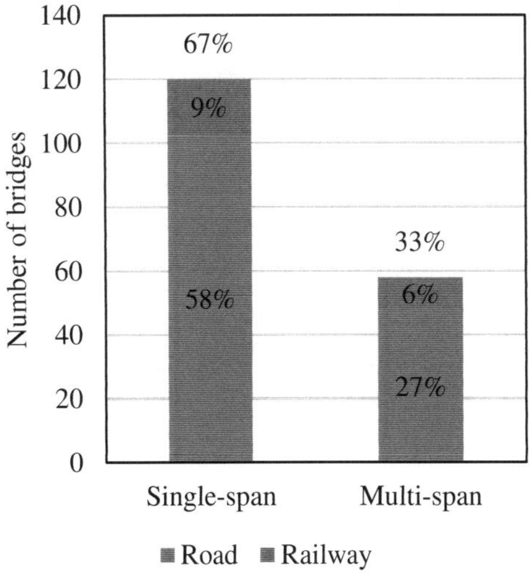

Figure 8. Distribution of bridges by number of spans.

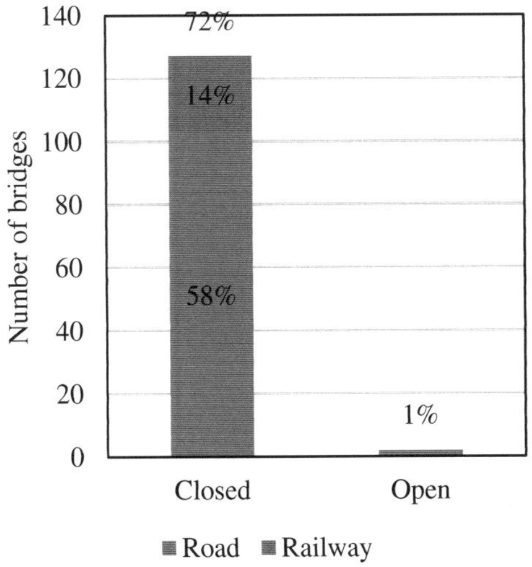

Figure 9. Distribution of bridges by superstructure type (closed or open spandrel walls).

The main statistical data, deduced from the distributions on Figures 5. – 9., can be summarized as:

1. Majority of existing bridges were built with natural stone (86%), while only a small fraction was built using clay bricks (13%)
2. Most represented types of arch geometry are segmental and circular arches, presenting 28% and 25% of bridge stock respectively.
3. Majority of bridges have a constant thickness of arch (69%)
4. Approximately 2/3 of bridge stock are single-span, and the remaining 1/3 are multi-span bridges
5. Only one bridge has open, while the rest have closed spandrel walls
6. The most common type of single-span bridge is a stone bridge with a circular arch of constant thickness and closed spandrel walls, presenting 15% of bridge stock
7. The most common type of multi-span bridge is a stone bridge with a segmental arch of constant thickness and closed spandrel walls, presenting 10% of bridge stock

4. Geometric analysis of brickwork arch bridges

A survey of geometrical properties of brickwork roadway arch bridges was carried out based on data acquired from official bodies, and visual in-situ inspection. Following parameters were collected: number of spans, clear span *s*, rise *r*, arch shape, arch thickness at crown d_c and springing d_s. Based on rise to span ratio and span length bridges were classified into three groups for each of those parameters [7]. Groups based on rise to span ratio (relative depth):

- shallow arch $0.00 < r/s \leq 0.25$
- semi-shallow arch $0.25 < r/s \leq 0.40$
- deep arch $0.40 < r/s$

Groups based on span length:

- short span $0.0 < s \leq 7.5$
- medium span $7.5 < s \leq 15.0$
- large span $15.0 < s$

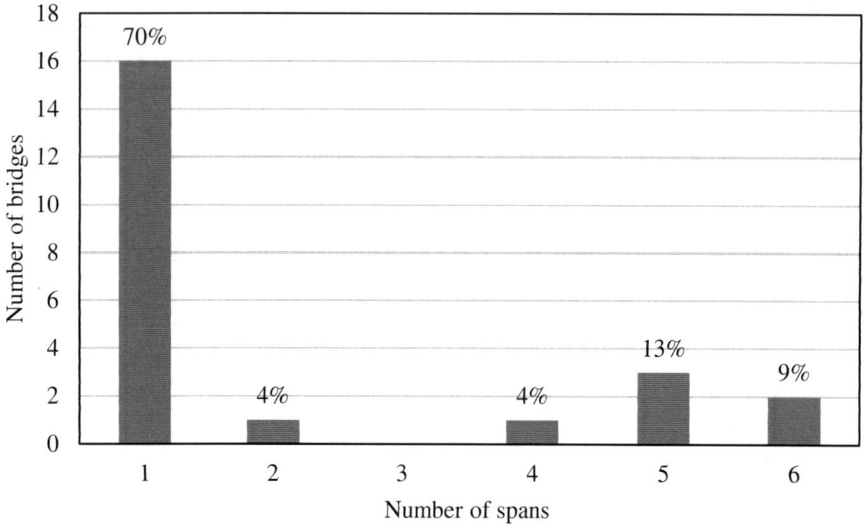

Figure 10. Number of bridges per number of spans.

The number of bridges in relation to the number of spans is shown in Figure 10. Single-span bridges present 70% of the stock, and the remaining 30% are multi-span bridges. The number of spans for multi-span bridges ranges from 2 to 6 spans, with an average of 5 spans.

Figures 11. and 12. show relationship between arch rise (*r*) and span length (*s*), and arch thickness at the crown (d_c) and span length (*s*), respectively. The ratio between arch thickness at the crown and springing (d_c/d_s) equals 1,00 for 83% of bridges and for the remaining 17% its mean value is 0.76.

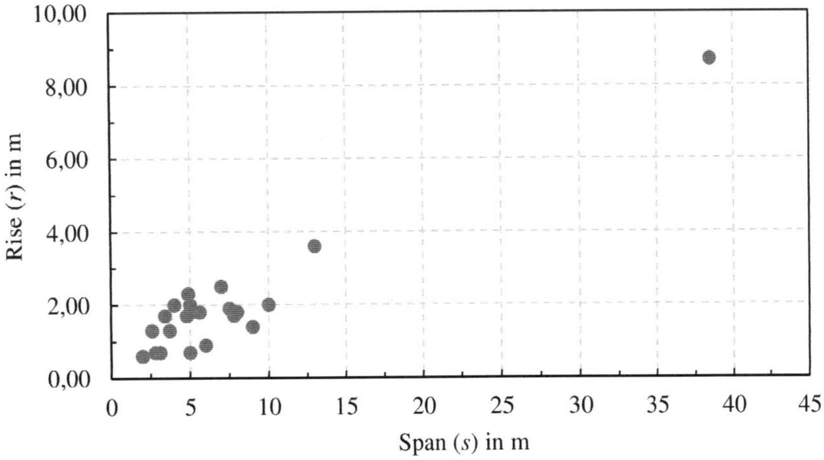

Figure 11. Relationship between rise (r) and span length (s).

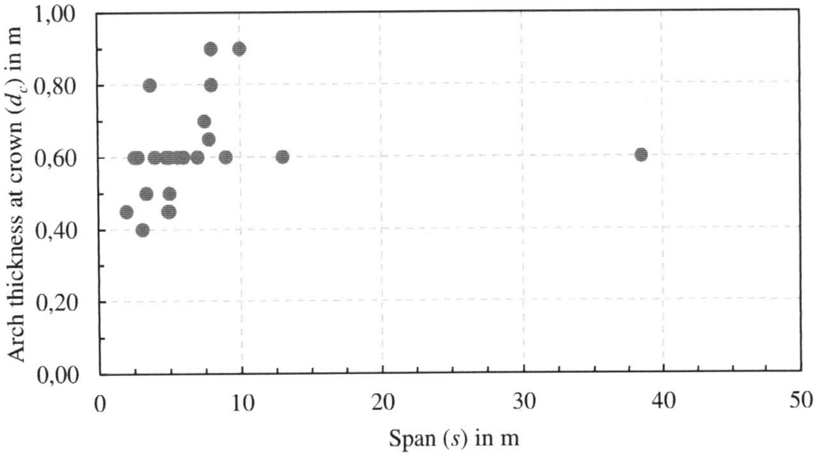

Figure 12. Relationship between arch thickness at crown (d_c) and span length (s).

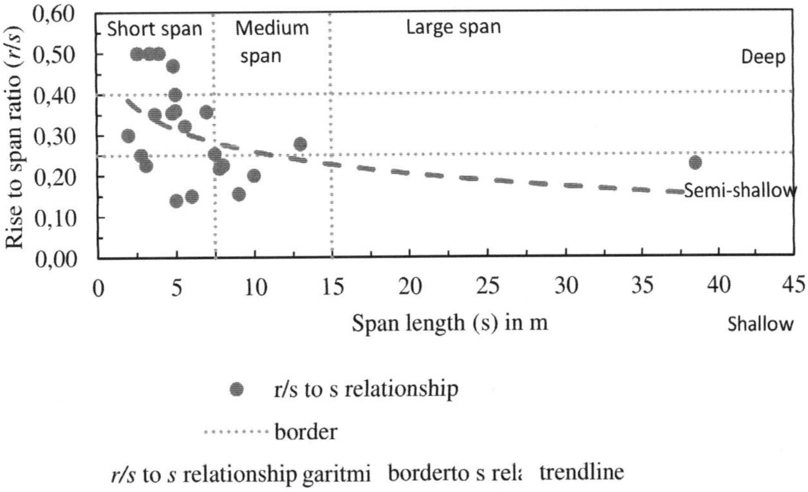

Figure 13. Relationship between rise to span ratio (r/s) and span length (s).

Figure 13. shows the relationship between rise to span ratio and span length, as well as corresponding areas of bridge groups based on rise to span ratio (*r/s*) and span length (*s*). The exact number of bridges in each of the groups and groups combined is presented in Table 1.

Table 1. Distribution of bridges by groups based on *r/s* and *s*.

	Short span	Medium span	Large span	$\sum$
Deep arch	**4 (18%)**	-	-	4 (18%)
Semi-shallow arch	**7 (30%)**	**1 (4%)**	-	8 (34%)
Shallow arch	**4 (18%)**	**6 (26%)**	**1 (4%)**	11 (48%)
$\sum$	15 (66%)	7 (30%)	1 (4%)	

From the data presented, three single-span reference bridges (i.e. idealized bridges) were defined as being geometrically representative of the whole sample, as shown in Table 2. Large span bridges and deep bridges were not considered due to their low presence in the sample. Values for span length, relative depth, and arch thickness are defined as mean values from corresponding groups. Defined reference bridges will be used for further study.

Table 2. Single-span reference bridges.

	Semi-shallow arch	Shallow arch
Short span	$s = 4.70$ m $r/s = 0.35$ $d_c = 0.60$ m (Bridge A)	$s = 4.20$ m $r/s = 0.20$ $d_c = 0.60$ m (Bridge B)
Medium span	-	$s = 8.40$ m $r/s = 0.20$ $d_c = 0.80$ m (Bridge C)

5. Damages on brickwork bridges

Masonry arch bridges suffer damages that decrease their load-bearing capacities, mainly caused by material degradation due to their advanced age and irregular maintenance. More often these are caused by environmental hazards than by human activity. Classification of six main types of damages proposed by [8] is listed below:

- **Contamination** – appearance of any type of plant vegetation or dirtiness
- **Deformation** – geometry changes (deflection of backfill, protection or masonry, and slipping and swelling of masonry)
- **Destruction** – the deterioration of physical and chemical structural features (such effects can be recognized for example by spalling and/or chipping of bricks)
- **Discontinuity** – a break in structure material continuity (cracks)
- **Displacement** – displacement of whole structure or its structural components (e.g. rotation or translation of spandrel walls)
- **Loss of material**

Inspection and categorization of damages was conducted on 22 brickwork masonry arch bridges using this classification. This was a rapid visual inspection and no major conclusions can be draw. In fact, further investigation should be carried out using destructive and non-destructive tests to understand the capacity of such structures which are old and deteriorating with time. This classification was selected because it offers a precise definition and categorization of damages on masonry bridges. Another list of damages can be found in older work by same authors [5], and [9] gives a classification of damage patterns for stone arch bridges of European railway organization.

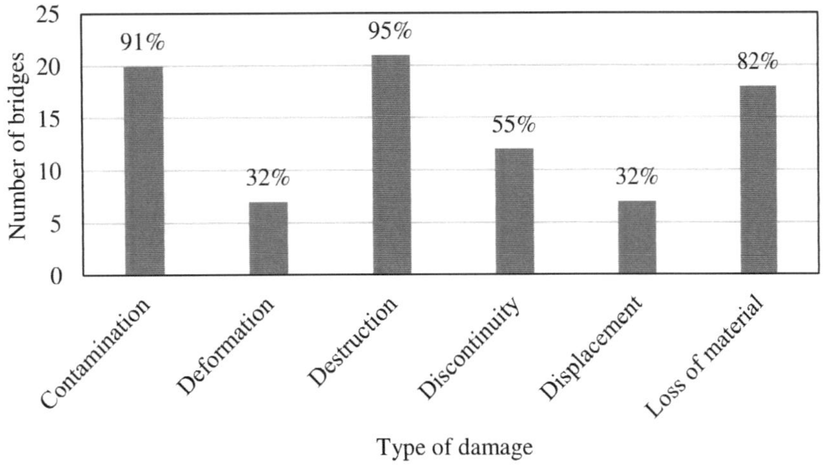

Figure 14. Frequency of damage types on bridges.

Figure 14. presents results of categorization of damages on 22 inspected bridges. The most common types of damages are contamination, destruction, and loss of material. Contamination is mostly in the form of salt attack and moss growth, and in few cases, water penetration and thick vegetation growth were observed. Destruction of material was observed on all bridges with contamination since destruction is often caused by contamination. However, it can be caused by physical effects like freeze-thaw actions, overloading, fatigue, etc. In few cases significant discontinuities (cracks) were observed, such as ring separation, and possible detachment of spandrel walls from arch. These types of damages cause a significant loss of load-bearing capacity and should be repaired as soon as possible. Figure 15. presents examples of observed damages.

Figure 15. Examples of observed damage a) contamination and material loss; b) destruction; c) discontinuity (longitudinal crack); d) material loss.

6. Conclusion

An overview of masonry arch bridges in Croatia is presented here. Data from 178 bridges was collected from official organisations, state archives, and conservation institutes, as well through visual in-situ inspection.

Collected data was statistically analysed to determine the most common bridge typologies in Croatia. Stone bridge with the circular arch of constant thickness and closed spandrel walls is the most common single-span bridge typology, presenting 15% of bridge stock. Stone bridge with segmental arch of constant thickness and closed spandrel walls is the most common multi-span bridge typology, presenting 10% of bridge stock. Collected data was also used to conduct a geometric analysis of brickwork arch bridges, to define single-span reference bridges. These reference bridges are geometrically representative of the whole sample and will be used for further study. Categorization of damages observed during the visual in-situ inspection of brickwork bridges was conducted. Damages were categorized into six main types. The most common damage types are contamination, destruction, and loss of material. If no measures are taken to stop their propagation, they could have serious consequences.

The number of masonry arch bridges in Croatia is modest in comparison to other European countries, nonetheless they are an important part of Croatian transportation infrastructure. Most of them are in poor condition and adequate maintenance and retrofitting are paramount for their intended use.

7. References

[1] Proske D and Gelder P 2009 *Safety of historical stone arch bridges* (Berlin, Heidelberg: Springer) pp 1–359
[2] Brencich A and Morbiducci R 2007 Masonry Arches: Historical Rules and Modern Mechanics *Int. J. Archit. Herit.* **1** 2 165–89
[3] De Santis S and de Felice G 2014 Overview of railway masonry bridges with a safety factor estimate *Int. J. Archit. Herit.* **8** 3 452–74
[4] Page J 1993 *Masonry arch bridges. State of the art review* (London : HMSO)
[5] Bień J and Kamiński T 2004 Masonry arch bridges in Poland *ARCH'04 - Proc. of the 4th International Conference on Arch Bridges (Barcelona)* p 10
[6] Sarhosis V, De Santis S and De Felice G 2016 A review of experimental investigations and assessment methods for masonry arch bridges *Struct. Infrastruct. Eng.* 50
[7] Oliveira D V, Lourenço P B and Lemos C 2010 Geometric issues and ultimate load capacity of masonry arch bridges from the northwest Iberian Peninsula *Eng. Struct.* **32** 12 3955–65
[8] Bień J and Kamiński T 2007 Damages to masonry arch bridges - proposal for reminology unification *ARCH'07 - Proc. of the 5th International Conference on Arch Bridges (Madeira)* pp 341–8
[9] Orbán Z 2004 Assessment, reliability and maintenance of masonry arch railway bridges in Europe *ARCH'04 - Proc. of 4th International Conference on Arch Bridges (Barcelona)* pp 1–10

Acknowledgments

This research has been conducted with the support of the Erasmus+ programme of the European Union project "Environmental Risk Assessment and Mitigation on Cultural Heritage Assets in Central Asia - ERAMCA" (Project Reference: 609574-EPP-1-2019-1-IT-EPPKA2-CBHE-JP) and Croatian Science Foundation (HRZZ) "Young Researchers' Career Development Project – Training of Doctoral Students" programme. The authors are also grateful to Croatian Railways, Croatian Roads, County Road Administrations, and Croatian State Archives for providing the data that contributed to this research.

Push-out tests of shear dowels

P Vanova[1], D Dubecky[1] and V Kvocak[1]

[1] Technical university of Kosice, Faculty of Civil Engineering, Institute of Structural Engineering, Vysokoskolska 4, 042 00 Kosice, Slovakia

E-mail: patricia.vanova@tuke.sk; vincent.kvocak@tuke.sk

Abstract. Continuous shear connectors in a shape of dowels are one of the newer shapes of composite steel-concrete bridges. In this article results of push-out tests of such a dowel with geometry designed at Faculty of Civil Engineering, Technical University of Kosice are presented and compared to the previous research.

1. Introduction

The paper deals with a type of a shear connector that is nowadays becoming more and more popular across Europe. It explains its geometrical attributes, material attributes and how they were obtained and an experimental study that was performed. Berthellemy et. al [1, 2] did a comparison of different dowel shapes following the research of Lorenc et. al [3, 4], who did an experimental as well as a numerical study of those dowels, using Abaqus FEM based software. Another research of continuous dowels was done by Classen and Hegger [5], who focused on pry-out failure of such dowels.

2. Geometrical Attributes

The continuous shear connector presented in this paper was designed in the shape of a puzzle tooth, so called dowel. As visible in the Figure 1, the shape is based on four circles of 32 mm diameter. The tooth itself was of 57 mm height and 85 mm width in the base, 58 mm on the top.

Each strip was made from IPE 220, by the longitudinal cut of the wall, and the strip also included the flanges. For the purposes of the push-out tests each strip was of 600 mm length.

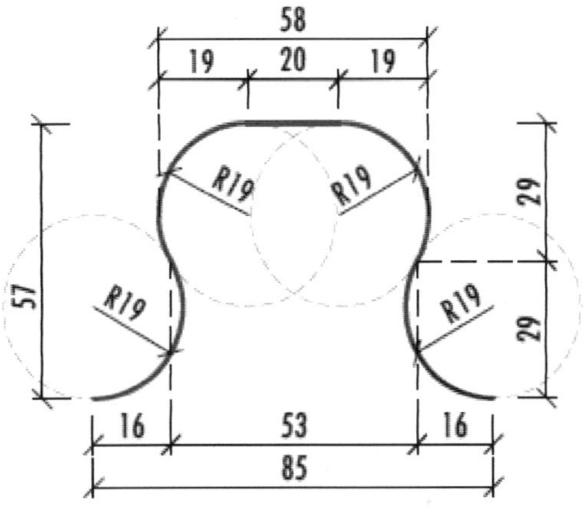

Figure 1. Puzzle tooth of the continuous shear connector [mm].

IOP Conf. Series: Materials Science and Engineering 1209 (2021) 012064 doi:10.1088/1757-899X/1209/1/012064

3. Material Characteristics

The composite tested is based on two materials – concrete and steel. All materials used were tested, except reinforcement, for which the expected normative values were taken into consideration.

3.1. Material Characteristic of Steel

The tensile tests of steel used were performed by Faculty of Metallurgy. The Yield point of the steel used for the connector was 315.3 MPa, as visible in the Figure 2. This value was also set in the FEM software for the middle steel plates of the push-out tests, as this does not affect the results. Other characteristics were not measured and therefore their respected normative values were used. [5]

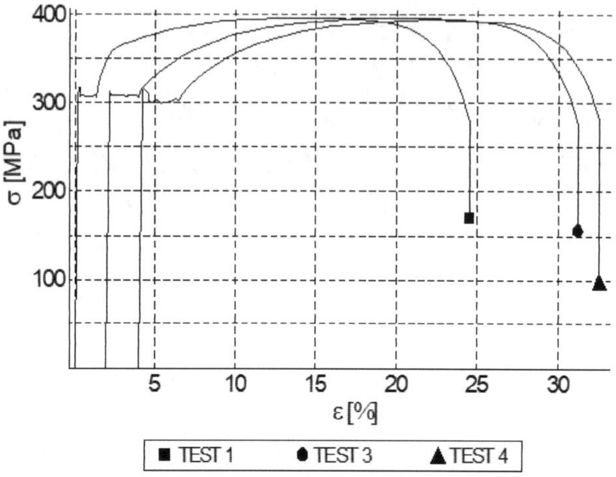

Figure 2. The tensile tests results provided by Faculty of Metallurgy.

3.2. Material Characteristics of Concrete

Material characteristics of concrete were tested in accordance with normative procedure. Several specimens – cubes, cylinders and beams, were made the same they as the specimens for push-out tests and then tested the same day as well. Compressive tests for finding compressive stress and Young's Modulus were done, and three point flexural test was done, from which the tensile stress was counted. The results of the experimental study of concrete properties are shown in the Table 1. [6, 7, 8]

Table 1. Material properties of tested concrete.

	Value	Unit
Mass Density	2.274E-09	tonne/mm^3
Young's Modulus	33 423.8	MPa
Tensile Stress	6.89	MPa
Conpressive Stress	66.31	MPa

4. Push-Out Tests

In order to find out the shear resistance of the shear connector it is necessary to perform at least three push-out tests. The strips, 600 mm long (see Figure 3), are poured into the two concrete blocks. 600x600x200 mm, 100 mm above the bottom, lying on the 10 mm thick polystyrene, which serves as the area for pushing between the steel and the concrete. A general layout of the push-out test performed is shown in the Figure 4. The specimens were then put into the hydraulic press, which pushed the middle steel part in the direction of axis Z (direction of gravitation). Loading of the specimens was done via loading cycles, 98 in total, which included 50 loading cycle during which the force went repeatedly from 400 kN to 100 kN and back in approximately one third of the expected maximum load to prove the elastic behaviour of the specimens at such loading. [9]

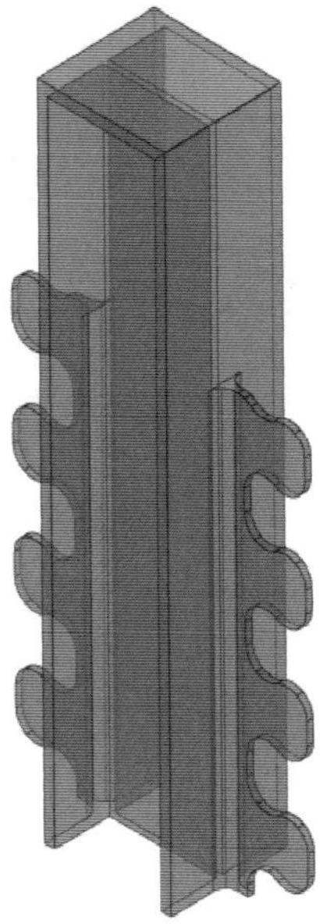

Figure 3. The connector.

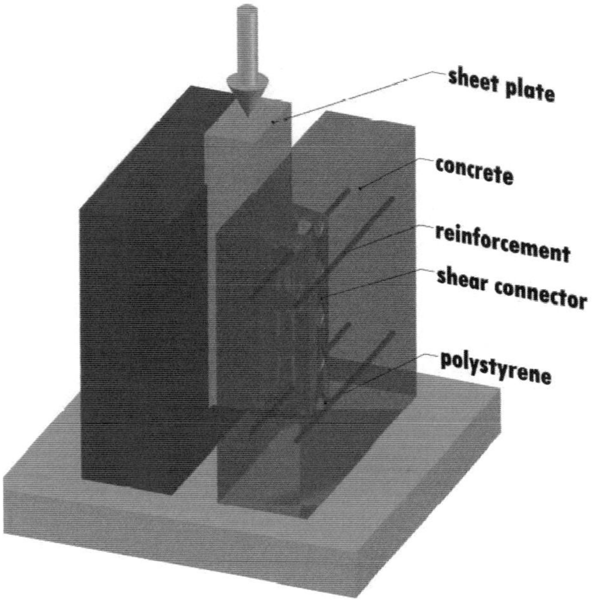

Figure 4. The layout of the general push-out tests.

5. Results

The shear resistance was measured at the point where the specimens stopped acting elastically and went into plastic-elastic behaviour. This is visible in the Figure 5 by the break in the curve – where the vertical part changes into horizontal. This occurred in the loading cycle of 700 kN value.

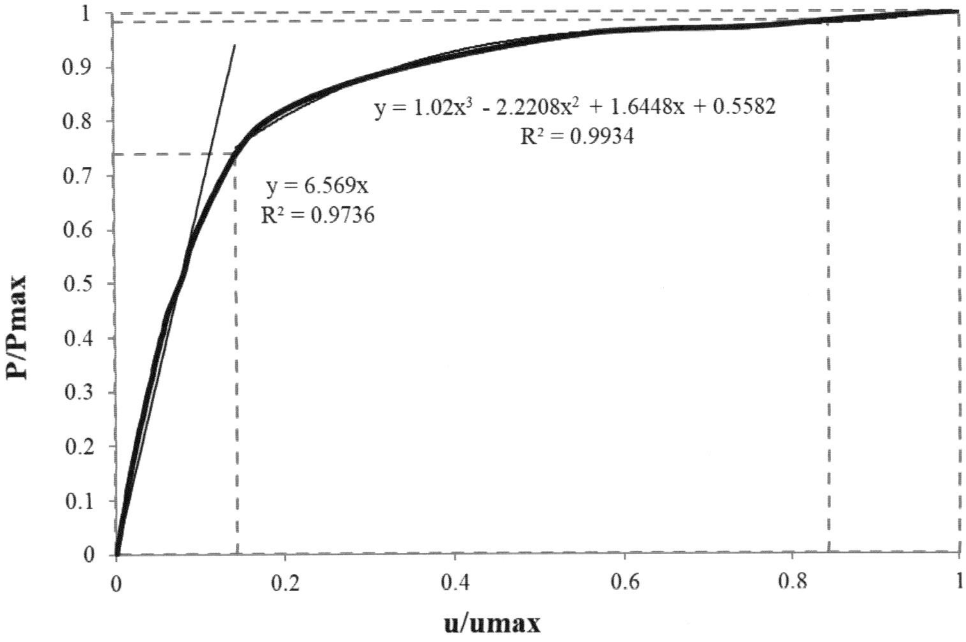

Figure 5. Composite action diagram of the shear tooth.

6. Conclusion

In conclusion, the experimental study went without any unexpected events, the behaviour of the specimens went through all the phases – elastic, plastic-elastic and plastic, after which the specimens were terminally damaged. The shear resistance of the proposed shear dowel was found, and its value showed to be high enough for sufficient shear connection in bridges.

However, if the dowel were to be used for composite bridge constructions, several additional tests need to be performed in order to prove its bending resistance, dynamic resistance and fatigue resistance.

7. References

[1] Berthellemy J, Hechler O, Lorenc W, Seidl G and Viefhues E 2009 *Premiers rèsultats du projet de recherche europèen precobeam de connexion par dècoupe d'une tôle* (Revue Construction Mètallique) vol 3 pp 3–26

[2] Berthellemy J, Zanon R, Seidl G and Lorenc W 2014 *An innovative solution for small span bridges - precobeam* (Toronto: Innovative steel bridges)

[3] Lorenc W 2009 *Boundary approach in shape study of composite dowel shear connector* (Archives of Civil and Mechanical Engineering) vol 9 pp 55–66

[4] Lorenc W 2016 *The design concept for the steel part of a composite dowel shear connection* (Steel Construction) vol 9(2) pp 89–97

[5] STN EN 10002-1 2002 *Skúška ťahom. časť 1: Skúška ťahom pri teplote okolia* (Bratislava: SÚTN)

[6] STN EN 12390-13 2014 *Skúšanie zatvrdnutého betónu. časť 13: Stanovenie sečnicového modulu pružnosti v tlaku* (Bratislava: SÚTN)

[7] STN EN 12390-3 2010 *Skúšanie zatvrdnutého betónu. časť 3: Pevnosť v tlaku skúšobných telies* (Bratislava: SÚTN)

[8] STN EN 12390-7 2011 *Skúšanie zatvrdnutého betónu. časť 7: Objemová hmotnosť zatvrdnutého betónu* (Bratislava: SÚTN)

[9] STN EN 1994-1-1 2006 *Eurokód 4: Navrhovanie spriahnutých oceľovobetónových konštrukcií. časť 1-1: Všeobecné pravidlá pre budov* (Bratislava: SÚTN)

Acknowledgments

The paper has been supported by The projects: VEGA 1/0172/20 "Stress and deformation analysis of load bearing components made out of steel, glass and composite materials" of the Scientific Grant Agency of the Ministry of Education, science, research and sport of the Slovak Republic and the Slovak Academy of Sciences and by the project Slovak Research and Development Agency under the contract No. APVV-15-0486 "Analysis of Shear Connection Influence in Bridges with Encased Beams".

Material characteristics of push-out tests

P Vanova[1], P Orolin[1] and D Dubecky[1]

[1] Technical University of Kosice, Faculty of Civil Engineering, Institute of Structural Engineering, Vysokoskolska 4, 042 00 Kosice, Slovakia

E-mail: patricia.vanova@tuke.sk

Abstract. Two types of push-out tests were carried out at the Centre of Research and Innovation in Construction, the Technical University of Košice – one at a composite based on steel continuous shear connector and the second one at the same connector, however made of glass-laminate material. For further research, the material characteristics of the material used needed to be found. In this article, the material tests performed as well as their results are presented.

1. Introduction

The composite structures and composite bridges are one of the newer additions to structural engineering. Any new geometrical or material addition to these structures has to be firstly severely experimentally tested in order to prove its reliability. Before any experiments are performed it is necessary to find out their material characteristics, which are also a platform for numerical study. The material characteristics and testing for analytical and software analysis was previously described by several authors. Both I. Rewers and M. Hafezolghorani et. Al. closely describe concrete damage plasticty and tensile behaviour of concrete in Abaqus software [1,2]. R. Kanishchev and V. Kvocak have a close description of material testing for different composite structures. [3,4]

2. Push-out Tests

The push-out test is an experiment intended for the determination of shear resistance in composite structures. It consists of two concrete blocks in which the composite element is built in, usually from steel, but other materials can be used as well. The inside element is placed on polystyrene; however, the material characteristics of this part play no role in the experiment results, it serves only as filling to prevent forcing trough of steel (or other material) into concrete, which would be counterproductive. [5]

3. Material Characteristics

Material characteristics of any experimental study are very important in structural engineering. The results serve for analytical as well as numerical study and can explain certain behaviour during the experiment and predict the change in results depending on the material used.

In this article, the material characteristics and the process of obtaining them of the two most commonly used structural materials – steel and concrete – are described.

3.1. Material Characteristic of Steel

For steel, only the tensile tests were performed (Figure 1). Three samples, cut out of the wall of IPE 160 (see Figure 2), which was later used as a continuous shear connector in push-out tests, were put into a universal testing machine. Each sample contains of two shoulders and one gauge. The shoulders are put

IOP Conf. Series: Materials Science and Engineering 1209 (2021) 012065 doi:10.1088/1757-899X/1209/1/012065

into the jaws of the machine and they serve as holders for enabling of pulling forces to go to the gauge. Onto the gauge, there is also a deformeter for capturing the smallest changes in length. [6]

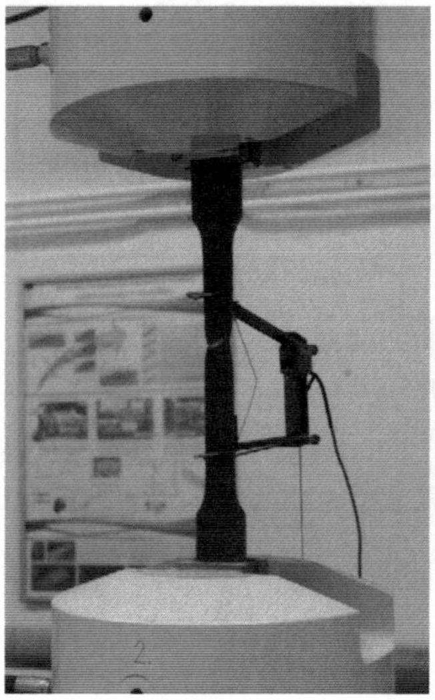

Figure 1. Results of tensile test.

Figure 2. The breached steel samples.

To adjust the results into the stress-strain diagram, the Eq.1 and Eq. 2 were used. [6]

$$\sigma = \frac{F_n}{A} \tag{1}$$

$$\varepsilon = \frac{\Delta L}{L_0} = \frac{L - L_0}{L_0} \tag{2}$$

Where σ and ε are the stress and strain, respectively; F_n and A are the force and cross-section are of the gauge, respectively, and L_0 and L are the initial gauge length and the length of the gauge under the influence of the load, respectively. [6]

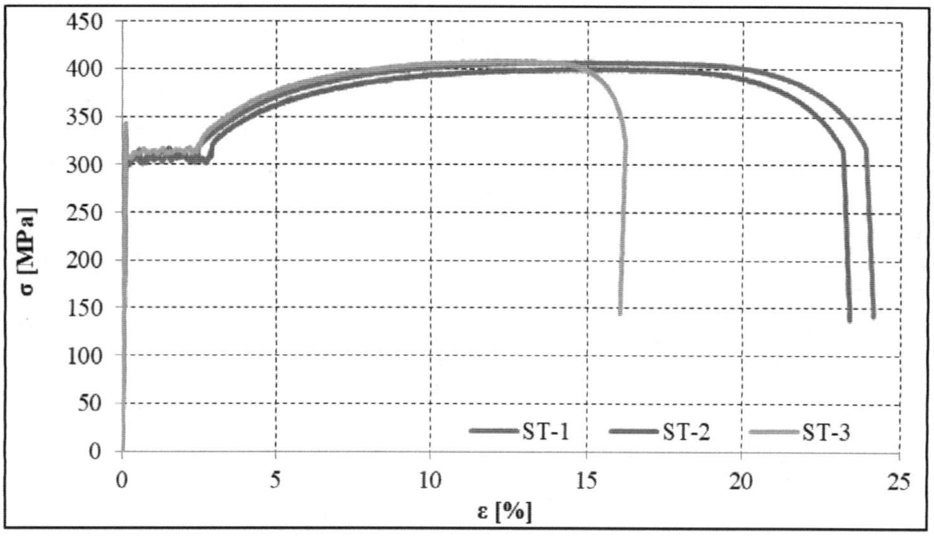

Figure 3. Results of tensile test.

IOP Conf. Series: Materials Science and Engineering 1209 (2021) 012065 doi:10.1088/1757-899X/1209/1/012065

From the experiments performed, the tensile strength of the steel used in push-out tests was determined to be 309.5 MPa (see Figure 3).

3.2. Material Characteristics of Concrete

The concrete used in the push-out test was marked as C30/37 by standard European norms, obtained from a concrete manufacturer. The push-out tests were performed approximately four months after the concreting. All experiments for the determination of concrete properties were done simultaneously with the push-out test. Compressive strength, Young's Modulus, and tensile strength of the concrete were all tested at three specimens each. [7,8,9,10]

Compressive strength was detected for cubes, cylinders, and beams. Each specimen was firstly cleaned of any possible impurities, then measured with Vernier caliper in order to determine its exact dimension with a 0.1 mm accuracy and finally weighted with 1 g accuracy. After the preparation process, the specimens were one by one put into a hydraulic press (see Figure 4, Figure 5 and Figure 6). Into the machine, information about its dimension was put and the press then determined the maximum load as well as the strength itself. The results of cubic, cylindric, and beam strength are presented in Table 1, Table 2 and Table 3, respectively. [7,8,9,10]

Table 1. Results of compressive strength of three concrete cubic specimens.

#	Sample Designation	Dimensions d_1 x d_2 x d_3 [mm]	Weight [kg]	Maximum load [kN]	Strength [MPa]
1	C-1	150.1 x 152.5 x 150.1	7.766	1091.582	47.688
2	C-2	151.3 x 150.5 x 150.2	7.732	1124.445	49.382
3	C-3	150.3 x 149.7 x 150.8	7.679	1145.681	50.919
Average				**1120.681**	**49.330**

Table 2. Results of compressive strength of three concrete cylindric specimens.

#	Sample Designation	Dimensions d_1 x d_2 x l [mm]	Weight [kg]	Maximum load [kN]	Strength [MPa]
1	CY-1	149.3 x 149.5 x 299.0	11.849	600.141	34.234
2	CY-2	149.3 x 149.3 x 299.8	11.866	641.545	36.645
3	CY-3	149.5 x 149.5 x 300.0	11.930	615.972	35.090
Average				**619.219**	**35.232**

Table 3. Results of compressive strength of three concrete beam specimens.

#	Sample Designation	Dimensions d_1 x d_2 x L [mm]	Weight [kg]	Maximum load [kN]	Strength [MPa]
1	B-1	97.4 x 95.5 x 399.9	8.939	306.241	30.624
2	B-2	97.5 x 93.8 x 399.7	8.825	331.062	33.755
3	B-3	100.1 x 99 x 400.1	9.064	323.960	32.691
Average				**321.960**	**32.357**

Tensile strength was tested as the transverse tensile strength of cubes (Figure 7) and flexural strength of concrete (Figure 8), from which the tension was counted via Eq. 3. Both cubes and cylinders were prepared the same way as for the compressive tests. The transverse tensile strength of cubes was done with help of a slim lamella plate, which was attached to the hydraulic press and placed just above the centre of the cube. The three-point flexural test was performed on the beams. Two metal cylinders served as supports, placed 50 mm from the specimen faces. The third metal cylinder was above the longitudinal centre of the beam and the force from the hydraulic press was applied to the specimen by it. The results are shown in Table 4 and Table 5, respectively. [7,8,9,10]

$$f_{cf} = \frac{F \cdot l_0}{d_1 \cdot d_2^2} \quad (3)$$

Where F is the maximum load and l_0, d_1 and d_2 are the dimensions from Table 6. [7,8,9,10]

Table 4. Results of transverse tensile strength of three concrete cubic specimens.

#	Sample Designation	Dimensions d_1 x d_2 x d_3 [mm]	Weight [kg]	Maximum load [kN]	Strength [MPa]
1	C-1-T	151.0 x 150.3 x 150.1	7.654	113.325	5.037
2	C-2-T	151.5 x 150.1 x 150.0	7.701	118.083	5.145
3	C-3-T	149.9 x 149.9 x 151.6	7.823	137.237	6.108
Average				**122.882**	**5.428**

Table 5. Results of flexural strength of three concrete beam specimens.

#	Sample Designation	Dimensions d_1 x d_2 x L [mm]	Weight [kg]	Maximum load [kN]	Strength [MPa]	Tensile strength [MPa]
1	B-1-B	100.2 x 98.5 x 400.0	8.961	18.089	5.586	7.443
2	B-2-B	102.0 x 100.2 x 399.4	9.283	17.532	5.142	6.838
3	B-3-B	99.9 x 99.9 x 399.5	8.885	17.591	5.293	7.049
Average				**17.737**	**5.340**	**7.110**

Figure 4. The breached concrete cube after the compression strength test.

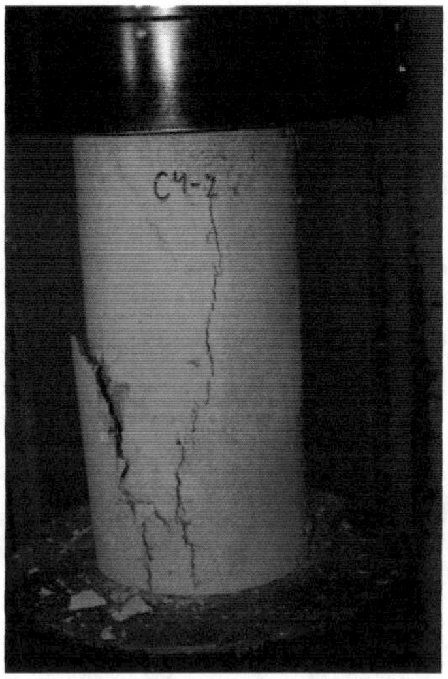

Figure 5. The breached concrete cylinder after the compression strength test.

Figure 6. The breached concrete beam after the compression strength test.

For finding Young's Modulus three beam samples were tested. Compression tests were performed in accordance with Eurocode. Three cycles between the loads 5 kN and 95 kN were made, where in each position the load was upheld for 60 seconds. The maximum and minimum of the loading cycle were counted out of the previously measured compressive strength of beams. After the cycling period, the hydraulic press was evenly increasing the load until the sample was breached and no more usable. Four deformeters (see Figure 9) were put onto the specimen from each side. The results can be seen in the following Table 6. [7,8,9,10]

Table 6. Results of Young's Modulus.

#	Sample Designation	Dimensions d_1 x d_2 x L [mm]	Weight [kg]	Maximum load [kN]	Young's Modulus [MPa]
1	B-1-E	97.1 x 94.5 x 399.8	8.975	100.180	32.016
2	B-2-E	97.2 x 98.0 x 399.8	9.127	100.478	29.373
3	B-3-E	97.8 x 97.3 x 399.9	9.110	105.440	30.548
Average				**102.033**	**30.646**

Figure 7. The breached cube sample after the transverse tensile strength test.

Figure 8. The breached beam sample after the three-point-flexural test.

Figure 9. The Young's Modulus test.

4. Conclusion
The material testing was performed successfully; all searched values were obtained and can be used for further investigation of steel-concrete composite structures. The materials met expectation based on their marking.

5. References
[1] Rewers I 2019 *Numerical Analysis of RC beam with High Strength Steel Reinforcement using CDP model* (Bristol: IOP Publishing) vol 571
[2] Hafezolghorani M, Hejazi F, Vaghei R, Jaafar M S B, Karimzade K 2017 *Simplified Damage Plasticity Model for Concrete* (Switzerland: Structural Engineering International) pp 68-78

[3] Kanishchev R, Kvočák V 2019 *Local buckling of rectangular steel tubes filled with concrete* (South Korea: Steel and Composite Structures) vol 2 pp 201-2016
[4] Kvočák V, Kožlejová V 2011 *Research into Filler-Beam Deck Bridges With Encased Beams of Various Sections* (Poland: Technical Gazette) vol 3 pp 385-392
[5] STN EN 1994-1-1 2006 *Eurokód 4: Navrhovanie spriahnutých oceľovobetónových konštrukcií. časť 1-1: Všeobecné pravidlá pre budovy* (Bratislava: SÚTN)
[6] STN EN 10002-1 2002 *Skúška ťahom. časť 1: Skúška ťahom pri teplote okolia* (Bratislava: SÚTN)
[7] STN EN 12390-13 2014 *Skúšanie zatvrdnutého betónu. časť 13: Stanovenie sečnicového modulu pružnosti v tlaku* (Bratislava: SÚTN)
[8] STN EN 12390-3 2010 *Skúšanie zatvrdnutého betónu. časť 3: Pevnosť v tlaku skúšobných telies* (Bratislava: SÚTN)
[9] STN EN 12390-4 2001 *Skúšanie zatvrdnutého betónu. časť 4: Požiadavky na skúšobné stroje* (Bratislava: SÚTN)
[10] STN EN 12390-5 2011 *Skúšanie zatvrdnutého betónu. časť 5: Pevnosť v ťahu priohybe skúšobných telies* (Bratislava: SÚTN)

Acknowledgments
The paper has been supported by The projects: VEGA 1/0172/20 "Stress and deformation analysis of load-bearing components made out of steel, glass and composite materials" of the Scientific Grant Agency of the Ministry of Education, science, research and sport of the Slovak Republic and the Slovak Academy of Sciences and by the project Slovak Research and Development Agency under the contract No. APVV-15-0486 "Analysis of Shear Connection Influence in Bridges with Encased Beams".

IOP Conf. Series: Materials Science and Engineering 1209 (2021) 012066 doi:10.1088/1757-899X/1209/1/012066

Determination of temperature gradient of slender structure

M Velešík[1], M Požár and R Nečas[1]

[1] Institute of concrete and masonry structure, Faculty of Civil Engineering, Brno University of Technology, Veveří 331/95, 602 00 Brno, Czech Republic

E-mail: velesik.m@fce.vutbr.cz

Abstract. This study investigates convenient methods of determination of temperature gradient of slender structures in time. Correct determination of temperature along the height of the cross-section takes a key role in defining the temperature loading of slender bridge structures. A proper definition of temperature loading is crucial for structures such as stress ribbon because of their sensitivity to temperature change in terms of geometrical changes of the structure. Correct determination of the temperature gradient is important during long-term geodetic monitoring, which might be used to prove the correctness of computational models of bridge structures. To approximate heat transfer in cross-section, a test specimen with temperature sensors installed along the height was formed and continuously monitored. The accuracy of the retrofitted temperature sensors was also investigated. The temperature at the surface of the specimen, solar radiation and wind velocity were used as input data for heat transfer analysis. The measured values from sensors situated along the height were subsequently used for verification of performed heat transfer analysis on volume computational model in the software Ansys Mechanical.

1. Introduction

Cable-supported structures like stress ribbon or suspension bridge are frequently used in modern architecture because of their clean design and the possibility of forming independently from existing terrain. For statically certain structures the thermal load is not crucial for the design because free deformation is allowed. However, for the structurally and statically more complex structures like pedestrian bridges with slender decks supported by cables, a combination of temperature loading and restraining of free deformation plays a key role for static analysis. For proper evaluation of cable-supported structures, non-linear analysis with large deformation effects needs to be performed.

2. Description of the examined stress ribbon structures

In order to verify the computational model of the structures such as stress ribbon structures or suspension pedestrian bridges, it is necessary to correctly evaluate the geodetic surveying of the bridge. Precision and accuracy in geodetic surveying can measure the exact sag of the structure, however, the geometry of cable-supported structures is very sensitive to the surrounding temperature and whether or not the solar rays reach its surface. These and other conditions affect the temperature along the height of the deck and can cause incorrect evaluation of geodetic surveying. It follows that without knowing the temperature load in the deck we cannot precisely determine the sag of the structure.

2.1. Structural design and Geometry of the structure

The proposed experiments examining the effect of thermal loading on the geometry of bridge structures with slender deck were designed in response to an actual structures of stress ribbon in Kroměříž (figure

1,2 and 3) and Radonice nad Ohří. The superstructure of both bridges is structurally composed of a prestressed concrete deck supported by cables, which is rigidly connected with outer monolithic abutments. The bridges deck is made of precast segments (type DS-L and DS-Lv) 0.30 m high, 3.80 m wide and 3.00 m long. The segments are carried by 2×5 tension cables type "A" from 3×2 strands and prestressed by 14 tension cables type "B" 3×2 strands and cables type "C" 4×2 strands. All tendons are formed by Lp 15.5 mm diameter strands [5] [6].

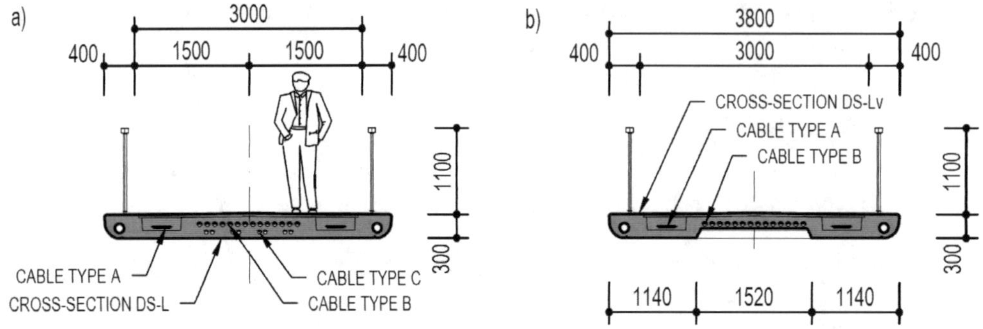

Figure 1. Pedestrian bridge in Kroměříž, Czech Republic – a) cross-section at the abutment, b) cross-section at the midspan.

The bridge is straight in plan and has a variable longitudinal slope. The length of the suspended stress ribbon is 63.36 m. The width between the railing is constant 3.00 m and the width of the bridge deck is constant 3.80 m. The variable longitudinal slope decreases towards the mid-span of the bridge to a value of zero. According to the original design documentation the transverse cross-slope is 1.00 % [5].

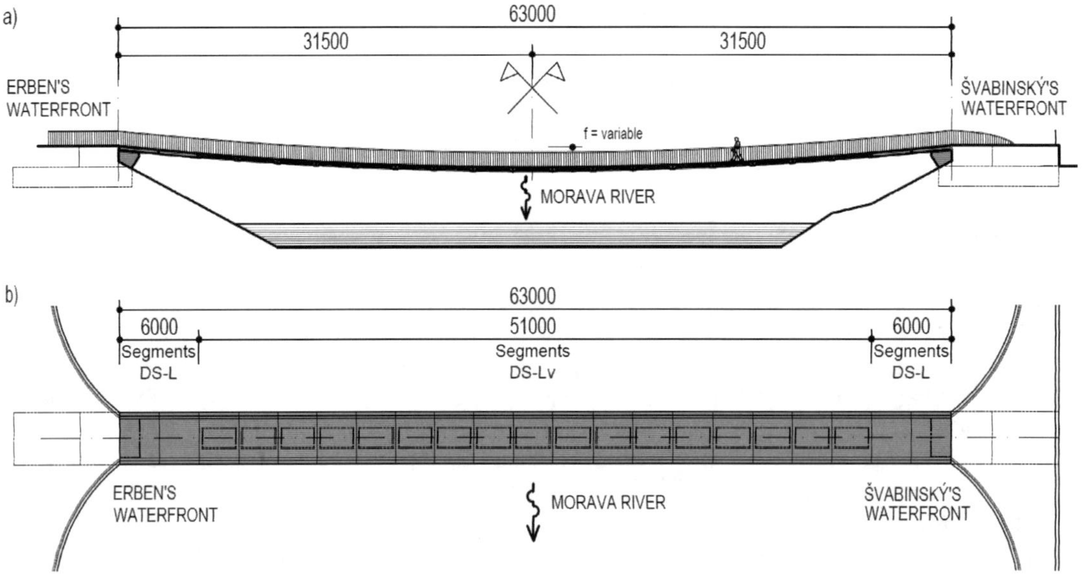

Figure 2. Pedestrian bridge in Kroměříž, Czech Republic – a) elevation, b) plan.

To reduce the local stresses at the connection of end segment and abutment, the outermost segments are bedded on the abutments on unreinforced elastomeric bearings. Bearings are not connected to the bridge structure therefore the structure may unwound from the bearings during construction or temperature drop and rewound under the pedestrian load or temperature rise. This results in a variable span of the superstructure in the range of 57.73 m to 63.36 m. The length of the span and the sag depends on the temperature and the magnitude of the load. During the frost in winter, the sag of the stress ribbon decreases, on the contrary at higher temperature level the sag increases, for example, a sag value of 1.56 m corresponds to the temperature of 10 °C and the difference in the sag is in the range of 0.30 m.

Figure 3. Stress ribbon pedestrian bridge in Kroměříž, Czech Republic.

2.2. Temperature loading of the structure

Structures can be thermally loaded by uniform temperature component, linearly varying temperature difference component and non-linear temperature difference component. According to Eurocode 1991-1-5 are bridge structures loaded by uniform temperature component, which depends on characteristic values of maximum and minimum shade temperature defined by national maps of isotherms and by temperature difference component, which can be considered as linear temperature difference component (Approach 1) or as nonlinear temperature difference component (Approach 2). Considering the thermal loads applied to structures according to Eurocode, it can be assumed that the actual temperature load cannot be simplified to a constant value of the ambient air temperature or the temperature at the surface of the bridge deck throughout the entire cross-section height.

Above mentioned assumptions were confirmed by geodetic measurements carried out on similar structures in the past. They showed that measuring air temperature and the temperature of the top and bottom surfaces of the bridge deck is not sufficient to determine the thermal load and therefore initial geometry of the structure. The differences between the surface and interior temperature of concrete segments can be up to 10 °C. Such a significant difference will impact the sag of the stress ribbon and invalidate the comparison of the geodetic measurement with the sags determined by the structural analysis. Distorted conclusions can be made and misinterpretation of the results leads to incorrect determination of stiffness of the structure, which may also explain different deflection from the measurement and analysis. Reduced stiffness might be caused by damage of prestressing cables by corrosion and leads to the wrongly justified reconstruction of the bridge. In worse cases, reduced stiffness may not be detected.

3. Experimental determination of heat transfer through a cross-section

In order to continuously measure and monitor the temperature along the height of the precast segment was created specimen from reinforced concrete with built-in temperature sensors. These sensors served for the verification of computational models investigating heat transfer and for verification of the correctness of the measured data by additionally fitted sensors.

The experimental model (figure 5) consists of a reinforced concrete block with dimensions of 2.00 × 2.00 × 0.30 m, placed outdoors in an open area on steel supports at a height of 1.00 m above the ground. The main parameters measured on the specimen are the temperature along the height of the cross-section in three plan positions - A, B and C (see figure 4), the ambient temperature above and below the specimen, and the solar radiation on the panel. There are 7 internal temperature sensors Pt100 along the height of the specimen at each positions A, B, and C with a spacing of 50 mm. The Pt100 temperature sensors by Sensit, type TG68-60 are a resistance sensor designed for measuring the temperature of gaseous and liquid substances. The temperature sensors placed on the top and bottom of the specimen are concreted to face with the surface with a level difference of up to 1 mm. Therefore, it can be considered that the temperature measured on the sensors corresponds to the surface temperature of the specimen, i.e., it considers the effect of solar radiation that heats the specimen. Considering the specimen size and the sun exposure, it can be assumed that the temperature is the same all over the top respectively

IOP Conf. Series: Materials Science and Engineering 1209 (2021) 012066 doi:10.1088/1757-899X/1209/1/012066

the bottom surfaces, but due to the placement of the temperature sensors, we can observe the temperature change near the specimen surface along the height (position A and C) in contrast to the data measured by sensors in the centre of the specimen (position B). The air temperature sensors are installed just above and below the surface of the specimen. Illumination is measured by a light sensor located in the middle of the element on the top surface. The sensor type is a photodiode with a spectral sensitivity of 600 nm and a measurement range from 0 to 100000 Lux. Measured data is collected by the dataTaker multi-channel measurement control unit DT80G with a CEM20 expansion module.

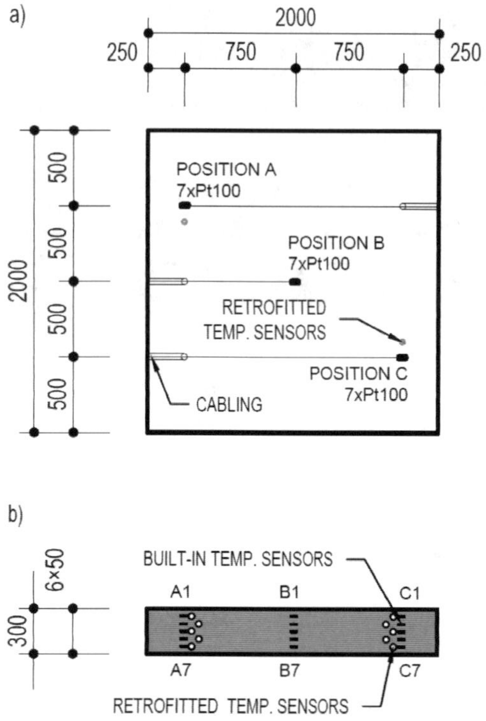

Figure 4. Location of the temperature sensors in the specimen – a) plan, b) cross-section.

Figure 5. Experimental model – reinforcement with built-in sensors and concreted specimen with additionally installed sensors.

In addition to the built-in temperature sensors, additional temperature sensors were installed near the built-in sensors. These sensors were at a sufficient distance from the concreted sensors to prevent their damage and to prevent affecting the measured data but reasonably close to allow the comparison of the measured values. To measure the deformation of the specimen, a strain gauge is placed on the upper and lower surfaces. Since deformation on the structure is not restrained, it should be possible to estimate the temperature of the specimen through the deformation.

4. Finite-Element Modeling of heat transfer

In addition to experimental determination of the temperature along the height of the cross-section, a simulation of the time-dependent heat transfer analysis in the finite element software Ansys Mechanical 2021 R1 was performed. The approach of modelling temperature constraints at the outer surface of the elements was chosen. Since the temperature on the surface of the structure is not difficult to measure, this modelling method is very easy to implement on any construction. Moreover, by measuring the temperature at the surface of the structure and not the ambient air temperature, the effect of solar radiation is also taken into account.

The early thermal analysis was performed on a two-dimensional element Plane 55, which has four nodes and it is suitable for solving steady-state and transient thermal analysis. Subsequently was the problem complexify into a three-dimensional problem modelled by element Solid70, which has eight nodes. Both types of elements have thermal conduction capability and only one degree of freedom at each node, and that is the temperature.

Since we investigate the heat transfer through an element that is loaded at its surfaces by external temperature, the thermal conductivity k is a key input parameter for correct analysis and its evaluation. In the Czech Republic, this variable can be determined according to ČSN EN 73 0540-3 Annex A [3], where is given the characteristic value of thermal conductivity for reinforced concrete with a density of $2500\ kg/m^3$ $k = 1.48\ W/m{\cdot}K$. This value cannot be sufficient for modelling the heat transfer for all type of concrete. The thermal conductivity of reinforced concrete depends on many parameters such as density of concrete, moisture content, type of aggregate or cementitious material and amount of concrete reinforcement [1]. The second problematic factor for accurately analyzing the heat transfer through cross-section is the flow of the surrounding medium and ambient humidity. The second factor can significantly complicate both the measurement and the subsequent analysis in the software, but its correct determination improves the results as will be shown in the following chapters.

4.1. Description of the model and input parameters

The geometry of the computational model (figure 6) was defined to match the actual geometry of the specimen 2.00×2.00×0.30 m. The meshing of the model is chosen so that the evaluated nodes correspond to the position of the sensors in the specimen. Thus, the model is divided by 0.05 m in height.

The specimen is made of concrete of strength class C30/37 but just like on the real structure we lack accurate information about the material properties such as modulus of elasticity, Poisson's ratio and density. The values are defined by Eurocode 2 [4]. An overview of the variables used in the computational model is given in Table 1. Differences in these properties from their actual values do not affect our analysis because these properties are not crucial for the thermal analysis. The specific heat capacity c can be considered for concrete $880\ J/kg{\cdot}K$ and concrete reinforcement (steel) $480\ J/kg{\cdot}K$. Concerning the amount of concrete reinforcement in the specimen, a value of specific heat capacity will be considered $c = 870\ J/kg{\cdot}K$.

For heat transfer analysis is important the correct determination of the thermal conductivity k. Thermal conductivity can be described very simply for a homogeneous and steady heat flux as the unit of heat [W] that passes through a unit of the area [m^2] per time unit [s]. According to [1] thermal conductivity can be calculated with respect to the density according to the equation:

$$k_r = 0{,}0013 \cdot \rho - 0{,}5502 \tag{1}$$

We can see that for a density of $2500\ kg/m^3$ the thermal conductivity calculated according to equation (1) is $k = 2.7\ W/m{\cdot}K$, which is a significantly greater value than the value declared according to standard [3]. In addition, the thermal conductivity value should be adjusted depending on the speed of the medium, i.e., the velocity of wind and the roughness of the concrete [2]. Based on the performed calculations and their calibration with the measured data, the method of increasing the thermal conductivity in the depth of 50 mm was found effective. The increase of the thermal conductivity was implemented as a function of the wind velocity $w\ [m/s]$, by increasing the conductivity by 19% on the top and by 9% on the bottom for every 1 m/s.

Table 1. Overview of material properties used in the calculation model.

Material properties	**Modulus of elasticity**	**Minor Poisson's ratio**	**Secant coefficient of thermal expansion**	**Density**	**Thermal conductivity**	**Specific heat**
Symbol *units*	E_{cm} $[GPa]$	ν $[-]$	α $[-]$	ρ $[kg/m^3]$	k $[W/m \cdot K]$	c $[J/kg \cdot K]$
Value	33	0.20	0.00001	2500	2.7	870

Just as important as the definition of material properties is the definition of the constraints. Thanks to the generated substeps we can describe the value change of the temperature constraint at every surface of the element at any time. In our case, due to the continuous measurement on the specimen, it is not a problem to specify temperature constraints condition accurately. In chapter 5.1, the problem of temperature constraints on the specimen and effect time step size will be discussed. If one surface of the specimen is warmed by 20 °C, which can be considered as the maximum temperature difference during a day-night cycle, the temperature of the specimen at the given material properties will stabilize throughout the entire height after less than 24 hours. Therefore, it is unnecessary to collect data in intervals longer than 24 hours before the evaluation of the structure.

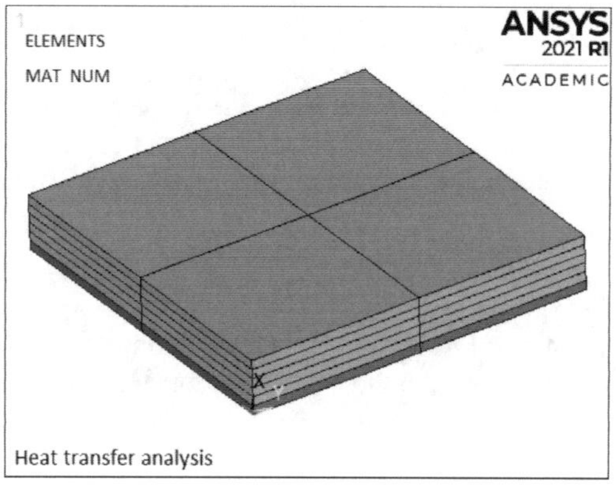

Figure 6. Axonometry of the computational model with the colour distinction of the materials.

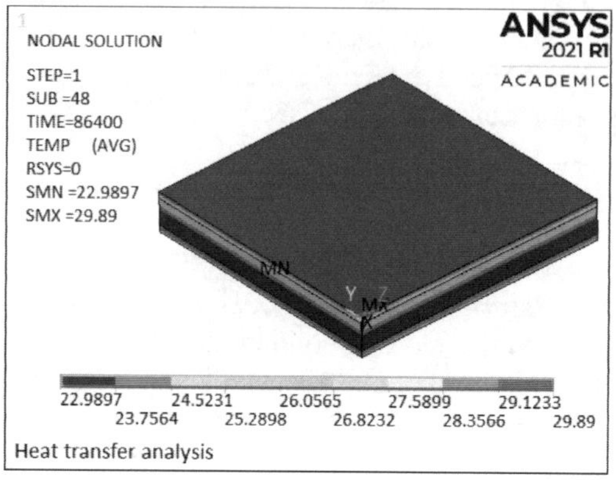

Figure 7. Graphical display of results on the computational model.

5. Evaluation of data obtained from measurements and calculations

The measurements and calculations aim to approximate as closely as possible the actual thermal loading of the slender structures. The influence of long-term temperature and climate conditions is evaluated together with short-term effects such as solar radiation and wind during the day.

The measured data during the winter months (figure 8) show that although the upper surface of the specimen is exposed to solar radiation, it does not warm up to exceed the temperature of the lower surface. From the measured data, the highest solar intensities were obtained on the dates from 12 February to 15 February. On these days the average solar intensities were 90-130 W/m^2. The fact that despite the solar radiation the temperature of the top surface is not higher than the temperature of the bottom surface may be caused by the flow of the surrounding medium. The specimen is placed only 1.0 m above the ground. As a result of the roughness of the terrain and obstacles cause that the air below the specimen does not circulate as freely as above the specimen. This effect is respected in computational models, where the thermal conductivity at the bottom is not increased as much as at the top.

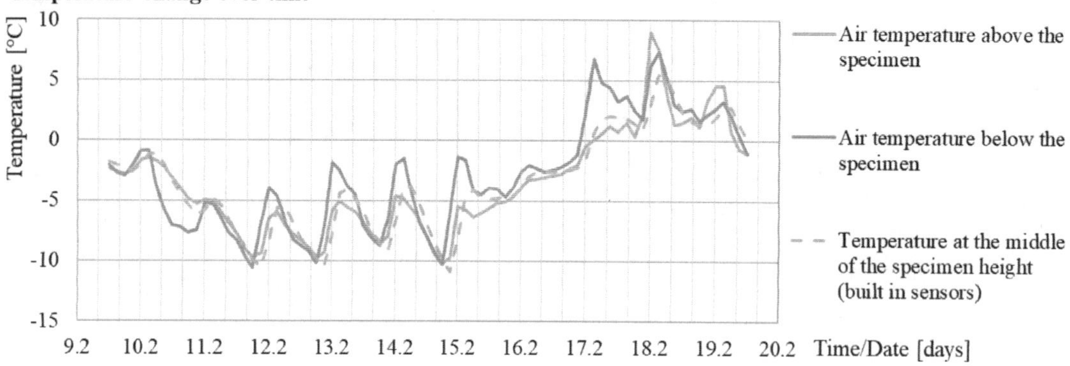

Figure 8. Chart of temperature change over time in winter.

For the data presented in the spring/summer months (figure 9), we can observe a significant temperature difference between the upper and lower surface during the afternoon hours. The average daily values of solar radiation intensity are approximately three times higher than in the winter months. Average values of daily solar intensity are around 300-350 W/m^2. It can also be observed that the top surface of the concrete sample does not retain the received solar energy until the next day. Overnight, the surface of the specimen manages to cool down to the same temperature as the bottom surface.

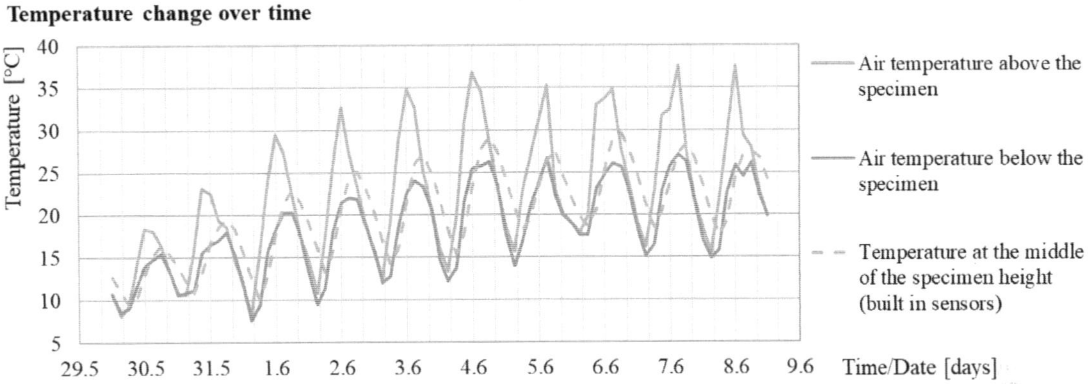

Figure 9. Chart of temperature change over time in spring/summer.

5.1. Evaluation of the size of the time step in the heat transfer analysis

To simplify as much as possible the amount of input data for the heat transfer analysis, the maximum possible size of the time step in which the temperature on the surfaces must be defined was investigated without affecting the accuracy of the results. Chart of the accuracy of heat transfer analysis depending on the time step size (figure 10) represents the temperature along the height of the cross-section (grey dashed curves) obtained from the computational models with time steps of 15, 30, 60, 120 and 240 minutes. The blue curve shows the actual temperature along the height of the cross-section measured by the built-in sensors. A relatively good match is obtained for time steps from 15 to 60 minutes. For further analyses, a time step of 30 minutes was chosen. To obtain input data for heat transfer analysis, it is sufficient to measure the temperature at the surfaces of the real structure every hour for 24 hours before the geodetic survey.

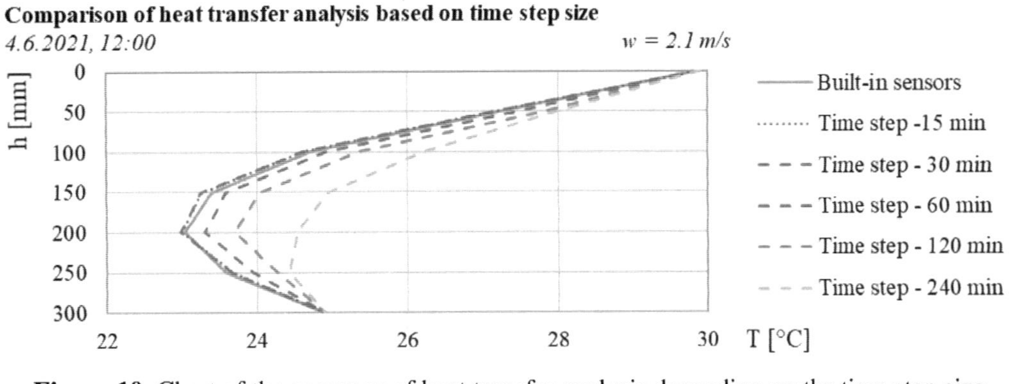

Figure 10. Chart of the accuracy of heat transfer analysis depending on the time step size.

5.2. Comparison of measurements and numerical solution

The following charts (figure 11,12,13 and 14) show the temperature along the height of the cross-section obtained by the built-in sensors (blue curve) and the retrofitted sensors (green curve) - only for the data measured in June. The graphs also show the results from the computational models, where the grey dashed curve represents the temperature determined by heat transfer analysis with unadjusted thermal conductivity, the dotted grey line shows the temperature along the height from the analysis, where the thermal conductivity is increased over the entire height with respect to the wind velocity. The outcome of the heat transfer analysis is shown by the dash-dotted orange curve, where the thermal conductivity is adjusted only at the top and bottom of the cross-section.

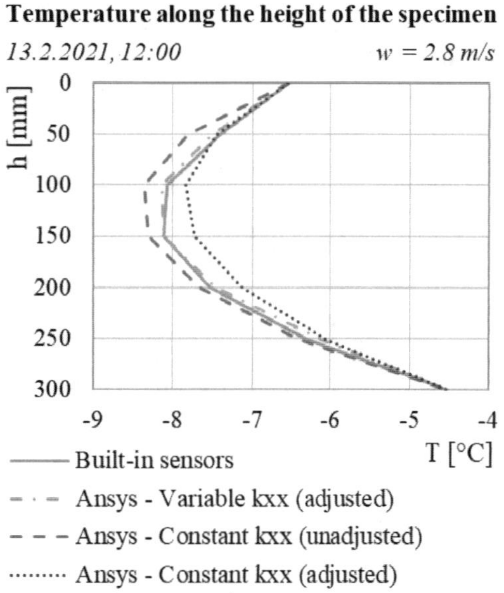

Figure 11. Comparison of data obtained by different methods (13.2. 12:00).

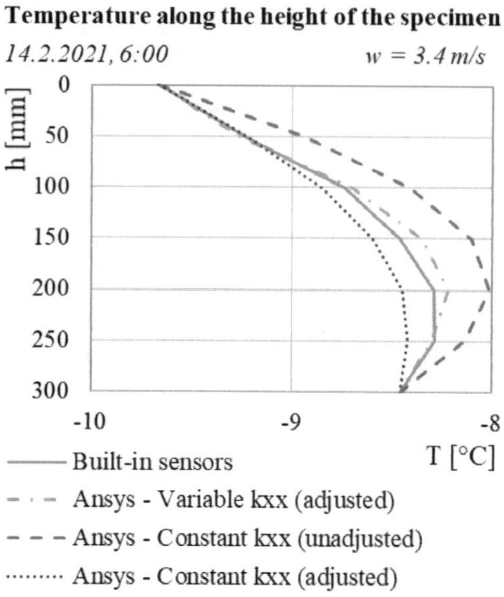

Figure 12. Comparison of data obtained by different methods (14.2. 6:00).

In the following graphs (figure 13 and 14), which track temperature in June, we can compare measured data from retrofitted sensors and the built-in sensors. The differences between the values are significant and this is mainly caused by the fact that the retrofitted sensors are only sealed on the outer surface. This is done on purpose to check that the sensors cannot be installed in the structure this way. In the future, it is planned to fill the holes with cement suspension. It can be assumed that then the measured data from the built-in and retrofitted sensors will match significantly better.

Based on the performed analyses, it is clear that the temperature obtained from the numerical models have a very good match with the measured data. It can be observed from the graphs that it is important to adjust the thermal conductivity appropriately concerning the wind velocity. This adjustment described in chapter 4.1 is only empirical based on the calibration of numerical models with measured data. Therefore, its application to structures may not always reflect the actual behaviour of the material. It is not possible to perform a local test on a real structure to verify the actual thermal conductivity, at the same time the effect of wind on heat transfer cannot be neglected.

In the graph from 5.6.21 (figure 14) we can see that although the cross-section shows signs of cooling on the surface, the cross-section holds heat in its core. This can also be seen in the Chart of temperature change over time (figure 9), where a delay in peak temperature can be seen for values measured at the centre of the specimen height. The graph from 4.6.21 (figure 13) shows that we cannot assume a linear temperature distribution along the height of the cross-section defined by the temperature at the top and bottom surfaces. Likewise in the winter months (figures 11 and 12), where the difference in the temperatures is not as significant and hence, we would make a smaller error.

IOP Conf. Series: Materials Science and Engineering 1209 (2021) 012066 doi:10.1088/1757-899X/1209/1/012066

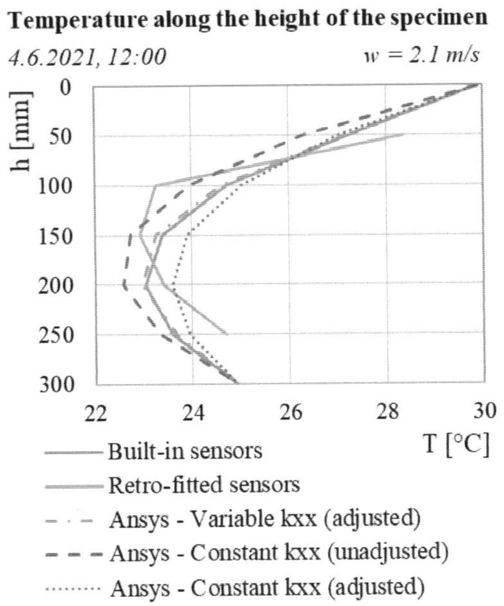

Figure 13. Comparison of data obtained by different methods (4.6. 12:00).

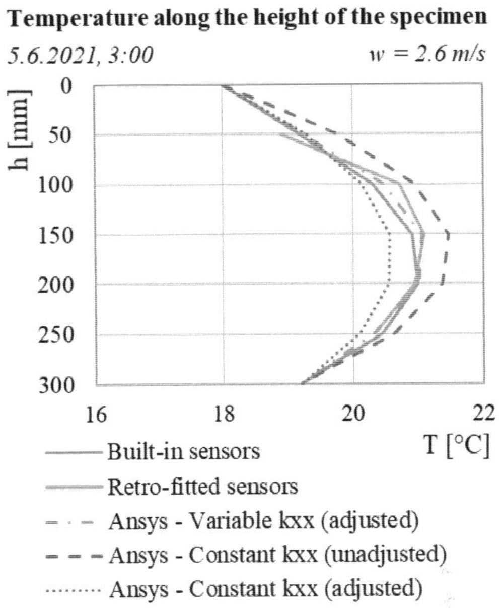

Figure 14. Comparison of data obtained by different methods (5.6. 3:00).

6. Conclusion

The paper outlined the importance and practical application of the correct determination of temperature along the cross-section height in cable-supported structures with slender concrete deck.

From the collected data, we get an idea of the temperature distribution along the height of the cross-section during the day and night during the year. During the daytime, the temperature curve along the cross-section height has a convex shape with a higher temperature on the top surface in summer and on the bottom surface in winter. Conversely, during the nighttime, the temperature curve along the cross-section height has a concave shape with comparable temperature on both surfaces. Furthermore, it can be stated that retrofitted temperature sensors cannot be considered as reliable data sources without proper hole filling by e.g., cement suspension.

As one of the main contributions of the research, we can evaluate the gained knowledge from the time-dependent heat transfer analysis in the finite element software Ansys Mechanical. The analyses show excellent consistency and matching with the values obtained by built-in sensors. As input data for the analysis on a real structure is sufficient to measure the surface temperature. The determination of thermal conductivity as a function of density and increasing its value in outer parts of the cross-section with respect to the wind velocity have been proven effective.

7. References

[1] Asadi I, Shafigh P, Abu Hassan Z F B and Mahyuddin N B 2018 Thermal conductivity of concrete – A review *Journal of Building Engineering* **20** p 81-93

[2] Guo L, Guo L, Zhong L and Zhu Y 2011 Thermal conductivity and heat transfer coefficient of concrete *Journal of Wuhan University of Technology-Mater. Sci. Ed* **26** p 791- 796

[3] ČSN 73 0540-3 2005 *Thermal protection of buildings – Part 3: Design value quantities* (Praha: Český normalizační institut) p 96

[4] ČSN EN 1992-1-1 ed. 2 2019 *Eurocode 2: Design of concrete structures – Part 1-1: General rules and rules for buildings* (Praha: Česká agentura pro standardizaci) p 208

[5] Klusáček L, Nečas R, Koláček J and Strnad J 2018 *Diagnostický průzkum lávky pro pěší přes řeku Moravu v Kroměříži* Brno

[6] Klusáček L, Svoboda A, Nečas R, Koláček J and Olšák M 2020 *Diagnostický průzkum předpjaté lávky pro pěší přes řeku Ohři v Radonicích nad Ohří* Brno

Acknowledgements
The authors gratefully acknowledge the financial support from the Faculty of Civil Engineering at Brno University of Technology project nos. FAST-S-21-7409 and FAST-J-21-7299.

IOP Conf. Series: Materials Science and Engineering 1209 (2021) 012067 doi:10.1088/1757-899X/1209/1/012067

Experimental analysis of shear connection effect composite beams with concrete

M Weissova[1], D Dubecky[1] and V Kvocak[1]

[1] Institute of Structural Engineering, Faculty of Civil Engineering, Technical university of Kosice, Vysokoskolska 4, 042 00 Kosice, Slovakia

E-mail: michala.weissova@student.tuke.sk

Abstract. The subject of this research is the analysis of shear connection effect on the fiber glass reinforced composite beams. The research contains the detailed analysis of materials like concrete and fiber glass reinforced composite. The shear connection has been tested by the experiment of push out test run in the Centre of research and innovation in construction of Technical University of Košice. The shear connection has been tested on two types of connectors used on composite beams. The main idea of this research is find out whether it is possible to use the alternative material, which should substitute the steel. The aim of research is to design a type of connector for fiber reinforced composite materials or another composite materials developping in the world of material engineering. The part od research is devoted to the FEM analysis.

1. Introduction – Composite materials

Composite materials represent a part of the market that is intended to replace currently used materials. Due to the rapid development of science, technology and industry, the technical requirements for the employed materials are also increasing. The currently used materials are gradually ceasing to meet the requirements of recent technologies. Therefore, composite materials come to the fore, which have higher strength properties but also higher resistance to external influences. The development of composite materials is necessary because the gradual increase in the requirements for strength and durability is inefficiently achieved by current methods, such as the addition of additives, heat treatment, or moulding. Especially for this reason, different materials have begun to be combined into one construction material, a composite material, to merge the positive properties of the individual materials. Brittle or high-strength materials have begun to be combined with plastic ones.

The development of composite materials is preceded not only by science and technology, but also by the gradual lack of used mineral materials, from which majority of today's materials are made. For the material industry, raw materials such as minerals and ores, fossil fuels (coal, natural gas oil) and biomass (wood, cotton, corn) are recently employed. The effort of the whole material industry is the ability to replace materials is the goal of the whole material industry.

Composite materials can be divided into four groups. The first includes composite materials with a plastic matrix, which is reinforced with fibers of low plasticity (boron fibers, carbon fibers). The second includes materials with a plastic matrix but also with relatively plastic fibers (steel, beryllium, molybdenum, tungsten). The third group includes composites with a brittle matrix reinforced with fibers characterized by considerable plasticity (metal fibers of highly fusible metals) and the fourth includes materials with a brittle matrix and reinforcing carbon fibers, refractory compounds, and some crystals with very low plasticity.

The advantage of fibre-reinforced composite materials is their anisotropy, because, as is well known, different elements are stressed differently in different directions. The strength of fibre-reinforced materials is increased by placing the fibres evenly in one direction. The reinforcing fibres can be metallic or non-metallic. For example, the metal fibres include steel fibres, whose strength is increasing along with the increasing plastic deformation, as well as tungsten or molybdenum fibres. Non-metallic fibres include carbon fibres, which have a high young's modulus, ultimate strength, considerable heat resistance and low weight. Glass fibres are another type of non-metallic fibres. They have low weight and thermal conductivity, but high chemical and thermal stability together with sufficient strength. Glass fibres are prepared in two basic types, continuous and staple. Continuous fibres are produced by pultrusion from a molten glass mass and staple fibres are made by pultrusion of glass fibres from a stream of molten glass and then its subsequent interruption on the desired length. Other types of fibres are boron, ceramic, or filament crystals [1].

The composite materials′ production by pultrusion consists in drawing the material through a heated steel mould by a pultrusion device. It is a continuous process of producing reinforced resins of various shapes and lengths, while the input material is a mixture of liquid resin and fibre reinforcement.

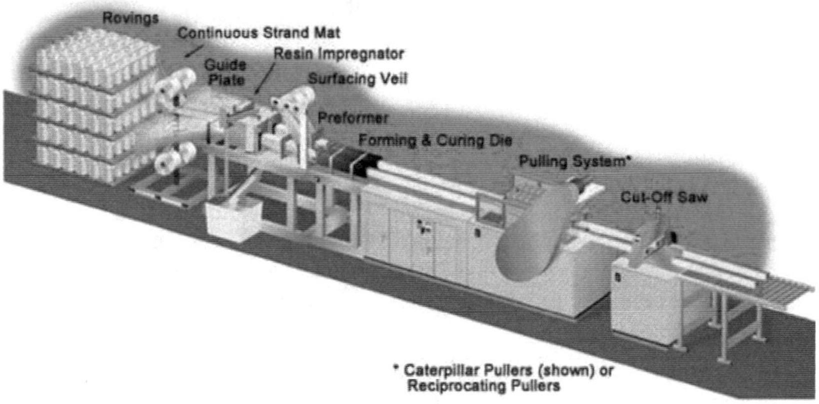

Figure 1. Pultrusion [2].

Most of the reinforcing material is in the form of glass fibres, also known as roving, and sheet mats. The fibres wound on the spools enter the leveller, which ensures the distribution of the fibres in the cross section together with the placement of the mats. On the surface of the profile, there is a polyester drape that wraps the glass reinforcement and forms a cover, which increases the resistance to the penetration of chemicals, radiation, and improves the appearance of the composite. In general, it prevents the glass fibres from protruding after the surface has been damaged. Subsequently, the fibres, mats and drapes are wetted in a mixture of resin, filler, catalyst dye and other additives, to improve the material properties of the resulting profile. Excess binder is extruded in the mould and the profile is gradually moulded and enters the mould to cure the final profile. At the end of the mould, the finished profile is pulled by the feeding device and divided into the required length.

The pultrusion permits to achieve a higher proportion of fibres in the volume in the resulting profile. In contrast to manual lamination, where this proportion is about 20%, and to pressing, where it is about 40%, it is more than 70% in profiles produced by pultrusion. The material passes through the production line at a speed of approximately 30 to 150 cm per minute. This production process creates a strong and light hollow profile with a minimum wall width of 1.5 mm [2].

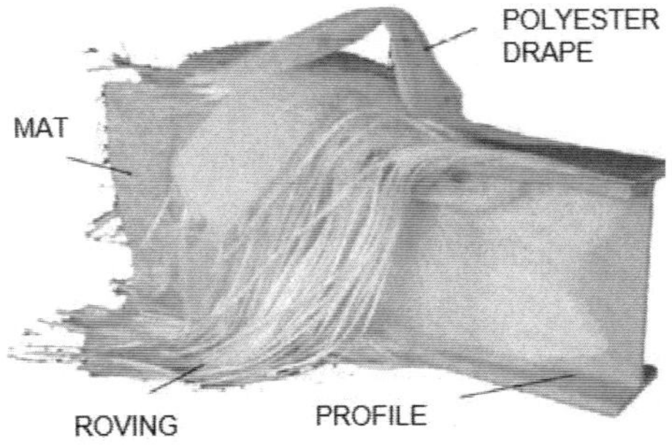

Figure 2. Composite profile [2].

2. Preparation of experiment

Experimental analysis of the shear resistance of an element with concrete is described in the standard EN 1994-1-1. Eurocode 4: Design of composite steel and concrete structures. This standard describes a push out test, which will result in the shear resistance itself. For the purposes of the experiment, it was necessary to make several geometric adjustments to the sample.

The standard test describes the dimensions of the tested specimen as well as the loading procedure. The dimensions are described in detail in the following figure. The loading of the sample begins with a continuous load of up to 40% of the value of the presumed failure of the shear resistance and should then be relieved 25 times and loaded between 5% and 40% of the expected value of the load at failure. At the end of the unloading and loading, the sample is loaded again with a continuous load until the sample breaks. The failure of the sample must not occur until after 15 minutes of continuous loading. The longitudinal slip of the sample should be measured either continuously throughout the loading process or continuously but at least 80% of the value of the assumed maximum load [3] [4] [5].

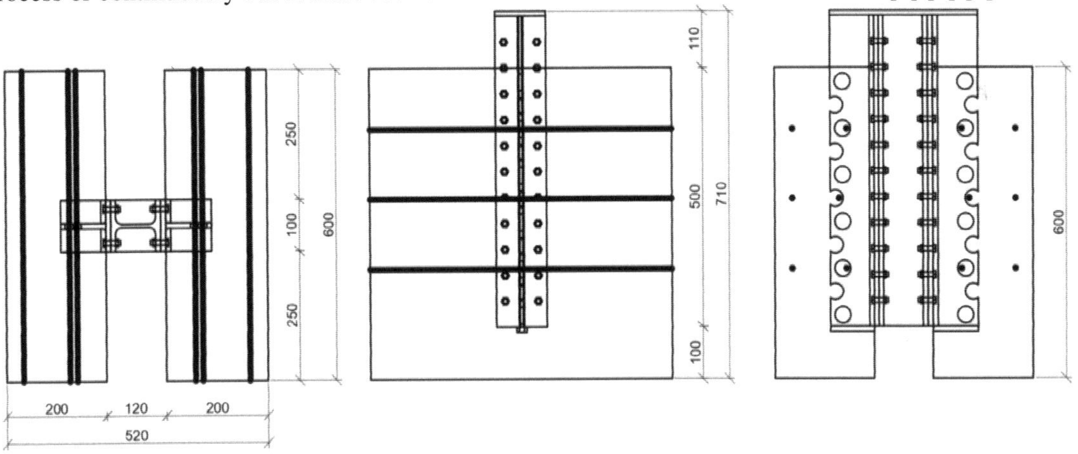

Figure 3. Geometry of push out test sample.

Due to previous experience with push out tests, the geometry of the sample was adjusted so that the coupling strip is up to the edge of the concrete and the flange of the composite beam reaches the edge of the steel part of the sample. The top plate, by which the force is transferred from the hydraulic press to the sample, reaches the edge of the composite wall.

To achieve a sufficiently rigid connection between the composite and the steel, the connection of the middle steel part with the composite beams will be made by using M12 screws at an axial distance of 50 mm, while the tightening force in the screw is 60 Nm. To avoid the coupling of the bolt nut with the concrete and the consequent deterioration of the test, the bolts will be protected from the inside with mounting foam.

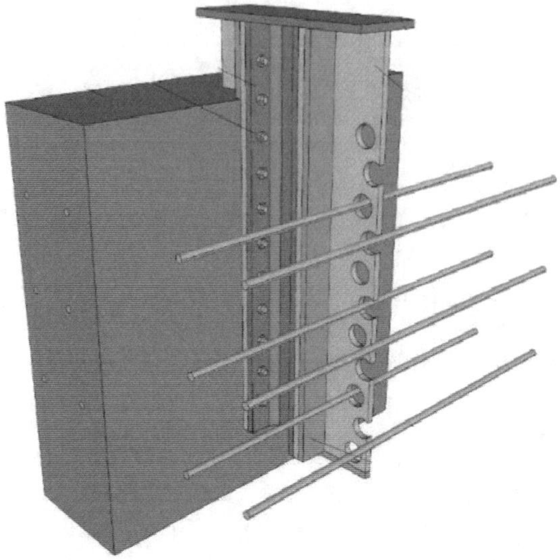

Figure 4. Arrangement of push out test sample.

Two types of coupling belt were selected for experimental verification of shear resistance. The first type of sample was chosen in the form of a ridge strip with a ridge height of 57 mm. A 10 mm diameter GFRP reinforcement rod was inserted into each wave.

The second type of sample was chosen with the shape of a perforated strip, with a hole diameter of 32 mm and reinforced with GFRP reinforcement according to the following figure.

3. Experiment

The main principle of the push out test is to determine the shear resistance of the connection belt. The push out test sample consists of two concrete blocks with a width of 200 mm, a length of 600 mm and a height of 600 mm. Connection strips with a connection length of 400 mm were concreted into these two concrete blocks. The connection strips form two beams cut from a composite profile with a length of 600 mm and have been concreted into the blocks so that their position is 100 mm from the lower edge of the block and 100 mm above the upper edge. To determine the actual connection resistance and to prevent the beam from being crushed in the concrete, a 10 mm thick polystyrene was inserted under the beam. The beams were attached to the steel profile so that it was possible to achieve the displacement of the entire middle part together with the composite beams. Therefore, it was necessary to ensure sufficient rigidity of the joint. This joint is usually made by welding, but since it is a test of the composite beam connection, this joint had to be made by screws. Thus, a central part with a total length of 600 mm was created, where the flanges of two composite beams were attached to the flanges of the steel beam by steel screws. As an excessive connection of the screws with the concrete could appear, it was necessary to secure the area around the screws with mounting foam. The concrete blocks were reinforced with GFRP reinforcement so that the reinforcing bars were part of the connection strip.

Figure 5. Push out test.

A total of 5 sensors were attached to the sample. One inductive sensor measured the horizontal displacement of the middle part of the element during loading. The remaining sensors were placed on the concrete surface, in areas where maximum stresses were expected. Two inductive sensors and two tensometers were used. The length of inductive sensors and tensometers was 150 mm.

Using an INSTRON hydraulic press with a maximum force of 2500 kN, the middle part of the test body was pushed over the distribution surface. Two types of connection belts were tested, a perforated belt and a ridge belt.

The loading of the sample took place in loading cycles in such a way that the sample was loaded to 40% of the assumed shear resistance, then drawn to 5% of the resistance. Subsequently, such 25 cycles were performed with the lowest value of 5% and the highest 40% of the assumed resistance. These cycles are described in the following figures.

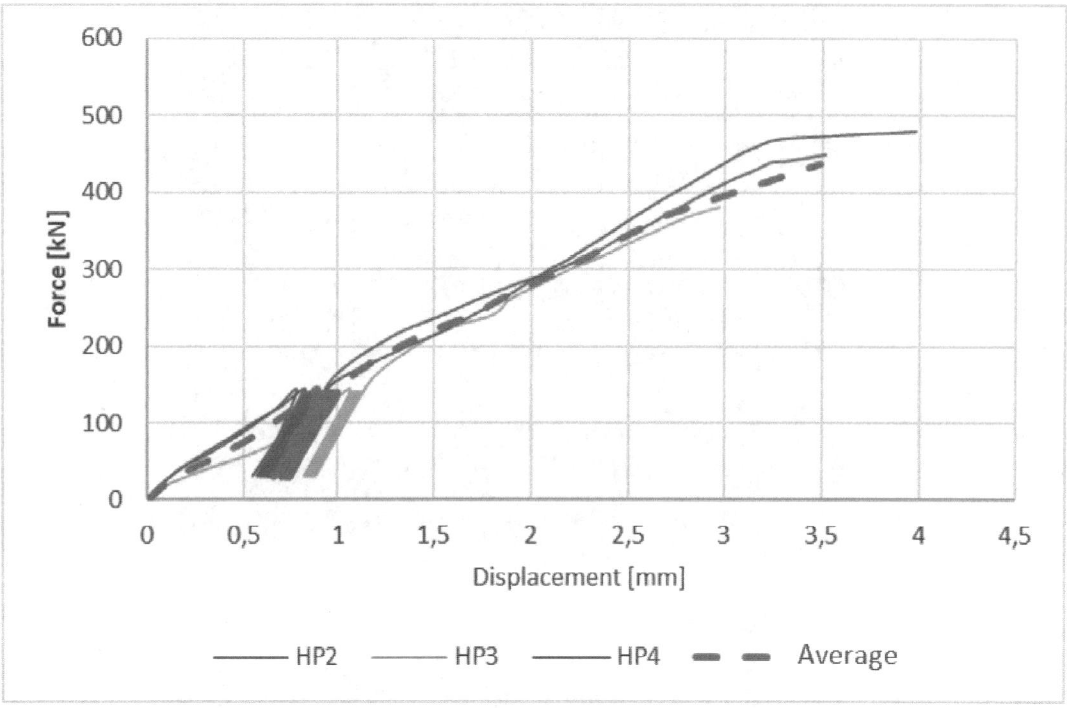

Figure 6. Vertical displacement of ridge belt.

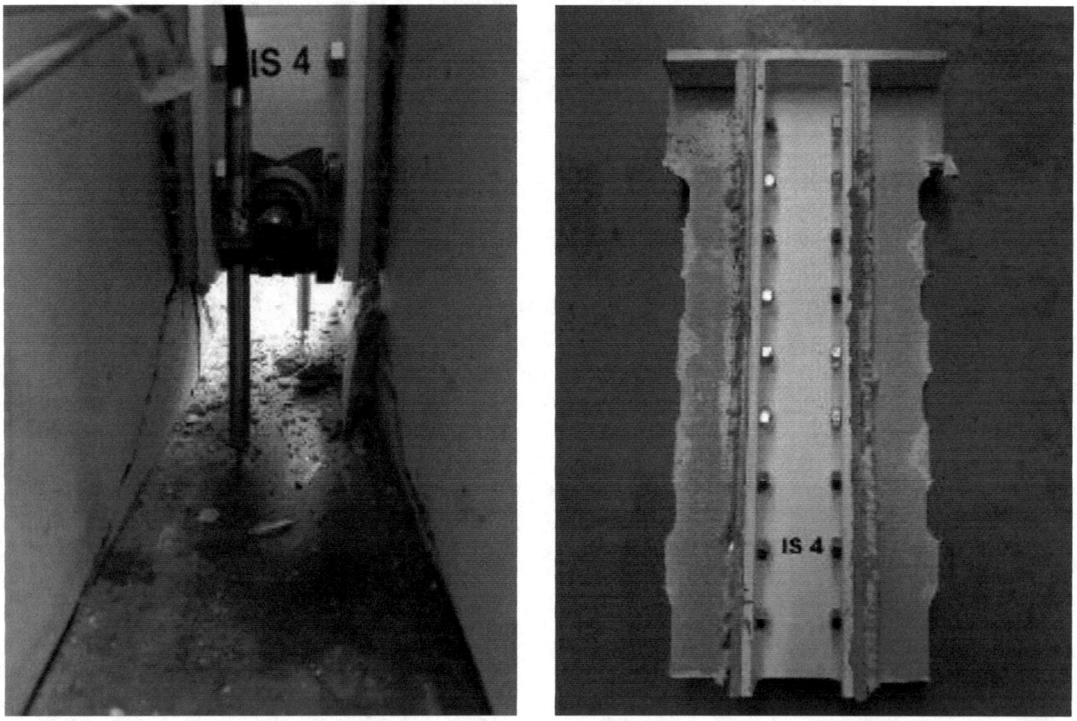

Figure 7. The result of experiment – ridge belt.

Due to the material characteristics and the manufacturing method of the composite beam, the matrix between the individual fibres was broken and thus the connection strip separated from the rest of the beam.

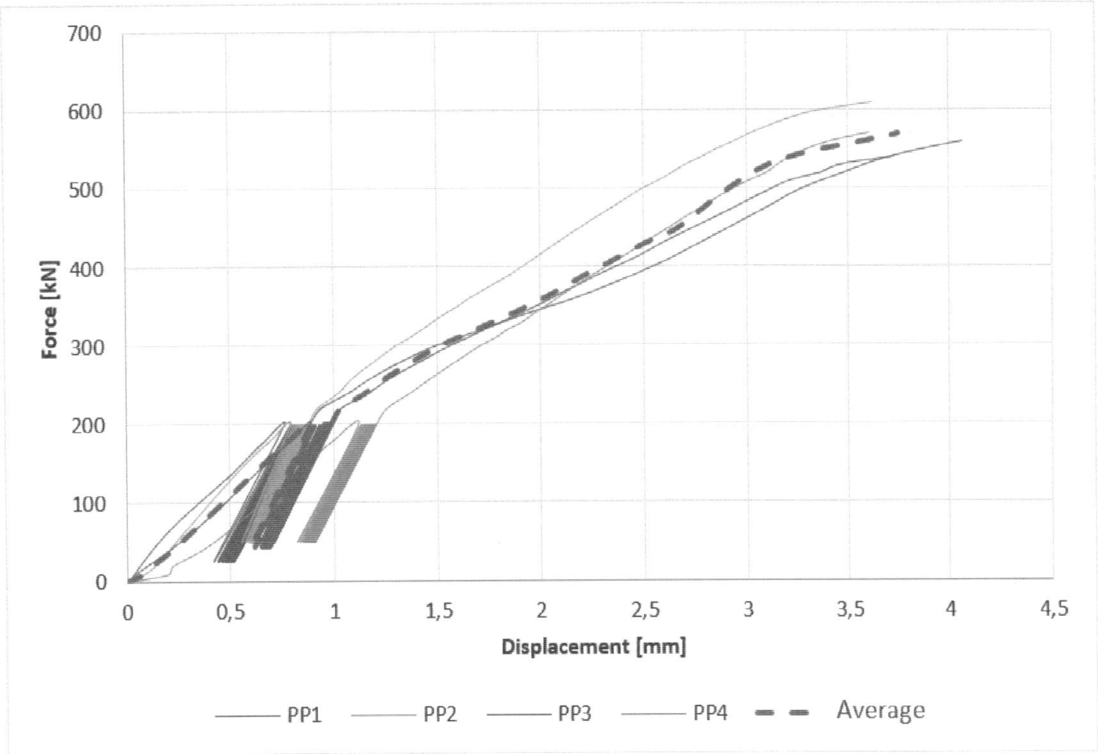

Figure 8. Vertical displacement - perforated belt.

Figure 9. The result of experiment – failure.

Based on these experiments, it is possible to determine the maximum force that represents the resistance of this type of connection at maximum deformation. The previous graphs point out to the fact that composite materials tend to behave elastically with a slight hint of plasticity until the failure. After breaking the connection, the middle part had hit the concrete, where the breakage occurred.

IOP Conf. Series: Materials Science and Engineering 1209 (2021) 012067 doi:10.1088/1757-899X/1209/1/012067

It is clear from the experiments that failures on the concrete surface did not occur until the connection failure. The failure on the concrete occurred at the upper edge, and at the lower after the impact in the shape of a load cone.

It is clear from the experiments that failures on the concrete surface did not occur until the connection failure. The failure on the concrete occurred at the upper edge, and at the lower after the impact in the shape of a load cone.

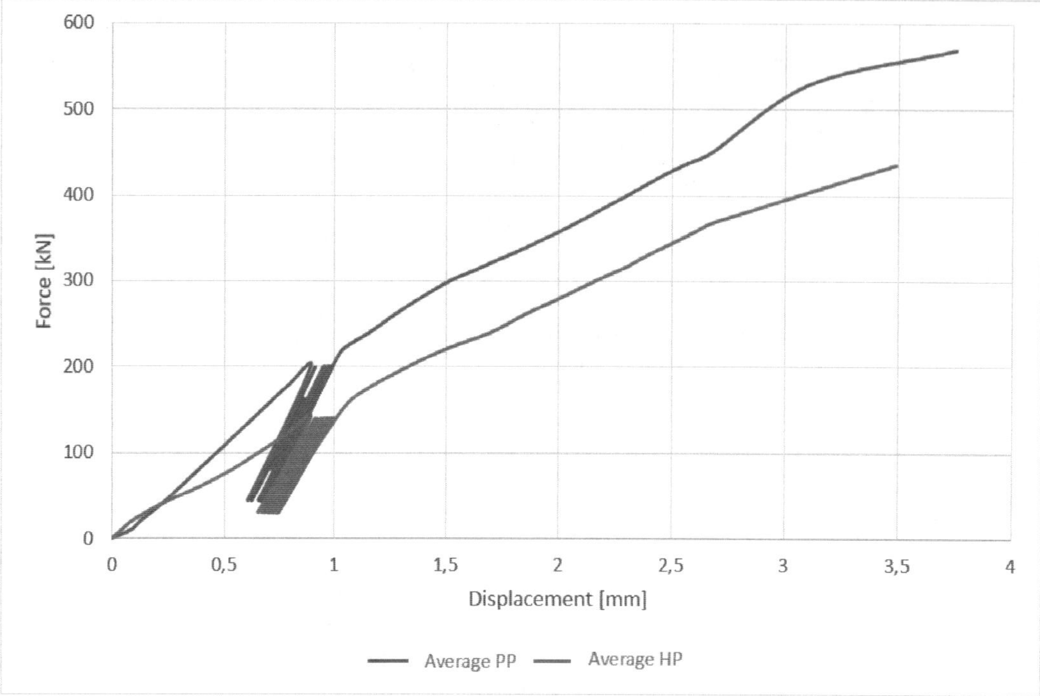

Figure 10. Comparison of resistance.

From these graphs and experimental verification, it is clear that the connection resistance with the ridge strip is 436.33 kN and with the perforated strip is 569.55 kN.

4. Conclusion

The result of the analysis can be evaluated from several points of view. First of all, it is clear that the shear resistance of the composite material is lower compared to steel. Specifically, according to the previous experimental verification of the shear resistance of the connection on steel elements of the same geometry, the resistance of the ridge strip is approximately 620 kN and the perforated strip approximately 725 kN. The resistance of the composite connection with the ridge strip is 436.33 kN and the perforated strip is 569.55 kN.

Based on these results, it can be concluded that the resistance of the ridge strip is lower by about 30%, which greatly affected the matrix failure during the experiment. The resistance of the perforated strip is lower compared to steel by about 20%.

Due to the positive properties of the composite material such as corrosion protection and low weight of the material, the composite beam as a connection element can be considered sufficient, but to use such beams on bridge structures, it is necessary to experimentally verify dynamic, fatigue, bending or long-term effects for this kind of coupling.

5. References

[1] Blaščík F and Pulc V 1994 *Kompozitné materiály* (Bratislava: Vydavateľstvo STU v Bratislave)
[2] Prefa Kompozity *Výroba a montáž kompozitních konstrukcí, Technologie výroby* Available online www.prefa-kompozity.cz

[3] Rovňák M, Ďuricová A, Kundrát K and Naď Ľ 2006 *Spriahnuté oceľovo-betónové mosty* (Košice: Elfa)
[4] Kvočák V, Vičan J and et al. 2013 *Navrhovanie oceľových mostov podľa európskych noriem* (Košice: Technická univerzita v Košiciach)
[5] Dubecký D. and Kvočák V 2018 *Analýza odolnosti doskových mostov so zabetónovanými oceľovými nosníkmi* (Košice: Technická univerzita)

Acknowledgments
The paper presented was supported by the projects: VEGA 1/0172/20 "Stress and deformation analysis of load bearing components made out of steel, glass and composite materials" of the Scientific Grant Agency of the Ministry of Education, Science, Research and Sport of the Slovak Republic

Sustainable concept of energy management of buildings as an effective tool for green building

J Domanický[1], Z Vranayová[1] and F Vranay[1]

[1] Technical university of Kosice, Faculty of Civil Engineering, Institute of Architectural Engineering, Vysokoskolska 4, 042 00 Kosice, Slovakia

E-mail: jan.domanicky@tuke.sk,zuzana.vranayova@tuke.sk,frantisek.vranay@tuke.sk

Abstract. Nowadays, due to high external temperatures, the topic of cooling is very actual. However, cooling buildings is energy intensive. Therefore, it is necessary to focus on more than one method of cooling. We can say that cooling can be provided in 2 ways. By building services and by architectural design. For existing buildings, we cannot change the shape and orientation of the building. But we can change the composition of structures. Green architecture is often mentioned in this area. Therefore, the aim of this research is to interconnect these two areas and determine the influence of vegetative roof on the radiant cooling and heating systems. This connection shows us how the design will affect the HVAC systems.

1. Introduction

Global warming is one of the most fundamental problems in the world. It significantly affects every one of us. The energy crisis and environmental pollution pose great challenges to human development. Traditional vapor-compression cooling consumes abundant energy and leads to a series of environmental problems. An example of its consequences is e.g. higher temperatures in summer and at the same time lower temperatures in winter, rising world ocean levels, etc. It leads to massive cooling demand [1].

One of the ways to contribute to solving this problem is to reduce the energy intensity of buildings. Whether it is family houses, apartment buildings, office buildings and much more.

The study always has to start from the smallest part. The representative could be the small house or office.

2. Cooling and heating systems

Every building has heat losses and heat gains in its operation. It depends on the orientation of the building, location, architectural composition and building structures. Cooling and heating systems must be used to maintain the correct indoor climate. The value in the interior in the summer easily exceeds the temperature even 30 ° C and more and in cold days is under 20°C. In this case, productivity and concentration of decrease. But what's worse that in the European Union is an increasing trend of human mortality due to high temperatures during the summer months. This fact forces us to think about the possibilities of how to reduce this number. One option is to use cooling system in a building [2].

2.1. Legislation

The Decree no. 99/2016 Coll. (Decree of the Ministry of Health of the Slovak Republic on details on health protection against heat and cold stress at work). describes in Annex 2, table No. 1,2 The range of optimal and permissible values of heat-humidity microclimate factors for the warm and cold period. It depends on the class of work.

Table 1 describes the work classes and its parameters. Tables 2 and 3 show the maximum operating temperatures according to these classes.

Table 1. Classes of work according to total energy expenditure (Translation from the decree) [3].

Class of work		Energy expenditure q_M [W.m^{-2}]	Examples of activities
1	1a	≤ 80	Seating: easy administrative work (writing, computer work, drawing, accounting, office work).
	1b	81-105	Seating: easy manual work (simple sewing, laboratory work); arm and shoulder work (small work tools, inspection work, assembly or sorting of light objects); arm and leg work (driving under normal conditions, operating the pedals).
	1c	106-130	Standing: drilling or milling of small parts, coil winding, threading of small fittings, machining with low-power tools, occasional walking (speed up to 2.5 km / h).
2	2a	131-165	Continuous work of hands and arms (nailing, filling); arm and leg work (driving trucks, tractors or construction machinery off-road or in the field); arm and torso work (working with a pneumatic hammer, tractor assembly,
	2b	166-200	plastering, intermittent handling of medium-heavy material, weeding, hoeing, picking fruit or vegetables, pushing or pulling light carts or furrows, walking at a speed of 2.5 to 5.5 km / h, forging).
3		201-260	Intensive work of arms and torso (carrying heavy material, working with a shovel, working with a heavy hammer, sawing, planing or working hardwood, walking at a speed of 5.5 to 7 km / h, pushing or pulling heavily loaded carts or blowers, beating castings, laying concrete panels).
4		>260	Very intensive activity at a fast to maximum pace (working with an ax, intensive work with a shovel or digging, climbing stairs, on a ramp or ladder, fast walking in small steps, walking at a speed greater than 7 km / h).

Table 2. The range of optimal and permissible values of heat-humidity microclimate factors for the warm period (Translation from the decree) [3]

Class of work	Operative temperature t_o [°C]			Permissible speed air flow v_a [m.s^{-1}]	Permissible relative humidity r_h [%]
	optimal	permissible			
		min.	max.		
1a	23-27	20	28	≤ 0,25	30 - 70
1b	22-25	19	27	≤ 0,3	
1c	20-24	17	26	≤ 0,3	
2a	18-21	15	25	0,1-0,3	
2b	17-20	12	25		
3	Not determined*	10	Not determined*	0,1-0,5	
4		10			

Table 3. The range of optimal and permissible values of heat-humidity microclimate factors for the cold period (Translation from the decree) [3]

Class of work	Operative temperature t_o [°C]			Permissible speed air flow v_a [m.s^{-1}]	Permissible relative humidity r_h [%]
	optimal	permissible			
		min.	max.		
1a	23-27	20	28	≤ 0,25	
1b	22-25	19	27	≤ 0,3	

IOP Conf. Series: Materials Science and Engineering 1209 (2021) 012068 doi:10.1088/1757-899X/1209/1/012068

1c	20-24	17	26	≤ 0,3	
2a	18-21	15	25	0,1-0,3	30 - 70
2b	17-20	12	25		
3	Not determined*	10	Not determined*	0,1-0,5	
4		10			

Based on these values, a cooling and heating system must be provided to ensure thermal comfort. By calculating the heat load and heat loss, it is possible to determine the required power. Then it is necessary to select a suitable system.

2.2. Cooling systems
Cooling is a very necessary system today. The indoor temperature is overheated due to the increase in outdoor temperatures. We can use one of the following methods for cooling (air conditioning, air flow, radiant cooling. Each of them has its advantages. These properties must be taken in the optimal design.

2.2.1. Properties for optimal design
- energy consumption
- indoor climate quality
- noise parameters
- design
- amount of fresh air
- requirements for air distribution
- self-regulation of the system
- economic aspect

2.3. Heating systems
Over the years, heating has been designed to a much more than cooling. However, this has changed today. It can be provided by conventional, radiant and hot air heating method. The same principles apply to design as for cooling (par. 2.2.1) In the past, convection heating has been used more. Today, due to heat pumps and low operating temperature, radiant heating is used.

2.4. Radiant cooling and heating systems
The best and most economical solution is if both cooling and heating are provided by one system. In this case, it is necessary to select a suitable source or a combination of both. A suitable solution is e.g. heat pump. A heat pump is a device that can generate heat as well as cold. Heat pumps and photovoltaics offer the most energy-efficient way to provide heating and cooling in many applications, as they can use renewable heat sources in our surroundings. The principle is based on obtaining cold and heat from external sources (air, ground, water) and transfers it to the cooling or heating system. There are many other options such as. combination of a separate source of heat and cold, but it is more economically demanding. Radiant cooling and heating system can be designed as a wall, ceiling and the most common floor method [4].

3. Green roofs
When it comes to thermal comfort with regard to the interior and exterior, the building envelope plays an important role. The main example is a green roof, in some cases a wall. Green roofs are constructed vegetated systems designed to perform ecosystem services such as water catchment, building insulation, air purification, noise reduction, and creation of animal habitat. Many studies have been prepared that deal with the combination of constructions with green elements. This aspect plays a big role in solving heat islands, lowering the ambient temperature and much more. If we disregard the economic, static demands and the issue of maintenance, it is a very effective solution. It has many advantages that can be divided into ecological and economic areas. Both are closely interconnected [5].

A roof receives solar radiation directly, which makes it an important component to reduce cooling energy consumption. Cool roofs have been studied and verified as an effective passive cooling strategy for diverse conditions and a green roof was studied from many perspectives, such as energy saving potential in buildings, reduction of GHG emissions, etc [6].

3.1. Ecological advantages of green roof
They retain water (retention) and relieve the sewer system, improve the air around the building (by evaporation), prevent the object from overheating or cooling down, act as a filter against dust and harmful substances, they also participate in the formation of oxygen and bind carbon dioxide (photosynthesis), reduce light smog by absorption, last but not least create a new living space for plants, animals and humans [7].

3.2. Economic advantages of green roof
Green roof insulates building in winter and cool in summer, reduce cooling and heating costs, expand the possibilities of use and the total value of the properties, reduce the cost of draining the building, protect roof layers from UV radiation and extend the service life of the structure by min. twice [7].

4. Impact of green architecture elements on HVAC systems
The object of research consists in a combination of two areas. Cooling and heating systems and impact of green roof on them. The research has 2 parts. Creation of a simulation model on which it will be possible to simulate structures under various boundary conditions (exterior, interior temperature, humidity, etc.). The second part consists of "in situ" measurements of structures at different boundary conditions and comparison with a simulation model.

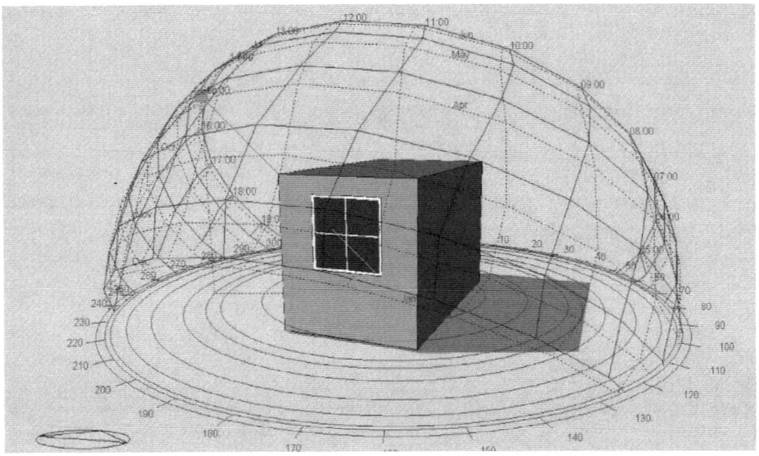

Figure 1. Experimental chamber – Simulation model.

The simulation model shown on the figure 1 is created using the DesignBuilder program. All packaging, interior and filling constructions are entered. We can model various exterior conditions as well as interior properties. It is possible to quantify the simulated values of temperatures in given time periods as well as the value of the annual energy demand for cooling and heating. We can compare these values with the real measured values in the sample house.

5. Description of structures
Figure 2 shows 3 experimental containers. Each container has dimensions 4200x2440x2680mm (l x w x h) with an area of 8 m2. For the research I chose the side containers as they have the same thermal technical properties, disposition as well as orientation. Construction of these containers consist of metal core, on the north-east side door and on the south-west side window is situated.

Figure 2. Experimental object – Laboratory of Technical university in Kosice.

5.1. Material composition

There are five types of constructions in each container. Floor, roof with added insulation and 3 types of walls. Two are basic (sheep panels and polyurethane foam). These constructions are identical for both objects. The difference is in the south wall, where in the left container (C3) there is a wall of OSB boards and insulating filling. In the right (C1), this wall is designed from a pair of masonry. Half aerated concrete and half ceramic brick.

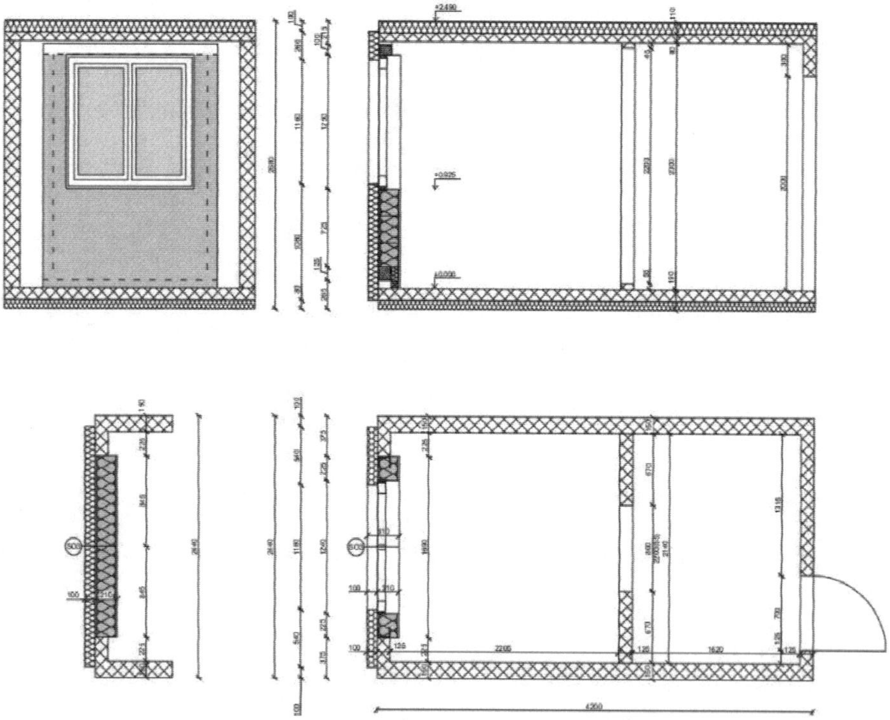

Figure 3. Plan of left container [8].

IOP Conf. Series: Materials Science and Engineering 1209 (2021) 012068 doi:10.1088/1757-899X/1209/1/012068

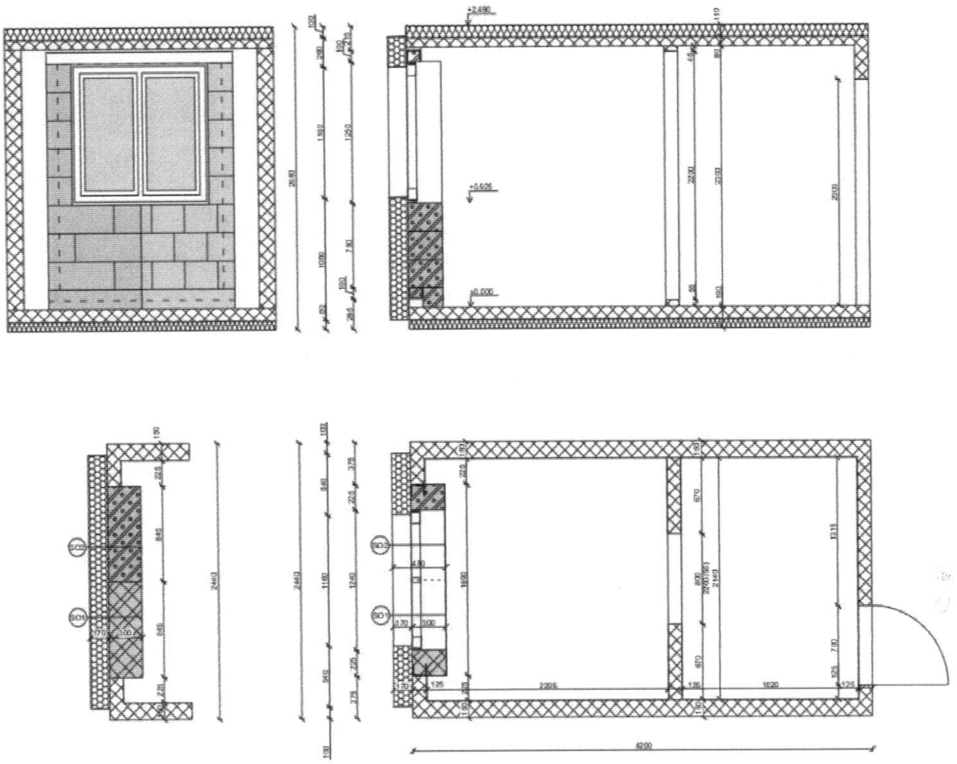

Figure 4. Plan of right container [8].

Construction characteristics:

SO0 - Perimeter construction of the measuring chamber - metal prefabricated panels, 2xprofiled sheet metal, thickness 3 mm, the filling is made of polyurethane foam - thickness 125 and 150 mm.

SO1 - Perimeter construction - brick block HELUZ STI (300/247/238), strength 8.0 N / mm2, for YTONG mortar hr. 3mm, foam graphite polystyrene GreyWall 033, hr.170 mm, Adhesive trowel and silicate plaster CERESIT hr. 15 mm.

SO2 - Perimeter construction - block YTONG P2-350 (300/249/599), strength 2.0 N / mm2, for YTONG mortar hr. 3mm, foam graphite polystyrene GreyWall 033, hr.170 mm, Adhesive trowel and silicate plaster CERESIT hr. 15 mm.

SO3 - Perimeter construction - OSB board hr.12mm, thermal insulation ISOVER UNIROL hr.180mm, OSB board hr.12mm, facade insulation system ISOVER TF hr.100mm, Adhesive drain and silicate plaster CERESIT hr.15 mm.

S1 – Roof construction - metal prefabricated panels, 2xprofiled sheet metal, thickness 3 mm, the filling is made of polyurethane foam - thickness 80 mm, Roof insulation thickness 110mm.

F1 - Floor construction - metal prefabricated panels, 2xprofiled sheet metal, thickness 3 mm, the filling is made of polyurethane foam - thickness 110 mm, Floor insulation thickness 80mm.

IOP Conf. Series: Materials Science and Engineering 1209 (2021) 012068 doi:10.1088/1757-899X/1209/1/012068

Table 4. R, U value of constructions at boundary conditions (Rsiw = 0.13 m2K/W, Rsir = 0.10m2K/W, Rsif= 0.17 m2K/W, Rse = 0.04 m2K/W, θe = -13.0° C, θai = 20.0° C)

Construction	SO0	SO1	SO2	S1
U (W/m2.K)	0,121	0,121	0,121	0,186

5.2. Design construction definition

Thanks to the almost identical chambers, there is a unique opportunity to determine the influence of the green structure on the energy intensity of the building. The design consists in the application of green structures to one of the chambers, thanks to which a pair of experimental chambers will be created. The adjustment has two stages. Intensive vegetation roof and exterior vegetation facade.

5.2.1. Intensive vegetation roof

Due to its properties, an intensive vegetation roof with was chosen. Expectations are based on the influence of the roof composition as well as its storage capacity on the overall energy consumption.

Figure 5. Roof characteristics [9].

S1+ - Intense greenery
- Vegetation layer (Bauder vegetation substrate, intensive (layer thickness 24 cm)
- Filter layer (Bauder filter fleece FV 125)
- Hydroaccumulation / drainage layer (Bauder hydroaccumulation board WSP 75)
- Protective layer (Bauder protective mat FSM 600)
- Separation and sliding layer (Bauder separation foil PE 02)
- S1 (metal prefabricated panels, 2xprofiled sheet metal, thickness 3 mm, the filling is made of polyurethane foam - thickness 80 mm, Roof insulation thickness 110mm).

5.2.2. Intensive vegetation facade

System solutions for the construction of vegetation vertical walls in the interior and on the exterior of buildings offer the possibility of creating diverse, aesthetic and architecturally interesting areas, permanently overgrown with vegetation.

The system of vertical gardens consists of overhanging modular aluminum panels (which can also be called cassettes or facade baskets) in production filled with a substrate into which plants are planted on site. The panels are attached to hanging profiles screwed to the wall or facade.
The classic size of the panel is 600x1000mm. The thickness of the construction is 60mm + 30mm aluminum construction. The substrate can be diverse (white - pumice, brown - lava, beige - tuff, or red - brick crumb).

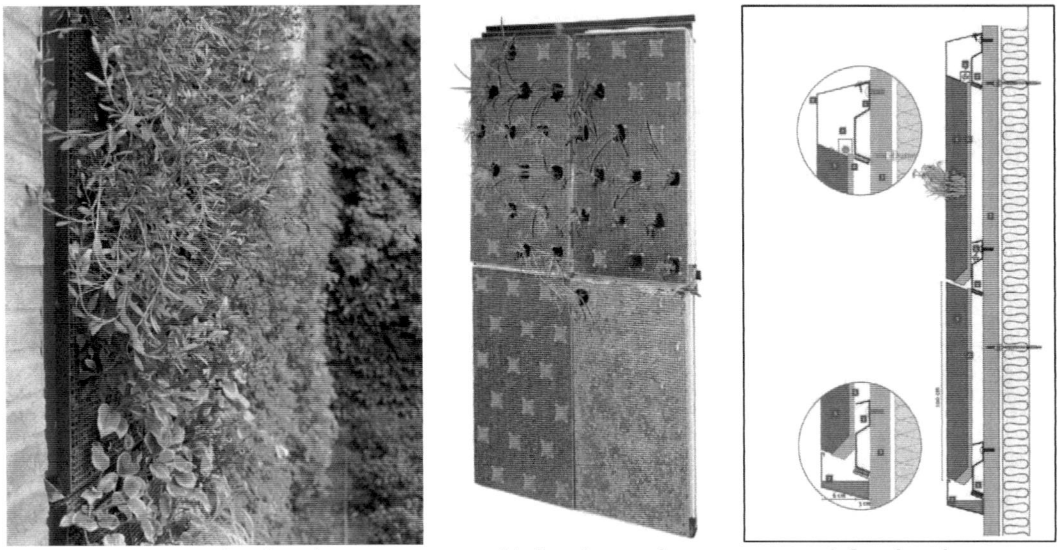

a) vegetative facade b) facade panel c) facade scheme

Figure 6. Vegetative fasade Fassadengarten Optigrün [10].

SO0+ - SO0
- Hanging profile
- Load-bearing profile
- drip irrigation
- absorbent capillary fabric
- absorbent capillary fabric
- facade panel
- flowerpot with planted plant
- at the bottom – collecting trough

The same construction will be applied to the other wall SO1.

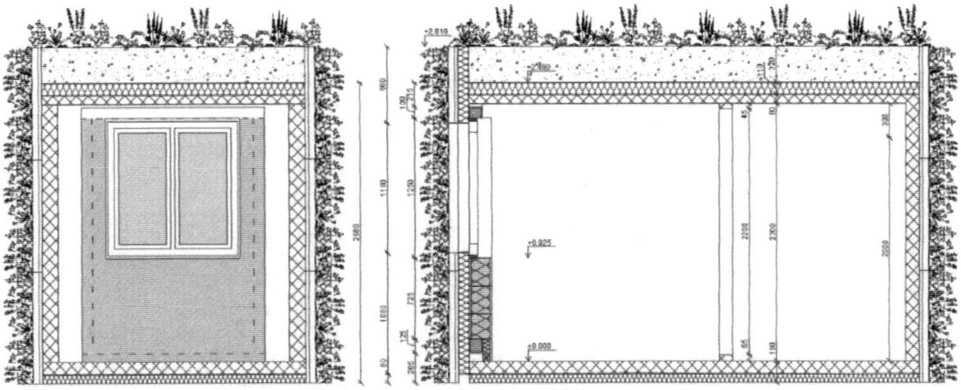

Figure 7. Vegetative facade and roof.

5.3. Description of window

The dimensions of the window are 1200x1200mm. The area of the glazing is 0,88 m2. Area of the fixed part is 0,56m2.

The glazing in chamber C3 is designed as double-leaf opening-tilting window, profile made of composite plastic REHAU GENEO MD plus with insulating triple glazing 4-12-4-12-4 Ar from the company NITRASKLO. The glazing in C1 is INTERNORMA EDITION double-leaf opening and folding window with wood-aluminum frame profile with highly insulating thermal foam in the filling of the aluminum part. The window is glazed with insulating triple glazing 4-16-4-16-4 Kr from the company NITRASKLO [8].

6. Measuring process and instrumentation

After the completion of construction modifications, the measurement follows. As measurements have already taken place in this chamber, it will be possible to use the existing instrumentation:

- humidity senors
- weather staton VAISALA
- temperature senors (temperature sensors based on resistance without Ntc and NiCr type tip)

Other devices will be needed for the research:

- energy measurement – calorimeter
- electricity consumption meter

The experiment involves measuring the temperature and relative humidity of the outside and inside air, the temperature and relative humidity inside the structure, the surface temperature and the heat flux.

7. Conclusion

In pursuit of energy savings, reducing production costs and securing heat to buildings, we are trying to find new suitable solutions. The intention is also supported by the requirements for energy efficiency of buildings [9].

Radiant systems as well as green architecture have a significant impact on tackling energy consumption and global warming. How their combination will affect the energy consumption and comfort of the indoor climate will only be demonstrated by this research. In conclusion, we would like to examine the economic impact of this combination, not only in new buildings but also in reconstructions.

8. References

[1] Liu J, Zhou Z, Zhang D, Jiao S, Zhang J, Gao F, Ling J, Feng W and Zuo J 2020 Research on the performance of radiative cooling and solar heating coupling module to direct control indoor temperature p. 01

[2] Kušnír M, Vranay F, Vranayová Z, Kapalo P and Košičanová D 2015 Synergy between progressive indoor environment cooling and photovoltaic system, *15th International Multidisciplinary Scientific Geoconference (SGEM)* p.01

[3] Decree no. 99/2016 Coll.Decree of the Ministry of Health of the Slovak Republic on details on health protection against heat and cold stress at work (Available 29.06.2021) https://www.zakonypreludi.sk/zz/2016-99

[4] Kušnír M, Vranay F, Kapalo P, Košičanová D and Vranayová Z 2014 The effective use of renewable energy sources in office building, 9th *International Conference on Environmental Engineering (ICEE)* p.01

[5] Droz A G Coffman R R Blackwood Ch B 2021 Plant diversity on green roofs in the wild: Testing practitioner and ecological predictions in three midwestern (USA) cities, *Urban Forestry & Urban Greening* **60** (2021) 127079 p. 01

[6] La Roche P, Yeom D J and Ponce A 2020 Passive cooling with a hybrid green roof for extreme climates p.01

[7] Jávorka T and Kovaly L 2010 (Available online) https://www.asb.sk/stavebnictvo/strecha/zelene-strechy-ekologicky-ekonomicky-esteticky

[8] Bizoňová S Katunský D and Bagoňa M Comparative analysis of selected glass systems by dynamic simulation using measured real environmental conditions, E3S Web of Conferences 17 2, 19009 (2020)
[9] Composition of an intense vegetation roof (Available online) https://www.bauder.sk/sk/gruendach-eu/systemove-riesenia-pre-vegetacne-strechy/intenzivne-ozelenenie-striech/hydroakumulacny-zazrak.html
[10] Dostálová J 2011 (Available online) https://www.asb.sk/stavebnictvo/fasada/vertikalne-zahrady-vegetacne-fasady
[11] Vranay F Vranayová Z 2020 Influence of heat source on building energy certification process and CO2 emissions, 17th International Scientific Conference on Current Issues of Civil and Environmental Engineering Lviv- Kosice - Rzeszow (CEE) p. 01

Acknowledgments
This work was supported by Development Agency under the contract (APVV APVV-18-0360 ACHIEve Active hybrid infrastructure for the sponge city).

IOP Conf. Series: Materials Science and Engineering 1209 (2021) 012069 doi:10.1088/1757-899X/1209/1/012069

The trend of using solar energy of a green intelligent building and thermal energy storage to reduce the energy intensity of the building

M Gorás[1], Z Vranayová[1] and F Vranay[1]

[1] Institute of Architectural Engineering, Faculty of Civil Engineering, Technical university of Kosice, Vysokoskolska 4, 042 00 Kosice, Slovakia

E-mail: michal.goras@tuke.sk; zuzana.vranayova@tuke.sk; frantisek.vranay@tuke.sk

Abstract. The trend is to reduce the energy intensity of buildings. Thermal energy storage (TES) is the biggest challenge for buildings. It is a technology that supplies thermal energy by heating or cooling a tank, which then serves for the system in the building. Comparison of hitherto known systems ATES, BTES, PTES and research TTES. The most important factors for the accumulation of thermal energy are capacity (the energy stored in the system - depends on the storage process, the medium, and the size of the system), power (how fast the energy stored in the system can be discharged and charged), efficiency (the ratio of the energy provided to the user to the energy needed to charge the storage system. It accounts for the energy loss during the storage period and the charging/discharging cycle), storage (how long the energy is stored and lasts hours to months), charging and discharging (how much time is needed to charge or discharge the system), and cost (refers to capacity (€/kWh) or power (€/kW) of the TES system and depends on the capital and operation costs of the storage equipment and its lifetime).

1. Introduction

Reducing the carbon footprint is a hot topic today. In civil engineering, there are as many demands on buildings as possible. The trend today is to build a building from ecological materials, energy-efficient in production, application, use and also in its removal. This is not only the norm and the European Union, but also social responsibility. The construction of a building is the beginning of the building's existence, but what is more energy-intensive is its operation during the period of its use. In the rough construction and its construction, humanity has come to a break where it is no longer as parameters to move to better numbers. And so the true energy performance of buildings is a major theme for green architecture.

We assess buildings according to energy efficiency. In terms of their energy efficiency, buildings are classified into energy classes, which allows them to be compared with each other within individual building categories. The building (for these purposes) is a roofed structure with walls, in which energy is used to modify the indoor environment. A building is a building as a whole or a part thereof that has been designed or modified for independent use [1].

2. Legislation

The Act is implemented by Decree no. 364/2012 Coll. It classifies buildings into energy classes. It distinguishes a total of 8 classes, while buildings in the "G" class have the highest energy

IOP Conf. Series: Materials Science and Engineering 1209 (2021) 012069 doi:10.1088/1757-899X/1209/1/012069

consumption. The upper limit of energy class "A0" for the global indicator is for buildings with almost zero energy demand.

Table 1. Global indicator energy class scale - primary energy.

Global indicator energy class scale - primary energy in kWh / (m^2.a)									
		Building energy efficiency classes							
	Building categories	A0	A1	B	C	D	E	F	G
Global indicator - primary energy	Family houses	≤54	55-108	190-216	217-324	325-432	433-540	541-648	>648
	Apartment houses	≤32	33-63	64-126	127-189	190-252	253-315	316-378	>378
	Administrative buildings	≤60	61-120	121-240	241-360	361-480	481-600	601-720	>720
	School and school facilities	≤34	35-68	69-136	137-204	205-272	273-340	341-408	>408
	Hospital buildings	≤96	97-192	193-328	385-576	577-769	770-961	962-1153	>1153
	Hotel and Restaurant buildings	≤82	83-164	165-328	329-492	493-656	657-820	821-984	>984
	Sports halls and other sports buildings	≤38	39-76	77-153	153-258	259-304	305-380	381-456	>456
	Wholesale buildings	≤85	86-170	171-340	341-510	511-680	681-850	851-1020	>1020

Decree 364/2012 Z.z., Annex no. 3 [2].

Global Energy Indicator; the amount of energy that must be consumed in order to obtain the necessary "supplied energy" from the system. It takes into account the type of heat source used, the type of fuel, the use of renewable energy sources [8].

The need for primary energy to meet energy class A1 = 55-108 kWh / (m^2 .a) for family houses with a building permit from 1.1.2016 is mandatory [2].

3. Thermal energy storage

Different storage systems are used for solar thermal systems depending on the application. The objective of heat storage systems is to provide desired heat even at locations with fluctuating solar irradiances. They can be divided into two groups:

- short-term storage systems (daily cycle)
- long-term storage (inter-seasonal storage) systems.

Long-term storage systems should compensate for seasonal fluctuations, whereas short-term storage systems generally are only required to store heat over periods of a few hours to a few days. Hence, long-term storage systems need significantly larger volumes [3].

3.1 The categories according to the principle (Figure 1).

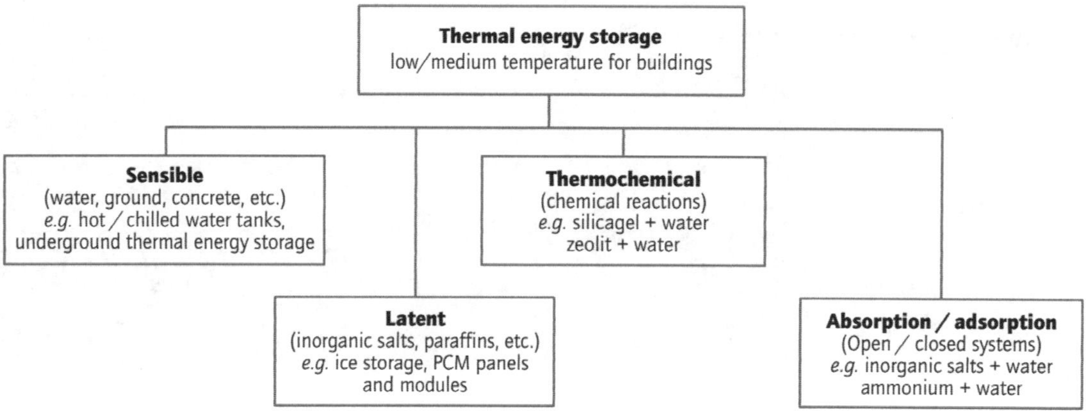

Figure 1. The categories according to the principle [4].

3.2 Thermal energy storage (TES) the main parameters

TES systems can be charged with heat or cold and hold this energy over time. The main parameters of TES are:

- **Capacity** (the energy stored in the system - depends on the storage process, the medium, and the size of the system);
- **Power** (how fast the energy stored in the system can be discharged and charged);
- **Efficiency** (the ratio of the energy provided to the user to the energy needed to charge the storage system. It accounts for the energy loss during the storage period and the charging/discharging cycle);
- **Storage period** (how long the energy is stored and lasts hours to months);
- **Charge and discharge** (how much time is needed to charge or discharge the system);
- **Cost** (refers to capacity (€/kWh) or power (€/kW) of the TES system and depends on the capital and operation costs of the storage equipment and its lifetime) [5].

3.3 Thermal energy storage - Comparison

Comparison of storage concepts with regard to the storage medium, energy density and equivalent storage volume (Figure 2).

	TTES	PTES	BTES	ATES
Storage medium	Water	Gravel-water [a]	Soil	Sand-water
Energy density [kWh/m³]	60–80	30–50	15–30	30–40
Storage volume for 1 m³ water equivalent	1 m³	1.3–2 m³	3–5 m³	2–3 m³

Figure 2. Comparison of thermal energy storage [6].

Energy capacities, power, efficiency and storage time of thermal energy storage technologies (Figure 3).

TES technology	Capacity (kWh/t)	Power (kW)	Efficiency (%)	Storage time
Hot water tank	20-80	1 - 10 000	50-90	day-year
Chilled water tank	10-20	1 - 2 000	70-90	hour-week
ATES low temp.	5-10	500 - 10 000	50-90	day-year
BTES low temp.	5-30	100 - 5 000	50-90	day-year
PCM-general	50-150	1 - 1 000	75-90	hour-week
Thermal-chemical	120-150	10 - 1 000	75-100	hour-day

Figure 3. Comparison of thermal energy storage according to properties [6].

4. Thermal energy storage system in Košice

Thermal energy storage in Košice uses techology of Hot water tank (Figure 4).

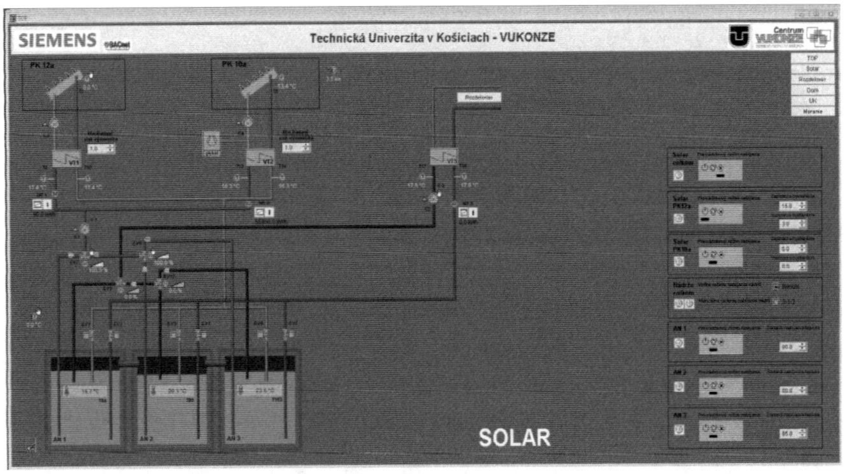

Figure 4. Scheme system of TES [7].

The system is made up of three different sizes of underground water reservoir heat. Different sizes of underground water reservoir heat are designed for charging according to the season and for discharging according to the energy consumption requirement. The entire storage with a volume of 160 m^3 is charged with an area of 225 m^2 of solar hot water collectors (Figure 5).

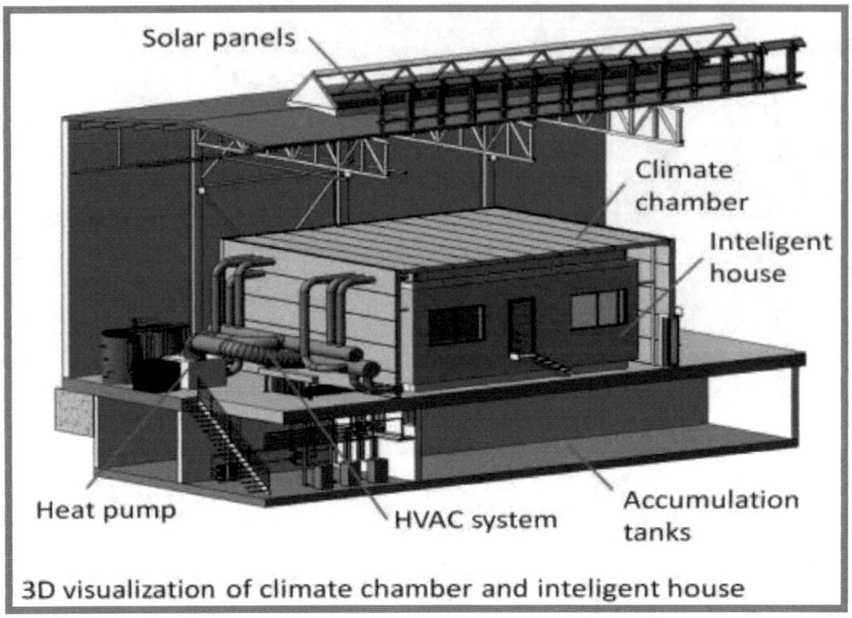

Figure 5. Experimental house for TES [7].

5. Solar thermal system

Solar thermal systems are an important use of solar energy. The use of solar thermal applications has a long history: Archimedes is reported to have boiled water using a concave mirror as early as 214 BC; thermal systems today still represent the most cost-effective use of solar energy.

Solar thermal signifies the thermal use of solar energy in general.

Therefore, a number of different technical applications exist. Besides building space heating, water heating or industrial processes, solar thermal systems can be used for cooling applications or electricity generation with solar thermal power plants.

The main operational areas are:

- solar swimming pool heating
- solar domestic water heating
- solar low-temperature heat for space heating in buildings
- solar process heat
- solar thermal electricity generation [3].

6. Conclusion

Accumulation is important in green architecture if we want to continue the trend of reducing the carbon footprint. Over the last 10 years, the requirements for buildings have increased so that even today it is possible to get into the A0 category only using green technology. If the requirements increase in the coming years, it will inevitably find an affordable and effective green solution. Accumulation today is known for heat pumps only on the secondary side. The use of hot water accumulated on the primary side of the heat pump will move us closer to this goal.

7. References

[1] Ústredný portál verejnej správy 2016 *Energetická hospodárnosť budov* Available on the internet: https://www.slovensko.sk/sk/agendy/agenda/_energeticka-hospodarnost-budov11/

[2] Ministerstvo dopravy, výstavby a regionálneho rozvoja Slovenskej republiky 2012 *Vyhláška č. 364/2012 Z. z.* Available on the internet: https://www.epi.sk/zz/2012-364#p1
[3] Quaschning V 2005 *Understanding Renewable Energy Systems* (Bath: Bath Press) pp 1 - 289
[4] IEA 2011 *Technology Roadmap Energy-efficient Buildings: Heating and Cooling Equipment* International Energy Agency – IEA 9 rue de la Fédération (France: Corlet)
[5] Sarbu I and Sebarchievici C A 2018 Comprehensive Review of Thermal Energy Storage *Sustainability* **10** 191
[6] Dahash A, Ochs F, Janetti M B and Streicher W 2019 Advances in seasonal thermal energy storage for solar district heating applications: A critical review on large-scale hot-water tank and pit thermal energy storage systems *Applied Energy* **239** 296-315
[7] Kapalo P 2021 *Nesica conf.* Presentation of Thermal energy storage
[8] Vranay F and Vranayova Z 2020 Influence of Heat Source Choice on Building Energy Certification Process and CO_2 Emissions *Proc. of CEE 2019 Lecture Notes in Civil Engineering (Lviv)* vol 47, ed Z Blikharskyy, P Koszelnik and P Mesaros (Cham: Springer) pp 541-548

Acknowledgments
This work was supported by the project of Slovak Research and Development Agency APVV-18-0360 Active hybrid infrastructure towards sponge city.

Sustainability of green walls

I Halaszova[1] and M Kozlovska[1]

[1] Institute of Construction Technology, Economic and Management, Faculty of Civil Engineering, Technical University of Kosice, Vysokoskolska 4, 042 00 Kosice, Slovakia

E-mail: ivana.halaszova@tuke.sk

Abstract. Not long ago, the sustainability of buildings was not taken as a design priority, although it was considered to some extent. The required level of sustainability has risen sharply in recent years, buildings are becoming more self-sufficient, environmental friendly and greener. Towards more attention to green sustainable alternatives in architecture, new solutions for sustainability began to be sought, such as the use of vertical and horizontal surfaces of the building for green facades, the use of roofs for roof greenery, as well as introduction of greenery into building in the form of green walls. Along with the introduction of greenery into the interior comes the new necessary technical solutions, thinking about the overall green wall with respect to the indoor environment and building itself as well as new possibilities of using rainwater, beautify the interior, improving the air etc. However, it is important to perceive a green wall as a living element that must be sustainable, and this is what this article is about. The article focuses on the green walls from the point of examining the available information on this subject in domestic as well as in foreign sources. The results of this article can help to improve understanding of various green walls from different perspectives and compare their view on the green wall in terms of sustainability. The paper helps designers and researchers to orientate in the given topic to create an overview of the different options of the current greenery systems.

1. Introduction

In the present building practice, it is common to perceive the building as a whole object and also focus on aspects such as economic sphere, environmental impact, construction massing, location of the building and necessary of HVAC system integration in the building. Although the environmental aspect also comes to the fore, it is not yet an established practice to think about green systems as a whole concept. Green systems should be seen in global during the entire construction process, which is used from creation to its liquidation and thus should be seen as a part of the entire life cycle of the building, not just as a short-term additional application.

During the life cycle, buildings consume a considerable amount of resources and contribute to the transformation of areas. This can result in significant economic consequences and impacts on the environment and human health. Therefore, we try to minimize the impact of buildings on the environment during their entire life cycle - its operation, restoration, demolition and waste management [1].

In today´s climatic conditions average annual temperatures are constantly rising [2], emphasis is placed on the need for HVAC systems and their correct setting during the design of the building (current location, rotation of the building, choice of materials, etc.). The role of cooling and heating systems is to improve the conditions in the building for the end costumer. In addition to the HVAC systems, which is artificially made to improve indoor environment, we also know how to help improve environment in a greener way, for example, by incorporating green walls. It is important to think about green walls throughout the whole life cycle if we want to achieve a functioning sustainable long-term system.

Simulation of various aspects during the life cycle of the building can be predicted using life cycle assessment (LCA) programs, which make it easier to predict possible problems in various stages of use of the building. As stays in [3,4] a LCA is used to analyse the similarity and differences in the environmental impacts in relation with benefits estimated for two climate types for building energy saving (reduction of electrical energy used for building cooling and heating). Renewable energy has emerged as an economically realistic alternative for supplying global energy needs to drive human production activities and modern comfort.

2. Adverse impacts arising during the life cycle of buildings

The global trend towards the creation of self-sufficient systems and zero-energy buildings is growing. Many times it happens that after a more thorough study we find that the construction which is labeled as *ecological* or *green* actually has the same impact on the environment as the classic variant or even worse. It is therefore necessary to know why to focus at different stages of the life cycle, if we want to distinguish this variant. Structures like the green wall must have a very diverse structure. First of all, the reduction of the production of greenhouse effects, for example CO_2, which must be eliminated already in the production phase.

The life cycle assessment of construction materials shows the most impact on the greenhouse is the manufacturing process. While the smallest impact is the disposal process. This is because the manufacturing process requires high energy resulting in high emissions as well. Based on information from the source [5] it can be easily seen in the Figure 1 in which part of the life cycle the largest CO_2 production occurs.

When it comes to exactly just how the advantages of Green Wall surfaces effect your company, that originates from the lifestyle of vegetations on their own. Plants normally absorb CO_2 as well as various other contaminants and after that remove pure oxygen.

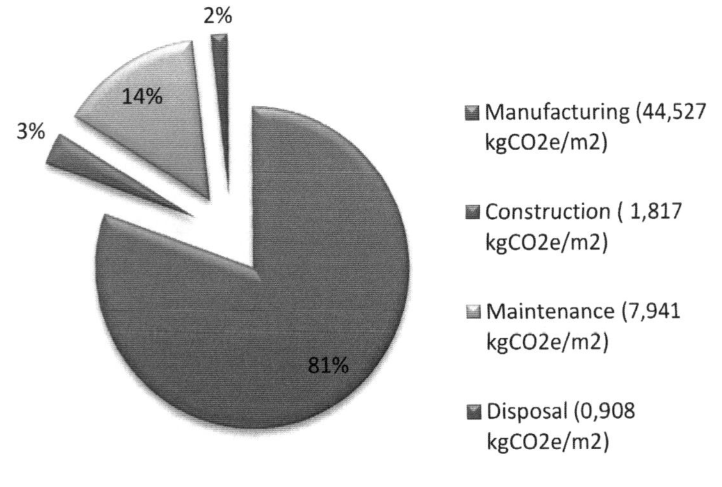

Figure 1. LCA impact of construction materials on the greenhouse [5].

In addition to the mentioned reduction of CO_2, there are some other factors that need to be considered when talking about side effects during the life cycle. From the study of source [7] it is possible to see the impact of individual aspects can be seen in Table 1 which is based on data from this ECOCHAIN source.

Table 1. Environmental impact indicators [7].

Impact category Indicator	Unit	Description
Global warming	kg CO_2-eq	Indicator of potential global warming due to emissions of greenhouse gases to air
Ozone depletion	kg CFC-11-eq	Indicator of emissions to air that cause the destruction of the stratospheric ozone layer
Acidification of soil and water	kg SO_2- eq	Indicator of the potential acidification of soils and water due to the release of gases such as nitrogen oxides and sulphur oxides
Eutrophication	kg PO_4^{3}-eq	indicator of the enrichment of the aquatic ecosystem with nutritional elements, due to the emission of nitrogen or phosphor containing compounds
Photochemical ozone creation	kg ethene-eq	Indicator of emissions of gases that affect the creation of photochemical ozone in the lower atmosphere (smog) catalysed by sunlight.
Depletion of abiotic resources- elements	kg Sb-eq	Indicator of the depletion of natural non-fossil resources
Depletion of abiotic resources- fossil fuels	MJ	Indicator of the depletion of natural fossil fuel resources
Human toxicity	1,4-DCB-eq	Impact on humans of toxic substances emitted to the environment (Dutch version of EN15804 only)
Fresh water aqatic ecotoxicity	1,4-DCB-eq	Impact on freshwater organisms of toxic substances emitted to the environment (Dutch version of EN15804 only)
Marine aquatic ecotoxicity	1,4-DCB-eq/sup>	Impact on sea water organisms of toxic substances emitted to the environment (Dutch version of EN15804 only)
Terrestrial ecotoxicity	1,4-DCB-eq	Impact on land organisms of toxic substances emitted to the environment (Dutch version of EN15804 only)
Water pollution	m^3	Indicator of the amount of water required to dilute toxic elements emitted into water or soil (french version of EN15804 only)
Air pollution	m^3	Indicator of the amount of air required to dilute toxic elements emitted into air (French version of EN15804 only)

3. Green Wall Life Cycle Assessment

LCA studies measure the environmental impact of a product. Starting from raw materials extracted, transportation, material manufacturing, installation, use and maintenance, to returning to the ground.

As it is written in [6], to understand the full environmental impact of a structure over decades of use, all phases, starting before construction and continuing through demolition, must be considered. Life cycle assessment (LCA) seeks to quantify the environmental impacts over the infrastructure life cycle.

Besides that, the green plant layer will also reduce the amount of UV light that will fall on building materials. Since UV light deteriorates the material and mechanical properties of coatings, plastic, etc. plants will also have positive effect on durability aspects and on maintenance costs [3].

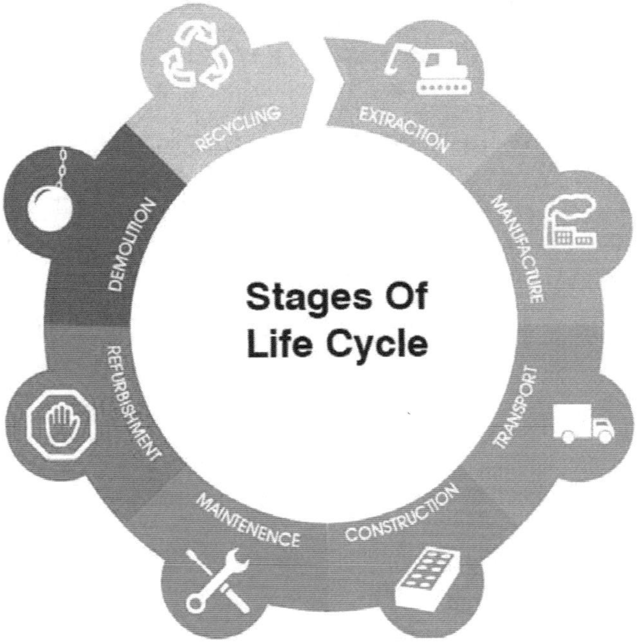

Figure 2. Stages of green wall LCA according to the general LCA of constructions from the source [6].

The life cycle of green walls begins with the actual mining of materials, as well as with the life cycle of other structures or the entire building. To better understand the life cycle, we could divide the stages of the life cycle from Figure 2 into 4 main categories according to ISO 14040 and ISO 14044:

- The project stage
- The construction process stage
- The stage of exploitation
- The closure stage

It is important to note that there is always a transport phase between individual stages.

3.1. The project stage

The project phase is at the very beginning of the life cycle and, in terms of greenhouse gas production, puts the most burden on the whole process. During the project phase, we start with the production of the material. Specifically, for green walls, we must distinguish between the living vegetation part and the part of the supporting structure and all wall layers. The vegetation part itself can enter the life cycle directly in the installation process, while the support system must be built much earlier. There is a wide range of support systems on the market for different variations of green walls. Some are based on plastic, others are made of steel frames. When choosing, it is necessary to take into account their recyclability and, if possible, choose a suspension system from recycled materials.

Another part of the project stage is the off-site production. Green walls can also be produced as modular, which greatly simplifies production, provides system variability and reduces production costs and production time. For example, in the newly created modular GEOGREEN system [8] the materials used in the production of the modular parts aim to minimize the environmental impact of this system. Each Geogreen modular element comprise 50% of industrial waste, namely mine waste mud from Panasqueira mines (43%) and milled waste glass (7%).

3.2. The construction process stage

The main part of this stage is the installation of the system. There are several ways to create a sustainable green wall. One of them is the water connection system. One of the most popular wall walling systems is the connection to rainwater. Based on the source [9] green walls can also be used as a tool for storm water management. It is possible and highly recommended to irrigate green walls with rainwater whenever possible. The naturally dissolved minerals found in rain provide plant nutrients, thus reducing the amount of fertilizer required. Integrating reservoirs, retention ponds, or rainwater harvesting techniques into a green wall system will increase onsite infiltration.

In addition to rainwater, the use of grey water is also gaining more and more prominence. This type of water, which would otherwise be removed from the building, can still be reused to irrigate the green wall. There are many publications on the use of grey water for a sustainable green wall [10,11,12], or [13] where research is talked about great potential of designing green walls as a greywater treating systems, but points to the current lack of scientific literature.

3.3. The stage of exploitation

The longer the wall lasts, the longer it will serve its purpose and we will be able to say that the construction is really sustainable. Because it could be said that this is the primary phase whose duration is measured in years and decades, it consists of several parts. According to information from source [14], figure 3 was created. The picture shows the phases that belong to this stage. These are the phases from the use of the green wall to its reuse via repair, replacement and refurbishment.

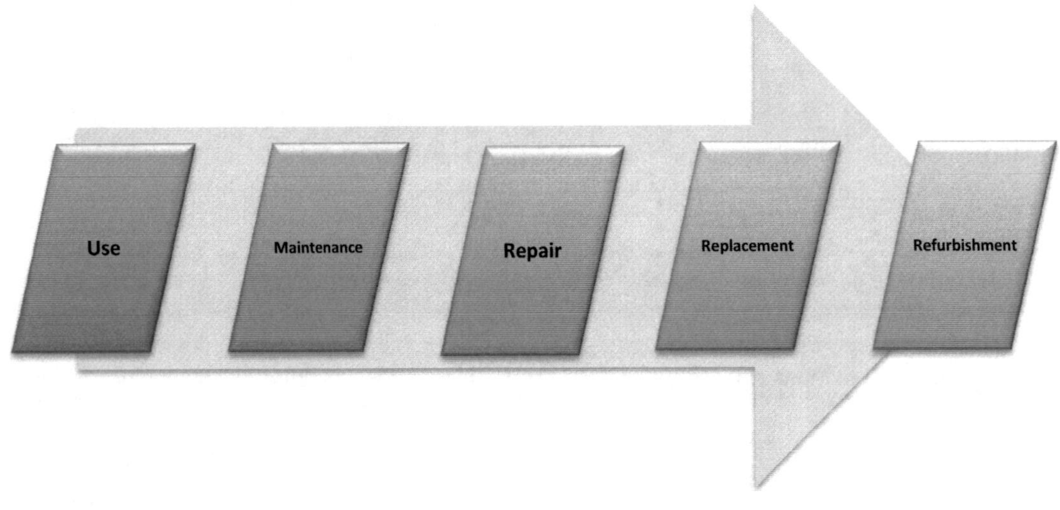

Figure 3. Phase of the green wall during the stage of exploitation.

The length of life depends primarily on the plants that require the most attention. As described by the GREENWALL [15], climbers have the longest life span, shrub type plants can live for several years with appropriate maintenance: The older parts need to be removed in order to provide room for the new ones. The annuals are best from spring to autumn, the bi-annuals from autumn to spring. The life span of perennials is 3-10 years, while the sedums continuously renew and change.

According to [16], the three stages of execution stage namely construction, operations and refurbishment approximately 15 % of the world's freshwater resources; 40 % of the world's energy; and produce approximately 23-40 % of the world's greenhouse gas emissions. Large quantities of solid waste is generated during maintenance and refurbishments, and as well large quantities of natural resources are again consumed for new materials and thereby all the prior stages to maintenance are repeated creating even more impacts on the environment.

3.4. The closure stage

Green walls as well as various other constructions have their lifespan. Although this service life can be delayed by proper maintenance and design, it cannot be avoided. Waste from dead plants can be used as biomass, as it is an ecological material. Waste from the rest of the green walls must be properly quarried.

For example, either the waste is going for incineration, the process during which large emissions of greenhouse gases are released affecting the air quality; or recycling, or waste generated is dumped in landfills where it becomes obsolete. Other environmental impacts associated with the demolition of a building is the release of greenhouse gas emissions through burning fossil fuel used for the demolition machinery as well as the transportation of waste to landfills, or materials to recovery sites [16]. As further discussed in this source, some components or materials can be reused directly after being dismantled. Some of these materials may for example, require to be re-painted and therefore they will impact the environment through paint. For materials that require re-processing, fuel will be required for transportation, energy and other resources to make up the new products and thus there is going to be environmental impacts involved in the re-processing.

4. Conclusion

The aim of this paper was to focus on green walls from a life cycle perspective. There are a number of publications that focus on the impact of the green wall on the indoor environment, the right choice of technology, the location of the wall, etc. However, there is little information on how the living room behaves throughout the life of the building itself. The article explained the basic parts of the life cycle with a focus on the green wall in each phase. It was about monitoring the green wall from the selection of materials to the demolition of the wall and the reuse of its components.

In a separate chapter, we focused on which aspects need to be observed in the design. The biggest enemy is the production of greenhouse gases, especially CO_2, which needs to be closely monitored and eliminated as much as possible. Not only green walls but also other green solutions that contribute to sustainability can do this.

The article used to obtain basic information on this issue. In further research, authors would like to pay more attention to the green walls in the life cycle of the building. Each of the mentioned phases can be given a separate analysis, because this issue is very extensive. It could also be interesting for further publication to focus on green walls from the point of view of UPD-united process data and environmental data EIOD-environmental input-output data, or analysis of the current state of green stein on the market (whether these walls are really sustainable).

5. References

[1] Pifko H and Krjcsovics 2017 *Hodnotenie udržateľnosti budov- metodika CESBA, časť 1:Aktuálny stav udržateľnosti v stavebníctve* (Bratislava: Vydavateľstvo STU) Available online https://www.archinfo.sk/diela/teoreticka-praca/hodnotenie-udrzatelnosti-budov-metodika-cesba-cast-1-aktualny-stav-udrzatelnosti-v-stavebnictve.html

[2] SHMU statistic data *Slovenský hydrometeorologický ústav* (Slovakia) Available online http://www.shmu.sk/sk/?page=1&id=klimat_operativneudaje1&identif=11934&rok=2021&obdobie=1981-2010&sub=2

[3] Ottéle M, Perini K, Fraaij A L A, Haas E M and Raiteri R 2011 Comparative life cycle analysis for green façades and living wall systems *Energy and Buildings* **43** 12 3419–29

[4] Martines E and Siew Ng K 2021 Life cycle Analysis of renewable energy *Reference Module in Earth Systems and Environmental Sciences* (Amsterdam: Elsevier)

[5] Small Cycle 2021 *Selasa Life cycle assessment of contruction materials living green wall* Available online https://www.smallcycle.com/2021/02/life-cycle-assessment-of construction-materials.html

[6] ENVIROLINK *Building life cycle assessment* Available online https://www.envirolink.me/building-life-cycle-assessment/
[7] Hillege L 2019 *Impact categories (LCA)- Overwiew* Available online https://ecochain.com/knowledge/impact-categories-lca/
[8] Manso M, Castro-Gomesab J, Pauloc B, Bentesac I and Teixeirad C A 2018 Life cycle analysis of a new modular greening system *Science of The Total Environment* **627** 15 1146–53
[9] Weinmaster M 2009 Are green walls as green as they look? An intoruction to the various and ecological benefits of green walls *Journal of Green Building* **4** 4 3–18
[10] Sitzenfrei R, Kleidorfer M, Bach P M and Kuzniecow Bacchin T 2020 Green infrastructures for urban water system: Balance between cities and nature *Water* **12** 5 1456
[11] Prodanovic V, Hatt B, McCarthy D, Zhang K and Deletic A 2017 Green walls for greywater reuse: Understanding the role of media on pollutant removal *Ecological Engineering* **102** 625–635
[12] Prodanovic V, McCarthy D, Hatt B and Deletic A 2019 Designing green walls for greywater treatment: The role of plants and operational factors on nutrient *Ecological Engineering* **130** 184–195
[13] Prodanovic V, Zhang K, Hatt B, McCarthy D and Deletic A 2018 Optimisation of lightweight green wall media for greywater treatment and reuse *Building and Environment* **131** 99–107
[14] Souza E 2021 *Why should architects understand and care about carbon and life assessment?* Available online https://www.archdaily.com/959800/how-embodied-carbon-and-life- cycle analysis-can-support-decisions-in-an-architectural-project?ad_medium=gallery
[15] GREENWALL *Green facades in general* Available online https://greenwall.pro/en/vegetation/
[16] Ngwepe L and Aigbavboa C A 2015 *theoretical review of building life cycle stages and their related environmental impact*

Acknowledgments
The activity presented in the article is part of the VEGA research grant No. 1/0557/18 “Research and development of technological and product innovations of modern construction methods in the context of the principles of Industry 4.0”

IOP Conf. Series: Materials Science and Engineering 1209 (2021) 012071 doi:10.1088/1757-899X/1209/1/012071

A method of shaping an energy-efficient building envelope based on natural patterns in order to achieve save of energy

I Wawrek[1] and R Staňa[2]

[1] Institute of Architectural Engineering, Faculty of Civil Engineering, Technical University of Košice, Vysokoškolská 4, 042 00 Kosice, Slovakia
[2] Faculty of science, Institute of Computer Science, Pavol Jozef Šafárik University of Košice, Jesenná 5, 040 01 Košice, Slovakia

E-mail: igor.wawrek@tuke.sk; Richard.stana@upjs.sk

Abstract. This article aims to present an innovative way of designing objects called A method of shaping an energy-efficient building envelope based on natural patterns. The method uses currently available technologies in an unusual way. Digital photogrammetry data are inserted into a process of designing envelope structure of future objects. In the process of construction, this method counts on the use of all the advantages of 3D printing technology. This combination of the use of relatively innovative technologies in the process of design and construction of buildings opens up new possibilities for the design of the building envelope. Consequently, new building envelope shaping options create space for more efficient use of the energy savings associated with the building envelope shape as a building shape factor. The final design solution of the proposed building takes inspiration directly from the environment in which it is to be situated. Final design takes inspiration of its shape from the environment in which it will be built as a future object. As a result, the building works harmoniously and in accordance with the environment in which it was designed. This creates a pleasant ecological environment for its inhabitants and the surrounding area.

1.Introduction

Each building in order to work properly, should respond adequately to the environment in which it is installed. If we build the same house within one state, for example Slovakia, in five different conditions in terms of settlement in the field and altitude, this house will have different energy requirements for its operation in terms of energy [6].

In the classic approach to building design, the architect or engineer gets to know the environment in which he starts projecting new building as much as possible. Architect evaluates input parameters such as: altitude, slope, global orientation, the direction of prevailing winds and many more. Architect usually determines value of input parameters given by the environment. It often depends on his/hers proper initial consideration of how well the object will work in the context of its environment in the future. Unfortunately, building energy specialists and other professions enter the project process in many cases only at a point where it is very difficult to change something fundamentally. In this respect, BIM technology can significantly influence the creation process. Experts, through individual professions acting in the project, enter the design process in the early stages of the design and are cooperating throughout the creation of the construction project.

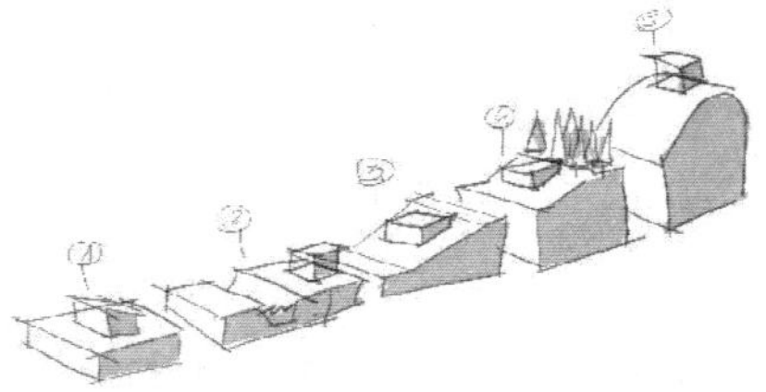

Figure 1. Types of different level installation of building.

BIM design even extends beyond the imaginary boundaries of traditional project documentation as we know it, as it is also helpful as a documentation of post-implementation modifications of the project, or documentation of operation and maintenance of the building.

This article aims to present an experimental method of designing objects, in which the creative process of designing the building envelope is taken from the hands of the architects. The final shape of the future building is formed only on the basis of models of living and inanimate nature observed directly at the place of origin of the future building.

2.Method

To verify the method, a model situation was created, in which the task was to design a family house. In a classic case, location of future building is characterised by borders of land, access road, connection at engineering networks, terrain, orientation and many others parameters of building site. These are all parameters that the architect normally works with and to which he responds in his design of the shape or disposition solution of the building.

In our case, this assignment was moved to an abstract level to a less specific location, which was characterized only by altitude. Such an assignment pushes us to think about the true essence of form, its natural effect and defensibility in its environment. Paradoxically, the model release of input parameters complicates the task of designing the building envelope.

A place in a sparse forest on a hill with an altitude of 894 meters above sea level was chosen for the experiment. The natural model was a rock formation near the top of the Veľký Milič hill in eastern Slovakia. The rock formation was captured by a mobile device camera. A total of 21 photographs were taken around the perimeter of the rock formation. The photos were processed by the open source software called Meshroom.

Figure 2. Founded stone object.

3. Photo processing

The photographs were processed using electronic photogammetry. GPL (General Public License) software called Mashroom was used for fully automated photo processing. Mashroom software is freely available on the alicevision website [1]. A series of images that were used as some input data for photogrammetry processing were taken with a standard Samsung SM-A750FN mobile camera, with the F1.7 1 / 265s 3.93mm ISO40 camera set without flashlight. The use of this available technology demonstrates how fast the software is evolving, and thus that it is no longer necessary to use a digital SLR camera for scanning, as described by Marčiš Marián in his book [2].

However, the condition of a sufficiently powerful computer still remains for the success of photo processing, because the processing time depends mainly on the performance of the Processor and the size of the RAM.

The complexity of the process of automated image processing into 3D spatial models can be divided into two basic parts.

The first part is 1. Reconstruction of the scene when the mutual orientation of the images is calculated, the camera is calibrated and a simplified cloud is created, the so-called connecting points that roughly describe the shape of the object.

The second part is the 2nd detailed reconstruction of the scanned object, which means the calculation of a dense cloud of points that cover the object.

3.1 Scene reconstruction

In the first step, natural functions are extracted. The aim of this step is to identify characteristic groups of pixels that are, to some extent, constant when rotating, scaling and angle the camera during image capture. The most common method of function detection is the SIFT algorithm, which is also used by the Mashroom software.

In the second step of the scene reconstruction, the images are matched. The purpose of this step is to find images that look at the same point in the scene being shot. Simply put, the program assigns a 128-dimensional vector to each significant point. Since each significant point is characterized by its own characteristic vector, it is possible to search in two images for points with the most similar characteristic vector.

The third step is to find a match of functions. In this step, the match of significant points from one image with the significant points in the other image is searched using an optimized algorithm. In the case of Meshroom software, the RANSAC algorithm is used, which finds application in a technology known in the field of computer vision as SfM (Structure from Motion). In general, it is a process in which a three-dimensional structure of an object is sought by analysing local motion signals across time. In computer vision, this is especially the case when the observed scene is stable and only the camera moves around the object. By analysing the changes between individual images, finding significant points and pairing them, it is possible to gradually work towards a three-dimensional reconstruction of the entire scene.

Figure 3. Reconstruction of photo scene by Mashroom software.

3.2 Detailed reconstruction of the scanned object

In the first step, to obtain a detailed reconstruction of the scanned object, the Structure from Motion (sfm) is obtained. In this step, the position of individual points in three-dimensional space is calculated based on the comparison of significant points on 2D images. In the same way, in this step, the position of the cameras relative to the subject is determined.

The next step is to obtain a depth map. In this step, the program obtains a depth value for each pixel. There are number of approaches. The Mashlab program uses the SGM method (Semi-Global Matching) [3].

The task of the last step is to get a polygonal network (Mesh) from a spatial cloud of points. We call this step Meshing. In this step, the individual points of the cloud are interconnected into a dense network of triangular polygons [4].

Figure 4. High detailed triangulation of model and texturing from photos.

The resulting network can be very detailed in some places. To simplify subsequent use, it is necessary to simplify the resulting network by omitting some vertices. Finally, the polygon model is assigned a texture, which is a combination of images from all cameras. Each polygon is assigned a section of the camera image that has the most orthogonal viewing angle for that polygon.

4. Conclusions

The obtained mesh model was subsequently processed in 3Ds Max software from Autodesk. The mesh model that resulted from the photogrammetric process was simplified and smoothed. To simplify the network, the ProOptimizer modifier was used, which reduces the percentage of vertices in the model.

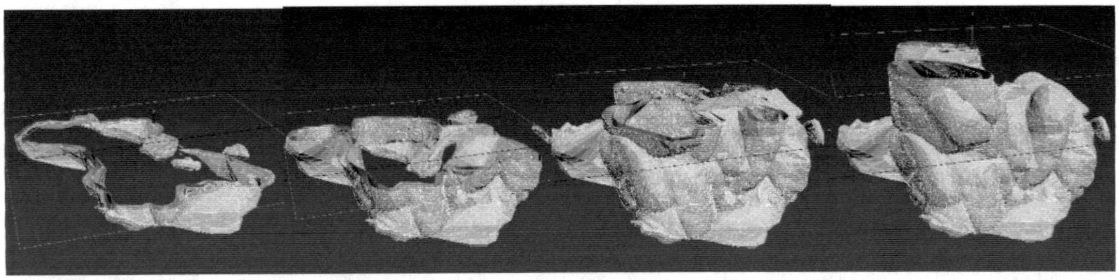

Figure 5. Horizontal section cuts of building envelope.

The thickness of the envelope of the perimeter shell was created by offsetting the outer surface with a Shell modifier. The result is an organic undulating interior of a family house, which is in harmony with the environment in which it is created. The obtained volume serves as an input design to the BIM process, replacing the design proposal of the architect in the introduction. At this point, all specialists enter the process of completing the building through professions simultaneously with the architect.

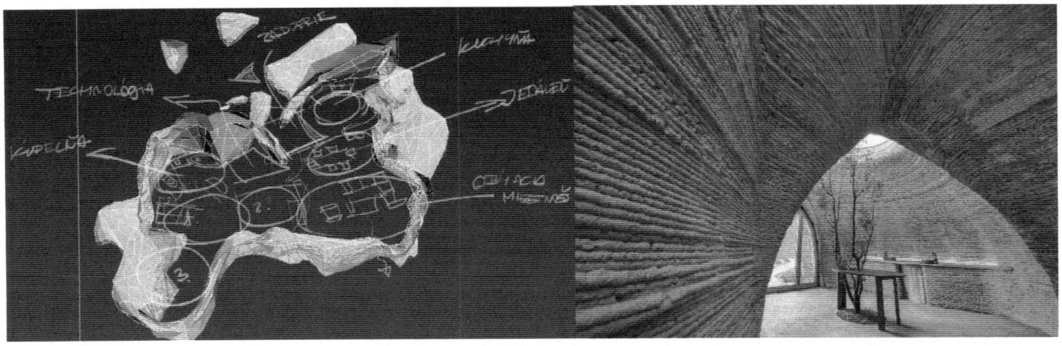

Figure 6. Disposition diagram of future house drawn in to shape of envelope structure.

The task of the architect in this case, is to breathe life into the given form as a dispositional solution, the correct placement of construction openings and the layout of the installation of technical equipment installations in coordination with other construction specialists. Such depersonalization from the form by detaching the design process from the author can contribute to a better incorporation of essential comments of specialists into the design and implementation of the construction. At the same time, the form of the building, which draws the morphology from the environment in which it is created, acts harmoniously with respect to its surroundings and does not appear as a foreign element.

5. References

[1] Alicevision Photogrammetric Computer Vision Framework Available online https://alicevision.org/#photogrammetry

[2] Marčiš M 2019 *Fotogrametrické metódy v procese digitalizácie kultúrneho dedičstva* (Bratislava: Slovenská Technická Univerzita v Bratislave)

[3] Hirschmüller H 2011 Semi-global matching-motivation, developments and applications *Photogrammetric Week* **11** 173–84

[4] Jančošek M 2014 Large Scale Surface Reconstruction based on Point Visibility

[5] Lowe D G 2004 Distinctive image features from scale-invariant keypoints *International journal of computer vision* **60** 2 91–110

[6] Pifko H 2005 Kritériá hodnotenia kvality architektúry *Život. Prostr.* **39** 6 304 – 9

[7] Pifko H 2004 Alternatívne stavebné materiály a technológie *Život. Prostr.* **38** 4 188 –93

[8] Buch M and Uhlík A 2020 Modifikovanie cementových kompozitov pre 3d tlač budov *Mladá veda* **8** 1 1–16

[9] Rogovská V 2015 3d tlač ako nová priemyselná revolúcia *Conf. Marketing Identity: Digital Life (Slovakia)* vol 12

[10] Špaček R, Peciar M and Šíp L 2016 Sketching and drawing in the new age-the role of sketching and drawing in architectural and technical education *World Trans. on Engng. and Technol. Educ.* **14** 1 8 –13

[11] Raspall F 2016 vMESH: How to print Architecture? *XX Congreso de la Sociedad Iberoamericana de Gráfica Digital* vol 3 1

[12] Hager I, Golonka A and Putanowicz R 2016 3D printing of buildings and building components as the future of sustainable construction? *Procedia Engineering* **151** 292-299

[13] Yossef M and Chen A 2015 Applicability and limitations of 3D printing for civil structures *Civil, Construction and Environmental Engineering Conference Presentations and Proceedings* **35** 1–11

[14] Marijnissen M and Van Der Zee A 2017 3D Concrete printing in architecture. A research on the potential benefits of 3D Concrete Printing in Architecture. *Material Studies–Methodologies* **2** 299-308

[15] Kreiger M A, MacAllister B A, Wilhoit J M and Case M P 2015 The current state of 3D printing for use in construction *Proc. of the 2015 Conference on Autonomous and Robotic Construction of Infrastructure (Ames)* pp 149-158

[16] De Schutter G, Lesage K, Mechtcherine V, Nerella V N, Habert G and Agusti-Juan I 2018 Vision of 3D printing with concrete—Technical, economic and environmental potentials *Cement and Concrete Research* **112** 25-36
[17] Sakin M and Kiroglu Y C 2017 3D Printing of Buildings: Construction of the Sustainable Houses of the Future by BIM *Energy Proc.* **134** 702-711

Acknowledgments
This paper was created with the support of grant project VEGA 1/0512/20 "Analysis of new approaches to the assessment and certification of sustainable office buildings in terms of employee satisfaction and performance".

Assessment of the effect of shapes of a shaft intake structure using the PIV method

L Bytčanková[1]

[1] Department of Hydraulic Engineering, Faculty of Civil Engineering, Slovak University of Technology in Bratislava, Radlinského 11, 810 05 Bratislava, Slovakia

E-mail: lucia.bytcankova@stuba.sk

Abstract. The construction of shaft intake structures in Slovakia has increased. The shaft intake structures overcome significant vertical height over short horizontal distance. In their front horizontal section, the water flows with free surface, then in the vertical section the flow changes its direction and character to a pressurized flow. The flow of water in these shaft intake structures is therefore very complicated. A hydraulically suitable design of the intake structure is associated with achieving the required parameters of the small hydropower plant (SHPP), but due to the reduction of project costs, the shapes of shaft intake structures of SHPP are often not correctly hydraulically designed. One of the important aspects is the distribution of flow velocity of these intake structures. Uneven distribution of flow velocity causes negative effects on turbine performance. Therefore, the investigation of the effects of shaft intake structure design on flow velocity distribution has been realized. The velocity field at a shaft intake of a small hydropower plant was investigated on a physical model in a hydraulic laboratory using the PIV (Particle Image Velocimetry) method. The PIV measurements were realized for different shaft heights and proved negative effects of the design on the flow homogeneity in the turbine intake.

1. Introduction

Intake structures connect hydropower plants with reservoirs and lead water to the turbine units. They therefore belong to the most important parts of Small HydroPower Plants (SHPPs) and are directly related to the whole function of the power plant. Proper construction of intake structures ensures sufficient water flow and should ensure minimum pressure losses of the flow [1], [2] and also prevents the formation of vortices in this area [3], [4]. Uneven inflow in the intake structures causes irregular load of the turbine impeller [5]. The homogeneity of the flow velocity fields at the intakes to the turbines has a great influence on the flow conditions, while the inhomogeneous distribution of the flow velocities has a negative effect on the power and efficiency of the turbine [6].

Shaft intake structures in Slovakia began to be built after 1990. Shaft intake structures are named after their shape. In the first part, the shape is horizontal and the flow can be both pressure and non-pressure here. Before entering the turbines, the flowing water overcomes the slope in the "shaft" and changes the flow to pressure. The flow is therefore very complex and the hydraulic design of these objects should be subjected to hydraulic research. For this reason, it is necessary to examine these structures and examine the velocity field at the intake to the turbine. Shaft intakes are usually built with additional use of hydropower potential on already built water structures. Despite their construction, many of them were in Slovakia built without research, only on the basis of a design proposal [7]. This has led to many operational problems of the SHPPs connected to hydraulic issues of the design. One of the main hydraulic problems in the shaft intakes is the inhomogeneous flow velocity distribution, which has negative effects on the performance of the turbines leading to the reduction in their efficiency and uneven mechanical load on the turbine parts, which results in a reduction in their lifetime. Therefore, a hydraulic research of the effects of the shaft intake shape on the flow was realized.

2. Materials and methods

To investigate the effects of the shaft shape on the flow velocity distribution a physical model of the SHPP Dobrohošť shaft intake structure was created in the flow channel of the hydraulic laboratory. The model was built as a 2D model of the axial cross section of the shaft intake in a scale of 1:30. The model was designed to investigate the effects of the shaft height on the flow velocity distribution in the turbine intake. The transverse flow was neglected and the research was focused only on longitudinal flow distribution.

For the velocity measurements on the model of the shaft intake, the PIV (Particle Image Velocimetry) method was used, as a very effective method. It is an optical method that allows to measure the instantaneous values of sizes and direction of velocity vectors in a given flow cross section. The so-called reflective particles are introduced into the liquid, which reflect the laser beam. The narrow light beam is guided directly from the laser. It is modified by a cylindrical lens in the shape of a light knife, by means of which the selected plane of the measuring space is illuminated. The illuminated area is bounded by the field of view of the camera and reflections from reflective particles are recorded by it. The PIV system records images in two pulses. The laser sends the first beam, which irradiates the particle. The particles shift in time between the transmission of the second laser pulse, but do not leave the pixel. The new location is also captured by the camera. It uses a Charge-Coupled Device (CCD) camera or Complementary Metal Oxide Semiconductor (CMOS) camera for processing, which is connected to a computer via a digital card. Information about the change of position and time is evaluated then the speed is calculated [8].

The PIV system was installed on a physical model of a shaft intake structure (Figure 1). Subsequently, the basic settings of the PIV system took place: camera settings, laser pulse power settings, laser flash frequencies and basic system calibration. The shooting time for the observed measurements was set to 6000 microseconds, pulses of 3 Hz and the number of measurements (frames) was set to 500. For each scenario, the flow velocity field in the shaft in axial cross section of the model was measured by the PIV method.

Figure 1. Physical model of a shaft intake structure.

To observe the velocity field in the shaft intake structure, 5 scenarios were proposed, which consisted of changes the depth of the shaft (Figure 2). The measuring cross-section was located in the axis of the 2D model of the shaft intake.

IOP Conf. Series: Materials Science and Engineering 1209 (2021) 012072 doi:10.1088/1757-899X/1209/1/012072

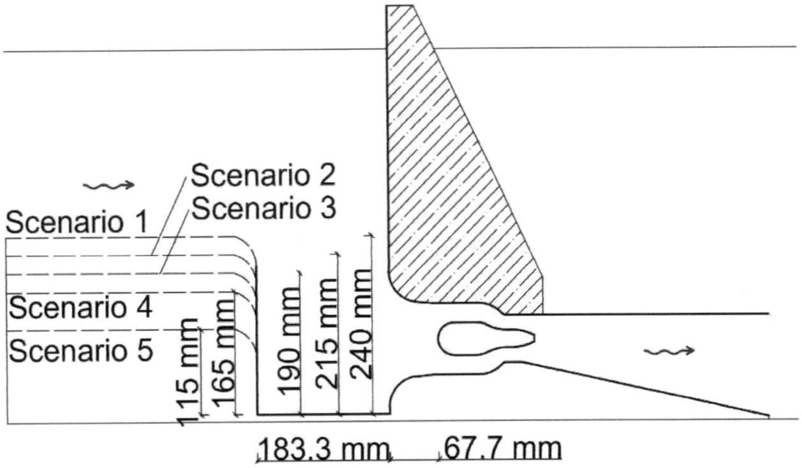

Figure 2. Scheme of the modelled scenarios.

3. Result and discussion

The results of PIV measurements were processed in DynamicStudio software. Figures 3-7 show velocity maps with flow vectors for all scenarios. These maps show the area of the shaft and the intake of the turbine. The areas with high and low flow velocities can be observed. A significant flow and no-flow area in the shaft can be observed. At the bottom of the shaft intake, the smallest flow velocities occur. The assessment of the measured flow maps also shows an uneven vertical flow distribution in the turbine intake, where the most of the flow and the highest velocities are in the upper part of the turbine intake cross section.

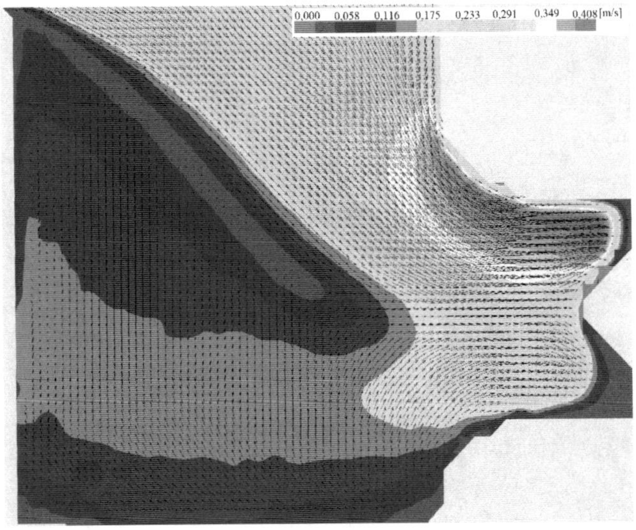

Figure 3. Scalar velocity map with flow vectors - Scenario 1 - shaft height 240 mm.

IOP Conf. Series: Materials Science and Engineering 1209 (2021) 012072 doi:10.1088/1757-899X/1209/1/012072

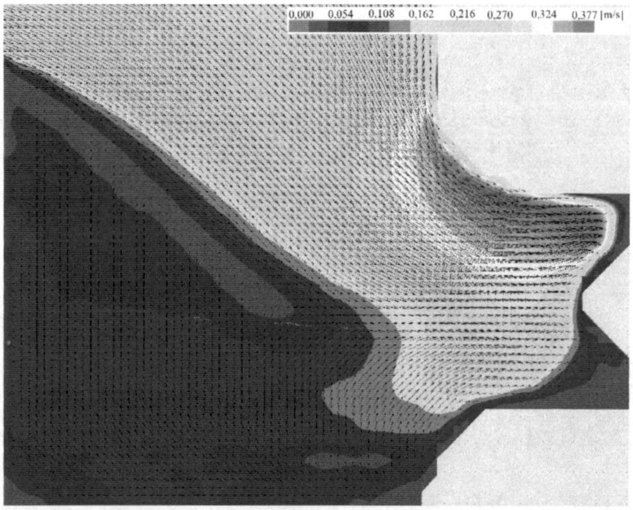

Figure 4. Scalar velocity map with flow vectors - Scenario 2 - shaft height 215 mm.

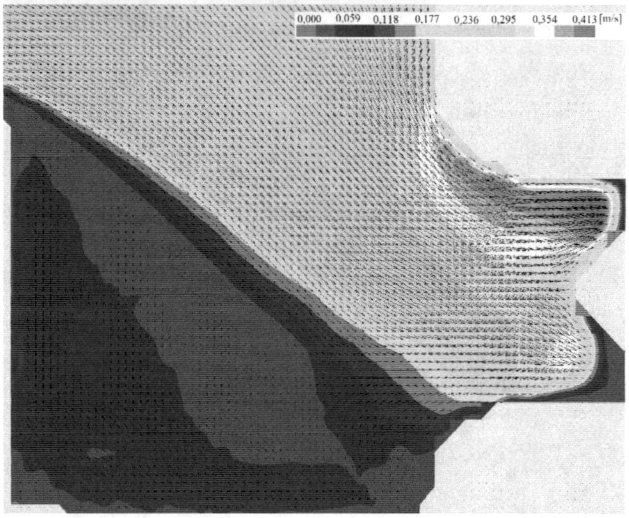

Figure 5. Scalar velocity map with flow vectors - Scenario 3 - shaft height 190 mm.

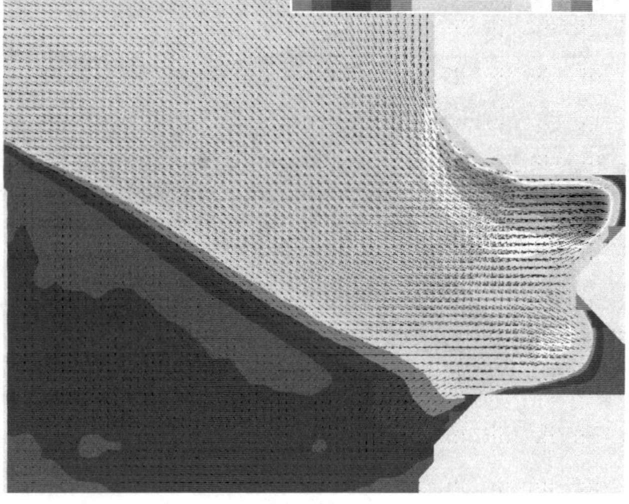

Figure 6. Scalar velocity map with flow vectors - Scenario 4 - shaft height 165 mm.

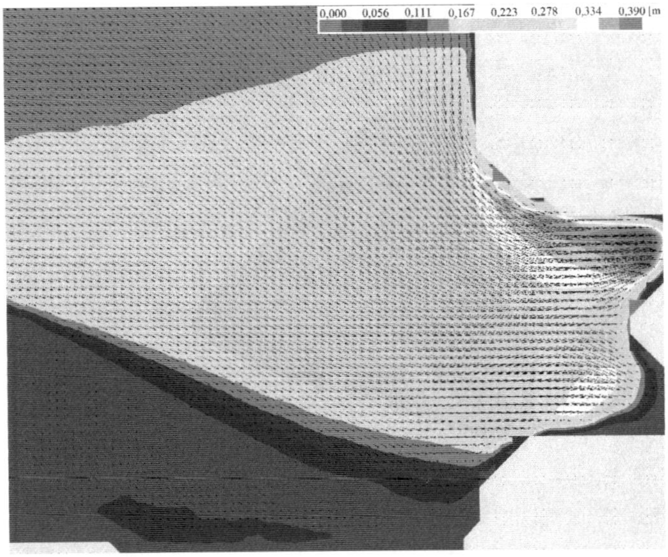

Figure 7. Scalar velocity map with flow vectors - Scenario 5 - shaft height 115 mm.

Figure 8 shows the distribution of flow velocity in the observed profile for each selected scenario and is compared to the average cross-sectional velocity. The average cross-sectional velocity was 0.24 $m.s^{-1}$. The most significant deviations from the average cross-sectional velocity are obtained in scenario 1, while in scenario 5 these deviations are the smallest. By decreasing the height of the shaft, the deviations of the flow velocities are reduced. It is also possible to observe the distribution of the flow between the turbines on the graph, where most of them are concentrated in the upper half of the observed profile.

The results show that the shaft height has a significant effect on the flow velocity distribution in a shaft intake. The increase in the shaft height increases the size of the ineffective flow area in the shaft intake and leads to significant uneven flow velocity distribution in the turbine intake. The decreasing shaft height increases the effective flow area and helps to distribute the flow velocities more evenly in the turbine intake.

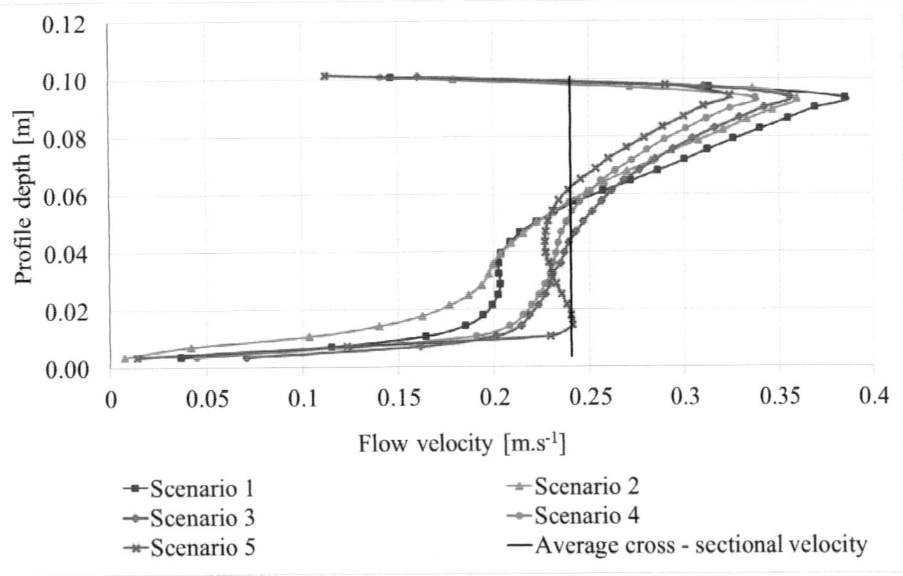

Figure 8. Distribution of flow velocities in the observed profile.

4. Conclusion

Research of the shaft height effects on the flow velocity distribution in the turbine intake was realized on a 2D physical model for 5 different shaft heights. The research helped to identify the problems and tools for assessing the homogeneity of the velocity field. Uneven distributions of flow velocities were observed in the observed shaft intake structure. The performed measurements proved the negative effect of the shaft intakes especially the shaft height, on the distribution of flow and the flow velocities in the turbine intakes. Increase of the shaft height leads also to the increase of ineffective flow area in the shaft, which can be avoided or minimize by modifications to the design of the intake structure itself.

5. References

[1] Fošumpaur P and Čihák F 2005 Design and optimization of a turbine intake structure *Acta Polytechnica* **45** 3 87–91
[2] Hodák T 1984 *Water Power Utilization* (Bratislava: University of Technology in Bratislava)
[3] Taştan K and Yildirim N 2018 Effect of intake geometry on the occurrence of a free-surface vortex *Journal of Hydraulic Engineering* **144** 4 04018009
[4] Sun H and Liu Y 2015 Theoretical and experimental study on the vortex at hydraulic intakes *Journal of Hydraulic Research* **53** 6 787–796
[5] Lichtneger P 2009 Intake flow problems at low-head hydropower *Wasserbaukolloquium, Wasserkraft im Zeichen des Klimawandels* (Dresden) 259–266
[6] Angulo M, Liscia S, Lopez A and Lucino C 2014 Experimental validation of a low-head turbine intake designed by CFD following Fisher and Franke guidelines *IOP Conf. Series: Earth and Environmental Science, Hydraulic Systems* **22** 042014
[7] Dušička M, Rumann J and Dušička P 2016 Shaft intake structures SHHP *Hydroturbo* (Brno) pp 1–4
[8] Seika Co 2008 *Particle Image Velocimetry (PIV) Overview & Principle*

Acknowledgments
This article was created with the support of the Ministry of Education, Science, Research and Sport of the Slovak Republic within the Slovak Research and Development Agency, project no. APVV-18-0472.

Managed Aquifer Recharge (MAR) in areas used for agricultural purposes: the Rye Island pilot study

M Červeňanská[1], J Mydla[1], A Šoltész[1], Z Danáčová[2] and E Kullman[2]

[1] Department of Hydraulic Engineering, Faculty of Civil Engineering, Slovak University of Technology in Bratislava, Radlinského 11, 810 05 Bratislava, Slovakia
[2] Slovak Hydrometeorological Institute, Jeséniova 17, 833 15 Bratislava, Slovakia

E-mail: michaela.cervenanska@stuba.sk; andrej.soltesz@stuba.sk

Abstract. For a long-term retention of water in aquifers and its subsequent use in drier or heavier demand periods, the Managed Aquifer Recharge (MAR) techniques are studied and implemented in 4 pilot areas of the DEEPWATER-CE project. In Slovakia, the pilot study is situated in the Rye Island. A calibrated MODFLOW model is used for a prediction of groundwater level changes caused by the Recharge Dam MAR. Results of the simulations showed that the increased groundwater level caused by the realization and operation of three proposed weirs affects the volume of water infiltrated to the aquifer but does not cause the flooding of the adjacent area.

1. Introduction

Fast water transfer to rivers during flood situations causes a water deficit during dry periods. For long-term retention of water in aquifers and its subsequent use in drier or heavier demand periods, the Managed Aquifer Recharge (MAR) techniques are studied and implemented in 4 pilot areas of the DEEPWATER-CE project funded by the Interreg CENTRAL EUROPE programme [1]. In Slovakia, the research is carried out by the Water Research Institute in cooperation with the Slovak University of Technology in Bratislava. Data required for the model design were provided by the Slovak Hydrometeorological Institute. The pilot study is situated in an area used for agricultural purposes bounded by irrigation channels – the Rye Island. The Rye Island is located in the Podunajska lowland, in the southwestern part of the country. There are 4 channels located in the investigated area – S VII (the Gabčíkovo-Topoľníky channel), A VII (Vojka-Kračany), B VII (Šuľany-Jurová) and C VII (Baka-Gabčíkovo). The last one mentioned was not taking into account in the simulation model because of the missing data.

The local geology is characterized by fluvial Quaternary sediments, which determine the hydrogeological conditions. The hydrological conditions are strongly affected by the construction of the Gabčíkovo water structure. After construction of the Gabčíkovo water structure, decrease of groundwater level in the adjacent area of the river Danube was observed. Groundwater regime as well as the water level regime in drainage channels changed. To improve and to control the groundwater regime in the floodplain area, a water supply structure situated on the left-hand side of the seepage channel was constructed. This structure allows an artificial water supply into the branch system and consequently into the groundwater [2] [3] [4].

For retention of surface water in aquifer and for recharge of groundwater and its subsequent using for irrigation during dry periods, the Recharge Dam MAR type was investigated using the numerical modelling technique.

IOP Conf. Series: Materials Science and Engineering 1209 (2021) 012073 doi:10.1088/1757-899X/1209/1/012073

2. Data availability and sources of data

Data (and their sources) needed for a model design are briefly presented in table 1.

During the period from October 2020 to May 2021, several field measurements were carried out to determine the course of water levels in drainage channels and the bottom elevation of these channels.

Table 1. Data type and sources of data.

Data type	Description	Source
Geomorphological	digital relief model (DRM)	Geodesy, Cartography and Cadastre Authority of the Slovak Republic (GCCA SR)
Geological and hydrogeological	bottom elevation of the aquifer hydraulic conductivity	State Geological Institute of Dionýz Štúr (SGIDŠ)
Hydrological	water levels in channel S VII daily precipitation groundwater levels in observation wells	Slovak Hydrometeorological Institute (SHMI)
	longitudinal profiles and cross sections of channels S VII, A VII and B VII	Slovak Water Management Enterprise (SWME)

Three hydrological yeas have been selected for the data analysis:

- 2008 as the year which can be characterized as the precipitation normal year for the regions of western Slovakia (100 to 109 % of the long-term normal);
- 2010 as the year extremely above normal in terms of precipitation (159 % of the long-term normal);
- and 2018 with extremely low precipitation totals recorded in April, May, July and October.

Based on the data analysis, the hydrological years 2010 and 2018 were selected for the purpose of the model design.

3. Numerical modelling

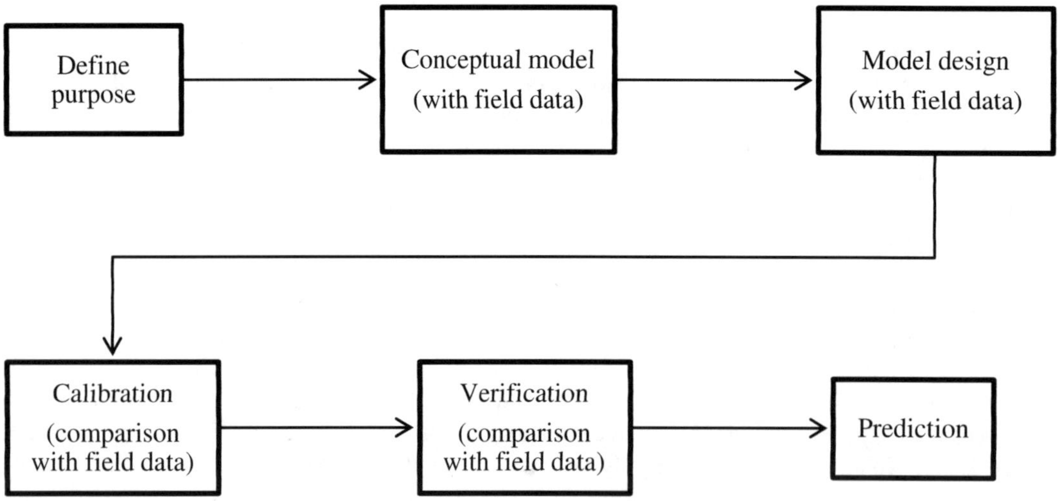

Figure 1. Modelling protocol (adapted from [5]).

The numerical model was developed using the MODFLOW program in the Groundwater Modeling System (GMS) environment, which allows us to use the conceptual model approach. A modelling protocol, which was used for creation of the numerical model, is described in figure 1.

3.1. Conceptual model and model design

During the conceptual model stage, the system boundaries were identified, field data describing geomorphological, geological, hydrogeological and hydrological conditions were assembled, and the model area was visited. After that, the conceptual model was mapped to the designed grid, the boundary conditions were set and preliminary selection of values for aquifer parameters and hydrologic stresses was performed (model design).

The modelled area is bounded by the left-hand side seepage channel of the supply channel of Gabčíkovo water structure and 13 groundwater level monitoring objects of the State Hydrological Network of groundwater quantity monitoring (or observation wells) of the Slovak Hydrometeorological Institute (figure 2). The Dirichlet (or first-type) boundary condition was specified on the entire edge of the model boundary. The value of the boundary condition was computed as the mean value of groundwater levels measured in every observation well during the modelled period.

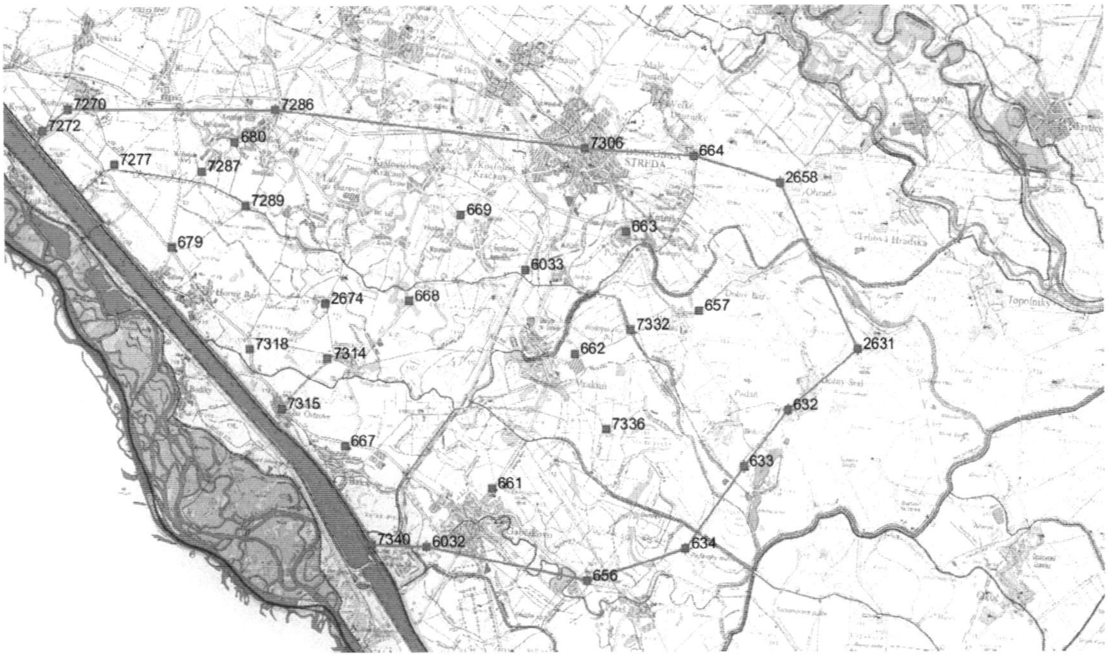

Figure 2. Groundwater level observation wells of the SHMI (red squares) located in the modelled area with the model boundary (orange arc).

Channels located in the area of interest represent the Cauchy boundary condition. Two elevations had to be specified – the elevation of the bottom of the river (channel) bed and the head in the channel. Also, the conductance had to be specified. The value of the conductance was one of the calibration parameters.

3.2. Calibration and verification

During the calibration stage, a set of values for aquifer parameters and stresses, that approximates field-measured heads and flows, was found. The calibration of the model was performed on a steady-state model using the period from May 26, 2010 to September 9, 2010 (figure 3) during the 2010 flood situation on the rivers Danube, Váh and Little Danube. Calibration of the model was done using:

- the automated parameter estimation code PEST [6] for the calibration of a hydraulic conductivity;

- and by trial-and-error adjustment of a conductance of the riverbed.

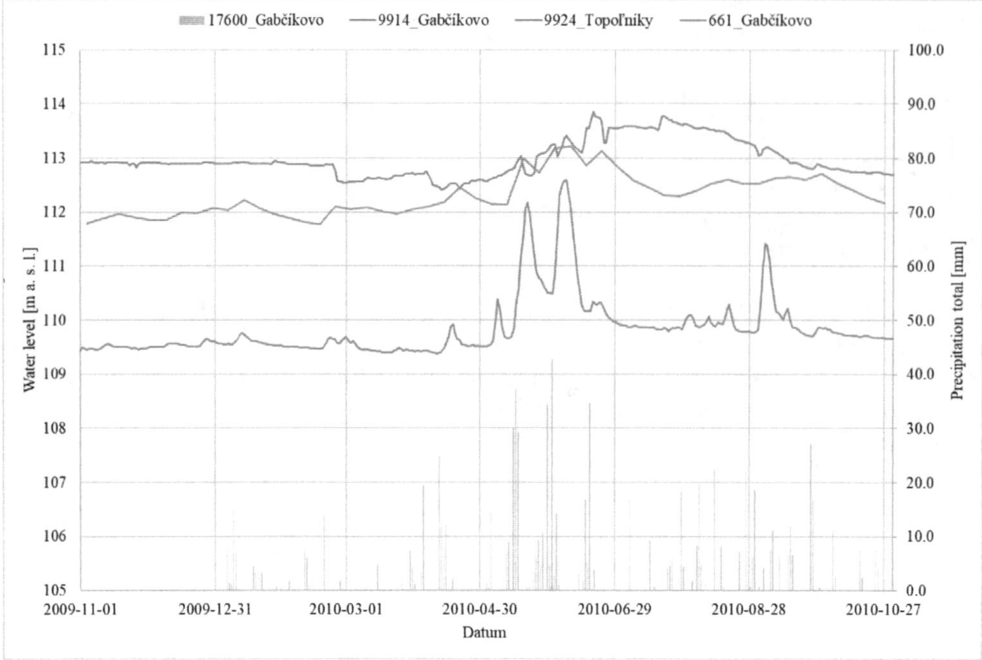

Figure 3. Course of water level at stage discharges 9914 Gabčíkovo and 9924 Topoľníky (the S VII channel) and the course of the groundwater level in well no. 661 in Gabčíkovo; daily total precipitation in the rain gauging station 17600 Gabčíkovo – hydrological year 2010 (SHMI, 2021).

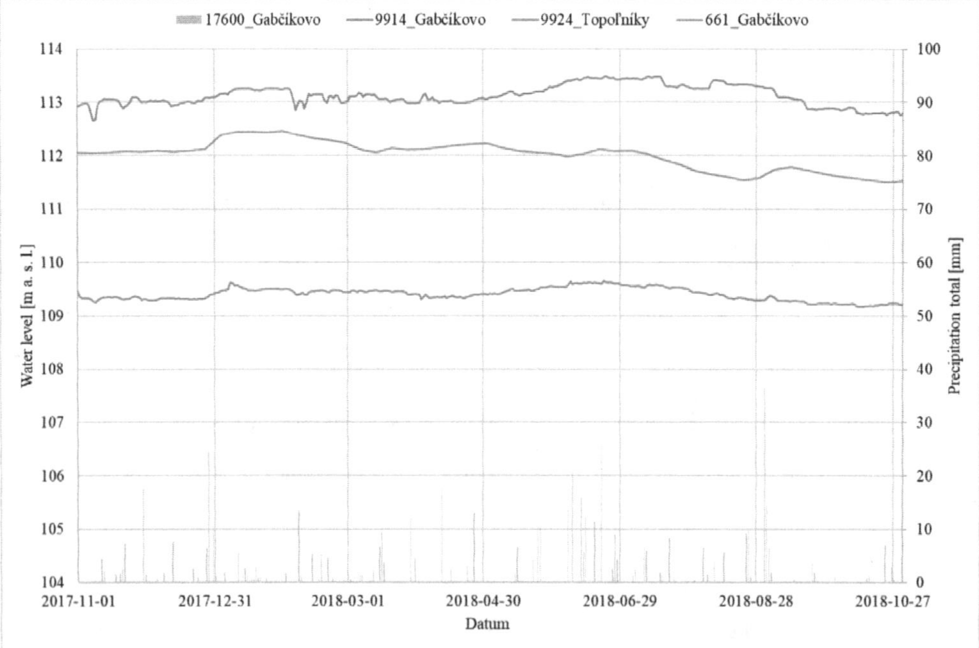

Figure 4. Course of water level at stage discharges 9914 Gabčíkovo and 9924 Topoľníky (the S VII channel) and the course of the groundwater level in well no. 661 in Gabčíkovo; daily total precipitation in the rain gauging station 17600 Gabčíkovo – hydrological year 2018 (SHMI, 2021).

Figure 5 captures the contours of the calculated groundwater head for the calibrated steady-state model. Calibration targets display a calibration error at each object. As can be seen from the figure 5, all targets (except from two targets) have the green colour, which means that the difference between the observed and computed head is less than 0.10 m. The difference between the observed and calculated head in the two yellow calibration targets is 0.17 (12.04 % of the difference between the maximum and minimum value in the observation well, or a variance) and -0.11 m (12.94 % of the variance). The model can be assumed calibrated in the accepted level of accuracy.

The verification of the model was done using conditions during the "dry" hydrological year 2018, especially the period from the March 13, 2018 to May 2, 2018 (figure 4).

Figure 6 captures the contours of the calculated groundwater head for the steady-state model of the verification period. As can be seen from the figure 6, several targets have the green colour. The rest of the calibration targets have the yellow and red colour, which means, that the difference between the observed and computed head is more than 0.10 m. Nine of these objects (table 2) have the proportion of the difference of the observed and computed head (D) to the variance, i.e. the difference between the maximum and minimum measured groundwater level (V), greater than 30 %. The reason is, on the one hand, that the influence of the left-hand side seepage channel is not taking into account because of the missing data – the water level on the seepage channel is not measured or observed. On the other hand, it could be caused by the estimation of the water level courses in channels A VII and B VII. There are not any records about the water levels on these channels, nor the records about the manipulation with gate slides of weirs on the channels. The course of water levels is therefore estimated based on the field measurements. However, if we look at the sum of the differences, the value itself is close to zero.

Table 2. Differences of the observed and computed head (D) compared to the difference between the maximum and minimum measured level (variance, V) in observation points.

Name	Variance (V)	Difference (D)	D/V*100
657	0.32	-0.08	-24.6
661	0.96	0.07	6.82
662	0.55	-0.04	-7.33
663	0.25	-0.09	-37.4
667	0.70	0.38	55.0
668	0.45	0.09	20.7
669	0.33	0.28	85.5
679	0.58	0.22	38.0
680	0.32	-0.07	-23.2
2674	0.37	0.13	34.2
6033	0.35	0.04	11.4
7277	0.59	-0.03	-5.69
7287	0.41	0.01	1.63
7289	0.35	-0.07	-19.5
7314	0.41	-0.13	-32.7
7315	0.53	0.36	68.3
7318	0.47	0.34	71.8
7332	0.39	-0.14	-36.8
7336	0.76	0.13	17.3
	Σ =	1.38	

IOP Conf. Series: Materials Science and Engineering 1209 (2021) 012073 doi:10.1088/1757-899X/1209/1/012073

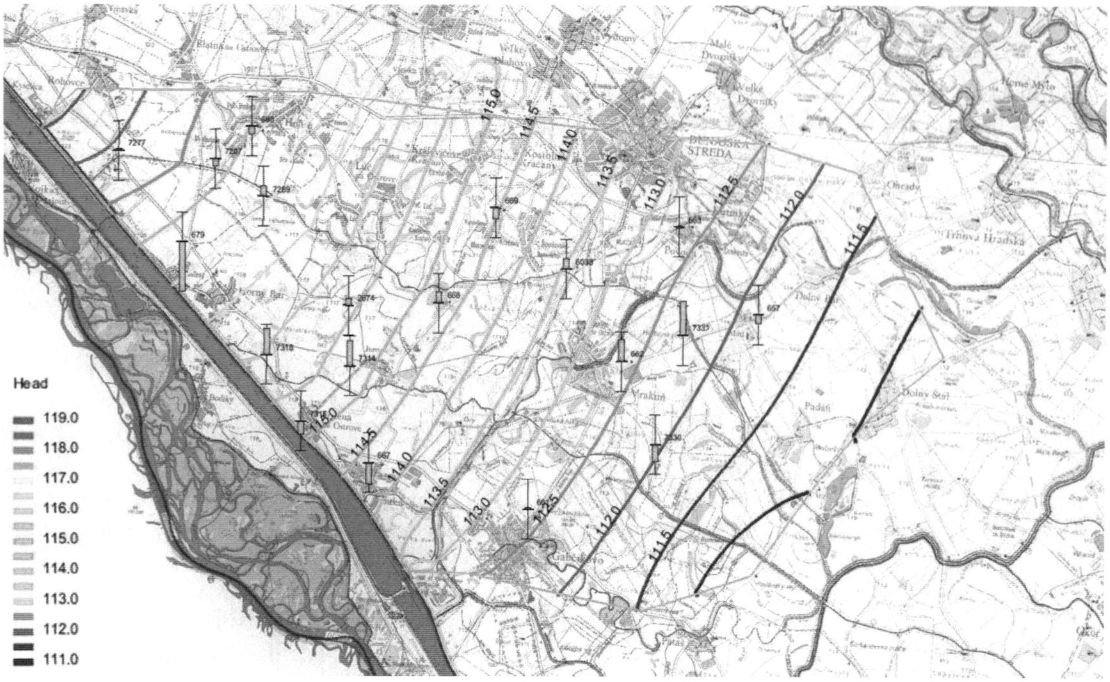

Figure 5. The calculated groundwater head for the period from May 26, 2010 to September 9, 2010 [m a. s. l.].

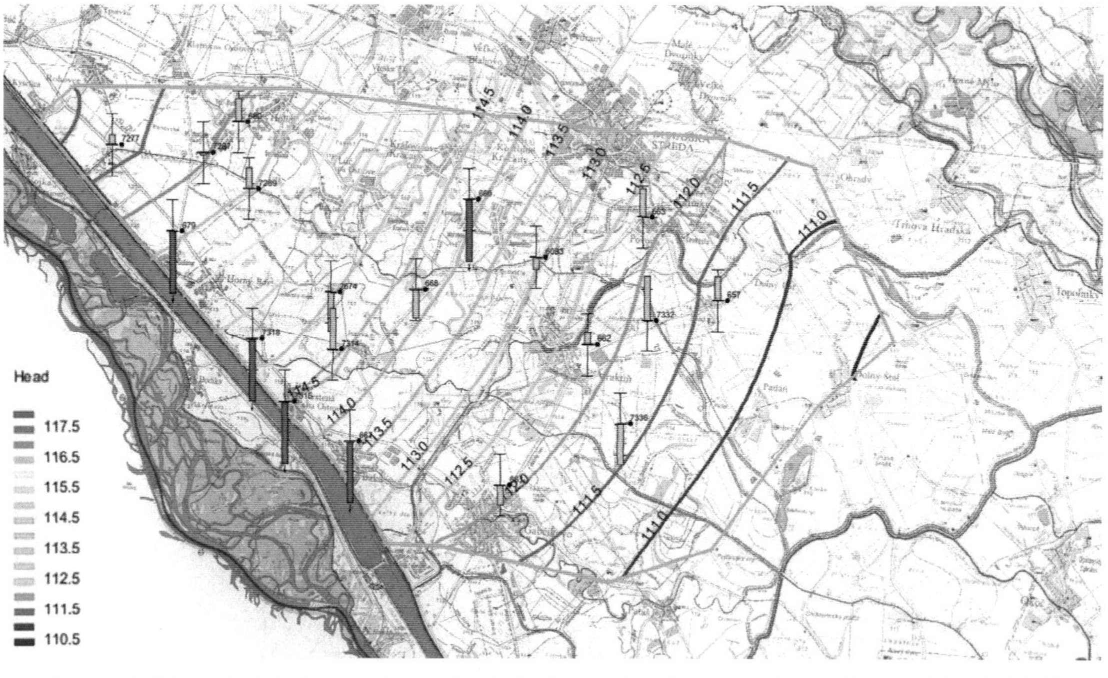

Figure 6. The calculated groundwater head for the period from March 13, 2018 to May 2, 2018 [m a. s. l.].

3.3. Prediction

A prediction represents a response of the system to future events. The model is run with calibrated values for parameters and stresses, except for those stresses that are expected to change in the future. The prognosis focused on the proposal of three weirs on the A VII channel at river kilometres 2.270, 7.060

and 12.530. The height of each gate slide is 2.0 m. Water level corresponds to the maximum level at each weir (for the period from 2018, figure 8) or to the maximum level during the flood situation (in 2010, figure 7). The calculated groundwater heads for the steady-state models capturing the prognosis for the period of 2010 is in figure 9 and for the period from 2018 in figure 11. Figure 10 and figure 12 show the differences between prognosis and the so called "zero variant".

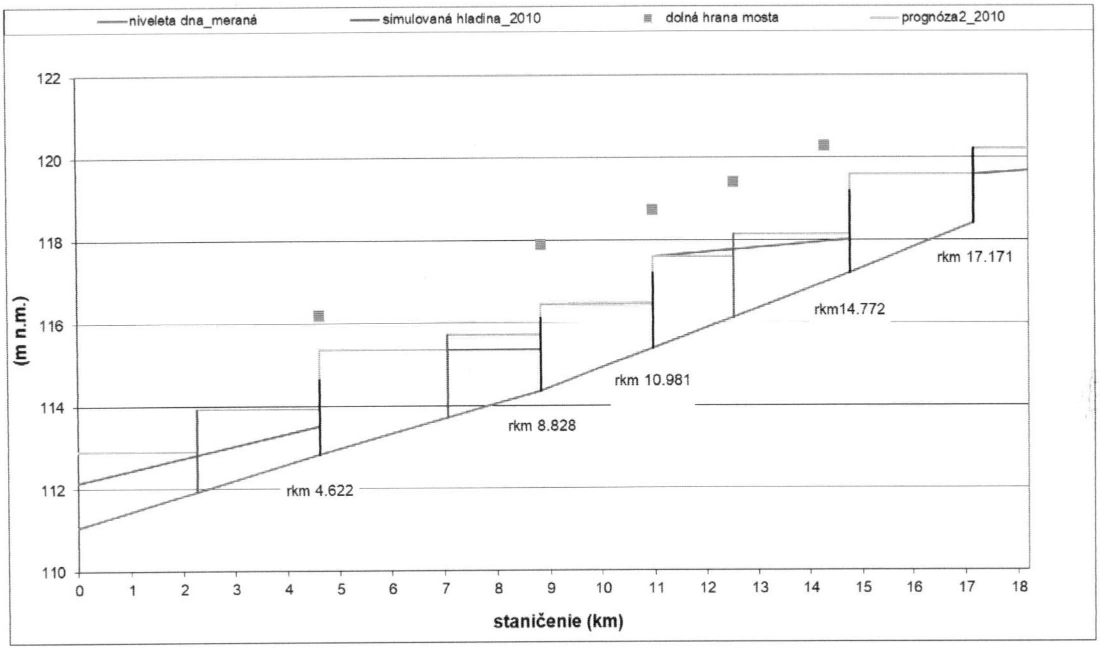

Figure 7. Longitudinal profile of the A VII drainage channel – simulated water level in 2010 and prognosis (black: existing weirs; red: proposed weirs).

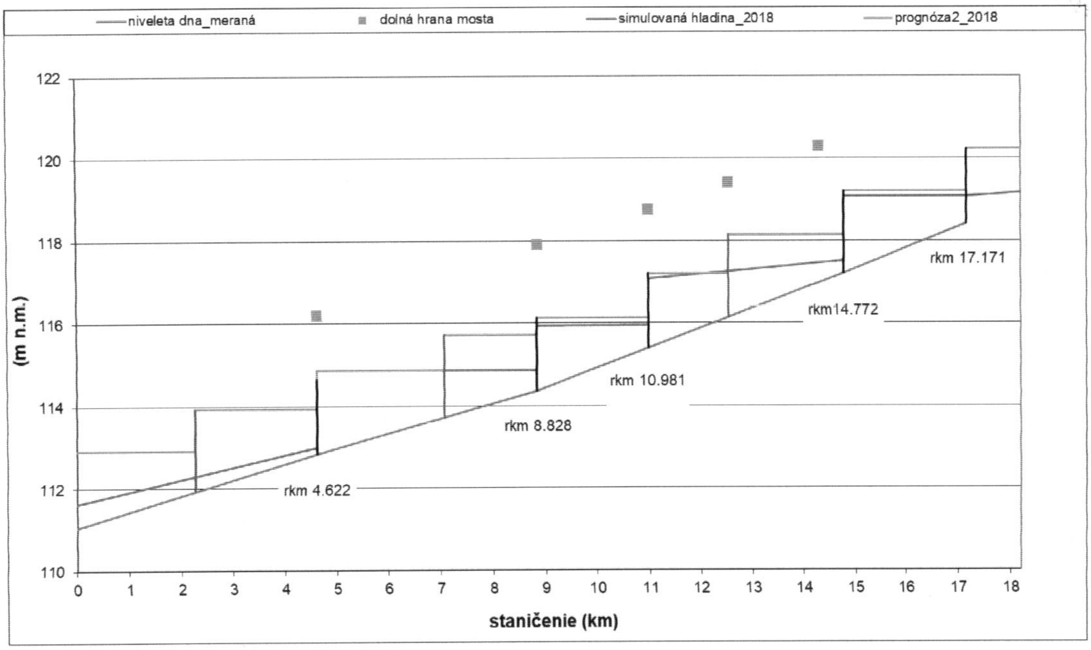

Figure 8. Longitudinal profile of the A VII drainage channel – simulated water level in 2018 and prognosis (black: existing weirs; red: proposed weirs).

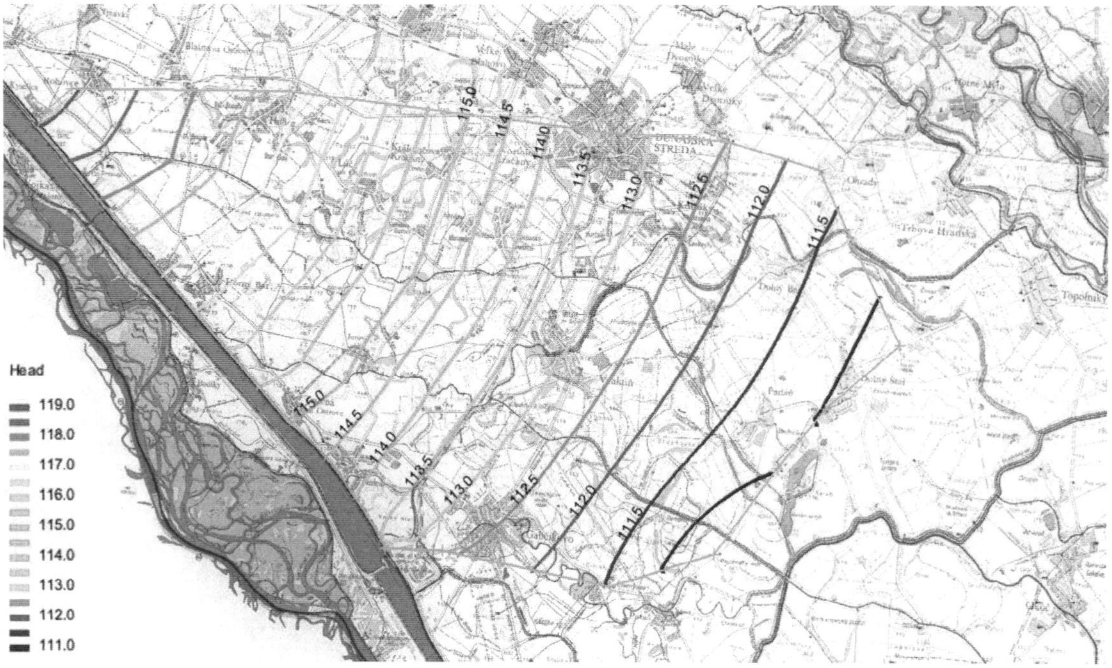

Figure 9. Prognosis: calculated groundwater head for the period from May 26, 2010 to September 9, 2010 [m a. s. l.].

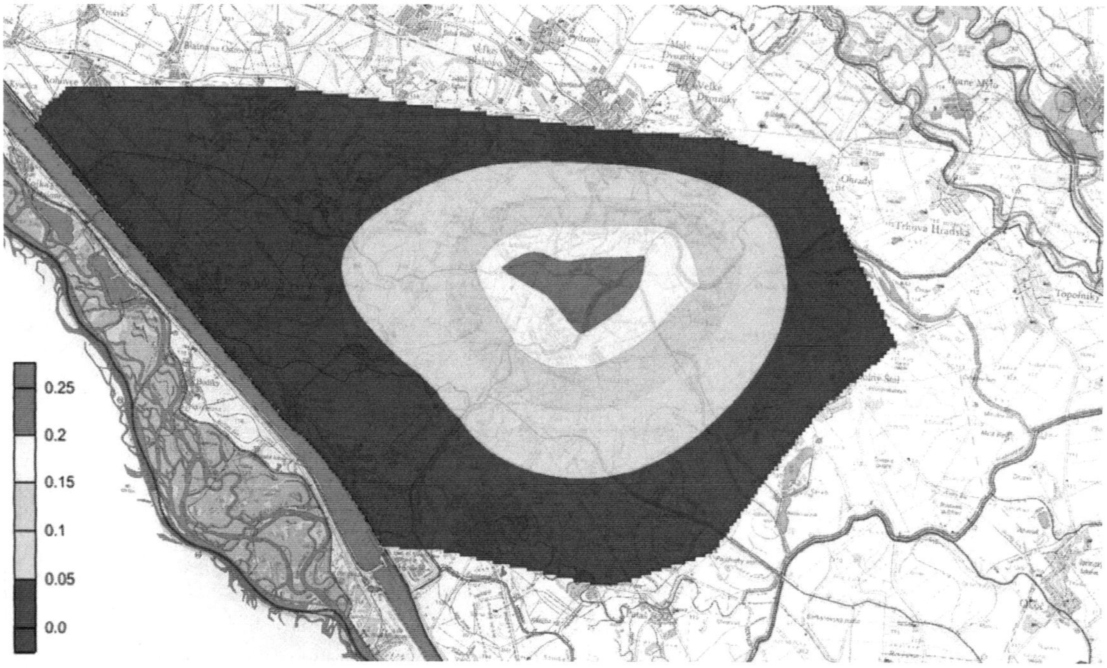

Figure 10. Difference between the prognosis 2010 and the „zero variant“ [m].

Figure 11. Prognosis: calculated groundwater head for the period from March 13, 2018 to May 2, 2018 [m a. s. l.].

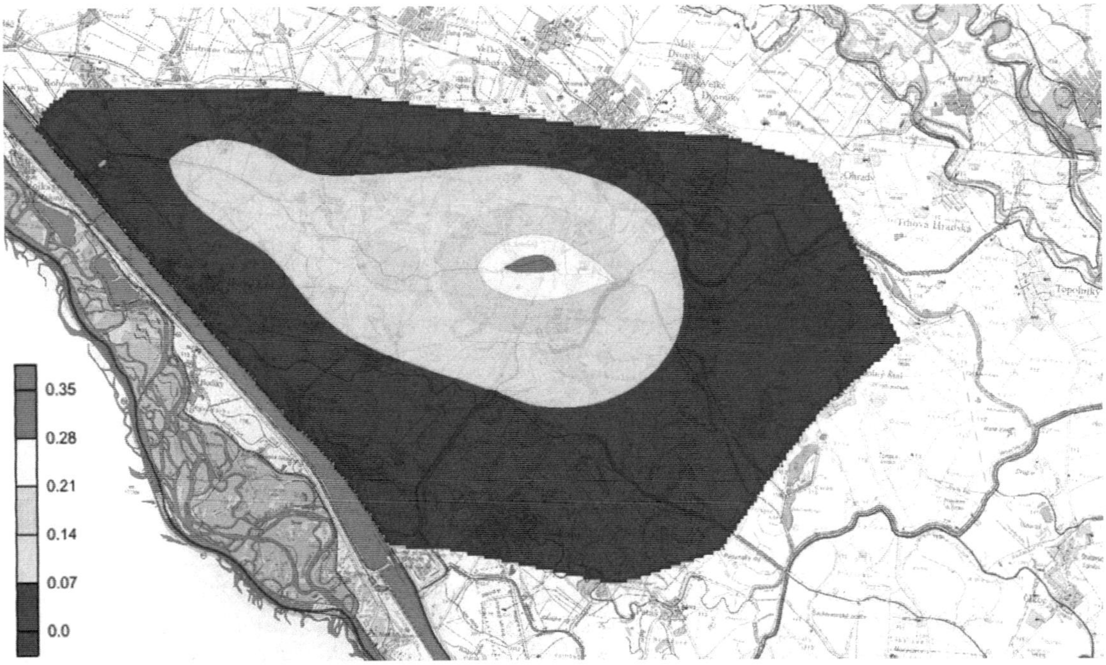

Figure 12. Difference between the prognosis 2018 and the „zero variant“ [m].

As can be seen from figure 10 and figure 12, the differences in groundwater levels vary from 0.0 to 0.25 m in the modelled period from 2010 and from 0.0 to 0.35 m in the modelled period from 2018. Increased groundwater level affects the volume of water infiltrated to the aquifer but does not cause the flooding of the adjacent area. Due to the slight increase in water level, the groundwater gets closer to the roots of cultivated crops.

4. Conclusion

For retention of surface water in aquifer and for recharge of groundwater and its subsequent using for irrigation during dry periods, the Recharge Dam MAR type was investigated using the numerical modelling technique. The numerical model was developed using the MODFLOW program in the Groundwater Modeling System (GMS) environment. The pilot study was situated in the Rye Island in the Podunajska lowland. Field measurements were performed to provide data for setting up boundary conditions for a numerical model of the surface-groundwater flow interaction. Three hydrological yeas have been selected for the data analysis, two of them, 2010 for the calibration of the model and 2018 for the verification, were selected for the purpose of the model design.

The calibration of the model was performed on a steady-state model using the period from May 26, 2010 to September 9, 2010 during the 2010 flood situation. Calibration of the model was done using the automated parameter estimation code PEST for the calibration of a hydraulic conductivity and by trial-and-error adjustment of a conductance of the riverbed. The calibrated model was verified using conditions during the period from the March 13, 2018 to May 2, 2018.

The prognosis focused on the proposal of three weirs at river kilometres 2.270, 7.060 and 12.530 in addition to six existing weirs on the A VII channel. The height of each gate slide is 2.0 m. The groundwater heads for steady-state models capturing the prognosis for the period of 2010 and 2018 were calculated and figures showing the differences between prognosis and the so called "zero variant" were created. These figures indicated, that increased groundwater level affects the volume of water infiltrated to the aquifer but does not cause the flooding of the adjacent area. Due to the slight increase in water level, the groundwater gets closer to the roots of cultivated crops.

5. References

[1] Stachniak K 2020 *DEEPWATER-CE. Workpackage WPC, Activity D.C.2.1. Press release. Final (summary from all PPS)* (Poland: University of Selesia in Katowice) p 8

[2] Sikora A and Slota R 1992 *Research of discharge and water level regime in left-hand side branch system of the Danube by method of physical modelling* (Bratislava: Water Research Institute) p 42

[3] Šoltész A 2002 Water Management in Regions Affected by Water Structure Construction and Operation *Advances in Hydro-Science and –Engineering. Abstracts. Proc. of the 5th Int. Conf. on Hydro Science and –Engineering* (Warsaw: University of Technology) p 127

[4] Červeňanská M, Šoltész A, Baroková D and Janík A 2016 Evaluation of Water Management in Regions Affected by Water Structure Construction and Operation *"Ovidius" University Annals – Constantza, Year XVIII – Issue 18, Series: Civil Engineering* (Constantza, Romania) pp 17-24

[5] Anderson M P and Woessner W 1992 *Applied Groundwater Modeling. Simulation of Flow and Advective Transport* (California: Academic press)

[6] Doherty J 2004 *Model-Independent Parameter Estimation. User Manual: 5th Edition* (Watermark Numerical Computing) p 336

Acknowledgments

The contribution was developed within the frame and based on the financial support of the APVV-16-0278 project "Use of hydromelioration structures for mitigation of the negative extreme hydrological phenomena effects and their impacts on the quality of water bodies in agricultural landscape" as well as the APVV-19-0383 project "Natural and technical measures oriented to water retention in sub-mountain watersheds of Slovakia".

IOP Conf. Series: Materials Science and Engineering 1209 (2021) 012074 doi:10.1088/1757-899X/1209/1/012074

Drought assessment and prediction for Gidra river, Slovakia

L Čubanová[1] and W Almikaeel[1]

[1] Slovak University of Technology in Bratislava, Faculty of Civil Engineering, Department of Hydraulic Engineering, Radlinského 11, 810 05 Bratislava, Slovakia

E-mail: lea.cubanova@stuba.sk; wael.almikaeel@stuba.sk

Abstract. Climate change is affecting every aspect of the world including water resources and water scarcity. Drought is one of many big problems associated with climate change that could occur all over the world. Moreover, hydrological drought is one form of drought that relates to decreased river discharges, below-normal groundwater level, declining the area of wetlands and low water level in lakes or reservoirs. In this study, an assessment of hydrological drought in Gidra river is conducted to characterize dry and normal hydrological years according to Slovak Hydrometeorological Institute (SHMI) Methodology. Furthermore, making benefit of machine learning and artificial intelligence in this field is applicable now, as data of many types are being recorded every day. Deploying machine learning algorithms for the purpose of drought prediction is one way to regulate many operations of water management to prevent irrigation problems. By catching patterns through historical data and deploying machines to learn from those patterns, it is possible to use the values of daily average discharges for January, February, March, and April to correctly predict the hydrological situation in Gidra river whether it is dry or normal, knowing that normal situation refers to wet or normal hydrologically assessed years as the optimal goal in this study is drought assessment and prediction of Gidra river.

1. Introduction

The definition of drought is dependent on the objective of a study, which is very important when quantifying drought [1]. Generally, drought could be defined as a prolonged dry period in the natural climate cycle that can occur anywhere in the world. It is a slow on-set phenomenon caused by a lack of rainfall (World Meteorological Organization WMO). Droughts are generally correlated with large-scale impacts, often driven by regional or even global-scale climate features. The historic classification of drought types has emerged mainly from the meteorological and hydrological sciences to tackle agricultural and socioeconomic impacts [2]. However, drought classification is usually divided into 4 major categories: Hydrological drought, soil moisture drought, Meteorological drought, and Socio-economic drought [3].

Hydrological drought mainly refers to negative anomalies in surface and subsurface water system. For example, hydrological drought relates to decreased river discharges, below-normal groundwater level, declining the area of wetlands and low water level in lakes or reservoirs [3]. Hydrological drought consists of low flow periods, but continuous seasonal appearance of low flow is not necessarily a hydrological drought, although many researchers define hydrological drought as a prolonged period of low flow in the river [4].

2. Gidra river location, intakes, and data

The area of Gidra river catchment above the Píla village on the southeastern slopes of the Small Carpathians are located in the southern part of Slovakia [5] Figure 1. The catchment is forested by over 95 %. These mountains belong to the southern corner of the inner Carpathian massive. In their upper parts, the streams flow through the more or less original beech forest environment.

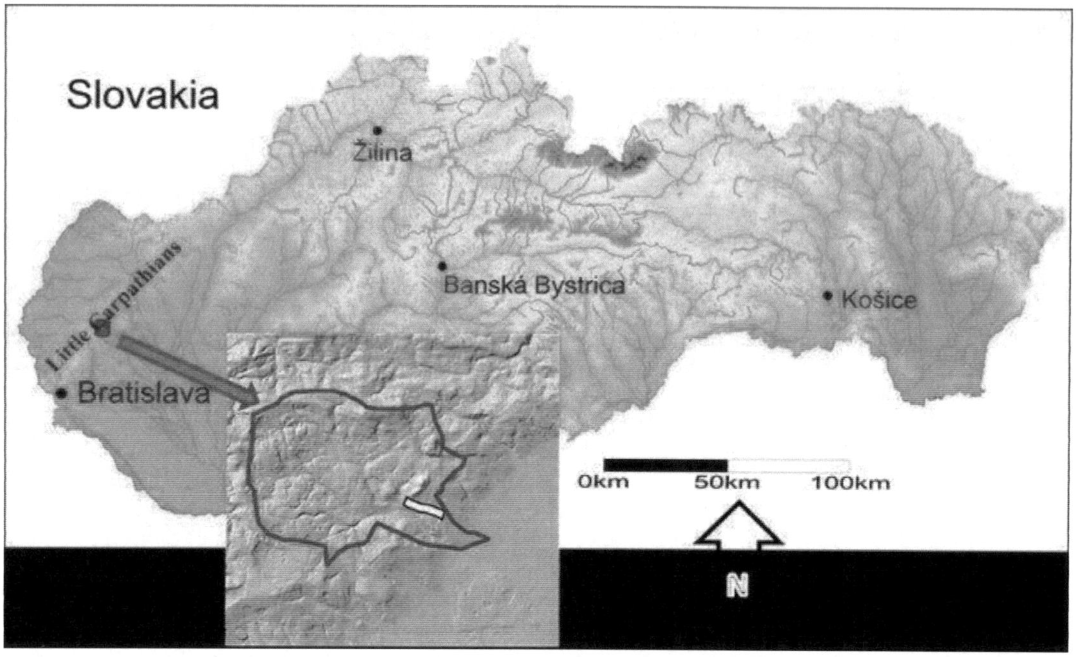

Figure 1. Positioning of the research area (upper Gidra river basin) inside the area of Slovakia. The white rectangle indicates the village Píla) [5].

Many water intakes exist along the catchment of Gidra river, and those intakes are of different types (private, Public). Any drought assessment needs to trace every source of water intake reservoirs, ponds, weirs, and fisheries to control every possibility of water loss along the river. Figure 2 shows Gidra river with all the existing intakes which are:

- Fishery Častá [6] – system of ponds – water is diverted by side take-off; water level is increased by the low sill built across riverbed.
- Water structure Budmerice and pond Hájiček [7]– water diversion structure upstream Budmerice, which diverts water into the reservoir Budmerice and into the stream Gidra, water from Budmerice reservoir is overflowing into the pond Hájiček and then back to Gidra (according to operation manual).
- Pond in Cífer [8] – through-flow pond (volume of water, which is inflowing into the pond is outflowing back to Gidra stream at the other side of pond).
- Upper Ronava [9] – this take-off serves for filling the Ronava reservoir in the neighbouring catchment (built in 1965 – 1967).
- Lower Ronava [10] – this take-off serves for filling the Ronava reservoir in the neighbouring catchment (built in 1988 – 1992).
- Side take-off upstream Voderady weir [11] – this take-off supplies small pond in the park of Voderady manor house, weir of Voderady is located in rkm 10.967 to increase water level for side take-off in rkm 11.040

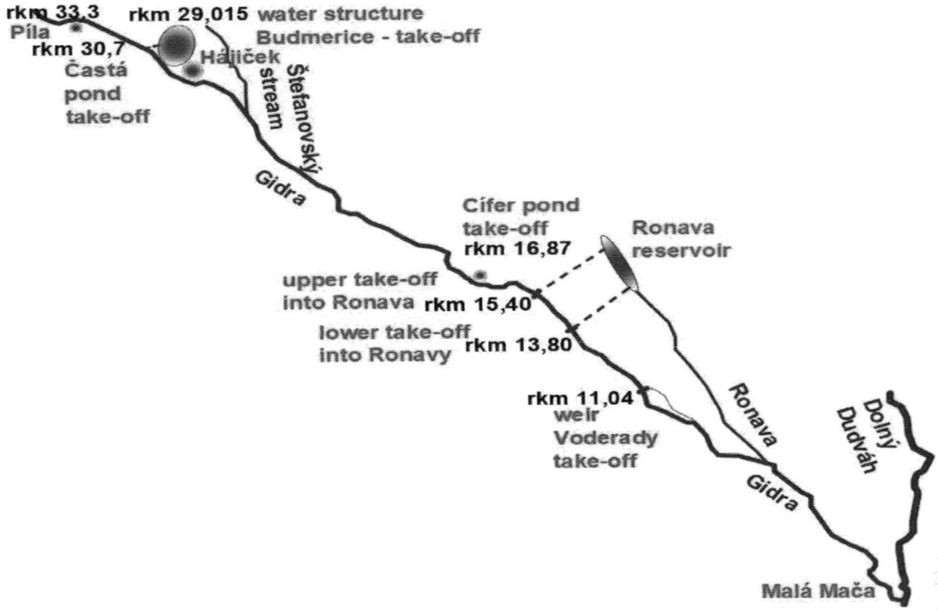

Figure 2. Gidra river and the intakes along the river stream.

In 2018, approximately the last quarter of Gidra's river long was completely dry with no water flow according to many local observations. The reason for such event was unclear and only two actions could have been taken. Firstly, trace back the volume of water used in every intake along the river, which is almost impossible because there is no existing record of water take-off in each intake. The second option was to assess the hydrological situation in Gidra river to check drought level in the same year and if the missing amount of water infiltrated to the ground due to low groundwater level. Therefore, this study aims to analyze historical data and deploys machine learning techniques in order to provide predictions of the hydrological situation in Gidra river based on the discharges of the first 4 months of each year.

In this study, the daily mean discharge over 57 years is collected from gauging stations called Píla, Gidra for the period (1961 – 2018). The data set contains date of discharge in the form of dd/mm/yyyy, code of gauging stations according to the SHMI, name of the stream/village, and daily mean discharges in $m^3.s^{-1}$.

3. Methods and indices

There are various indices used to measure and identify drought of different types. The most practical and widely used indices to evaluate hydrological drought situations are: Standardized Precipitation Index (SPI), the Standardized Streamflow Index (SSI), a threshold-based index, Palmer Drought Severity Index and Snow Water Supply Index [12]. The Standardized Precipitation Evaporation Index and the Standardized Snow Melt and Rain Index use precipitation and temperature to estimate Stream flow [12].

However, most of these indices require more information than only average daily discharge to give the right hydrological assessment. Therefore, in such cases, water bearing coefficient is used to assess hydrological status of a given river by comparing the average discharge value of any year with long-term average value (which represent the normal status) [13]. Exploratory data analysis is performed on daily average discharges by extracting one of basic low-flow statistics (Mean flow) monthly and annually, then compare the results on monthly basis over the study period (1961 – 2018). For predictions, there are several ways to model hydrological events (flood, drought). Physically-based models have been used to forecast hydrological phenomena like storms, rainfall/runoff, shallow water conditions, hydraulic flow models, Physical models have been shown to be capable of forecasting a wide range of flooding/drought situations, but they frequently need a variety of hydro-geomorphological monitoring information, which necessitates extensive computing and prevents short-term prediction [14]. Machine

Learning (ML) is a branch of artificial intelligence (AI) that is used to detect regularities and patterns. providing easier implementation with low computation cost, as well as faster training, validation, testing, and assessment when compared to physical models [14]. Therefore, drought forecasting in this paper is conducted using ML algorithms to pre-assess each year hydrologically for Gidra river.

4. Drought assessment

As mentioned before, the data obtained for this study is for 57 years over the period (1961 – 2018). Moreover, the hydrological situation of each year in the dataset is evaluated using water bearing coefficient. Thus, the ratio between annual average discharge value and long-term average discharge is compared to standard values which define the normal, wet, and dry hydrological status of the river. In Table 1 the standard values of water bearing coefficient are obtained to differentiate between several statuses of the hydrological situation of the river. Standard values of water bearing coefficient represent the percentage of average discharge in a given period compared to the long-term average, which is demonstrated in standard intervals. The standard intervals are divided into three main categories (determine whether the year is dry, normal, or wet) and more subcategories (to measure the severity of drought and wet years) [13]. Based on those intervals, it is possible to assess the hydrological status by calculating water bearing coefficient for each year and compare it with the standard intervals then assess the situation accordingly.

Table 1. Standard values of Water Bearing Coefficient [13].

10 - 29	**DRY YEARS**
30 - 49	
50 - 69	
70 - 89	
90 - 110	**NORMAL YEARS**
111 - 130	**WET YEARS**
131 - 150	
151 - 170	
171 - 180	
More	

However, the evaluation of hydrological situation using water bearing coefficient for the period between 2005 – 2018 is demonstrated in table 2 as an illustrative example, Furthermore, table 2 consists of 5 columns, the first column contains year index, while the second column has only one value (long-term average discharge Qa (1961 – 2000)), the third column contains values of annual average discharge of each year in the study period, the fourth column contains calculated water bearing coefficient for each year, and the last column contains the hydrological status of Gidra river for each year in the study period

Table 2. Assessment of Hydrological status of Gidra river using water bearing coefficient values.

Year	Q_a (1961-2000) ($m^3.s^{-1}$)	Q_{avg} ($m^3.s^{-1}$)	Q_{avg}/ Q_a (%)	Status
2005	0.297	0.195	66	Dry
2006	0.297	0.462	155	Normal or wet
2007	0.297	0.187	63	Dry
2008	0.297	0.110	37	Dry
2009	0.297	0.372	125	Normal or wet
2010	0.297	0.553	186	Normal or wet
2011	0.297	0.379	128	Normal or wet
2012	0.297	0.151	51	Dry
2013	0.297	0.348	117	Normal or wet
2014	0.297	0.239	81	Dry
2015	0.297	0.304	102	Normal or wet
2016	0.297	0.220	74	Dry
2017	0.297	0.105	35	Dry
2018	0.297	0.149	50	Dry

5. Normal and dry years characterization

Preparing information of drought conditions for modelling requires the data to be divided into two separate parts. The first part is the data of all daily and monthly average discharges for dry assessed years, while the second part contains daily and monthly average discharges of all wet and normal assessed years and starting from this point the notation of normal year refers to both normal and wet years. Using this notation, it is possible to detect any existing pattern in both dry and normal parts of data. Therefore, it is possible now to extract the average monthly discharges of the long-term normal and dry hydrological years and compare them in order to find the major differences in monthly average discharges. Figure 3 illustrates the differences in monthly average discharges between normal and dry assessed years.

Differences between monthly mean discharges in dry and normal years are changing in value each month. However, the differences in January, February, March, and April are the highest, while the differences in mean discharges in summer months are smaller. So, the difference between monthly average discharges in the first 4 months is significant. Moreover, the higher the difference in monthly mean discharges between normal and dry years for each month, the more effective mean discharge value of the same month in determining the hydrological situation of Gidra river. Therefore, the monthly mean discharge in April is the most effective value to assess the hydrological situation and monthly mean discharge in March is considered the second effective value in this approach..

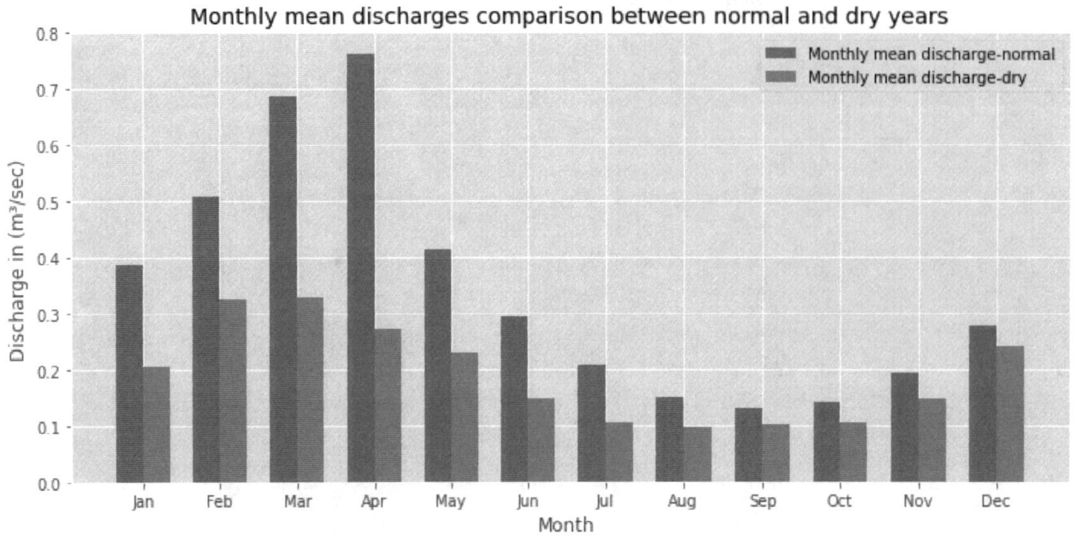

Figure 3. Monthly mean discharges for dry and normal years for period 1961 – 2018.

6. Monthly mean discharge's ranges comparison

Average discharges of January, February, March, and April provided basic indicators to differentiate normal assessed year from the dry one. This comparison aimed to determine the overlapping rate of monthly average discharges values between normal and dry years using box plots in January, February, March, and April. Figure 4 illustrates the distribution of monthly average discharges in February and January in both dry and normal assessed years. Boxes represent interquartile ranges IQR, which contain 50% of data (the discharges between 25% and 75% of the ordered data) while the horizontal line in the middle of the box represents the median. The whiskers contain the rest of data point, which their values are between (Q1 – 1.5*IQR) and (Q3 + 1.5*IQR), where Q1 is the point of 25% of the data and Q3 represent the point with 75th percentage of the ordered data.

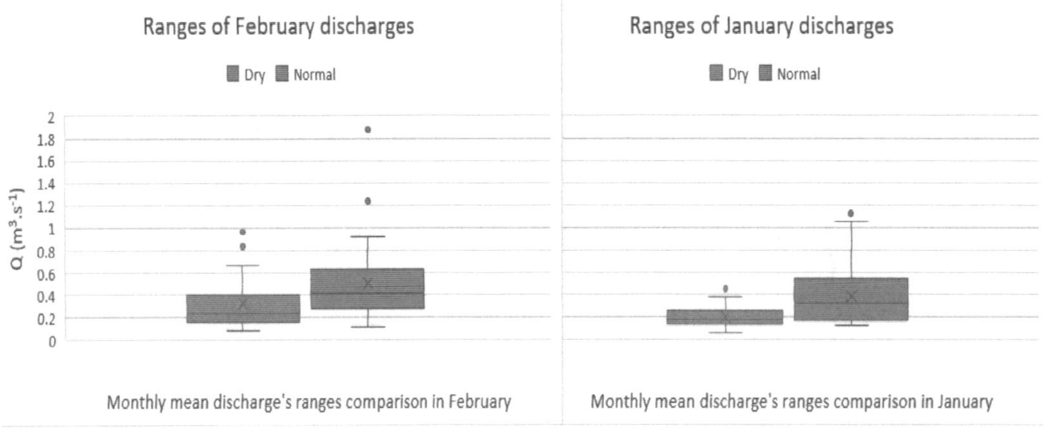

Figure 4. Monthly mean discharge´s ranges comparison in January and February.

By comparing the ranges of January average discharge, a high rate of overlapping is observed, where over 50% of average discharges in normal years contain similar values as the mean average discharges in dry years, and over 75% percent of average discharges in dry years overlap with less than 50% of average discharge values in normal years. However, more than 25% of discharges in normal years did not occur in any previous dry assessed year.

In February, the situation is different, and the rate of overlapping is even higher than what is observed in January. For better understanding of monthly mean discharges distribution in normal and dry years, it is observed that 75% of average discharges in dry years overlap with 50% of average discharges in normal years, and over 50% of discharges in dry years overlap with less than 25% of discharges in normal years.

In Figure 5, the ranges of monthly mean discharges of March and April for both dry and normal years are displayed in box plots.

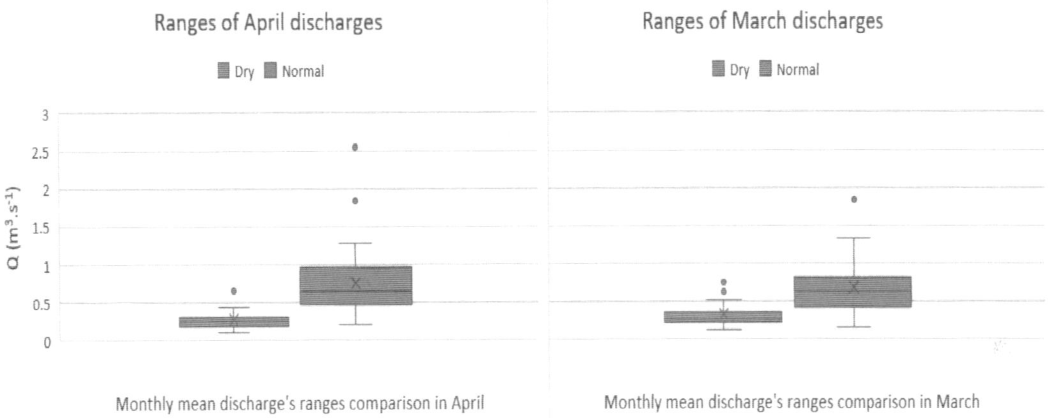

Figure 5. Monthly mean discharge´s ranges comparison in March and April.

In March, over 90% of average discharges in dry assessed years are distributed in the range [0 – 0.5 $m^3.s^{-1}$], while less than 30% of average discharges in normal years are located in the same range. Therefore, over 60% of average discharges values in normal years have never occurred in any dry assessed year, and more than 75% of average discharges in dry years are overlapping with less than 25% of discharges values in normal years. Moreover, the overlapping rate in average discharges for normal and dry years is significant (around 30% excluding outliers) over the period between 1961 – 2018, and this rate also provide valuable information in drought assessment process as an input of a combination of multiple inputs could be used for this purpose.

On the other hand, the ranges of average discharge in April for both dry and normal years are located in different positions. Over 95% of average discharges in dry years are in the range [0 – 0.5 $m^3.s^{-1}$], while less than 30% of average discharges in normal years are in the same range. Only 1 extreme value of average discharges, which is classified as an outlier, is overlapping with 75% of average discharges in normal years. Therefore, the probability of correctly assessing the hydrological situation of Gidra river is more than 70% depending only on average discharges in April. Thus, the hydrological situation of Gidra river in a given year could be assessed as normal year, if the value of April average discharge is greater than 0.5 $m^3.s^{-1}$in general (excluding the outlier discharge value).

7. Pattern detection and hypothesis building

Two separate patterns were extracted for dry and normal hydrological years of Gidra river's discharges. However, the discharges of the first 4 months (January, February, March, and April) of each year vary the most comparing dry and normal average hydrological year. Furthermore, discharges distribution for both dry and normal year has low overlapping rate, which could be useful to differentiate between dry and normal years.

Although there are differences in monthly average discharges of years in the first 4 months between dry and normal assessed years, the level of overlapping in monthly average discharge distributions is complicated, especially in January and February. The main point is to find any combination of monthly or daily average discharges that indicates the hydrological situation of Gidra river based on the study of historical data over the period (1961 – 2018).

Therefore, the study results conduct a hypothesis based on the following facts:

- Monthly average discharges of January, February, March, and April vary and differ between normal and dry hydrological situation.
- The difference in monthly average discharges of the first 4 months is the highest compared to any sequence of monthly discharges.
- Monthly average discharges distribution has low overlapping rate in March, and April and relatively moderate overlapping rate in January, February.
- The overlapping rate of monthly discharges distribution in the first 4 months is the lowest compared any sequence of monthly discharges.

Hypothesis statement: 'For every hydrological year with no unprecedented extreme level of discharges due to climate change crisis or unnatural level of discharges that never happened in the past for 60 years, it is possible to predict the hydrological situation of Gidra river only depending on average daily discharges of the first 4 months (January, February, March, and April) of the same year using machine learning and deep learning algorithms'.

8. Modelling and hypothesis testing

To test the hypothesis of this study, Machine learning, specifically, Support vector machine algorithm (SVM) and deep learning (fully connected neural networks FcNN) models were used. The main task of both models is to classify the hydrological situation of Gidra river based on the data of daily average discharge of the first 4 months.

Support vector machines (SVM) are a popular machine learning method to analyse data and recognize patterns. SVM performs classification by constructing an N-dimensional hyperplane (a plane generalized into N dimensions), that optimally separates the data into two categories. They are used for classification and regression analysis, among other tasks [15].

First, the data must be split into two groups. The first group is for training the model and contains daily average discharges of the first 4 months for 40 years in the period (1961 – 2000) and the hydrological situation of Gidra river for every year over the same period. The training data set is used to feed the model with historical information about daily average discharges and the resulting hydrological situation of each year, therefore, the model will be able to recognize an empirical pattern (N-dimensional hyperplane) to separate the values of the daily discharge into a dry and normal situation.

Afterwards, the model obtained a separable N-dimensional hyperplane, that can separate daily average discharges into two positions (dry or normal) and it should be applicable for testing data set, which contains daily average discharges of Gidra river over the period (2001 – 2018) without providing the hydrological situation of those years. Now, SVM model must produce results regarding the hydrological situation of Gidra river over the testing period using the separable N-dimensional hyperplane in the format "0" for dry hydrological status and "1" for normal hydrological status.

As the real hydrological situations over the testing period (2001 – 2018) are evaluated in this study, therefore, it is possible to measure the accuracy of model's results (predicted- status) by comparing them with the real hydrological situations over the testing period (real- status). Table 3 compares predicted and real status over the testing period, where the first row represents the real hydrological status, and the second row represents the predicted values of hydrological status.

As shown in Table 3 the real and predicted hydrological status are identical, which indicates, that SVM model with a linear kernel can correctly predict the drought situation of Gidra river over the period (2001 – 2018).

Table 3. Predicted and real hydrological status comparison over testing period (2001 – 2018) using *SVM* model.

year	2001	2002	2003	2004	2005	2006	2007	2008	2009	2010	2011	2012	2013	2014	2015	2016	2017	2018
Real-status	0.0	0.0	0.0	1.0	0.0	1.0	0.0	0.0	1.0	1.0	1.0	0.0	1.0	0.0	1.0	0.0	0.0	0.0
Predicted-status	0.0	0.0	0.0	1.0	0.0	1.0	0.0	0.0	1.0	1.0	1.0	0.0	1.0	0.0	1.0	0.0	0.0	0.0

Deep learning algorithm also provides the same results but using a complicated technique, which is called fully connected neural network. Although the definition of deep learning and neural network is not the object of this study, however, for better understanding of the concept a brief definition is provided: neural network is a parametric supervised deep learning algorithm that contains N layers and takes inputs as N-dimensional array and apply nonlinear transformation (by using multiple activation functions) on the input values by multiplying them by learnable parameters and constantly tuning those parameters (this process called backpropagation and it requires three steps: loss functions to measure difference between predicted outputs and the real outputs, metric function to measure the accuracy of the model, and an optimization algorithm to estimate the parameter values depending on the values of loss function) until the output of this model satisfy the required outputs in the learning phase [15], [16], [17].

The neural network model used to classify the hydrological situation of Gidra river contains 6 layers (including the input layer), each layer has a defined number of neurons (120, 90, 60, 30, 4, 1) respectively and used the following activation functions (selu, tanh, and softplus). For parameters optimization, cosine similarity function is used as loss function and binary cross entropy function is used as metric function. In addition, stochastic gradient descent algorithm used as optimization algorithm to tune the parameters with learning rate is set to 0.1.

In addition, one specific step has been used before feeding training data to FcNN model, which is scaling input data using robust scaler to transform the input values into values that are robust to outliers. This method removes the median and scales the data in the range between 1st quartile and 3rd quartile by subtracting the median and then dividing by the inter-quartile range (75 % value — 25 % value), i.e., in between 25th quartile and 75th quartile range. If outliers are present in the data-set, then the median and the interquartile range provide better results and outperform the sample mean and variance [17]. Table 4 shows the predicted results of the hydrological situation over the testing period (2001 – 2018), which are identical to the real assessed values.

Table 4. Predicted and real hydrological status comparison over testing period (2001 – 2018) using *FcNN* model.

year	2001	2002	2003	2004	2005	2006	2007	2008	2009	2010	2011	2012	2013	2014	2015	2016	2017	2018
Real-status	0.0	0.0	0.0	1.0	0.0	1.0	0.0	0.0	1.0	1.0	1.0	0.0	1.0	0.0	1.0	0.0	0.0	0.0
Predicted-status	0.0	0.0	0.0	1.0	0.0	1.0	0.0	0.0	1.0	1.0	1.0	0.0	1.0	0.0	1.0	0.0	0.0	0.0

Both models succeeded to correctly predict drought hydrological situations in Gidra river using only daily average discharges of the first 4 months of each year over training and testing period. Those results could provide useful information for any operation project of Gidra river to manage water distribution for agriculture or any other purposes during possible dry periods in the future. Precisely due to the number of water structures or take-offs along the river. The forecast of the hydrological status of the Gidra river based on daily discharges from the first four months of the year can be immediately included in the changes of operation manuals of water structures on Gidra to prevent it from drying out at the lower part of the stream.

Finally, to succeed in building both models, it is necessary to split the data in a homogeneous way, which means to make sure, that training data obtains variety of hydrological situations (dry, normal) with different water bearing coefficient values (especially the values located around the limits of determining new hydrological situation). In this regard and after performing multiple experiences with the different training sets, FcNN turns out to obtain better results and in many cases, these results obtain information about drought severity.

9. Conclusion

Analysing historical data of different types (hydrological, metrological, climatological and many others) for drought assessment is one of many possible ways to characterize drought and dry period. However, performing broad statistical analysis on hydrological data for any river or streamflow reveals valuable information about average discharges, ranges of average discharges, the typical average year of dry and normal hydrological situations.

In Gidra river, after performing detailed statistical analysis according to the SHMI methods, the following results were obtained:

- the discharges of January, February, March, and April play a major role in the final hydrological situation of Gidra river,
- there is a significant difference in average discharge values for the first 4 months between dry and normal assessed years,
- drought has significant impact on monthly average discharges, mainly in April and March, perhaps because it is stream of under mountain region influenced by the snow melting,
- it is possible to use historical data and deploy machine learning models to predict the hydrological situation,
- to obtain more accurate results on hydrological situation, precipitation and snow melting data are recommended to consider, analyse, and combine with discharge values to track every possible pattern over time.

At the end of the study, daily average discharges for 2019 and 2020 were provided by SHMI. The discharges of the first 4 months were extracted and used to predict the hydrological situation for 2019 and 2020. Both SVM and FcNN models produce results of a dry hydrological situation for 2019, 2020 which correspond with the results obtained from the analysis using SHMI methodology (water bearing coefficient values for 2019, 2020: 72, 85 respectively) [18, 19].

10. References

[1] Lloyd-Hughes B 2013 *The impracticality of a universal drought definition: Theoretical and Applied Climatology* (United Kingdom: University of Reading) p 3
[2] European Environment Agency 2007 *Climate change and water adaptation issues* (Copenhagen) p 9
[3] Sheffield J and Wood E F 2012 *Drought: past problems and future scenarios* (United State) 12
[4] Zelenhasić E and Salvai A 1987 A method of streamflow drought analysis *Water Resources Research* **23** 158
[5] Orfánus T, Jenčo M, Bebej J, and Benko M 2017 Simulation of the effects of forest roads ON Stormflow generation using GIS and 2D vadose Zone hydrological model *Ekológia* **36** 26
[6] Simanová I 2019 *Operational manual for water structure "Častá – completion of ponds" on the Gidra River at km 30.7* (Proma Ltd: Bratislava, Slovakia) p 97
[7] Magula P 2012 *Operational manual for water structure Budmerice on the Gidra River at km 29.015 (intake)* (Piešťany: Slovak Water Management Company) p 49
[8] Magula P 2007 *Operational manual for permanent operation of Hájiček Pond in Budmerice (side reservoir on the Gidra River)* (Bratislava: Geohydro) p 28
[9] Ďuriš V 2019 *Technical report. Measurement after the implementation of the sluice weir for the pond in Cífer and the supply channel to the pond* (Trnava: CS Ltd) p 7

[10] Matulík Š 2019 *Operational manual for water structure Water reservoir Ronava, update 2019* (Nitra: AGROPROJEKT Nitra Ltd) p 36
[11] Potis F 2018 *Operational manual for water structure Weir on the Gidra River in the river kilometer 10.967* (Bratislava) p 36
[12] Svoboda M and Fuchs B A 2016 *Handbook of Drought Indicators and Indices Integrated Drought Management Programme (IDMP)* (Geneva: World Meteorological Organization (WMO) and Global Water Partnership (GWP)) p 55
[13] Fatulova E and GWP Slovakia 2013 *Guidelines for Drought Management Plans (Slovak Study Report)* pp 71 – 4
[14] Mosavi A, Ozturk P, and K.-wing Chau 2018 Flood Prediction Using Machine Learning Models: Literature Review *Water* **10** 11 p 2
[15] Farber R and Kaufmann M 2011 *CUDA for real problems (CUDA Application Design and Development)* pp 265 – 75
[16] Buitinck L, Louppe G, Blondel M, Pedregosa F, Mueller A, Grisel O, and Varoquaux G 2013 *API design for machine learning software: Experiences from the scikit-learn project*
[17] Brunton S L and Kutz J N 2019 *Data-driven science and engineering machine learning, dynamical systems, and control* (Cambridge: Cambridge University Press)
[18] Jeneiová K, Janečková L, Blaškovičová L, Podolinska J, Síčová B, and Liová S 2020 *Zhodnotenie hydrologického roka 2019* (Slovakia: Vodohospodársky spravodajca) pp 20 – 4
[19] Jeneiová K, Blaškovičová L, Podolinska J, Slivková K, Síčová B, and Liová S 2021 *Zhodnotenie hydrologického roka 2020* (Slovakia: Vodohospodársky spravodajca) pp 20 – 4

Acknowledgments
This work was supported by The Slovak Research and Development Agency under the contract No. APVV-15-0489 and also by The Slovak Scientific Grant Agency under the contract No.VEGA 1/0728/21.

Analysis of anomalies in the control probes of the Rozgrund dam

F Koval'[1]

[1] Department of Hydraulic Engineering, Faculty of Civil Engineering, Slovak University of Technology in Bratislava, Radlinského 2766/11, 810 05 Bratislava, Slovakia

E-mail: frederik.koval@stuba.sk

Abstract. The subject of the contribution is to clarify the causes of anomalous development of water levels in some observation objects of the Rozgrund dam. To clarify the anomalous development of the water levels, a detailed analysis of the development of water levels in all observation probes built on the dam, the water levels in the reservoir and the daily total precipitation was performed. An important knowledge is the amplitudes of fluctuations in water levels in observation probes and in the water levels in the reservoir. The calculations of correlation coefficient, expressing the relationship between water level in individual boreholes and the water level in the reservoir are another step in assessment of anomalies. Based on the knowledge obtained, it was possible to assess the degree of the influence of the water level in the reservoir on the development of the water level regime in observation objects. At the same time, it was also possible to detect the existence of other effects influenced the water levels in the probes, such as the leakage into their surroundings caused by precipitation or the impact of waters flowing from hillslopes.

1. Introduction

Dam Rozgrund was built between 1743 and 1744 (Figure 1). The main object of the Rozgrund reservoir is a dam, which is built in geologically and morphologically good conditions. Local materials were used for its construction: alumina-stone and stone-clay and clay. They were built into the dam to form a heterogeneous dam with a central seal, bound to the rock bedrock by a tooth. The elevation of the dam crown above the maximum level was designed to minimize the risk of overflowing [3].

Table 1. Characteristic parameters.

River		Vyhniansky
River km		12,8
Hydrological condition		
Drainage-basin area	km^2	2,7
Average year precipitation	mm	950
Q_{364}	$m^3.s^{-1}$	0,002
Q_{355}	$m^3.s^{-1}$	0,003
Q_{100}	$m^3.s^{-1}$	11
Characteristic parameters of the reservoir		
Max. ground elevation of water level	m n. m.	705,00
Min. ground water elevation of water level	m n. m.	693,00
Volume reservoir	mil. m^3	0,515 103
Characteristic dam parameters		
Ground elevation of dam crown	m n. m.	706,20
Length dam in crown	m	125,0
Max. high dam	m	27,0
Width dam in crown	m	6,0
Capacity dyke drain		
Dyke drain	mm	2 x DN 250
- Capacity by max. water level	$m^3.s^{-1}$	2 x 0,33
Safety spillway		
Type of spillway		Non controlled
Width spillway	m	4,2
Capacity spillway	$m^3.s^{-1}$	7,5

The dam consists of five parts:

1. Sealing core, beaten from the soil, width in the crown of 2 m and in the valley riveting 14 m, binding with a tooth depth of about 4 m.
2. Guide stabilization prism – alumina-stone and stone-clay, fortified from broken stone of flat shape, slope 1:1,5 - 1:1,7.
3. Aerial stabilization prism – alumina-stone and stone-clay, fortified by thickening, grassing and since 2004 reinforced by nailing, slope at the top 1:1,22,
4. Steepness gradually decreases from 1:1,35 to 1:2,00. Fortifications of the directional slope by stone straightening.
5. Stone foot.

Figure 1. Dam Rozgrund (photo Kovaľ, F.).

1.1. Significant changes and reconstructions

In 1987, the reconstruction of the safety level was completed, in which the safety fallout was deepened and its shape was adjusted. In 1991, an attempt was made to rebuild the adage and reconstruct the bottom outlet. The ad was not fully restored, the work was stopped as it was carried out at a nearly full reservoir. This caused increased tributary to the stampede. On the air slope of the dam above the "portal" of the tunnel, dangerous deformations occurred, accompanied by a landslide. The tunnel so built (in 1991 – Figure 2a) with a length of 19.8 m showed signs of instability due to increased humidity and unprofessionally realized works, therefore in 1994 its remediation with stone vault reinforcement made of rhyolite masonry (Figure 2b) was necessary. The bottom was sprinkled with sorted aggregates of hr. 8-16 mm. Drainage was built in the bottom with a result in front of the ad, where there was a threshold and an object for measuring seeps [1].

a) b)

Figure 2. Tunnel in year 1991 (a) and in year 1994 (b) (photo Vodohospodárska výstavba (VV)).

In 2004 and 2005, dam was extensively reconstructed. The purpose of which was to build a stage of the bottom outlet, inlet object, bottom outlet caps and their control, adjustment of the trough under the dam, stabilization of subsoil parts of the air slope by nailing method (Figure 3), construction of a new staircase on the directional slope of the dam and installation of a new watertight lath. The bottom drain adage was designed as excavated through of the dam in the route of the original wooden pipe of the bottom outlets. The route of the athele took advantage of the stamped part from 1991 with a length of 19.5 m. Tunnel was excavated in a mining way with primary lining secured by mining TH – reinforcement deposited after 0.5 m and steel profile Union used as casings. The netting was attached to the lining thus secured, followed by fixation with sprayed concrete. After sections of max. 10 m length, filling of over-the-counter low-pressure injection with cement-ash mixture was carried out. The total length of the ad is 69.5 m, of which the newly stamped 52 m, the length of the concreted part is 18.0 m, the passage part 51.5 m. The most significant fact, as established during the reconstruction works, is the confirmation of the existence of a clay central nuclear seal of relatively high quality. The aerial slope of the Rozgrund dam was stabilized with the help of earth nails, the anchoring depth of which was 4 m. During implementation, comments occurred (inappropriate inclination and insufficient depth of nails). Nevertheless, after the implementation of the stabilisation measure, there was a significant reduction of local surface deformations (creaps) on the air slope of the dam [4].

Figure 3. Stabilization of the aerial slope of Rozgrund earth nails in 2006 with the detail of the nature nail (photo VV).

1.2. Dam monitoring

Between 1982 and 2005, dam was gradually equipped with devices for measuring and observing deformations and development of the movement of groundwater and leachate. Elevation changes of the dam and its objects are monitored by geodetic measurement by a very precise levelling method at the points of gravity and at the observed elevation points. On the water construction site there are points of contact (PVB-1 to PVB-3), observation elevation points (KVB-1 to 10) and observation probes (PS-1 to PS-6, MG-1 to MG-6, HPS-1 to HPS-3, Figure 4).

In total, facilities for measuring and observing the following phenomena are built on the water construction site:

- Weather conditions (air temperature, water temperature, atmospheric precipitation).
- Operating conditions (inflow into the reservoir, water level in the reservoir, water abstraction from the reservoir - currently not carried out, level in the treated trough under the dam).
- Deformations (height changes of the dyke). Leaks – measured in stone beads and in the admiss of bottom outlets.
- Surface mode of groundwater and leachate – using observation probes built into the dam, adjacent valley slopes and the area below the dam.
- Accompanying signs of suffo-like phenomena – by measuring the encracing of observation probes.
- Risk of violation of filtration stability – geophysical measurements of the flow rate of groundwater and leachate in observation probes.

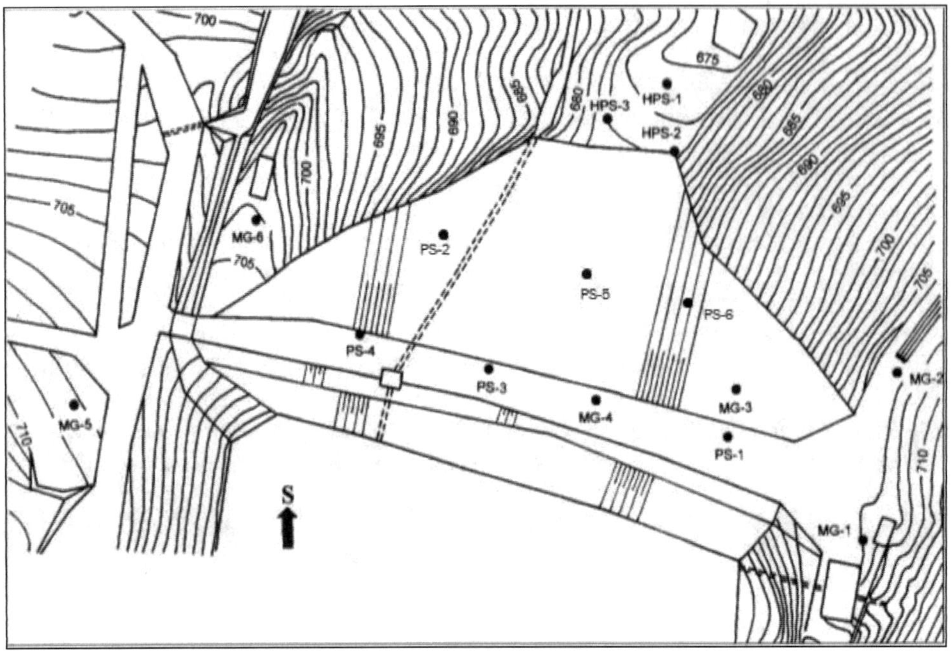

Figure 4. Situation with marked observation probes.

2. Development of the level regime in observation probes

The effect of the water level in the reservoir on changes in the level regime of groundwater and leachate can be expressed through correlation dependencies [2]. I have quantified the correlation components, expressing the link between the surface in the reservoir and the surface regime in individual probes, presenting the development of groundwater and leachate in the dam, in its bedside, in the subsoil and in adjacent valley slopes. The analysis also monitored amplitudes of level fluctuations in the reservoir and observation probes in two separate time periods before and after reconstruction in 2004 (1991-2004 and 2006-2019).

For a summary analysis of the development of groundwater and leachate levels, calculated correlation factors were used, expressing the links between the observed levels as well as amplitudes of fluctuations in the levels in the probes and the water level in the reservoir. After analysing the results obtained, expressing the correlation of the levels in the probes to the level in the reservoir (Table 2, Figure 5). It can be concluded that a more pronounced correlation between groundwater and leachate levels and the level in the reservoir has been demonstrated in observation probe MG-1, MG-3, MG-6 and PS-1, indicated by a red ring in Figure 5. The positions of these observation objects indicate the process of wrapping the dam with water from the reservoir in areas bound to valley slopes. The mean correlation to the level in the reservoir (blue ring marking in Figure 5) can be recorded in observation objects MG-4, PS-4, PS-2 and HPS-3 located in the dam and in MG-2 – in the right-hand valley slope. Low correlation to reservoir level (green ring marking in Figure 5) and other observation objects PS-5, PS-6 and HPS-1 was observed in observation objects PS-3, MG-5 and HPS-1 (marking with a yellow ring in Figure 5).

When evaluating the amplitudes of the splitting levels in the probes and the level in the reservoir (Table 2), no consideration was made for the reconstruction period when the reservoir was discharged. A higher amplitude of water level fluctuations in the probe, compared to the water level in the reservoir, indicates the impact of post-reservoir phenomena, such as the leakage of the probe by precipitation waters, or the impact of slope waters. Such anomalies have been reported in observational objects PS-1, PS-4, MG-1, MG-2, MG-3 and MG-4. In Figure 5, these phenomena are highlighted by a blue double-stitch. Of these, PS-1, PS-4 and MG-4 are on the crown of the dam, with the risk of leakage by precipitation waters and MG-1, MG-2 and MG-3 in areas of adjacent slopes with possible impact of slope waters, or on the forest road (MG-2). The MG-5 probe is predestined to monitor the level of slope waters. Based on the above knowledge and summary information on the evolution of the level regimen, anomalies in probes PS-1, PS-4 and MG-4 can be recorded.

Table 2. Degree of correlation bonds and amplitude of splitting of levels.

Probe	Correlation	Amplitude unbalance: Before reconstruction	Amplitude unbalance: After reconstruction	Side effects
PS-1	High	**6,59**	**3,95**	Impact of slope water?
PS-2	Middle	-	**2,44**	
PS-3	Low	**4,20**	**3,73**	
PS-4	Middle	12,83	**12,40**	Leakage?
PS-5	Without	-	**1,51**	
PS-6	Without	-	**1,07**	
MG-1	High	5,96	3,82	Impact of slope water?
MG-2	Middle	**5,64**	6,65	Slope water? Leakage?
MG-3	High	**5,70**	6,17	Slope water? Leakage?
MG-4	Middle	**8,74**	**5,71/6,12**	Leakage? Slope water?
MG-5	Low	**5,74**	**5,01**	Slope water – dominant influence
MG-6	High	**3,31**	**3,90**	
HPS-1	Without	**1,67**	**0,62**	
HPS-2	Low	**1,45**	**1,58**	
HPS-3	Middle	**1,80**	**2,32**	
Reservoir		**5,31**	**2,10**	

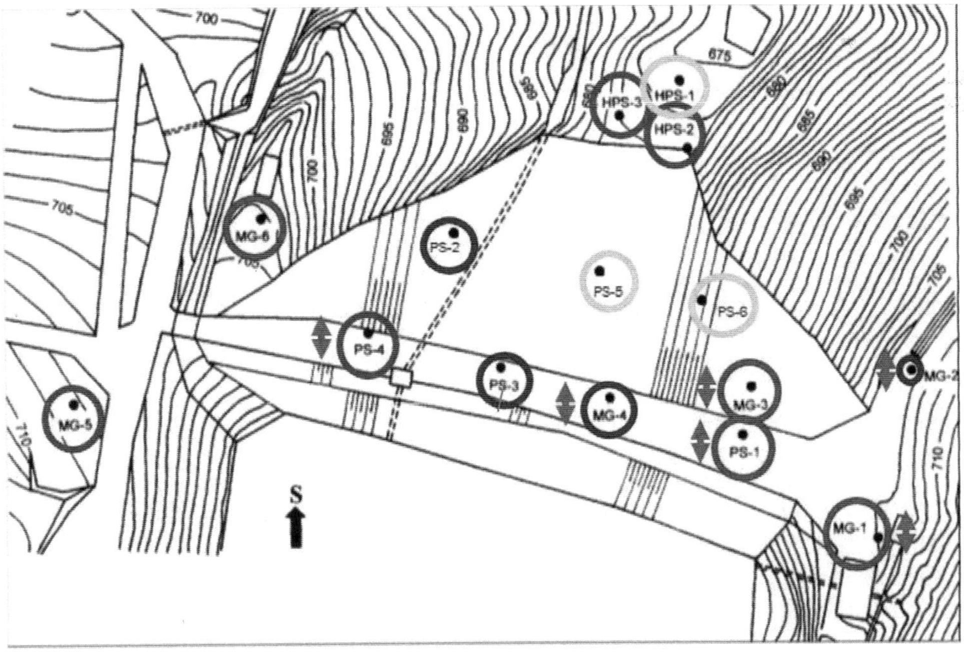

Figure 5. Situation of correlation bands in measuring probes.

2.1. Analysis of anomalies in observation probe PS-1 a MG-4

The probes are located on the crown of the dam on the right, in the area of binding to the valley slope. Since they show anomalies in the development of the level regime for a long time, an analysis of the development of water temperature was carried out here. On 26.08.2019, automated water temperature sensors were installed in the probes in various depth horizons. The water temperature in the reservoir was measured, at a depth of 3 m. The results of the measurements are documented in Figures 6 and 7. The fluctuations in the temperature of the water in the probes range from 8°C to 11°C (for the period from August 2019 to February 2020), the water temperature in the reservoir gradually decreases from about 20°C in August to 2°C in February. Measurements show that the effect of seasonal water temperature in the reservoir does not occur in the water temperatures in the probes. The results of the measurements seem to indicate that there is no link between the temperature of the water in the reservoir and in the probes. However, it should be taken into account that the water leaking through the rocky environment or the environment of the dam will be cooled. The link between the water temperature in the PS-1 probe and the water temperature in the reservoir can be slightly registered in the change in its temperature regime. The change in water temperature in the PS-1 probe decreases in depth (Figure 6). The temperature difference is most pronounced in the autumn months, most in October – about 2°C, with a culmination of 11°C at the level of 701.35 m.s.l. Then the temperature difference decreases and in the winter period, in February the temperature after the probe depth is stabilized at 9.5°C.

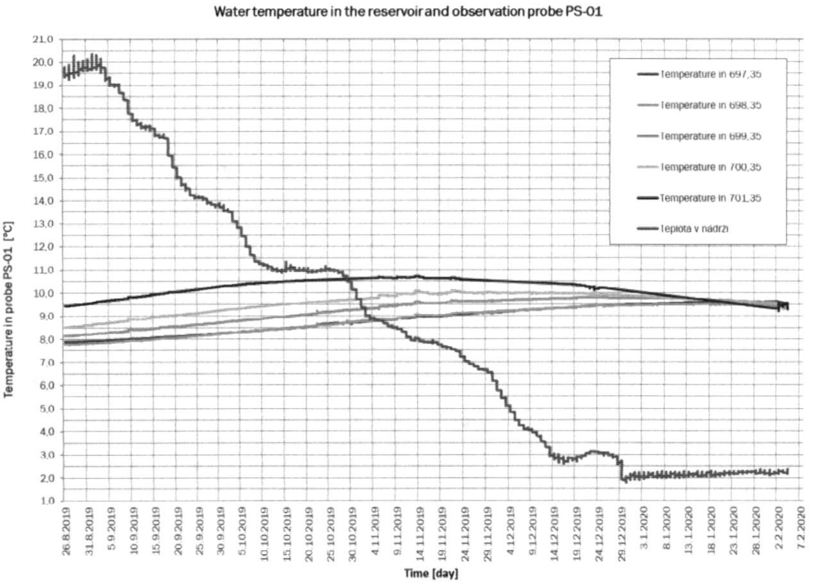

Figure 6. Temperature mode in different groundwater level horizons in observation probe PS-1.

Due to the proven high correlation between the surface in probe PS-1 and in the reservoir, the impact of the reservoir (preferred see-through roads) in this area of binding of the dam to the valley slope cannot be excluded. At the same time, the correlation between the level in the PS-1 probe and the precipitation activity signals the existence of drainage bypass leakage through precipitation waters and slope waters (Figure 7).

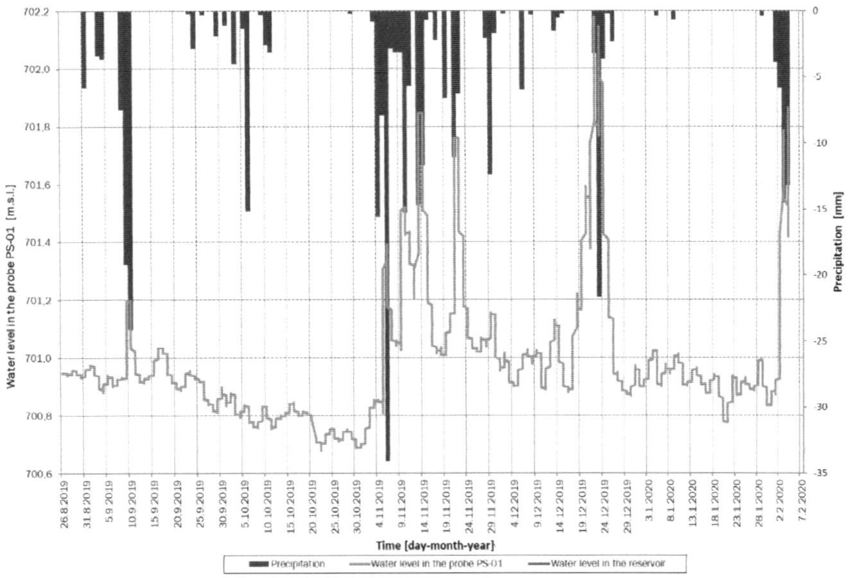

Figure 7. Hydrological and climatological data in observation probe MG-4 for the period 26.8.2019 – 4.2.2020.

Its growing trend cannot be ignored in the evolution of water levels in the PS-1 probe. For the years 1991 – 2020 there is a registered increase of about 3.5 m, which can mean not only the previously mentioned accumulation of preferred seepage roads in the area of right-hand binding of the dam to the valley slope, but also a local change in the permeability of the affected environment.

In the MG-4 probe, the variation in water temperature at depth is between 9°C and 11°C, depending on the time the evolution of the water temperature in the probe is marked by sudden changes, possibly caused by other post-effects (Figure 8). The water temperature in the probes is determined by the rock environment itself, seepage waters become groundwater. In the case of the MG-4 probe, due to the reported sudden changes in temperature development (slight increase), partial mixing with precipitation waters from around the borehole (bypass) due to leakage cannot be excluded. The impact of slope waters cannot also be excluded. Sudden changes in the temperature mode of the water in the MG-4 probe indicate constant (pulsating) filter movement in the affected environment, which can also be registered at its level mode (Figure 9).

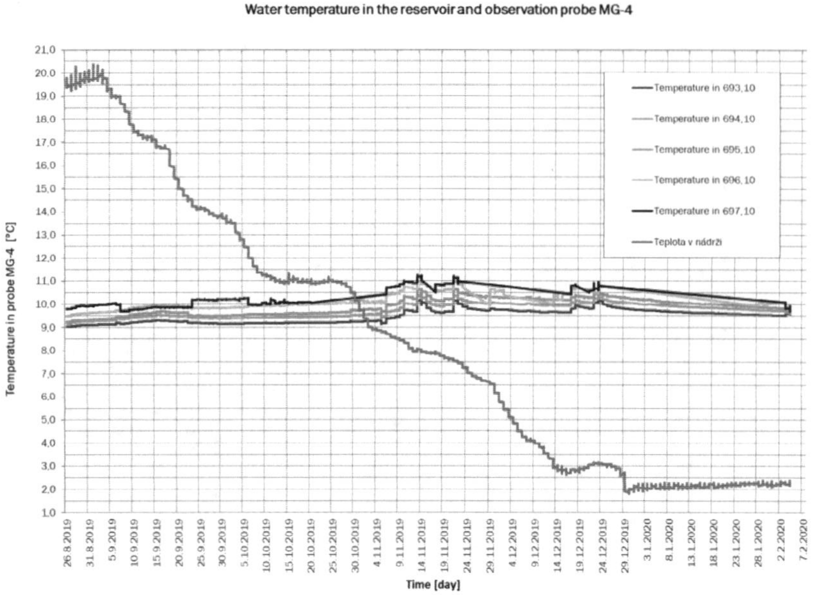

Figure 8. Temperature mode in different groundwater level horizons in observation probe MG-4.

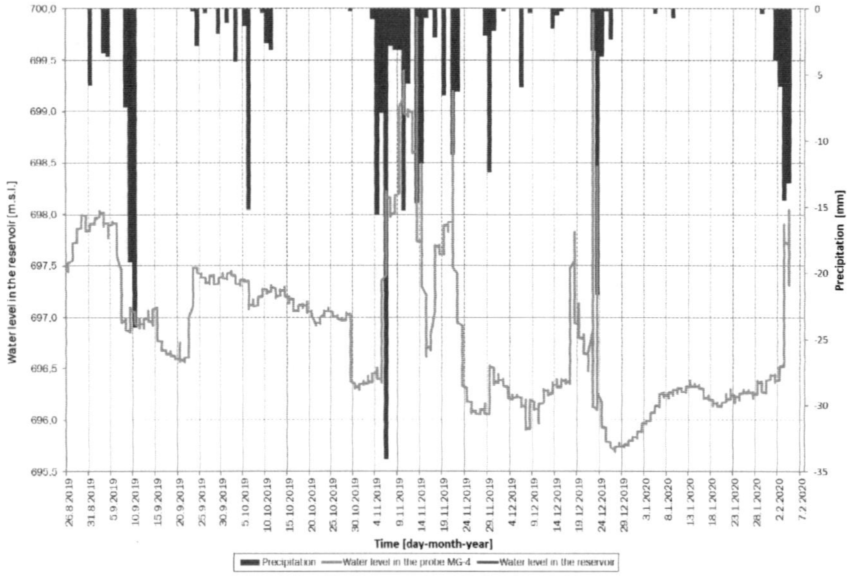

Figure 9. Hydrological and climatological data in observation probe PS-1 for the period 26.8.2019 – 4.2.2020.

The filtration movement may be due to the nature of the geological composition of the surroundings of the MG-4 well, as a result of which it can be assumed that changes in the level mode are due to short-term changes in pressure ratios. This means that if a certain value (pressure height) is exceeded, the water level in the borehole may suddenly decrease. These anomalies phenomena in MG-4 are associated, as with the PS-1 probe, with an increasing trend in level development. For the period 1995-2020, this increase is up to 8 m. It would therefore be appropriate to closely monitor and analyse the causes of these anomalies other than by analysing the level and development of precipitation. E.g., a pouring test is considered when measuring the filtration speeds at the same time or at the same time monitoring the movement of water in the borehole by the camera system.

2.2. Analysis of anomalies in observation probe PS-4

Graphical processing of the development of water level in the reservoir and in probe PS-4, as well as supporting calculations of correlation factories, shows that a medium correlation bond was recorded between the observed phenomena (in addition to the reconstruction work period) (Figure 10). The trend in the development of water levels in the probe is stable, with minimum levels around 692 m.s.l. in recent years marked by a drop in the level below this level (2018-2019). Now the probe is dry. Noteworthy, however, are the amplitudes of level fluctuations in this probe (12.83 m and 12.40 m, respectively), which are extremely high compared to the amplitudes of fluctuations in levels in other probes. The high amplitude of fluctuations in the period before 2018 is related to both the development of the water level in the reservoir and the development of precipitation activity (Figure 10). The sudden change in the level regime after 2018, or even its current decrease below the level of the probe's bottom, may be related to a decrease in the water level in the reservoir below 704 m.s.l., as well as a long-term decrease in rainfall totals in 2018-2019. It can be concluded from the above that either there are preferred leak roads in the dam or the probe is leaking through precipitation waters, and a combination of these effects cannot be excluded either. Such a presumption is signalled by the fact that as the water level in the reservoir decreases, the level in the probe decreases (Figure 10), which may indicate the occurrence of local preferred sees in the upper areas of the dam.

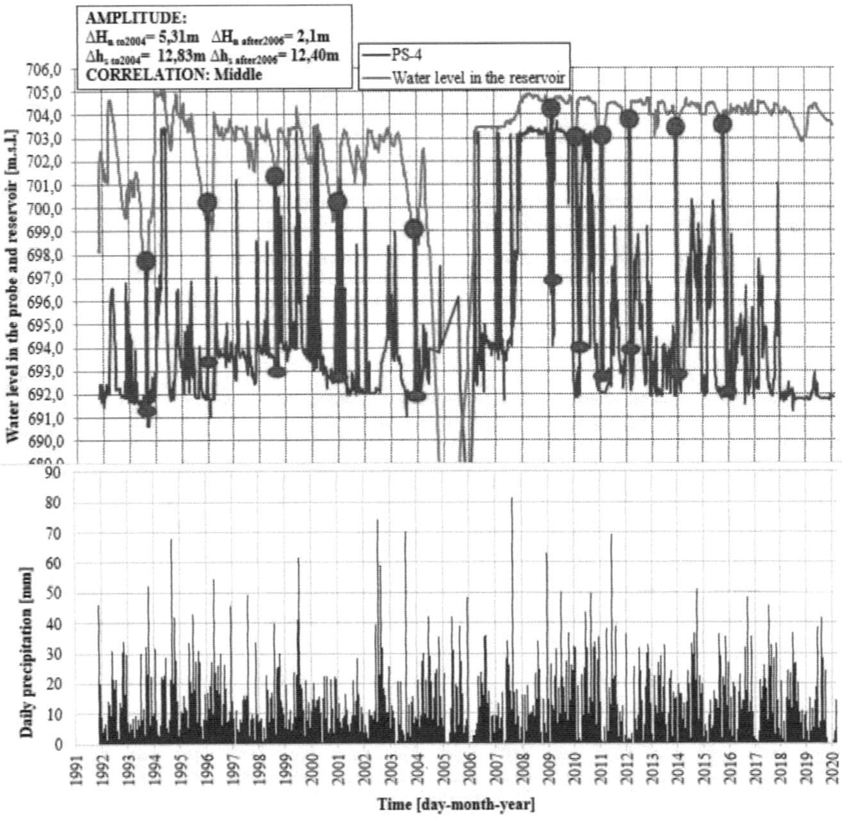

Figure 10. Situation of correlation bands in measuring probes.

3. Conclusion

The development of water levels in observation objects and the water level in the reservoir formed a substantial part of the evidence to investigate the causes of existing anomalies. These were supplemented by amplitudes of fluctuations in these levels. Their analysis was able to analyse in which observation objects, in addition to the water level in the reservoir, other influences are involved in the surface mode, e.g. leakage of their surroundings by precipitation water, or slope waters. The additional documents were information obtained by calculating correlation activities, expressing the link between the level in the reservoir and the level regime in individual probes, presenting the development of groundwater and leachate of the dam, it's in the sub-dam and in adjacent valley slopes. From summary information on the development of the level regimen, it is clear that anomalies occur for a long time in observation objects PS-1, PS-4 and MG-4. Their detailed analysis concludes that in addition to the water level in the reservoir, precipitation is a significant contributor to their development. This is evidenced by significantly higher amplitudes of their fluctuations, which are generally cumulated into periods of increased precipitation totals. Due to their situational arrangement (on the crown of the dam where the collector is located), it is considered that the surroundings of observation objects are leaked through precipitation waters. However, in the case of probes PS-1 and MG-4, the impact of slope waters cannot be excluded either. A negative phenomenon in the development of the level regime in probes PS-1 and MG-4 is their growing trend. In the case of the PS-4 probe, the occurrence of preferred see-through roads in the upper positions of the dam is also considered.

4. References

[1] Bednárová E and Lukáč M 2010 *Dam construction in Slovakia. Originals – milestones curiosities* (Bratislava: Kuskus) p 207

[2] Bednárová E, Gramblicková D and Bakaljarová M 2002 Use of geophysical methods in clarifying filter mode anomalies on dam Rozgrund *International seminar. Field geotechnical method 12.-13.9 (Ústí nad Labem)* vol 22 p 88-94

[3] Lukáč M, Bednárová E and Gramblicková D 2003 Historical importance of Banska Štiavnica reservoir in the context of the construction of reservoirs and dam in Slovakia and the world *Report BEVEX- mining research* vol 43 2-4 (Prievidza: BEVEX-Banský výskum) pp 136-40

[4] Vodohospodárska výstavby 2012 *Dam: ROZGRUND – Summary report for the period of dam existence* (Bratislava)

Acknowledgments

This work was supported by The Slovak Research and Development Agency under the contract No. APVV-15-0489 and also by The Slovak Scientific Grant Agency under the contract No.VEGA 1/0728/21.

Designing the sewerage network of the city in purpose to reduce impact to the environment and water bodies

M Portnov[1], A Popov[2], J Hrudka[3] and S Stanko[3]

[1] Department of Technological Processes, Apparatus and Technosphere Safety, Faculty of Technology, Tambov State Technical University, Sovetskaya 106, 392000 Tambov, Russia
[2] Department of Engineering and Technology of Nanoproduct Production, Faculty of Technology, Tambov State Technical University, Sovetskaya 106, 392000 Tambov, Russia
[3] Department of Sanitary and Environmental Engineering Slovak, Faculty of Civil Engineering, University of Technology in Bratislava, Radlinského 11, 810 05 Bratislava, Slovakia

E-mail: maksportnov1997@mail.ru, olimp_popov@mail.ru, stefan.stanko@stuba.sk, jaroslav.hrudka@stuba.sk.

Abstract. Today, the study of the city's sewerage network is an important part of the life of the city and the environment. Ensuring optimal living conditions not only for oneself, but also for other participants in the environment should be a priority for a person. Today we can face various problems that complicate the operation of the city sewer network and can cause detrimental effects on the environment. To prevent this impact on the environment, you first need to understand the cause of this impact, and then look for a solution. Wastewater is a pollutant of rivers, lakes and the water system in general. Wastewater is formed in connection with human activity, the appearance of rainwater, as well as under the influence of industrial enterprises. The purpose of the article is to develop the management of wastewater flows within the city sewer network to ensure the highest quality water clarification and minimize environmental harm.

1. Introduction

Sewerage is one of the types of engineering equipment and improvement of settlements, residential, public and industrial buildings that provide the necessary sanitary and hygienic conditions and a high level of convenience for work, life and recreation of the population. sewer network hydraulic manifold

Sewerage refers to a set of equipment, networks and structures designed for the organized reception and removal of contaminated wastewater through pipelines outside settlements or industrial enterprises, as well as for their purification and neutralization before disposal and discharge into a water body.

The sewerage network is one of the constituent parts of sewerage facilities. The cost of the network often accounts for 60-70% of the cost of the entire sewage system, and therefore special attention is paid to its calculation, design and installation.

When designing a sewer network, the following tasks are solved:

- sewerage basins are determined, the route of the network is planned,

- costs for calculated sections of the network are determined,
- hydraulic calculation and design of the network is carried out,
- longitudinal profiles are drawn up,
- constructions on the sewerage network are being designed.

The sewerage network operates in a gravity mode, but this cannot be achieved with deep collectors. In the lower sections of the network, pumping stations are arranged to raise water to higher elevations. The section of the network from the pumping station to the gravity collector or treatment facilities is called a pressure header. The outlet pipe is a pipeline for draining water into a body of water. The channel is a large diameter collector.

The design of the sewerage network begins with the division of the city's territory into sewerage basins, the choice of the location of the treatment facilities and the release of treated wastewater into the reservoir. The boundaries of the sewerage basins are drawn on the city plan with horizontals, they are determined by the terrain and the draft of the city's vertical planning. Determine the direction of movement of water and areas where pumping of water is required (if gravity is not possible). Then they begin to trace the external and internal sewerage.

2. Principles of tracing the sewer network

Tracing is a graphical display of the sewer network, main and main collectors on the plan of the sewage facility. This stage is of fundamental importance for the entire design. The labor intensity, cost and timing of the creation of the sewerage system depend on how the collector lines will be laid over the terrain.

When tracing the sewer network, the following principles should be followed:

1. Maximum provision of the service area with a gravity drainage regime.
2. Economic rationality of the depths of the pipes and collector channels.

Optimal division of the route into gravity sections with the location of sewage pumping stations (SPS) between them for pumping wastewater. This may be required if the ground conditions are unsatisfactory, the relief is heavily indented, with large differences in elevation and for other reasons.

The final element of the main collector in the schematic diagram of the sewage system of a populated area is a treatment plant with a place where treated effluents are discharged into a reservoir. You should also choose sites for the installation of sewage pumping stations.

Several options for tracing the sewer network should be offered. Of these, the best is selected in terms of technical and economic indicators.

The network routing begins with the two most important collectors, the main and the branch. The possibility of mechanization of construction and installation work should be taken into account to the maximum extent. All geometrical parameters of reservoir filling: depths, slopes, cross-sections of pipes and channels are determined according to the calculated flow rates. The sewerage network is tied to the location of treatment facilities and the release of treated water.

The tracing lines of the main collectors are mainly the embankments of streams and rivers or the thalweg of the valleys. If the terrain is flat, then it is preferable to route the collectors along the centerline of the sewage basin with the device of several treatment plants.

The sewer network will gain greater reliability if it is looped back with a connecting pipeline. At the same time, it will become easier to operate it. This solution must be facilitated by local conditions or specific requirements must be placed on the network.

It is advisable to lay collectors with a large cross-section in places with the least interference for construction and maintenance. These are areas of the territory with minimal underground structures and low development and load of transport lines. The crossing of water channels, railways, tunnels, heating mains and other structures underground by sewer networks increases the cost and complicates both the creation of the sewage system and its future operation.

If there are drinking water supply lines on the tracing line of the sewer network, it is necessary to provide for laying the collector below them at least 0.4 m.This is the minimum distance between the sewage pipes and the water pipes. If it is impossible to withstand it, as well as, if necessary, for technical reasons, laying sewer pipes above water conduits at a vertical distance of less than 40 cm, protective measures should be taken:

1. Use of steel pipes for water supply,
2. Use of cast iron for sewage;
3. Installation for a water pipe of a protective case (the minimum length of the case on each side of the intersection: in filter soils - 10 m, in clayey - 5 m).

Sewerage and water supply lines at the point of intersection must be strictly perpendicular to each other.

In residential areas, it is prohibited to make above-ground or above-ground laying of sewer pipelines. Open-type sewerage canals with gravity flow of treated wastewater can be installed only outside the boundaries of a populated area. The local conditions must always be taken into account.

It is also important to trace the sewerage system inside buildings in order to ensure high-quality and correct output of the internal network to the external network.

Figure 1. External network tracing.

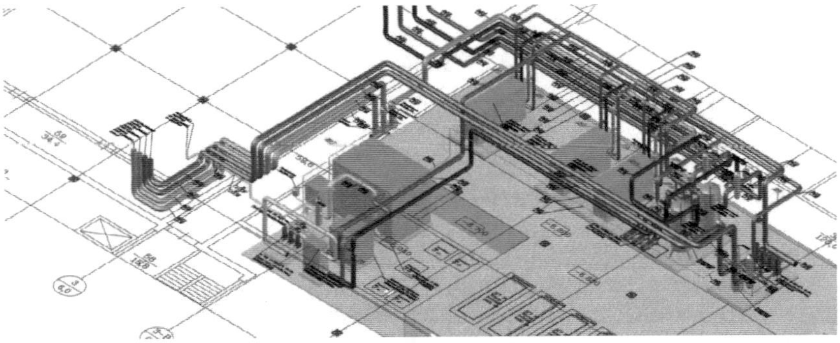

Figure 2. Internal network tracing.

3. Schemes of sewerage networks and laying of pipes

The scheme of sewer networks is the relative position in terms of the main elements of the external sewage system, the cost and quality of the network as a whole depends on their correct solution. The choice of the scheme is influenced by many factors, the decisive factor is the relief.

The types of circuits are as follows:

- with a smooth lowering of the relief - a crossed pattern;
- in the absence of reverse slopes - perpendicular;
- with a sharp slope of the terrain to the reservoir - parallel;
- in case of complex - radial;
- in case of uneven - zonal.

The perpendicular scheme is used only to drain clean atmospheric precipitation into the river. It is envisaged that sewer pool collectors are routed along the shortest distance to the reservoir, perpendicular to it.

The intersected scheme, the sewer pool collectors going to the reservoir are intercepted by the main one, which transports the wastewater to the pumping station, from where it is fed to the treatment plant via the pressure line. The main collector is laid parallel to the reservoir. This scheme is used for perpendicular reconstruction.

Parallel circuit. Basin collectors are laid parallel or with a slight slope in the direction of water movement in the river and are intercepted by the main collector, which diverts wastewater to the treatment plant.

The zone scheme (belt) provides for the presence of 2 main collectors in the upper and lower zones of the city. From the lower zone, the wastewater is pumped by the pumping station into the waste collector.

Radial scheme. The collectors have a radial direction from the center to its periphery. Each district has its own main sewer, sewage treatment plant and pumping station. This scheme is convenient when expanding cities; it does not require rebuilding existing collectors.

Methods for tracing street networks. Correct routing of street sewer networks is important.

There are types:

- encompassing - networks describe each quarter from 4 sides along the thoroughfares;
- on the lowered side of the quarter - the network only on the lowered sides of the quarter;
- over-the-quarter - is located within the block in a straight line, the length of the network is reduced, but operation becomes more difficult.

Depth of pipe laying. The depth of placement is influenced by the parameters:

- ensuring the reception of wastewater from the intra-quarter networks of buildings;
- protection against mechanical damage;
- freeze protection.

The minimum depth of laying is taken depending on the depth of freezing, taking into account that the temperature of the drains is 10-15 ° C and not less than 0.7 m, which keeps the pipe from mechanical damage:

$$h = H - (0.3\text{-}0.5) \geq 0.7 + D_{\text{pipes}} \tag{1}$$

The initial depth of laying is determined by the formula:

$$H_{initial} = h + i\,(L+L_1) - (Z_1\text{-}Z_2) + \Delta d, \tag{2}$$

where h - the smallest depth of the network pipes from the ground surface to the pipe chute in the most distant well of the intra-quarter network;

i - slope of the intra-quarter network;

$L+L_1$ - the length of the intra-block network from the most distant well to the place of its connection to the street network;

Z_1; Z_2 - marks of the ground surface at the most distant well and at the place of connection to the street network;

$\Delta\ d$ - the difference in the diameters of the pipelines of the street and intra-quarter network at the junction.

The maximum laying depth depends on the way they are laid. With an open method of work, the depth of the collectors is 10-15m – in dry soils; 5-7 m in wet soil. With closed methods, usually up to 20 m.

It is recommended to lay the network within the carriageway and in areas of green spaces. At the intersection with railways, near unique buildings, subway routes, duplicate lines with cameras for communication are provided.

It is rational to provide for laying pipelines for various purposes together. Laying is carried out parallel to the red building line or the axis of the road. When crossing sewer pipes with water pipes, the first ones are laid below, with a clear distance between the pipes of at least 0.4 m, or the water supply system is enclosed in a casing with a length of at least 5 m, in filter soils - 10 meters on both sides of the intersection.

4. Conclusion

Thanks to high-quality tracing of sewerage networks, we get a properly designed sewerage network. The network is capable of passing much more streaming wastewater and rainwater through itself. In this regard, more water will reach the sewage treatment plant through the sewer network, which means that more water can be purified and rendered harmless before the water enters the water sources. Thanks to this, we will reduce the risks of harm to the designer for the past environment and water bodies.

5. References

[1] Lukinykh A and Lukinykh N 2009 *Tables for hydraulic calculation of sewer networks and siphons according to the formula of academician N. Pavlovsky* (Russia) p 156

[2] Galperin E M 2014 On the procedure for determining the reliability of the functioning of objects of water supply and water disposal systems *Urban planning and architecture* (Russia: Bulletin of SGASU) pp 52-67

[3] Abramov L I and Manaenkova E A 2010 *Organization and planning of construction production. Construction organization management. Textbook for universities* (Moscow: Stroyizdat) p 400

[4] Zubareva G 2020 *Water supply and water disposal with the basics of hydraulics* (Perm: IPC Prokrost) p 106

[5] Galperin E M and Strelkov A K 2015 On the reliability of water supply and sanitation systems *Water supply and sanitary engineering* pp 39-46

Acknowledgments

This work was supported by the Scientific Grant Agency of the Ministry of Education, Youth and Sports of the Slovak Republic and the Slovak Academy of Sciences within the project VEGA 1/0727/20, co-funded by the Slovak Research and Development Agency under contract No. APVV-18-0203.

IOP Conf. Series: Materials Science and Engineering 1209 (2021) 012077 doi:10.1088/1757-899X/1209/1/012077

Restoration of a wetland Čéne in eastern Slovakia

A Repel[1]

[1] Technical university of Kosice, Faculty of Civil Engineering, Institute of Environmental Engineering, Vysokoskolska 4, 042 00 Kosice, Slovakia

E-mail: adam.repel@tuke.sk

Abstract. The aim of this paper is to present the possibility of revitalizing the disappearing wetland of Čéne in eastern Slovakia. Due to climate change and anthropogenic interventions in the natural landscape, natural habitats are currently extinct. Wetlands are one such habitat, of which the Čéne wetland is a very good example. Wetland Čéne is a place of occurrence of rare species of fauna and flora. The paper proposes and evaluates the method of revitalization of the wetland by subsidizing the water in the wetland from the river Tisa through the proposed dam culvert. The paper also evaluates the technical measures that need to be implemented and the economic aspects of this solution.

1. Introduction

Many new environmental problems are currently emerging, some of which result from anthropogenic activities and from changes in use and anthropogenic interventions in the landscape. One such problem is the lack of water, drought and rising average temperatures. In the territory of the Slovak Republic, the lack of water is manifested mainly in the lowlands, where large anthropogenic interventions took place in the 50s and 60s of the 20th century. The runoff conditions in the vast East Slovakian lowland were most significantly adjusted, especially in the years 1956 - 1968. Changes of land use and impact of anthropogenic activities to fauna and flora was presented in many studies [1-3].

Nowadays, the renewal of the water regime and the assessment of the possibilities of retaining water in the country as much as possible are becoming more and more important. There are several studies focused on the restoration of the water regime in the East Slovak lowlands [4].

Changes in runoff conditions are also significant in the East Slovakian lowlands, where an extensive network of hydromelioration structures was built in the 1960s, especially drainage and irrigation canals with several pumping stations, as well as flood protection water structures. Such interventions in the landscape have caused that the water that should be retained in the country and thus create suitable conditions not only for the life of many protected animals and plants, but also humans are quickly drained away from the entire territory of the East Slovak lowlands. As a result, there is a lack of water, a drop in the groundwater level and thus problems caused by the more frequent hydrological drought [5, 6, 7].

One of the many examples of the negative consequences of anthropogenic activity on the environment is the wetland Čéne on the Slovak-Ukrainian border, which was in the past a naturally meandering branch of the river Tisa. At present, it dries out and disappears, as it is not regularly supplied with water.

2. Study area

The paper is focused on the restoration of the Čéne wetland in eastern Slovakia in the area called Medzibodrožie. The location of the wetland Čéne is shown in the Figure 1. Medzibodrožie is an area bordered by the rivers Tisa, Bodrog and Latorica with a total area of 945 km^2. It is crossed by the Slovak-Hungarian border, which divides it into Horné Medzibodrožie with an area of 389 km^2 on the Slovak side and Dolné Medzibodrožie with an area of 556 km^2 on the Hungarian side. The flat landscape is predominant, but in Horné Medzibodrožie there are several volcanic hills, the highest of which is Tarbucka (277.5 m above sea level). The lowest part of the Slovak part of Medzibodrožie is located in places where the river Bodrog crosses the Slovak-Hungarian border near the town of Streda nad Bodrogom and the altitude in these places is 94.3 m above sea level. From a geographical point of view, the Medzibodrožie area is located in the East Slovakian lowlands. The average annual total precipitation at this station for the observed period is 557.7 mm, the maximum recorded annual total precipitation was 1029.8 mm and the minimum 379.2 mm, which proves significant differences in annual totals between individual years. The rainiest months are June and July, but also the months of May and August. In the summer months, the amount of precipitation is about half as high as in the winter months. Rainfall in summer is mainly caused by short-term intense showers arising from heat [8].

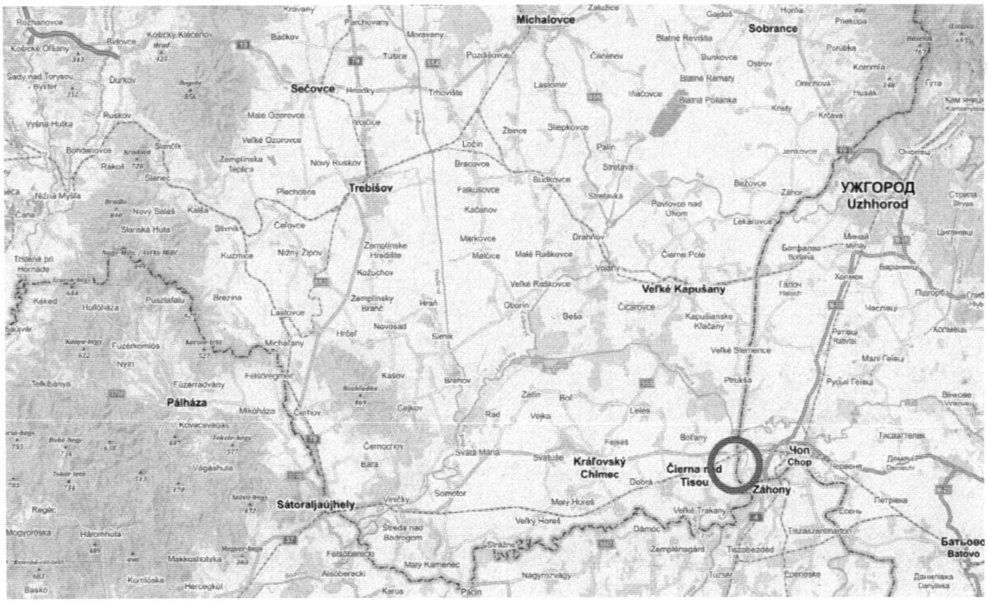

Figure 1. Location of Čéne wetland.

The water level in the Čéne wetland fluctuates significantly over time. Based on field observations, it can be said that the wetland has dried up almost completely in recent years. At present, the wetland serves as a natural habitat for water-loving fauna and flora and also as a fishing area. The wetland is divided into a northern part with an open water surface (surface) and a southern part, which has the character of a waterlogged area, overgrown mainly with reeds and shrubs.
Basic information about Čéne wetland:

- Maximum water level: 99.25 m above sea level.
- Terrain altitude in the southern, muddy part of the wetland: 99.80 m above sea level.
- Open water area: 12.4 ha
- Total wetland area in the Slovak Republic: 14.2 ha
- Terrain altitude around the wetland: 102.50 - 103.00 m above sea level.

The maximum altitude of the water level in the wetland was 99.25 m above sea level, the area of the open water area at the level at this level is 12.4 ha. Around the wetland there is mostly agricultural land. The surrounding terrain is at an altitude of 102.50 to 103.00 m above sea level. The southern part of the arm is covered with mud sediments, has the character of a wetland without an open water surface and is mostly overgrown with reeds and shrubs.

3. Restoration of wetland Čéne

This paper deals with the restoration of the wetland by providing water from the Tisza River during high flood conditions. This will be provided by a newly designed dam culvert in the left bank of the Tisza in the immediate vicinity of the border with Ukraine. The left bank dam is currently without any culvert. For this reason, it would be necessary to build a new pipe dam with a sluice, including a supply channel on both sides of the dam. For the purpose of subsidizing water from the Tisza to the Čené wetland, it was necessary to assess the hydrological regime of the Tisza River at the site of the proposed culvert. The culvert has to be designed with a bottom at such a height that it could be used at least once a year. Data obtained from the Tisza watercourse administrator were used for the hydrological assessment.

Nowadays, there is no dam culvert in the left bank dam of the Tisza near the wetland of Čené, which would be usable for the inflow of water from the Tisza into the wetland of Čené. The left-bank dam is 2.0 m wide in the crest and the elevation of the dam crest is at the level of 106.60 m above sea level. Dam height is 3.5 m. The elevation of the foot of the dam is on the water-side from the river Tisza at the elevation of 103.20 m above sea level and on the aerial side of the dam at a height of 102.40 m above sea level. The total width of the dam, including the water-side and aerial slope, is 30.0 m.

At a distance of approx. 90.0 m from the foot of the water-side, there is a abandoned meander in the inter-dam culvert of the Tisza, which is subsidized with water from the Tisza at higher water levels and is flooded all year round. The length of this meander to the main course of the river Tisza is approximately 1700 m. The abandoned meander is not directly connected to the main flow, the inlet of the meander into the flow is partialy dammed with soil. The lowest point between the meander and the main stream of the Tisza is at 101.23 m above sea level, which means that the meander is subsidized with water from the Tisza after reaching this level. There is no canal on the aerial side of the dam, between the Čéne wetland and the dam, the Čéne wetland near the dam is significantly grounded and overgrown with shrubs and trees. The distance from the left bank dam of Tisza from the permanently flooded part of the wetland Čéne is approx. 575 m.

Proposed solution, within alternative no. 2, consists of the construction of a new dam culvert with the sluice in the left bank dam of the Tisza and landscaping in the inter-dam area and also in the grounded parts of the wetland Čéne on the aerial side of the left bank. The proposed landscaping consists of deepening parts of the Čené wetland, deepening the dam connection of the Tisza arm in the inter-dam area to the main Tisza flow and also the construction of a supply channel to the proposed culvert, both on the aerial and on the water-side of the dam.

The most important object, which is necessary for the proper functioning of water supply to the Čené wetland under alternative 2, is the construction of a dam culvert in the left bank of the Tisza dam. The site of the proposed dam is in the immediate vicinity of the border with Ukraine. The dam is grassed, without any objects and also without damages in this location.

A pipe dam culvert is designed from the DN800 pipeline, which will provide sufficient capacity to infuse water into the Čené wetland. The pipe culvert could be realized by pushing the pipe and thus there would be no interference with the dam body itself.

The pipe culvert is designed with a total length of 30.0 m. The level of the bottom of the pipe culvert is designed at a height of 100.00 m above sea level. Concrete faces of the culvert are designed at the beginning and end of the pipe culvert. A sluice gate would be installed on the water-side of the dam, which, if opened, would be used to fill the water into the wetland of Čené. The depth of the bottom of the pipeline from the terrain under the crest of the dam is 6.60 m.

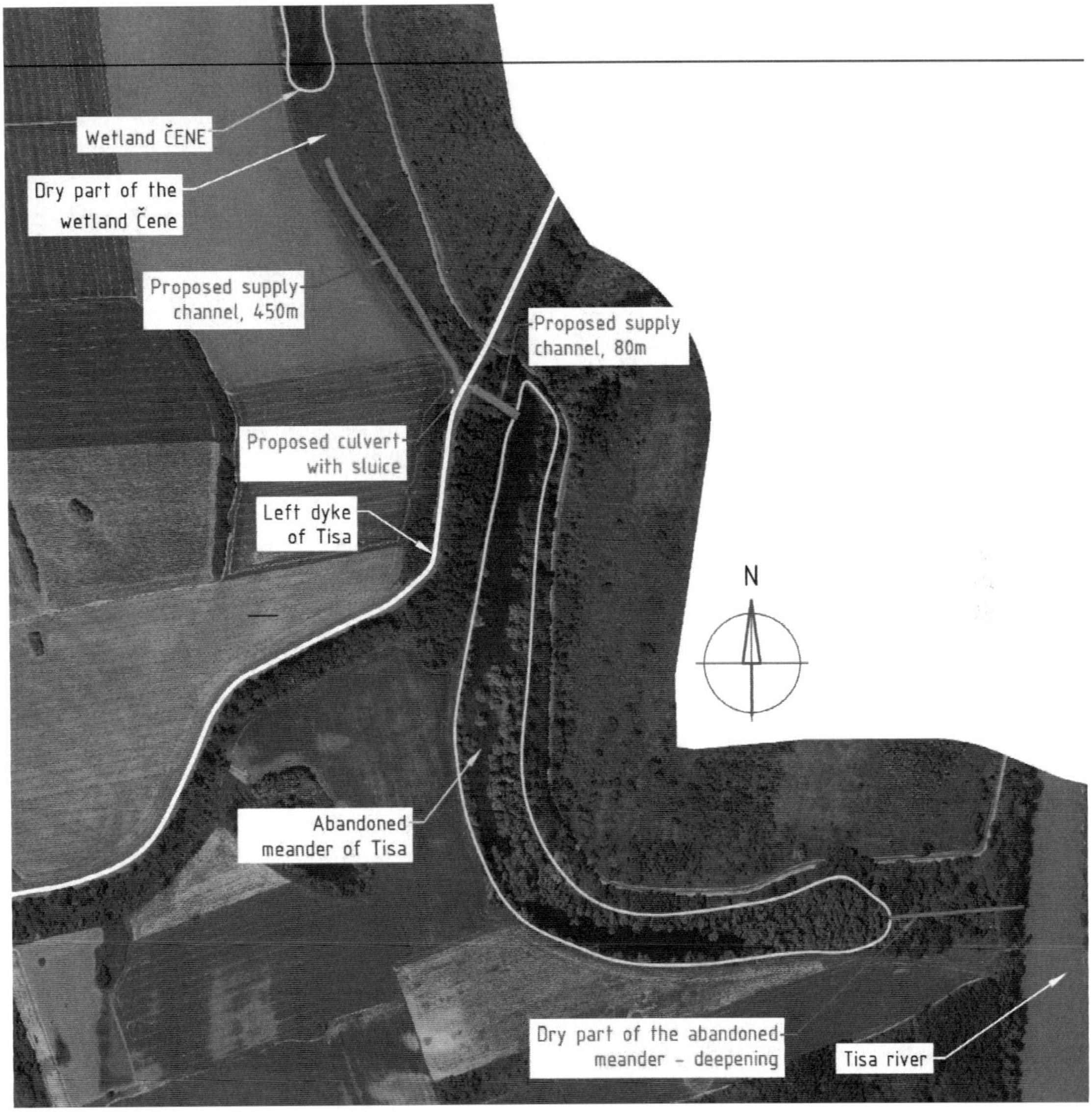

Figure 2. Proposal of wetland restoration.

The bottom of the dam culvert is designed at a height of 100 m above sea level. It follows from the above that the minimum water level in the Tisza River required for the inflow of water into the wetland of Čené is 100 m above sea level. Historical observations and data available from the watercourse administrator have shown that such a level is reached in the Tisza River for at least two months of the year, which is sufficient for the flooding of the Čené wetland. The estimated cost of the construction is 172,000€.

4. Conclusion

The presented paper deals with the proposal for the restoration of the disappearing wetland of Čéne in Eastern Slovakia. This wetland is a remnant of the former Tisa riverbed, but it has been endangered by anthropogenic interventions, especially by the construction of flood protection structures and changes in agriculture. Wetland Čéne is a place of occurrence of rare species of fauna and flora. This paper proposes a way to restore the wetland by building a dam and a supply channel to the wetland. The investment cost of the construction was set at approximately € 172,000. With such a measure, it is possible to restore natural habitats and preserve the function of the wetland.

5. References

[1] Vavrovič J, Čorba J and Šiška, B 2013 Impacts of climate change on plant production in Eastern Slovak Lowland *Environmental changes and adaptation strategies. Conf. proceedings Abstracts (Slovakia)*

[2] Toth S 1998 The trend of selected climatic characteristics in the East Slovak Lowland *Rostlinna Vyroba-UZPI (Czech Republic)*

[3] Pálinkášová Z and Šoltész A 2012 Hydrologic and hydraulic evaluation of drainage system in eastern Slovak Lowland *Pollack Periodica* ***7*** 3 91-8

[4] Repel A and Zeleňáková M 2021 Proposal of technical solutions for water regime renewal in the eastern Slovakian lowland *Resumos do 15º Congresso da Agua APRH (Lisbon)* ISBN 978-989-8509-27-7

[5] Kubiak-Wojcicka K, Nagy P, Zeleňáková M, Hlavatá H and Abd-Elhamid H F 2021 Identification of Extreme Weather Events Using Meteorological and Hydrological Indicators in the Laborec River Catchment *Water* **10** 1413

[6] Zelenakova M, Fijko R, Findoráková L, Gaňová L 2016 Flood occurrence evaluation based on the heights of water level in the rivers *3rd International Conference Water Resources and Wetlands 8-10 September Tulcea (Romania)* pp 153-57

[7] Repel A, Zeleňáková M, Jothiprakash V, Hlavatá H, Blišťan P, Gargar I and Purcz P 2021 Long-Term Analysis of Precipitation in Slovakia *Water* ***13*** 7 952

[8] Terek J and Dobos E 2008 *Život Medzi Riekami—Monografia Krajinného Manažmentu Medzibodrožia* (University of Miskolc: Miskolc, Hungary) ISBN 978 9630642651

Acknowledgments

This work was supported by the Slovak Research and Development Agency under the Contract no. APVV-20-0281. This work was supported by project of the Ministry of Education of the Slovak Republic VEGA 1/0308/20 Mitigation of hydrological hazards, floods and droughts by exploring extreme hydroclimatic phenomena in river basins.

Effect of percent air voids content on the deformation properties of the asphalt mixture

J Bokomlasko[1] and J Mandula[1]

[1] Technical University of Košice, Faculty of Civil Engineering, Vysokoškolská 4

E-mail: jakub.bokomlasko@tuke.sk

Abstract. Asphalt mixture is a building material with many advantages. Therefore, it is most used in road construction. If the asphalt mixture is laid with the prescribed technology, it can withstand load effects to long-term. It is necessary to take samples that will be subjected to laboratory measurements. There are several laboratory test, for example measurement thickness of the asphalt mixture layers, the aggregate fraction, quantity of binder in the mixture, determination of air void in asphalt mixture layers. Samples taken directly from the construction site are subjected to laboratory tests. This article focuses on one of the laboratory tests and it is determination of air void in asphalt mixture layers. The determination of air void in asphalt mixture layers is test in detail, because this effect has influence on the deformation properties of asphalt mixture layers. Therefore, it was necessary to model of air void in asphalt mixture layers with different degrees air void. On this purpose was use program Abaqus. The results were plotted. This graphs showed that increasing the air void in asphalt mixture layers has effect on the expansion of deformations. This can lead to faster pavement degradation.

1. Introduction

At present, the dominant material in Slovakia in the construction of roads is an asphalt mixture. This is justified, because asphalt mixture has many advantages. These include the speed of laying, laying without joints, ease of laying and the like. Slovakia has a wide network of roads of various classes of traffic load, which are divided into six categories. Where the first class of traffic load represents the most important roads (motorways, expressways) and the sixth class of traffic load represents the roads of the lower importance (roads III., parking and traffic areas). These traffic load classes require its attention, either in the actual implementation or in its maintenance. One of the prerequisites for the long life of the road is the quality laying of layers of the road and adherence to the technological process. Therefore, it is necessary to control the individual layers, even during their laying. There are several laboratory tests to verify the quality of laying asphalt mixture layers. These include, for example, the grain size test, the thickness of the layers, the amount of binder, air void content and so on.

Specifically, this article focuses on the air void content. In terms of significance, it can be said that it has the greatest informative value for the quality of laying [11, 12]. When, on the basis of an unacceptable air void content, the negative factors that have affected the quality of the laying can be precisely defined. Also the grain size test, the amount of asphalt binder in the mixture is certified by the manufacturer of the mixture and so on. Air void content may indicate non-compliance with the technological procedure for laying asphalt mixture layers. Factors include for example, low temperature of the asphalt mixture during laying, poor condition of the paver finisher, lack of compaction rollers, compaction of the cold asphalt mixture, thickness of the laid layer and so on.

2. Test samples

The test samples consisted of boreholes taken from different consumption points. The samples depended on whether it was wearing course or binder course. There were a total of ten consumption points. Three samples were taken from nine consumption points and two samples were taken from one consumption points. Samples were taken of the road from the left edge of the road, from the middle of the road and from the right edge of the road. The test samples different consumption points and therefore it was necessary to determine which asphalt mixture it was. Based on the extraction of the samples taken, a sieve analysis was performed and it was subsequently possible to classify the individual asphalt mixtures sample. From the result of the sieve analysis, three types of asphalt mixtures were finally determined, namely AC 8, AC 11 and AC16, which had the same aggregate size line. The resulting relative proportion of the individual asphalt mixtures can be seen in figure 1.

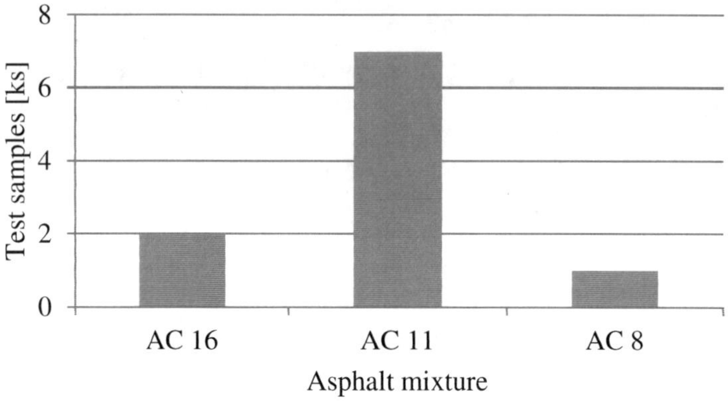

Figure 1. Representation of individual types of asphalt mixtures.

The thickness of the samples taken was variable (figure 2). It depended on the consumption points and the thickness of the laying of the asphalt mixture layers in the given place. The only dimension that was constant was the diameter of the samples taken, namely 100 mm. This was because the same drilling rig was used for sampling, which used a 100 mm diameter drill bit.

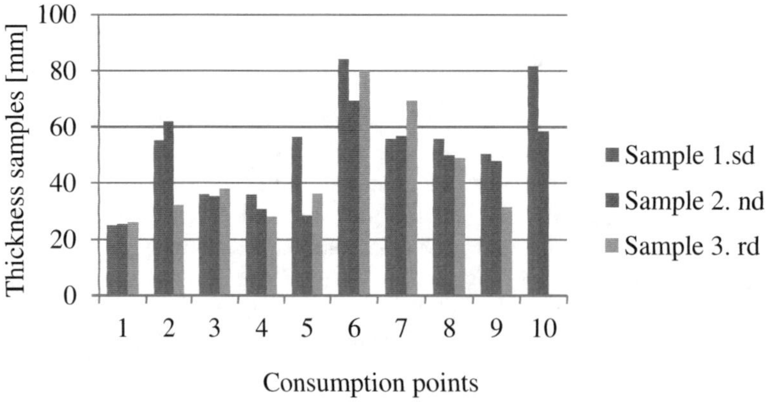

Figure 2. Representation of individual types of asphalt mixtures.

3. Air voids content

Air voids content is one of the verification criteria for determining the quality of the laying of the asphalt mixture. This verification criteria has a great influence on the resulting life of the pavement [11]. Too high air voids content has an adverse effect on the pavement. With high air voids content Longitudinal as well as transverse inequalities occur. Water also enters the air voids content, and in the winter, due to frost, the water expands and thus disrupts the layer. Especially in summer, this effect is greater because the asphalt mixture has a lower modulus of stiffness and the viscoelastic behavior of the asphalt mixture is beginning to prevail [6]. With a high air voids content in the summer, compaction occurs faster and affects this phenomenon the formation of longitudinal and transverse inequalities. Conversely, if the asphalt mixture is laid in the colder period, it becomes stiff and the stiffness modulus is higher and the elastic behavior of the asphalt mixture is more prevalent [6]. As a result, the mixture becomes brittle and more sensitive to the effects of deformation, when cracks or protrusions can occur due to deformation.

Therefore, the test samples taken were subjected to a air voids content test. The test consists of determining the bulk density by immersion in water and testing the maximum bulk density. Based on these tests, the percentage air voids content was determined using the formula given in STN EN 12697-8 [4]. The resulting air voids content values from the samples taken are shown in figure 3.

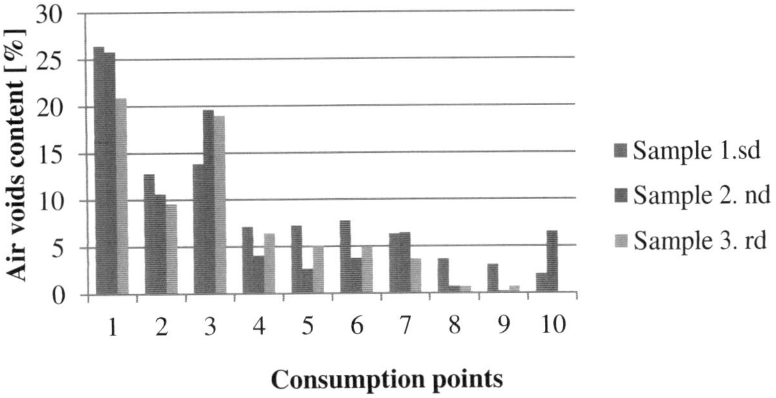

Figure 3. Determination of air voids content in percent.

The x-axis shows the consumption points (one to ten) and also shows the number of samples taken from the consumption points. The y-axis shows the percentage of air voids content. From these results, a high air voids content was found in some samples, where the average maximum allowed value for unmodified asphalt mixtures is 5% [1].

This leads to the idea of how the air voids content affects the resulting deformation properties of a asphalt mixture layer. For this reason, the pavement structure was modelled in the Abaqus software.

4. Modelling in the program Abaqus

The pavement structure was modelled in the ABAQUS program. It is an analytical model of the pavement structure. The material characteristics were taken from TP 033 due to the fact that these values are also considered in the current design methodology on Slovakia [2]. Values for spring and autumn, were taken to the modelling because this temperature is considered the average annual temperature in Slovakia. Average temperature on Slovakia is 11 °C. Modelling was done in 2D space to simplify calculations and computer requirements [5, 7, 8]. The load area was considered as square and was determined on the basis of TP 098 [3]. The load value was represented by a design axle weighing 100 kN, which represents a contact pressure of 0.6 MPa. The air voids content was modelled as a void in the asphalt mixture. Air voids content was considered in the wearing course. Different percentages of air voids content were considered and that 0, 5, 10, 15, 20, 25, 30 and 50%. The reference value was taken at 0% air voids content. Vertical deformation and vertical strain were observed at individual percent of air voids content.

5. Evaluation of a numerical experiment

The evaluation of a numerical experiment consisted of plotting the vertical deformation and the vertical strain in graphs. A regression curve and a correlation coefficient were also plotted, which indicates the relationship between the plotted values.

Figure 4. plots the relationship between air voids content and vertical deformation. On the x-axis the percent of air voids content are plotted from 0 to 50% and on the y-axis the vertical deformation in mm.

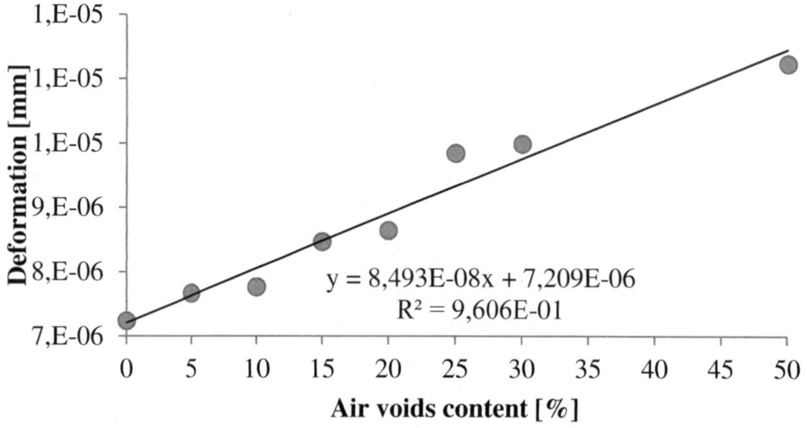

Figure 4. Percentage increase in vertical deformation.

As can be seen in figure 4, the increase in vertical deformation with increasing air voids content has a linear course. The correlation coefficient R reached the value of 0.972, which indicates a high dependence between the plotted values.

Figure 5. plots the relationship between air voids content and vertical strain. Where the air voids content was plotted on the x-axis and the vertical strain was plotted on the y-axis. Also in this case, a regression curve and a correlation coefficient were plotted.

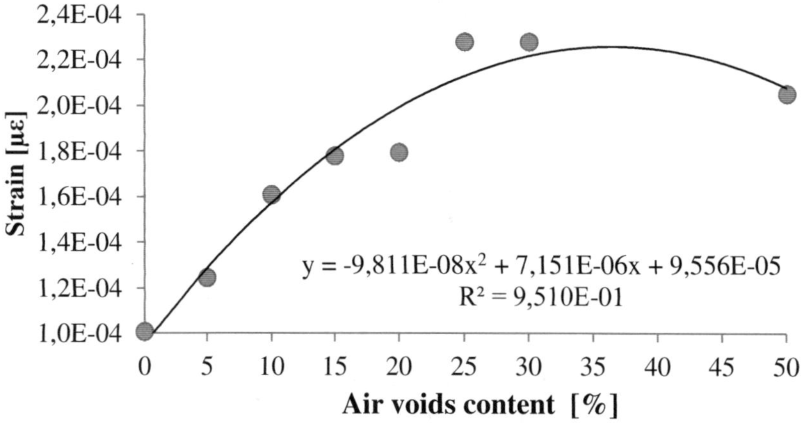

Figure 5. Percentage increase in vertical strain.

The largest increase is right at the beginning of the air voids content, when the peak was recorded at the level of air voids content of 25 to 30%. After reaching the peak, the strain has a decreasing tendency. The correlation coefficient R reached the value of 0.975, which indicates a high dependence between the plotted values.

6. Discussion

From the obtained results it can be shown that the air voids content has an effect on increasing the deformation. Figure 4 shows that the increase in vertical deformation is evident, which is reflected in the occurrence of transverse and longitudinal deformations in the pavement. With an average air voids content of 5% allowed by the standard as well as KLAZ 01/2019, the percentage increase is small and amounts to only a 6% increase in deformation compared to the reference value. Also, at 10% air voids content, the percentage increase in vertical deformation is only 7%. But, at 26.4% of the air voids content, which was also measured for one sample from the consumption point marked number one, the percentage increase in deformation is 36% compared to the reference value. Such an increase will be reflected in the emergence of greater pavement inequalities. Over time, it will compaction and the air voids content will decrease, but irreversible damage will occur which will lead to:

- the creation of longitudinal deformation which have an impact on the quality and safety driving,
- the thickness of the layer at a given point is reduced and this has an effect on the load distribution to the lower layers,
- if the asphalt mixture is laid in the winter and unfavourable air voids content is formed, this can result in the formation of cracks and these lead to the formation of hole.

At vertical strain (figure 5), a clear increase in values was noted at the beginning. In this case, it has the shape of a curve, which indicates the fact that at a certain degree of air voids content, a critical limit is reached. At 5% air voids content, the increase in vertical strain is only 20%. But, at a air voids content of 26.4% (the value obtained for the sample from the consumption point one), the percentage increase is about 100%. With a vertical strain, a boundary air voids content was detected at the level of 25 to 30% (figure 5.). After exceeding the value of 30%, the vertical strain begins to decrease. This phenomenon was to be expected. The decreasing vertical strain also points to the fact that the asphalt mixture begins to transmit less stress with increasing deformation. After exceeding this limit, the vertical strain begins to decrease. This leads to the conclusion that the given asphalt layer ceases to transfer the load and the load begins to carry the layer beneath it. This results in earlier wear of the layer and the pavement as a whole.

7. Conclusion

From the obtained results it can be stated that the air voids content has a high influence on the final life of the layers of asphalt bonded materials. This effect will later be reflected in the greater load of the lower layers in the pavement structure. Therefore, it is necessary to monitor the quality of laying asphalt bonded pavement layers. This will translate into longer pavement life, as it will not be necessary to change the wearing course or other road layers for a long time. Longer service life also leads to reduced consumption of aggregates and asphalt binder. It also has an impact on the environment, as a low-repair road produces less emissions, either due to smoother traffic or a reduction in repair intensity. Also, a good pavement surface leads to lower fuel consumption.

8. References

[1] KLAZ 1/2019 *Catalogue list of asphalt mixtures (supplement to valid TKP)*
[2] TP 033 *8/2009 Technical condition. Design of non-rigid and semi-rigid pavements, MDPT SR.*
[3] TP 098 *2/2015 Technical conditions. Design of cement-concrete pavement on roads*
[4] STN EN 12697-8 8/2020 *Bituminous mixtures. Test methods. Part 8: Determination of void characteristics of bituminous specimens*
[5] Králik J 2009 *Modelovanie konštrukcií v metóde konečných prvkov* STU Bratislava Stavebná fakulta, Bratislava
[6] Komba J Maina J W Anochie-Boateng J K and O´connell J 2012 Abstracts of the 31st Southern African Transport Conference *Analytical modelling of visco-elastic behaviour of hot-mix asphalt* (SATC 2012)
[7] Kudrna J 2010-2012 *Podkladní vrstvy a navrhování vozovek* TÉMA II – Realizace staveb pozemních komunikací (2. SEMINAŘ) Studijní materiál pro seminář
[8] Šoltýs R 2013 *Nelineárna analýza vynúteného kmitania lanových sústav* / Róbert Šoltýs - 1. vyd. - Košice: Technická univerzita v Košiciach p 237 [print]
[9] Zexin M Liping L and Lijun S 2018 Construction and Building Materials, *Investigation of top-down cracking performance of in-situ asphalt mixtures based on accelerated pavement testing and laboratory tests*, Volume 179, Pages pp 277-284
[10] Russo F Salvatore A B Formisano A and Dell'acqua G 2018 Construction and Building Materials *Predicting percent air voids content in compacted bituminous hot*

mixture specimens by varying the energy laboratory compaction and the bulk density assessment method Volume 164 pp 508-524
[11] Qingguo Y and Jincheng N 2011 Energy Procedia 5 *The environmental influence of asphalt pavement and Countermeasures*, 2432–2436, IACEED2010, 1876–6102 © 2011 Published by Elsevier Ltd.

Acknowledgments
This work was supported by the Slovak Research and Development Agency under the contract No. APVV-15-0777.
This work was also supported by the Scientific Grant Agency of the Ministry of Education of Slovak Republic and the Slovak Academy of Sciences under Project VEGA 1/0796/21.

Young Scientist 2021 IOP Publishing
IOP Conf. Series: Materials Science and Engineering 1209 (2021) 012079 doi:10.1088/1757-899X/1209/1/012079

Analysis of available shared transport solutions in Krakow

A Grabarz[1] and W Sawicki[1]

[1] Doctoral School of Cracow University of Technology, Civil Engineering and Transport, 24 Warszawska Street, 31-155 Cracow, Poland

E-mail: w.sawicki@hotmail.com

Abstract. Sharing modes of transport in Kraków: in the context of the transport policy of large cities, it is worth noting sharing as an increasingly common method of transport. The policy of sharing: cars, scooters, bicycles and motor scooter has become an important way to complement public transport. The aim of this paper is to analyse the available shared transport modes used by Krakow residents and their impact on the environment, over the last few years. The increase in popularity of available shared transport solutions is due, among other things, to their availability and relatively low cost of use. This solution is a response to the needs of modern cities and has a positive impact on the environment.

1. Introduction

The development of transport systems has a very large impact on the urban planning of cities and its constant growth causes major changes in the transport policy of cities. Analysing the impact of shared transport is extremely important in terms of urban planning and transport policy, which is also really important for inhabitants as it directly affects their comfort of living. Often, decisions made in life take into account transport possibilities in cities, and also have an impact on the choice of workplace or place of residence. Proper planning and regulation of transport have a huge impact on the life of the city and the traffic jams in it. Providing an appropriate transport infrastructure, parking lots, a coherent network of connections, and well-functioning transfer gives the city - character.

For many people, the advent of shared means of transport has radically changed the way how people move around the city. The possibility of supplementing the already functioning communication system of the city of Krakow perfectly matched the decision-making process of passengers as to the choice of means of transport. There are motor scooter, scooters, bicycles and also shared cars. Each of the above-mentioned offers the possibility of transport in Krakow, and also gives the possibility of free parking for the end user. The parking policy is on side of company providing each transport system. In addition, this transport solution is characterised by its availability and optimal usage costs. These connections support the removal of white transport spots, where it is often necessary to drive from point A to point B on a long, inefficient road. Shared transport also allows us to take shortcuts. Maintaining such a growing trend in the popularity of this mode of transport means that it has a major impact on the environment. It is not only bicycles, motor scooters and scooters that are shared by users that are good for the environment. It is worth noting that shared car fleets are new cars that meet environmental requirements at the highest level. Because of the economics and use of this type of transport, companies can afford to use the latest available vehicles and can therefore emit less harmful substances into the environment.

2. Sharing means of transport

In describing shared transport it is necessary to explain the definition. Sharing means of transport - it is a transport system that allows the user to share the vehicle with other people. This is done on a private rental basis, which allows you to share the cost of the trip, and at the same time fills the gap in the transport system between the private car and public transport. Shared mobility is a strategy that allows you to use the transport system when you need it. It could be said that shared means of transport work in a similar way to taxis - they are a solution that you get exactly when you need it, in addition to giving you the opportunity to move by yourself from point A to point B in the city. Of course, transport systems have additional advantages, for example some of them do not require the age of majority or a driving licence. Sharing also aims to economise on travel costs and eliminate the downtime associated with transport by individual means. The aim, or actually the effect of shared transport, is also to reduce harmful emissions from the transport sector through more efficient use of vehicles on the road. The progress of civilization indicating an increasing sharing policy is aimed at optimizing costs, using unused resources, while increasing the comfort of travelers. Current solutions allowing different people to use the same means of transport are extremely innovative. Such solutions require IT systems and access to the Internet, which enables people to choose and hire the means of transport themselves. This approach has brought a new meaning of transport to cities.

The most popular means of shared transport in Poland is the bicycle. Such a system is called bikesharing. In this case, two solutions are available: a standard bike or an electric bike. Bikesharing is operated by the so-called self-service rental station, i.e. a stand from which bicycles are collected or returned. The system is operated remotely by a mobile application [1].

Scooters are also available for shared use. This is a new but very popular means of transport for getting around the city on a daily basis. Like bicycles, scooters are rented using an app. Operators who provide equipment in cities most often offer electric scooters so-called e-scooters. However, there is a difference when it comes to where the scooter is taken, as they are available wherever someone leaves them. Companies that provide scooters inform users that once the journey is over, they should leave the scooter in any but a safe place so as not to endanger other road users [1].

The car is a very popular means of transport, but its use involves costs: purchase, maintenance and exploitation. The introduction of shared cars is a solution in which people pay only for the use, i.e. the period during which they drive. Carsharing is a system where people rent a car by themselves for a selected period of time using an application on their phone [1].

Riding a scooter is one way of getting around town. Using shared scooters is an optimal solution for the user, who will be able to rent one through the app for the duration of the journey.

The availability of different means of shared transport allows the potential user to make a choice tailored to their needs. The use of each of them is similar because it uses an application on the phone. The advantages are also similar: flexible time and optimal rental costs [1].

3. Description of shared transport

Implementing shared mobility solutions is attractive because of independent and accessibility and effectiveness. The benefits of shared-transport services require the creation of appropriate market conditions and operational frameworks. To be effective, solutions of this type must be implemented on a large scale in the entire area, not just in part of it. The appropriate scale of implementation of this type of solutions is also very important to be able to properly manage the costs of the entire project. The challenge faced by cities is very big, because it is not easy to predict the exact behavior of customers, because the transit does not take place along fixed routes, but almost in the door-2-door concept. The possibility for some means of transport to enter restricted traffic zones is also an interesting advantage. In order to guarantee the success of this type of service, it is necessary to ensure its availability: both territorial and temporal. Keep in mind that it is very easy to discourage the consumer if the service is not available in their area or has time availability restrictions (for example, not available late at night). The feature without which this type of service is doomed to failure is also the reliability of the means of transport. Solutions of this type must be 100% reliable and provide the passenger with a complete and comfortable solution and safe while traveling. The availability of the service system in the form of an app is also an interesting and easy experience for the passenger. Ordering or booking a ride must be easy, pleasant and at the same time smooth and hassle-free. Appropriate promotion and range of services are crucial for building user awareness. For the system administrator and creator, the ability to adapt in almost real time to new circumstances is required, with particular emphasis on the quality of passenger service. Equally crucial is the appropriate combination of micro-transport with the already existing alternative transport models. The possibility to get to interchanges, railway stations, metro stations should be much more convenient for the passenger than reaching the same destination by own means of transport or on foot. Creating an appropriate system that will enable efficient and easy planning as well as connecting communication in cities is equally important. The size and availability of shared transport services must be large enough to be both convenient and profitable for the user. The introduction of a clear fee system will also contribute to social acceptance for the new solution. If the system is consistent with other means of communication - the willingness to use such a solution will be even greater. Consumers are becoming more and more demanding, expecting comprehensive, personalized and convenient solutions. Shared transport is becoming the answer to the current transport needs of travelers by offering tailor-made solutions to the consumer. It is, inter alia, the lack of the need to travel "to the stop" and "from the stop" is to translate into an increase in the share of shared transport in large cities.

Shared mobility is a solution that affects the area in which it is used. In this case, the urban environment and its inhabitants should be seen in this way. The basic idea is to see that a defined group of users share the same vehicles in the same area. In addition, it should be pointed out that the same demand for mobility in a city is met. Furthermore, space in car parks, for example, is also used more optimally. These benefits result from the fact that the user only uses the vehicle when he needs it. After the journey, users can leave the vehicle and another person can use it. Moreover, traffic on the roads and CO2 emissions are reduced. The positive environmental aspects are one of the main objectives of this system in cities.

It is also worth adding that the policy of sharing transport also touches upon the aspect of the sharing economy, which is a social phenomenon that uses a large change in organizational models aimed at combining dispersed networks into individuals and communities co-creating, co-using and co-purchasing. These trends make it possible to increase the efficiency of the use of resources. Many of sharing companies were founded between 2008 and 2010 after global financial crisis [2].

4. The popularity of shared transport means

According to the report "Transport-Sharing in Poland" published at the end of 2018, there were as many as 21 companies providing sharing solutions in Poland. There were 11 900 bikes and 3 800 cars. As many as 86% of providers said they wanted to expand their current fleet, and 64% planned to expand to other cities in Poland. At that time, there were 0,3 vehicles per 100 inhabitants in Cracow and 6,1 vehicles per 1 km2. These numbers show the availability of transport vehicles in Kraków in 2018.

When it comes to the transport policy of large cities, including Krakow, it is clear here that modal sharing policy has a growing and extremely promising trend. There are speculations estimating the growth of car-sharing users worldwide from 7 million people and 112 000 cars to 36 million users and 427 000 cars by 2025. It can therefore be predicted that we will move away from the habit of owning one or more cars, replacing the ownership model with a sharing model. When it comes to Krakow and the data from the "MOBILE CITY" association, about 467 shared cars are mentioned in Krakow. According to data from the Central Statistical Office, Kraków is a city with a population of 780,000, of which about 60 percent are of working age. We can therefore conclude that the persons interested and authorized to use are about 468 thousand people. Clearly, it is not possible to calculate the exact number of potential customers for sharing platforms, as there is a problem with accurately estimating the number of people staying in Kraków (we are talking about students, tourists). However, if numbers are presented, it is possible to speak (disregarding the pandemic) about 14 million tourists who will potentially use shared means of transport and about 145 000 students. Krakow can be called the city of millions [3].

It is also worth adding that the number of companies dealing with sharing transport is constantly growing. In addition, it is worth remembering that the service is available 24/7. This is again a big advantage over other public means of transport.

When it comes to the bicycle system, WAWELO in Krakow had 170 stations and 1500 bicycles. According to the calculations of the "MOBILE CITY" association, each bike was rented on average 2.2 times a day, and about 70 000 users were registered in the entire system. The supplier terminated the contract with the city of Krakow, citing unprofitability. The city authorities are planning to introduce about 1 000 bicycles and 100 stations to Krakow, forecasting that the bicycles will be used to cover distances of 10-15 km. The bikes are supposed to be electric this time. An important disadvantage for the bikes which could reduce their attractiveness for users was the requirement to return the bikes to the rental station. This is a property that scooters do not have, the number of which seems to be growing incredibly on the streets of Krakow. In 2019, there were about 137 of them, in 2020 the Lime company itself has already delivered 400 of them. It is also said that the Blinkee company introduced another 500 units in Krakow and Warsaw, so we can estimate, taking into account several suppliers - that the number of scooters in Krakow is certainly greater than 1000 units. By observing the streets, we can see that the scooters have flooded the city, making it possible to travel from point A to point B without having to return it to the designated place. Scooters are everywhere, which is obviously a big problem for the authorities and residents, because they can be found on every corner, on every street, on the sidewalks. You can see that it is a large shared means of transport and cities do not know yet how to treat scooters, which have appeared in large agglomerations with a huge hit [4].

4.1. *Shared transport use in the City of Krakow*

In order to identify basic information about shared transport in Krakow, a survey research was conducted. The survey was prepared to be completed online. A total of 150 randomly selected people residing in Kraków joined the research. The first stage was to check the age of the respondents. Most people answering the questionnaire were between 19 and 26 years old, they constituted 65% of all respondents. The remaining results are presented in figure 1. In the next question respondents were asked if they had ever used shared means of transport such as bikes, motor scooters, scooters or cars. The majority, 84%, confirmed that they had used such means of transport. Only 16% said they had not used such a transport solution. The results are shown in figure 2.

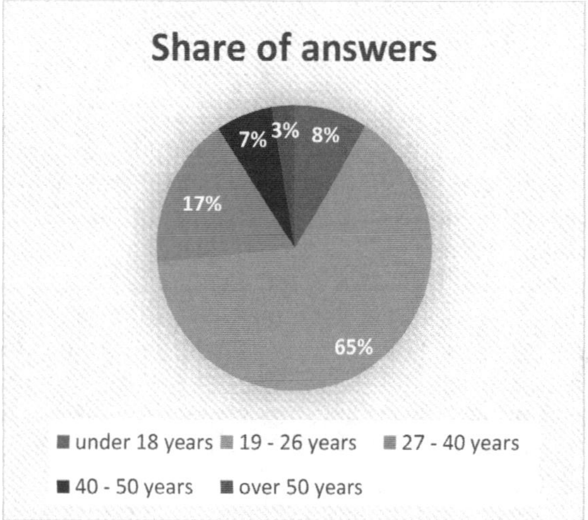

Figure 1. Age of respondents [own study].

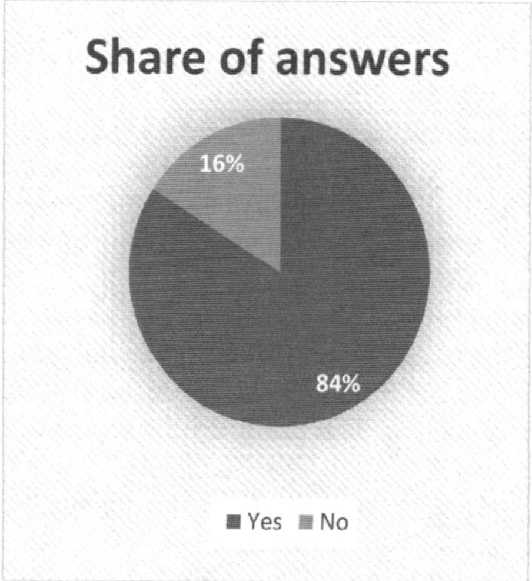

Figure 2. Percentage of people using shared transport. [own study].

A very important question in the survey was which means of shared transport are most commonly used by users. The answers made it possible to prepare the graph presented in figure 3, which shows a significant share of people using car-sharing - as much as 38%. The smallest number of people admitted to using scooters. The question on why people use shared transport was relevant for this paper. The percentage results are shown in figure 4. In this question 29% of the people declared that they use this type of transport to get to work, school and university. This is a very important answer, which indicates that shared modes of transport may to some extent start to replace the previously used own vehicles. 21% of respondents indicated that they choose this mode of transport to get to a meeting with friends. Only about 1% of respondents answered that they use this mode of transport to get to a bus, tram or train stop to use public transport.

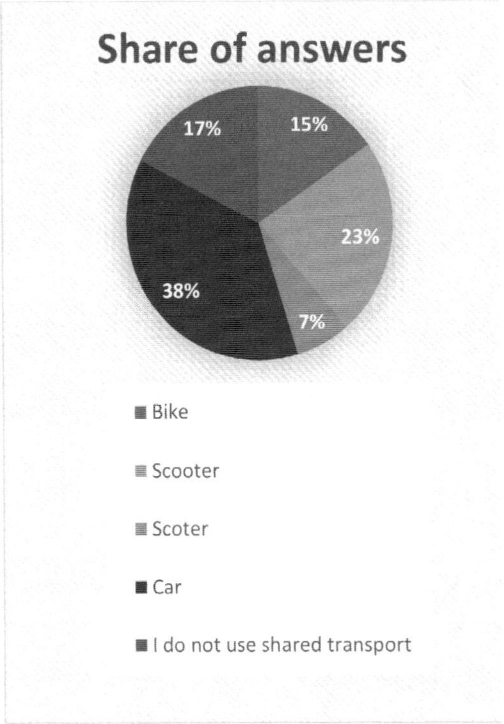

Figure 3. Means of transport [own study].

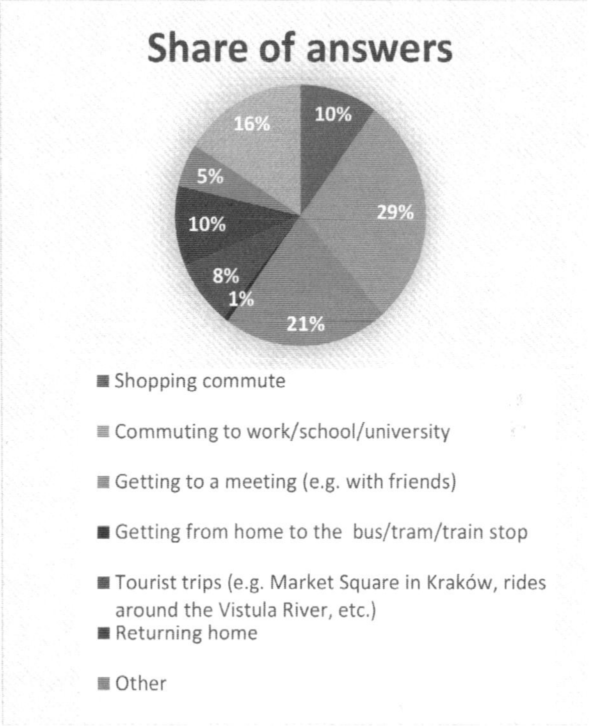

Figure 4. Purpose of the trip [own study].

The final stage was a question about the possible replacement of one's own means of transport with the use of shared transport. Here the large majority of respondents, 74%, said they would not want to trade in their vehicle. However, 26% said they would be willing to change their vehicle for shared transport in the city. The graph which shows the responses can be found in figure 5.

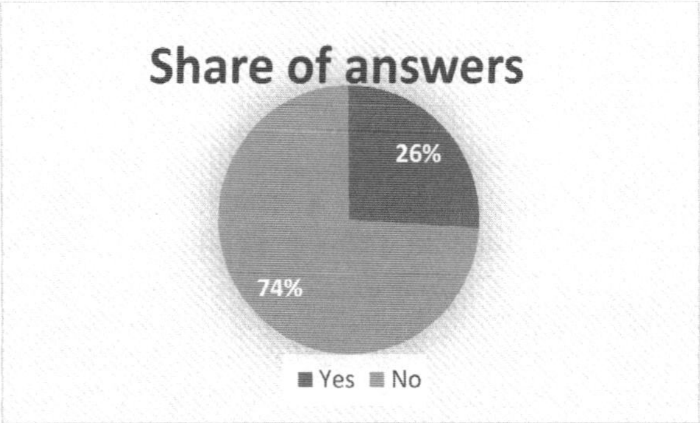

Figure 5. Percentage of people who would change their own vehicle for shared transport[own study].

The results show that shared transport solutions, by using different means of transport, work well in different situations and respond to the needs of potential users. The answers obtained may be a consequence of the predominant group of interviewees aged 19 - 26, who are young people studying or working. However, such results show the real occurrence of shared vehicles in daily life, and this has an impact on the environment. The use of shared transport in daily trips replaces other means of transport, with a positive impact on the environment by reducing the number of vehicles on the road and lowering CO2 emissions.

5. Availability of shared transport means in Poland

Shared transport is an increasingly popular solution both among inhabitants of Krakow and tourists. In the next few years experts expect a more than twofold increase in the number of rented cars and bicycles. Much higher growth is predicted for scooters and motor scooters, where a six-fold increase in the number of means of transport is estimated, as shown in Figure 6.

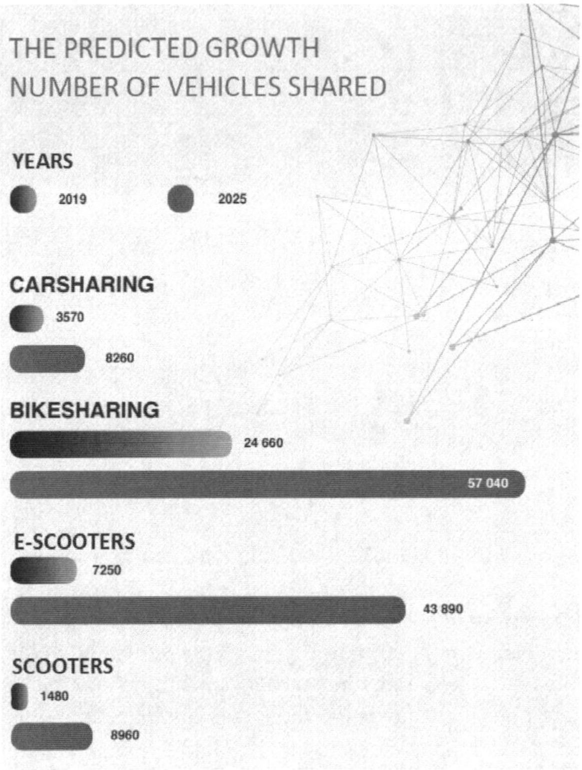

Figure 6. Projected increase in the number of shared vehicles in Poland [own study based on 4].

Scooters fall into the category of "UTO" so personal transport. As seen, the growth and number of scooters will revolutionise urban transport. A factor that may have halted the growth of this mode of transport was the Covid-19 pandemic. However, according to the latest report: An Analysis of the Mobile City in Poland the scooter sharing system has grown significantly. The report informs of 12 operators that offered 18 800 vehicles for use in 39 cities. A new phenomenon in 2020 was the increased availability of this solution in smaller cities [5].

The pandemic has negatively affected other forms of shared transport such as bikesharing. The number of available bikes has decreased to 19 800 in the country. A smaller decrease has also been recorded for scooters, where availability is 1 700 (figures quoted are for the fourth quarter of the year, but before the winter modes were withdrawn). The data on the sudden decrease in numbers for selected shared transport modes is related to the so-called Stress Test phenomenon represented by the pandemic. It should also be noted that a decrease has also been noted in the data for the use of individual means of transport. At the time of the greatest restrictions (estimated for April), the decrease is as much as 55% compared to the typical mobility of that period [6].

6. Conclusion

From the above work, it is clear that the city's transport policy has been supplemented by a very interesting model - transport sharing. Certainly, changes and new transport solutions are not as popular as the individual means of transport that have existed for centuries. We are still attached to having our own means of transport. However, the decision to use shared ones is gradually becoming more popular. The survey shows quite clearly that shared means of transport are often chosen by young people, who are often more open to new solutions and places. It will probably still take time, promotion and the development of companies providing such services for these transport models to become even more popular and accessible to a wider group of users. Constantly improving applications that support shared transport will certainly influence the experience of travellers and make shared transport even more accessible and comfortable for the user.

The availability and variety of shared vehicles fits the needs of people in cities. With relatively low running costs and a simple rental service, it is a solution that fits current needs and standards. The increase in the number of users of shared vehicles has an impact on their number on the roads. As a result, traffic jams on the roads are also reduced. This choice has a positive impact on the environment, because the use of shared vehicles affects air quality. In addition, shared transport improves the mobility of citizens in cities.

7. References

[1] Mobilne Miasto 2019 Na Przełomie Progu *Raport Współdzielolna Mobilność w Polsce* p 75
[2] Carlton C 2017 *The Sharing Economy in our Modern Society* (Stanford: Stanford University)
[3] Urząd Statystyczny 2020 *Report Statystyczne Vademecum Samorządowca (Cracow)*
[4] DataArt 2019 Analiza dostępności usług transport-sharingu w największych osrodkach miejskich na koniec 2018 roku *Transport-sharing w Polsce*
[5] Mobilne Miasto 2021 Jak współdzielone rowery, elektryczne hulajnogi i skutery przetrwały pandemię? *Stress test zaliczony. Mikromobilność w Polsce 2020* p 17
[6] Farsewicz P 2021 Jak wyglądał sharing w 2020 roku? Hulajnogi dogoniły rynek rowerów *Transport Publiczny* (Poland) [online] https://www.transport-publiczny.pl/wiadomosci/jak-wygladal-sharing-w-2020-roku-hulajnogi-dogonily-rynek-rowerow-67707.html

Acknowledgments

The choice of the topic of the paper was supported by Professor Andrzej Szarata. Acknowledgements for substantive support provided.

IOP Conf. Series: Materials Science and Engineering 1209 (2021) 012080 doi:10.1088/1757-899X/1209/1/012080

Introduction of cold-in-place recycling technology in practice as an innovative tool

R Kovács[1] and J Mandula[1]

[1] Department of Geotechnics and Traffic Engineering, Technical University of Košice Faculty of Civil Engineering, Vysokoskolska 4, 042 00 Kosice, Slovakia

E-mail: kovacs.robert1@gmail.com

Abstract. Roads exposed to continuous and heavy traffic often shown sight of damage which particularly extend to the underlying layers of roads. Due to the advantages of less raw materials and fossil fuel consumption, lower carbon footprint and the capability of pavement performance improvement enormous effect have been done in the western countries over the past decades. Thanks to the efforts of some construction companies in Slovakia our focus in this paper on the innovation technology of cold-in-place recycling is given. The presumption of success of this innovative technology is the introduction and use of the set of a powerful and interconnected machines. This technology allows the repair of deep defects, rutting and fatigue failure, which cannot be done by surface treatment with significant reducing the impact on the environment, material, and economical savings. Moreover, it enables the repair of deep defects, rutting and fatigue failure, which cannot be done by surface treatment. To promote the widespread application of cold-in-place recycling technology in practice and to monitor the progress of developments, this article has examined and discussed the basic aspects of the use of this technology based on laboratory analyzes of recycled material.

1. Introduction

So far known „linear economy“ which is based on the „take-produce-throw“ model is gradually replacing with the so called „circular economy“. This type of model – often referred – to green economy is intended to ensure the competitiveness of companies, stable economic growth, and a healthy environment, respectively. The benefits are based on the efficient use of natural resources through the efficient recovery of the materials and products used. Their constant return to the technological cycle closes the flow of the materials, this radically minimalize waste, material input costs and energy required for the production of new products [1]. Road infrastructure is an area that integrates very effectively into this model. An example should be the innovative technology of cold-in-place recycling of asphalt roads. Although this technology has been known in Slovakia for a few decades; the initial huge financial investments in technological and machinery have pushed its use in practice to the background. Thanks to the enormous efforts of construction companies and contractors in Slovakia, the most modern type of WIRGTEN W380 CRi was imported and successfully put into practice from in international group of companies operating in the construction product industry (Figure 1).

The main parts of the cold-in-place recycling train are the follows:

1. Water tanker truck
2. Binder tanker track (foamed asphalt or asphalt emulsion)
3. Cold Recycler

4. Injection system for the hydraulic binder (cement etc.)
5. Grader (in case of use of recycler without leveling bar or without laying by finisher)
6. Pneumatic-tyred roller.

Figure 1. Recycling train WIRTGEN W380 CRi (image taken from the official website of the company www.wirtgen-group.com).

Basic principle of cold-in-place recycling:

Cold-in-place recycling is a process where in one continuous cycle the existing structural layer is disconnected, and the resulting mixture is crushed. It´s properties are improved by adding the necessary materials for „rejuvenating/strengthening" of aged asphalt (binders should consist of cement, foamed asphalt, asphalt emulsion, etc.) and placed on the road again. This layer is then flattened and compacted creating a new recycled road structure. The initial type test of RAP (recycled asphalt pavement) material needs to be performed before the recycling. In case of non-optimal aggregate grain size line of the resulting mixture it is necessary to apply the missing amount of aggregate to the original pavement layer. The hydraulic binder is directly applied on the road. A recycling machine emerges on the surface thus prepared, the structural layer in the milling drum is crushed and the prescribed amount of water is added continuously according to the prescribed recipe. If the type test prescribes an asphalt-based binder the foamed asphalt or asphalt emulsion is added. All components are mixed in the mixing device and then the prepared recycled mixture is placed back on the road. Recycler equipped with a leveling bar, or with the finisher evenly spread the mixture and the levels the surface and pre-compacts the recycled mixture. Finally, using of the compaction roller the compaction of the new structural layer takes place [3, 4].

Advantages of this technology:

1. Fast and high-quality road repairs, which will be reflected in lower financial costs for the entire process.
2. The machine set can efficiently cut, disrupt, and continuously mix the individual components. It has a high performance and precise injection system, which results in increase the efficiency of the construction system, better homogenization of the base layer across the entire width and prevents the formation of transverse cracks. These effects will have a positive effect on the quality of the new recycled layer and faster road renewal, respectively.
3. This technology enables the use of 100 % of the material from the old road on site. In addition, there are savings when adding asphalt binder. This is then reflected in the efficient, multiple use of the material and does not waste natural resources and burden the environment.
4. Material from the construction is not removed, not stored, but is used immediately. There is less load and damage to adjacent roads and other objects, in addition, there is no burden on the environment with exhaust fumes.
5. The technology can also be implemented in the case of a partial closure of the road and the work can be carried out on individual lanes.

The main goal of this work was to assess based on the valid Technical Conditions the suitability of milled RAP material for reprocessing into the construction layers of roads using the machine set WIRTGEN W380 CRi. Based on the measured data, three series of test mixtures with cement and asphalt binder with different contents were designed and mixed in the laboratory. According to the valid test standard, the prescribed empirical properties of individual series of mixtures were determined. Based on the verified properties of hydraulically bonded mixtures, a mixture that complied with the relevant standard was selected and the design of the mixture was submitted to the construction company for the implementation of the test section using a machine set. This part is not a subject of this study.

2. Materials and methods

Type test of mixed R-material taken from the construction site:

Test samples were taken from the place of production, from the construction: Road No. III / 2413 Hronsek – Vlkanová. Exact designation of the object is: 3,102 - 5,092 km. From the construction sample point of view all asphalt layers were designed for recycling, with the classified type of material as mixed RAP material. Trade name of the sample according to the standard is: ZRM 32 CEM III / A 32.5 N 180 mm. The sample was taken after milling the layer with the WIRTGEN W 380 CRi recycling machine set, the climatic conditions at the time of milling were 22 ° C and clear. The ZRM type test was performed in accordance with the technical conditions of TP 046/2017 "Cold processing of non-rigid road on spot layers, Ministry of Transport and Construction of Slovak republic, 2017."
During the verification of the input materials, the following properties were determined: the content of the asphalt binder, according to the test standard STN EN 12697-1 [6] and the grain size according to the STN EN 933-1 [7].

Preparation and type testing of hydraulically bound mixtures produced at the laboratory conditions:

The following building materials were used to design the composition of the mixture in the laboratory:

Hydraulic binder CEM III / A 32.5 N was obtained from the manufacturer Duna-Dráva Cement Kft., Vác. Manufacturer on the basis of product tests performed according to the harmonized standard EN 197-1: 2011 "Cement. Part 1: Composition, specification and conformity criteria for common cements." issued the "Declaration of parameters, no. 1414-CPR-T0071-01 [8]. According to the above-mentioned declaration, the product meets the intended use of the construction product.

Used road asphalt CA 50/70 was obtained from MOL Plc., Százhalombatta, Hungary. Based on the "Declaration of Parameters" issued (No. PB50 / 70-DR) by the manufacturer, the product meets the criteria for use in road construction and maintenance.

The water used in the production of the mixture met the requirements of STN EN 1008 [9].

The following methods and procedures were used in the production of samples: Compaction according to STN EN 13286-2 (Proctor modified test, 5 x 56 strokes) [10], Determination of water content according to STN EN 1097-5 (in a ventilated oven) [11]. In all procedures, the treatment regime was set according to TP 046/2017 [5].

To verify the properties of the hydraulically bound mixture produced in the laboratory, the determination of the grain size line according to STN EN 933-1 [7], the determination of the bulk density according to STN EN 13286-2 [6] and the compressive strength R_{c7} according to TP 046/2017 [5] were used.

3. Results and discussion

At the beginning of the discussion on the achieved results, it should be noted that from the legislation of the Slovak Republik point of view that the whole process - production, laying and quality control of recycled cold mixture, is governed by technical regulation TP 046 „Cold processing of non-rigid road on spot layers, Ministry of Transport and Construction of Slovak republic, 2017."

Asphalt road milling, sampling:

Based on a mutual agreement with the construction company and the contractor of the construction work, the place of construction, time and conditions of road recycling were determined at the beginning of the works. The road No. III / 2413 between the villages of Hronsek and Vlkanová was designated as the construction site. In the first stage of the scientific work, the project documentation for the construciton has been examined (it is not a part of this work). The marked section was selected based on a diagnostic survey of the road, which basically determines the suitability of the technology. Sampling for initial type testing was performed after milling of the original road structure by a WIRTGEN W380 CRi cold recycler. This type of recycling cutter allows repairs of even deep defects, the maximum possible depth of the milled layer is up to 350 mm. It should be done thanks to a high-performance engine (775 kW) with huge torque. The robust milling and mixing rotor granulates the damaged asphalt road. Rotor speed at maximum possible operating speed reaches 100 rpm [2, 4].

It is important to realize that this whole part of the cold recycling process – milling and crushing of the layers – is essentially a technological matter. The recycler must have a sufficient working width and must be characterized by the greatest possible engagement depth. The most important component is the milling roller with special tips, which ensure sufficient excavation and crushing of damaged layers (Figure 2). The milling drum promotes mixing in the vertical plane, i.e. the recovered material is not thrown to the side or longitudinally from its original location within the roadway. In practice, this means that the materials in the recycled length of the section return approximately to the same location after recycling, which reflects their original location. An important role is played by the milling tips of the roller, which can vary from model to model. The layers of unbound material disintegrate easily with the tip and return to their original uncompressed state. The degree of pulverization of the layers of bound material (e.g. asphalt) is influenced mainly by the speed of the recycler, but also by the speed of rotation of the drum. Generally speaking the faster the speed of the recycler and the lower the rotation speed, the coarser the product [4].

Figure 2. Milling drum of the recycler WIRTGEN W380 CRi (image taken from the official website of the company www.wirtgen-group.com).

To achieve the required road quality, the materials used should meet the technical requirements set out in the relevant standards. The quality of the used RAP material for reprocessing must be checked before the works start and during production, respectively. To assess the suitability of the RAP material for reprocessing, need to have a suitable composition. Based on the survey, the depth of the milled road was determined to be 180 mm. Due to the evaluation of aggregate properties, focus is basically given on the determination of the content of asphalt binder and the physical and structural properties of the aggregate. The results of the type tests are summarized in Table 1 and Figure 3. The content of the asphalt binder is an important criterion for determination of the suitability of the RAP material for reprocessing.

Our representative sample from the investigated section has a soluble binder content of 3.1 weight percent. This data was used in the calculation of the hydraulic binder added and water to the mixing drum to obtain the ideal mixture of the new road structure.

For the determination of the properties of RAP material (fraction size 0/32) the sieve analysis method was used. Figure 3 shows the grain size line of the studied aggregate. In addition to the grain size line, the values of fine grain content and moisture were also determined (Table 1). From the definition of hydraulically bound aggregate it is evident that the resulting aggregate mixture must have a continuous grain line. If the grain size of the RAP material does not correspond to the prescribed boundary lines it is necessary to add the amount of the missing fraction aggregate. The amount of added binders must be in accordance with the relevant material data sheets, such as emulsions – Emulsion and sealing catalogue lists (KLEaZ) asphalt [12] – Hydraulic adhesives catalogue lists (KLHS) [13].

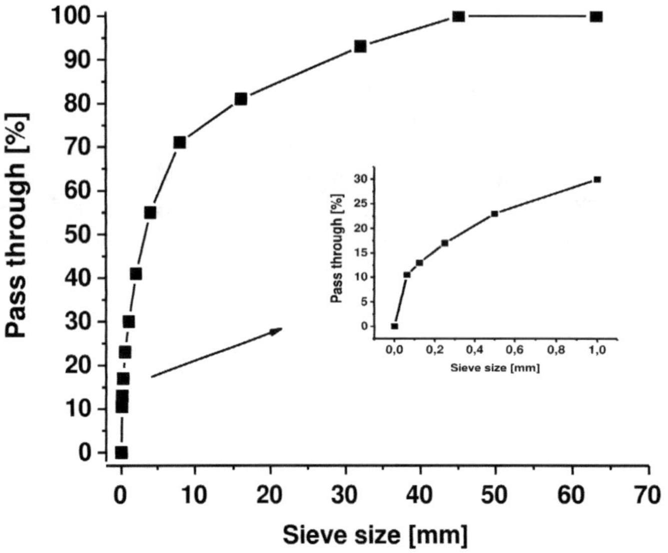

Figure 3. Aggregate gradation of RAP material.

Table 1. Measured values of RAP material.

Test	Measured value	Units	Test method
Asphalt content	3.1	w %	STN EN 12697-1
Content of fine grains	10.5	w %	STN EN 933-1
Humidity	1.7	%	STN EN 1097-5
Bulk density of the soil	2008	kg m^{-3}	STN EN 13286-2

The results of the milled RAP material and the assessed input materials show that they meet all the requirements of the relevant Catalogue sheets and are in accordance with the requirements for use in the mixture for the intended use, for the transported load class II - IV, determined according to STN 73 6114 [14].

Hydraulically compacted mixtures prepared in the laboratory:

In the first part of the work, the focus was given on the assessing the suitability of a real sample of RAP material milled from the roadway and added initial materials. Based on these measured data, three various hydraulically bonded mixtures with different contents of asphalt binder were designed. In the design a real milled RAP material from the construction site was used. The mixtures prepared and tested in this part are the model samples and based on the most important parameters measured, the most ideal mixture was selected, the design of which was submitted to the construction contractor, who can adjust the addition of components to prepare the recycled layer according to the laboratory tested of this study (this part has not been reviewed, nor is it the subject of this study).

The second part of the cold recycling process, i. e. - the mixing of the binder and other components - is more of a scientific matter. Based on the properties of the milled RAP material, the correct choice and amount of rejuvenating agent needs to be adjusted. In our case, a hydraulic binder - cement - with water and asphalt binder is used. The determination of the amount of rejuvenating additives must be in full accordance with the several worn binders in the existing layer to achieve the properties of the new construction layer. It should be noted that during the use of the road at higher temperatures, its´ rigidity decreases and due to wheel pressure vehicles create rails and deformations. These defects are mainly due to the increasing total weight of trucks. During the road use, as well as due to climatic changes (higher / lower temperature during the changing seasons, UV radiation, exposure of various materials to the road surface (sanding salts, acid rain, spilled chemical compounds, water, etc.) and asphalt loses its´ original physical and chemical properties [15].

The most striking factor is the loss of light, oily fractions, the so-called maltenes. They mainly consist of resins, aromatic oils and saturated components whose are responsible for creating a dispersive environment in the binder. They carry the plastic-elastic and adhesive properties of the binder. In practice they are highly responsible and affect the adhesion of the asphalt binder to the aggregate. During aging, these volatile components are lost from the binder. The remaining heavy fractions, the so-called asphaltenes, which are responsible for the rheological properties without this disperse-oily maltene medium, will reduce the viscosity of the binder. Such a reduction subsequently influences the surface activity of the asphalt binder and the binder can lose its´ flexibility. Therefore, it is very important to determine the asphalt binder content correctly and as accurately as possible, because according to this value we can subsequently determine how much maltene in the form of various additives (cationic asphalt emulsion, foamed asphalt) to add to "rejuvenate" functional properties of the original aged binder [16].

Figure 4. The injection system (left) and material mixing process (right) of a recycler WIRTGEN W380 CRi. (image taken from the official website of the company www.wirtgen-group.com).

With the right choice of binders and additives, as well as their amount, it is important to know what technology will be used. In our case, the recycler allows the dosing of a cement, water, and asphalt binder continuously directly into the mixing drum [Figure 4]. The first component is cement, as a hydraulic binder. The type of cement used affects the initial waterproofing ability of the road. The water used in this process helps to create a uniform cement / asphalt consistency. Although physical sorption occurs between the particles, this mixture must eventually split in the mixing drum and total adsorption must occur on the surface of the aggregate [17]. At the required time (a few seconds) between the recycled aggregate in this mixture, flocculation must take place, where the binders are partially joined, the aqueous phase is depleted and subsequently coalescence occurs [18]. In this process, droplets of asphalt binder are bonded to the cement on the surface of the aggregate, the aqueous phase is separated and a layer of asphalt / cement covers the surface of the recycled material. In the final phase, coalescence occurs when the aged binder is combined with the new binder and the individual phases are combined into larger units and wrap the recycled material. In order for this process to be stable and satisfactory, it is necessary to determine the exact amount of binders and water under laboratory conditions. If an unsatisfactory amount of adhesion to the aggregate is used, it will be weak and the entire material will not be coated. In Figure 5 is a schematic illustration of the described process

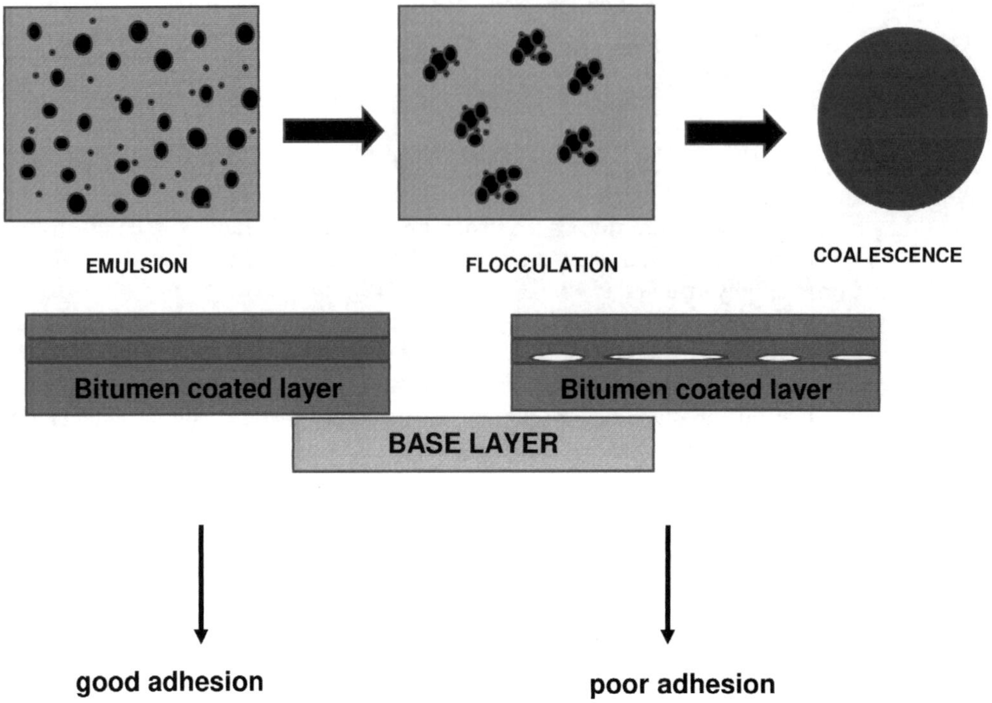

Figure 5. Schematic illustration of the emulsification process (a) and emulsion coatings of aggregates resulting from good adhesion (left) and poor adhesion (right).

According to the data described on Table 2 (hydraulic and asphalt binder content), three series of test rollers were prepared. For each series, the optimal water content in the building mix was calculated. The basic parameters were determined according to TP 046/2017 [5]. After evaluating of all results, a mixture with 2.2 % by weight of asphalt binder was selected. However, it is evident that the measured value of the strength of this mixture does not meet the basic requirements set out in the relevant technical regulations. After extrusion from the mold, the samples were stored at laboratory conditions at 20 ° C ± 5 ° C at 30 % to 40 % relative humidity for 24 hours. Then they were placed in a climate chamber (> 95% relative humidity) for 6 days. The compressive strength values of R_{c7} are approximately 1 MPa higher as it is given in relevant technical standards. The explanation for this deviation should be based on the used RAP material. Recycling of RAP material requires careful consideration of the following factors: nature and the composition of the original structural layer (type of the mixture, aggregate content, binder aging), type and cause of deformation, thermal and fatigue regime of road, cracking of tracks, damage extent, purpose of recycling (complete renovation or partial repair of the road). In addition, it should be noted that this type of machine set is just starting to be applied on Slovak roads. In the future, it will be necessary to look comprehensively at the entire technological process and also on the basis of our measured values and research in this area to re-evaluate some parameters so that they are in line with current advanced technology and take greater account of the construction contractors´needs.

4. Conclusion

Innovative technology cold-in-place recycling can be considered as a green technology from a technological, economic and especially from environmental point of view. However, the results obtained during this study showed that an efficient machine assembly is needed for such a technology to be feasible. On the other hand, the technological process must also take into account the conditions of implementation, the type and defects of the milled layer and the properties of the material. Only after a whole diagnostic examination and setting of the parameters of all components will be possible to achieve results that will be absolutely relevant. Determination of functional and structural properties of RAP material, choice of rejuvenating agent will not avoid the aged layer without quality laboratory research. During the elaboration of this work, the requirements of practice were fully taken into account, all procedures and the chosen methodology were consulted by experts in the relevant technology field.

5. References

[1] Cséfalvayová P, Maleš I and Ľuptáková M 2017 *Slovensko a cirkulárna ekonomika* (Bratislava: INCIEN)
[2] Wirtgen Group 2018 *High-quality in-place recycling of asphalt layers. Cold recyclers – CR – model series, W 240 CR / W 240 CRi / W 380 CR / W 380 CRi* (Germany: Wirtgen GmbH)
[3] Gschwendt I, Staňo R, Polakovič Ľ, Loveček Z and Sekera M 2001 *Vozovky Materiály a technológie* (Bratislava: Jaga group)
[4] Wirtgen Group 2012 *Wirtgen Cold Recycling Technology* 1ed (Germany: Windhagen)
[5] Ministry of Transport and Construction of Slovak republic 2017 *TP 046 Cold processing of non-rigid road on spot layers*
[6] STN EN 12697-1 *Bituminous mixture – Test method – Part 1: Soluble binder content*
[7] STN EN 933-1 *Tests for geometrical properties of aggregates. Determination of particle size distribution – Sieving method*
[8] STN EN 197-1 *Cement. Part 1: Composition, specification and conformity criteria for common cements*
[9] STN EN 1008 *Mixing water for concrete – Specification for sampling, testing and assessing the suitability for water recovered from processes in the concrete industry, as mixing water for concrete*
[10] STN EN 13286-2 *Unbound and hydraulically bound mixtures - Part 2: Test methods for laboratory reference density and water content - Proctor*
[11] STN EN 1097-5 *Tests for mechanical and physical properties of aggregates. Determination of the water content by drying in a ventilated oven*
[12] Ministry of Transportation and Construction of Slovak republic KLEaZ 1/*2014 Emulsion and sealing catalogue lists: Annex No. 1/2016 to KLEaZ 1/2014.*
[13] Ministry of Transportation and Construction of Slovak republic KLHS 1/0216 *Hydraulic adhesives catalogue list*
[14] STN 73 6114 *Road pavement. Basic provisions for design.*
[15] Willis J R, Turner P, Julian G, Taylor A J, De G and Padula F 2012 Effect on Changing Virgin Binder Grade and Content on RAP Mixture Properties *NACT Report No. 12-03*
[16] Aleksandrov A S, Zaev A A and Kurmanov D H 2021 The use of cold recycling technology to increase the resistance of pavements to rutting *IOP Conf. Series: Mat. Sc. and Eng.* **1079** 052017
[17] Rossi C O, Teltayev B and Angelico R 2017 Adhesion promoters in bituminous road materials: A review *Applied Sciences* **7** 524
[18] Khan A, Redelius N and Kringos N *Effect of surfactants and adhesion promoters on the* bitumen-minerals interfacial bound during breaking of bitumen emulsions Available online www.diva-portal.org.

Acknowledgements
This work was supported with the financial support of Roads Slovensko, Ltd., the exclusive distributor of the WIRTGEN W380 CRi recycling machine set in Slovakia. The authors of the article also thank the company VIAKORP, Ltd., to provide the possibility of implementing this contribution on the construction work. This work was supported by the Slovak Research and Development Agency under

the contract No. APVV-19-0520. This work was also supported by the Scientific Grant Agency of the Ministry of Education of Slovak Republic and the Slovak Academy of Sciences under Projects VEGA 1/0796/21.

IOP Conf. Series: Materials Science and Engineering 1209 (2021) 012081 doi:10.1088/1757-899X/1209/1/012081

Evaluation of the mechanical properties of fluvial coarse-grained soils from dynamic penetration test

E Kučová[1] and M Kuvik[1]

[1] Department of Geotechnics, Faculty of Civil Engineering, Slovak University of Technology in Bratislava, Radlinského 11, 810 05 Bratisava, Slovakia

E-mail: eliska.kucova@stuba.sk

Abstract. Soil and rock properties are essential inputs for the design of geotechnical structures. But unlike man-made materials, the determination of the characteristic value is not simple, mostly because of the large space variability of the soils and rocks. In many cases, the large spatial variability makes it difficult to collect sufficient ground samples for further laboratory testing so that many tests are performed directly in the field. In Slovakia, the dynamic penetration test (DP) is, among others, a quite popular field test for soil investigation. From the measured data, it is possible to derive relative density (I_D) shear (ϕ_{ef}) and deformation (E_{def}) parameters of coarse-grained soils. There are many available correlations between the DP results and geotechnical parameters. Therefore, it is necessary to choose the most appropriate one as it considerably affects the final evaluation. This paper shows the assessment of geotechnical parameters of coarse-grained fluvial soil from the DP results and presents the statistical determination of its characteristic values, which are furthermore compared with the nominal values used in Slovakia.

1. Introduction

The determination of characteristic value of geotechnical parameters is a necessary input for the design of geotechnical structures. However, unlike man-made materials (e.g., steel or concrete), there is more significant uncertainty in the determination of soil properties, mostly due to spatial variability and local dispersions of soils [1].

The characteristic value of soil properties is determined from the set of data obtained from the field and laboratory tests. Schnaid [2] describes the most common field tests and explores new methods for interpreting the measured properties of various soil formations. In addition to the common tests, the dynamic penetration test (DP) is another field test often used in Slovakia to measure the soil properties. It is a relatively simple test, where the dynamic penetration resistance (q_{dyn}), representing the soil resistance against the rod penetration, is measured. Variations of this test, categorized according to hammer mass, are closely specified in European standards [3,4]. DP is recommended to test coarse-grained soils due to less friction on the rod, but it can also be applied to investigate fine-grained soils. For example, the dynamic penetration was used to monitor the compaction of soils (including clays) in the UK [5], or it was applied for analysis of deformation parameters of collapsible loess in Slovakia [6]. Using appropriate correlations, it is possible to derive the relative density (I_D), the angle of shear strength (ϕ_{ef}) or the deformation modulus (E_{def}) of coarse-grained soils from the results of DP. Available correlations are listed in Eurocode 7 (EC 7) [3], various national standards [7,8,9] and other Slovak literature [10]. Most correlations are derived for relative density as it directly affects the soil resistance, thus other parameters. Tran, Chevalier and Breul [11] found out that the soil resistance is also affected by the grain size distribution. They supported their theory with the spectral analysis modelled with DEM assessing the effect of grain size variation on the DP results. They found out that resistance decrease with the increasing content of smaller particles.

The correlations between DP results and geotechnical parameters of coarse-grained soils reflects various aspects. EC 7 [3] and national standards [7,8,9] describe correlations for I_D derived from the number of blows N (measured for 10 or 20 cm of penetration depth). The correlations are based on the laboratory measurements distinguishing two situations: above and below groundwater level. Vrettos, Papamichael [12] discussed the applicability and reliability of these correlations. They stated that the correlations are reliable for the determination of medium density, but the determination of other geotechnical parameters from the results of I_D is uncertain. For that reason, they recommended using direct correlations.

Direct correlations between the DP results and mechanical parameters (ϕ_{ef}, E_{def}) of coarse-grained soil are published in Slovak literature [9,10]. According to Matys et al. [10], N's values are only indicative data as they do not count rod friction and change of the rod's weight with the increasing depth. Therefore, in Slovakia, the correlations derived from the results of q_{dyn} [9,10] are used more frequently. If there is no direct correlation with dynamic penetration, it is possible to correlate with another test. Czado, B., Pietras, J.S., [13] studied the correlations between the cone penetration test and dynamic penetration, concluding that the correlations should be applied with caution, because in some cases, it may lead to overestimation of the results.

Another important aspect influencing the correctness and appropriate selection of the correlations is the soil characterization (i.e., composition, grain size/ shape, and genesis). While standards like EC 7 [3], DIN 4094-3:2002 [7] and PN-B-04552:2002 [8] classify correlations according to soil type and uniformity coefficients C_u, Slovak literature [9,10] mentions the correlations which also consider the genesis of the soil. Overall, the literature (both international and national) offers a wide range of correlations, so it is necessary to choose the appropriate correlation wisely and, if possible, verify its suitability to the local conditions.

The challenging step in the determination of the characteristic value of geotechnical parameters is the statistical evaluation. Despite EC 7 defines a statistical approach for determining characteristic value, some problems were specified and discussed by various authors. Frankovská, J. [14] describes the statistical evaluation of the characteristic values of geotechnical parameters and indicates that the assessment of geotechnical parameters with natural origin cannot follow the same rules as made materials and that the genesis of the soil/rocks should be taken into consideration. Schneider H. R. [15] provides guidelines for assessing the characteristic values bearing in mind uncertainties in measurement and empirical correlations. A.J. Bond [16] observed that many practicing engineers use various statistical approaches but didn't have a conservative estimate of the characteristic value, mainly when data are scattered. C. Pohl [1] commented on the effectiveness of the statistical approach from EC 7 and listed some recommendations for practical applications. He pointed out that the statistical method can be an effective tool when experience, indirect site investigation, and field and laboratory tests are available.

2. Site Description – Geology

The tested locality belongs to Podunajská nížina (lowland). The region is rich in coarse-grained soils which were in the past deposited and accumulated in layers by the Danube River [17]. The most represented coarse-grained soils in the area are well-graded (G1/GW) and poor-graded (G2/GP) gravels. These layers alternate with the positions of fine-grained soils (G3/G-F or G4/GM). The layer thickness of fluvial gravels varies between 3 to 13 meters. Below the gravels fine-grained soils with low permeability are located [18]. Gravel grains are well-rounded with a smooth spherical or cylindrical shape. Most of the gravel deposits have a fraction of 1-5 centimetres, but there are few locations with fractions of 5-12 centimetres.

The fluvial coarse-grained soils were investigated by the heavy dynamic penetration test (DPH) in the 3-kilometer-long section near Bratislava. The depth of penetrations varied from 7 to 25 meters. In addition to the penetration test, exploratory boreholes were drilled for soil classification purposes. A steady underground water level was observed in the depths of 0.5 to 4 meters below the terrain. Based on the collected data, the most predominant soils in the section are medium dense to dense poor-graded gravels (G2/GP) and medium dense gravels with fine particles (G3/G-F). In some places, medium dense fine-grained soils (G4/GM, G5/GC) intersect these layers. In addition, some sandy soils were detected (S3/S-F, S4/SM), but they occurred less often than gravels.

3. Methodology

Fluvial sediments of the area were tested with a heavy dynamic penetration test (DPH). The parameters of the device are standardized in [3]. The data of dynamic penetration resistance recorded for every 10 centimeters of penetration were collected. Furthermore, these data were used for derivation of relative density (I_D) and mechanical properties of coarse-grained soil (ϕ_{ef}, E_{def}).

In this article, the analyzed input data consist of dynamic penetration resistance (q_{dyn}) and thus, the correlations for determining geotechnical parameters were selected respectively (Table 1). The chosen correlations are used in Slovakia and can be found in Slovak literature [9,10]. Other correlations from EC 7 [3] and other national standards [7,8] were excluded as they are derived from the number of blows *N*.

The selected correlations for determination of relative density (I_D), angle of shear strength (ϕ_{ef}) and deformation modulus (E_{def}), listed in Table 1, reflect the soil type and soil composition. But none of these correlations considers the genesis of the soil (e.g., shape of the grains), which is a crucial aspect influencing the soil's resistance and so the determination of other geotechnical parameters.

Table 1. Selected correlations for determination of geotechnical parameters of coarse-grained soils.

Reference	Soil parameter	Correlation	Soil	Coefficients	
Švasta [10]	**Relative density (I_D)**	$I_D = a \times (q_{dyn})^b$ [-]	Silty, Clayey Sand	***a***	0.16
				b	0.7
			Fine-grained Sand	***a***	0.15
				b	0.67
			Medium and Coarse-grained Sand	***a***	0.14
				b	0.63
			Gravel soils, Gravely Sands	***a***	0.13
				b	0.6
	Angle of shear resistance (ef)	$\varphi_{ef} = p \times (q_{dyn})^r$ [°]	Fine-grained Sand, Gravel Soils	***p***	24
				r	0.16
Švasta [10]	**Deformation modulus**	$E_{def} = n \times (q_{dyn})^m$ [MPa]	Silty Sand	***n***	2.0
				m	0.83
			Fine-grained Sand	***n***	3.5
				m	0.83
			Coarse and Medium-grained Gravel	***n***	5.5
				m	0.83
			Sandy Gravel	***n***	8.8
				m	0.83
Obert [10]		$E_{def} = n \times q_{dyn}$ [MPa]	Silty, Clayey Sand	***n***	2
			Sandy Soils		3
			Gravel Soils		10

4. Results and discussion

In this paper, the results of the dynamic penetration resistance q_{dyn} [MPa] are analysed and used for the determination of the relative density (I_D), the angle of shear strength (ϕ_{ef}) and the deformation modulus (E_{def}) using correlation in table 1.

During the ground investigation, there were detected medium dense to dense gravel and sand layers. The range of dynamic penetration resistance for dense gravels was 14.9 MPa – 36.0 MPa for poor-graded gravel (G2/GP) and 28.3 – 61.8 MPa for well-graded gravel (G1/GW). For dense sands, the ranges of dynamic penetration resistance were determined as followed: 28.5 – 43.8 MPa for well-graded sand (S1/SW), 11.9 – 25.6 MPa for poor-graded sands (S2/SP) and 8.1 – 11.9 MPa for sands with fine particles (S3/S-F). For the medium dense gravels and sands, the dynamic penetration resistance was measured in this way: 12.2 MPa – 14.8 MPa for poor-graded gravel, 9.3 – 14.8 MPa for gravels with fine-grained particles (G3/G-F), 7.1 – 14.6 MPa for silty gravels (G4/GM), 5.2 – 9.2 MPa for clayey gravels (G5/GC), 4.7 – 6.5 MPa for S3/S-F, 3.8 – 6.9 MPa for silty sands (S4/SM) and 3.3 – 5.5 MPa for clayey sands (S5/SC). The results of q_{dyn} are graphically presented in Figure 1. Similarly, the ranges of values for mechanical properties of sands and gravels (e.g., angle of shear strength and deformation modulus) were specified and are displayed in Figure 2 and Figure 3.

As a next step, the characteristic values for q_{dyn}, ϕ_{ef} and E_{def} were determined using the statistical approach given in Eurocode 7 [14]. The characteristic and mean values for the mentioned parameters are listed in Table 2 and Table 3. From the statistical perspective, we can observe that difference between the mean and characteristic values is not significant, therefore in this case it does not matter whether we use mean or characteristic value for further comparison. But the mean value may appear more convenient as its statistical evaluation is more straightforward than that of the characteristic value.

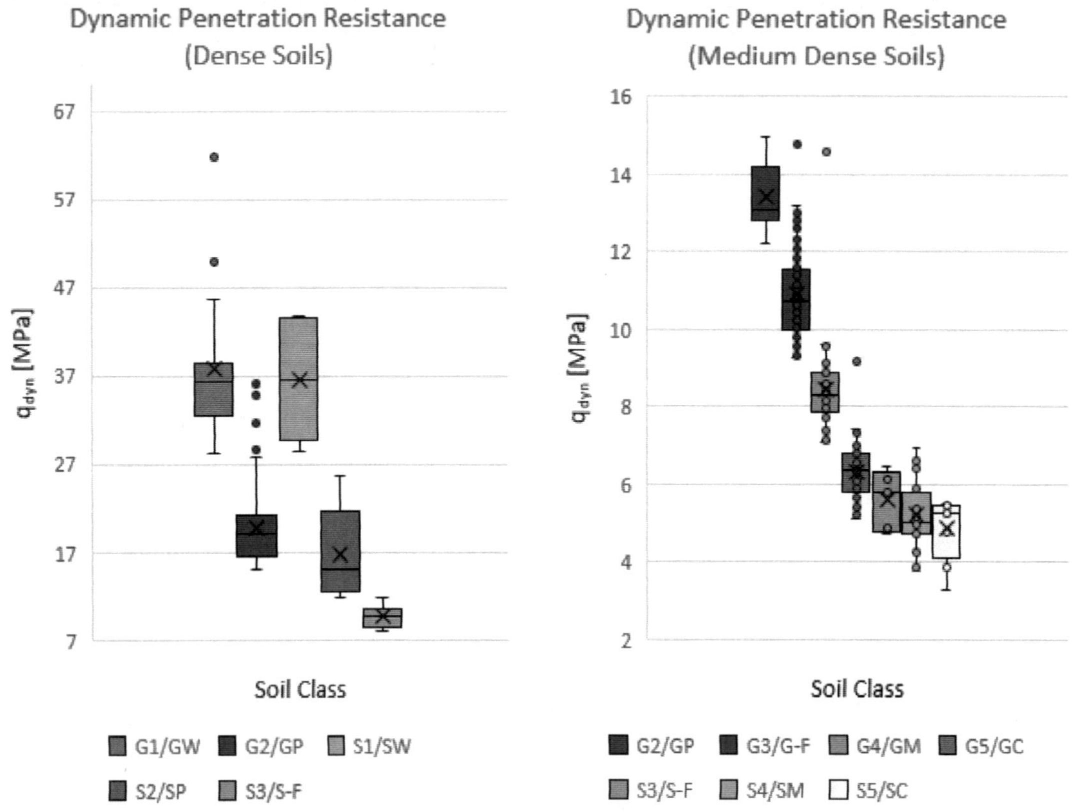

Figure 1. Dynamic penetration resistance of medium dense and dense coarse-grained soils.

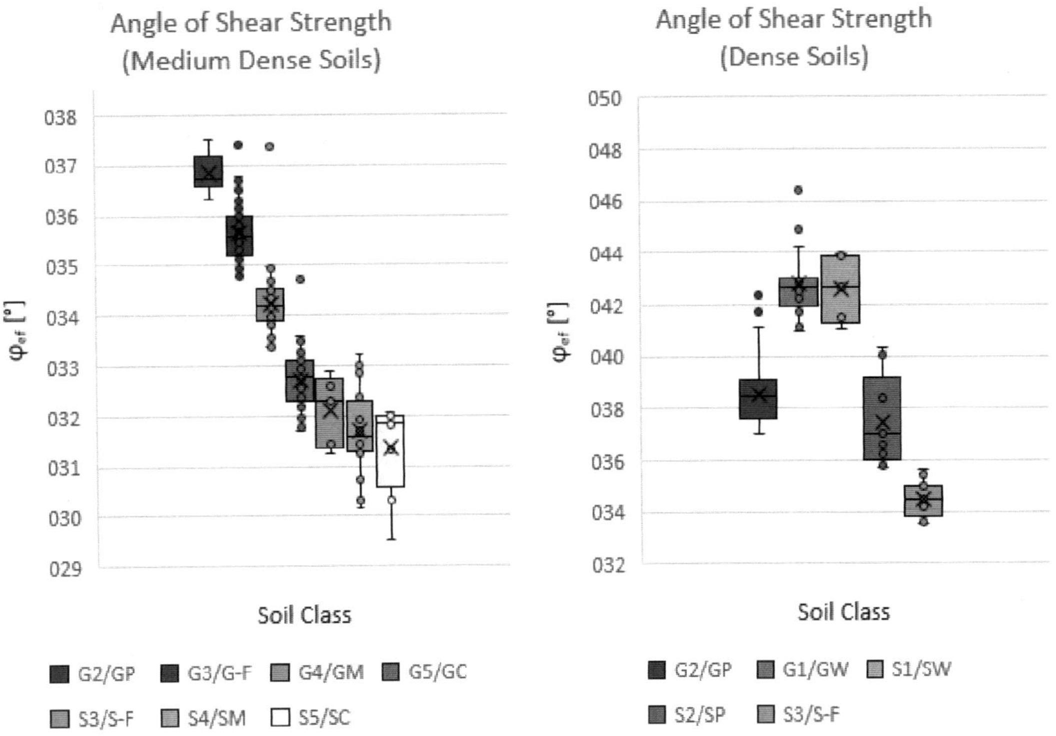

Figure 2. Angle of shear strength of medium dense and dense coarse-grained soils.

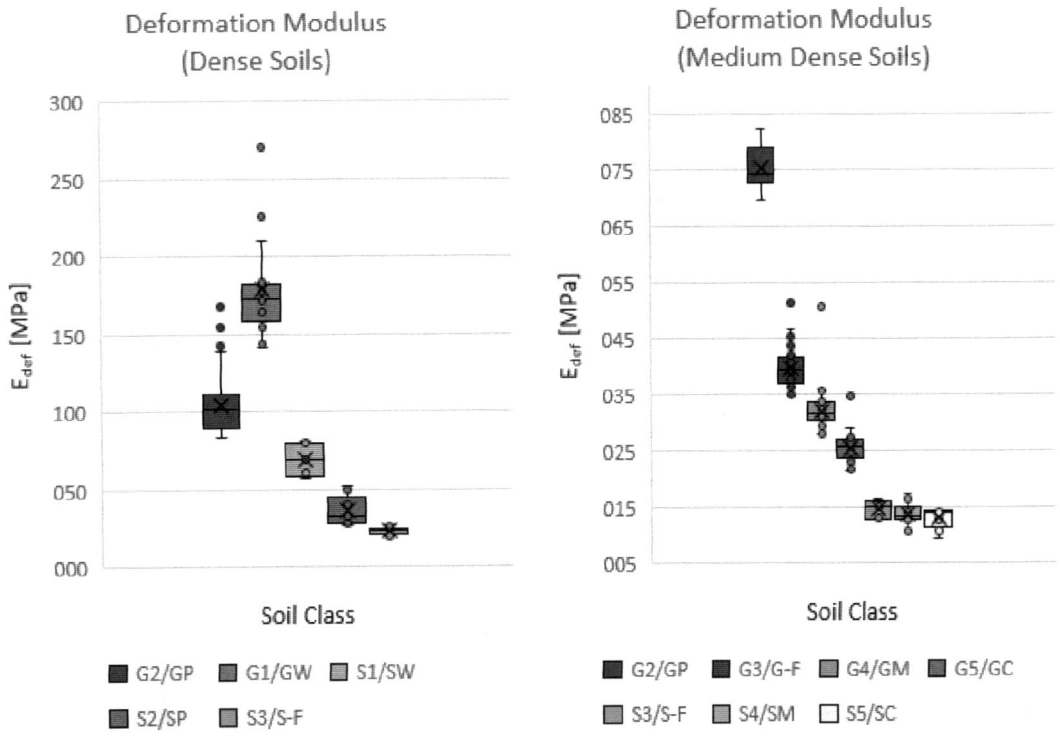

Figure 3. Deformation modulus of medium dense and dense coarse-grained soils.

IOP Conf. Series: Materials Science and Engineering 1209 (2021) 012081 doi:10.1088/1757-899X/1209/1/012081

Table 2 summarizes the results of q_{dyn} and ϕ_{ef} for coarse-grained fluvial soils and shows their comparison with nominal values specified in cancelled Slovak standard STN 73 1001:1989 [19]. Comparison of the results ϕ_{ef} with nominal values shows slight inconsistency. While the majority of ϕ_{ef} characteristic values (sand S1 – S3, all classes of gravels) are following suggested intervals of nominal values, the mean ϕ_{ef} values of soils S3, S4, S5 exceed these intervals by more than 1°.

Despite some exceeded values, we can conclude that the results of ϕ_{ef} for fluvial sands and gravels, derived from the results of dynamic penetration resistance, are in accordance with nominal values suggested in STN 73 1001:1989 [19]. This indicates that the chosen correlation (Švasta, Table 1) could be used for the determination of the angle of shear strength for coarse-grained fluvial soils.

Table 2. Comparison of the angle of shear strength results with nominal values.

Density	*Soil*		*Results* q_{dyn} [MPa] Mean	*Results* q_{dyn} [MPa] Charact.	*Results* ϕ_{ef} [°] Mean	*Results* ϕ_{ef} [°] Charact.	*Nominal values (STN 73 1001:1989)* ϕ_{ef} [°]
Dense (I_D = 0.67-1.0)	*Gravel*	*G1*	37.9	34.2	42.8	42.2	39 - 44
		G2	19.6	19.0	38.5	38.4	36 - 41
	Sand	*S1*	36.5	29.9	42.6	41.3	37 - 42
		S2	16.7	13.5	37.5	36.4	34 - 37
		S3	9.7	9.0	34.5	34.1	30 - 33
Medium Dense (I_D = 0.33-0.67)	*Gravel*	*G2*	13.3	13.1	36.3	36.2	33 - 38
		G3	10.9	10.7	35.1	35.0	30 - 35
		G4	8.4	8.2	33.7	33.6	30 - 35
		G5	6.3	6.1	32.2	32.0	28 - 32
	Sand	*S3*	5.6	4.9	31.6	30,9	28 - 31
		S4	5.2	4.8	31.5	31.2	28 - 30
		S5	4.9	4.3	30.8	30.2	26 - 28

Table 3 presents the results of q_{dyn} and E_{def} for tested coarse-grained soils. The E_{def} was derived from the results of dynamic penetration resistance using two correlations - E_{def}^1 (Švasta) and E_{def}^2 (Obert). Simultaneously, the nominal values published in cancelled Slovak standard STN 73 1001:1989 [19] and recommended by Frankovská et al. in [20] are listed in the same table. The results of deformation modulus (E_{def}^1, E_{def}^2) were compared with these nominal values.

From the comparison, we can see that derived values are in many cases out of the interval range of recommended nominal values. For instance, results of E_{def}^1 of well-graded gravels fall into the intervals recommended by Frankovská et al. but are far below the nominal values given in STN 73 1001:1989. On the other hand, the results of E_{def}^2 belong to intervals given by STN 73 1001:1989 and fall out of the interval recommended by Frankovská et al. The results of dense poor-graded gravels do not belong to any intervals of nominal values. For medium dense gravels, only results of E_{def}^2 for poor-graded gravels are within the intervals given by STN 73 1001:1989. The deformation modulus estimated for other classes of medium dense gravels (i.e., G3, G4, G5) does not comply with any of the respective intervals given in Table 3.

Comparison of results for sands shows a better match with nominal values. Regarding dense sands, only the results of E_{def}^2 for well-graded sand (S1) exceed the given intervals of nominal values. A comparison of the results of medium dense sands with nominal values shows that in some cases the results are slightly underestimated or overestimated (S3, S5). A comparison of the deformation modulus of medium dense gravels shows that the results hardly fall within the intervals of nominal values.

Despite that the deformation modulus was derived using two different correlations - E_{def}^1 (Švasta) and E_{def}^2 (Obert), most of the results do not fit into the intervals of nominal values. Grater offset between the E_{def} results and nominal values can be observed in gravels. The results of E_{def}^1, which were derived by Švasta's correlations, tend to undergo the intervals of nominal values. In contrary, the results of E_{def}^2, derived by Obert's correlation, reach significantly higher values. Comparing them with the nominal values, they exceed given intervals.

In summary, there are two cases considering E_{def} results and their comparison with nominal values. Either way, the results undergo or exceed the intervals, which points out obvious discrepancy between the results and the nominal values. This inconsistency may be caused by choice of unsuitable correlations or by the set intervals of nominal values. Regarding the correlations, it can be seen in Table 1 that both empiric relations consider the type and composition of the coarse-grained soils. But as none of the correlations mentions the genesis of the soil, we cannot be certain if the selected correlations are suitable for E_{def} determination of coarse-grained soils with fluvial origin (i.e., well-rounded soil grains shaped by the river activity). In general, the soils with rounded grains have smaller deformation modulus than the soils with semi angular or angular grains. The second uncertainty in the evaluation of deformation modulus is the nominal values. For example, Table 3 shows that STN 73 1001:1989 [19] presents higher values of E_{def} than Frankovská et al. [20]. But considering the genesis, we cannot assess for which soils are the values derived. For this reason, it would be appropriate to add the information of soil genesis or reconsider these intervals of nominal values with respect to the genesis of the coarse-grained soils.

Table 3. Comparison of the deformation modulus results with nominal values.

Density	*Soil*		q_{dyn} [MPa]		*Results* E_{def}^1 [MPa]		E_{def}^2 [MPa]		*Nominal values* E_{def}^3 [MPa]	E_{def}^4 [MPa]
			Mean	*Charact.*	*Mean*	*Charact.*	*Mean*	*Charact.*		
Dense (I_D = 0.67-1.0)	*Gravel*	*G1*	37.9	34.2	179.2	165.0	378.8	341.9	360 - 500	140 - 200
		G2	19.6	19.0	103.8	101.1	196.1	189.9	170 - 250	110 - 170
	Sand	*S1*	36.5	29.9	69.2	58.7	109.6	89.7	50 - 100	
		S2	16.7	13.5	36.0	30.3	50.0	40.4	30 - 50	
		S3	9.7	9.0	23.0	21.7	19.4	18.0	17 - 25	
Medium Dense (I_D = 0.33-0.67)	*Gravel*	*G2*	13.3	13.1	75.6	75.4	133.4	131.3	100 - 190	100 - 120
		G3	10.9	10.7	39.9	39.8	108.9	106.9	80 - 90	70 - 90
		G4	8.4	8.2	32.2	31.5	84.4	81.9	60 – 80	50 - 70
		G5	6.3	6.1	20.5	20.3	63.1	61.2	40 - 60	40 - 50
	Sand	*S3*	5.6	4.9	14.6	13.0	11.2	9.7	12 - 19	
		S4	5.2	4.8	13.7	12.9	10.4	9.7	5 – 15	
		S5	4.9	4.3	13.0	11.7	9.7	8.6	4 – 12	

E_{def}^1 - The deformation modulus was derived by Švasta´s correlation (Table 1)
E_{def}^2 - The deformation modulus was derived by Obert´s correlation (Table 1)
E_{def}^3 – Nominal values of deformation modulus listed in Slovak standard STN 73 1001:1989.
E_{def}^4 – Nominal values of deformation modulus recommended by Frankovská et al.

5. Conclusion

The knowledge of ground geotechnical parameters is fundamental for the design of geotechnical structures. Geotechnical parameters can be determined by field or laboratory tests. In Slovakia, dynamic penetration (DP) is one of the frequently used field tests for soil investigation. This test is especially suitable for testing coarse-grained soils due to less friction. The test measures the soil resistance against the rod penetration expressed by dynamic penetration resistance (q_{dyn}). Using suitable correlations, it is possible to derive physical (I_D) and mechanical (ϕ_{ef}, E_{def}) properties of coarse-grained soils. Other uses of dynamic penetration include DEM modelling to validate the methods of analysis [21] or to predict the grading of the coarse-grained soil [22].

This article focuses on the derivation of mechanical properties of coarse-grained fluvial soils from the results of q_{dyn} measured by heavy dynamic penetration test (DPH). The angle of shear strength (ϕ_{ef}) and deformation modulus (E_{def}) were derived using correlations established by Švasta and Obert [10]. These correlations consider the soil type and its composition but not the soil genesis. Results of q_{dyn}, ϕ_{ef} and E_{def} were furthermore statistically analyzed, expressing the mean and characteristic values. After comparing the mean and characteristic results of each parameter, we concluded that the difference between values is not so significant. Therefore, it is not important whether we take mean or characteristic values for further analysis.

Derived mechanical parameters (ϕ_{ef}, E_{def}) were compared with the nominal values specified in STN 73 1001:1989 [19] and recommended by Frankovská et al. [20]. The angle of shear strength results shows a good fit with the given intervals of nominal values indicating that chosen correlation by Švasta is suitable for coarse-grained fluvial soils. A similar comparison was performed with the results of E_{def}. Two sets of results, derived by Švasta´s and Obert´s correlations [10], were compared with two sets of nominal values specified by STN 73 1001:1989 [19] and Frankovská et al. [20]. The worst match was observed in gravels, where many of the results did not fit within any of the intervals. The irregularities between the results and nominal values can be caused either by the unsuitability of the used correlations or by the overestimated nominal values.

Regarding coarse-grained fluvial soils, to critically assess the applicability of correlations and the correctness of the nominal values, we need to supplement the survey with the results of laboratory tests.

6. References

[1] Pohl C 2011 Determination of characteristic soil values by statistical methods *Geotechnical Safety and Risk ISGSR* pp 427–434
[2] Schnaid F 2009 *In Situ Testing in Geomechanics, The main tests* (London, New York: Taylor & Francis) p 328
[3] EN 1997-2:2007 Eurocode 7 Geotechnical design Part 2: Ground investigation and testing
[4] EN ISO 22476-2 Geotechnical investigation and testing - Field testing Part 2: Dynamic Probing
[5] Langton D 1999 The Panda lightweight penetrometer for soil investigation and monitoring material compaction *Ground Engineering* no. September p 33– 37
[6] Frankovská J 2016 Collapsibility of loess evaluated from field and laboratory tests *16th International Multidisciplinary Scientific Geoconference (SGEM 2016: Science and Technologies in Geology, Exploration and Mining* vol I) p 731–7
[7] DIN 4094-3:2002 *Subsoil Field testing Part – 3: Dynamic probing*
[8] PN-B-04552:2002 *Geotechnics Field testing*
[9] STN 72 1032:1997 *Dynamic penetration test*
[10] Matys M, Ťavoda O and Cuninka M 1990 *Poľné skúšky zemín* (Bratislava: Alfa)
[11] Tran Q A, Chevalier B and Breul P 2018 Spectral analysis of the response of coarse granular material to dynamic penetration test modelled with DEM *Int. J. of Geosynth. and Ground Eng.* **4** 22
[12] Vrettos C and Papamichael S 2018 Relative density of cohesionless soils from dynamic probe and cone penetration test – An update assessment of empirical relationship *Geotechnik* **41** 186–96
[13] Czado B and Pietras J S 2012 Comparison of the cone penetration resistance obtained in static and dynamic field tests *Journal of Mining and Geography* p 97–105
[14] Frankovská J 2016 Statistical approach for evaluation of characteristic values of geotechnical parameters *Geotechnika* **19** 2 p 3–7
[15] Schneider H R 2010 Characteristic Soil Properties for EC 7: Influence of quality of test results and soil volume involved *Proc. 14th Danube-European Conference on Geotechnical Engineering*
[16] Bond A J 2011 A procedure for determining the characteristic value of a geotechnical parameter *Geotechnical Safety and Risk ISGSR 2011* p 420–426
[17] Panuska J and Frankovská J 2016 Effect of a void ratio on the small strain shear modulus G(max) for coarse - grained soils *World Multidisciplinary Civil Engineering-Architecture-Urban Planning Symposium, WMCAUS 2016: Proc. Engineering* **161** p 1239–1235
[18] Frankovská J, Kopecký M and Turček P 2014 The correlation of surface area with hydraulic conductivity of fine-grained soils *SGEM 2014 GeoConference on Science and Technologies in Geology, Exploration and Mining* (Sofia: STEF 92 Technology) p 893–900
[19] STN 73 1001:1989 Geotechnical structures Foundation
[20] Frankovská J, Súľovská M and Turček P 2011 *Foundations – Introduction to geotechnical design,*

spread and deep foundations (Bratislava: STU Press) p 162
[21] Sastre C, Benz M, Gourvès R, Breul P and Bacconnet C 2016 Automatic methodology to predict grain size class from dynamic penetration test using neural networks *5th International Conference on Geotechnical and Geological Site Characterization, ISC'5 2016* p 1465 – 1470
[22] Tran Q A, Chevalier B, Benz-Navarrete, Breul P and Gourve`s R 2017 Modeling of light dynamic cone penetration test Panda 3R in granular material by using 3D discrete element method *EPJ Web of Conferences Powders & Grains (Montpellier)* **140** 16003

Acknowledgment
This paper was supported by Grant Agency of the Ministry of Education, Science, Research and Sport of the Slovak Republic, VEGA, grant project No. 1/0129/21.

AUTHOR INDEX

Abrahamczyk, L ... 411
Al Ali, M ... 312
Albu, D C ... 48
Almikaeel, W ... 495
Bajzecerová, V ... 194
Baran, J ... 356
Baroš, T ... 181
Bartoň, V ... 228
Baránek, Š ... 221
Baskova, R ... 7
Beckovsky, D ... 330
Begić, H ... 59
Bečkovský, D ... 215
Blaho, S ... 363
Bokomlasko, J ... 527
Borzovic, V ... 390
Borzovič, V ... 356, 396
Brandis, A ... 344
Brandtner, M ... 1
Brecka, J ... 373
Budajová, J ... 98
Burzio, C ... 104
Bytčanková, L ... 479
Charvát, J ... 255
Chlebková, L ... 255
Cikrle, P ... 228
Csóka, M ... 108, 119, 162
Danek, P ... 270
Danáčová, Z ... 485
Dasović, B ... 65
Domanický, J ... 450
Doubek, P ... 324
Doubek, R ... 71
Drochytka, R ... 221, 234, 241
Dubecky, D ... 421, 425, 441
Eštoková, A ... 248
Figala, P ... 234, 241
Frantík, P ... 270
Freiherrova, N ... 188
Gajdošová, K ... 363, 368
Garaj, M ... 114
Gorás, M ... 460
Grabarz, A ... 533
Halaszova, I ... 466
Hejl, M ... 76
Hermann, R ... 234, 241
Hodovanets, E ... 312
Hološová, M ... 248
Hornakova, M ... 188
Hrabovska, K ... 373
Hrudka, J ... 119, 132, 150, 517
Husovská, A ... 82
Hutkai, K ... 13
Jaroš, P ... 23
Jun, D ... 28
Juracka, D ... 188
Juračka, D ... 263
Jílek, J ... 76
Kanócz, J ... 194
Kapolka, L ... 188, 318, 339
Karla, V ... 194
Katunský, D ... 13, 181
Kawulok, M ... 263
Kersner, Z ... 270
Klanšek, U ... 65
Klosova, D ... 201
Kmet, S ... 339
Kodymová, J ... 167
Kovac, M ... 380
Koval, F ... 506
Kovács, R ... 540
Kozlovska, M ... 466
Kozlovská, M ... 88, 201
Kraus, I ... 344
Krupík, P ... 43
Król, P A ... 335
Kubínová, T ... 126
Kullman, E ... 485
Kumpova, I ... 270
Kusnir, S ... 380
Kusák, I ... 285
Kuvik, M ... 549
Kučová, E ... 549
Kvapilová, V ... 386
Kvocak, V ... 312, 421, 441
Kyncl, M ... 126
Lapčík, V ... 167
Lesan, A ... 48
Lisztwan, D ... 270
Luptáková, A ... 248
Malíková, L ... 324
Mandula, J ... 527, 540
Marko, I ... 108, 132, 175
Mattsson, A ... 104
Miarka, P ... 324
Mizerová, C ... 285

Modin, O.......104
Mohapl, M.......76
Mokrenko, D.......88
Mydla, J.......485
Máčalová, K.......255
Mészárosová, L.......241
Nagy, P.......138
Nechyporchuk, Y.......7
Nespesny, O.......330
Nespešný, O.......215
Nespěšný, O.......28
Nečas, R.......431
Nivert, E.......104
Orolin, P.......425
Papesch, R.......255
Penava, D.......411
Pencik, J.......330
Perić, A.......291
Persson, F.......104
Peráčková, J.......34
Petrík, A.......209
Pełczyński, J.......335
Popov, A.......517
Portnov, M.......517
Požár, M.......431
Predajnianska, A.......142
Raczková, A.......119
Ravaszová, S.......298
Repel, A.......522
Rinaudo, F.......411
Rovnaník, P.......285
Rózsa, G.......108, 150, 162
Sarhosis, V.......411
Sarvaicova, S.......390
Sawicki, W.......533
Schmid, P.......285
Seitl, S.......324
Selník, P.......215
Sokol, M.......34
Stanko, S.......517
Stanko, Š.......108, 132, 162, 175
Staňa, R.......473
Struková, Z.......82
Stulerova, L.......188, 318, 339
Svahn, O.......104
Svoboda, J.......255
Takacs, J.......142
Taušan, D.......344
Trajber, D.......411
Vanova, P.......421, 425
Velešík, M.......431
Venkrbec, V.......1
Vertal, M.......23
Vranay, F.......450, 460
Vranayová, Z.......450, 460
Vrdoljak, I.......304
Vystrcil, J.......330
Vystrčil, J.......215
Vácha, M.......167
Wawrek, I.......473
Weissova, M.......441
Wilén, B M.......104
Wittmanová, R.......132, 150, 175
Yakovlev, G.......221
Zeleňaková, M.......114
Čereš, D.......368
Černý, V.......221
Červeňanská, M.......485
Čubanová, L.......495
Šenšelová, Ž.......396
Škultétyová, I.......119, 150, 162
Šnédar, J.......228
Šoltész, A.......485
Štecák, R.......402
Šugareková, M.......156
Šuhajda, K.......215
Šutúš, M.......119, 162, 175